NONLINEAR AND ADAPTIVE CONTROL DESIGN

Adaptive and Learning Systems for Signal Processing, Communications, and Control

Editor Simon Haykin

Werbos / THE ROOTS OF BACKPROPAGATION: FROM ORDERED DERIVATIVES TO NEURAL NETWORKS AND POLITICAL FORECASTING

Krstić, Kanellakopoulos, and Kokotović / NONLINEAR AND ADAPTIVE CONTROL DESIGN

Nikias and Shao / SIGNAL PROCESSING WITH ALPHA-STABLE DISTRIBUTIONS AND APPLICATIONS

NONLINEAR AND ADAPTIVE CONTROL DESIGN

Miroslav Krstić
Ioannis Kanellakopoulos
Petar Kokotović

A WILEY-INTERSCIENCE PUBLICATION
JOHN WILEY & SONS, INC.
New York / Chichester / Brisbane / Toronto / Singapore

> **A NOTE TO THE READER**
> This book has been electronically reproduced from digital information stored at John Wiley & Sons, Inc. We are pleased that the use of this new technology will enable us to keep works of enduring scholarly value in print as long as there is a reasonable demand for them. The content of this book is identical to previous printings.

This text is printed on acid-free paper.

Copyright © 1995 by John Wiley & Sons, Inc.

All rights reserved. Published simultaneously in Canada.

Reproduction or translation of any part of this work beyond that permitted in Section 107 or 108 of the 1976 United States Copyright Act without the permission of the copyright owner is unlawful. Requests for permission or further information should be addressed to the Permissions Department, John Wiley & Sons, Inc., 605 Third Avenue, New York, NY 10158-0012.

Library of Congress Cataloging in Publication Data

Krstić, Miroslav.
    Nonlinear and adaptive control design / Miroslav Krstić, Ioannis Kanellakopoulos, Petar Kokotović.
       p. cm. — (Adaptive and learning systems for signal processing, communications, and control)
    "A Wiley-Interscience publication."
    Includes bibliographical references and index.
    ISBN 0-471-12732-9
      1. Automatic control. 2. Nonlinear control theory. 3. Adaptive control systems. I. Kanellakopoulos, Ioannis. II. Kokotović, Petar III. Title. IV. Series.
TJ213.K748 1995
629.8—dc20
                                              95-10082
                                                 CIP

Printed in the United States of America

10 9 8 7

Preface

This book opens a view to the largely unexplored landscape of nonlinear systems with uncertainties. Its main subject is feedback design for such systems. New design tools and systematic design procedures are developed which guarantee that the designed feedback systems will possess desired properties not only locally, but also globally or in a specified region of the state space.

Backstepping. Compared with other books on nonlinear control, the major novelty of this book is a recursive design methodology: *backstepping*. With this methodology the construction of both feedback control laws and associated Lyapunov functions is systematic. Strong properties of global or regional stability and tracking are built into the nonlinear system in a number of steps, which is never higher than the system order. While feedback linearization methods require precise models and often cancel some useful nonlinearities, backstepping designs offer a choice of design tools for accommodation of uncertain nonlinearities and can avoid wasteful cancellations.

Classes of systems. By the very nature of the modeling process, nonlinear models are more structured than their linear offspring. Many nonlinearities introduced by physical laws (e.g., centrifugal forces and chemical kinetics) can not be confined to a linear sector, because, with increasing magnitudes, their growth is polynomial or even exponential. Systems with such nonlinearities are prone to explosive instabilities which must be anticipated and prevented by feedback control.

Nonlinearities usually appear multiplied with physical constants, often poorly known or dependent on the slowly changing environment. This common form of uncertainty is captured by nonlinear models with unknown constant parameters. In "parametric pure-feedback" systems studied in this book, the growth of nonlinearities is unrestricted, while the unknown parameters appear linearly. When only an output is measured, the nonlinearities are assumed to be functions of the output. The considered class of nonlinear systems is broad and encompasses linear systems as a special case.

Adaptive control. The control of nonlinear systems with unknown parameters is traditionally approached as an adaptive control problem, separate from the rest of nonlinear control theory. Here we bridge this gap and let adaptive control be what it is—a paradigm for constructing nonlinear dynamic feedback for both nonlinear and linear systems. We depart from the traditional "certainty equivalence" approach and design stronger nonlinear control laws that achieve the desired objectives either by interacting with parameter update laws or by attenuating the effect of parameter estimation errors. In this way we achieve not only stronger stability properties, but also quantifiable improvements of transient performance.

Applications. Strong regional or global properties achieved by the design methods presented in this book have the potential to expand the operating range of feedback controllers and make them applicable to physical plants for which linear or other local controllers are inadequate. In this text, application examples of this type include automotive suspensions, jet engine stall and surge control, biochemical processes, aircraft wing rock control, induction motors, robotic manipulators, and magnetic levitation.

Organization of the book. Although most of the results in this book are new, they are presented at a level accessible to audiences with a standard undergraduate background in control theory and a basic knowledge of stability concepts. Each chapter is written in a pedagogical style, with detailed proofs and illustrative examples. More involved extensions and generalizations are only outlined. Basic tools for nonadaptive backstepping design with state feedback are given in Chapter 2, and with output feedback in Chapter 7, while adaptive backstepping is introduced in Chapter 3. Chapters 1, 2, 3, and 7 are written as a short text for a part of either a course on nonlinear systems or on adaptive control. The remaining chapters, although more advanced, are accessible without Chapters 2, 3, and 7. The core of the controller design is the tuning functions method in Chapters 4 and 8. The modular designs in Chapters 5, 6, and 9 achieve the independence of the controller and the identifier. The new adaptive nonlinear designs are applied to linear systems in Chapter 10 and are shown to outperform the traditional linear designs. This chapter can be read independently from the rest of the book. Thus, depending on the reader's interest, the book can be read in several ways:

	Chapters
• Nonlinear stabilization without adaptation	2, 7
• Introduction to adaptive backstepping	2, 3, 7
• Adaptive Lyapunov design with tuning functions	2, 4, 8, 10
• Estimation-based design with modularity	2, 5, 6, 9, 10
• State feedback adaptive designs	2, 4, 5, 6
• Adaptive control of linear systems	10

Preface

This book is a joint effort based on recent research of its authors. Chapters 4, 5, 6, 8, 9, and 10 and Appendices A–F were written by the first author, and Chapters 2, 3, and 7 and Appendix G were written by the second author. Their labor was arduous due to frequent revisions by the third author, their former Ph.D. advisor, who also wrote Chapter 1.

Acknowledgments. At the initial stage of this research, Riccardo Marino introduced us to nonlinear geometric methods and collaborated with us and David Taylor on the early adaptive nonlinear schemes, contributing a key extended matching idea. A separate line of research with Hector Sussmann and Ali Saberi led us to a nonadaptive recursive design. About a year later, in a joint effort with Steve Morse, we developed the first adaptive recursive design procedure, which we named adaptive backstepping. We express our deepest gratitude to these colleagues. We are also thankful to Randy Freeman, Petros Ioannou, Laurent Praly, and Andy Teel for continuous exchange and discussion of current results. Among many other colleagues who helped us with frequent debates on broader issues of nonlinear and adaptive control are Brian Anderson, Karl Åström, Tamer Başar, Gil Blankenship, Bill Boothby, Mohammed Dahleh, Reza Ghanadan, Graham Goodwin, Jessy Grizzle, Alberto Isidori, Mrdjan Janković, Hassan Khalil, Art Krener, P. R. Kumar, Ioan Landau, David Mayne, Rick Middleton, Carl Nett, Shankar Sastry, Rodolphe Sepulchre, Eduardo Sontag, Mark Spong, Jing Sun, and Gang Tao. And above all, our warmest thanks go to the most generous contributors to this project, our wives, Angela, Georgia, and Anna.

The writing of this book was supported in part by the National Science Foundation under Grants ECS-9203491 and RIA ECS-9309402, by the Air Force Office of Scientific Research under Grant F-49620-92-J-0495, by UCLA through the SEAS Dean's Fund, and also by grants from Ford Motor Company and Rockwell International.

Miroslav Krstić
Ioannis Kanellakopoulos
Petar Kokotović

Santa Barbara, California

Contents

1 Introduction — 1
  1.1 Adaptive Linear Control — 1
    1.1.1 Emergence of adaptive control — 1
    1.1.2 Achievements of adaptive linear control — 2
    1.1.3 Adaptive control as dynamic nonlinear feedback — 3
    1.1.4 Lyapunov-based design — 4
    1.1.5 Estimation-based design — 5
  1.2 Adaptive Nonlinear Control — 7
    1.2.1 A nonlinear challenge — 7
    1.2.2 A structural obstacle — 9
    1.2.3 Early results — 10
  1.3 Preview of the Main Topics — 11
    1.3.1 Classes of nonlinear systems — 11
    1.3.2 Adaptive backstepping and tuning functions — 12
    1.3.3 Modular designs — 14
    1.3.4 Output-feedback designs — 15
    1.3.5 Linear systems — 15
    1.3.6 Transient performance — 16
    1.3.7 Caveat: robustness — 17
  1.4 Notation — 17
  Notes and References — 17

I State Feedback — 19

2 Design Tools for Stabilization — 21
  2.1 Stability — 22
    2.1.1 Main stability theorems — 22
    2.1.2 Control Lyapunov functions (clf) — 25
  2.2 Backstepping — 29
    2.2.1 Integrator backstepping — 29
    2.2.2 Example: active suspension (parallel) — 37
    2.2.3 Feedback linearization and zero dynamics — 39

	2.2.4	Stabilization of cascade systems	42
	2.2.5	Block backstepping with zero dynamics	49
	2.2.6	Example: active suspension (series)	55
2.3	Recursive Design Procedures		58
	2.3.1	Strict-feedback systems	58
	2.3.2	Pure-feedback systems	61
	2.3.3	Block-strict-feedback systems	64
2.4	Design Flexibility: Jet Engine Example		66
	2.4.1	Jet engine stall and surge	67
	2.4.2	A two-step design	68
	2.4.3	Avoiding cancellations	70
2.5	Stabilization with Uncertainty		72
	2.5.1	Nonlinear damping	73
	2.5.2	Backstepping with uncertainty	80
	2.5.3	Robust strict-feedback systems	84
Notes and References			86

3 Adaptive Backstepping Design — 87

3.1	Adaptation as Dynamic Feedback		88
3.2	Adaptive Backstepping		92
	3.2.1	Adaptive integrator backstepping	92
	3.2.2	Adaptive block backstepping	98
3.3	Recursive Design Procedures		99
	3.3.1	Parametric strict-feedback systems	99
	3.3.2	Multi-input systems	103
	3.3.3	Parametric block-strict-feedback systems	105
3.4	Extended Matching Design		110
	3.4.1	Reducing the overparametrization	110
	3.4.2	Example: biochemical process	113
	3.4.3	Transient performance improvement	115
Notes and References			121

4 Tuning Functions Design — 123

4.1	Adaptive Control Lyapunov Functions		124
	4.1.1	Departure from certainty equivalence	124
	4.1.2	Certainty equivalence for a modified system	128
	4.1.3	Adaptive backstepping via aclf	134
4.2	Set-Point Regulation		139
	4.2.1	Design procedure	140
	4.2.2	Stability and convergence	151
	4.2.3	Passivity	154
4.3	Tracking		156
	4.3.1	Design procedure	157
	4.3.2	Trajectory initialization	162

4.4	Transient Performance	165
	4.4.1 $\mathcal{L}_2$ performance	165
	4.4.2 $\mathcal{L}_\infty$ performance	166
4.5	Extensions	168
	4.5.1 Unknown virtual control coefficients	168
	4.5.2 Block-strict-feedback systems	173
	4.5.3 Pure-feedback systems	175
4.6	Example: Aircraft Wing Rock	180
	Notes and References	183

5 Modular Design with Passive Identifiers — 185

5.1	Weakness of Certainty Equivalence	186
5.2	ISS-Control Lyapunov Functions	189
5.3	ISS-Controller Design	198
5.4	Observers for Strict Passivity	206
5.5	z-Passive Scheme	209
5.6	x-Passive Scheme	212
5.7	Transient Performance	218
5.8	SG-Scheme (Weak Modularity)	222
	5.8.1 Controller design	223
	5.8.2 Scheme with strengthened identifier	225
5.9	Unknown Virtual Control Coefficients	229
	5.9.1 Controller design	229
	5.9.2 Passive adaptive scheme	232
	Notes and References	233

6 Modular Design with Swapping Identifiers — 235

6.1	ISS-Controller	235
6.2	Swapping and Static Parametric Models	237
6.3	z-Swapping Scheme	239
6.4	x-Swapping Scheme	248
6.5	Transient Performance	254
	6.5.1 Simulation examples	259
6.6	SG-Scheme	265
6.7	Schemes with Weak ISS-Controller	271
6.8	Unknown Virtual Control Coefficients	277
	Notes and References	282

II Output Feedback — 283

7 Output-Feedback Design Tools — 285

7.1	Observer Backstepping	285
	7.1.1 Unmeasured states	285

	7.1.2	Output-feedback systems	291
7.2		MIMO Design: Induction Motor	294
7.3		Adaptive Observer Backstepping	301
	7.3.1	An introductory example	301
	7.3.2	Parametric output-feedback systems	307
	7.3.3	Example: single-link flexible robot	313
7.4		Extensions	315
	7.4.1	Interlaced controller-observer design	315
	7.4.2	Design with partial-state feedback	320
		Notes and References	324

8 Tuning Functions Designs — 327

8.1		Design with K-Filters	329
	8.1.1	Filters and observer	329
	8.1.2	Adaptive controller design	332
	8.1.3	Stability	340
	8.1.4	Transient performance	346
8.2		Design with MT-Filters	348
	8.2.1	Filtered transformations and observer	348
	8.2.2	Adaptive controller design	351
	8.2.3	Stability	357
	8.2.4	Transient performance	363
8.3		Example: Levitated Ball	365
		Notes and References	369

9 Modular Designs — 371

9.1		ISS-Controller Design	372
9.2		y-Passive Scheme	380
9.3		y-Swapping Scheme	384
9.4		x-Swapping Scheme	388
9.5		Schemes with Parametric z-Model	392
9.6		Transient Performance	394
	9.6.1	Passive schemes	394
	9.6.2	Swapping schemes	397
9.7		Swapping Schemes with Weak ISS-Controller	400
9.8		Schemes with MT-Filters	403
	9.8.1	ISS-observer	404
	9.8.2	ISS-controller	405
	9.8.3	χ-Passive scheme	407
	9.8.4	ε-Swapping scheme	409
	9.8.5	Transient performance	412
		Notes and References	416

10 Linear Systems — 417
- 10.1 State Estimation Filters — 418
- 10.2 Tuning Functions Design — 422
  - 10.2.1 Design procedure — 422
  - 10.2.2 Stability analysis — 434
  - 10.2.3 Passivity — 438
  - 10.2.4 Design example — 440
- 10.3 Properties of the Nonadaptive System — 443
  - 10.3.1 Underlying linear controller — 443
  - 10.3.2 Parametric robustness and nonadaptive performance — 448
- 10.4 Transient Performance with Tuning Functions — 453
  - 10.4.1 Transient performance of the adaptive system — 454
  - 10.4.2 Performance improvement due to adaptation — 459
- 10.5 Comparison with a Traditional Scheme — 464
  - 10.5.1 Choice of a traditional scheme — 464
  - 10.5.2 Comparison of the schemes — 465
- 10.6 Modular Designs — 470
  - 10.6.1 SG-controller — 470
  - 10.6.2 y-Passive scheme — 473
  - 10.6.3 x-Swapping scheme — 477
  - 10.6.4 Comparison: modular vs. tuning functions design — 482
- 10.7 Summary — 484
- Notes and References — 484

Appendices — 487

A Lyapunov Stability and Convergence — 489

B Input-Output Stability — 493

C Input-to-State Stability — 501

D Passivity — 507

E Parameter Projection — 511

F Nonlinear Swapping — 515

G Differential Geometric Conditions — 521
- G.1 Partial-State-Feedback Forms — 521
- G.2 Output-Feedback Forms — 533
- G.3 Full-State-Feedback Forms — 535

Bibliography — 541

Index — 559

NONLINEAR AND ADAPTIVE CONTROL DESIGN

Chapter 1

Introduction

1.1 Adaptive Linear Control

1.1.1 Emergence of adaptive control

Attempts to invent, design, and build systems capable of controlling unknown plants or adapting to unpredictable changes in the environment have a long and rich history. This history could justifiably go back to early inventions of feedback, because even the most elementary feedback loops can tolerate significant uncertainties.

Major advances of control theory in the 1950's and 1960's encouraged the thinking about more sophisticated forms of feedback systems. Many appealing concepts were proposed in which the notion of feedback was often invoked through the prefix "self," such as "self-learning," "self-optimizing," "self-organizing," and "self-tuning" or "adaptive" control. More specific proposals included on-line identification or pattern recognition inside the feedback control loop. On the theoretical end of the spectrum were stochastic and dual control. Under the influence of the artificial intelligence community, many other concepts were born: "fuzzy," "neural," and, more recently, "intelligent" control. Most of these concepts have widened our intellectual horizon, even if some have since disappeared from our technical vocabulary.

Adaptive control is one of the ideas conceived in the 1950's which has firmly remained in the mainstream of research activity with hundreds of papers and several books published every year. In the almost forty years of its existence, adaptive control theory has been steadily growing into a well-formed scientific discipline: from inventions to rigorous problem formulations; from solutions of basic problems to more demanding tasks for broader classes of systems; from questions of existence and solvability to application-oriented issues of robustness and performance.

One of the reasons for the rapid growth and continuing popularity of adaptive control is its clearly defined goal: to control plants with unknown parameters. Adaptive control has been most successful for plant models in which the unknown parameters appear linearly. Fortunately, such a "linear parametrization" can be achieved in most situations of practical interest.

1.1.2 Achievements of adaptive linear control

While in this book we will be preoccupied with nonlinear systems, we must not forget that the control of linear plants with unknown parameters was a formidable problem which took almost twenty years to solve. The adaptive control community deserves full credit for providing not only one, but several solutions to this fundamental problem [5, 31, 44, 109, 136, 138, 142, 165]. Each of these solutions was a breakthrough in the development of adaptive control. By the early 1980's, several types of adaptive schemes were proven to provide stable operation and asymptotic tracking. We refer to the results from that period as *adaptive linear control* or *traditional adaptive control*. Traditional adaptive schemes are classified as "direct" and "indirect" and as "Lyapunov-based" and "estimation-based." They involve parameter identification with "parameter estimators" or "identifiers." The vital part of the identifier is the parameter adaptation algorithm, commonly referred to as the "parameter update law." The direct-indirect classification reflects the fact that the updated parameters are either those of the controller (direct) or those of the plant (indirect). According to this classification, all the schemes in this book are indirect.

The distinction between Lyapunov-based and estimation-based schemes is more substantial and is dictated in part by the type of parameter update law and the corresponding proof of stability and convergence. Lyapunov-based design is one of the oldest results of adaptive control. Until recently, however, its applicability was restricted to linear plants with relative degree one or two. This limitation has been removed by the recursive design procedures presented in this book, commonly referred to as *backstepping*.

Estimation-based designs are more broadly applicable and allow a choice of parameter update laws from a wide repertoire of gradient and least-squares optimization algorithms. This flexibility is achieved by treating the identifier as a separate module and guaranteeing its properties independent of the controller module. The estimation-based designs presented in this book achieve complete modularity. For this reason they are called *modular designs*.

An important feature of traditional adaptive control is its reliance on "certainty equivalence" controllers. This means that a controller is first designed as if all the plant parameters were known. The controller parameters are determined as functions of the plant parameters. Given the true values of the plant parameters, the controller parameters are calculated by solving design equations for model-matching, pole-zero placement, or optimality. When the true plant parameters are unknown, the controller parameters are either estimated

1.1 ADAPTIVE LINEAR CONTROL

directly (direct schemes) or computed by solving the same design equations with plant parameter estimates (indirect schemes). The resulting controller, which is either estimated (direct) or designed for the estimated plant (indirect), is called a *certainty equivalence* controller.

It is not at all obvious that a certainty equivalence controller will work inside an adaptive feedback loop and achieve stabilization and tracking. Even when the plant is stable, bad parameter estimates may yield a destabilizing controller. The situation is more critical when the plant is unstable, because then the controller must achieve stabilization in addition to its tracking task. It is therefore significant that certainty equivalence controllers have been proven to be satisfactory for adaptive control of linear systems.

In spite of major advances in the development of adaptive control schemes for linear systems, they have not yet become tools for systematic engineering design. Each adaptive scheme leaves up to the designer the choice of various filters, design coefficients, initialization rules, and so on. It is still unclear how the adaptive system's performance, especially its transient performance, depends on these design choices. Current research activity is aimed at providing the designer with clearer choices and trade-offs between transient performance and robustness. Transient performance improvement is one of the prominent features of this book.

1.1.3 Adaptive control as dynamic nonlinear feedback

If a linear plant contains unknown parameters without any information about their bounds, then, in general, it cannot be stabilized by a linear controller. This is true for the simplest scalar plant

$$\dot{x} = u + \theta x, \qquad (1.1)$$

where u is the control and θ is an unknown constant. If an a priori bound $\bar{\theta}$ on $|\theta|$ were known, $|\theta| \leq \bar{\theta}$, then $u = -2\bar{\theta}x$ would be a linear stabilizing controller. If such a bound is not known, no linear controller can be designed to guarantee stability of (1.1).

To examine whether a static nonlinear controller can help, let us try the controller

$$u = -k_1 x - k_2 x^3, \qquad (1.2)$$

where $k_1 > 0$, $k_2 > 0$. The resulting feedback system is

$$\dot{x} = (\theta - k_1)x - k_2 x^3. \qquad (1.3)$$

For $\theta > k_1$, the equilibrium $x = 0$ is unstable, but the nonlinear term $-k_2 x^3$ prevents $x(t)$ from growing unbounded. It is easy to see that $x(t)$ will converge to one of the two new equilibria $\pm\sqrt{\frac{\theta-k_1}{k_2}}$. Thus, the static nonlinear controller

(1.2) has achieved boundedness of $x(t)$ without any knowledge of a bound on θ.

Our goal is more ambitious than just boundedness of $x(t)$. We also want to achieve its regulation: $\lim_{t\to\infty} x(t) = 0$. Can this be accomplished by a dynamic nonlinear controller? The answer is affirmative: One such controller is

$$u = -(p+\xi)x, \quad \dot{\xi} = x^2, \qquad (1.4)$$

where $p > 0$ is a design parameter. The resulting feedback system is of second order:

$$\dot{x} = -(p+\xi)x + \theta x \qquad (1.5a)$$
$$\dot{\xi} = x^2. \qquad (1.5b)$$

Its stability properties can be checked by examining the derivative of the Lyapunov function

$$V(x,\xi) = \frac{1}{2}x^2 + \frac{1}{2}(\xi - \theta)^2, \qquad (1.6)$$

which turns out to be nonpositive:

$$\dot{V} = -px^2 - \xi x^2 + \theta x^2 + (\xi - \theta)x^2 = -px^2. \qquad (1.7)$$

Thus, $V(x(t), \xi(t))$ evaluated along the solutions of (1.5) is a nonincreasing function of time. This proves that $x(t)$ and $\xi(t)$ remain bounded for all $t \geq 0$. The proof that $\lim_{t\to\infty} x(t) = 0$ is also achieved can be given using Theorem 2.1 in the next chapter.

How was the dynamic nonlinear controller (1.4) conceived? Not as a nonlinear controller, but rather as a parameter adaptation scheme! Its dynamic part $\dot{\xi} = x^2$ is, in fact, an update law for ξ as an estimate of θ. Consequently, the estimation error $\xi - \theta$ is penalized in the Lyapunov function (1.6).

1.1.4 Lyapunov-based design

The controller (1.4) is an outcome of a systematic Lyapunov design procedure. In this procedure we seek a parameter update law for the estimate $\hat{\theta}(t)$,

$$\dot{\hat{\theta}} = \tau(x, \hat{\theta}), \qquad (1.8)$$

which, along with a control law $u = \alpha(x, \hat{\theta})$, will make the Lyapunov function

$$V(x, \hat{\theta}) = \frac{1}{2}x^2 + \frac{1}{2}(\hat{\theta} - \theta)^2 \qquad (1.9)$$

a nonincreasing function of time:

$$V(x(t), \hat{\theta}(t)) \leq V(x(t_0), \hat{\theta}(t_0)), \quad \forall t \geq t_0, \; \forall t_0 \geq 0. \qquad (1.10)$$

1.1 ADAPTIVE LINEAR CONTROL

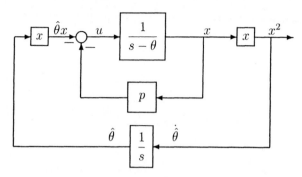

Figure 1.1: Lyapunov-based adaptive scheme for the scalar system $\dot{x} = u + x\theta$.

To this end, we express $\dot{V}$ as a function of u and $\dot{\hat{\theta}}$ and seek $\alpha(x, \hat{\theta})$ and $\tau(x, \hat{\theta})$ to guarantee that $\dot{V} \leq -px^2$ with $p > 0$, namely

$$\dot{V} = x(u + \theta x) + (\hat{\theta} - \theta)\dot{\hat{\theta}} \leq -px^2. \tag{1.11}$$

Rearranging terms we get

$$xu + \hat{\theta}\dot{\hat{\theta}} + \theta\left(x^2 - \dot{\hat{\theta}}\right) \leq -px^2. \tag{1.12}$$

Since neither $\alpha(x, \hat{\theta})$ nor $\tau(x, \hat{\theta})$ is allowed to depend on the unknown θ, we must take $\tau(x, \hat{\theta}) = x^2$, that is,

$$\dot{\hat{\theta}} = x^2. \tag{1.13}$$

The remaining condition

$$xu + \hat{\theta}x^2 \leq -px^2 \tag{1.14}$$

allows us to select $\alpha(x, \hat{\theta})$ in various ways. The choice which results in the dynamic nonlinear controller (1.4) is

$$u = -(p + \hat{\theta})x. \tag{1.15}$$

We have thus designed our first Lyapunov-based adaptive scheme shown in Figure 1.1, where s is the complex variable of the Laplace transform. This scheme already exhibits some features of more general schemes to be designed in this book.

1.1.5 Estimation-based design

Another dynamic nonlinear controller for the same linear plant (1.1) will now be designed starting with the design of a parameter identifier. Since the signal

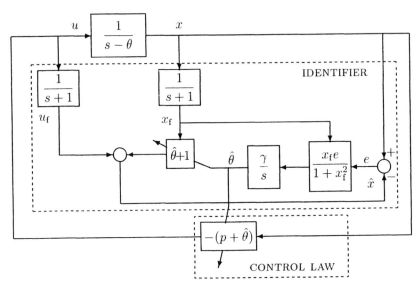

Figure 1.2: Estimation-based adaptive scheme for the scalar system $\dot{x} = u + x\theta$.

$\dot{x}$ is not available for measurement, we cannot solve (1.1) for the unknown θ. To overcome this difficulty, we resort to filtering both sides of (1.1) by $\frac{1}{s+1}$:

$$\frac{s}{s+1}x = \frac{1}{s+1}u + \theta\frac{1}{s+1}x. \qquad (1.16)$$

Denoting the filtered versions of the known signals $x(t)$ and $u(t)$ by

$$x_\mathrm{f} = \frac{1}{s+1}x, \quad u_\mathrm{f} = \frac{1}{s+1}u, \qquad (1.17)$$

we can rewrite (1.16) as

$$x(t) = (\theta + 1)x_\mathrm{f}(t) + u_\mathrm{f}(t). \qquad (1.18)$$

If instead of the unknown θ we use its estimate $\hat{\theta}$, the corresponding predicted value of x is

$$\hat{x}(t) = (\hat{\theta}(t) + 1)x_\mathrm{f}(t) + u_\mathrm{f}(t), \qquad (1.19)$$

and the prediction error e is related to the estimation error $\tilde{\theta}$ as follows:

$$e = x - \hat{x} = (\theta - \hat{\theta})x_\mathrm{f} = \tilde{\theta}x_\mathrm{f}. \qquad (1.20)$$

A parameter update law for $\hat{\theta}$ can now be derived to aim at a minimum of e^2. To this end, the rate of change of $\hat{\theta}$ is set to be proportional to the negative gradient of e^2 with respect to $\hat{\theta}$:

$$\dot{\hat{\theta}} = -\frac{\gamma}{2}\frac{\partial(e^2)}{\partial\hat{\theta}} = \gamma e x_\mathrm{f}, \qquad (1.21)$$

where $\gamma > 0$ is the "adaptation gain."

More commonly used are "normalized" versions of (1.21). One of these is

$$\dot{\hat{\theta}} = \frac{\gamma}{1 + x_f^2} e x_f. \tag{1.22}$$

Once $\hat{\theta}(t)$ is available, it can be used in a certainty equivalence control law. For example, the same control law as in (1.15) would result from the specification that the closed-loop pole be placed at $-p$. This task would indeed be achieved if the estimate $\hat{\theta}$ were correct, $\hat{\theta} \equiv \theta$.

The closed-loop adaptive system with the normalized update law (1.22) is shown on Figure 1.2. The dynamic order of the adaptive controller is three due to the identifier module which contains two first-order filters and the first-order gradient-type update law. The stability analysis of estimation-based adaptive schemes is intricate. For gradient-type schemes, this analysis is conclusive only in the case of normalized update laws. One of its key results establishes the boundedness properties of the identifier module without any restrictions on $x_f(t)$ and hence on $u(t)$. This means that the identifier module can be connected not only with the pole-placement controller used in this example, but also with a wide variety of other controllers.

1.2 Adaptive Nonlinear Control

1.2.1 A nonlinear challenge

To introduce the topic of this book, let us apply the estimation-based approach to the adaptive control problem for the nonlinear plant

$$\dot{x} = u + \theta x^2. \tag{1.23}$$

We first select a certainty equivalence control law

$$u = -px - \hat{\theta} x^2, \tag{1.24}$$

which, if $\hat{\theta}$ were exact, $\hat{\theta} \equiv \theta$, would result in the closed-loop system $\dot{x} = -px$, as in the preceding linear example.

The construction of an identifier is analogous to (1.16)–(1.20). The resulting estimation error equation is the same, except for $x_f(t)$, which is now the filtered version of $x^2(t)$:

$$e = \tilde{\theta} x_f, \quad x_f = \frac{1}{s+1} x^2. \tag{1.25}$$

We proceed with the normalized update law (1.22) and, upon the substitution of $\dot{\hat{\theta}} = -\dot{\tilde{\theta}}$ and $e = \tilde{\theta} x_f$, we obtain

$$\dot{\tilde{\theta}} = -\gamma \frac{x_f^2}{1 + x_f^2} \tilde{\theta}. \tag{1.26}$$

This differential equation is linear. It clearly shows that the parameter error $\tilde{\theta}(t)$ cannot converge to zero faster than exponentially. Let us consider the most favorable case where

$$\tilde{\theta}(t) = e^{-\gamma t}\tilde{\theta}(0). \tag{1.27}$$

This rate would have been satisfactory for the certainty equivalence control (1.15) of the linear plant (1.1). Is this so with the control (1.24) of the nonlinear plant (1.23)? The resulting nonlinear closed-loop system is

$$\dot{x} = -x + \tilde{\theta}x^2, \tag{1.28}$$

where, for simplicity, $p = 1$. The substitution of (1.27) with $\gamma = 1$ into (1.28) yields the equation

$$\dot{x} = -x + x^2 e^{-t}\tilde{\theta}(0), \tag{1.29}$$

whose explicit solution is

$$x(t) = \frac{2x(0)}{x(0)\tilde{\theta}(0)e^{-t} + [2 - x(0)\tilde{\theta}(0)]e^{t}}. \tag{1.30}$$

It is easy to see that if $x(0)\tilde{\theta}(0) < 2$, then $x(t)$ will converge to zero as $t \to \infty$. However, if $x(0)\tilde{\theta}(0) > 2$, then at the time

$$t_{\text{esc}} = \frac{1}{2}\ln\frac{x(0)\tilde{\theta}(0)}{x(0)\tilde{\theta}(0) - 2} \tag{1.31}$$

the difference of the two exponential terms in the denominator becomes zero, that is,

$$|x(t)| \to \infty \text{ as } t \to t_{\text{esc}}. \tag{1.32}$$

The adaptive closed-loop system is not only unstable, but worse than that: Its state $x(t)$ escapes to infinity in finite time t_{esc}. The escape time becomes shorter as the difference $x(0)\tilde{\theta}(0) - 2$ grows, that is, as the initial conditions become larger.

This simple example clearly demonstrates why a traditional estimation-based design cannot be applied to nonlinear systems. Normalized update laws are too slow in providing the estimates $\hat{\theta}(t)$ with which certainty equivalence control would be able to prevent catastrophic forms of instability. We need either stronger controllers which will be able to achieve stabilization with standard identifiers, or faster identifiers, or a combination of both. Lyapunov-based designs in Chapters 3 and 4 provide faster identifiers, while in modular designs in Chapters 5 and 6 we construct stronger controllers.

1.2.2 A structural obstacle

If the estimation-based approach failed because of the slowness of normalized estimation, will a Lyapunov-based design be more successful? Let us start with the same Lyapunov function (1.9) as for the linear plant (1.1). Its derivative for the nonlinear plant (1.28) is

$$\dot{V} = x(u + \theta x^2) + (\hat{\theta} - \theta)\dot{\hat{\theta}}. \tag{1.33}$$

The requirement $\dot{V} \leq -px^2$ imposes the following condition on the choice of an update law for $\hat{\theta}$ and a control law for u:

$$xu + \hat{\theta}\dot{\hat{\theta}} + \theta\left(x^3 - \dot{\hat{\theta}}\right) \leq -px^2. \tag{1.34}$$

To eliminate the unknown θ, the update law must be

$$\dot{\hat{\theta}} = x^3, \tag{1.35}$$

so that (1.34) reduces to

$$xu + x^3\hat{\theta} \leq -px^2. \tag{1.36}$$

A control law which satisfies this condition is the certainty equivalence controller

$$u = -px - \hat{\theta}x^2. \tag{1.37}$$

Since the update law (1.35) and the control law (1.37) yield $\dot{V} = -px^2$, we achieve the same stability properties and state regulation as in the adaptive control of the linear plant (1.1). Comparing this result with the estimation-based design, we observe that the control laws (1.24) and (1.37) are the same, while the Lyapunov update law (1.35) is much faster than the normalized gradient update law (1.26).

It would appear that a Lyapunov approach to adaptive nonlinear control is more promising than the estimation-based approach. However, in the linear case the Lyapunov-based design has been restricted to plants with transfer functions of relative degree one and two. In the nonlinear state-feedback case, this structural restriction is translated into the "level of uncertainty," that is, the number of integrators between the control input and the unknown parameter. In the plant (1.23) the level of uncertainty is zero. The uncertainty and the control are "matched," because they appear in the same equation. In the plant

$$\begin{aligned} \dot{x}_1 &= x_2 + \theta x_1^2 \\ \dot{x}_2 &= u \end{aligned} \tag{1.38}$$

the level of uncertainty is one, which corresponds to the so-called "extended-matching" case. As we shall see, an extension of the Lyapunov design to this case is relatively straightforward, because $\dot{\hat{\theta}}$, which appears in the control law, can be substituted from the update law. For the system with uncertainty level two

$$\begin{aligned} \dot{x}_1 &= x_2 + \theta x_1^2 \\ \dot{x}_2 &= x_3 \\ \dot{x}_3 &= u \end{aligned} \qquad (1.39)$$

this is no longer so, because in this case the control law would have to include $\ddot{\hat{\theta}}$, which is not available. Backstepping designs developed in Chapter 2, 3, and 4 remove this structural obstacle and allow the Lyapunov-based designs to be applied to wide classes of uncertain nonlinear systems.

1.2.3 Early results

Interest in adaptive control of nonlinear systems was stimulated by major advances in the differential-geometric theory of nonlinear feedback control in the mid-1980's. A thorough treatment of this theory was given by Isidori in his seminal book [53] which unified a decade of results by many researchers. Particularly popular were the results on "feedback linearization," that is, the state and feedback transformation of nonlinear systems into linear ones [27, 48, 56, 117, 118]. This methodology helped convert many previously intractable nonlinear problems into much simpler problems solvable by familiar linear methods.

It soon became clear, however, that along with their many advantages, the nonlinear geometric methods have some shortcomings. One of them is their inability to handle the presence of unknown parameters. This motivated the first series of adaptive nonlinear control schemes. They were all restricted to systems satisfying the *matching condition*. Examples of such systems are rigid models of robotic manipulators. While the first robotic adaptive scheme by Craig [22] required measurement of joint accelerations, this impractical assumption was soon removed by Slotine and Li [169, 170], Middleton and Goodwin [130], and Ortega and Spong [148], among others. A more general treatment of adaptive nonlinear regulation under the matching condition was given by Taylor, Kokotović, Marino and Kanellakopoulos [186], including unmodeled dynamics which violated the matching condition.

The matching condition was relaxed to the *extended matching* condition by Kanellakopoulos, Kokotović and Marino [65] and Campion and Bastin [8, 15]. For a period, the extended matching condition was the frontier which could not be crossed by Lyapunov-based designs. This redirected researchers to estimation-based designs. Nam and Arapostathis [141] and Sastry and Isidori [166] combined feedback linearization with adaptation techniques from adaptive linear control. However, to achieve global stability, these schemes required that the nonlinearities be restricted by linear growth conditions. Similar restrictions on system nonlinearities were imposed by Kanellakopoulos, Kokotović, and Middleton [67, 68] and Teel, Kadiyala, Kokotović, and Sastry [190].

The only nonlinear estimation-based results which went beyond the linear growth constraints were obtained by Pomet and Praly [152, 153, 154], who used Lyapunov functions to characterize relationships between nonlinear growth constraints and controller stabilizing properties. In the absence of matching conditions, their schemes still involved some growth restrictions but were able to handle the benchmark third-order example (1.39).

The state-of-the-art of adaptive control, including adaptive nonlinear control, was reviewed in the 1990 Grainger lectures [86]. One of these lectures presented the result of Kanellakopoulos, Kokotović, and Morse [69, 87], which finally broke the extended matching barrier. This was achieved with a new recursive design procedure called *adaptive backstepping*. Adaptive backstepping, developed by Ioannis Kanellakopoulos [63] in collaboration with Petar Kokotović and Steve Morse, emerged as a confluence of the adaptive estimation idea, on one side, and, on the other side, nonlinear control ideas expressed in works of Tsinias [193], Byrnes and Isidori [12], Sontag and Sussmann [175], Kokotović and Sussmann [85], and Saberi, Kokotović, and Sussmann [163]. Adaptive backstepping was also strongly influenced by the properties of an early adaptive scheme by Feuer and Morse [36], which, although designed for linear systems, preserved global stability under output feedback for a class of output nonlinearities, as shown by Kanellakopoulos, Kokotović, and Morse [70, 71]. Adaptive backstepping influenced further developments in adaptive nonlinear control. Marino and Tomei [122, 123, 124] combined it with their filtered transformations [120, 121] to solve the adaptive output-feedback problem for a class of nonlinear systems that has not since been enlarged. Adaptive backstepping also stimulated efforts to reduce its overparametrization. A partial reduction was achieved by Jiang and Praly [59]. With the invention of *tuning functions*, Miroslav Krstić [92, 94] introduced a new design which completely removed the overparametrization.

1.3 Preview of the Main Topics

1.3.1 Classes of nonlinear systems

The main topic of this book is the design of feedback controllers for nonlinear systems with unknown constant parameters. The most important design specification is to achieve asymptotic tracking of a known reference trajectory with the strongest possible form of stability. Another key requirement is that the designed controller should provide effective means for shaping the transient performance and thus allow different performance-robustness trade-offs.

The largest classes of nonlinear systems for which the stated design problem is solvable with either state-feedback or output-feedback controllers are not known at this time. The largest classes for which solutions have been obtained are those considered in this book.

State feedback solutions are given for the so-called class of "parametric pure-feedback systems." They are first presented for the subclass of "parametric strict-feedback systems," for which the achieved stability and tracking properties are global. By analogy with linear systems, strict-feedback systems are also called "triangular."

Output-feedback solutions are restricted to a narrower class of minimum phase systems in which the nonlinearities depend only on the output variable.

The class of pure-feedback systems with unknown parameters is well represented by the third order system

$$\begin{aligned} \dot{x}_1 &= x_2 + \varphi_1^T(x_1, x_2)\theta \\ \dot{x}_2 &= x_3 + \varphi_2^T(x_1, x_2, x_3)\theta \\ \dot{x}_3 &= u + \varphi_3^T(x_1, x_2, x_3)\theta, \end{aligned} \quad (1.40)$$

where the $p \times 1$ vector θ is constant and unknown. (In a more general case, the terms x_2, x_3, and u can be multiplied by unknown constant parameters provided that the signs of these parameters are known.) Apart from the requirement that the dependence of the right-hand side of (1.40) on θ be linear, or, to be precise, affine, pure-feedback systems are characterized by the structure of the known nonlinearities φ_1, φ_2, and φ_3. The function φ_1 must not depend on x_3, and a further implicit function restriction is imposed on the dependence of φ_1 on x_2, and of φ_2 on x_3. This restriction is automatically satisfied if φ_1 does not depend on x_2 and φ_2 does not depend on x_3, that is, if we have $\varphi_1(x_1)$ and $\varphi_2(x_1, x_2)$. In this "strict-feedback" case the results are global.

1.3.2 Adaptive backstepping and tuning functions

When the parameter vector θ is known, the pure-feedback restriction essentially amounts to feedback linearizability. However, even when achievable, the goal of feedback linearization is not pursued in this book, because it often leads to cancellation of useful nonlinearities. *Backstepping designs* are more flexible and do not force the designed system to appear linear. They can avoid cancellations of useful nonlinearities and often introduce additional nonlinear terms to improve transient performance.

The idea of backstepping is to design a controller for (1.40) recursively by considering some of the state variables as "virtual controls" and designing for them intermediate control laws. In (1.40) the first virtual control is x_2. It is used to stabilize the first equation as a separate system. Since θ is unknown, this task is solved with an adaptive controller consisting of the control law $\alpha_1(x_1)$ and the update law $\dot{\hat{\theta}} = \tau(x_1)$, as in the Lyapunov-based design in

1.3 Preview of the Main Topics

Section 1.2.2. In the next step the state x_3 is the "virtual control" which is used to stabilize the subsystem consisting of the first two equations of (1.40). This is again an adaptive control task, and a new update law is to be designed. However, an update law $\dot{\hat{\theta}} = \tau(x_1)$ has already been designed in the first step and this does not seem to allow any freedom to proceed further. Chapters 3 and 4 provide two different cuts of this Gordian knot: *adaptive backstepping* and *tuning functions*.

Adaptive backstepping treats the parameter θ in the second equation of (1.40) as a new parameter and assigns to it a new estimate with a new update law. As a result, there are several estimates for the same parameter. This *overparametrization* is avoided in Chapter 4 by considering that in the first step $\dot{\hat{\theta}} = \tau(x_1)$ is not an update law but only a function $\tau(x_1)$. This "tuning function" is used in subsequent recursive steps and the discrepancy $\dot{\hat{\theta}} - \tau(x_1)$ is compensated with additional terms in the controller. Whenever the second derivative $\ddot{\hat{\theta}}$ would appear, it is replaced by the analytic expression for the first derivative of $\tau(x_1)$.

Both adaptive backstepping and tuning functions achieve the goals of stabilization and tracking. The proof of these properties is a direct consequence of the recursive procedure during which a Lyapunov function is constructed for the entire system, including the parameter estimates. For strict-feedback systems, this Lyapunov function provides the proof of global uniform stability and, if $x_1(t)$ is required to follow a trajectory $x_{1\,\text{ref}}(t)$, also the proof of asymptotic tracking $x_1(t) - x_{1\,\text{ref}}(t) \to 0$. For pure-feedback systems the solution is not global, but the same Lyapunov function is used for estimating the region of attraction within the feasible set.

The tuning functions approach is an advanced form of adaptive backstepping. It has the advantage that the dynamic order of the adaptive controller is minimal. The dimension of the set to which the states and parameter estimates converge is also minimal.

Both adaptive backstepping and tuning functions have crossed the "extended matching" barrier which blocked the traditional Lyapunov-based design. They have achieved this by designing controllers "more intelligent" and "stronger" than certainty equivalence controllers.

The departure from certainty equivalence designs is epitomized by the "adaptive control Lyapunov function" (aclf) in Chapter 4. This function $V(x,\theta)$ is a control Lyapunov function (clf) in the sense of Artstein [4] and Sontag [171], but for a modified system in which the term $\frac{\partial V}{\partial \theta}$ is added to the unknown parameter θ. A certainty equivalence controller for this modified system is used for adaptive control of the original system. It is equipped with terms which counteract the effects of parameter estimate transients on the original system.

1.3.3 Modular designs

Adaptive backstepping and tuning functions procedures are Lyapunov-based designs. These recursive procedures have removed the critical relative degree (i.e., extended matching) restriction of the traditional Lyapunov design and provided the simplest proofs of the strongest stability properties. Along with their advantages, they have certain drawbacks. One of them is that they do not offer freedom of choice of parameter update laws. For systems with many unknown parameters, a further drawback of adaptive backstepping is that the dynamic order of its overparametrized controller is high. On the other hand, the order of the tuning functions controller is minimal, but for high-order systems its nonlinear expressions become increasingly complex. The main source of complexity is the built-in interaction between the identifier and the control law.

Both of these drawbacks are removed by modular designs in Chapters 5 and 6 which are inspired by traditional estimation-based designs. In adaptive linear control the estimation-based designs achieve a significant level of *modularity* of the controller-identifier pair: Any stabilizing controller can be combined with any identifier. Parameter update laws can be of either gradient or least-squares type. The controller module is capable of stabilizing the plant when all the parameters are known. This is its *certainty equivalence* property. The identifier module, in turn, guarantees certain boundedness properties independently of the controller module. The modularity of the estimation-based designs makes them much more versatile than the Lyapunov-based designs.

Chapters 5 and 6 present the development of a new modular approach to adaptive nonlinear control. They first show that a major obstacle to earlier attempts to apply estimation-based designs to nonlinear systems was the weakness of the certainty equivalence controllers. Such controllers cannot achieve any controller-identifier separation without severely restricting the system nonlinearities.

To overcome this weakness of certainty equivalence controllers, a new controller is developed with strong parametric robustness properties: *It achieves boundedness without adaptation*. Furthermore, the new controller, called the *ISS-controller*, guarantees boundedness not only in the presence of constant parameter errors, but also in the presence of time-varying parameter estimates. These input-to-state stability (ISS) properties make the ISS-controller suitable for modular adaptive nonlinear designs. In addition to the ISS-controller, another type of controller is also designed, called the *SG-controller*, with a small gain (SG) property analogous to that used in adaptive linear control. This weaker controller reduces some of the nonlinear complexity of the ISS-controller, but its performance is not as high.

The adaptive schemes in Chapter 5 employ two "observer-like" *passive identifiers* which are still restricted to unnormalized gradient-type update laws.

The first identifier is based on the open-loop plant model, while the second is based on the the closed-loop model.

Adaptive schemes designed in Chapter 6 achieve complete modularity. They can employ identifiers with any standard gradient or least-squares update law. This is accomplished by extending the well-known linear swapping technique to nonlinear systems. This technique converts dynamic parametric models into a static form to which the standard update laws are applicable.

1.3.4 Output-feedback designs

In Part I of the book it is assumed that the full state of the system is available for measurement. Part II addresses a class of problems in which only the plant output is measured. There are many difficulties that stand in the way to extending the state-feedback results to output-feedback problems.

A common approach to output-feedback control of linear systems is to use the certainty equivalence principle and combine state-feedback controllers with state estimators (observers). However, the separation principle does not hold for nonlinear systems, where even an exponentially decaying state estimation error can lead to instability and even finite escape time.

The results in Part II solve the adaptive control problem for a class of nonlinear systems in the *output-feedback form*. It seems that this class of systems is not much narrower than as yet unknown class of nonlinear systems which are globally stabilizable by output feedback.

Backstepping design of output-feedback controllers in Chapter 7 is performed on systems enlarged by filters, and the filter states are used for feedback. Chapter 7 is introductory and can be read immediately after Chapters 2 and 3. After this introductory chapter, two groups of adaptive designs are developed as in the case of full-state feedback: the tuning functions designs in Chapter 8 and the modular designs in Chapter 9. These designs employ the *K-filters* originally introduced for adaptive linear observers by Kreisselmeier [91] and modified for nonlinear systems in [72, 80, 95, 103]. For completeness, Chapters 8 and 9 also present new designs with the *MT-filters* introduced by Marino and Tomei [121, 123].

The output-feedback modular designs in Chapter 9 result in separation of three design modules: the control law, the identifier, and the state estimator. Both the schemes with passive identifiers and with swapping identifiers use parametric models which require only output measurement.

1.3.5 Linear systems

For linear systems with unknown parameters the output-feedback adaptive control problem was solved in the 1970's and 1980's with several well-known adaptive schemes. A question of interest here is whether any of the traditional schemes reemerge from the new output-feedback designs in Chapters 8 and 9. This question is answered in Chapter 10, where the tuning functions design is applied to linear minimum phase systems. The new scheme is radically different from the traditional schemes in that its controller is nonlinear and has stronger stability properties. With the new controller, boundedness can be preserved even when the adaptation is switched off. This means that the adaptive part of the controller does not have to carry the burden of preserving boundedness, but only of achieving asymptotic tracking. The dynamic order of the new adaptive scheme is no higher (and in most cases is significantly lower) than in traditional adaptive schemes. Thus, even for linear systems, this book offers entirely new adaptive control solutions. Analytical bounds and simulation studies show that the new adaptive designs achieve better transient performance than the traditional schemes.

1.3.6 Transient performance

The tracking error of most adaptive control schemes converges to zero, that is, they satisfy the asymptotic performance requirement. In applications, however, the system's transient performance is often more important.

Analytical quantification and systematic improvement of transient performance have been open problems in adaptive control. For traditional adaptive linear controllers there are virtually no results which allow the designer to compute bounds on the transient behavior.

One of the novelties of this book is that it addresses the problem of transient performance in an analytical framework. Computable transient performance bounds are derived for most of the designed controllers. The dependence of these bounds on design parameters and initial conditions is clearly displayed and serves as a guide for transient performance improvement.

It is well known that during the initial period, while the parameter estimates are poor, the states of traditional adaptive schemes exhibit large initial swings. The designs in this book improve the transients following two different routes. The first route is a rapid exchange of information between the controller and the identifier. This is the case with the tuning functions controller in Chapters 4 and 8, which takes into account the effect of the parameter estimation transients by incorporating, as one of its parts, the parameter update law. The second route uses a stronger controller which suppresses the destabilizing effects of poor parameter estimates. This is how transient performance improvement is achieved in modular designs in Chapters 5, 6, and 9.

The performance analysis for the tuning functions scheme shows that the adaptive controller outperforms its nonadaptive counterpart. As discussed in Chapter 10, this result is new even for adaptive control of linear systems.

1.3.7 Caveat: robustness

Robust adaptation is a current topic of adaptive control research motivated by the practical question raised in the early 1980's: What happens if some assumptions in the stability proofs are shattered by unmodeled dynamics and disturbances? Major advances have been made in robust redesigns of adaptive schemes for linear systems [2, 50, 51]. They employ various modifications of parameter update laws.

Adaptive nonlinear control theory is new and is still focused on the development of the basic schemes. Their robust modifications are yet to be developed and are not discussed in this book. Based on the results of adaptive linear control, particularly on those presented in the recent book by Ioannou and Sun [51], we expect that similar robust modifications will be effective for adaptive nonlinear control schemes developed in this book.

1.4 Notation

For vectors we use

$$|x|_P \triangleq \left(x^\mathrm{T} P x\right)^{1/2}$$

to denote the weighted Euclidean norm of x. For matrices,

$$|X|_\mathcal{F} \triangleq \left(\mathrm{tr}\{X^\mathrm{T} X\}\right)^{1/2} = \left(\mathrm{tr}\{X X^\mathrm{T}\}\right)^{1/2}$$

denotes the Frobenius, and $|X|_2$ denotes the induced 2-norm of X.

The $\mathcal{L}_\infty$, $\mathcal{L}_2$ and $\mathcal{L}_1$ norms for signals are denoted by $\|\cdot\|_\infty$, $\|\cdot\|_2$, and $\|\cdot\|_1$, respectively.

The spaces of all signals which are globally bounded, locally bounded, and square-integrable on $[0, t_f)$, $t_f > 0$, are denoted by $\mathcal{L}_\infty[0, t_f)$, $\mathcal{L}_{\infty e}[0, t_f)$, and $\mathcal{L}_2[0, t_f)$, respectively. By saying that a signal belongs to $\mathcal{L}_\infty[0, t_f)$ or to $\mathcal{L}_2[0, t_f)$ we mean that the corresponding bound is independent of t_f.

By referring to a matrix $A(t)$ as exponentially stable we mean that the corresponding LTV system $\dot{x} = A(t)x$ is exponentially stable.

The ith coordinate vector in $\mathbb{R}^n$ is denoted by e_i. The same symbol e_i will be used in spaces of different dimension, and the dimension of e_i will be clear from the context. Symbols $X_{(i)}$ and X_j, respectively, denote the ith row and the jth column of matrix X.

Notes and References

Basic texts on adaptive linear control are Goodwin and Sin [44] and Sastry and Bodson [165] for estimation-based approaches, and Narendra and Annaswamy [142] for the model-reference approach. An applications-oriented text is Åström and Wittenmark [5]. The most recent in-depth treatment of adaptive linear control, including robustness analysis and redesign, is Ioannou and Sun [51]. Familiarity with parts of these references would be helpful, but is not required for the understanding of this book. Background needed in nonlinear stability is at the level of a few introductory chapters in Khalil [83] or Vidyasagar [195]. It is summarized in Appendices A–D. The book makes use of special forms of nonlinear systems. The necessary and sufficient conditions for the existence of these forms are given in Appendix G and are based on the differential geometric theory given in Isidori [53] and Nijmeijer and van der Schaft [144].

Part I

State Feedback

Chapter 2

Design Tools for Stabilization

Recursive designs in this book are composed of simple basic steps. They are referred to as "backstepping designs" because they "step back" toward the control input starting with the scalar equation which is separated from it by the largest number of integrations.

After a brief review of Lyapunov stability, this chapter introduces basic backstepping tools, first for systems without uncertainty, and then for those with uncertainty. The most elementary integrator backstepping tool evolves into a recursive design for the class of "strict feedback" systems. An extension to block backstepping is made via a result for stabilization of cascade systems.

In the absence of uncertainties, backstepping can be used to force a nonlinear system to behave like a linear system in a new set of coordinates. However, this and other forms of "feedback linearization" require cancellation of nonlinearities, even those which are helpful for stabilization and tracking. A major advantage of backstepping is that it has the flexibility to avoid cancellations of useful nonlinearities and pursue the objectives of stabilization and tracking, rather than that of linearization.

The task of nonlinear design is much more challenging in the presence of uncertainty. When the uncertainty is *matched*, that is, when it appears in the same equation as the control, the design with "nonlinear damping" guarantees boundedness even when the bound on the uncertainty is not known. A more advanced backstepping tool is used to achieve boundedness in the case when the uncertainty is not matched, that is, when it appears before the control input. Backstepping with such a general form of bounded uncertainties with unknown bounds is a key tool used to achieve boundedness without adaptation. When the uncertainty is in the form of constant but unknown parameters, then a more suitable form of backstepping is adaptive backstepping, developed in Chapter 3.

2.1 Stability

2.1.1 Main stability theorems

For all control systems, and for adaptive control systems in particular, stability is the primary requirement. Stability concepts that are widely used in control theory are *Lyapunov stability* and *input-output stability*. Tools for analysis of both of these types of stability are summarized in Appendices A through D. The initial chapters of this book deal mostly with Lyapunov stability, which we now briefly review.[1]

Lyapunov Stability. To begin with, we remind the reader that Lyapunov stability, asymptotic stability, uniform stability, uniform asymptotic stability, etc., are properties not of a dynamic system as a whole, but rather of its individual solutions. Consider the time-varying system

$$\dot{x} = f(x, t), \qquad (2.1)$$

where $x \in \mathbb{R}^n$, and $f : \mathbb{R}^n \times \mathbb{R}_+ \to \mathbb{R}^n$ is piecewise continuous in t and locally Lipschitz in x. The solution of (2.1) which starts from the point x_0 at time $t_0 \geq 0$ is denoted as $x(t; x_0, t_0)$ with $x(t_0; x_0, t_0) = x_0$. Lyapunov stability concepts describe continuity properties of $x(t; x_0, t_0)$ with respect to x_0. If the initial condition x_0 is perturbed to $\tilde{x}_0$, then, for stability, the resulting perturbed solution $x(t; \tilde{x}_0, t_0)$ is required to stay close to $x(t; x_0, t_0)$ for all $t \geq t_0$. In addition, for asymptotic stability, the error $x(t; \tilde{x}_0, t_0) - x(t; x_0, t_0)$ is required to vanish as $t \to \infty$. So, the solution $x(t; x_0, t_0)$ of (2.1) is

- *bounded*, if there exists a constant $B(x_0, t_0) > 0$ such that

$$|x(t; x_0, t_0)| < B(x_0, t_0), \quad \forall t \geq t_0; \qquad (2.2)$$

- *stable*, if for each $\varepsilon > 0$ there exists a $\delta(\varepsilon, t_0) > 0$ such that

$$|\tilde{x}_0 - x_0| < \delta \implies |x(t; \tilde{x}_0, t_0) - x(t; x_0, t_0)| < \varepsilon, \quad \forall t \geq t_0; \qquad (2.3)$$

- *attractive*, if there exist an $r(t_0) > 0$ and, for each $\varepsilon > 0$, a $T(\varepsilon, t_0) > 0$ such that

$$|\tilde{x}_0 - x_0| < r \implies |x(t; \tilde{x}_0, t_0) - x(t; x_0, t_0)| < \varepsilon, \quad \forall t \geq t_0 + T; \qquad (2.4)$$

- *asymptotically stable*, if it is stable and attractive; and

- *unstable*, if it is not stable.

---
[1] For a detailed treatment of the subject, the reader is referred to the book by Khalil [81].

2.1 Stability

The stability properties of $x(t;x_0,t_0)$ in general depend on the initial time t_0. For different t_0, different values of $B(x_0,t_0), \delta(\varepsilon,t_0), r(t_0)$, and $T(\varepsilon,t_0)$ may be needed to satisfy (2.2), (2.3) and (2.4). When these constants are independent of t_0, the corresponding properties are *uniform*.[2] For adaptive systems, *uniform stability* is more desirable than just stability. Even more desirable is *uniform asymptotic stability*, often shortened to UAS. The solution $x(t;x_0,t_0)$ is UAS if it is *uniformly stable and uniformly attractive*, that is, if $\delta(\varepsilon,t_0) = \delta(\varepsilon)$, $r(t_0) = r$, and $T(\varepsilon,t_0) = T(\varepsilon)$ do not depend on t_0.

Some solutions of a given system may be stable and others unstable. In particular, (2.1) may have stable and unstable *equilibria*, that is, constant solutions $x(t;x_e,t_0) \equiv x_e$ satisfying $f(x_e,t) \equiv 0$. If an equilibrium x_e is asymptotically stable, then it has a *region of attraction* — a set Ω of initial states x_0 such that $x(t;x_0,t_0) \to x_e$ as $t \to \infty$ for all $x_0 \in \Omega$.[3] In this book, the stability properties for which an estimate of the region of attraction is given are referred to as *regional*. Otherwise they are called *local*. When the region of attraction is the whole space $\mathbb{R}^n$, then the stability properties are *global*.

Any equilibrium under investigation can be translated to the origin by redefining the state x as $z = x - x_e$. Such a translation $z = x - x(t;x_0,t_0)$ can be defined for any solution $x(t;x_0,t_0)$ so that the solution under investigation can always be considered to be an equilibrium at the origin with a corresponding redefinition of $f(x,t)$ into $\bar{f}(z,t)$ such that $\bar{f}(0,t) \equiv 0$, namely:

$$\dot{z} = f(z + x(t;x_0,t_0),t) - f(x(t;x_0,t_0),t) \triangleq \bar{f}(z,t). \qquad (2.5)$$

Therefore, there is no loss of generality in standardizing the stability results for the zero solution $z(t;0,t_0) \equiv 0$. In adaptive tracking problems, this zero solution is particularly meaningful when the state z represents the tracking error and its derivatives.

To be of practical interest, stability conditions must not require that we explicitly solve (2.1). The direct method of Lyapunov aims at determining the stability properties of $x(t;x_0,t_0)$ from the properties of $f(x,t)$ and its relationship with a positive definite function $V(x,t)$. For global results, this function must be radially unbounded, that is, $V(x,t) \to \infty$ as $|x| \to \infty$ uniformly in t. For simplicity, we will assume that the translation to the origin has been performed, that is, $f(0,t) \equiv 0$, and thus the solution under investigation is $x \equiv 0$.

Uniform asymptotic stability is a desirable property, because systems that possess it can deal better with perturbations and disturbances. We shall see that, in general, adaptive designs achieve less than uniform asymptotic stability. However, they achieve more than uniform stability because they force the tracking error to converge to zero. This key property is referred to as *regulation*

---

[2]Clearly, all properties are uniform if the system is time-invariant: $\dot{x} = f(x)$.
[3]When x_e is only stable, then the solutions starting in Ω remain close to x_e in the sense of (2.3).

when the reference signal is constant, and *tracking* when it is a time-varying signal. For convergence analysis, a powerful tool is the following theorem due to LaSalle [110] and Yoshizawa [201]:

Theorem 2.1 (LaSalle-Yoshizawa) *Let $x = 0$ be an equilibrium point of (2.1) and suppose f is locally Lipschitz in x uniformly in t. Let $V : \mathbb{R}^n \to \mathbb{R}_+$ be a continuously differentiable, positive definite and radially unbounded function $V(x)$ such that*

$$\dot{V} = \frac{\partial V}{\partial x}(x)f(x,t) \leq -W(x) \leq 0, \quad \forall t \geq 0, \, \forall x \in \mathbb{R}^n, \qquad (2.6)$$

where W is a continuous function. Then, all solutions of (2.1) are globally uniformly bounded and satisfy

$$\lim_{t \to \infty} W(x(t)) = 0. \qquad (2.7)$$

In addition, if $W(x)$ is positive definite, then the equilibrium $x = 0$ is globally uniformly asymptotically stable (GUAS).

Because of its importance, a more general version of this theorem and its proof are included in Appendix A (Theorem A.8), along with a frequently used technical lemma due to Barbalat [155] (Lemma A.6). The LaSalle-Yoshizawa theorem is applicable to time-varying systems and allows us to establish convergence to the set E where $W(x) = 0$. For most of our design tasks, we will construct $V(x)$ such that the set E consists solely of the trajectories which meet the tracking objective, that is, along which the tracking error is zero.

For the regulation task, the designed system is usually time-invariant,

$$\dot{x} = f(x), \qquad (2.8)$$

in which case we are interested in its *invariant sets*. A set M is called an invariant set of (2.8) if any solution $x(t)$ that belongs to M at some time instant t_1 must belong to M for all future and past time:

$$x(t_1) \in M \implies x(t) \in M, \quad \forall t \in \mathbb{R}. \qquad (2.9)$$

A set Ω is *positively invariant* if this is true for all future time only:

$$x(t_1) \in \Omega \implies x(t) \in \Omega, \quad \forall t \geq t_1. \qquad (2.10)$$

Can we guarantee convergence to a desired invariant set? A rewarding answer to this question is provided by LaSalle's Invariance Theorem and its asymptotic stability corollary:

2.1 STABILITY

Theorem 2.2 (LaSalle) *Let Ω be a positively invariant set of (2.8). Let $V : \Omega \to \mathbb{R}_+$ be a continuously differentiable function $V(x)$ such that $\dot{V}(x) \leq 0$, $\forall x \in \Omega$. Let $E = \{x \in \Omega \mid \dot{V}(x) = 0\}$, and let M be the largest invariant set contained in E. Then, every bounded solution $x(t)$ starting in Ω converges to M as $t \to \infty$.*

Corollary 2.3 (Asymptotic Stability) *Let $x = 0$ be the only equilibrium of (2.8). Let $V : \mathbb{R}^n \to \mathbb{R}_+$ be a continuously differentiable, positive definite, radially unbounded function $V(x)$ such that $\dot{V}(x) \leq 0$, $\forall x \in \mathbb{R}^n$. Let $E = \{x \in \mathbb{R}^n \mid \dot{V}(x) = 0\}$, and suppose that no solution other than $x(t) \equiv 0$ can stay forever in E. Then the origin is globally asymptotically stable (GAS).*

These invariance results will motivate us to closely examine the invariant subsets of E. As we shall see, the convergence properties of the designed system are stronger if the dimension of M is lower. In the most favorable case of asymptotic stability, the largest invariant subset M of E is just the origin $x = 0$. Our aim will thus be to render the dimension of M as low as possible.

Input-to-State Stability. Another stability concept which is used throughout the book is that of input-to-state stability (ISS), introduced by Sontag [173]. The system

$$\dot{x} = f(x, u) \tag{2.11}$$

is said to be *input-to-state stable (ISS)* if for any $x(0)$ and for any input $u(\cdot)$ continuous and bounded on $[0, \infty)$ the solution exists for all $t \geq 0$ and satisfies

$$|x(t)| \leq \beta(|x(0)|, t) + \gamma\left(\sup_{0 \leq \tau \leq t} |u(\tau)|\right), \quad \forall t \geq 0, \tag{2.12}$$

where $\beta(s, t)$ and $\gamma(s)$ are strictly increasing functions of $s \in \mathbb{R}_+$ with $\beta(0, t) = 0$, $\gamma(0) = 0$, while β is a decreasing function of t with $\lim_{t \to \infty} \beta(s, t) = 0$, $\forall s \in \mathbb{R}_+$.

This definition of input-to-state stability is appropriate for nonlinear systems since it explicitly incorporates the effect of the initial conditions $x(0)$: (2.12) shows that the norm of the state $x(t)$ depends not only on the input $u(\tau)$, but also includes an asymptotically decaying contribution from $x(0)$. A more extensive treatment of ISS is given in Appendix C.

2.1.2 Control Lyapunov functions (clf)

This book is about control *design*: Our objective is to create closed-loop systems with desirable stability properties, rather than analyze the properties of a given system. For this reason, we are interested in an extension of the Lyapunov function concept, called a *control Lyapunov function* (clf).

Suppose that our problem for the time-invariant system

$$\dot{x} = f(x, u), \quad x \in \mathbb{R}^n, \quad u \in \mathbb{R}, \quad f(0,0) = 0, \qquad (2.13)$$

is to design a feedback control law $\alpha(x)$ for the control variable u such that the equilibrium $x = 0$ of the closed-loop system

$$\dot{x} = f(x, \alpha(x)) \qquad (2.14)$$

is globally asymptotically stable. We can pick a function $V(x)$ as a Lyapunov candidate, and require that its derivative along the solutions of (2.14) satisfy $\dot{V}(x) \leq -W(x)$, where $W(x)$ is a positive definite function. We therefore need to find $\alpha(x)$ to guarantee that for all $x \in \mathbb{R}^n$

$$\frac{\partial V}{\partial x}(x) f(x, \alpha(x)) \leq -W(x). \qquad (2.15)$$

This is a difficult task. A stabilizing control law for (2.13) may exist but we may fail to satisfy (2.15) because of a poor choice of $V(x)$ and $W(x)$. A system for which a good choice of $V(x)$ and $W(x)$ exists is said to possess a clf. Let us make this notion precise.

Definition 2.4 *A smooth positive definite and radially unbounded function* $V : \mathbb{R}^n \to \mathbb{R}_+$ *is called a* control Lyapunov function (clf) *for (2.13) if*

$$\inf_{u \in \mathbb{R}} \left\{ \frac{\partial V}{\partial x}(x) f(x, u) \right\} < 0, \quad \forall x \neq 0. \qquad (2.16)$$

The clf concept of Artstein [4] and Sontag [172] is a generalization of Lyapunov design results by Jacobson [54] and Judjevic and Quinn [62]. Artstein [4] showed that (2.16) is not only necessary, but also sufficient for the existence of a control law satisfying (2.15), that is, the existence of a clf is equivalent to global asymptotic stabilizability.

For systems affine in the control,

$$\dot{x} = f(x) + g(x)u, \quad f(0) = 0, \qquad (2.17)$$

the clf inequality (2.15) becomes

$$\frac{\partial V}{\partial x} f(x) + \frac{\partial V}{\partial x} g(x)\alpha(x) \leq -W(x). \qquad (2.18)$$

If $V(x)$ is a clf for (2.17), then a particular stabilizing control law $\alpha(x)$, smooth for all $x \neq 0$, is given by Sontag's formula [171]

$$u = \alpha_s(x) = \begin{cases} -\dfrac{\frac{\partial V}{\partial x} f + \sqrt{\left(\frac{\partial V}{\partial x} f\right)^2 + \left(\frac{\partial V}{\partial x} g\right)^4}}{\frac{\partial V}{\partial x} g}, & \frac{\partial V}{\partial x} g \neq 0 \\ 0, & \frac{\partial V}{\partial x} g = 0. \end{cases} \qquad (2.19)$$

2.1 STABILITY

It should be noted that (2.18) can be satisfied only if

$$\frac{\partial V}{\partial x} g(x) = 0 \Rightarrow \frac{\partial V}{\partial x} f(x) < 0, \quad \forall x \neq 0, \tag{2.20}$$

and that in this case (2.19) results in

$$W(x) = \sqrt{\left(\frac{\partial V}{\partial x} f\right)^2 + \left(\frac{\partial V}{\partial x} g\right)^4} > 0, \quad \forall x \neq 0. \tag{2.21}$$

A further characterization of a stabilizing control law $\alpha(x)$ for (2.17) with a given clf V is that $\alpha(x)$ is continuous at $x = 0$ if and only if the clf satisfies the *small control property*: For each $\varepsilon > 0$ there is a $\delta(\varepsilon) > 0$ such that, if $x \neq 0$ satisfies $|x| < \delta$, then there is some u with $|u| < \varepsilon$ such that

$$\frac{\partial V}{\partial x} [f(x) + g(x)u] < 0. \tag{2.22}$$

The main deficiency of the clf concept as a design tool is that for most nonlinear systems a clf is not known. The task of finding an appropriate clf may be as complex as that of designing a stabilizing feedback law. For several important classes of nonlinear systems, we will solve these two tasks simultaneously using a *backstepping* procedure. To initiate this procedure we need to be able to find $V(x)$ and $\alpha(x)$ at least for scalar systems. Fortunately, for scalar systems, $V(x) = \frac{1}{2}x^2$ is always a reasonable clf and the inequality (2.18) is easy to satisfy. This is illustrated by an example which also issues a warning that some designs may lead to a waste of control effort.

Example 2.5 For the scalar system shown in Figure 2.1,

$$\dot{x} = \cos x - x^3 + u, \tag{2.23}$$

our task is to design a feedback control law which creates and globally stabilizes the equilibrium at $x = 0$. We will compare three different designs.

In a *feedback linearization* design, the control law

$$u = -\cos x + x^3 - x \tag{2.24}$$

cancels both nonlinearities ($\cos x$ and $-x^3$) and replaces them by $-x$ so that the resulting feedback system is linear: $\dot{x} = -x$. Taking

$$V(x) = \frac{1}{2}x^2 \tag{2.25}$$

as a clf for (2.23), we see that the control law (2.24) satisfies the requirement (2.18) with $W(x) = x^2$, that is, $\dot{V}(x) \leq -x^2$. However, there is an obvious irrationality of this control law: It cancels not only $\cos x$, but also $-x^3$. For

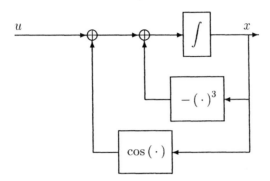

Figure 2.1: The block diagram of system (2.23).

stabilization at $x = 0$, the negative feedback term $-x^3$ is helpful, especially for large values of x. On the other hand, the presence of x^3 in the control law (2.24) is harmful: It leads to large magnitudes of u and may cause nonrobustness.

A more reasonable design is not to cancel $-x^3$. With $V(x) = \frac{1}{2}x^2$ as before, we take $W(x) = x^2 + x^4$, so that the control law satisfying (2.18) becomes

$$u = -\cos x - x \triangleq \alpha(x). \qquad (2.26)$$

In this case, the magnitude of u grows only linearly with $|x|$.

Finally, as our third control law we employ Sontag's formula (2.19). Since this formula is based on the assumption that $f(0) = 0$, we first cancel $\cos x$ by introducing $u = -\cos x + u_s$. We again use $V(x) = \frac{1}{2}x^2$ as our clf and evaluate $\alpha_s(x)$ from (2.19) with $f = -x^3$ and $g = 1$:

$$u_s = \alpha_s(x) = x^3 - x\sqrt{x^4 + 1}. \qquad (2.27)$$

A remarkable property of (2.27) is that $\alpha_s(x) \to 0$ as $|x| \to \infty$, which means that for large $|x|$ the control law for u reduces to the term $-\cos x$ required to place the equilibrium at $x = 0$. The rationale is clear: Except for the cancellation of $\cos x$, the control is inactive for large $|x|$ because then the internal nonlinear feedback $-x^3$ takes over and forces x towards zero. In this way the control effort is not wasted to achieve a property already present in the system. On the other hand, for small $|x|$ we have $\alpha_s \approx -x$, which is the same as in the previous two control laws. It is easy to check that $u = -\cos x + \alpha_s(x)$ satisfies (2.18) with $W(x) = x^2\sqrt{x^4 + 1}$. This control law is superior because it requires less control effort than the other two. $\diamond$

Example 2.6 The scalar system shown in Figure 2.2,

$$\dot{x} = x^3 + x^2 u, \qquad (2.28)$$

is of interest because it is smoothly stabilizable in spite of the singularity at $x = 0$. We proceed with $V(x) = \frac{1}{2}x^2$ and, because of the term x^3, we choose

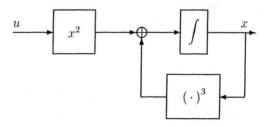

Figure 2.2: The block diagram of system (2.28).

$W(x) = c_1 x^4$ where $c_1 > 0$. Solving

$$\dot{V}(x, u) = x\left[x^3 + x^2 u\right] = -c_1 x^4 \tag{2.29}$$

for u, we obtain the control law

$$u = -(1 + c_1)x \triangleq \alpha(x), \tag{2.30}$$

which yields the globally asymptotically stable closed-loop system $\dot{x} = -c_1 x^3$. In this case, Sontag's formula yields

$$u = \alpha_s(x) = -x\left(1 + \sqrt{1 + x^4}\right), \tag{2.31}$$

which satisfies (2.18) with $W(x) = x^4 \sqrt{1 + x^4}$. Clearly, this control law requires more control effort than (2.30). ◇

For scalar systems the choice of a quadratic clf is obvious and the nonlinear design is straightforward: The harmful nonlinearities are cancelled by the control.

2.2 Backstepping

2.2.1 Integrator backstepping

The simplicity of scalar designs motivates us to use them as starting points of recursive designs for higher-order systems. Let us first construct clf's for second-order systems. We begin by augmenting the system (2.23) with an integrator:

$$\dot{x} = \cos x - x^3 + \xi \tag{2.32a}$$
$$\dot{\xi} = u. \tag{2.32b}$$

Let the design objective be the regulation of $x(t)$, that is, $x(t) \to 0$ as $t \to \infty$, for all $x(0)$, $\xi(0)$. Of course, $\xi(t)$ must remain bounded. From (2.32a), the only equilibrium with $x = 0$ is at $(x, \xi) = (0, -1)$. We will meet our design

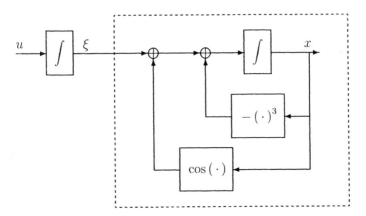

Figure 2.3: The block diagram of system (2.32).

objective by rendering this equilibrium GAS. In the block diagram in Figure 2.3 the scalar system (2.23) of Figure 2.1 appears in the dashed box. To construct a clf for (2.32) we will exploit the fact that a clf is known for its subsystem in the dashed box. Indeed, if ξ were the control input, then (2.32a) would be identical to (2.23), and the corresponding clf and control law would be $V(x) = \frac{1}{2}x^2$ and $\xi = -c_1 x - \cos x$ (cf. (2.26)). Of course ξ is just a state variable and not the control. Nevertheless, as its "desired value" we prescribe

$$\xi_{\text{des}} = -c_1 x - \cos x \triangleq \alpha(x). \tag{2.33}$$

Let z be the deviation of ξ from its desired value:

$$z = \xi - \xi_{\text{des}} = \xi - \alpha(x) = \xi + c_1 x + \cos x. \tag{2.34}$$

We call ξ a *virtual control*, and its desired value $\alpha(x)$ a *stabilizing function*. The variable z is the corresponding *error variable*. Now we rewrite the system (2.32) in the (x, z)-coordinates in which it takes on a more convenient form, as illustrated in Figures 2.4 and 2.5. Starting from (2.32) and Figure 2.3, we add and subtract the stabilizing function $\alpha(x)$ to the $\dot{x}_1$-equation as shown in Figure 2.4. Then we use $\alpha(x)$ as the feedback control inside the dashed box and "backstep" $-\alpha(x)$ through the integrator, as in Figure 2.5. In the new coordinates (x, z) the system is expressed as

$$\dot{x} = \cos x - x^3 + [\xi + c_1 x + \cos x] - c_1 x - \cos x = -c_1 x - x^3 + z \tag{2.35a}$$
$$\dot{z} = \dot{\xi} - \dot{\alpha} = \dot{\xi} + (c_1 - \sin x)\dot{x} = u + (c_1 - \sin x)\left(-c_1 x - x^3 + z\right). \tag{2.35b}$$

The first key feature of backstepping is that we don't use a differentiator to implement the time derivative $\dot{\alpha}$ in (2.35b); since $\alpha(x)$ is a known function, it is easy to compute its time derivative analytically as

$$\dot{\alpha} = \frac{\partial \alpha}{\partial x}\dot{x} = -(c_1 - \sin x)\left(-c_1 x - x^3 + z\right). \tag{2.36}$$

2.2 BACKSTEPPING

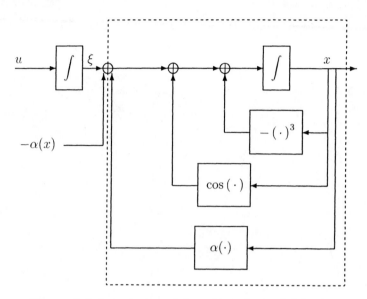

Figure 2.4: Introducing $\alpha(x)$ as the desired value for ξ.

We now need to select a clf V_a for the system (2.32). Let us try to construct it by augmenting $V(x)$ with a quadratic term in the error variable z:

$$V_a(x,\xi) = V(x) + \frac{1}{2}z^2 = \frac{1}{2}x^2 + \frac{1}{2}(\xi + c_1 x + \cos x)^2. \quad (2.37)$$

The derivative of V_a along the solutions of (2.35) is computed as

$$\begin{aligned}\dot{V}_a(x,z,u) &= x\left[-c_1 x - x^3 + z\right] + z\left[u + (c_1 - \sin x)\left(-c_1 x - x^3 + z\right)\right] \\ &= -c_1 x^2 - x^4 + z\left[x + u + (c_1 - \sin x)\left(-c_1 x - x^3 + z\right)\right]. \quad (2.38)\end{aligned}$$

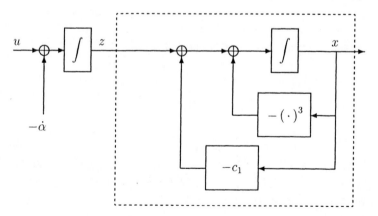

Figure 2.5: Closing the feedback loop in the dashed box with $+\alpha$ and "backstepping" $-\alpha$ through the integrator.

As always, we let $\dot{V}_a$ be an explicit function of u and design u to satisfy the clf inequality (2.18). For this reason, the cross-term xz, which is due to the presence of z in (2.35a), is grouped together with u. This is possible because u is also multiplied by z due to the chosen form of V_a. This is the second key feature of backstepping. Now we choose the control u to make $\dot{V}_a$ negative definite in x and z. The simplest way to achieve this is to make the bracketed term in (2.38) equal to $-c_2 z$, where $c_2 > 0$:

$$\begin{aligned} u &= -c_2 z - x - (c_1 - \sin x)\left(-c_1 x - x^3 + z\right) \\ &= -c_2(\xi + c_1 x + \cos x) - x - (c_1 - \sin x)\left(\xi + \cos x - x^3\right). \end{aligned} \quad (2.39)$$

With this control, the clf derivative is

$$\dot{V}_a = -c_1 x^2 - c_2 z^2, \quad (2.40)$$

which proves that in the (x, z) coordinates the equilibrium $(0, 0)$ is GAS. In view of (2.34), the equilibrium $(0, -1)$ in the (x, ξ) coordinates has the same property.

The resulting closed-loop system in the (x, z)-coordinates is

$$\begin{bmatrix} \dot{x} \\ \dot{z} \end{bmatrix} = \begin{bmatrix} -c_1 - x^2 & 1 \\ -1 & -c_2 \end{bmatrix} \begin{bmatrix} x \\ z \end{bmatrix}. \quad (2.41)$$

Although written in a linear-like form, this system is nonlinear. An important structural property of this system is that its nonlinear "system matrix" is the sum of a *negative diagonal* and a *skew-symmetric* matrix function of x. This is the third key feature of backstepping, which will be extremely useful in other designs.

Avoiding cancellations. The above control law is not the best way to achieve negativity of $\dot{V}_a$, because it involves at least one unnecessary cancellation. A closer examination of (2.38) reveals that the term $-z^2 \sin x$ need not be cancelled because it can be dominated by $-c_2 z^2$. A control law which avoids this cancellation is

$$u = -c_2 z - x - (c_1 - \sin x)\left(-c_1 x - x^3\right), \quad c_2 > c_1 + 1. \quad (2.42)$$

With this control, the clf derivative is

$$\dot{V}_a = -c_1 x^2 - x^4 - (c_2 - c_1 + \sin x) z^2. \quad (2.43)$$

Although more complicated than (2.40), this function is easily rendered negative definite by the choice $c_2 > c_1 + 1$. The resulting system in the (x, z) coordinates preserves its skew-symmetric form

$$\begin{bmatrix} \dot{x} \\ \dot{z} \end{bmatrix} = \begin{bmatrix} -c_1 - x^2 & 1 \\ -1 & -c_2 + c_1 - \sin x \end{bmatrix} \begin{bmatrix} x \\ z \end{bmatrix}. \quad (2.44)$$

2.2 BACKSTEPPING

The simplified control law (2.42) is an illustration of design flexibilities in satisfying the clf inequality $\dot{V}_a \leq 0$ and at the same time avoiding unnecessary cancellations. In fact, more detailed calculations show that the control law can be further simplified to

$$u = -k_1 z - k_2 x^2 z, \qquad (2.45)$$

with

$$k_1 > c_2 + c_1 + 1 + \frac{(c_1^2 + c_1 + 1)^2}{2c_1}, \quad k_2 \geq \frac{(c_1 + 1)^2}{4}. \qquad (2.46)$$

Using this control we obtain

$$\dot{V}_a \leq -\frac{1}{2} c_1 x^2 - c_2 z^2. \qquad (2.47)$$

Integrator backstepping as a general design tool is based on the following assumption:

Assumption 2.7 *Consider the system*

$$\dot{x} = f(x) + g(x)u, \quad f(0) = 0, \qquad (2.48)$$

where $x \in \mathbb{R}^n$ is the state and $u \in \mathbb{R}$ is the control input. There exist a continuously differentiable feedback control law

$$u = \alpha(x), \quad \alpha(0) = 0, \qquad (2.49)$$

and a smooth, positive definite, radially unbounded function $V : \mathbb{R}^n \to \mathbb{R}$ such that

$$\frac{\partial V}{\partial x}(x)\left[f(x) + g(x)\alpha(x)\right] \leq -W(x) \leq 0, \quad \forall x \in \mathbb{R}^n, \qquad (2.50)$$

where $W : \mathbb{R}^n \to \mathbb{R}$ is positive semidefinite.

Under this assumption, the control (2.49), applied to the system (2.48), guarantees global boundedness of $x(t)$, and, via the LaSalle-Yoshizawa theorem (Theorem 2.1), the regulation of $W(x(t))$:

$$\lim_{t \to \infty} W(x(t)) = 0. \qquad (2.51)$$

A stronger convergence result is obtained using LaSalle's theorem (Theorem 2.2) with $\Omega = \mathbb{R}^n$: $x(t)$ converges to the largest invariant set M contained in the set $E = \{x \in \mathbb{R}^n \mid W(x) = 0\}$. Clearly, if $W(x)$ is positive definite, the control (2.49) renders $x = 0$ the GAS equilibrium of (2.48).

Lemma 2.8 (Integrator Backstepping) *Let the system (2.48) be augmented by an integrator:*

$$\dot{x} = f(x) + g(x)\xi \qquad (2.52a)$$
$$\dot{\xi} = u, \qquad (2.52b)$$

and suppose that (2.52a) satisfies Assumption 2.7 with $\xi \in \mathbb{R}$ as its control.

(i) If $W(x)$ is positive definite, then

$$V_a(x,\xi) = V(x) + \frac{1}{2}[\xi - \alpha(x)]^2 \qquad (2.53)$$

is a clf for the full system (2.52), that is, there exists a feedback control $u = \alpha_a(x,\xi)$ which renders $x = 0$, $\xi = 0$ the GAS equilibrium of (2.52). One such control is

$$u = -c(\xi - \alpha(x)) + \frac{\partial \alpha}{\partial x}(x)[f(x) + g(x)\xi] - \frac{\partial V}{\partial x}(x)g(x), \quad c > 0. \qquad (2.54)$$

(ii) If $W(x)$ is only positive semidefinite, then there exists a feedback control which renders $\dot{V}_a \leq -W_a(x,\xi) \leq 0$, such that $W_a(x,\xi) > 0$ whenever $W(x) > 0$ or $\xi \neq \alpha(x)$. This guarantees global boundedness and convergence of $\begin{bmatrix} x(t) \\ \xi(t) \end{bmatrix}$ to the largest invariant set M_a contained in the set

$$E_a = \left\{ \begin{bmatrix} x \\ \xi \end{bmatrix} \in \mathbb{R}^{n+1} \ \Big| \ W(x) = 0, \ \xi = \alpha(x) \right\}.$$

Proof. Introducing the error variable

$$z = \xi - \alpha(x), \qquad (2.55)$$

and differentiating[4] with respect to time, (2.52) is rewritten as

$$\dot{x} = f(x) + g(x)[\alpha(x) + z] \qquad (2.56a)$$

$$\dot{z} = u - \frac{\partial \alpha}{\partial x}(x)[f(x) + g(x)(\alpha(x) + z)]. \qquad (2.56b)$$

Using (2.50), the derivative of (2.53) along the solutions of (2.56) is

$$\begin{aligned}
\dot{V}_a &= \frac{\partial V}{\partial x}(f + g\alpha + gz) + z\left[u - \frac{\partial \alpha}{\partial x}(f + g(\alpha + z))\right] \\
&= \frac{\partial V}{\partial x}(f + g\alpha) + z\left[u - \frac{\partial \alpha}{\partial x}(f + g(\alpha + z)) + \frac{\partial V}{\partial x}g\right] \\
&\leq -W(x) + z\left[u - \frac{\partial \alpha}{\partial x}(f + g(\alpha + z)) + \frac{\partial V}{\partial x}g\right], \qquad (2.57)
\end{aligned}$$

where the terms containing z as a factor have been grouped together. By the LaSalle-Yoshizawa theorem (Theorem 2.1), any choice of the control u which renders $\dot{V}_a \leq -W_a(x,\xi) \leq -W(x)$, with W_a positive definite in $z = \xi - \alpha(x)$, guarantees global boundedness of x, z, and $\xi = z + \alpha(x)$, and regulation of $W(x(t))$ and $z(t)$. Furthermore, LaSalle's theorem (Theorem 2.2)

---

[4]Once again, note that the time derivative $\dot{\alpha}$ in (2.56b) is implemented analytically without the need for a differentiator.

2.2 BACKSTEPPING

guarantees convergence of $\begin{bmatrix} x(t) \\ z(t) \end{bmatrix}$ to the largest invariant set contained in the set $\left\{ \begin{bmatrix} x \\ z \end{bmatrix} \in \mathbb{R}^{n+1} \;\middle|\; W(x) = 0, z = 0 \right\}$. Again, the simplest way to make $\dot{V}_a$ negative definite in z is to choose the control (2.54), which renders the bracketed term in (2.57) equal to $-cz$ and yields

$$\dot{V}_a \leq -W(x) - cz^2 \triangleq -W_a(x,\xi) \leq 0. \tag{2.58}$$

Clearly, if $W(x)$ is positive definite, Theorem 2.1 guarantees the global asymptotic stability of $x = 0, z = 0$, which in turn implies that $V_a(x,\xi)$ is a clf and $x = 0, \xi = 0$ is the GAS equilibrium of (2.52). □

While the choice of control (2.54) is simple, this control may not be desirable because it involves cancellation of nonlinearities, some of which may be useful. As illustrated by (2.39) and (2.40), the requirement that $\dot{V}_a$ in (2.57) be made negative by u allows considerable freedom in the choice of control law $u = \alpha_a(x,\xi)$ such that

$$\dot{V}_a \leq -W(x) + z\left[\alpha_a(x,\xi) - \frac{\partial \alpha}{\partial x}(f + g(\alpha + z)) + \frac{\partial V}{\partial x}g\right] = -W_a(x,\xi) \leq 0. \tag{2.59}$$

We stress that the main result of backstepping is not the specific form of the control law (2.54), but rather the construction of a Lyapunov function whose derivative can be made negative by a wide variety of control laws. In this way, the design of a stabilizing state-feedback controller is effectively reduced to satisfying the scalar inequality (2.59).

Example 2.9 As a design tool, backstepping is less restrictive than feedback linearization. In some situations it can overcome singularities such as lack of controllability. This is illustrated by the system

$$\dot{x} = x\xi \tag{2.60a}$$
$$\dot{\xi} = u, \tag{2.60b}$$

which is uncontrollable at $x = 0$. Comparing with (2.52), we see that $f(x) = 0$, $g(x) = x$. Applying Lemma 2.8 with $V(x) = \frac{1}{2}x^2$ we can choose

$$\alpha(x) = -x^2, \quad z = \xi - \alpha(x) = \xi + x^2, \tag{2.61}$$

so that $W(x)$ in (2.50) is positive definite: $W(x) = x^4$. The substitution of (2.61) into (2.60) yields

$$\dot{x} = -x^3 + xz \tag{2.62a}$$
$$\dot{z} = u + 2x^2(z - x^2), \tag{2.62b}$$

and the derivative of $V_a = \frac{1}{2}x^2 + \frac{1}{2}z^2$ is

$$\dot{V}_a = -x^4 + z\left(u + x^2 + 2x^2 z - 2x^4\right). \tag{2.63}$$

The control (2.54) which renders $\dot{V}_a = -x^4 - z^2$ is

$$u = -z - x^2 - 2x^2 z + 2x^4 = -\xi - 2x^2 - 2x^2\xi. \tag{2.64}$$

The resulting system in the (x, ξ) coordinates is

$$\dot{x} = x\xi \tag{2.65a}$$
$$\dot{\xi} = -\xi - 2x^2 - 2x^2\xi, \tag{2.65b}$$

and its equilibrium $(0, 0)$ is GAS.

A significant design flexibility of backstepping is in the choice of $\alpha(x)$. For the system (2.60), instead of (2.61) we can choose

$$\alpha(x) \equiv 0, \quad z = \xi, \tag{2.66}$$

so that $W(x) \equiv 0$ is semidefinite and

$$V_a = \frac{1}{2}x^2 + \frac{1}{2}\xi^2. \tag{2.67}$$

The derivative of V_a along the solutions of (2.60) is

$$\dot{V}_a = x^2\xi + \xi u = \xi\left(u + x^2\right). \tag{2.68}$$

In this case the best we can do is to render $\dot{V}_a$ negative semidefinite: The control

$$u = -\xi - x^2 \tag{2.69}$$

yields the closed-loop system

$$\dot{x} = x\xi \tag{2.70a}$$
$$\dot{\xi} = -\xi - x^2 \tag{2.70b}$$

and the Lyapunov derivative $\dot{V}_a = -\xi^2$. Then, Lemma 2.8(ii) guarantees that $(x(t), \xi(t))$ is bounded and converges to the largest invariant set M_a of (2.70) contained in the set E_a where $\xi = 0$. But $\xi(t) \equiv 0$ implies $x(t) \equiv 0$. Applying Corollary 2.3, we conclude that the equilibrium $(0, 0)$ is GAS.

Comparing the two control laws (2.64) and (2.69) we see that the choice $\alpha(x) \equiv 0$ simplified the control by eliminating the x^4-term. The design flexibility of backstepping will be further explored and exploited in the jet engine example of Section 2.4. ◇

2.2 BACKSTEPPING

Lemma 2.8 shows how to add a single integrator. This lemma can be repeatedly applied to add a whole chain of integrators.

Corollary 2.10 (Chain of Integrators) *Let the system (2.48) satisfying Assumption 2.7 with $\alpha(x) = \alpha_0(x)$ be augmented by a chain of k integrators so that u is replaced by ξ_1, the state of the last integrator in the chain:*

$$\begin{aligned} \dot{x} &= f(x) + g(x)\xi_1 \\ \dot{\xi}_1 &= \xi_2 \\ &\vdots \\ \dot{\xi}_{k-1} &= \xi_k \\ \dot{\xi}_k &= u\,. \end{aligned} \qquad (2.71)$$

For this system, repeated application of Lemma 2.8 with $\xi_1, \ldots, \xi_k$ as virtual controls, results in the Lyapunov function

$$V_a(x, \xi_1, \ldots, \xi_k) = V(x) + \frac{1}{2}\sum_{i=1}^{k}[\xi_i - \alpha_{i-1}(x, \xi_1, \ldots, \xi_{i-1})]^2\,. \qquad (2.72)$$

Any choice of feedback control which renders $\dot{V}_a \leq -W_a(x, \xi_1, \ldots, \xi_k) \leq 0$, with $W_a(x, \xi_1, \ldots, \xi_k) = 0$ only if $W(x) = 0$ and $\xi_i \neq \alpha_{i-1}(x, \xi_1, \ldots, \xi_{i-1})$, $i = 1, \ldots, k$, guarantees that $\left[x^T(t), \xi_1(t), \ldots, \xi_k(t)\right]^T$ is globally bounded and converges to the largest invariant set M_a contained in the set $E_a = \left\{\left[x^T, \xi_1, \ldots, \xi_k\right]^T \in \mathbb{R}^{n+k} \,\big|\, W(x) = 0, \xi_i = \alpha_{i-1}(x, \xi_1, \ldots, \xi_{i-1}), i = 1, \ldots, k\right\}$. Furthermore, if $W(x)$ is positive definite, that is, if $x = 0$ can be rendered GAS through ξ_1, then (2.72) is a clf for (2.71) and the equilibrium $x = 0, \xi_1 = \cdots = \xi_k = 0$ can be rendered GAS through u.

2.2.2 Example: active suspension (parallel)

The previous section presented integrator backstepping as a design tool for nonlinear systems. However, even when dealing with *linear* systems, one may encounter situations where a *nonlinear* closed-loop response is desired. In those cases, backstepping can be used to design the corresponding nonlinear controller. We now illustrate this with an active suspension example.[5]

When designing vehicle suspensions, the dual objective is to minimize the vertical acceleration of the car body (for passenger comfort), and, at the same time, to maximize tire contact with the road surface (for handling). In recent years, manufacturers of passenger cars and off-road vehicles have been developing *active* suspension systems, with hydraulic actuators. Feedback control

---

[5]The active suspension examples of Sections 2.2.2 and 2.2.6 were suggested to us by Jim Winkelman and Doug Rhode of Ford Motor Company.

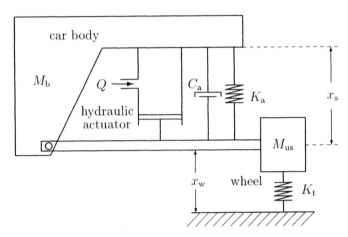

Figure 2.6: Quarter-car model for active suspension design with *parallel* connection of hydraulic actuator with passive spring/damper.

of the actuators improves both ride quality and handling performance, with the important secondary benefits of better braking and cornering because of reduced weight transfer.

Active suspension designs trade off ride quality and suspension travel. Hitting the suspension travel limits causes passenger discomfort as well as increased wear and tear on vehicle components. Hence, active suspensions should be designed to behave differently on smooth and rough roads. This can be achieved by introducing nonlinearities in the controller which make the suspension stiffer near its travel limits.

To see how backstepping can be used to design such a nonlinear controller, we consider the simplified quarter-car model in Figure 2.6, where Q is the fluid flow into the hydraulic actuator, x_s is the suspension travel, and the wheel is modeled as an unsprung mass M_{us} with a spring K_t. In the configuration of Figure 2.6, the hydraulic actuator is connected in parallel with a spring K_a and a damper C_a.

The fluid flow Q is adjusted by opening a current-controlled solenoid valve which can be modeled as a first-order linear system with the current i_v as input and the valve opening d_v as output:

$$\dot{d}_v = -c_v d_v + k_v i_v. \tag{2.73}$$

While the resulting flow Q is usually proportional to the product of the valve opening with the square root of the pressure differential across the valve, we will consider a more advanced valve, which effectively cancels the square-root nonlinearity and renders the current-to-flow dynamics linear:

$$\dot{Q} = -c_f Q + k_f i_v. \tag{2.74}$$

2.2 BACKSTEPPING

In the parallel configuration of Figure 2.6, neglecting leakage and compressibility, the flow Q is related to the suspension travel x_s through the equation

$$\dot{x}_s = \frac{1}{A} Q, \qquad (2.75)$$

where A is the effective surface of the piston. To apply backstepping, we view the flow Q as the virtual control, and design for it a nonlinear stabilizing function $\alpha(x_s)$ which will stiffen the suspension near its travel limits:

$$Q_{\text{des}} = \alpha(x_s) = -A \left[c_1 x_s + \kappa_1 x_s^3 \right]. \qquad (2.76)$$

The nonlinear term $\kappa_1 x_s^3$ is negligible when $|x_s|$ is small, but grows very fast when $|x_s| > \kappa_1^{1/3}$. With $z = Q - \alpha(x_s)$ and (2.76), the equation (2.75) becomes

$$\dot{x}_s = -c_1 x_s - \kappa_1 x_s^3 + \frac{1}{A} z. \qquad (2.77)$$

Using (2.74), we obtain

$$\dot{z} = -c_f Q + k_f i_v + \left(c_1 + 3\kappa_1 x_s^2 \right) Q. \qquad (2.78)$$

Applying Lemma 2.8, we obtain the control law for the current:

$$i_v = \frac{1}{k_f} \left[c_f \alpha(x_s) - \frac{1}{A} x_s - \left(c_1 + 3\kappa_1 x_s^2 \right) Q \right], \qquad (2.79)$$

which results in the error system

$$\begin{aligned} \dot{x}_s &= -c_1 x_s - \kappa_1 x_s^3 + \frac{1}{A} z \\ \dot{z} &= -c_f z - \frac{1}{A} x_s. \end{aligned} \qquad (2.80)$$

Clearly, the derivative of $V(x_s, z) = \frac{1}{2}(x_s^2 + z^2)$ is negative definite: $\dot{V} = -c_1 x_s^2 - \kappa_1 x_s^4 - c_f z^2$. The nonlinear closed-loop system (2.80) is GAS and possesses the desired property of becoming "stiffer" as $|x_s|$ becomes larger.

2.2.3 Feedback linearization and zero dynamics

One of the popular methods for nonlinear control design is *feedback linearization*, which employs a change of coordinates and feedback control to transform a nonlinear system into a system whose dynamics are linear (at least partially). A great deal of research has been devoted to this subject over the last two decades, as evidenced by the comprehensive books of Isidori [53] and Nijmeier and Van der Schaft [144] and the references therein. Since feedback linearization is not a goal pursued in this book, we only briefly review some concepts needed for the remainder of the chapter. For maximum accessibility, we avoid the use of differential geometric notation.

Let us consider the nonlinear system

$$\dot{x} = f(x) + g(x)u, \quad x \in \mathbb{R}^n, \ u \in \mathbb{R}$$
$$y = h(x), \quad y \in \mathbb{R} \tag{2.81}$$

where f, g, h are smooth (that is, infinitely differentiable) vector functions. The derivative of the output $y = h(x)$ is given by:

$$\dot{y} = \frac{\partial h}{\partial x}(x)f(x) + \frac{\partial h}{\partial x}(x)g(x)u. \tag{2.82}$$

If $\frac{\partial h}{\partial x}(x_0)g(x_0) \neq 0$, then the system (2.81) is said to have *relative degree one* at x_0.[6] In our terminology, this implies that the output y is separated from the input u by one integrator only.

If $\frac{\partial h}{\partial x}(x_0)g(x_0) = 0$, there are two cases:

(i) If there exist points x arbitrarily close to x_0 such that $\frac{\partial h}{\partial x}(x)g(x) \neq 0$, then (2.81) does not have a well-defined relative degree at x_0.

(ii) If there exists a neighborhood B_0 of x_0 such that $\frac{\partial h}{\partial x}(x)g(x) = 0$ for all $x \in B_0$, then the relative degree of (2.81) at x_0 may be well-defined.

In case (ii), we define

$$\psi_1(x) = h(x), \quad \psi_2(x) = \frac{\partial h}{\partial x}(x)f(x) \tag{2.83}$$

and compute the second derivative of y:

$$\ddot{y} = \frac{\partial}{\partial x}\left(\frac{\partial h}{\partial x}f\right)f + \frac{\partial}{\partial x}\left(\frac{\partial h}{\partial x}f\right)gu = \frac{\partial \psi_2}{\partial x}(x)f(x) + \frac{\partial \psi_2}{\partial x}(x)g(x)u. \tag{2.84}$$

If $\frac{\partial \psi_2}{\partial x}(x_0)g(x_0) \neq 0$, then (2.81) is said to have relative degree *two* at x_0.

If $\frac{\partial \psi_2}{\partial x}(x)g(x) = 0$ in a neighborhood of x_0, then we continue the differentiation procedure.

Definition 2.11 *The system (2.81) is said to have* relative degree ρ *at the point x_0 if there exists a neighborhood B_0 of x_0 on which*

$$\frac{\partial \psi_1}{\partial x}(x)g(x) = \frac{\partial \psi_2}{\partial x}(x)g(x) = \cdots = \frac{\partial \psi_{\rho-1}}{\partial x}(x)g(x) = 0 \tag{2.85}$$

$$\frac{\partial \psi_\rho}{\partial x}(x)g(x) \neq 0, \tag{2.86}$$

where

$$\psi_1(x) = h(x), \quad \psi_i(x) = \frac{\partial \psi_{i-1}}{\partial x}(x)f(x), \ i = 2, \ldots, \rho. \tag{2.87}$$

If (2.85) and (2.86) are valid for all $x \in \mathbb{R}^n$, then the relative degree of (2.81) is said to be globally defined.

---

[6]Note that since the functions are smooth, $\frac{\partial h}{\partial x}(x_0)g(x_0) \neq 0$ implies that there exists a neighborhood of x_0 on which $\frac{\partial h}{\partial x}g(x) \neq 0$.

2.2 BACKSTEPPING

Suppose now that (2.81) has relative degree ρ at x_0. Then we can use a change of coordinates and feedback control to locally transform this system into the *cascade connection* of a ρ-dimensional linear system and an $(n-\rho)$-dimensional nonlinear system. In particular, after differentiating ρ times the output $y = h(x)$, the control u appears:

$$\begin{aligned} y^{(\rho)} &= \underbrace{\frac{\partial}{\partial x}\left(\frac{\partial}{\partial x}\underbrace{\left(\cdots \frac{\partial}{\partial x}\left(\frac{\partial h}{\partial x}f\right)f\cdots\right)}_{\rho-1}f\right)}_{\rho-1}f(x) \\ &+ \underbrace{\frac{\partial}{\partial x}\left(\frac{\partial}{\partial x}\underbrace{\left(\cdots \frac{\partial}{\partial x}\left(\frac{\partial h}{\partial x}f\right)f\cdots\right)}_{\rho-1}f\right)}_{\rho-1}g(x)u \\ &= \frac{\partial \psi_\rho}{\partial x}(x)f(x) + \frac{\partial \psi_\rho}{\partial x}(x)g(x)u. \end{aligned} \qquad (2.88)$$

Since $\frac{\partial \psi_\rho}{\partial x} g \neq 0$ in a neighborhood of x_0, we can linearize the input-output description of the system (2.81) using feedback to cancel the nonlinearities in (2.88):

$$u = \frac{1}{\frac{\partial \psi_\rho}{\partial x}(x)g(x)}\left[-\frac{\partial \psi_\rho}{\partial x}(x)f(x) + v\right]. \qquad (2.89)$$

Then the dynamics of y and its derivatives are governed by a chain of ρ integrators: $y^{(\rho)} = v$. Since our original system (2.81) has dimension n, we need to account for the remaining $n-\rho$ states. Using differential geometric tools, it is easy to show that it is always possible to find $n-\rho$ functions $\psi_{\rho+1}(x), \ldots, \psi_n(x)$ with $\frac{\partial \psi_i}{\partial x}(x)g(x) = 0$, $i = \rho+1, \ldots, n$ such that the change of coordinates

$$\begin{aligned} \zeta_1 = y = h(x) = \psi_1(x), \; \zeta_2 = \dot{y} = \psi_2(x), \ldots, \zeta_\rho = y^{(\rho-1)} = \psi_\rho(x), \\ \zeta_{\rho+1} = \psi_{\rho+1}(x), \ldots, \zeta_n = \psi_n(x) \end{aligned} \qquad (2.90)$$

is locally invertible and transforms, along with the feedback (2.89), the nonlinear system (2.81) into

$$\begin{aligned} \dot{\zeta}_1 &= \zeta_2 \\ &\vdots \\ \dot{\zeta}_{\rho-1} &= \zeta_\rho \\ \dot{\zeta}_\rho &= v \\ \dot{\zeta}_{\rho+1} &= \frac{\partial \psi_{\rho+1}}{\partial x}(x)f(x) = \phi_{\rho+1}(\zeta) \\ &\vdots \\ \dot{\zeta}_n &= \frac{\partial \psi_n}{\partial x}(x)f(x) = \phi_n(\zeta) \\ y &= \zeta_1. \end{aligned} \qquad (2.91)$$

As a cascade connection of a chain of ρ integrators with an $(n-\rho)$-dimensional nonlinear system, this system is a special case of the cascade systems to which we will apply backstepping in the following sections.

The states $\zeta_{\rho+1}, \ldots, \zeta_n$ of the nonlinear subsystem in (2.91) have been rendered *unobservable* from the output y by the control (2.89). Hence, feedback linearization in this case is the nonlinear equivalent of placing $n - \rho$ poles of a linear system at the origin and cancelling the ρ zeros with the remaining poles. Of course, to guarantee stability, the cancelled zeros must be stable. In the nonlinear case, using the new control input v to stabilize the linear subsystem of (2.91) does not guarantee stability of the whole system, unless the stability of the nonlinear part of (2.91) has been established separately.

When v is used to keep the output y equal to zero for all $t > 0$, that is, when $\zeta_1 \equiv \cdots \equiv \zeta_\rho \equiv 0$, the dynamics of $\zeta_{\rho+1}, \ldots, \zeta_n$ are described by

$$\begin{aligned}
\dot{\zeta}_{\rho+1} &= \phi_{\rho+1}(0, \ldots, 0, \zeta_{\rho+1}, \ldots, \zeta_n) \\
&\vdots \\
\dot{\zeta}_n &= \phi_n(0, \ldots, 0, \zeta_{\rho+1}, \ldots, \zeta_n).
\end{aligned} \quad (2.92)$$

They are called *the zero dynamics* of (2.81), because they evolve on the subset of the state space on which the output of the system is identically zero. If the equilibrium at $\zeta_{\rho+1} = \cdots = \zeta_n = 0$ of the zero dynamics (2.92) is asymptotically stable, the system (2.81) is said to be *minimum phase*. With a slight abuse of notation, we will refer to the $(\zeta_{\rho+1}, \ldots, \zeta_n)$-subsystem as the *zero dynamics subsystem* of (2.81), even when $\zeta_1, \ldots, \zeta_\rho$ are not zero.

In (2.81) the output $y = h(x)$ is prespecified, possibly from a tracking objective, and the resulting cascade system is linear from the input v to the output y. This linearization process is usually called *input-output feedback linearization* [53]. If our goal is only to design a stabilizing controller, we may attempt to find an output with respect to which the relative degree is $\rho = n$. If such an output exists, the whole system is linearized without zero dynamics. This process is referred to as *full-state feedback linearization* [48, 49, 56, 178]. If such an output cannot be found, then we may look for an output which yields the highest relative degree, and thus results in a cascade whose linear subsystem has the highest dimension [117]. It is desirable that with respect to the chosen output the system be minimum phase. The importance of this property will be clear in the following sections which address problems of stabilization of cascade systems.

2.2.4 Stabilization of cascade systems

The Integrator Backstepping Lemma (Lemma 2.8) is a stabilization result for the cascade connection of a nonlinear system with an integrator. The nonlinear system, which is allowed to be unstable, is stabilized through the integrator.

2.2 BACKSTEPPING

We now consider cascade connections in which the nonlinear system is globally stable, but the input subsystem is more complex than just an integrator. We begin with the case where the input subsystem is linear:

$$\dot{x} = f(x) + g(x)y, \quad f(0) = 0, \quad x \in \mathbb{R}^n, \quad y \in \mathbb{R} \tag{2.93a}$$
$$\dot{\xi} = A\xi + bu, \quad y = h\xi, \quad \xi \in \mathbb{R}^q, \quad u \in \mathbb{R}. \tag{2.93b}$$

We assume that when $y = 0$ the nonlinear subsystem (2.93a) has a globally stable equilibrium at $x = 0$, and that an appropriate Lyapunov function $V(x)$ is known:

$$\frac{\partial V}{\partial x}(x) f(x) \le -W(x) \le 0. \tag{2.94}$$

The problem is to stabilize the linear subsystem (2.93b) without destabilizing the nonlinear subsystem (2.93a), and, if possible, to achieve GAS of the equilibrium of (2.93) at $(0,0)$. This problem is not solvable in general. Here it will be solved by requiring the input subsystem (2.93b) to have the following passivity property:

Assumption 2.12 *The triple (A, b, h) is feedback positive real (FPR), that is, there exists a linear feedback transformation $u = K\xi + v$ such that $A + bK$ is Hurwitz and there are matrices $P > 0$, $Q \ge 0$ which satisfy*

$$(A + bK)^T P + P(A + bK) = -Q \tag{2.95a}$$
$$Pb = h^T. \tag{2.95b}$$

A sufficient condition for FPR is that there exists a feedback gain row vector K such that $A+bK$ is Hurwitz, the transfer function $Z(s) = h(sI - A - bK)^{-1}b$ is positive real (PR) (see Appendix D, Definition D.6), and the pair $(A+bK, h)$ is observable. It should be noted from (2.95b) that the relative degree of PR transfer functions is one because $b^T Pb = hb > 0$.

Lemma 2.13 (Stabilization with FPR) *Let $V(x)$ be a Lyapunov function for (2.93a) satisfying (2.94). If the triple (A, b, h) is FPR, then a Lyapunov function for the cascade system (2.93) is*

$$V_a(x, \xi) = V(x) + \xi^T P \xi, \tag{2.96}$$

and the corresponding control law

$$u = \alpha_a(x, \xi) = K\xi - \frac{1}{2}\frac{\partial V}{\partial x}g(x) \tag{2.97}$$

guarantees that $\begin{bmatrix} x(t) \\ \xi(t) \end{bmatrix}$ is globally bounded and converges to the largest invariant set M_a contained in the set $E_a = \left\{ \begin{bmatrix} x \\ \xi \end{bmatrix} \in \mathbb{R}^{n+q} \;\middle|\; W(x) = 0, \; Q^{\frac{1}{2}}\xi = 0 \right\}$.

If $W(x)$ is positive definite, that is, if the nonlinear subsystem (2.93a) with $y = 0$ has a globally asymptotically stable equilibrium at $x = 0$, then the equilibrium $x = 0$, $\xi = 0$ is also GAS.

Proof. Using (2.94) and (2.95) and denoting $u = K\xi + v$ with $v = -\frac{1}{2}\frac{\partial V}{\partial x}g(x)$ from (2.97), the derivative of $V_a(x,\xi)$ is

$$\begin{aligned}
\dot{V}_a &= \frac{\partial V}{\partial x}(x)\left[f(x) + g(x)y\right] \\
&\quad + \xi^T P\left[(A+bK)\xi + bv\right] + \left[(A+bK)\xi + bv\right]^T P\xi \\
\text{by (2.94) and (2.95a)} &\leq -W(x) + \frac{\partial V}{\partial x}(x)g(x)y - \xi^T Q\xi + 2\xi^T Pbv \\
\text{by (2.95b)} &= -W(x) + \frac{\partial V}{\partial x}(x)g(x)y - \xi^T Q\xi + 2y\left[-\frac{1}{2}\frac{\partial V}{\partial x}g(x)\right] \\
&= -W(x) - \xi^T Q\xi \leq 0. \quad (2.98)
\end{aligned}$$

Since V_a is positive definite, radially unbounded and has a negative semidefinite derivative, $x(t)$ and $\xi(t)$ are globally bounded. Furthermore LaSalle's theorem (Theorem 2.2) guarantees convergence to the largest invariant set M_a in the set E_a.

If, in addition, $W(x)$ is positive definite, then the global asymptotic stability of $x = 0, \xi = 0$ is shown using Corollary 2.3. From the positive definiteness of $W(x)$, the set E_a, on which $\dot{V}_a = 0$, is given by $E_a = \left\{(x,\xi)\,\middle|\, x = 0,\ Q^{\frac{1}{2}}\xi = 0\right\}$. Since $V(x)$ is positive definite, it has a minimum at $x = 0$, and thus $\frac{\partial V}{\partial x}(0) = 0$. This implies that on the set E_a the control term $v = -\frac{1}{2}\frac{\partial V}{\partial x}g(x)$ vanishes. Hence, on the set E_a the state $\xi(t)$ satisfies

$$\dot{\xi} = (A+bK)\xi, \quad V_a(x,\xi) = \xi^T P\xi. \quad (2.99)$$

But V_a is constant on E_a, which means that $\xi^T P\xi$ must be constant on E_a. Since $A + bK$ is Hurwitz, $\xi = 0$ is the only solution of $\dot{\xi} = (A+bK)\xi$ that satisfies $\xi^T P\xi = \text{constant}$. Thus, $\xi = 0$ on the largest invariant set contained in E_a. This implies that this invariant set M_a is just the equilibrium $x = 0$, $\xi = 0$, which, by Corollary 2.3, is GAS. □

The stabilizing control law (2.97) consists of two terms, one linear and one nonlinear. The purpose of the latter is to preserve the stability of the nonlinear subsystem.

Example 2.14 For a comparison with backstepping, let us first reexamine the second-order system stabilized in Example 2.9:

$$\begin{aligned}
\dot{x} &= x\xi & (2.100a) \\
\dot{\xi} &= u. & (2.100b)
\end{aligned}$$

2.2 BACKSTEPPING

In this system we have $f(x) \equiv 0$, $g(x) = x$, $A = 0$, $B = 1$, and $y = \xi$. Using $V(x) = x^2$ we see from (2.94) that $W(x) \equiv 0$. The FPR condition is trivially satisfied and the stabilizing control (2.97) is

$$u = -k\xi - x^2, \quad k > 0. \tag{2.101}$$

With $k = 1$ this is the same control law as (2.69) obtained by backstepping with $\alpha(x) \equiv 0$. We know from Example 2.9 that this control law achieves GAS of the equilibrium $(x, \xi) = (0, 0)$.

Next we consider a third-order system to which backstepping is not directly applicable:

$$\dot{x} = x(\xi_1 + \xi_2) \triangleq xy \tag{2.102a}$$
$$\dot{\xi}_1 = \xi_2 \tag{2.102b}$$
$$\dot{\xi}_2 = u. \tag{2.102c}$$

In this case $b^T = [0\ 1]$ and $h = [1\ 1]$, so that the condition (2.95b) yields

$$\begin{bmatrix} p_{11} & p_{12} \\ p_{12} & p_{22} \end{bmatrix} \begin{bmatrix} 0 \\ 1 \end{bmatrix} = \begin{bmatrix} 1 \\ 1 \end{bmatrix} \Rightarrow \begin{matrix} p_{12} = 1 \\ p_{22} = 1. \end{matrix} \quad p_{11} \neq ? \tag{2.103}$$

With this restriction on P and with $K = [-k_1\ -k_2]$, (2.95a) results in

$$\begin{bmatrix} -2k_1 & p_{11} - k_1 - k_2 \\ p_{11} - k_1 - k_2 & 2 - 2k_2 \end{bmatrix} = -\begin{bmatrix} q_{11} & q_{12} \\ q_{12} & q_{22} \end{bmatrix}. \tag{2.104}$$

For the simplest choice $q_{11} = q_{22} = 1$ and $q_{12} = 0$ we get $k_1 = 0.5$, $k_2 = 1.5$, $p_{11} = 2$. Then the control law (2.97) is

$$u = -\frac{1}{2}\xi_1 - \frac{3}{2}\xi_2 - x^2. \tag{2.105}$$

The equilibrium $(x, \xi_1, \xi_2) = (0, 0, 0)$ is GAS because $Q = I$ is positive definite.
◇

Example 2.15 Let us now consider a system in which $\dot{x} = f(x)$ has a GAS equilibrium at $x = 0$:

$$\dot{x} = -x^3 - x^3(h_1\xi_1 + h_2\xi_2) \tag{2.106a}$$
$$\dot{\xi}_1 = \xi_2 \tag{2.106b}$$
$$\dot{\xi}_2 = u. \tag{2.106c}$$

When $h_2 = 0$ this system is stabilizable by two steps of integrator backstepping as in Corollary 2.10. Thus, the case of interest is when $h_2 \neq 0$ and $h_1 h_2 \geq 0$. This includes the case $h_1 = 0$ when the transfer function $\frac{h_2 s + h_1}{s^2}$ of the linear part is only *weak minimum phase* [163] because it has a zero at $s = 0$.

Choose a feedback $u = -k_1\xi_1 - k_2\xi_2 + v$ with $k_1, k_2 > 0$ which makes the polynomial $q(s) = s^2 + k_2 s + k_1$ Hurwitz, and denote $p(s) = h_1 + h_2 s$. We can choose $k_1 = a^2$ and $k_2 = 2a$, with $a > \frac{h_1}{2h_2}$, so that the transfer function $Z(s) = p(s)/q(s)$ is positive real. We can then write (2.106a) as $\dot{x} = f + gy$, $f = g = -x^3$. Clearly, $x = 0$ is a GAS equilibrium for $\dot{x} = f$, so all the conditions of Lemma 2.13 are satisfied. Using $V(x) = x^2$ in (2.97), we obtain the control law

$$u = -a^2\xi_1 - 2a\xi_2 + x^4. \qquad (2.107)$$

The situation is quite different when $h_1 h_2 < 0$, that is, when the transfer function $Z(s) = p(s)/q(s)$ is nonminimum phase. Then, Lemma 2.13 does not apply. In fact, a detailed calculation given in [85, Example 3.4] shows that in this case the system *cannot* be globally stabilized. ◇

The FPR property is a passivity property. Its nonlinear counterpart will be employed in the stabilization of the nonlinear cascade

$$\dot{x} = f(x,\xi) + g(x,\xi)y, \quad f(0,\xi) = 0, \quad \forall \xi \in \mathbb{R}^q, \quad x \in \mathbb{R}^n, \quad y \in \mathbb{R} \quad (2.108a)$$
$$\dot{\xi} = m(\xi) + \beta(\xi)u, \quad y = h(\xi), \quad h(0) = 0, \quad \xi \in \mathbb{R}^q, \quad u \in \mathbb{R}. \quad (2.108b)$$

Our key assumption is that (2.108b) can be rendered *passive* or *strictly passive* (cf. Appendix D) via a feedback transformation $u = k(\xi) + r(\xi)v$.

Definition 2.16 *The system*

$$\dot{\xi} = m(\xi) + \beta(\xi)u, \quad y = h(\xi), \quad h(0) = 0, \quad \xi \in \mathbb{R}^q, \quad u \in \mathbb{R} \qquad (2.109)$$

is said to be feedback passive (FP) *if there exists a feedback transformation*

$$u = k(\xi) + r(\xi)v \qquad (2.110)$$

such that the resulting system $\dot{\xi} = m(\xi) + \beta(\xi)k(\xi) + \beta(\xi)r(\xi)v$, $y = h(\xi)$ *is passive with a storage function $U(\xi)$ which is positive definite and radially unbounded:*

$$\int_0^t y(\sigma)v(\sigma)d\sigma \geq U(\xi(t)) - U(\xi(0)). \qquad (2.111)$$

The system (2.109) is said to be feedback strictly passive (FSP) *if the feedback (2.110) renders it strictly passive:*

$$\int_0^t y(\sigma)v(\sigma)d\sigma \geq U(\xi(t)) - U(\xi(0)) + \int_0^t \psi(\xi(\sigma))d\sigma, \qquad (2.112)$$

where $\psi(\cdot)$ is the positive definite dissipation rate.

As in the linear case, FP systems of the form (2.109) must have relative degree one.

2.2 BACKSTEPPING

Lemma 2.17 (Stabilization with Passivity) *Let $V(x)$ be a radially unbounded Lyapunov function for $\dot{x} = f(x,\xi)$ satisfying*

$$\frac{\partial V}{\partial x}(x)f(x,\xi) \leq -W(x) \leq 0, \quad \forall x \in \mathbb{R}^n, \ \forall \xi \in \mathbb{R}, \tag{2.113}$$

and let (2.108b) be FP as in Definition 2.16. Then, a Lyapunov function for the cascade system (2.108) is

$$V_a(x,\xi) = V(x) + U(\xi), \tag{2.114}$$

and the corresponding control law

$$u = \alpha_a(x,\xi) = k(\xi) - r(\xi)\frac{\partial V}{\partial x}(x)g(x,\xi) \tag{2.115}$$

guarantees that $\begin{bmatrix} x(t) \\ \xi(t) \end{bmatrix}$ *is globally bounded and converges to the largest invariant set $\bar{M}_a$ contained in the set $\bar{E}_a = \left\{ \begin{bmatrix} x \\ \xi \end{bmatrix} \in \mathbb{R}^{n+q} \mid W(x) = 0 \right\}$. If (2.108b) is FSP, then (2.115) guarantees convergence to the largest invariant set M_a contained in the set $E_a = \left\{ \begin{bmatrix} x \\ \xi \end{bmatrix} \in \mathbb{R}^{n+q} \mid W(x) = 0, \ \xi = 0 \right\}$. Finally, if (2.108b) is FSP and $W(x)$ is positive definite, that is, if $\dot{x} = f(x,\xi)$ has a GAS equilibrium at $x = 0$ uniformly in ξ, then the equilibrium $x = 0$, $\xi = 0$ of (2.108) is also GAS.*

Proof. We first recall Theorem D.4 in Appendix D, which states that the negative feedback interconnection of two (strictly) passive systems is (strictly) passive, and of Lemma D.3, which states that a (strictly) passive system with a positive definite and radially unbounded storage function has a G(A)S equilibrium at $x = 0$ when its external input is set to zero.

The closed-loop system (2.108) with the control (2.115) is

$$\begin{aligned} \dot{x} &= f(x,\xi) + g(x,\xi)y \\ \dot{\xi} &= m(\xi) + \beta(\xi)k(\xi) + \beta(\xi)r(\xi)v \\ y &= h(\xi), \quad v = -\frac{\partial V}{\partial x}(x)g(x,\xi). \end{aligned} \tag{2.116}$$

Let us now express (2.116) as the feedback interconnection of two passive systems Σ_1 and Σ_2:

$$\Sigma_1 \begin{cases} \dot{x} = f(x,\xi) + g(x,\xi)y \\ \eta = \frac{\partial V}{\partial x}(x)g(x,\xi) \end{cases} \tag{2.117a}$$

$$\Sigma_2 \begin{cases} \dot{\xi} = m(\xi) + \beta(\xi)k(\xi) + \beta(\xi)r(\xi)v \\ y = h(\xi) \end{cases} \tag{2.117b}$$

$$v = -\eta. \tag{2.117c}$$

We already know that Σ_2 is passive, since (2.111) is satisfied. To show that Σ_1 is passive with storage function $V(x)$, we use (2.113):

$$\dot{V} = \frac{\partial V}{\partial x}(f + gy) \leq -W(x) + \frac{\partial V}{\partial x} gy = -W(x) + \eta y. \tag{2.118}$$

Integrating (2.118) on $[0, t]$ we obtain

$$\int_0^t \eta(\sigma) y(\sigma) d\sigma \geq V(x(t)) - V(x(0)) + \int_0^t W(x(\sigma)) d\sigma, \tag{2.119}$$

which shows that Σ_1 is passive since $W(x) \geq 0$. From Theorem D.4 we conclude that (2.117) is passive with the positive definite and radially unbounded storage function $V_a(x, \xi) = V(x) + U(\xi)$. Lemma D.3 then states that $x = 0$, $\xi = 0$ is a globally stable equilibrium of (2.116). To see that $W(x) \to 0$ as $t \to \infty$, we differentiate (2.111) and combine the result with (2.113):

$$\begin{aligned} \dot{V}_a &= \dot{V} + \dot{U} \leq \frac{\partial V}{\partial x}(f + gy) + yv \\ &\leq -W(x) + \frac{\partial V}{\partial x} gy + yv = -W(x). \end{aligned} \tag{2.120}$$

Then, LaSalle's theorem (Theorem 2.2) guarantees convergence to the set $\bar{M}_a$. If (2.108b) is FSP, we replace (2.111) by (2.112). Then (2.120) becomes

$$\dot{V} \leq -W(x) - \psi(\xi), \tag{2.121}$$

which, since $\psi(\xi)$ is positive definite, guarantees convergence to the set M_a. Finally, if $W(x)$ is also positive definite, we conclude from (2.121) and Theorem 2.1 that $x = 0$, $\xi = 0$ is GAS. □

Example 2.18 Consider the cascade system:

$$\begin{aligned} \dot{x} &= -x(1 + e^\xi) + x^3 \xi^3 & (2.122a) \\ \dot{\xi} &= \xi^2 u. & (2.122b) \end{aligned}$$

The choice of output $y = \xi^3$ satisfies all the conditions of Lemma 2.17. First, (2.122b) is FSP: The feedback

$$u = -\xi + v \tag{2.123}$$

results in $\dot{\xi} = -\xi^3 + \xi^2 v$, $y = \xi^3$, which is strictly passive with storage function $U(\xi) = \frac{1}{2}\xi^2$, since

$$\dot{U} = -\xi^4 + \xi^3 v = -\xi^4 + yv \tag{2.124}$$

implies that

$$\int_0^t y(\sigma) v(\sigma) d\sigma \geq U(\xi(t)) - U(\xi(0)) + \int_0^t \xi^4(\sigma) d\sigma. \tag{2.125}$$

2.2 BACKSTEPPING

Furthermore, (2.122a) can be represented in the form (2.108a) with

$$f(x,\xi) = -x(1+e^\xi), \quad g(x,\xi) = x^3, \quad y = \xi^3, \qquad (2.126)$$

and (2.113) is satisfied with $V(x) = \frac{1}{2}x^2$, $W(x) = -x^2$.

Applying Lemma 2.17, we conclude that the control

$$u = -\xi - x^4 \qquad (2.127)$$

guarantees GAS of $x = 0$, $\xi = 0$. Indeed, the derivative of the clf $V_a(x,\xi) = \frac{1}{2}(x^2 + \xi^2)$ is negative definite:

$$\dot{V}_a = -x^2(1+e^\xi) + x^4\xi^3 - \xi^4 - x^4\xi^3 \leq -x^2 - \xi^4. \qquad (2.128)$$

◇

Remark 2.19 Lemmas 2.13 and 2.17 allow the zero dynamics of the input subsystem to be only stable rather than asymptotically stable. Such systems are said to be *weak minimum phase* [163]. This is illustrated in Example 2.15, where the cascade system (2.106) is stabilized even with $h_1 = 0$, that is, when the input subsystem (2.106a) has a simple zero at $s = 0$. ◇

2.2.5 Block backstepping with zero dynamics

Integrator backstepping (Lemma 2.8) is a recursive design tool. Now we want to develop a similar tool for feedback stabilization of a system augmented by a dynamic block more complicated than just an integrator. At first glance, it may appear that the cascade design in the preceding subsection provides us with such a tool. Not quite! The achievement of the cascade design is in being able to stabilize the input subsystem (2.93b) or (2.108b) *without destabilizing the original system*. What if the original system is not stable? Can we cascade it with a complicated input subsystem and still stabilize it in one step? We first show that this can be done with a linear input subsystem that is a minimum phase system with relative degree one. We then give a nonlinear extension of that result.

Example 2.20 Let us start with an example in which we cascade the system (2.122) of Example 2.18 with a linear minimum phase system:

$$x\text{-subsystem} \begin{cases} \dot{x}_1 = -x_1(1+e^{x_2}) + x_1^3 x_2^3 \\ \dot{x}_2 = x_2^2 y \end{cases}$$

$$\xi\text{-subsystem} \begin{cases} \dot{\xi}_1 = \xi_2 \\ \dot{\xi}_2 = u \\ y = \xi_1 + \xi_2. \end{cases} \qquad (2.129)$$

The transfer function of the input subsystem is $\frac{s+1}{s^2}$ and its zero is at $s = -1$. One of its minimal realizations is

$$\dot{y} = y - \xi_1 + u \qquad (2.130a)$$
$$\dot{\xi}_1 = -\xi_1 + y. \qquad (2.130b)$$

Its zero dynamics, that is, the dynamics constrained by $y(t) \equiv 0$, are described by $\dot{\xi}_1 = -\xi_1$.

The cascade design of the preceding subsection is not applicable to (2.129) because the equilibrium $x = 0$ of the x-subsystem with $y = 0$ is unstable. To circumvent this obstacle, we first convert (2.130a) into an integrator via the feedback transformation

$$u = -y + \xi_1 + v, \qquad (2.131)$$

where v is our new control variable. The system (2.129) is then rewritten as

$$\begin{aligned} \dot{x}_1 &= -x_1(1 + e^{x_2}) + x_1^3 x_2^3 \\ \dot{x}_2 &= x_2^2 y \\ \dot{y} &= v \\ \dot{\xi}_1 &= -\xi_1 + y. \end{aligned} \qquad (2.132)$$

Now the subsystem consisting of the first three equations in (2.132) is in a form convenient for integrator backstepping. From Example 2.18 we already know that the x-subsystem can be stabilized with y as its virtual control (cf. (2.127)):

$$y_{\text{des}} = \alpha(x) = -x_1^4 - x_2. \qquad (2.133)$$

The corresponding clf is $V(x, \xi) = \frac{1}{2}(x_1^2 + x_2^2)$. Hence, we can achieve stabilization and regulation of x_1, x_2, y by a direct application of Lemma 2.8. The resulting control law is

$$u = -\left(y + x_1^4 + x_2\right) - y + \xi_1 - x_2^2 y + 4x_1^4\left(1 + e^{x_2} - x_1^2 x_2^3\right) - x_2^3. \qquad (2.134)$$

This design ignored the presence of the zero dynamics subsystem $\dot{\xi}_1 = -\xi_1 + y$. However, this subsystem is input-to-state stable (ISS) with respect to y, so that ξ_1 is bounded because y is bounded, and moreover $\lim_{t \to \infty} \xi_1(t) \to 0$ since $\lim_{t \to \infty} y(t) \to 0$. ◇

We now want to generalize the above example and formulate design tools which allow the original system to be unstable when $y = 0$ and let us backstep more than a simple integrator at a time. Since we want to be able to apply these tools repeatedly, each lemma we formulate must guarantee for the cascade system all the properties assumed for the original system. As we will see, the constructed $V_a(x, \xi)$ for the cascade system does not include the zero dynamics variables, but their boundedness is guaranteed by the boundedness of V_a. Hence, we must reformulate Assumption 2.7 to assume the same properties for the original system, by including the case when $V(x)$ is not positive definite:

2.2 BACKSTEPPING

Assumption 2.21 *Suppose Assumption 2.7 is valid with $V(x)$ positive semidefinite, and the closed-loop system (2.48) with the control (2.49) has the property that $x(t)$ is bounded if $V(x(t))$ is bounded.*

Under this assumption, the control (2.49), applied to the system (2.48), guarantees not only global boundedness of $x(t)$, but also regulation of $W(x(t))$: From (2.50) we conclude that $W(x(t))$ is integrable on $[0, \infty)$ and uniformly continuous, and hence converges to zero by Lemma A.6. Furthermore, since all solutions $x(t)$ are bounded, we can apply LaSalle's theorem (Theorem 2.2) to conclude that $x(t)$ convergences to the largest invariant set M contained in the set $E = \{x \in \mathbb{R}^n \,|\, W(x) = 0\}$. The following fact is easy to prove:

Corollary 2.22 *When Assumption 2.7 is replaced by Assumption 2.21, then the boundedness and convergence properties in part (ii) of Lemma 2.8 still hold.*

Lemma 2.23 (Linear Block Backstepping) *Consider the cascade system*

$$\dot{x} = f(x) + g(x)y, \quad f(0) = 0, \quad x \in \mathbb{R}^n, \quad y \in \mathbb{R} \quad (2.135a)$$
$$\dot{\xi} = A\xi + bu, \quad y = h\xi, \quad \xi \in \mathbb{R}^q, \quad u \in \mathbb{R}, \quad (2.135b)$$

where (2.135b) is a minimum phase system of relative degree one ($hb \neq 0$). If (2.135a) satisfies Assumption 2.21 with y as its input, then there exists a feedback control which guarantees global boundedness and convergence of $\begin{bmatrix} x(t) \\ \xi(t) \end{bmatrix}$ to the largest invariant set M_a contained in the set $E_a = \left\{ \begin{bmatrix} x \\ \xi \end{bmatrix} \in \mathbb{R}^{n+q} \,\middle|\, W(x) = 0, y = \alpha(x) \right\}$. One choice for this control is

$$u = \frac{1}{hb} \left\{ -c(y - \alpha(x)) - hA\xi + \frac{\partial \alpha}{\partial x}(x)\left[f(x) + g(x)y\right] - \frac{\partial V}{\partial x}(x)g(x) \right\}, \quad c > 0. \quad (2.136)$$

Moreover, if $V(x)$ and $W(x)$ are positive definite, then the equilibrium $x = 0, \xi = 0$ is GAS.

Proof. We recall from [164] that the relative-degree-one SISO linear system (2.135b) can be represented in the form

$$\dot{y} = hA\xi + hbu \quad (2.137a)$$
$$\dot{\zeta} = A_0\zeta + b_0 y, \quad (2.137b)$$

where the eigenvalues of A_0 are the (stable) zeros of the transfer function $H(s) = h(sI - A)^{-1}b$ of the minimum phase system (2.135b). Using (2.137) and the feedback transformation

$$u = \frac{1}{hb}(v - hA\xi), \quad (2.138)$$

we rewrite (2.135) as follows:

$$\dot{x} = f(x) + g(x)y \tag{2.139a}$$
$$\dot{y} = v \tag{2.139b}$$
$$\dot{\zeta} = A_0\zeta + b_0 y. \tag{2.139c}$$

We first ignore the zero dynamics (2.139c) and, using Corollary 2.22, apply Lemma 2.8 to (2.139a)–(2.139b) to achieve global boundedness of x and y and regulation of $W(x(t))$ and $y(t) - \alpha(x(t))$. In view of (2.138) and (2.54), one choice of control is given by (2.136). Returning to (2.139c), we note that ζ is bounded because y is bounded and A_0 is Hurwitz. Thus, ξ is bounded. Since all solutions of (2.135) are bounded, we can apply LaSalle's theorem (Theorem 2.2) with $\Omega = \mathbb{R}^{n+q}$ to conclude convergence to the set M_a.

From Lemma 2.8 we also know that if $V(x)$ and $W(x)$ are positive definite, then the equilibrium $x = 0$, $y = 0$ of (2.139a)–(2.139b), which is completely decoupled from (2.139c), is GAS. The fact that in this case the equilibrium $x = 0, \xi = 0$ of the cascade system (2.135) is also GAS follows immediately from the following lemma:

Lemma 2.24 *Consider the cascade system with $\zeta \in \mathbb{R}^m$, $x \in \mathbb{R}^n$:*

$$\dot{\zeta} = A_0\zeta + b_0 y \tag{2.140a}$$
$$\dot{x} = f(x), \quad f(0) = 0 \tag{2.140b}$$
$$y = h(x), \quad h(0) = 0. $$

If (2.140b) is GAS and A_0 is Hurwitz, then the equilibrium $\zeta = 0$, $x = 0$ of the cascade (2.140) is GAS.

Proof. From the definition of GAS (Definition A.4 in Appendix A) we know that the GAS property of (2.140b) implies the existence of class $\mathcal{KL}_\infty$ functions β and β_1 such that

$$|x(t)| \leq \beta(|x(0)|, t), \quad |y(t)| \leq \beta_1(|x(0)|, t). \tag{2.141}$$

The solutions of (2.140a), on the other hand, are given by

$$\zeta(t) = e^{A_0 t}\zeta(0) + \int_0^t e^{A_0(t-\tau)} b_0\, y(\tau)\, d\tau. \tag{2.142}$$

Since A_0 is Hurwitz, we know that $\left|e^{A_0 t}\right| \leq k_1 e^{-\alpha t}$. Using this with (2.141) in (2.142), we obtain

$$|\zeta(t)| \leq \left|e^{A_0 t}\right| |\zeta(0)| + \int_0^t \left|e^{A_0(t-\tau)} b_0\right| |y(\tau)|\, d\tau$$

2.2 BACKSTEPPING

$$\leq k_1 e^{-\alpha t}|\zeta(0)| + k_2 \int_0^t e^{-\alpha(t-\tau)} \beta_1(|x(0)|, \tau) d\tau$$

$$\leq k_1 e^{-\alpha t}|\zeta(0)| + k_2 \sup_{0 \leq \tau \leq t/2} \beta_1(|x(0)|, \tau) \int_0^{t/2} e^{-\alpha(t-\tau)} d\tau$$

$$+ k_2 \sup_{t/2 \leq \tau \leq t} \beta_1(|x(0)|, \tau) \int_{t/2}^t e^{-\alpha(t-\tau)} d\tau$$

$$\leq k_1 e^{-\alpha t}|\zeta(0)| + k_2 \beta_1(|x(0)|, 0) \int_0^{t/2} e^{-\alpha(t-\tau)} d\tau$$

$$+ k_2 \beta_1(|x(0)|, t/2) \int_{t/2}^t e^{-\alpha(t-\tau)} d\tau$$

$$= k_1 e^{-\alpha t}|\zeta(0)| + \frac{k_2}{\alpha} \beta_1(|x(0)|, 0) e^{-\alpha t/2} \left(1 - e^{-\alpha t/2}\right)$$

$$+ \frac{k_2}{\alpha} \beta_1(|x(0)|, t/2) \left(1 - e^{-\alpha t/2}\right)$$

$$\leq k_1 e^{-\alpha t}|\zeta(0)| + \frac{k_2}{\alpha} \beta_1(|x(0)|, 0) e^{-\alpha t/2} + \frac{k_2}{\alpha} \beta_1(|x(0)|, t/2)$$

$$\stackrel{\triangle}{=} \beta_2 \left(\left| \begin{matrix} \zeta(0) \\ x(0) \end{matrix} \right|, t \right), \quad (2.143)$$

where β_2 is a class $\mathcal{KL}_\infty$ function. Combining (2.141) with (2.143) proves that $\zeta = 0, x = 0$ is GAS:

$$\left| \begin{matrix} \zeta(t) \\ x(t) \end{matrix} \right| \leq \beta_3 \left(\left| \begin{matrix} \zeta(0) \\ x(0) \end{matrix} \right|, t \right), \quad \beta_3 \in \mathcal{KL}_\infty. \quad (2.144)$$

□

Comparing Lemmas 2.13 and 2.23 we see that, instead of assuming global stability of $x = 0$ when $y = 0$, Lemma 2.23 assumes only global stabilizability of $x = 0$ through y. The corresponding assumptions on the input subsystems, however, reveal the price paid for this generalization: The minimum phase assumption of Lemma 2.23 is stronger than the FPR assumption of Lemma 2.13, which, as noted in Remark 2.19, allows some zeros to be on the imaginary axis, that is, to be weak minimum phase.

Let us now examine the cascade system

$$\dot{x} = -x(1 + e^{\xi_1}) + x^3 \xi_1^3 \quad (2.145a)$$
$$\dot{\xi}_1 = \xi_1^2 \xi_2^3 \quad (2.145b)$$
$$\dot{\xi}_2 = \xi_2^2 u. \quad (2.145c)$$

As we have already shown in Example 2.18, (2.145a)–(2.145b) is stabilizable through $y = \xi_2^3$, while (2.145c) with this output is FSP. However, if we try to stabilize the cascade (2.145), we run into difficulties because the relative degree of (2.145c) is not defined at $\xi_2 = 0$.

This example shows that we need to assume that the input subsystem

$$\dot{\xi} = m(\xi) + \beta(\xi)u, \quad y = h(\xi), \tag{2.146}$$

has a globally defined constant relative degree. For a nonlinear analog of Lemma 2.23, we also assume that the zero dynamics subsystem of (2.146) is ISS.

Lemma 2.25 (Nonlinear Block Backstepping) *Consider the cascade system:*

$$\dot{x} = f(x) + g(x)y, \quad f(0) = 0, \quad x \in \mathbb{R}^n, \ y \in \mathbb{R} \tag{2.147a}$$
$$\dot{\xi} = m(x,\xi) + \beta(x,\xi)u, \quad y = h(\xi), \ h(0) = 0, \ \xi \in \mathbb{R}^q, \ u \in \mathbb{R}. \tag{2.147b}$$

Assume that (2.147b) has globally defined and constant relative degree one uniformly in x, and that its zero dynamics subsystem is ISS with respect to x and y as its inputs. If (2.147a) satisfies Assumption 2.21 with y as its input, then there exists a feedback control which guarantees global boundedness and convergence of $\begin{bmatrix} x(t) \\ \xi(t) \end{bmatrix}$ to the largest invariant set M_a contained in the set $E_a = \left\{ \begin{bmatrix} x \\ \xi \end{bmatrix} \in \mathbb{R}^{n+q} \ \Big| \ W(x) = 0, \ y = \alpha(x) \right\}$. One particular choice is

$$u = \left(\frac{\partial h}{\partial \xi}(\xi)\beta(x,\xi) \right)^{-1} \left\{ -c(y - \alpha(x)) - \frac{\partial h}{\partial \xi}(\xi)m(x,\xi) \right.$$
$$\left. + \frac{\partial \alpha}{\partial x}(x)\left[f(x) + g(x)y \right] - \frac{\partial V}{\partial x}(x)g(x) \right\}, \quad c > 0. \tag{2.148}$$

Moreover, if $V(x)$ and $W(x)$ are positive definite, then the equilibrium $x = 0, \xi = 0$ is GAS.

Proof. Since the relative degree of the subsystem (2.147b) is globally defined and equal to one uniformly in x, there exists a global[7] change of coordinates of the form (2.90), in particular $(y, \zeta) = (y, \phi(x, \xi))$ with $\frac{\partial \phi}{\partial \xi}\beta \equiv 0$, which transforms (2.147b) into

$$\dot{y} = \frac{\partial h}{\partial \xi}(\xi)m(x,\xi) + \frac{\partial h}{\partial \xi}(\xi)\beta(x,\xi)u \triangleq f_1(x,y,\zeta) + g_1(x,y,\zeta)u \tag{2.149a}$$
$$\dot{\zeta} = \frac{\partial \phi}{\partial x}(x,\xi)\left[f(x) + g(x)y\right] + \frac{\partial \phi}{\partial \xi}(x,\xi)m(x,\xi) \triangleq \Phi(\zeta, x, y). \tag{2.149b}$$

---

[7]This change of coordinates is a global diffeomorphism under additional conditions of connectedness and completeness [13].

2.2 BACKSTEPPING

We now consider the cascade system consisting of (2.147a) and (2.149a). If we linearize (2.149a) with the feedback given by (2.89),

$$u = \left(\frac{\partial h}{\partial \xi}\beta\right)^{-1}\left(v - \frac{\partial h}{\partial \xi}m\right), \qquad (2.150)$$

we obtain $\dot{y} = v$. Then we can apply Lemma 2.8, with v as the new control input, to guarantee global boundedness of x and y and regulation of $W(x(t))$ and $y(t) - \alpha(x(t))$. From (2.149b) and the ISS assumption on the zero dynamics, ζ is also bounded, and thus ξ and u are bounded. Since all solutions of (2.147) are bounded, we can apply LaSalle's theorem (Theorem 2.2) with $\Omega = \mathbb{R}^{n+q}$ to conclude convergence to the set M_a. Combining (2.150) with (2.54), we see that a particular choice of control is given by (2.148).

From Lemma 2.8 we also know that if $V(x)$ and $W(x)$ are positive definite, then the equilibrium $x = 0$, $y = 0$ of (2.147a) and (2.149a), which is completely decoupled from (2.149b), is GAS. The fact that in this case the equilibrium $x = 0, \xi = 0$ of the cascade system (2.147) is also GAS follows from Lemma C.4 by noting that the state (x, y) of the GAS system (2.147a) and (2.149a) is the input of the ISS system (2.149b). □

Lemma 2.25 relaxes the global stability assumption of Lemma 2.17 to global stabilizability of $x = 0$ through y. As in the case of Lemmas 2.13 and 2.23, however, the price paid for this generalization is the strengthening of the FP assumption of Lemma 2.17 to the ISS assumption of Lemma 2.25.

The following example illustrates the use of block backstepping as a design tool.

2.2.6 Example: active suspension (series)

We return to the active suspension example, but in contrast to the parallel configuration of Figure 2.6, we now work with the quarter-car model of Figure 2.7, where the hydraulic actuator is connected in series with the spring/damper system.[8]

In this configuration, the piston position x_a is determined from the equation

$$\dot{x}_\mathrm{a} = \frac{1}{A}Q. \qquad (2.151)$$

Again, the control objective is to stiffen the suspension near its travel limits. Neglecting the wheel acceleration $\ddot{x}_\mathrm{w}$, the acceleration of the car body is equal to the suspension acceleration $\ddot{x}_\mathrm{s}$ and is given by

$$M_\mathrm{b}\ddot{x}_\mathrm{s} = -K_\mathrm{a}(x_\mathrm{s} - x_\mathrm{a}) - C_\mathrm{a}(\dot{x}_\mathrm{s} - \dot{x}_\mathrm{a}). \qquad (2.152)$$

---

[8]Both of these configurations are currently used in active suspension research and design.

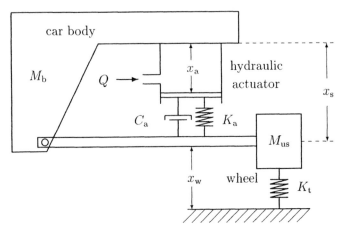

Figure 2.7: Quarter-car model with *series* connection of hydraulic actuator with passive spring/damper.

Combining (2.151) and (2.152), we obtain the following suspension equation:

$$\ddot{x}_s = -\frac{K_a}{M_b}(x_s - x_a) - \frac{C_a}{M_b}\left(\dot{x}_s - \frac{1}{A}Q\right). \tag{2.153}$$

For the flow Q, we consider again the linear equation (2.74):

$$\dot{Q} = -c_f Q + k_f i_v. \tag{2.154}$$

The system composed of (2.151), (2.153), and (2.154) is a linear system which can be rewritten in the form of (2.135) with $x_1 = x_s$, $x_2 = \dot{x}_s$, $\xi_1 = x_a$, $\xi_2 = Q$, $u = i_v$:

$$\begin{aligned}
\dot{x}_1 &= x_2 \\
\dot{x}_2 &= -\frac{K_a}{M_b}x_1 - \frac{C_a}{M_b}x_2 + y \\
\dot{\xi}_1 &= \frac{1}{A}\xi_2 \\
\dot{\xi}_2 &= -c_f\xi_2 + k_f u \\
y &= \frac{K_a}{M_b}\xi_1 + \frac{C_a}{M_b A}\xi_2.
\end{aligned} \tag{2.155}$$

Clearly, the assumptions of Lemma 2.23 are satisfied, since the (ξ_1, ξ_2)-subsystem is minimum phase and its relative degree is one. It should be noted that the assumptions of Lemma 2.13 are also satisfied, since the (x_1, x_2)-subsystem is GAS when $y = 0$, and the (ξ_1, ξ_2)-subsystem is FPR. Thus, if the objective were just stabilization, Lemma 2.13 would be applicable.

2.2 BACKSTEPPING

We first rewrite the system (2.155) in the form (2.139) with $\zeta = \xi_1$:

$$\begin{aligned}
\dot{x}_1 &= x_2 \\
\dot{x}_2 &= -\frac{K_a}{M_b}x_1 - \frac{C_a}{M_b}x_2 + y \\
\dot{y} &= \left(\frac{K_a}{M_b A} - \frac{C_a c_f}{M_b A}\right)\xi_2 + \frac{C_a k_f}{M_b A}u \\
\dot{\zeta} &= -\frac{K_a}{C_a}\zeta + \frac{M_b}{C_a}y.
\end{aligned} \qquad (2.156)$$

Next, we apply Lemma 2.8 to the (x_1, x_2)-subsystem: We view x_2 as the virtual control and design for it the stabilizing function

$$\alpha_1(x_1) = -c_1 x_1 - \kappa_1 x_1^3, \qquad (2.157)$$

which, with $z_1 = x_2 - \alpha_1(x_1)$, yields

$$\dot{x}_1 = -c_1 x_1 - \kappa_1 x_1^3 + z_1. \qquad (2.158)$$

Using (2.155), we obtain

$$\dot{z}_1 = -\frac{K_a}{M_b}x_1 - \frac{C_a}{M_b}x_2 + y + \left(c_1 + 3\kappa_1 x_1^2\right)x_2. \qquad (2.159)$$

Applying Lemma 2.23, we view y as the virtual control in (2.159) and design for it the stabilizing function

$$\alpha_2(x_1, x_2, \xi_1, \xi_2) = -c_2 z_1 - x_1 - \left(c_1 + 3\kappa_1 x_1^2\right)x_2 + \frac{K_a}{M_b}x_1 + \frac{C_a}{M_b}x_2, \qquad (2.160)$$

which results in

$$\begin{aligned}
\dot{x}_1 &= -c_1 x_1 - \kappa_1 x_1^3 + z_1 \\
\dot{z}_1 &= -c_2 z_1 - x_1 + z_2,
\end{aligned} \qquad (2.161)$$

where $z_2 = y - \alpha_2$. Using (2.160) and (2.155), we obtain

$$\dot{z}_2 = \left(\frac{K_a}{M_b A} - \frac{C_a c_f}{M_b A}\right)\xi_2 + \frac{C_a k_f}{M_b A}u + \left(1 - \frac{K_a}{M_b} + 6\kappa_1 x_1 x_2\right)x_2 \\
+ c_2\left(-c_2 z_1 - x_1 + z_2\right) + \left(c_1 + 3\kappa_1 x_1^2 - \frac{C_a}{M_b}\right)\left(-\frac{K_a}{M_b}x_1 - \frac{C_a}{M_b}x_2 + y\right). \qquad (2.162)$$

The control law for the current i_v then becomes

$$u = \frac{M_b A}{C_a k_f}\left[-c_3 z_2 - z_1 - c_2\left(-c_2 z_1 - x_1 + z_2\right) - \left(1 - \frac{K_a}{M_b} + 6\kappa_1 x_1 x_2\right)x_2 \right. \\
\left. - \left(c_1 + 3\kappa_1 x_1^2 - \frac{C_a}{M_b}\right)\left(-\frac{K_a}{M_b}x_1 - \frac{C_a}{M_b}x_2 + y\right)\right] - \left(\frac{K_a}{C_a k_f} - \frac{c_f}{k_f}\right)\xi_2. \qquad (2.163)$$

58 DESIGN TOOLS FOR STABILIZATION

The resulting closed-loop system is nonlinear and is guaranteed to be GAS by Lemma 2.23:

$$\begin{aligned}
\dot{x}_1 &= -c_1 x_1 - \kappa_1 x_1^3 + z_1 \\
\dot{z}_1 &= -c_2 z_1 - x_1 + z_2 \\
\dot{z}_2 &= -c_3 z_2 - z_1 \\
\dot{\zeta} &= -\frac{K_a}{C_a}\zeta + \frac{M_b}{C_a}(z_2 + \alpha_2)\,.
\end{aligned} \quad (2.164)$$

The design objective of preventing large excursions of x_1 is achieved by the presence of the term $-\kappa_1 x_1^3$ in the first equation.

2.3 Recursive Design Procedures

Backstepping tools will now be employed to form systematic design procedures for general classes of nonlinear systems. In increasing order of complexity, the classes considered are strict-feedback systems, pure-feedback systems, and block-strict-feedback systems.

2.3.1 Strict-feedback systems

Nonlinear *strict-feedback systems* are of the form

$$\begin{aligned}
\dot{x} &= f(x) + g(x)\xi_1 \\
\dot{\xi}_1 &= f_1(x,\xi_1) + g_1(x,\xi_1)\xi_2 \\
\dot{\xi}_2 &= f_2(x,\xi_1,\xi_2) + g_2(x,\xi_1,\xi_2)\xi_3 \\
&\vdots \\
\dot{\xi}_{k-1} &= f_{k-1}(x,\xi_1,\ldots,\xi_{k-1}) + g_{k-1}(x,\xi_1,\ldots,\xi_{k-1})\xi_k \\
\dot{\xi}_k &= f_k(x,\xi_1,\ldots,\xi_k) + g_k(x,\xi_1,\ldots,\xi_k)u\,,
\end{aligned} \quad (2.165)$$

where $x \in \mathbb{R}^n$ and $\xi_1,\ldots,\xi_k$ are scalars. The reason for referring to the ξ-subsystem as "strict-feedback" is that the nonlinearities f_i, g_i in the ξ_i-equation ($i = 1,\ldots,k$) depend only on $x, \xi_1,\ldots,\xi_i$, that is, on state variables that are "fed back."

The x-subsystem satisfies Assumption 2.7 with ξ_1 as its control input. Our recursive design starts with the subsystem

$$\begin{aligned}
\dot{x} &= f(x) + g(x)\xi_1 \\
\dot{\xi}_1 &= f_1(x,\xi_1) + g_1(x,\xi_1)\xi_2\,.
\end{aligned} \quad (2.166)$$

If $f_1 \equiv 0$ and $g_1 \equiv 1$, the Integrator Backstepping Lemma 2.8 would be directly applicable to (2.166), treating ξ_2 as the control. In the presence of $f_1(x,\xi_1)$ and

2.3 RECURSIVE DESIGN PROCEDURES

$g_1(x, \xi_1)$, we proceed in the same way by constructing $V_1(x, \xi_1)$ for (2.166) as

$$V_1(x, \xi_1) = V(x) + \frac{1}{2}[\xi_1 - \alpha(x)]^2, \qquad (2.167)$$

where $\alpha(x)$ is a stabilizing feedback that satisfies (2.50) for the x-subsystem. Such intermediate control laws will be called *stabilizing functions*. To find a stabilizing function $\alpha_1(x, \xi_1)$ for ξ_2, the virtual control in (2.166), we need to make the derivative of $\dot{V}_1$ nonpositive when $\xi_2 = \alpha_1$:

$$\begin{aligned}
\dot{V}_1 &\leq -W(x) + [\xi_1 - \alpha(x)]\left\{\frac{\partial V}{\partial x}(x)g(x) + f_1(x, \xi_1) + g_1(x, \xi_1)\xi_2 \right. \\
&\qquad \left. -\frac{\partial \alpha}{\partial x}(x)\left[f(x) + g(x)\xi_1\right]\right\} \\
&= -W(x) + [\xi_1 - \alpha(x)]\left\{\frac{\partial V}{\partial x}(x)g(x) + f_1(x, \xi_1) + g_1(x, \xi_1)\alpha_1(x, \xi_1) \right. \\
&\qquad \left. + g_1(x, \xi_1)[\xi_2 - \alpha_1(x, \xi_1)] - \frac{\partial \alpha}{\partial x}(x)\left[f(x) + g(x)\xi_1\right]\right\} \\
&= -W_1(x, \xi_1) + \frac{\partial V_1}{\partial \xi_1}(x, \xi_1)g_1(x, \xi_1)[\xi_2 - \alpha_1(x, \xi_1)], \qquad (2.168)
\end{aligned}$$

where $W_1(x, \xi_1) > 0$ when $W(x) > 0$ or $\xi_1 \neq \alpha(x)$. If $g_1(x, \xi_1) \neq 0$ for all x and ξ_1, one choice for α_1 is

$$\begin{aligned}
\alpha_1(x, \xi_1) &= \frac{1}{g_1(x, \xi_1)}\left\{-c_1[\xi_1 - \alpha(x)] - \frac{\partial V}{\partial x}(x)g(x) - f_1(x, \xi_1) \right. \\
&\qquad \left. +\frac{\partial \alpha}{\partial x}(x)\left[f(x) + g(x)\xi_1\right]\right\}, \qquad (2.169)
\end{aligned}$$

with $c_1 > 0$, which yields $W_1(x, \xi_1) = W(x) + c_1[\xi_1 - \alpha(x)]^2$. However, as we pointed out before, many other, possibly better, choices for α_1 are available, even if $g_1(x, \xi_1) = 0$ at some points.

With $\alpha_1(x, \xi_1)$ determined, our next step is to augment (2.166) with the ξ_2-equation from (2.165). In a compact notation, we obtain

$$\begin{aligned}
\dot{X}_1 &= F_1(X_1) + G_1(X_1)\xi_2 \\
\dot{\xi}_2 &= f_2(X_1, \xi_2) + g_2(X_1, \xi_2)\xi_3,
\end{aligned} \qquad (2.170)$$

where $f_2(X_1, \xi_2), g_2(X_1, \xi_2)$ stand for $f_2(x, \xi_1, \xi_2), g_2(x, \xi_1, \xi_2)$ and

$$X_1 = \begin{bmatrix} x \\ \xi_1 \end{bmatrix}, \quad F_1(X_1) = \begin{bmatrix} f(x) + g(x)\xi_1 \\ f_1(x, \xi_1) \end{bmatrix}, \quad G_1(X_1) = \begin{bmatrix} 0 \\ g_1(x, \xi_1) \end{bmatrix}. \qquad (2.171)$$

The structure of (2.170) is identical to that of (2.166). We therefore repeat the same step by introducing

$$\begin{aligned}V_2(X_1,\xi_2) &= V_1(X_1) + \frac{1}{2}[\xi_2 - \alpha_1(X_1)]^2 \\ &= V(x) + \frac{1}{2}\sum_{i=1}^{2}[\xi_i - \alpha_{i-1}(X_{i-1})]^2,\end{aligned} \quad (2.172)$$

where, for notational convenience, we use $X_0 = x$ and $\alpha_0(X_0) = \alpha(x)$. The stabilizing function $\alpha_2(X_2)$, $X_2^T = [X_1^T \ \xi_2]^T$, for the virtual control ξ_3 is then determined to render

$$\dot{V}_2 \le -W_2(X_1,\xi_2) + \frac{\partial V_2}{\partial \xi_2}(X_1,\xi_2)g_2(X_1,\xi_2)[\xi_3 - \alpha_2(X_2)], \quad (2.173)$$

with $W_2(X_1,\xi_2) > 0$ when $W_1(x,\xi_1) > 0$ or $\xi_2 \ne \alpha_1(X_1)$.

It is clear that this procedure will terminate at the kth step, at which the whole system (2.165) is to be stabilized by the actual control u. In our compact notation, (2.165) is rewritten as

$$\begin{aligned}\dot{X}_{k-1} &= F_{k-1}(X_{k-1}) + G_{k-1}(X_{k-1})\xi_k \\ \dot{\xi}_k &= f_k(X_{k-1},\xi_k) + g_k(X_{k-1},\xi_k)u,\end{aligned} \quad (2.174)$$

where

$$X_{k-1} = \begin{bmatrix} X_{k-2} \\ \xi_{k-1} \end{bmatrix}, \ F_{k-1}(X_{k-1}) = \begin{bmatrix} F_{k-2}(X_{k-2}) + G_{k-2}(X_{k-2})\xi_{k-1} \\ f_{k-1}(X_{k-2},\xi_{k-1}) \end{bmatrix} \\ G_{k-1}(X_{k-1}) = \begin{bmatrix} 0 \\ g_{k-1}(X_{k-2},\xi_{k-1}) \end{bmatrix}. \quad (2.175)$$

Once again, this is in the form of (2.166) and (2.170), and the Lyapunov function for (2.174) is

$$\begin{aligned}V_k(x,\xi_1,\ldots,\xi_k) &= V_{k-1}(X_{k-1}) + \frac{1}{2}[\xi_k - \alpha_{k-1}(X_{k-1})]^2 \\ &= V(x) + \frac{1}{2}\sum_{i=1}^{k}[\xi_i - \alpha_{i-1}(X_{i-1})]^2.\end{aligned} \quad (2.176)$$

The meaning of the stabilizing functions $\alpha_i(X_i)$ designed as intermediate control laws, is now clearer from (2.176). These would-be control laws are, in fact, the tools to construct a Lyapunov function for the strict-feedback system (2.165). The function V_k in (2.176) is indeed a Lyapunov function, because u can be chosen to make $\dot{V}_k \le -W_k \le 0$, with $W_k > 0$ when $W_{k-1} > 0$ or

2.3 Recursive Design Procedures

$\xi_k \neq \alpha_{k-1}$:

$$\begin{aligned}
\dot{V}_k &= \dot{V}_{k-1} + (\xi_k - \alpha_{k-1})\left[f_k + g_k u - \frac{\partial \alpha_{k-1}}{\partial X_{k-1}}(F_{k-1} + G_{k-1}\xi_k)\right] \\
&\leq -W_{k-1}(X_{k-2}, \xi_{k-1}) + \frac{\partial V_{k-1}}{\partial \xi_{k-1}} g_{k-1}(\xi_k - \alpha_{k-1}) \\
&\quad + (\xi_k - \alpha_{k-1})\left[f_k + g_k u - \frac{\partial \alpha_{k-1}}{\partial X_{k-1}}(F_{k-1} + G_{k-1}\xi_k)\right] \\
&= -W_{k-1}(X_{k-2}, \xi_{k-1}) + (\xi_k - \alpha_{k-1})\left[\frac{\partial V_{k-1}}{\partial \xi_{k-1}} g_{k-1} + f_k + g_k u \right. \\
&\quad \left. - \frac{\partial \alpha_{k-1}}{\partial X_{k-1}}(F_{k-1} + G_{k-1}\xi_k)\right] \\
&\leq -W_k(X_{k-1}, \xi_k) \leq 0.
\end{aligned} \quad (2.177)$$

If the nonsingularity condition

$$g_k(x, \xi_1, \ldots, \xi_k) \neq 0, \quad \forall x \in \mathbb{R}^n, \ \forall \xi_i \in \mathbb{R}, \ i = 1, \ldots, k, \quad (2.178)$$

is satisfied, then the simplest choice for u is

$$u = \frac{1}{g_k}\left[-c_k(\xi_k - \alpha_{k-1}) - \frac{\partial V_{k-1}}{\partial \xi_{k-1}} g_{k-1} - f_k + \frac{\partial \alpha_{k-1}}{\partial X_{k-1}}(F_{k-1} + G_{k-1}\xi_k)\right], \quad (2.179)$$

with $c_k > 0$, which yields $W_k = W_{k-1} + c_k(\xi_k - \alpha_{k-1})^2$. We reiterate that a more desirable control law with less cancellations may also be found, even if the condition (2.178) is violated at some points.

2.3.2 Pure-feedback systems

A more general class of "triangular" systems comprises *pure-feedback systems*:

$$\begin{aligned}
\dot{x} &= f(x) + g(x)\xi_1 \\
\dot{\xi}_1 &= f_1(x, \xi_1, \xi_2) \\
\dot{\xi}_2 &= f_2(x, \xi_1, \xi_2, \xi_3) \\
&\vdots \\
\dot{\xi}_{k-1} &= f_{k-1}(x, \xi_1, \ldots, \xi_k) \\
\dot{\xi}_k &= f_k(x, \xi_1, \ldots, \xi_k, u),
\end{aligned} \quad (2.180)$$

where $\xi_i \in \mathbb{R}$. The x-subsystem again satisfies Assumption 2.7.

Compared with the strict-feedback systems of (2.165), pure-feedback systems lack the affine appearance of the variables ξ_k, to be used as virtual controls, and of the actual control u itself. Our recursive procedure starts with

the subsystem
$$\begin{aligned} \dot{x} &= f(x) + g(x)\xi_1 \\ \dot{\xi}_1 &= f_1(x, \xi_1, \xi_2), \end{aligned} \qquad (2.181)$$

in which ξ_2 is the virtual control. Our Lyapunov function for (2.181) is again

$$V_1(x, \xi_1) = V(x) + \frac{1}{2}[\xi_1 - \alpha(x)]^2, \qquad (2.182)$$

and its derivative $\dot{V}_1$ is to be rendered nonpositive by $\xi_2 = \alpha_1(x, \xi_1)$:

$$\begin{aligned} \dot{V}_1 &\leq -W(x) + [\xi_1 - \alpha(x)]\left\{\frac{\partial V}{\partial x}(x)g(x) + f_1(x, \xi_1, \xi_2)\right. \\ &\quad \left. -\frac{\partial \alpha}{\partial x}(x)[f(x) + g(x)\xi_1]\right\} \\ &= -W(x) + [\xi_1 - \alpha(x)]\left\{\frac{\partial V}{\partial x}(x)g(x) + f_1(x, \xi_1, \alpha_1(x, \xi_1))\right. \\ &\quad \left. +\bar{f}_1(x, \xi_1, \xi_2)[\xi_2 - \alpha_1(x, \xi_1)] - \frac{\partial \alpha}{\partial x}(x)[f(x) + g(x)\xi_1]\right\} \\ &= -W_1(x, \xi_1) + \frac{\partial V_1}{\partial \xi_1}(x, \xi_1)\bar{f}_1(x, \xi_1, \xi_2)[\xi_2 - \alpha_1(x, \xi_1)], \qquad (2.183) \end{aligned}$$

where $W_1(x, \xi_1) > 0$ when $W(x) > 0$ or $\xi_1 \neq \alpha(x)$, and the function $\bar{f}_1$ is smooth if f_1 is smooth:

$$f_1(x, \xi_1, \xi_2) - f_1(x, \xi_1, \alpha_1(x, \xi_1)) = \bar{f}_1(x, \xi_1, \xi_2)[\xi_2 - \alpha_1(x, \xi_1)]. \qquad (2.184)$$

Previously, when $g_1 \neq 0$, a simple choice for α_1 was to render $W_1(x, \xi_1) = W(x) + c_1[\xi_1 - \alpha(x)]^2$. If we attempt to do the same in (2.183), we would have to solve for α_1 the following equation:

$$f_1(x, \xi_1, \alpha_1(x, \xi_1)) = -c_1[\xi_1 - \alpha(x)] + \frac{\partial \alpha}{\partial x}(x)[f(x) + g(x)\xi_1]. \qquad (2.185)$$

By the Implicit Function Theorem (see, e.g., [81]), a necessary condition for the solvability of (2.185) with respect to α_1 is

$$\frac{\partial f_1}{\partial \xi_2}(x, \xi_1, \xi_2) \neq 0, \quad \forall (x, \xi_1, \xi_2) \in \mathbb{R}^{n+2}. \qquad (2.186)$$

This is quite restrictive and unnecessary for our ability to find an α_1 that satisfies (2.183). Even when (2.186) is violated at a set of points, such a smooth stabilizing function $\alpha_1(x, \xi_1)$ may exist. We therefore proceed assuming that such an α_1 has been found.

2.3 RECURSIVE DESIGN PROCEDURES

At the next step we augment the subsystem (2.181) by one more equation from (2.180)

$$\begin{aligned} \dot{X}_1 &= F_1(X_1, \xi_2) \\ \dot{\xi}_2 &= f_2(X_1, \xi_2, \xi_3) \,, \end{aligned} \qquad (2.187)$$

where $f_2(X_1, \xi_2, \xi_3)$ stands for $f_2(x, \xi_1, \xi_2, \xi_3)$ and

$$X_1 = \begin{bmatrix} x \\ \xi_1 \end{bmatrix}, \quad F_1(X_1, \xi_2) = \begin{bmatrix} f(x) + g(x)\xi_1 \\ f_1(x, \xi_1, \xi_2) \end{bmatrix}. \qquad (2.188)$$

We use ξ_3 as the virtual control to stabilize (2.187) with respect to the Lyapunov function

$$\begin{aligned} V_2(X_1, \xi_2) &= V_1(X_1) + \frac{1}{2}[\xi_2 - \alpha_1(X_1)]^2 \\ &= V(x) + \frac{1}{2}\sum_{i=1}^{2}[\xi_i - \alpha_{i-1}(X_{i-1})]^2 \,, \end{aligned} \qquad (2.189)$$

with $X_0 = x$ and $\alpha_0(X_0) = \alpha(x)$. Our task is to find a stabilizing function $\alpha_2(X_2)$, $X_2^T = [X_1^T \ \xi_2]^T$, to yield

$$\dot{V}_2 \leq -W_2(X_1, \xi_2) + \frac{\partial V_2}{\partial \xi_2}(X_1, \xi_2)\bar{f}_2(X_1, \xi_2, \xi_3)[\xi_3 - \alpha_2(X_2)], \qquad (2.190)$$

with $\bar{f}_2(X_1, \xi_2, \xi_3) = f_2(X_1, \xi_2, \xi_3) - f_2(X_1, \xi_2, \alpha_2(X_1, \xi_2))$, and $W_2(X_1, \xi_2) > 0$ when $W_1(x, \xi_1) > 0$ or $\xi_2 \neq \alpha_1(X_1)$. Once again, if we want $W_2(X_1, \xi_2) = W_1(x, \xi_1) + c_2[\xi_2 - \alpha_1(X_1)]^2$, we need

$$\frac{\partial f_2}{\partial \xi_3}(X_1, \xi_2, \xi_3) \neq 0, \quad \forall X_1 \in \mathbb{R}^{n+1}, \ \forall \xi_2 \in \mathbb{R}, \ \forall \xi_3 \in \mathbb{R}. \qquad (2.191)$$

In most situations we would avoid this requirement by directly finding $\alpha_2(X_2)$ to satisfy the inequality (2.190).

Proceeding in the same fashion, in the kth step we arrive at the actual control u in

$$\begin{aligned} \dot{X}_{k-1} &= F_{k-1}(X_{k-1}, \xi_k) \\ \dot{\xi}_k &= f_k(X_{k-1}, \xi_k, u) \,, \end{aligned} \qquad (2.192)$$

where

$$X_{k-1} = \begin{bmatrix} X_{k-2} \\ \xi_{k-1} \end{bmatrix}, \quad F_{k-1}(X_{k-1}, \xi_k) = \begin{bmatrix} F_{k-2}(X_{k-2}, \xi_{k-1}) \\ f_{k-1}(X_{k-2}, \xi_{k-1}, \xi_k) \end{bmatrix}. \qquad (2.193)$$

For the system (2.192) we use the Lyapunov function

$$\begin{aligned} V_k(x, \xi_1, \ldots, \xi_k) &= V_{k-1}(X_{k-1}) + \frac{1}{2}[\xi_k - \alpha_{k-1}(X_{k-1})]^2 \\ &= V(x) + \frac{1}{2}\sum_{i=1}^{k}[\xi_i - \alpha_{i-1}(X_{i-1})]^2 \,. \end{aligned} \qquad (2.194)$$

The design is completed by finding a control law

$$u = \alpha_k(x, \xi_1, \ldots, \xi_k) \qquad (2.195)$$

which makes $\dot{V}_k \leq -W_k \leq 0$, with $W_k > 0$ when $W_{k-1} > 0$ or $\xi_k \neq \alpha_{k-1}$:

$$\begin{aligned}
\dot{V}_k &= \dot{V}_{k-1} + (\xi_k - \alpha_{k-1}) \left[f_k - \frac{\partial \alpha_{k-1}}{\partial X_{k-1}} F_{k-1} \right] \\
&\leq -W_{k-1}(X_{k-2}, \xi_{k-1}) + \frac{\partial V_{k-1}}{\partial \xi_{k-1}} \bar{f}_{k-1}(\xi_k - \alpha_{k-1}) \\
&\quad + (\xi_k - \alpha_{k-1}) \left[f_k(X_{k-1}, \xi_k, u) - \frac{\partial \alpha_{k-1}}{\partial X_{k-1}} F_{k-1} \right] \\
&= -W_k(X_{k-1}, \xi_k) \leq 0.
\end{aligned} \qquad (2.196)$$

Once again, under the condition

$$\frac{\partial f_k}{\partial u}(X_{k-1}, \xi_k, u) \neq 0, \quad \forall X_{k-1} \in \mathbb{R}^{n+k-1}, \ \forall \xi_k \in \mathbb{R}, \ \forall u \in \mathbb{R}, \qquad (2.197)$$

we would be able to find $u = \alpha_k$ to yield $W_k = W_{k-1} + c_k \left[\xi_k - \alpha_{k-1} \right]^2$. However, not only can (2.196) often be satisfied even if (2.197) is violated, but even when this condition is satisfied, we may prefer a different choice of W_k.

2.3.3 Block-strict-feedback systems

Lemma 2.25 can also be applied repeatedly to design controllers for nonlinear systems which can be transformed, by a change of coordinates, into the *block-strict-feedback form*:

$$\begin{aligned}
\dot{x} &= f(x) + g(x) y_1 \\
\dot{\chi}_1 &= \tilde{f}_1(x, \chi_1) + \tilde{g}_1(x, \chi_1) y_2 \\
y_1 &= h_1(\chi_1) \\
\dot{\chi}_2 &= \tilde{f}_2(x, \chi_1, \chi_2) + \tilde{g}_2(x, \chi_1, \chi_2) y_3 \\
y_2 &= h_2(\chi_2) \\
&\vdots \\
\dot{\chi}_i &= \tilde{f}_i(x, \chi_1, \ldots, \chi_i) + \tilde{g}_i(x, \chi_1, \ldots, \chi_i) y_{i+1} \\
y_i &= h_i(\chi_i) \\
&\vdots \\
\dot{\chi}_{k-1} &= \tilde{f}_{\rho-1}(x, \chi_1, \ldots, \chi_{k-1}) + \tilde{g}_{k-1}(x, \chi_1, \ldots, \chi_{k-1}) y_\rho \\
y_{k-1} &= h_{k-1}(\chi_{k-1}) \\
\dot{\chi}_k &= \tilde{f}_k(x, \chi) + \tilde{g}_k(x, \chi) u \\
y_k &= h_k(\chi_k),
\end{aligned} \qquad (2.198)$$

2.3 RECURSIVE DESIGN PROCEDURES

where each of the k subsystems with state $\chi_i \in \mathbb{R}^{n_i}$, output $y_i \in \mathbb{R}$, and input y_{i+1} (for convenience we denote $y_{k+1} \equiv u$) satisfies the following conditions:

(BSF-1) its relative degree is one uniformly in $x, \chi_1, \ldots, \chi_{i-1}$, and

(BSF-2) its zero dynamics subsystem is ISS with respect to $x, \chi_1, \ldots, \chi_{i-1}, y_i$.

Under conditions (BSF-1) and (BSF-2), the system (2.198) can be transformed into a form reminiscent of the strict-feedback form (2.165). In particular, (BSF-1) is equivalent to

$$\frac{\partial h_i}{\partial \chi_i} \tilde{g}_i \neq 0, \quad \forall \chi_1 \in \mathbb{R}^{n_1}, \ldots, \forall \chi_i \in \mathbb{R}^{n_i}, \quad i = 1, \ldots, k. \tag{2.199}$$

This means that for each χ_i-subsystem in (2.198) there exists a global change of coordinates $(y_i, \zeta_i) = (h_i(\chi_i), \phi_i(x, \chi_1, \ldots, \chi_i))$, with $\frac{\partial \phi_i}{\partial \chi_i} \tilde{g}_i \equiv 0$, which transforms it into the normal form (2.149):

$$\begin{aligned}
\dot{y}_i &= \frac{\partial h_i}{\partial \chi_i}(\chi_i) \left[\tilde{f}_i(x, \chi_1, \ldots, \chi_i) + \tilde{g}_i(x, \chi_1, \ldots, \chi_i) y_{i+1} \right] \\
&\stackrel{\triangle}{=} f_i(x, y_1, \zeta_1 \ldots, y_i, \zeta_i) + g_i(x, y_1, \zeta_1 \ldots, y_i, \zeta_i) y_{i+1} \qquad (2.200\text{a}) \\
\dot{\zeta}_i &= \sum_{j=1}^{i-1} \frac{\partial \phi_i}{\partial \chi_j}(x, \chi_1, \ldots, \chi_i) \left[\tilde{f}_j(x, \chi_1, \ldots, \chi_j) + \tilde{g}_j(x, \chi_1, \ldots, \chi_j) y_{j+1} \right] \\
&\quad + \frac{\partial \phi_i}{\partial \chi_i}(x, \chi_1, \ldots, \chi_i) \tilde{f}_i(x, \chi_1, \ldots, \chi_i) \\
&\stackrel{\triangle}{=} \bar{\Phi}_i(x, \chi_1, \ldots, \chi_{i-1}, y_i, \zeta_i) \\
&\stackrel{\triangle}{=} \bar{\Phi}_i(x, y_1, \zeta_1, \ldots, y_{i-1}, \zeta_{i-1}, y_i, \zeta_i). \qquad (2.200\text{b})
\end{aligned}$$

With this change of coordinates, (2.198) is transformed into

$$\begin{aligned}
\dot{x} &= f(x) + g(x) y_1 \\
\dot{y}_1 &= f_1(x, y_1, \zeta_1) + g_1(x, y_1, \zeta_1) y_2 \\
\dot{y}_2 &= f_2(x, y_1, \zeta_1, y_2, \zeta_2) + g_2(x, y_1, \zeta_1, y_2, \zeta_2) y_3 \\
&\vdots \qquad\qquad\qquad\qquad\qquad\qquad\qquad\qquad\qquad (2.201) \\
\dot{y}_{k-1} &= f_{k-1}(x, y_1, \zeta_1, \ldots, y_{k-1}, \zeta_{k-1}) + g_{k-1}(x, y_1, \zeta_1, \ldots, y_{k-1}, \zeta_{k-1}) y_k \\
\dot{y}_k &= f_k(x, y_1, \zeta_1, \ldots, y_k, \zeta_k) + g_k(x, y_1, \zeta_1, \ldots, y_k, \zeta_k) u \\
\dot{\zeta}_1 &= \bar{\Phi}_1(x, y_1, \zeta_1) \\
&\vdots \\
\dot{\zeta}_k &= \bar{\Phi}_k(x, y_1, \zeta_1, \ldots, y_{k-1}, \zeta_{k-1}, y_k, \zeta_k).
\end{aligned}$$

If the zero dynamics variables $\zeta_1,\ldots,\zeta_k$ were not present, (2.201) would be identical to the strict-feedback form (2.165) with ξ_i replaced by y_i. Hence, the design procedure of Section 2.3.1 can be applied *mutatis mutandis* to (2.201). The presence of ζ_i generates a few new terms and requires a modification of the proof of boundedness as well. First the boundedness of $x(t), y_1(t),\ldots, y_k(t)$ and the regulation of $W(x(t)), y_1(t) - \alpha(x(t)),\ldots,y_k(t) - \alpha_{k-1}(x(t), y_1(t), \zeta_1(t),\ldots,y_k(t), \zeta_k(t))$ is established via a Lyapunov-like argument using the function $V_k(x, y_1,\ldots, y_k) = V(x) + \frac{1}{2}\sum_{i=1}^{k}[y_i - \alpha_{i-1}(x, y_1, \zeta_1,\ldots, y_{i-1}, \zeta_{i-1})]^2$. This implies that y_1 is bounded. Then, the boundedness of $y_2,\ldots, y_k$, $\zeta_1,\ldots,\zeta_k$ and u (and, consequently, the boundedness of $\chi_1,\ldots,\chi_k$) is established via an induction argument for $i = 2,\ldots, k+1$ (with $y_{k+1} \equiv u$): If $y_1,\ldots, y_{i-1}$ and $\zeta_1,\ldots,\zeta_{i-2}$ are bounded, the ISS stability assumption on the ζ_{i-1}-subsystem guarantees that ζ_{i-1} is bounded. This implies that $\alpha_i(x, y_1, \zeta_1,\ldots, y_{i-1}, \zeta_{i-1})$ is bounded, which implies that y_i is also bounded. This completes the induction argument and shows that $\chi_1,\ldots,\chi_k, u$ are bounded, since they can be expressed as smooth functions of $x, y_1,\ldots, y_k, \zeta_1,\ldots,\zeta_k$.

2.4 Design Flexibility: Jet Engine Example

We have presented several backstepping, cascade, and block-backstepping design tools and procedures which make the design of nonlinear controllers systematic. We stress, however, that 'systematic' means neither rigid nor dogmatic. Following the same principles, various modifications of the tools and procedures are possible. The recursive construction of control Lyapunov functions is flexible, and so is the choice of stabilizing functions.

At present there are no specific optimality criteria to help us select the best member of the backstepping controller family. However, there are certain applications-oriented guidelines which in most cases will lead to a simpler and more robust controller.

It is clear from the design procedures that the complexity of the controller increases with the number of recursive steps. Much can be gained if the number of steps can be reduced. It is also desirable to satisfy the $\dot{V}$-inequalities with as few cancellations as possible. Exact cancellations can rarely be implemented and cancellation errors may lead to nonrobustness. Additional analysis may be required to identify useful nonlinearities and avoid their cancellation. For this purpose, a more flexible construction of control Lyapunov functions can be employed. Some of these guidelines are now illustrated on a design example of major practical interest.[9]

---

[9]This section is based on Krstić and Kokotović [105].

2.4.1 Jet engine stall and surge

Jet engine compression systems (Figure 2.8) have recently become the subject of intensive control studies aimed at understanding and preventing two types of instability: *rotating stall* and *surge*. Rotating stall manifests itself as a region of severely reduced flow that rotates at a fraction of the rotor speed. Surge is an axisymmetric pumping oscillation which can cause flameout and engine damage.

The simplest model[10] that describes these instabilities is a three-state Galerkin approximation of the nonlinear PDE model by Moore and Greitzer [137]. This model exhibits bifurcations analyzed by McCaughan [128], and was used by Liaw and Abed [113] for a nonlinear feedback control design. Control designs with experimental verifications are reported in Paduano et al. [149] and Eveker and Nett [34].

We will design a feedback controller for the three-state model

$$\dot{\Phi} = -\Psi + \Psi_C(\Phi) - 3\Phi R \tag{2.202}$$

$$\dot{\Psi} = \frac{1}{\beta^2}(\Phi - \Phi_T) \tag{2.203}$$

$$\dot{R} = \sigma R\left(1 - \Phi^2 - R\right), \quad R(0) \geq 0, \tag{2.204}$$

where Φ is the mass flow, Ψ is the pressure rise, $R \geq 0$ is the normalized stall cell squared amplitude, Φ_T is the mass flow through the throttle, and σ and β are constant positive parameters. The compressor and throttle characteristics, $\Psi_C(\Phi)$ and $\Phi_T(\Psi)$, analyzed in [128], are:

$$\Psi_C(\Phi) = \Psi_{C0} + 1 + \frac{3}{2}\Phi - \frac{1}{2}\Phi^3 \tag{2.205}$$

$$\Psi = \frac{1}{\gamma}(1 + \Phi_T(\Psi))^2, \tag{2.206}$$

where Ψ_{C0} is a constant, and γ can be changed by varying the throttle opening. The characteristic $\Psi_C(\Phi)$ has its maximum at $\Phi = 1$. The no-stall equilibria have $R = 0$ and they cannot be stable for $\Phi^2 < 1$, as can be easily seen from (2.204). The pairs of no-stall equilibrium values Φ^e, Ψ^e are given by the intersections of the characteristics $\Psi_C(\Phi)$ and $\Phi_T(\Psi)$ in (2.205) and (2.206).

Using the flow through the throttle Φ_T as a control input, our objective is to stabilize the equilibrium $R^e = 0$, $\Phi^e = 1$, $\Psi^e = \Psi_C(\Phi^e) = \Psi_{C0} + 2$. We translate the origin to the desired equilibrium, $\phi = \Phi - 1$, $\psi = \Psi - \Psi_{C0} - 2$, and let the control variable be

$$u = \frac{1}{\beta^2}(\Phi_T - 1 - \phi). \tag{2.207}$$

---

[10]The control problem for this jet engine model has been brought to our attention by Carl Nett of United Technologies Research Center and Jim Paduano of MIT.

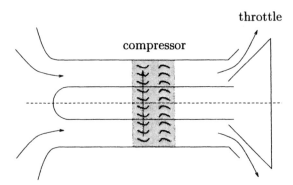

Figure 2.8: Compression system.

The model (2.202)–(2.204) is then rewritten as

$$\dot{R} = -\sigma R^2 - \sigma R \left(2\phi + \phi^2\right) \quad (2.208)$$

$$\dot{\phi} = -\psi - \frac{3}{2}\phi^2 - \frac{1}{2}\phi^3 - 3R\phi - 3R \quad (2.209)$$

$$\dot{\psi} = -u. \quad (2.210)$$

These equations are reordered to reveal the similarity with the pure-feedback form (2.180). The only discrepancy is in the first equation which is not affine in the second variable ϕ. Here is our chance to avoid being dogmatic. The only reason the $\dot{x}$-equation in (2.180) was assumed to be affine in ξ_1 was to simplify the statement of the stabilizability condition (2.50) in Assumption 2.7. The stabilizability of the $\dot{R}$-equation (2.208) is obvious by inspection. A stabilizing virtual control is simply $\phi = \alpha(R) = 0$ which yields $\dot{R} = -\sigma R^2$. Since physically $R \geq 0$, this means that $R(t) \to 0$. We can now use $V_2 = \frac{1}{2}R^2 + \frac{1}{2}\phi^2$ and proceed to the second step of backstepping. This design, which is left to the reader, is completed in three steps.

2.4.2 A two-step design

When at the first step of a three-step design the stabilizing function α_1 is zero, this suggests that a simpler two-step design is possible. We therefore consider that the initial subsystem consists of the first two equations (2.208) and (2.209). This subsystem is cascaded with the integrator (2.210) and the whole system can be viewed as being in the cascade form (2.147) where (R, ϕ) is x and ψ is ξ. Following Lemma 2.25, we need to satisfy Assumption 2.21 using ψ to stabilize the (R, ϕ)-subsystem. This can be done using the semidefinite $V_1 = \phi^2$ because the R-equation is ISS as can be seen from (2.208):

$$\begin{aligned}\dot{R} &\leq -\sigma R^2 + 2\sigma R|\phi| - \sigma\phi^2 R \\ &\leq -\frac{\sigma}{2}R^2 - \frac{\sigma}{2}R(R - 4|\phi|),\end{aligned} \quad (2.211)$$

2.4 DESIGN FLEXIBILITY: JET ENGINE EXAMPLE

which implies that when $R(t) > 4|\phi(t)|$, $R(t)$ decays faster than the solution of $\dot{w} = -\frac{\sigma}{2}w^2$. Hence, an upper bound for $R(t)$ is

$$R(t) \leq \frac{R(0)}{1 + R(0)\sigma t/2} + 4 \sup_{0 \leq \tau \leq t} |\phi(\tau)|. \tag{2.212}$$

Clearly, R is bounded if ϕ is bounded, and $R \to 0$ if $\phi \to 0$.

For ψ as the virtual control of the $\dot{\phi}$-equation we choose

$$\alpha(\phi, R) = c_1\phi - \frac{3}{2}\phi^2 - 3R. \tag{2.213}$$

With this choice we have avoided cancellation of the useful nonlinearities $-\frac{1}{2}\phi^3$ and $-3R\phi$. Substituting $\tilde{\psi} = \psi - \alpha(\phi, R)$ in (2.209), we get

$$\dot{\phi} = -c_1\phi - \frac{1}{2}\phi^3 - 3R\phi - \tilde{\psi}. \tag{2.214}$$

At the second step we differentiate $\dot{\tilde{\psi}} = \dot{\psi} - \dot{\alpha}(\phi, R) = -u - \frac{\partial \alpha}{\partial \phi}\dot{\phi} - \frac{\partial \alpha}{\partial R}\dot{R}$ and obtain

$$\dot{\tilde{\psi}} = -u - (c_1 - 3\phi)\left(-\psi - \frac{3}{2}\phi^2 - \frac{1}{2}\phi^3 - 3\phi R - 3R\right) + 3\sigma R\left(-2\phi - \phi^2 - R\right). \tag{2.215}$$

The control law is now chosen to render the derivative of $V_2 = \phi^2 + \tilde{\psi}^2$ negative definite:

$$\begin{aligned} u &= c_2\tilde{\psi} - \phi - (c_1 - 3\phi)\left(-\psi - \frac{3}{2}\phi^2 - \frac{1}{2}\phi^3 - 3\phi R - 3R\right) \\ &\quad + 3\sigma R\left(-2\phi - \phi^2 - R\right). \end{aligned} \tag{2.216}$$

Note that in the $(R, \phi, \tilde{\psi})$-space the Lyapunov function V_2 is only positive semidefinite, which is allowed by Assumption 2.21 and Lemma 2.25. In contrast, for the three-step design outlined in the preceding subsection, the final clf would be $V_2 = R^2 + \phi^2 + \tilde{\psi}^2$ and its derivative would involve terms from the $\dot{R}$-equation. This would make the control law more complicated than (2.216).

With the control (2.216) the resulting feedback system is

$$\begin{aligned} \dot{R} &= -\sigma R^2 - \sigma R\left(2\phi + \phi^2\right) \\ \dot{\phi} &= -\left(c_1 + \tfrac{1}{2}\phi^2 + 3R\right)\phi - \tilde{\psi} \\ \dot{\tilde{\psi}} &= \phi - c_2\tilde{\psi}. \end{aligned} \tag{2.217}$$

The equilibrium $(\phi, \tilde{\psi}) = 0$ of the $(\phi, \tilde{\psi})$-subsystem is GAS for all $R \geq 0$. In addition, $R(t) \to 0$ because of the ISS-property (2.212). This means that surge and stall are suppressed within the region of validity of this jet engine model.

2.4.3 Avoiding cancellations

Although we have already avoided cancellation of several useful nonlinearities, a further systematic simplification of the above controller is possible by a better choice of α and a more flexible construction of the control Lyapunov function V_2. We illustrate this possibility on the no-stall ($R = 0$) part of the jet engine model (2.208)–(2.210), rewritten here as

$$\dot{\phi} = -\psi - \frac{3}{2}\phi^2 - \frac{1}{2}\phi^3 \tag{2.218}$$

$$\dot{\psi} = -u. \tag{2.219}$$

To design a stabilizing function $\alpha(\phi)$ for ψ in (2.218) with respect to $V_1(\phi) = \frac{1}{2}\phi^2$, we examine the inequality $\dot{V}_1 < 0$, that is,

$$\phi\left[-\alpha(\phi) - \frac{3}{2}\phi^2 - \frac{1}{2}\phi^3\right] < 0, \quad \forall \phi \neq 0. \tag{2.220}$$

We already noticed that $-\frac{1}{2}\phi^3$ enhances this inequality and we did not cancel it in (2.213). But if we go further and rewrite (2.220) as

$$\phi\left[-\alpha_1(\phi) - \frac{1}{2}\left(\phi + \frac{3}{2}\right)^2\phi + \frac{9}{8}\phi\right] < 0, \quad \forall \phi \neq 0, \tag{2.221}$$

we recognize that $-\frac{1}{2}\left(\phi + \frac{3}{2}\right)^2\phi$ is also useful and should not be cancelled. Therefore, $\alpha(\phi)$ is chosen to be linear

$$\alpha(\phi) = \left(c_1 + \frac{9}{8}\right)\phi, \quad c_1 > 0. \tag{2.222}$$

The derivative of $V_1(\phi) = \frac{1}{2}\phi^2$ is then

$$\begin{aligned}\dot{V}_1 &= -c_1\phi^2 - \frac{1}{2}\left(\phi + \frac{3}{2}\right)^2\phi^2 - \phi\tilde{\psi} \\ &\leq -c_1\phi^2 - \phi\tilde{\psi},\end{aligned} \tag{2.223}$$

where $\tilde{\psi} = \psi - \alpha(\phi)$.

In the second step we denote $c_0 = c_1 + \frac{9}{8}$, and by differentiating $\tilde{\psi} = \psi - c_0\phi$ we get

$$\dot{\tilde{\psi}} = -u + c_0\left(\psi + \frac{3}{2}\phi^2 + \frac{1}{2}\phi^3\right). \tag{2.224}$$

Proceeding as usual, our choice for the Lyapunov function V_2 would be quadratic, $V_2(\phi, \psi) = V_1(\phi) + \frac{1}{2}\tilde{\psi}^2 = \frac{1}{2}\phi^2 + \frac{1}{2}\tilde{\psi}^2$, resulting in $\dot{V}_2 \leq -c_1\phi^2 +$

2.4 DESIGN FLEXIBILITY: JET ENGINE EXAMPLE

$\tilde{\psi}\left[-u + c_0\left(\psi + \frac{3}{2}\phi^2 + \frac{1}{2}\phi^3\right) - \phi\right]$. To satisfy the inequality $\dot{V}_2 < 0$, the control law would have to cancel the nonlinearities $c_0\frac{3}{2}\phi^2$ and $c_0\frac{1}{2}\phi^3$. We will now illustrate the construction of a more flexible Lyapunov function

$$V_2 = V_1 + F(V_1) + \frac{1}{2}\tilde{\psi}^2, \qquad (2.225)$$

where $F(\cdot)$ is yet to be selected as a continuously differentiable, nonnegative, and increasing function, $\frac{dF(V_1)}{dV_1} \geq 0$. In view of (2.223) and (2.224), the derivative of (2.225) satisfies

$$\dot{V}_2 \leq -c_1\phi^2 - \phi\tilde{\psi} + \frac{dF(V_1)}{dV_1}\left(-c_1\phi^2 - \phi\tilde{\psi}\right)$$
$$+ \tilde{\psi}\left(-u + c_0\psi + \frac{3c_0}{2}\phi^2 + \frac{c_0}{2}\phi^3\right). \qquad (2.226)$$

After collecting all the terms with $\tilde{\psi}$, we get

$$\dot{V}_2 \leq -c_1\phi^2 - c_1\phi^2\frac{dF(V_1)}{dV_1}$$
$$+ \tilde{\psi}\left(-u - \phi - \frac{dF(V_1)}{dV_1}\phi + c_0\psi + \frac{3c_0}{2}\phi^2 + \frac{c_0}{2}\phi^3\right). \qquad (2.227)$$

In addition to the choice of a control law for u, we now have the freedom to choose $\frac{dF(V_1)}{dV_1}$. With this choice we will avoid the cancellation of the cubic term $\frac{c_0}{2}\phi^3$ by u. We simply select $\frac{dF(V_1)}{dV_1}$ to eliminate $\frac{c_0}{2}\phi^3$:

$$\frac{dF(V_1)}{dV_1} = \frac{c_0}{2}\phi^2 = c_0V_1. \qquad (2.228)$$

This yields

$$F(V_1) = \frac{c_0}{2}V_1^2 = \frac{c_0}{8}\phi^4, \qquad (2.229)$$

and the resulting Lyapunov function (2.225) is nonquadratic:

$$V_2(\phi, \psi) = \frac{1}{2}\phi^2 + \frac{c_0}{8}\phi^4 + \frac{1}{2}(\psi - c_0\phi)^2. \qquad (2.230)$$

Substituting (2.228) into (2.227), we arrive at

$$\dot{V}_2 \leq -c_1\phi^2 - c_1\frac{c_0}{2}\phi^4 + \tilde{\psi}\left(-u - \phi + c_0\psi + \frac{3c_0}{2}\phi^2\right). \qquad (2.231)$$

The design could now be finished by selecting a control u which cancels $\frac{3c_0}{2}\phi^2$. However, even this cancellation can be avoided because of the strong stabilizing term $c_1\frac{c_0}{2}\phi^4$ in (2.231). Completing squares, $\frac{3c_0}{2}\phi^2\tilde{\psi} \leq c_1\frac{c_0}{2}\phi^4 + \frac{9c_0}{8c_1}\tilde{\psi}^2$, we get

$$\dot{V}_2 \leq -c_1\phi^2 + \tilde{\psi}\left(-u - \phi + c_0\psi + \frac{9c_0}{8c_1}\tilde{\psi}\right). \qquad (2.232)$$

Hence our control law can be selected to be linear,

$$u = -\phi + c_0\psi + \left(c_2 + \frac{9c_0}{8c_1}\right)\tilde{\psi}, \qquad c_2 > 0, \qquad (2.233)$$

and yield

$$\dot{V}_2 \leq -c_1\phi^2 - c_2\tilde{\psi}^2. \qquad (2.234)$$

This proves that the equilibrium $\phi = 0, \psi = 0$ is globally asymptotically stable. Denoting

$$k_1 = 1 + c_2 c_0 + \frac{9c_0^2}{8c_1}, \qquad k_2 = c_2 + c_0 + \frac{9c_0}{8c_1}, \qquad (2.235)$$

we rewrite (2.233) in the more compact form

$$u = -k_1\phi + k_2\psi \qquad (2.236)$$

and obtain the closed-loop system

$$\dot{\phi} = -\frac{1}{2}\phi^3 - \frac{3}{2}\phi^2 - \psi \qquad (2.237)$$

$$\dot{\psi} = k_1\phi - k_2\psi. \qquad (2.238)$$

For comparison, we also derive a feedback linearizing controller,

$$u = -k_1\phi + \left(k_2 - 3\phi - \frac{3}{2}\phi^2\right)\left(\psi + \frac{3}{2}\phi^2 + \frac{1}{2}\phi^3\right), \qquad (2.239)$$

which makes the system (2.218), (2.219) appear linear in the coordinates $\chi_1 = \phi$ and $\chi_2 = \dot{\phi}$. The controller simplification achieved with backstepping is impressive: While the linearizing control (2.239) grows as ϕ^5 and $\psi\phi^2$, the backstepping controller (2.236) is linear. The improvement over the control law (2.216) in Section 2.4 is also significant: (2.216) grows as ϕ^4 and $\psi\phi$ because the quadratic nonlinearity was cancelled at the first step, so the cancellation could not be avoided at the second step.

In the remainder of the book we will not assume the presence of useful nonlinearities. However, it should always be understood that whenever such additional information is available, backstepping designs should incorporate it.

2.5 Stabilization with Uncertainty

The full power of backstepping is exhibited in the presence of uncertain nonlinearities and unknown parameters, because for such applications no other systematic design procedure exists. We now begin the study of such design problems which are the main subject of this book. The first of the design tools that will be used to counteract uncertainty is *nonlinear damping*.

2.5 Stabilization with Uncertainty

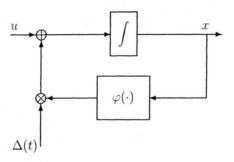

Figure 2.9: A system with "matched" uncertainty $\Delta(t)$.

2.5.1 Nonlinear damping

We introduce nonlinear damping for systems with "matched" uncertainty, in which both the uncertainty and the control appear in the same equation. The simplest example is the scalar nonlinear system depicted in Figure 2.9:

$$\dot{x} = u + \varphi(x)\Delta(t), \qquad (2.240)$$

where $\varphi(x)$ is a known smooth nonlinearity, and $\Delta(t)$ is a bounded function of t. Let us first examine the case when $\Delta(t)$ is an exponentially decaying disturbance:

$$\Delta(t) = \Delta(0)e^{-kt}. \qquad (2.241)$$

Can such an innocent-looking uncertainty cause harm? One might be tempted to ignore it and use the linear control $u = -cx$, which results in the closed-loop system

$$\dot{x} = -cx + \varphi(x)\Delta(0)e^{-kt}. \qquad (2.242)$$

While this design may be satisfactory when $\varphi(x)$ is bounded by a constant or a linear function of x, it is inadequate if $\varphi(x)$ is allowed to be any smooth nonlinear function. For example, when $\varphi(x) = x^2$ we have

$$\dot{x} = -cx + x^2\Delta(0)e^{-kt}. \qquad (2.243)$$

As we saw in Chapter 1, equations (1.29)–(1.32), the solution $x(t)$ of this system can be calculated explicitly using the change of variable $w = 1/x$:

$$\dot{w} = -\frac{1}{x^2}\dot{x} = c\frac{1}{x} - \Delta(0)e^{-kt} = cw - \Delta(0)e^{-kt}, \qquad (2.244)$$

which yields

$$w(t) = \left[w(0) - \frac{\Delta(0)}{c+k}\right]e^{ct} + \frac{\Delta(0)}{c+k}e^{-kt}. \qquad (2.245)$$

The substitution $w = 1/x$ gives

$$x(t) = \frac{x(0)(c+k)}{[c+k-\Delta(0)x(0)]e^{ct} + \Delta(0)x(0)e^{-kt}}. \qquad (2.246)$$

From (2.246) we see that the behavior of the closed-loop system (2.243) depends critically on the initial conditions $\Delta(0), x(0)$:

(i) If $\Delta(0)x(0) < c + k$, the solutions $x(t)$ are bounded and converge asymptotically to zero.

(ii) The situation changes dramatically when $\Delta(0)x(0) > c + k > 0$. The solutions $x(t)$ which start from such initial conditions not only diverge to infinity, but do so in *finite time*:

$$x(t) \to \infty \quad \text{as } t \to t_f = \frac{1}{c+k} \ln\left\{\frac{\Delta(0)x(0)}{\Delta(0)x(0) - (c+k)}\right\}. \quad (2.247)$$

Note that this finite escape cannot be eliminated by making c larger: For any values of c and k and for any nonzero value of $\Delta(0)$ there exist initial conditions $x(0)$ which satisfy the inequality $\Delta(0)x(0) > c + k$. This example shows that in a nonlinear system, neglecting the effects of exponentially decaying disturbances or nonzero initial conditions can be catastrophic.

To overcome this problem and guarantee that $x(t)$ will remain bounded for all bounded $\Delta(t)$ and for all $x(0)$, we augment the control law $u = -cx$ with a *nonlinear damping term* $-s(x)x$:

$$u = -cx - s(x)x. \quad (2.248)$$

We design $s(x)$ for the system (2.240), using the quadratic function $V(x) = \frac{1}{2}x^2$ whose derivative is

$$\begin{aligned}\dot{V} &= xu + x\varphi(x)\Delta(t) \\ &= -cx^2 - x^2 s(x) + x\varphi(x)\Delta(t). \end{aligned} \quad (2.249)$$

The objective of guaranteeing global boundedness of solutions can be equivalently expressed as rendering $\dot{V}$ negative outside a compact region. This is achieved with the choice

$$s(x) = \kappa \varphi^2(x), \quad \kappa > 0, \quad (2.250)$$

which yields the control

$$u = -cx - \kappa x \varphi^2(x) \quad (2.251)$$

and the derivative

$$\begin{aligned}\dot{V} &= -cx^2 - \kappa x^2 \varphi^2(x) + x\varphi(x)\Delta(t) \\ &= -cx^2 - \kappa \left[x\varphi(x) - \frac{\Delta(t)}{2\kappa}\right]^2 + \frac{\Delta^2(t)}{4\kappa} \\ &\leq -cx^2 + \frac{\Delta^2(t)}{4\kappa}. \end{aligned} \quad (2.252)$$

2.5 STABILIZATION WITH UNCERTAINTY

Comparing (2.249) with (2.252) we see that the nonlinear damping term (2.250) is chosen to allow the completion of squares in (2.252). In more complicated situations we can use *Young's Inequality*, which, in a simplified form, states that if the constants $p > 1$ and $q > 1$ are such that $(p-1)(q-1) = 1$, then for all $\varepsilon > 0$ and all $(x, y) \in \mathbb{R}^2$ we have

$$xy \leq \frac{\varepsilon^p}{p}|x|^p + \frac{1}{q\varepsilon^q}|y|^q. \tag{2.253}$$

Choosing $p = q = 2$ and $\varepsilon^2 = 2\kappa$, (2.253) becomes

$$xy \leq \kappa x^2 + \frac{1}{4\kappa}y^2, \tag{2.254}$$

which is the inequality used in (2.252):

$$x\varphi(x)\Delta(t) \leq \kappa x^2 \varphi^2(x) + \frac{\Delta^2(t)}{4\kappa}. \tag{2.255}$$

Global boundedness and convergence. Returning to (2.252), we see that $\dot{V}$ is negative whenever $|x(t)| \geq \frac{\Delta(t)}{2\sqrt{\kappa c}}$. Since $\Delta(t)$ is a bounded disturbance, we conclude that $\dot{V}$ is negative outside the compact residual set

$$\mathcal{R} = \left\{ x : |x| \leq \frac{\|\Delta\|_\infty}{2\sqrt{\kappa c}} \right\}. \tag{2.256}$$

Recalling that $V(x) = \frac{1}{2}x^2$, we conclude that $|x(t)|$ decreases whenever $x(t)$ is outside the set $\mathcal{R}$, and hence $x(t)$ is bounded:

$$\|x\|_\infty \leq \max\left\{ |x(0)|, \frac{\|\Delta\|_\infty}{2\sqrt{\kappa c}} \right\}. \tag{2.257}$$

Moreover, we can draw some conclusions about the asymptotic behavior of $x(t)$. Let us rewrite (2.252) as

$$\frac{d}{dt}\left(\frac{1}{2}x^2\right) \leq -cx^2 + \frac{\Delta^2(t)}{4\kappa}. \tag{2.258}$$

To obtain explicit bounds on $x(t)$, we consider the signal $x(t)e^{ct}$. Using (2.258) we get

$$\begin{aligned}
\frac{d}{dt}\left(x^2 e^{2ct}\right) &= \frac{d}{dt}\left(x^2\right)e^{2ct} + 2cx^2 e^{2ct} \\
&\leq -2cx^2 e^{2ct} + \frac{\Delta^2(t)}{2\kappa}e^{2ct} + 2cx^2 e^{2ct} \\
&= \frac{\Delta^2(t)}{2\kappa}e^{2ct}.
\end{aligned} \tag{2.259}$$

Integrating both sides over the interval $[0, t]$ yields

$$\begin{aligned}
x^2(t)e^{2ct} &\leq x^2(0) + \int_0^t \frac{1}{2\kappa}\Delta^2(\tau)e^{2c\tau}d\tau \\
&\leq x^2(0) + \frac{1}{2\kappa}\left[\sup_{0\leq\tau\leq t}\Delta^2(\tau)\right]\int_0^t e^{2c\tau}d\tau \\
&= x^2(0) + \frac{1}{4\kappa c}\left[\sup_{0\leq\tau\leq t}\Delta^2(\tau)\right]\left(e^{2ct}-1\right).
\end{aligned} \quad (2.260)$$

Multiplying both sides with e^{-2ct} and using the fact that $\sqrt{b^2+c^2} \leq |b|+|c|$, we obtain an explicit bound for $x(t)$:

$$\begin{aligned}
|x(t)| &\leq |x(0)|e^{-ct} + \frac{1}{2\sqrt{\kappa c}}\left[\sup_{0\leq\tau\leq t}|\Delta(\tau)|\right]\left(1-e^{-2ct}\right)^{\frac{1}{2}} \\
&\leq |x(0)|e^{-ct} + \frac{1}{2\sqrt{\kappa c}}\left[\sup_{0\leq\tau\leq t}|\Delta(\tau)|\right].
\end{aligned} \quad (2.261)$$

Since $\sup_{0\leq\tau\leq t}|\Delta(\tau)| \leq \sup_{0\leq\tau<\infty}|\Delta(\tau)| \triangleq \|\Delta\|_\infty$, (2.261) leads to

$$|x(t)| \leq |x(0)|e^{-ct} + \frac{\|\Delta\|_\infty}{2\sqrt{\kappa c}}, \quad (2.262)$$

which shows that $x(t)$ converges to the compact set $\mathcal{R}$ defined in (2.256):

$$\lim_{t\to\infty}\text{dist}\{x(t),\mathcal{R}\} = 0. \quad (2.263)$$

We reiterate that these properties of boundedness (cf. (2.257)) and convergence (cf. (2.263)) are guaranteed for any bounded disturbance $\Delta(t)$ and for any smooth nonlinearity $\varphi(x)$, including $\varphi(x) = x^2$. Furthermore, the nonlinear control law (2.251) does not assume knowledge of a bound on the disturbance, nor does it have to use large values for the gains κ and c. Indeed, the residual set $\mathcal{R}$ defined in (2.256) is compact for any finite value of $\|\Delta\|_\infty$ and for any positive values of κ and c. Hence, *global boundedness is guaranteed in the presence of bounded disturbances with unknown bounds, regardless of how small the gains κ and c are chosen*. While the size of $\mathcal{R}$ cannot be estimated *a priori* if no bound for $\|\Delta\|_\infty$ is given, it can be reduced *a posteriori* by increasing the values of κ and c.

This property is achieved by the "nonlinear damping" term $-\kappa x \varphi^2(x)$ in (2.251), which renders the effective gain of (2.251) "selectively high:" When κ and c are chosen to be small, the gain is low around the origin, but it becomes high when x is in a region where $\varphi(x)$ is large enough to make the term $\kappa\varphi^2(x)$ large. If we interpret the nonlinearity $\varphi(x)$, which multiplies the disturbance $\Delta(t)$, as the "disturbance gain," we see that the term $-\kappa\varphi^2(x)$ causes the control gain to become large when the disturbance gain is large.

2.5 STABILIZATION WITH UNCERTAINTY

Finally, we should note that, if the disturbance $\Delta(t)$ converges to zero in addition to being bounded, then the control (2.251) guarantees convergence of $x(t)$ to zero in addition to global boundedness. To show this, let $\bar{\Delta}(t)$ be a continuous nonnegative *monotonically decreasing* function such that $|\Delta(t)| \leq \bar{\Delta}(t)$ and $\lim_{t \to \infty} \bar{\Delta}(t) = 0$. Then, starting with the first inequality from (2.260) and using an argument almost identical to the proof of Lemma 2.24, we obtain

$$|x(t)| \leq |x(0)|e^{-ct} + \frac{1}{2\sqrt{\kappa c}} \left(\bar{\Delta}(0)e^{-\frac{c}{2}t} + \bar{\Delta}(t/2) \right). \quad (2.264)$$

Since $\lim_{t \to \infty} \bar{\Delta}(t/2) = 0$, we see that $\lim_{t \to \infty} x(t) = 0$.

ISS interpretation. For interpreting the effect of the nonlinear damping term $-\kappa x \varphi^2(x)$ in (2.251) from an input-output point of view, it is very convenient to use the concept of input-to-state stability: (cf. Appendix C) This κ-term renders the closed-loop system ISS with respect to the disturbance input $\Delta(t)$. To show that the ISS inequality (2.12) holds for our closed-loop system with $u(\tau)$ replaced by the disturbance $\Delta(\tau)$, we repeat the argument that led from (2.259) to (2.261), this time integrating over the interval $[t_0, t]$. The result is

$$|x(t)| \leq |x(t_0)|e^{-c(t-t_0)} + \frac{1}{2\sqrt{\kappa c}} \left[\sup_{t_0 \leq \tau \leq t} |\Delta(\tau)| \right], \quad (2.265)$$

which is identical to (2.12) with $\beta(r, s) = re^{-cs}$, $\gamma(r) = \frac{1}{2\sqrt{\kappa c}} r$, $r = |x(t_0)|$ and $s = t - t_0$.

Operator gain interpretation. It is also convenient to interpret the effect of nonlinear damping from an operator point of view on the basis of (2.257) and Figure 2.10. For all initial conditions $x(0)$ such that $|x(0)| < \frac{\|\Delta\|_\infty}{2\sqrt{\kappa c}}$, we obtain

$$\|x\|_\infty \leq \frac{1}{2\sqrt{\kappa c}} \|\Delta\|_\infty, \quad (2.266)$$

which shows that the nonlinear operator K mapping the disturbance $\Delta(t)$ to the output $x(t)$ is bounded, and its $\mathcal{L}_\infty$-induced gain is

$$\|K\|_{\infty\,\text{ind}} \leq \frac{1}{2\sqrt{\kappa c}}. \quad (2.267)$$

The nonlinear damping term renders the operator K bounded for *any* positive values of κ and c. Note, however, that (2.266) does not provide a complete description of this operator because, unlike (2.257), it hides the effect of initial conditions, which can be quite dangerous for nonlinear systems.

The following lemma recapitulates the properties achieved with nonlinear damping as a design tool.

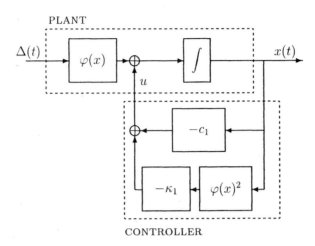

Figure 2.10: The bounded nonlinear operator $K: \Delta(t) \to x(t)$.

Lemma 2.26 (Nonlinear Damping) *Let the system (2.48) be perturbed as in*

$$\dot{x} = f(x) + g(x)\left[u + \varphi(x)^{\mathrm{T}}\Delta(x,u,t)\right], \qquad (2.268)$$

where $\varphi(x)$ is a $(p \times 1)$ vector of known smooth nonlinear functions, and $\Delta(x,u,t)$ is a $(p \times 1)$ vector of uncertain nonlinearities which are uniformly bounded for all values of x,u,t.

If Assumption 2.7 is satisfied with $W(x)$ positive definite and radially unbounded, then the control

$$u = \alpha(x) - \kappa \frac{\partial V}{\partial x}(x)g(x)|\varphi(x)|^2, \quad \kappa > 0, \qquad (2.269)$$

when applied to (2.268), renders the closed-loop system ISS with respect to the disturbance input $\Delta(x,u,t)$ and hence guarantees global uniform boundedness of $x(t)$ and convergence to the residual set

$$\mathcal{R} = \left\{ x \ : \ |x| \le \gamma_1^{-1} \circ \gamma_2 \circ \gamma_3^{-1} \left(\frac{\|\Delta\|_\infty^2}{4\kappa} \right) \right\}, \qquad (2.270)$$

where $\gamma_1, \gamma_2, \gamma_3$ are class-$\mathcal{K}_\infty$ functions such that [11]

$$\gamma_1(|x|) \le V(x) \le \gamma_2(|x|) \qquad (2.271)$$
$$\gamma_3(|x|) \le W(x). \qquad (2.272)$$

---

[11] Since $V(x)$ and $W(x)$ are positive definite and radially unbounded and $V(x)$ is smooth, there exist class-$\mathcal{K}_\infty$ functions $\gamma_1, \gamma_2, \gamma_3$ satisfying (2.271) and (2.272).

2.5 Stabilization with Uncertainty

Proof. The derivative of $V(x)$ is

$$\begin{aligned}
\dot{V} &= \frac{\partial V}{\partial x}[f+gu] + \frac{\partial V}{\partial x}g\varphi^T \Delta \\
\text{by (2.269)} &= \frac{\partial V}{\partial x}[f+g\alpha] - \kappa\left(\frac{\partial V}{\partial x}g\right)^2 |\varphi|^2 + \frac{\partial V}{\partial x}g\varphi^T \Delta \\
\text{by (2.50)} &\leq -W(x) - \kappa\left(\frac{\partial V}{\partial x}g\right)^2 |\varphi|^2 + \frac{\partial V}{\partial x}g\varphi^T \Delta \\
&\leq -W(x) - \kappa\left(\frac{\partial V}{\partial x}g\right)^2 |\varphi|^2 + \left|\frac{\partial V}{\partial x}g\right| |\varphi| \|\Delta\|_\infty \\
\text{by (2.254)} &\leq -W(x) + \frac{\|\Delta\|_\infty^2}{4\kappa}.
\end{aligned} \qquad (2.273)$$

From (2.273) it follows that $\dot{V}$ is negative whenever $W(x) > \frac{\|\Delta\|_\infty^2}{4\kappa}$. Combining this with (2.272) we conclude that

$$|x(t)| > \gamma_3^{-1}\left(\frac{\|\Delta\|_\infty^2}{4\kappa}\right) \Rightarrow \dot{V} < 0. \qquad (2.274)$$

This means that if $|x(0)| \leq \gamma_3^{-1}\left(\frac{\|\Delta\|_\infty^2}{4\kappa}\right)$, then

$$V(x(t)) \leq \gamma_2 \circ \gamma_3^{-1}\left(\frac{\|\Delta\|_\infty^2}{4\kappa}\right), \qquad (2.275)$$

which in turn implies that

$$|x(t)| \leq \gamma_1^{-1} \circ \gamma_2 \circ \gamma_3^{-1}\left(\frac{\|\Delta\|_\infty^2}{4\kappa}\right). \qquad (2.276)$$

If, on the other hand, $|x(0)| > \gamma_3^{-1}\left(\frac{\|\Delta\|_\infty^2}{4\kappa}\right)$, then $V(x(t)) \leq V(x(0))$, which implies

$$|x(t)| \leq \gamma_1^{-1} \circ \gamma_2 (|x(0)|). \qquad (2.277)$$

Combining (2.276) and (2.277) leads to the global uniform boundedness of $x(t)$:

$$\|x\|_\infty \leq \max\left\{\gamma_1^{-1} \circ \gamma_2 \circ \gamma_3^{-1}\left(\frac{\|\Delta\|_\infty^2}{4\kappa}\right),\ \gamma_1^{-1} \circ \gamma_2(|x(0)|)\right\}, \qquad (2.278)$$

while (2.274) and (2.271) prove the convergence of $x(t)$ to the residual set defined in (2.270). Finally, the ISS property of the closed-loop system with respect to the disturbance input $\Delta(x, u, t)$ follows from Theorem C.2. $\square$

2.5.2 Backstepping with uncertainty

Lemma 2.26 deals with the case where the uncertainty is in the span of the control u, i.e, the *matching condition* is satisfied. Combining Lemma 2.26 with Lemma 2.8 allows us to go beyond the matching case, as the following example illustrates.

Example 2.27 Consider the system

$$\dot{x} = \xi + x^2 \arctan \xi \Delta_0(t) \qquad (2.279a)$$
$$\dot{\xi} = (1+\xi^2)u + e^{x\xi}\Delta_0(t), \qquad (2.279b)$$

where $\Delta_0(t)$ is a bounded time-varying disturbance. Clearly, the uncertain terms in (2.279) are not in the span of the control u. Therefore, we will design a static nonlinear controller in two steps, combining nonlinear damping and backstepping.

Step 1. The starting point is equation (2.279a) and the choice of a virtual control variable. Clearly, ξ is the only choice. The fact that ξ is also present in the uncertain term does not present a problem, since it enters that term through the bounded function $\arctan(\cdot)$. In the notation of (2.268), we have

$$x^2 \arctan \xi \Delta_0(t) \triangleq x^2 \Delta_1(\xi, t) = \varphi_1(x)\Delta_1(\xi, t). \qquad (2.280)$$

The uncertain nonlinearity $\Delta_1(\xi, t)$ is bounded:

$$\|\Delta_1(\xi, t)\|_\infty = \|\Delta_0 \arctan \xi\|_\infty \leq \frac{\pi}{2}\|\Delta_0\|_\infty. \qquad (2.281)$$

Hence, Lemma 2.26 can be used to design a stabilizing function for ξ. The unperturbed system in this case would be the integrator $\dot{x} = \xi$, for which a clf is given by $V(x) = \frac{1}{2}x^2$ and the corresponding control is $\alpha(x) = -c_1 x$. From (2.269) we have

$$\alpha_1(x) = -c_1 x - \kappa_1 x \varphi_1^2(x), \qquad (2.282)$$

which results in

$$\dot{x} = -c_1 x + z - \kappa_1 x \varphi_1^2(x) + \varphi_1(x)\Delta_1(\xi, t), \qquad (2.283)$$

with the error variable z defined as in Lemma 2.8:

$$z = \xi - \alpha_1(x). \qquad (2.284)$$

The derivative of $V(x)$ along (2.283) is

$$\begin{aligned}
\dot{V} &= zx - c_1 x^2 - \kappa_1 x^2 \varphi_1^2 + x^3 \arctan \xi \Delta_0(t) \\
\text{by (2.280)} &\leq zx - c_1 x^2 - \kappa_1 x^2 \varphi_1^2 + |x\varphi_1(x)|\|\Delta_1\|_\infty \\
\text{by (2.254)} &= zx - c_1 x^2 + \frac{\|\Delta_1\|_\infty^2}{4\kappa_1},
\end{aligned} \qquad (2.285)$$

2.5 STABILIZATION WITH UNCERTAINTY

which confirms that if $z \equiv 0$, that is, if ξ were the actual control, then (2.282) would guarantee global uniform boundedness of x.

Step 2. Using the error variable z from (2.284), the system (2.279) is rewritten as

$$\dot{x} = -c_1 x + z - \kappa_1 x \varphi_1^2(x) + \varphi_1(x) \Delta_1(\xi, t) \qquad (2.286a)$$

$$\dot{z} = (1 + \xi^2) u + e^{x\xi} \Delta_0(t) - \frac{\partial \alpha_1}{\partial x} \left[\xi + x^2 \arctan \xi \Delta_0(t) \right]$$

$$= (1 + \xi^2) u - \frac{\partial \alpha_1}{\partial x} \xi + \left[e^{x\xi} - \frac{\partial \alpha_1}{\partial x} x^2 \arctan \xi \right] \Delta_0(t), \qquad (2.286b)$$

where the partial $\frac{\partial \alpha_1}{\partial x}$ is computed from (2.280) and (2.282):

$$\frac{\partial \alpha_1}{\partial x_1} = -c_1 - \kappa_1 \frac{\partial}{\partial x} \left[x \varphi_1^2(x) \right] = -c_1 - 5\kappa_1 x_1^4. \qquad (2.287)$$

If the $\Delta_0(t)$-term were not present in (2.286b), then Lemma 2.8 would dictate the Lyapunov function

$$V_2(x, \xi) = \frac{1}{2} x^2 + \frac{1}{2} z^2 = \frac{1}{2} x^2 + \frac{1}{2} \left[\xi - \alpha_1(x) \right]^2 \qquad (2.288)$$

and the following choice of control:

$$u = \bar{\alpha}(x, \xi) = \frac{1}{1 + \xi^2} \left[-c_2 z + \frac{\partial \alpha_1}{\partial x} \xi - x \right]. \qquad (2.289)$$

To compensate for the presence of the $\Delta_0(t)$-term in (2.286b), Lemma 2.26 is used again. From (2.269) we obtain

$$u = \frac{1}{1 + \xi^2} \left\{ -c_2 z + \frac{\partial \alpha_1}{\partial x} \xi - x - \kappa_2 z \left[e^{x\xi} - \frac{\partial \alpha_1}{\partial x} x^2 \arctan \xi \right]^2 \right\}, \qquad (2.290)$$

which renders the derivative of $V_2(x, \xi)$ negative outside a compact set, thus guaranteeing boundedness of $x(t)$ and $\xi(t)$:

$$\dot{V}_2 = \dot{V} + z\dot{z}$$

$$\text{by (2.285)} \quad \leq \quad zx - c_1 x^2 + \frac{\|\Delta_1\|_\infty^2}{4\kappa_1} + z\dot{z}$$

$$\text{by (2.286b) and (2.290)} \quad = \quad -c_1 x^2 + \frac{\|\Delta_1\|_\infty^2}{4\kappa_1}$$

$$+ z \left\{ -c_2 z - \kappa_2 z \left[e^{x\xi} - \frac{\partial \alpha_1}{\partial x} x^2 \arctan \xi \right]^2 \right.$$

$$\left. + \left[e^{x\xi} - \frac{\partial \alpha_1}{\partial x} x^2 \arctan \xi \right] \Delta_0(t) \right\}$$

$$\leq -c_1 x^2 - c_2 z^2 + \frac{\|\Delta_1\|_\infty^2}{4\kappa_1}$$

$$-\kappa_2 z^2 \left[e^{x\xi} - \frac{\partial \alpha_1}{\partial x} x^2 \arctan \xi \right]^2$$

$$+|z|\left| e^{x\xi} - \frac{\partial \alpha_1}{\partial x} x^2 \arctan \xi \right| \|\Delta_0\|_\infty$$

by (2.254) $\leq -c_1 x^2 - c_2 z^2 + \dfrac{\|\Delta_1\|_\infty^2}{4\kappa_1} + \dfrac{\|\Delta_0\|_\infty^2}{4\kappa_2}.$ (2.291)

$\diamond$

The combination of Lemmas 2.8 and 2.26, illustrated in the above example, is now formulated as another design tool.

Lemma 2.28 (Boundedness via Backstepping) *Consider the system*

$$\dot{x} = f(x) + g(x)u + F(x)\Delta_1(x, u, t), \qquad (2.292)$$

where $x \in \mathbb{R}^n$, $u \in \mathbb{R}$, $F(x)$ is an $(n \times q)$ matrix of known smooth nonlinear functions, and $\Delta_1(x, u, t)$ is a $(q \times 1)$ vector of uncertain nonlinearities which are uniformly bounded *for all values of x, u, t. Suppose that there exists a feedback control $u = \alpha(x)$ that renders $x(t)$ globally uniformly bounded, and that this is established via positive definite and radially unbounded functions $V(x), W(x)$ and a constant b, such that*

$$\frac{\partial V}{\partial x}(x)\left[f(x) + g(x)\alpha(x) + F(x)\Delta_1(x, u, t)\right] \leq -W(x) + b. \qquad (2.293)$$

Now consider the augmented system

$$\dot{x} = f(x) + g(x)\xi + F(x)\Delta_1(x, u, t) \qquad (2.294a)$$
$$\dot{\xi} = u + \varphi(x, \xi)^{\mathrm{T}} \Delta_2(x, \xi, u, t), \qquad (2.294b)$$

where $\varphi(x, \xi)$ is a $(p \times 1)$ vector of known smooth nonlinear functions, and $\Delta_2(x, \xi, u, t)$ is a $(p \times 1)$ vector of uncertain nonlinearities which are uniformly bounded *for all values of x, ξ, u, t. For this system, the feedback control*

$$u = -c\left[\xi - \alpha(x)\right] + \frac{\partial \alpha}{\partial x}(x)\left[f(x) + g(x)\xi\right] - \frac{\partial V}{\partial x}(x)g(x)$$
$$-\kappa\left[\xi - \alpha(x)\right]\left\{ |\varphi(x,\xi)|^2 + \left|\frac{\partial \alpha}{\partial x}(x) F(x)\right|^2 \right\} \qquad (2.295)$$

guarantees global uniform boundedness of $x(t)$ and $\xi(t)$ with any $c > 0$ and $\kappa > 0$.

2.5 STABILIZATION WITH UNCERTAINTY

Proof. Using the error variable

$$z = \xi - \alpha(x), \tag{2.296}$$

the system (2.294) is rewritten as

$$\begin{aligned} \dot{x} &= f(x) + g(x)[\alpha(x) + z] + F(x)\Delta_1(x, u, t) \tag{2.297a} \\ \dot{z} &= u + \varphi(x, \xi)^T \Delta_2(x, \xi, u, t) \\ &\quad - \frac{\partial \alpha}{\partial x}(x)[f(x) + g(x)\xi + F(x)\Delta_1(x, u, t)]. \tag{2.297b} \end{aligned}$$

The derivative of

$$V_2(x, \xi) = V(x) + \frac{1}{2}[\xi - \alpha(x)]^2 = V(x) + \frac{1}{2}z^2 \tag{2.298}$$

along the solutions of (2.297) with the control (2.295) is

$$\begin{aligned} \dot{V}_2 &= \frac{\partial V}{\partial x}(f + g\alpha + F\Delta_1) + \frac{\partial V}{\partial x}gz \\ &\quad + z\left[u + \varphi^T \Delta_2 - \frac{\partial \alpha}{\partial x}(f + g\xi + F\Delta_1)\right] \\ &\leq \frac{\partial V}{\partial x}(f + g\alpha + F\Delta_1) + z\left[u - \frac{\partial \alpha}{\partial x}(f + g\xi) + \frac{\partial V}{\partial x}g\right] \\ &\quad + z\left[\varphi^T \Delta_2 - \frac{\partial \alpha}{\partial x}F\Delta_1^T\right] \\ \text{by (2.293)} &\leq -W(x) + b + z\left[u - \frac{\partial \alpha}{\partial x}(f + g\xi) + \frac{\partial V}{\partial x}g\right] \\ &\quad + z\left[\varphi^T \Delta_2 - \frac{\partial \alpha}{\partial x}F\Delta_1\right] \\ \text{by (2.295)} &= -W(x) + b - cz^2 - \kappa z^2 \left[|\varphi|^2 + \left|\frac{\partial \alpha}{\partial x}F\right|^2\right] \\ &\quad + |z||\varphi|\|\Delta_2\|_\infty + |z|\left|\frac{\partial \alpha}{\partial x}F\right|\|\Delta_1\|_\infty \\ \text{by (2.254)} &= -W(x) - cz^2 + b + \frac{\|\Delta_1\|_\infty^2}{4\kappa} + \frac{\|\Delta_2\|_\infty^2}{4\kappa}. \tag{2.299} \end{aligned}$$

The radial unboundedness of $W(x)$ combined with (2.299) implies that $\dot{V}_2$ is negative outside a compact set, which in turn implies that $x(t)$ and $\xi(t)$ are globally uniformly bounded. □

2.5.3 Robust strict-feedback systems

Just as we generalized Lemma 2.8 to strict-feedback systems in Section 2.3.1 and Lemma 2.25 to block-strict-feedback systems in Section 2.3.3, we can generalize Lemma 2.28 to broader classes of uncertain nonlinear systems.

We consider systems in the *robust strict-feedback form*:

$$\begin{aligned}
\dot{x}_1 &= x_2 + \varphi_1^T(x_1)\Delta(x, u, t) \\
\dot{x}_2 &= x_3 + \varphi_2^T(x_1, x_2)\Delta(x, u, t) \\
&\vdots \\
\dot{x}_{n-1} &= x_n + \varphi_{n-1}^T(x_1, \ldots, x_{n-1})\Delta(x, u, t) \\
\dot{x}_n &= \beta(x)u + \varphi_n^T(x)\Delta(x, u, t),
\end{aligned} \qquad (2.300)$$

where $\beta(x) \neq 0$, $\forall x \in \mathbb{R}^n$, $\varphi_i(x_1, \ldots, x_i)$ is a $(p \times 1)$ vector of known smooth nonlinear functions, and $\Delta(x, u, t)$ is a $(p \times 1)$ vector of uncertain nonlinearities which are *uniformly bounded* for all values of x, u, t.

Corollary 2.29 (Robust Strict-Feedback Systems) *The state $x(t)$ of the system (2.300) will be globally uniformly bounded if the control is chosen as*

$$u = \frac{1}{\beta(x)}\alpha_n(x), \qquad (2.301)$$

where the function $\alpha_n(x)$ is defined by the following recursive expressions for $i = 1, \ldots, n$ (where we denote $z_0 \equiv \alpha_0 \equiv 0$):

$$z_i = x_i - \alpha_{i-1}(x_1, \ldots, x_{i-1}) \qquad (2.302)$$

$$\alpha_i(x_1, \ldots, x_i) = -c_i z_i - z_{i-1} + \sum_{j=1}^{i-1} \frac{\partial \alpha_{i-1}}{\partial x_j} x_{j+1} - \kappa_i z_i \left| \varphi_i - \sum_{j=1}^{i-1} \frac{\partial \alpha_{i-1}}{\partial x_j} \varphi_j \right|^2, \qquad (2.303)$$

with c_i, κ_i, $i = 1, \ldots, n$ positive design constants.

Proof. Using the definitions (2.302) and (2.303) and denoting $x_0 \equiv \alpha_0 \equiv 0$, $x_{n+1} \equiv \beta(x)u$, the derivative of the error variable z_i, $i = 1, \ldots, n$, becomes

$$\begin{aligned}
\dot{z}_i &= \dot{x}_i - \dot{\alpha}_{i-1}(x_1, \ldots, x_{i-1}) \\
&= x_{i+1} + \varphi_i^T \Delta - \sum_{j=1}^{i-1} \frac{\partial \alpha_{i-1}}{\partial x_j} \left(x_{j+1} + \varphi_j^T \Delta \right) \\
&= \alpha_i + z_{i+1} + \sum_{j=1}^{i-1} \frac{\partial \alpha_{i-1}}{\partial x_j} x_{j+1} + \left(\varphi_i - \sum_{j=1}^{i-1} \frac{\partial \alpha_{i-1}}{\partial x_j} \varphi_j \right)^T \Delta \\
&= -c_i z_i - z_{i-1} + z_{i+1} + \left(\varphi_i - \sum_{j=1}^{i-1} \frac{\partial \alpha_{i-1}}{\partial x_j} \varphi_j \right)^T \Delta - \kappa_i z_i \left| \varphi_i - \sum_{j=1}^{i-1} \frac{\partial \alpha_{i-1}}{\partial x_j} \varphi_j \right|^2.
\end{aligned} \qquad (2.304)$$

2.5 STABILIZATION WITH UNCERTAINTY

The choice of control (2.301) guarantees that $z_{n+1} \equiv 0$. The closed-loop error system can therefore be expressed as

$$\begin{aligned}
\dot{z}_1 &= -c_1 z_1 + z_2 + \varphi_1^T \Delta - \kappa_1 z_1 |\varphi_1|^2 \\
\dot{z}_2 &= -c_2 z_2 - z_1 + z_3 + \left(\varphi_2 - \frac{\partial \alpha_1}{\partial x_1}\varphi_1\right)^T \Delta - \kappa_2 z_2 \left|\varphi_2 - \frac{\partial \alpha_1}{\partial x_1}\varphi_1\right|^2 \\
&\vdots \\
\dot{z}_{n-1} &= -c_{n-1} z_{n-1} - z_{n-2} + z_n + \left(\varphi_{n-1} - \sum_{j=1}^{n-2} \frac{\partial \alpha_{n-2}}{\partial x_j}\varphi_j\right)^T \Delta \\
&\quad - \kappa_{n-1} z_{n-1} \left|\varphi_{n-1} - \sum_{j=1}^{n-2} \frac{\partial \alpha_{n-2}}{\partial x_j}\varphi_j\right|^2 \\
\dot{z}_n &= -c_n z_n - z_{n-1} + \left(\varphi_n - \sum_{j=1}^{n-1} \frac{\partial \alpha_{n-1}}{\partial x_j}\varphi_j\right)^T \Delta - \kappa_n z_n \left|\varphi_n - \sum_{j=1}^{n-1} \frac{\partial \alpha_{n-1}}{\partial x_j}\varphi_j\right|^2.
\end{aligned} \quad (2.305)$$

Now we can use the quadratic Lyapunov function

$$V_n(z_1, \ldots, z_n) = \frac{1}{2} \sum_{i=1}^n z_i^2 \quad (2.306)$$

to prove global uniform boundedness. Indeed, the derivative of (2.306) along the solutions of (2.305) is

$$\begin{aligned}
\dot{V}_n &= \sum_{i=1}^n \left[-c_i z_i^2 + z_i \left(\varphi_i - \sum_{j=1}^{i-1} \frac{\partial \alpha_{i-1}}{\partial x_j}\varphi_j\right)^T \Delta - \kappa_i z_i^2 \left|\varphi_i - \sum_{j=1}^{i-1} \frac{\partial \alpha_{i-1}}{\partial x_j}\varphi_j\right|^2 \right] \\
&\leq \sum_{i=1}^n \left[-c_i z_i^2 + |z_i| \left|\varphi_i - \sum_{j=1}^{i-1} \frac{\partial \alpha_{i-1}}{\partial x_j}\varphi_j\right| \|\Delta\|_\infty - \kappa_i z_i^2 \left|\varphi_i - \sum_{j=1}^{i-1} \frac{\partial \alpha_{i-1}}{\partial x_j}\varphi_j\right|^2 \right] \\
&\leq \sum_{i=1}^n \left[-c_i z_i^2 + \frac{\|\Delta\|_\infty^2}{4\kappa_i} \right]. \quad (2.307)
\end{aligned}$$

The last inequality implies that $z(t)$ is globally uniformly bounded. But from (2.303) we see that, since the α_i's are smooth functions, x_i can be expressed as a smooth function of $z_1, \ldots, z_i$:

$$x_1 = z_1, \quad x_i = z_i + \bar{\alpha}_{i-1}(z_1, \ldots, z_{i-1}), \quad i = 2, \ldots, n. \quad (2.308)$$

Hence, $x(t)$ is globally uniformly bounded and, furthermore, converges to the compact residual set

$$\mathcal{R} = \left\{ x \in \mathbb{R}^n : \sum_{i=1}^n c_i z_i^2 \leq \sum_{i=1}^n \frac{\|\Delta\|_\infty^2}{4\kappa_i} \right\}, \quad (2.309)$$

whose size is unknown since the bound $\|\Delta\|_\infty$ is unknown. □

Notes and References

Integrator backstepping is an idea whose origins are difficult to trace, because of its simultaneous appearance, often implicit, in the works of Tsinias [193], Koditschek [84], Byrnes and Isidori [12], and Sontag and Sussmann [175]. Kokotović and Sussmann [85] viewed the stabilization through an integrator as a special case of stabilization through an SPR transfer function, as in the early adaptive designs by Parks [150], Landau [109], and Narendra et al. [142]. This passivity view was extended to nonlinear cascades by Ortega [145] and Byrnes, Isidori, and Willems [14]. Integrator backstepping as a recursive design tool [85, Corollary 3.2] was employed in the cascade design of Saberi, Kokotović, and Sussmann [163] and further developed by Kanellakopoulos, Kokotović, and Morse [73]. The passivity aspect of this design was pointed out by Lozano, Brogliato, and Landau [116]. A tutorial overview of backstepping was given in the 1991 Bode lecture by Kokotović [88]. Among the current applications of backstepping are electric machines in the monograph of Dawson, Hu, and Burg [26] and steering and braking control by Chen and Tomizuka [17].

An important question, not addressed in the above references, is whether backstepping designs can incorporate optimization with respect to some meaningful cost functionals. It is clear that minimizing a partial cost functional at each step does not imply, and may in fact contradict, the overall optimality. A framework for a backstepping-like recursive optimization was proposed in a Russian language paper by Kolesnikov [89], which remained unnoticed in the English language literature.

All the above references apply backstepping to systems without uncertainty. For systems with uncertainty a robust nonlinear design was introduced for the matched case in the works of Corless and Leitmann [20] and Barmish, Corless, and Leitmann [6], and and extended by various forms of "generalized matching conditions" [19, 20]. However, these extensions haven't led to a discovery of backstepping with uncertainty. After the development of adaptive backstepping by Kanellakopoulos, Kokotović, and Morse [69, 87], backstepping with uncertainty was pursued in the works of Qu [160], Marino and Tomei [125], Freeman and Kokotović [37, 38], and Slotine and Hedrick [168]. Backstepping designs for systems with unmeasured or unmodeled dynamics have been developed by Praly and Jiang [159], Jiang, Teel, and Praly [61], Khalil [82], and Krstić, Sun, and Kokotović [104]. Recent advances in backstepping, which are beyond the scope of this book, are due to Coron and Praly [21] and Praly, d'Andréa-Novel, and Coron [158]. The restriction of backstepping to pure-feedback systems has motivated alternative designs applicable to other classes of nonlinear systems, such as those by Teel [187], Teel and Praly [192], Qu [161], and Mazenc and Praly [126].

Chapter 3

Adaptive Backstepping Design

The controllers designed in the preceding chapter guarantee that in the presence of uncertain bounded nonlinearities the closed-loop state remains bounded. In this chapter, and in the remainder of the book, the uncertainties are more specific. They consist of unknown constant parameters which appear linearly in the system equations. In the presence of such parametric uncertainties we will be able to achieve both boundedness of the closed-loop states and convergence of the tracking error to zero.

While all the controllers designed in Chapter 2 employ *static* feedback, the controllers in this chapter will, in addition, employ a form of nonlinear integral feedback. The underlying idea in the design of this *dynamic* part of feedback is *parameter estimation*. The dynamic part of the controller is designed as a *parameter update law* with which the static part is continuously *adapted* to new parameter estimates, hence its name: *Adaptive control law*. In using this traditional terminology, however, we should keep in mind that so conceived adaptive controllers are but one type of nonlinear dynamic feedback.

Adaptive controllers are dynamic and therefore more complex than the static controllers designed in Chapter 2. What is achieved with this additional complexity? As we will show in Section 3.1, an adaptive controller guarantees not only that the plant state x remains bounded, but also that it tends to a desired constant value ("regulation") or asymptotically tracks a reference signal ("tracking").

The first results leading to a new systematic design of adaptive controllers are presented in Section 3.2, which introduces *adaptive backstepping*. The recursive design procedure for *parametric strict-feedback systems* is then developed in Section 3.3.

In its basic form, the adaptive backstepping design employs *overparametrization*, that is, more than one estimate per unknown parameter. This means that the dynamic part of the controller is not of minimal order. In Chapter 4, a more intricate backstepping procedure is developed—the tuning functions method—which employs the minimal number of parameter estimates. The extended-matching design, presented in Section 3.4, is of interest as a transition between overparametrized and minimal-order designs. It also contains the first adaptive performance results.

3.1 Adaptation as Dynamic Feedback

The difference between a static and a dynamic (that is, adaptive) design will first be illustrated on the simplest nonlinear system:

$$\dot{x} = u + \theta\varphi(x). \tag{3.1}$$

This is the special case of the system (2.240), where the uncertainty $\Delta(t)$ is the unknown constant parameter θ.

Even if we do not know a bound for θ, we can use Lemma 2.26 to design a static nonlinear controller which guarantees global boundedness of $x(t)$. The nonlinear damping design (2.251) applies also here. The corresponding static controller is

$$u = -cx - \kappa x \varphi^2(x), \tag{3.2}$$

and the resulting closed-loop system is of first order:

$$\dot{x} = -cx - \kappa x \varphi^2(x) + \theta\varphi(x). \tag{3.3}$$

According to (2.252), the derivative of $V = \frac{1}{2}x^2$ satisfies

$$\dot{V} \leq -cx^2 + \frac{\theta^2}{4\kappa}, \tag{3.4}$$

which means that $x(t)$ converges to the interval

$$|x| \leq \frac{|\theta|}{2\sqrt{\kappa c}}. \tag{3.5}$$

This interval can be reduced by increasing the gains κ and c, but $x(t)$ will not converge to zero if θ is a nonzero constant. Excessive increase of these gains enlarges the system bandwidth, which is undesirable. Our task is therefore to achieve $\lim_{t\to\infty} x(t) = 0$ without increasing κ and c. In fact, we will first accomplish this task with $\kappa = 0$, and then use $\kappa > 0$ for further improvement of transients.

To achieve regulation of $x(t)$, we design a dynamic feedback controller, that is, we employ adaptation.

If θ were known, the control

$$u = -\theta\varphi(x) - c_1 x, \quad c_1 > 0 \tag{3.6}$$

would render the derivative of $V_0(x) = \frac{1}{2}x^2$ negative definite: $\dot{V}_0 = -c_1 x^2$. Of course the control law (3.6) can not be implemented, since θ is unknown.

3.1 ADAPTATION AS DYNAMIC FEEDBACK

Instead, one can employ its *certainty-equivalence* form in which θ is replaced by an estimate $\hat{\theta}$:
$$u = -\hat{\theta}\varphi(x) - c_1 x. \tag{3.7}$$
Substituting (3.7) into (3.6), we obtain
$$\dot{x} = -c_1 x + \tilde{\theta}\varphi(x), \tag{3.8}$$
where $\tilde{\theta}$ is the *parameter error*:
$$\tilde{\theta} = \theta - \hat{\theta}. \tag{3.9}$$
The derivative of $V_0(x) = \frac{1}{2}x^2$ becomes
$$\dot{V}_0 = -c_1 x^2 + \tilde{\theta} x \varphi(x). \tag{3.10}$$
Since the second term is indefinite and contains the unknown parameter error $\tilde{\theta}$, we can not conclude anything about the stability of (3.6). We make the controller dynamic with an update law for $\hat{\theta}$. To design this update law, we augment V_0 with a quadratic term in the parameter error $\tilde{\theta}$,
$$V_1(x, \tilde{\theta}) = \frac{1}{2}x^2 + \frac{1}{2\gamma}\tilde{\theta}^2, \tag{3.11}$$
where $\gamma > 0$ is the *adaptation gain*. The derivative of this function is
$$\begin{aligned}\dot{V}_1 &= x\dot{x} + \frac{1}{\gamma}\tilde{\theta}\dot{\tilde{\theta}} \\ &= -c_1 x^2 + \tilde{\theta} x \varphi(x) + \frac{1}{\gamma}\tilde{\theta}\dot{\tilde{\theta}} \\ &= -c_1 x^2 + \tilde{\theta}\left[x\varphi(x) + \frac{1}{\gamma}\dot{\tilde{\theta}}\right].\end{aligned} \tag{3.12}$$

The second term is still indefinite and contains $\tilde{\theta}$ as a factor. However, the situation is much better than in (3.10), because we now have the dynamics of $\dot{\tilde{\theta}} = -\dot{\hat{\theta}}$ at our disposal. With the appropriate choice of $\dot{\hat{\theta}}$ we can cancel the indefinite term. Thus, we choose the update law
$$\dot{\hat{\theta}} = -\dot{\tilde{\theta}} = \gamma x \varphi(x), \tag{3.13}$$
which yields
$$\dot{V}_1 = -c_1 x^2 \leq 0. \tag{3.14}$$
The resulting adaptive system consists of (3.1) with the control (3.7) and the update law (3.13), and is shown in Figure 3.1. In Figure 3.2, this system is redrawn in its closed-loop form consisting of (3.8) and (3.13), namely
$$\begin{aligned}\dot{x} &= -c_1 x + \tilde{\theta}\varphi(x) \\ \dot{\tilde{\theta}} &= -\gamma x \varphi(x).\end{aligned} \tag{3.15}$$

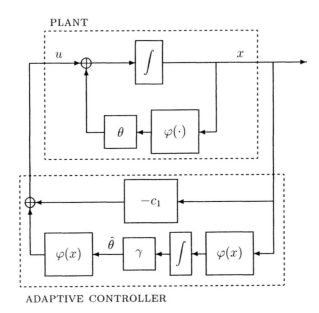

Figure 3.1: The closed-loop adaptive system (3.15).

Because $\dot{V}_1 \leq 0$, the equilibrium $x = 0$, $\tilde{\theta} = 0$ of (3.15) is globally stable. In addition, the desired regulation property $\lim_{t \to \infty} x(t) = 0$ follows from the LaSalle-Yoshizawa theorem (Theorem 2.1). The adaptive nonlinear controller which guarantees these properties is given by (3.8) and (3.13):

$$\begin{aligned} u &= -c_1 x - \hat{\theta}\varphi(x) \\ \dot{\hat{\theta}} &= \gamma x \varphi(x). \end{aligned} \quad (3.16)$$

One may think that the above adaptive design is so straightforward because (3.1) is a first-order system. In fact, this is due to the *matching condition*: The terms containing unknown parameters in (3.1) are in the span of the control, that is, they can be directly cancelled by u when θ is known. To illustrate this point, let us consider the following second-order system, where again the uncertain term is "matched" by the control u:

$$\begin{aligned} \dot{x}_1 &= x_2 + \varphi_1(x_1) \\ \dot{x}_2 &= \theta \varphi_2(x) + u. \end{aligned} \quad (3.17)$$

If θ were known, we would be able to apply Lemma 2.8: First view x_2 as the virtual control, design the stabilizing function

$$\alpha_1(x_1) = -c_1 x_1 - \varphi_1(x_1), \quad (3.18)$$

3.1 ADAPTATION AS DYNAMIC FEEDBACK

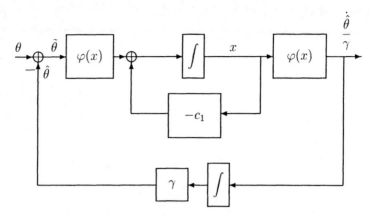

Figure 3.2: An equivalent representation of (3.15).

and then form the Lyapunov function

$$V_c(x) = \frac{1}{2}x_1^2 + \frac{1}{2}(x_2 - \alpha_1(x_1))^2, \qquad (3.19)$$

whose derivative is rendered negative definite

$$\dot{V}_c(x) = -c_1 x_1^2 - c_2 (x_2 - \alpha_1)^2 \qquad (3.20)$$

by the control

$$u = -c_2(x_2 - \alpha_1) - x_1 + \frac{\partial \alpha_1}{\partial x_1}(x_2 + \varphi_1) - \theta \varphi_2(x). \qquad (3.21)$$

Since θ is unknown, we again replace it with its estimate $\hat{\theta}$ in (3.21) to obtain the adaptive control law:

$$u = -c_2(x_2 - \alpha_1) - x_1 + \frac{\partial \alpha_1}{\partial x_1}(x_2 + \varphi_1) - \hat{\theta} \varphi_2(x). \qquad (3.22)$$

This results in the error system ($z_1 = x_1$, $z_2 = x_2 - \alpha_1$):

$$\frac{d}{dt}\begin{bmatrix} z_1 \\ z_2 \end{bmatrix} = \begin{bmatrix} -c_1 & 1 \\ -1 & -c_2 \end{bmatrix}\begin{bmatrix} z_1 \\ z_2 \end{bmatrix} + \begin{bmatrix} 0 \\ \varphi_2(x) \end{bmatrix}\tilde{\theta}. \qquad (3.23)$$

Then we augment (3.20) with a quadratic term in the parameter error $\tilde{\theta}$ to obtain the Lyapunov function:

$$V_1(z, \tilde{\theta}) = V_c + \frac{1}{2\gamma}\tilde{\theta}^2 = \frac{1}{2}z_1^2 + \frac{1}{2}z_2^2 + \frac{1}{2\gamma}\tilde{\theta}^2. \qquad (3.24)$$

Its derivative is

$$\dot{V}_1 = -c_1 z_1^2 - c_2 z_2^2 + \tilde{\theta}\left[z_2\varphi_2 - \frac{1}{\gamma}\dot{\hat{\theta}}\right]. \qquad (3.25)$$

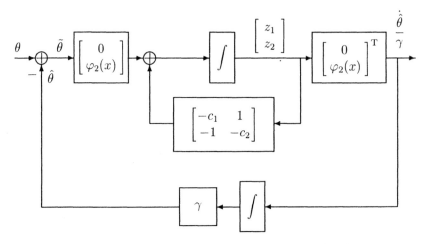

Figure 3.3: The closed-loop adaptive system (3.28).

The choice of update law

$$\dot{\hat{\theta}} = \gamma \varphi_2 z_2 \qquad (3.26)$$

eliminates the $\tilde{\theta}$-term in (3.25) and renders the derivative of the Lyapunov function (3.24) nonpositive:

$$\dot{V}_1 = -c_1 z_1^2 - c_2 z_2^2 \leq 0. \qquad (3.27)$$

This implies that the $z = 0, \tilde{\theta} = 0$ equilibrium point of the closed-loop adaptive system consisting of (3.23) and (3.26) (see block diagram in Figure 3.3)

$$\frac{d}{dt}\begin{bmatrix} z_1 \\ z_2 \end{bmatrix} = \begin{bmatrix} -c_1 & 1 \\ -1 & -c_2 \end{bmatrix}\begin{bmatrix} z_1 \\ z_2 \end{bmatrix} + \begin{bmatrix} 0 \\ \varphi_2(x) \end{bmatrix}\tilde{\theta} \qquad (3.28)$$

$$\dot{\tilde{\theta}} = -\gamma \begin{bmatrix} 0 & \varphi_2 \end{bmatrix}\begin{bmatrix} z_1 \\ z_2 \end{bmatrix}$$

is globally stable and, in addition, $x(t) \to 0$ as $t \to \infty$.

3.2 Adaptive Backstepping

3.2.1 Adaptive integrator backstepping

The adaptive design in the above examples was simple because of the matching: The parametric uncertainty was in the span of the control. We now move to the more general case of *extended matching*, where the parametric uncertainty enters the system one integrator before the control does:

$$\dot{x}_1 = x_2 + \theta \varphi(x_1) \qquad (3.29a)$$
$$\dot{x}_2 = u. \qquad (3.29b)$$

3.2 ADAPTIVE BACKSTEPPING

We use this example to introduce *adaptive backstepping*.

If θ were known, we would apply Lemma 2.8 to design the stabilizing function for x_2

$$\alpha_1(x_1, \theta) = -c_1 x_1 - \theta \varphi(x_1), \qquad (3.30)$$

with the Lyapunov function

$$V_c(x, \theta) = \frac{1}{2} x_1^2 + \frac{1}{2}(x_2 - \alpha_1(x_1, \theta))^2, \qquad (3.31)$$

whose derivative is rendered negative definite

$$\dot{V}_c(x, \theta) = -c_1 x_1^2 - c_2 (x_2 - \alpha_1(x, \theta))^2 \qquad (3.32)$$

by the control

$$u = -c_2 (x_2 - \alpha_1(x_1, \theta)) - x_1 + \frac{\partial \alpha_1}{\partial x_1}(x_2 + \theta \varphi). \qquad (3.33)$$

Since θ is unknown and appears one equation before the control does, we can not apply Lemma 2.8 because the dependence of $\alpha_1(x_1) = -c_1 x_1 - \theta \varphi(x_1)$ on the unknown parameter makes it impossible to continue the procedure. However, we can utilize the idea of integrator backstepping.

Step 1. If x_2 were the control, an adaptive controller for (3.29a) would be given by (3.16):

$$\alpha_1(x_1, \vartheta_1) = -c_1 z_1 - \vartheta_1 \varphi(x_1) \qquad (3.34)$$

$$\dot{\vartheta}_1 = \gamma z_1 \varphi(x_1). \qquad (3.35)$$

In the above equations we have replaced the parameter estimate $\hat{\theta}$ with the estimate ϑ_1, which denotes the estimate generated in this design step. As we will see, there will be another estimate generated in the next step. With (3.34) and the new error variable $z_2 = x_2 - \alpha_1$, the $\dot{z}_1$-equation becomes

$$\dot{z}_1 = -c_1 z_1 + z_2 + (\theta - \vartheta_1)\varphi. \qquad (3.36)$$

The derivative of the Lyapunov function

$$V_1(x, \vartheta_1) = \frac{1}{2} z_1^2 + \frac{1}{2\gamma}(\theta - \vartheta_1)^2 \qquad (3.37)$$

along the solutions of (3.36) is

$$\begin{aligned}
\dot{V}_1 &= z_1 \dot{z}_1 - \frac{1}{\gamma}(\theta - \vartheta_1)\dot{\vartheta}_1 \\
&= z_1 z_2 - c_1 z_1^2 + (\theta - \vartheta_1)\left(\varphi_1 z_1 - \frac{1}{\gamma}\dot{\vartheta}_1\right) \\
&= z_1 z_2 - c_1 z_1^2.
\end{aligned} \qquad (3.38)$$

Step 2. The derivative of z_2 is now expressed as

$$\begin{aligned} \dot{z}_2 &= \dot{x}_2 - \dot{\alpha}_1 \\ &= u - \frac{\partial \alpha_1}{\partial x_1}\dot{x}_1 - \frac{\partial \alpha_1}{\partial \vartheta_1}\dot{\vartheta}_1. \end{aligned}$$

Substituting (3.29a) and the update law (3.35) results in

$$\begin{aligned} \dot{z}_2 &= u - \frac{\partial \alpha_1}{\partial x_1}(x_2 + \theta\varphi) - \frac{\partial \alpha_1}{\partial \vartheta_1}\gamma\varphi z_1 \\ &= u - \frac{\partial \alpha_1}{\partial x_1}x_2 - \frac{\partial \alpha_1}{\partial \vartheta_1}\gamma\varphi z_1 - \theta\frac{\partial \alpha_1}{\partial x_1}\varphi. \end{aligned} \quad (3.39)$$

At this point we need to select a Lyapunov function and design u to render its derivative nonpositive. Our first attempt is the augmented Lyapunov function

$$V_2(z_1, z_2, \vartheta_1) = V_1(z_1, \vartheta_1) + \frac{1}{2}z_2^2,$$

whose derivative, using (3.38) and (3.39), is

$$\begin{aligned} \dot{V}_2 &= \dot{V}_1 + z_2 \dot{z}_2 \\ &= -c_1 z_1^2 + z_2 \left[z_1 + u - \frac{\partial \alpha_1}{\partial x_1}x_2 - \frac{\partial \alpha_1}{\partial \vartheta_1}\gamma\varphi z_1 - \theta\frac{\partial \alpha_1}{\partial x_1}\varphi \right]. \end{aligned}$$

The control u should now be able to cancel the indefinite terms in $\dot{V}_2$. To deal with the terms containing the unknown parameter θ, we will try to employ the existing estimate ϑ_1:

$$u = -z_1 - c_2 z_2 + \frac{\partial \alpha_1}{\partial x_1}x_2 + \frac{\partial \alpha_1}{\partial \vartheta_1}\gamma\varphi z_1 + \vartheta_1\frac{\partial \alpha_1}{\partial x_1}\varphi.$$

From the resulting derivative

$$\dot{V}_2 = -c_1 z_1^2 - c_2 z_2^2 - (\theta - \vartheta_1)\frac{\partial \alpha_1}{\partial x_1}\varphi_1 z_2,$$

we see that we have no design freedom left to cancel the $(\theta - \vartheta_1)$-term. To overcome this difficulty, we replace ϑ_1 in the expression for u with a *new* estimate ϑ_2:

$$u = -z_1 - c_2 z_2 + \frac{\partial \alpha_1}{\partial x_1}x_2 + \frac{\partial \alpha_1}{\partial \vartheta_1}\gamma\varphi z_1 + \vartheta_2\frac{\partial \alpha_1}{\partial x_1}\varphi. \quad (3.40)$$

With the choice (3.40), the $\dot{z}_2$-equation becomes

$$\dot{z}_2 = -c_2 z_2 - z_1 - (\theta - \vartheta_2)\frac{\partial \alpha_1}{\partial x_1}\varphi. \quad (3.41)$$

3.2 ADAPTIVE BACKSTEPPING

The presence of the new parameter estimate ϑ_2 suggests the following augmentation of the Lyapunov function:

$$\begin{aligned} V_2(z_1, z_2, \vartheta_1, \vartheta_2) &= V_1 + \frac{1}{2}z_2^2 + \frac{1}{2\gamma}(\theta - \vartheta_2)^2 \\ &= \frac{1}{2}\left(z_1^2 + z_2^2\right) + \frac{1}{2\gamma}\left[(\theta - \vartheta_1)^2 + (\theta - \vartheta_2)^2\right]. \end{aligned} \quad (3.42)$$

The derivative of V_2 is

$$\begin{aligned} \dot{V}_2 &= \dot{V}_1 + z_2 \dot{z}_2 - \frac{1}{\gamma}(\theta - \vartheta_2)\dot{\vartheta}_2 \\ &= z_1 z_2 - c_1 z_1^2 + z_2 \left[-c_2 z_2 - z_1 - (\theta - \vartheta_2)\frac{\partial \alpha_1}{\partial x_1}\varphi\right] - \frac{1}{\gamma}(\theta - \vartheta_2)\dot{\vartheta}_2 \\ &= -c_1 z_1^2 - c_2 z_2^2 - (\theta - \vartheta_2)\left(\frac{\partial \alpha_1}{\partial x_1}\varphi + \frac{1}{\gamma}\dot{\vartheta}_2\right). \end{aligned} \quad (3.43)$$

Now the $(\theta - \vartheta_2)$-term can be eliminated with the update law

$$\dot{\vartheta}_2 = -\gamma \frac{\partial \alpha_1}{\partial x_1}\varphi z_2, \quad (3.44)$$

which yields

$$\dot{V}_2 = -c_1 z_1^2 - c_2 z_2^2. \quad (3.45)$$

The equations (3.41) and (3.44) along with (3.36) and (3.35) form the error system representation of the resulting closed-loop adaptive system:

$$\begin{aligned} \dot{z}_1 &= -c_1 z_1 + z_2 + (\theta - \vartheta_1)\varphi \\ \dot{z}_2 &= -c_2 z_2 - z_1 - (\theta - \vartheta_2)\frac{\partial \alpha_1}{\partial x_1}\varphi \\ \dot{\vartheta}_1 &= \gamma \varphi z_1 \\ \dot{\vartheta}_2 &= -\gamma \frac{\partial \alpha_1}{\partial x_1}\varphi z_2. \end{aligned} \quad (3.46)$$

The matrix form of this system,

$$\begin{aligned} \frac{d}{dt}\begin{bmatrix} z_1 \\ z_2 \end{bmatrix} &= \begin{bmatrix} -c_1 & 1 \\ -1 & -c_2 \end{bmatrix}\begin{bmatrix} z_1 \\ z_2 \end{bmatrix} + \begin{bmatrix} \varphi & 0 \\ 0 & -\frac{\partial \alpha_1}{\partial x_1}\varphi \end{bmatrix}\begin{bmatrix} \theta - \vartheta_1 \\ \theta - \vartheta_2 \end{bmatrix} \\ \frac{d}{dt}\begin{bmatrix} \vartheta_1 \\ \vartheta_2 \end{bmatrix} &= \gamma \begin{bmatrix} \varphi & 0 \\ 0 & -\frac{\partial \alpha_1}{\partial x_1}\varphi \end{bmatrix}\begin{bmatrix} z_1 \\ z_2 \end{bmatrix}, \end{aligned} \quad (3.47)$$

makes its properties more visible:

- The constant system matrix has negative terms along its diagonal, while its off-diagonal terms are skew-symmetric, and

- the matrix that multiplies the parameter errors in the $\dot{z}$-equation is used in the update laws for the parameter estimates.

The stability properties of (3.47) follow from (3.42) and (3.45): The LaSalle-Yoshizawa theorem (Theorem 2.1) establishes that $z_1, z_2, \vartheta_1, \vartheta_2$ are bounded, and $z \to 0$ as $t \to \infty$. Since $z_1 = x_1$, x_1 is also bounded and converges to zero. The boundedness of x_2 then follows from the boundedness of α_1 (defined in (3.34)) and the fact that $x_2 = z_2 + \alpha_1$. Using (3.40) we conclude that the control u is also bounded. Finally, we note that the regulation of z and x_1 does *not* imply the regulation of x_2: From $z_2 = x_2 - \alpha_1$ and (3.34) we see that $x_2 + \vartheta_1 \varphi(0)$ will converge to zero. Thus, x_2 is not guaranteed to converge to zero unless $\varphi(0) = 0$. However, x_2 will converge to a constant value:

$$\lim_{t \to \infty} = -\theta \varphi(0) \triangleq x_2^e. \tag{3.48}$$

This can be seen from (3.29a): Since x_1 and $\dot{x}_1$ converge to zero, so does $x_2 + \theta \varphi(0)$.

With the above example we have illustrated the idea of adaptive backstepping. To formulate it as a design tool analogous to Lemma 2.8, we start with the assumption that an adaptive controller is known for an initial system.

Assumption 3.1 *Consider the system*

$$\dot{x} = f(x) + F(x)\theta + g(x)u, \tag{3.49}$$

where $x \in \mathbb{R}^n$ is the state, $\theta \in \mathbb{R}^p$ is a vector of unknown constant parameters, and $u \in \mathbb{R}$ is the control input. There exist an adaptive controller

$$\begin{aligned} u &= \alpha(x, \vartheta) \\ \dot{\vartheta} &= T(x, \vartheta), \end{aligned} \tag{3.50}$$

with parameter estimate $\vartheta \in \mathbb{R}^q$, and a smooth function $V(x, \vartheta) : \mathbb{R}^{(n+q)} \to \mathbb{R}$ which is positive definite and radially unbounded in the variables $(x, \vartheta - \theta)$ such that for all $(x, \vartheta) \in \mathbb{R}^{(n+q)}$:

$$\frac{\partial V}{\partial x}(x, \vartheta) \left[f(x) + F(x)\theta + g(x)\alpha(x, \vartheta) \right] + \frac{\partial V}{\partial \vartheta}(x, \vartheta) T(x, \vartheta) \leq -W(x, \vartheta) \leq 0, \tag{3.51}$$

where $W : \mathbb{R}^{n+q} \to \mathbb{R}$ is positive semidefinite. □

Under this assumption, the control (3.50), applied to the system (3.49), guarantees global boundedness of $x(t), \vartheta(t)$ and, by the LaSalle-Yoshizawa theorem (Theorem 2.1), regulation of $W(x(t), \vartheta(t))$. Adaptive backstepping allows us to achieve the same properties for the augmented system.

Lemma 3.2 (Adaptive Backstepping) *Let the system (3.49) be augmented by an integrator,*

$$\begin{aligned} \dot{x} &= f(x) + F(x)\theta + g(x)\xi & (3.52a) \\ \dot{\xi} &= u, & (3.52b) \end{aligned}$$

3.2 ADAPTIVE BACKSTEPPING

where $\xi \in \mathbb{R}$. Consider for this system the dynamic feedback controller

$$u = -c(\xi - \alpha(x,\vartheta)) + \frac{\partial \alpha}{\partial x}(x,\vartheta)\left[f(x) + F(x)\bar{\vartheta} + g(x)\xi\right]$$
$$+ \frac{\partial \alpha}{\partial \vartheta}T(x,\vartheta) - \frac{\partial V}{\partial x}(x,\vartheta)g(x), \quad c > 0 \tag{3.53}$$

$$\dot{\vartheta} = T(x,\vartheta) \tag{3.54}$$

$$\dot{\bar{\vartheta}} = -\Gamma\left[\frac{\partial \alpha}{\partial x}(x,\vartheta)F(x)\right]^{\mathrm{T}}(\xi - \alpha(x,\vartheta)), \tag{3.55}$$

where $\bar{\vartheta}$ is a new estimate of θ, $\Gamma = \Gamma^{\mathrm{T}} > 0$ is the adaptation gain matrix. Under Assumption 3.1, this adaptive controller guarantees global boundedness of $x(t)$, $\xi(t)$, $\vartheta(t)$, $\bar{\vartheta}(t)$ and regulation of $W(x(t),\vartheta(t))$ and $\xi(t) - \alpha(x(t),\vartheta(t))$. These properties can be established with the Lyapunov function

$$V_{\mathrm{a}}(x,\xi,\vartheta,\bar{\vartheta}) = V(x,\vartheta) + \frac{1}{2}[\xi - \alpha(x,\vartheta)]^2 + \frac{1}{2}(\theta - \bar{\vartheta})^{\mathrm{T}}\Gamma^{-1}(\theta - \bar{\vartheta}). \tag{3.56}$$

Proof. With the error variable $z = \xi - \alpha(x,\vartheta)$, (3.52) is rewritten as

$$\dot{x} = f(x) + F(x)\theta + g(x)\left[\alpha(x,\vartheta) + z\right] \tag{3.57a}$$
$$\dot{z} = u - \frac{\partial \alpha}{\partial x}(x,\vartheta)\left[f(x) + F(x)\theta + g(x)(\alpha(x,\vartheta) + z)\right] - \frac{\partial \alpha}{\partial \vartheta}T(x,\vartheta). \tag{3.57b}$$

Note that in (3.57b) the derivative of ϑ was replaced by the update law (3.54). Introducing a new parameter estimate $\bar{\vartheta}$, we augment the Lyapunov function:

$$V_{\mathrm{a}}(x,\xi,\vartheta,\bar{\vartheta}) = V(x,\vartheta) + \frac{1}{2}z^2 + \frac{1}{2}(\theta - \bar{\vartheta})^{\mathrm{T}}\Gamma^{-1}(\theta - \bar{\vartheta}). \tag{3.58}$$

Using (3.51), it is easy to show that the derivative of (3.58) satisfies

$$\begin{aligned}
\dot{V}_{\mathrm{a}} &= \frac{\partial V}{\partial x}(f + F\theta + g\alpha + gz) + \frac{\partial V}{\partial \vartheta}T \\
&\quad + z\left[u - \frac{\partial \alpha}{\partial x}(f + F\theta + g(\alpha + z)) - \frac{\partial \alpha}{\partial \vartheta}T\right] - \dot{\bar{\vartheta}}^{\mathrm{T}}\Gamma^{-1}(\theta - \bar{\vartheta}) \\
&= \frac{\partial V}{\partial x}(f + F\theta + g\alpha) + \frac{\partial V}{\partial \vartheta}T \\
&\quad + z\left[u - \frac{\partial \alpha}{\partial x}(f + F\theta + g(\alpha + z)) - \frac{\partial \alpha}{\partial \vartheta}T + \frac{\partial V}{\partial x}g\right] - \dot{\bar{\vartheta}}^{\mathrm{T}}\Gamma^{-1}(\theta - \bar{\vartheta}) \\
&\leq -W(x,\vartheta) + z\left[u - \frac{\partial \alpha}{\partial x}\left(f + F\bar{\vartheta} + g(\alpha + z)\right) - \frac{\partial \alpha}{\partial \vartheta}T + \frac{\partial V}{\partial x}g\right] \\
&\quad - \left[\frac{\partial \alpha}{\partial x}Fz + \dot{\bar{\vartheta}}^{\mathrm{T}}\Gamma^{-1}\right](\theta - \bar{\vartheta}).
\end{aligned} \tag{3.59}$$

The $(\theta - \bar{\vartheta})$-term is now eliminated with the update law (cf. (3.55))

$$\dot{\bar{\vartheta}} = -\Gamma\left(\frac{\partial \alpha}{\partial x}F\right)^{\mathrm{T}}z, \tag{3.60}$$

and the control (3.53) is chosen to make the bracketed term multiplying z in (3.59) equal to $-cz$ (cf. (2.54)):

$$u = -cz + \frac{\partial \alpha}{\partial x}\left(f + F\bar{\vartheta} + g(\alpha + z)\right) + \frac{\partial \alpha}{\partial \vartheta}T - \frac{\partial V}{\partial x}g. \tag{3.61}$$

This results in the desired nonpositivity of $\dot{V}_a$:

$$\dot{V}_a \leq -W(x, \vartheta) - cz^2 \leq 0. \tag{3.62}$$

From (3.56) and (3.62) we conclude that $V(x, \vartheta)$, $\bar{\vartheta}$ and z are bounded. By Assumption 3.1, this means that $x(t)$ and $\vartheta(t)$ are bounded. Hence, $\xi = z + \alpha(x, \vartheta)$ and u are bounded. By Thorem 2.1, the boundedness of all the signals combined with (3.62) proves the regulation of $W(x(t), \vartheta(t))$ and $z(t)$. □

3.2.2 Adaptive block backstepping

We now extend the Adaptive Backstepping Lemma (Lemma 3.2) by augmenting the initial system with a relative-degree-one nonlinear system whose zero dynamics subsystem is ISS, just like we did in Chapter 2, Lemmas 2.8 and 2.25. The adaptive counterpart of Assumption 2.7 was Assumption 3.1. We now formulate the adaptive counterpart of Assumption 2.21, with analogous changes in the properties of $V(x, \vartheta)$ from Assumption 3.1.

Assumption 3.3 *Suppose Assumption 3.1 is valid, but $V(x, \vartheta)$ is only positive semidefinite, and the closed-loop system (3.49) with the adaptive controller (3.50) has the property that $x(t)$ and $\vartheta(t)$ are bounded if $V(x(t), \vartheta(t))$ is bounded.* □

Under this assumption, the control (3.50), applied to the system (3.49), guarantees global boundedness of $x(t), \vartheta(t)$ and, by Lemma A.6, regulation of $W(x(t), \vartheta(t))$.

Lemma 3.4 (Adaptive Block Backstepping) *Let the system (3.49) be augmented by a nonlinear system which is linear in the unknown parameter vector θ,*

$$\dot{x} = f(x) + F(x)\theta + g(x)y \tag{3.63a}$$
$$\dot{\xi} = m(x, \xi) + M(x, \xi)\theta + \beta(x, \xi)u, \quad y = h(\xi), \tag{3.63b}$$

where $\xi \in \mathbb{R}^q$, and suppose that (3.63b) has relative degree one uniformly in x and that its zero dynamics subsystem is ISS with respect to y and x. Under Assumption 3.3, the feedback control

$$u = \left[\frac{\partial h}{\partial \xi}(\xi)\beta(x, \xi)\right]^{-1}\left\{-c(y - \alpha(x, \vartheta)) - \frac{\partial h}{\partial \xi}(\xi)\left[m(x, \xi) + M(x, \xi)\bar{\vartheta}\right]\right.$$
$$\left. + \frac{\partial \alpha}{\partial x}(x, \vartheta)\left[f(x) + F(x)\bar{\vartheta} + g(x)y\right] + \frac{\partial \alpha}{\partial \vartheta}T(x, \vartheta) - \frac{\partial V}{\partial x}(x, \vartheta)g(x)\right\}, \tag{3.64}$$

with $c > 0$ and $\bar{\vartheta}$ a new estimate of θ, along with the update laws

$$\dot{\vartheta} = T(x, \vartheta) \tag{3.65}$$

$$\dot{\bar{\vartheta}} = \Gamma \left[\frac{\partial h}{\partial \xi}(\xi) M(x, \xi) - \frac{\partial \alpha}{\partial x}(x, \vartheta) F(x) \right]^{\mathrm{T}} (y - \alpha(x, \vartheta)), \tag{3.66}$$

with the adaptation gain matrix $\Gamma = \Gamma^{\mathrm{T}} > 0$, guarantees global boundedness of $x(t)$, $\xi(t)$, $\vartheta(t)$, $\bar{\vartheta}(t)$ and regulation of $W(x(t), \vartheta(t))$ and $\xi(t) - \alpha(x(t), \vartheta(t))$.

Proof. As in Lemma 2.25, we employ the change of coordinates $(y, \zeta) = (h(\xi), \phi(x, \xi))$, with $\frac{\partial \phi}{\partial \xi} \beta \equiv 0$, to transform (3.63b) into the normal form

$$\dot{y} = \frac{\partial h}{\partial \xi}(\xi) \left[m(x, \xi) + M(x, \xi)\theta + \beta(x, \xi) u \right] \tag{3.67a}$$

$$\dot{\zeta} = \frac{\partial \phi}{\partial x}(x, \xi) \left[f(x) + F(x)\theta + g(x)y \right] + \frac{\partial \phi}{\partial \xi}(x, \xi) \left[m(x, \xi) + M(x, \xi)\theta \right]$$

$$\stackrel{\Delta}{=} \Phi_0(x, y, \zeta) + \Phi(x, y, \zeta)\theta. \tag{3.67b}$$

Introducing a new parameter estimate $\bar{\vartheta}$, we use the feedback transformation

$$u = \left(\frac{\partial h}{\partial \xi} \beta \right)^{-1} \left\{ v - \frac{\partial h}{\partial \xi} \left[m + M \bar{\vartheta} \right] \right\} \tag{3.68}$$

to rewrite (3.63a) and (3.67a) as

$$\dot{x} = f(x) + F(x)\theta + g(x)y \tag{3.69a}$$

$$\dot{y} = v + \frac{\partial h}{\partial \xi}(\xi) M(x, \xi)(\theta - \bar{\vartheta}). \tag{3.69b}$$

We now apply Lemma 3.1 to (3.69). The only difference between (3.69) and (3.52) is the presence of the additional parameter error term $\frac{\partial h}{\partial \xi} M(\theta - \bar{\vartheta})$ in (3.69b). This term can be eliminated in $\dot{V}_a$ by adding the term $-\Gamma[\frac{\partial h}{\partial \xi} M]^{\mathrm{T}}(y - \alpha)$ to the update law (3.55). Combining this modification with (3.68), we see that the resulting adaptive controller is given by (3.64)–(3.66). This guarantees the boundedness of x, ϑ, $\bar{\vartheta}$, z and the regulation of $W(x, \vartheta)$ and z. Hence, $y = z + \alpha(x, \vartheta)$ is bounded. Then, from (3.67b) and the ISS property of the zero dynamics, ζ is also bounded, and thus ξ and u are bounded. □

3.3 Recursive Design Procedures

3.3.1 Parametric strict-feedback systems

Through repeated application of Lemma 3.2, the backstepping design procedure is now generalized to nonlinear systems which can be transformed[1] into

---
[1] The coordinate-free characterization of these systems in terms of differential geometric conditions is given in Appendix G, Corollary G.15.

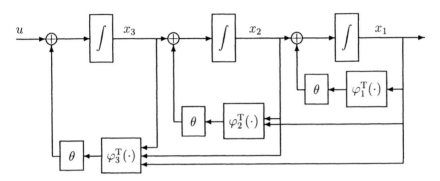

Figure 3.4: Block diagram of a third-order parametric strict-feedback system with $\beta(x) = 1$. The nonlinearities depend only on variables which are "fed back."

the *parametric strict-feedback form*

$$\begin{aligned}
\dot{x}_1 &= x_2 + \varphi_1^T(x_1)\theta \\
\dot{x}_2 &= x_3 + \varphi_2^T(x_1, x_2)\theta \\
&\vdots \\
\dot{x}_{n-1} &= x_n + \varphi_{n-1}^T(x_1, \ldots, x_{n-1})\theta \\
\dot{x}_n &= \beta(x)u + \varphi_n^T(x)\theta,
\end{aligned} \tag{3.70}$$

where $\beta(x) \neq 0$ for all $x \in \mathbb{R}^n$. The reason for the name "parametric strict-feedback" can be deduced from the block diagram in Figure 3.4, where, except for the integrators, there are only feedback paths.

For systems in the form (3.70), the number of design steps required is equal to the degree n of the system. At each step, an error variable z_i, a stabilizing function α_i, and a parameter estimate ϑ_i are generated. As a result, if a system contains p unknown parameters, the overparametrized adaptive controller may employ as many as pn parameter estimates. A schematic representation of this design procedure is given in Figure 3.5, and the resulting expressions are summarized in the following theorem:

Theorem 3.5 (Parametric Strict-Feedback Systems) *For the system (3.70) with $\beta(x) \neq 0$ for all $x \in \mathbb{R}^n$, consider the adaptive controller*

$$u = \frac{1}{\beta(x)} \alpha_n(x, \vartheta_1, \ldots, \vartheta_n) \tag{3.71}$$

$$\dot{\vartheta}_i = \Gamma \left(\varphi_i - \sum_{j=1}^{i-1} \frac{\partial \alpha_{i-1}}{\partial x_j} \varphi_j \right) z_i, \quad i = 1, \ldots, n, \tag{3.72}$$

where $\vartheta_i \in \mathbb{R}^p$ are multiple estimates of θ, $\Gamma = \Gamma^T > 0$ is the adaptation gain matrix, and the variables z_i and the stabilizing functions α_i, $i = 1, \ldots, n$,

3.3 RECURSIVE DESIGN PROCEDURES

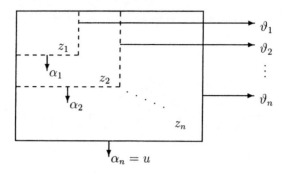

Figure 3.5: The design procedure for overparametrized schemes. Each step generates an error variable z_i, a stabilizing function α_i, and a *new* estimate ϑ_i of the unknown parameter vector θ.

are defined by the following recursive expressions (with $c_i > 0$ being design constants, and $z_0 \equiv \alpha_0 \equiv 0$ used for notational convenience):

$$z_i = x_i - \alpha_{i-1}(x_1, \ldots, x_{i-1}, \vartheta_1, \ldots, \vartheta_{i-1}) \tag{3.73}$$

$$\begin{aligned}\alpha_i = {}& -c_i z_i - z_{i-1} - \left(\varphi_i - \sum_{j=1}^{i-1} \frac{\partial \alpha_{i-1}}{\partial x_j} \varphi_j\right)^{\mathrm{T}} \vartheta_i \\ & + \sum_{j=1}^{i-1}\left[\frac{\partial \alpha_{i-1}}{\partial x_j} x_{j+1} + \frac{\partial \alpha_{i-1}}{\partial \vartheta_j}\Gamma\left(\varphi_j - \sum_{k=1}^{j-1}\frac{\partial \alpha_{j-1}}{\partial x_k}\varphi_k\right) z_j\right].\end{aligned} \tag{3.74}$$

This overparametrized adaptive controller guarantees global boundedness of $x(t)$, $\vartheta_1(t), \ldots, \vartheta_n(t)$, and regulation of $x_1(t)$ and $x_i(t) - x_i^{\mathrm{e}}$, $i = 2, \ldots, n$, where $x_i^{\mathrm{e}} = -\theta^{\mathrm{T}} \varphi_{i-1}(0, x_2^{\mathrm{e}}, \ldots, x_{i-1}^{\mathrm{e}})$.

Proof. Using the definitions (3.73), (3.74) and denoting $x_0 \equiv \alpha_0 \equiv 0$, $x_{n+1} \equiv \beta(x)u$, the derivative of the error variable z_i, $i = 1, \ldots, n$, becomes

$$\begin{aligned}\dot{z}_i ={}& \dot{x}_i - \dot{\alpha}_{i-1}(x_1, \ldots, x_{i-1}, \vartheta_1, \ldots, \vartheta_{i-1}) \\ ={}& x_{i+1} + \varphi_i^{\mathrm{T}}\theta - \sum_{j=1}^{i-1}\frac{\partial\alpha_{i-1}}{\partial x_j}\underbrace{\left(x_{j+1} + \varphi_j^{\mathrm{T}}\theta\right)}_{\dot{x}_j} \\ & - \sum_{j=1}^{i-1}\frac{\partial\alpha_{i-1}}{\partial \vartheta_j}\underbrace{\Gamma\left(\varphi_j - \sum_{k=1}^{j-1}\frac{\partial\alpha_{j-1}}{\partial x_k}\varphi_k\right)z_j}_{\dot{\vartheta}_j} \\ ={}& \alpha_i + z_{i+1} + \left(\varphi_i - \sum_{j=1}^{i-1}\frac{\partial\alpha_{i-1}}{\partial x_j}\varphi_j\right)^{\mathrm{T}}\theta\end{aligned}$$

$$-\sum_{j=1}^{i-1}\left[\frac{\partial\alpha_{i-1}}{\partial x_j}x_{j+1} + \frac{\partial\alpha_{i-1}}{\partial\vartheta_j}\Gamma\left(\varphi_j - \sum_{k=1}^{j-1}\frac{\partial\alpha_{j-1}}{\partial x_k}\varphi_k\right)z_j\right]$$

$$= -c_i z_i - z_{i-1} + z_{i+1} + \left(\varphi_i - \sum_{j=1}^{i-1}\frac{\partial\alpha_{i-1}}{\partial x_j}\varphi_j\right)^{\mathrm{T}}(\theta - \vartheta_i). \quad (3.75)$$

The choice of control (3.71) guarantees that $z_{n+1} \equiv 0$. The closed-loop error system can therefore be expressed as

$$\begin{aligned}
\dot{z}_1 &= -c_1 z_1 + z_2 + \varphi_1^{\mathrm{T}}(\theta - \vartheta_1) \\
\dot{z}_2 &= -c_2 z_2 - z_1 + z_3 + \left(\varphi_2 - \frac{\partial\alpha_1}{\partial x_1}\varphi_1\right)^{\mathrm{T}}(\theta - \vartheta_2) \\
&\vdots \\
\dot{z}_{n-1} &= -c_{n-1} z_{n-1} - z_{n-2} + z_n + \left(\varphi_{n-1} - \sum_{j=1}^{n-2}\frac{\partial\alpha_{n-2}}{\partial x_j}\varphi_j\right)^{\mathrm{T}}(\theta - \vartheta_{n-1}) \\
\dot{z}_n &= -c_n z_n - z_{n-1} + \left(\varphi_n - \sum_{j=1}^{n-1}\frac{\partial\alpha_{n-1}}{\partial x_j}\varphi_j\right)^{\mathrm{T}}(\theta - \vartheta_n) \\
\dot{\vartheta}_i &= -\Gamma\left(\varphi_i - \sum_{j=1}^{i-1}\frac{\partial\alpha_{i-1}}{\partial x_j}\varphi_j\right)z_i,
\end{aligned} \quad (3.76)$$

or, equivalently, in the matrix form

$$\begin{aligned}
\frac{d}{dt}\begin{bmatrix} z_1 \\ z_2 \\ \vdots \\ z_{n-1} \\ z_n \end{bmatrix} &= \begin{bmatrix} -c_1 & 1 & 0 & \cdots & 0 \\ -1 & -c_2 & 1 & \ddots & \vdots \\ 0 & \ddots & \ddots & \ddots & 0 \\ \vdots & \ddots & -1 & -c_{n-1} & 1 \\ 0 & \cdots & 0 & -1 & -c_n \end{bmatrix} \begin{bmatrix} z_1 \\ z_2 \\ \vdots \\ z_{n-1} \\ z_n \end{bmatrix} \\
&\quad + \begin{bmatrix} w_1 & 0 & \cdots & 0 \\ 0 & w_2 & \ddots & \vdots \\ \vdots & \ddots & \ddots & 0 \\ 0 & \cdots & 0 & w_n \end{bmatrix}^{\mathrm{T}} \begin{bmatrix} \theta - \vartheta_1 \\ \theta - \vartheta_2 \\ \vdots \\ \theta - \vartheta_n \end{bmatrix} \\
\frac{d}{dt}\begin{bmatrix} \vartheta_1 \\ \vartheta_2 \\ \vdots \\ \vartheta_n \end{bmatrix} &= \begin{bmatrix} \Gamma & 0 & \cdots & 0 \\ 0 & \Gamma & \ddots & \vdots \\ \vdots & \ddots & \ddots & 0 \\ 0 & \cdots & 0 & \Gamma \end{bmatrix} \begin{bmatrix} w_1 & 0 & \cdots & 0 \\ 0 & w_2 & \ddots & \vdots \\ \vdots & \ddots & \ddots & 0 \\ 0 & \cdots & 0 & w_n \end{bmatrix} \begin{bmatrix} z_1 \\ z_2 \\ \vdots \\ z_n \end{bmatrix},
\end{aligned} \quad (3.77)$$

3.3 RECURSIVE DESIGN PROCEDURES

where we have used the convenient notation

$$w_1 = \varphi_1, \quad w_i = \varphi_i - \sum_{j=1}^{i-1} \frac{\partial \alpha_{i-1}}{\partial x_j} \varphi_j, \quad i = 2, \ldots, n. \qquad (3.78)$$

This system has two important properties: (i) The z-system matrix in (3.77) has negative diagonal and skew-symmetric off-diagonal terms, and (ii) the transpose of the matrix that multiplies the parameter errors in the $\dot{z}$-equation appears in the update law. This structure is a result of the design procedure, and it allows us to use the simple quadratic Lyapunov function

$$V_n(z_1, \ldots, z_n, \vartheta_1, \ldots, \vartheta_n) = \frac{1}{2} \sum_{i=1}^{n} \left[z_i^2 + (\theta - \vartheta_i)^{\mathrm{T}} \Gamma^{-1} (\theta - \vartheta_i) \right] \qquad (3.79)$$

to prove stability and regulation. Its derivative along the solutions of (3.77) is

$$\begin{aligned}
\dot{V}_n &= z^{\mathrm{T}} \dot{z} - \sum_{i=1}^{n} \dot{\vartheta}_i^{\mathrm{T}} \Gamma^{-1} (\theta - \vartheta_i) \\
&= \sum_{i=1}^{n} \left[-c_i z_i^2 + z_i w_i^{\mathrm{T}} (\theta - \vartheta_i) - z_i w_i^{\mathrm{T}} (\theta - \vartheta_i) \right] \\
&= -\sum_{i=1}^{n} c_i z_i^2. \qquad (3.80)
\end{aligned}$$

The LaSalle-Yoshizawa theorem (Theorem 2.1) now guarantees the global uniform boundedness of $z(t)$, $\vartheta_1(t), \ldots, \vartheta_n(t)$, as well as the regulation of $z(t)$. Since $z_1 = x_1$, we see that x_1 is also bounded and regulated. The boundedness of $x_2, \ldots, x_n$ then follows from the boundedness of α_i (defined in (3.74)) and the fact that $x_i = z_i + \alpha_{i-1}$. Since x is bounded, $\beta(x)$ is bounded away from zero. Combining this with (3.71) we conclude that the control u is also bounded. Finally, the regulation of $x_i - x_i^e$ is concluded as follows: Since $z_i(t)$, $i = 1, \ldots, n$ converge to zero, $\dot{\vartheta}_i(t)$, $i = 1, \ldots, n$ also converge to zero and $\dot{z}_i(t)$ is integrable over $[0, \infty)$. Furthermore, the boundedness of all the signals and their derivatives guarantees the boundedness of $\ddot{z}_i(t)$ and hence the uniform continuity of $\dot{z}_i(t)$. From Lemma A.6, we conclude that $\lim_{t \to \infty} \dot{z}_i(t) = 0$, $i = 1, \ldots, n$. Since x can be expressed as a smooth vector function of $z_1, \ldots, z_n$ and $\vartheta_1, \ldots, \vartheta_n$, we can express $\dot{x}$ as a linear combination of $\dot{z}_i$ and $\dot{\vartheta}_i$ with coefficients which are bounded because they are smooth vector functions of the bounded signals $z_1, \ldots, z_n$ and $\vartheta_1, \ldots, \vartheta_n$. Hence, the convergence of $\dot{z}$ and $\dot{\vartheta}_i$ to zero implies that $\dot{x}$ converges to zero. Combining this with (3.70) and the regulation of x_1 leads to the desired result. □

3.3.2 Multi-input systems

The adaptive backstepping design procedure of Theorem 3.5 can be easily extended to nonlinear systems which have been transformed into the *multi-input parametric strict-feedback form*

$$\dot{x}_{1,1} = x_{1,2} + \varphi_{1,1}^{\mathrm{T}}(x_{1,1}, x_{2,1}, \ldots x_{2,\rho_2-\rho_1+1}, \ldots, x_{m,1}, \ldots, x_{m,\rho_m-\rho_1+1})\theta$$
$$\dot{x}_{1,2} = x_{1,3} + \varphi_{1,2}^{\mathrm{T}}(x_{1,1}, x_{1,2}, x_{2,1}, \ldots x_{2,\rho_2-\rho_1+2},$$
$$\ldots, x_{m,1}, \ldots, x_{m,\rho_m-\rho_1+2})\theta$$
$$\vdots$$
$$\dot{x}_{1,\rho_1-1} = x_{1,\rho_1} + \varphi_{1,\rho_1-1}^{\mathrm{T}}(x_{1,1}, \ldots, x_{1,\rho_1-1}, x_{2,1}, \ldots x_{2,\rho_2-1},$$
$$\ldots, x_{m,1}, \ldots, x_{m,\rho_m-1})\theta$$
$$\dot{x}_{1,\rho_1} = \sum_{j=1}^{m} \beta_{1,j}(x) u_j + \varphi_{1,\rho_1}^{\mathrm{T}}(x)\theta$$
$$\vdots \qquad (3.81)$$
$$\dot{x}_{i,j} = x_{i,j+1} + \varphi_{i,j}^{\mathrm{T}}(x_{1,1}, \ldots, x_{1,\rho_1-\rho_i+j}, \ldots, x_{i,1}, \ldots, x_{i,j},$$
$$\ldots, x_{m,1}, \ldots, x_{m,\rho_m-\rho_i+j})\theta$$
$$\vdots$$
$$\dot{x}_{m,1} = x_{m,2} + \varphi_{m,1}^{\mathrm{T}}(x_{1,1}, \ldots, x_{1,\rho_1-\rho_m+1}, x_{2,1}, \ldots x_{2,\rho_2-\rho_m+1}, \ldots, x_{m,1})\theta$$
$$\dot{x}_{m,2} = x_{m,3} + \varphi_{m,2}^{\mathrm{T}}(x_{1,1}, \ldots, x_{1,\rho_1-\rho_m+2}, x_{2,1}, \ldots x_{2,\rho_2-\rho_m+2},$$
$$\ldots, x_{m,1}, x_{m,2})\theta$$
$$\vdots$$
$$\dot{x}_{m,\rho_m-1} = x_{m,\rho_m} + \varphi_{m,\rho_m-1}^{\mathrm{T}}(x_{1,1}, \ldots, x_{1,\rho_1-1}, x_{2,1}, \ldots x_{2,\rho_2-1},$$
$$\ldots, x_{m,1}, \ldots, x_{m,\rho_m-1})\theta$$
$$\dot{x}_{m,\rho_m} = \sum_{j=1}^{m} \beta_{m,j}(x) u_j + \varphi_{m,\rho_1}^{\mathrm{T}}(x)\theta,$$

where $u_1, \ldots, u_m$ are the inputs, and the input matrix is nonsingular $\forall\, x \in \mathbb{R}^n$ ($n = \rho_1 + \cdots + \rho_m$):

$$\det B(x) \neq 0, \quad \forall\, x \in \mathbb{R}^n, \quad B(x) = \begin{bmatrix} \beta_{1,1}(x) & \cdots & \beta_{1,m}(x) \\ \vdots & & \vdots \\ \beta_{m,1}(x) & \cdots & \beta_{m,m}(x) \end{bmatrix}. \qquad (3.82)$$

The design procedure for this class of systems consists of applying the design procedure of Theorem 3.5 to the first $\rho_i - 1$ equations of each of the m subsystems of (3.81), to obtain the system

$$\dot{z}_{i,j} = -c_{i,j} z_{i,j} - z_{i,j-1} + z_{i,j+1} + w_{i,j}^{\mathrm{T}}(x, \vartheta_1, \ldots, \vartheta_{\ell-1})(\theta - \vartheta_\ell)$$
$$\ell = \sum_{k=1}^{i-1}(\rho_k - 1) + j, \quad 1 \leq j \leq \rho_i - 1, \quad 1 \leq i \leq m$$
$$\dot{\vartheta}_\ell = \Gamma w_{i,j}(x, \vartheta_1, \ldots, \vartheta_\ell) z_{i,j}, \quad 1 \leq \ell \leq n - m \qquad (3.83)$$
$$\frac{d}{dt}\begin{bmatrix} z_{1,\rho_1} \\ \vdots \\ z_{m,\rho_m} \end{bmatrix} = B(x) u + \Phi(x, \vartheta_1, \ldots, \vartheta_{n-m}) + W_{n-m+1}^{\mathrm{T}}(x, \vartheta_1, \ldots, \vartheta_{n-m})\theta,$$

3.3 RECURSIVE DESIGN PROCEDURES

where the functions $w_{i,j}$, Φ and W_{n-m+1} are defined appropriately. Now let ϑ_{n-m+1} be a new estimate of θ and define the control u as

$$u = B^{-1}(x) \left\{ - \begin{bmatrix} c_{1,\rho_1} z_{1,\rho_1} + z_{1,\rho_1 - 1} \\ \vdots \\ c_{m,\rho_m} z_{m,\rho_m} + z_{m,\rho_m - 1} \end{bmatrix} - \Phi(x, \vartheta_1, \ldots, \vartheta_{n-m}) \right.$$

$$\left. - W_{n-m+1}^T(x, \vartheta_1, \ldots, \vartheta_{n-m}) \vartheta_{n-m+1} \right\}, \quad (3.84)$$

and the update law for ϑ_{n-m+1} as

$$\dot{\vartheta}_{n-m+1} = \Gamma W_{n-m+1}(x, \vartheta_1, \ldots, \vartheta_{n-m+1}) \begin{bmatrix} z_{1,\rho_1} \\ \vdots \\ z_{m,\rho_m} \end{bmatrix}. \quad (3.85)$$

The stability properties of the resulting closed-loop system are analogous to those listed in Theorem 3.5, and can be similarly established using the Lyapunov function

$$V(z, \vartheta_1, \ldots, \vartheta_{n-m+1}) = \frac{1}{2} z^T z + \frac{1}{2} \sum_{i=1}^{n-m+1} (\theta - \vartheta_i)^T \Gamma^{-1} (\theta - \vartheta_i). \quad (3.86)$$

3.3.3 Parametric block-strict-feedback systems

Lemma 3.4 can be applied repeatedly to design adaptive controllers for nonlinear systems which can be transformed, after a change of coordinates, into the *parametric block-strict-feedback form*

$$\begin{aligned}
\dot{\chi}_1 &= \tilde{f}_1(\chi_1) + \tilde{F}_1(\chi_1)\theta + \tilde{g}_1(\chi_1) y_2 \\
y_1 &= h_1(\chi_1) \\
\dot{\chi}_2 &= \tilde{f}_2(\chi_1, \chi_2) + \tilde{F}_2(\chi_1, \chi_2)\theta + \tilde{g}_2(\chi_1, \chi_2) y_3 \\
y_2 &= h_2(\chi_2) \\
&\vdots \\
\dot{\chi}_i &= \tilde{f}_i(\chi_1, \ldots, \chi_i) + \tilde{F}_i(\chi_1, \ldots, \chi_i)\theta + \tilde{g}_i(\chi_1, \ldots, \chi_i) y_{i+1} \quad (3.87) \\
y_i &= h_i(\chi_i) \\
&\vdots \\
\dot{\chi}_{\rho-1} &= \tilde{f}_{\rho-1}(\chi_1, \ldots, \chi_{\rho-1}) + \tilde{F}_{\rho-1}(\chi_1, \ldots, \chi_{\rho-1})\theta + \tilde{g}_{\rho-1}(\chi_1, \ldots, \chi_{\rho-1}) y_\rho \\
y_{\rho-1} &= h_{\rho-1}(\chi_{\rho-1}) \\
\dot{\chi}_\rho &= \tilde{f}_\rho(\chi) + \tilde{F}_\rho(\chi)\theta + \tilde{g}_\rho(\chi) u \\
y_\rho &= h_\rho(\chi_\rho),
\end{aligned}$$

where each of the ρ subsystems with state $\chi_i \in \mathbb{R}^{n_i}$, output $y_i \in \mathbb{R}$, and input y_{i+1} (for convenience we denote $y_{\rho+1} \equiv u$) satisfies conditions (BSF-1) and (BSF-2) (see Chapter 2, equation (2.198)), that is, it has relative degree one uniformly in $\chi_1, \ldots, \chi_{i-1}$, and its zero dynamics subsystem is ISS with respect to $\chi_1, \ldots, \chi_{i-1}, y_i$.

Using the change of coordinates which transformed (2.198) into (2.201) in Section 2.3.3, we can now transform the system (3.87) into

$$\begin{aligned}
\dot{y}_1 &= f_1(y_1, \zeta_1) + \bar{\varphi}_1^T(y_1, \zeta_1)\theta + g_1(y_1, \zeta_1)y_2 \\
\dot{y}_2 &= f_2(y_1, \zeta_1, y_2, \zeta_2) + \bar{\varphi}_2^T(y_1, \zeta_1, y_2, \zeta_2)\theta + g_2(y_1, \zeta_1, y_2, \zeta_2)y_3 \\
&\vdots \\
\dot{y}_{\rho-1} &= f_{\rho-1}(y_1, \zeta_1, \ldots, y_{\rho-1}, \zeta_{\rho-1}) + \bar{\varphi}_{\rho-1}^T(y_1, \zeta_1, \ldots, y_{\rho-1}, \zeta_{\rho-1})\theta \\
&\quad + g_{\rho-1}(y_1, \zeta_1, \ldots, y_{\rho-1}, \zeta_{\rho-1})y_\rho \quad (3.88) \\
\dot{y}_\rho &= f_\rho(y_1, \zeta_1, \ldots, y_\rho, \zeta_\rho) + \bar{\varphi}_\rho^T(y_1, \zeta_1, \ldots, y_\rho, \zeta_\rho)\theta + g_\rho(y_1, \zeta_1, \ldots, y_\rho, \zeta_\rho)u \\
\dot{\zeta}_1 &= \bar{\Phi}_{1,0}(y_1, \zeta_1) + \bar{\Phi}_1(y_1, \zeta_1)\theta \\
&\vdots \\
\dot{\zeta}_\rho &= \bar{\Phi}_{\rho,0}(y_1, \zeta_1, \ldots, y_{\rho-1}, \zeta_{\rho-1}, y_\rho, \zeta_\rho) + \bar{\Phi}_\rho(y_1, \zeta_1, \ldots, y_{\rho-1}, \zeta_{\rho-1}, y_\rho, \zeta_\rho)\theta.
\end{aligned}$$

Then we employ another change of coordinates which replaces y_i by $x_i = \psi_i(y_1, \zeta_1, \ldots, y_{i-1}, \zeta_{i-1}, y_i)$, where

$$\begin{aligned}
x_1 &= y_1 \stackrel{\Delta}{=} \psi_1(y_1) \\
x_2 &= \frac{\partial \psi_1}{\partial y_1}(f_1 + g_1 y_2) = f_1(y_1, \zeta_1) + g_1(y_1, \zeta_1)y_2 \stackrel{\Delta}{=} \psi_2(y_1, \zeta_1, y_2) \quad (3.89) \\
x_{i+1} &= \sum_{j=1}^{i-1} \frac{\partial \psi_i}{\partial y_j}(f_j + g_j y_{j+1}) + \sum_{j=1}^{i-1} \frac{\partial \psi_i}{\partial \zeta_j}\bar{\Phi}_{j,0} + g_1 \cdots g_{i-1}f_i + g_1 \cdots g_i y_{i+1} \\
&\stackrel{\Delta}{=} \psi_{i+1}(y_1, \zeta_1, \ldots, y_i, \zeta_i, y_{i+1}), \quad i = 2, \ldots, \rho-1.
\end{aligned}$$

Finally, we use the feedback transformation

$$v = \sum_{j=1}^{\rho-1} \frac{\partial \psi_\rho}{\partial y_j}(f_j + g_j y_{j+1}) + \sum_{j=1}^{\rho-1} \frac{\partial \psi_\rho}{\partial \zeta_j}\bar{\Phi}_{j,0} + g_1 \cdots g_{\rho-1}f_\rho + g_1 \cdots g_\rho u. \quad (3.90)$$

Condition (BSF-1) guarantees that $g_1, \ldots, g_\rho \neq 0$ everywhere. Hence, the change of coordinates (3.89) relating $[y_1, \ldots, y_\rho, \zeta_1^T, \ldots, \zeta_\rho^T]^T$ to $[x_1, \ldots, x_\rho, \zeta_1^T, \ldots, \zeta_\rho^T]^T$ is a global diffeomorphism, and the feedback transformation (3.90) relating u to v is nonsingular. It is now straightforward to verify that (3.89) and (3.90) transform (3.88) into a form reminiscent of the

3.3 Recursive Design Procedures

parametric strict-feedback form (3.70):

$$\begin{aligned}
\dot{x}_1 &= x_2 + \varphi_1^T(x_1, \zeta_1)\theta \\
\dot{x}_2 &= x_3 + \varphi_2^T(x_1, x_2, \zeta_1, \zeta_2) \\
&\vdots \\
\dot{x}_{\rho-1} &= x_\rho + \varphi_{\rho-1}^T(x_1, \ldots, x_{\rho-1}, \zeta_1, \ldots, \zeta_{\rho-1})\theta \\
\dot{x}_\rho &= v + \varphi_\rho^T(x, \zeta)\theta \\
\dot{\zeta}_1 &= \Phi_{1,0}(x_1, \zeta_1) + \Phi_1(x_1, \zeta_1)\theta \\
&\vdots \\
\dot{\zeta}_\rho &= \Phi_{\rho,0}(x_1, \ldots, x_\rho, \zeta_1, \ldots, \zeta_{\rho-1}, \zeta_\rho) + \Phi_\rho(x_1, \ldots, x_\rho, \zeta_1, \ldots, \zeta_{\rho-1}, \zeta_\rho)\theta \\
y &= x_1.
\end{aligned} \qquad (3.91)$$

In (3.91) each ζ_i-subsystem is ISS with respect to $x_1, \ldots, x_i, \zeta_1, \ldots, \zeta_{i-1}$ as its inputs, and φ_i, $\Phi_{i,0}$, Φ_i, $i = 1, \ldots, \rho$ are defined as

$$\begin{aligned}
\varphi_i^T(x_1, \ldots, x_i, \zeta_1, \ldots, \zeta_i) &\triangleq \sum_{j=1}^{i} \frac{\partial \psi_i}{\partial y_j}(y_1, \zeta_1, \ldots, y_{i-1}, \zeta_{i-1}, y_i) \\
&\quad \bar{\varphi}_j^T(y_1, \zeta_1, \ldots, y_j, \zeta_j) \\
&\quad + \sum_{j=1}^{i-1} \frac{\partial \psi_i}{\partial \zeta_j}(y_1, \zeta_1, \ldots, y_{i-1}, \zeta_{i-1}, y_i) \\
&\quad \bar{\Phi}_j(y_1, \zeta_1, \ldots, y_j, \zeta_j) \qquad (3.92)
\end{aligned}$$

$$\Phi_{i,0}(x_1, \ldots, x_i, \zeta_1, \ldots, \zeta_i) \triangleq \bar{\Phi}_{i,0}(y_1, \zeta_1, \ldots, y_i, \zeta_i) \qquad (3.93)$$

$$\Phi_i(x_1, \ldots, x_i, \zeta_1, \ldots, \zeta_i) \triangleq \bar{\Phi}_i(y_1, \zeta_1, \ldots, y_i, \zeta_i). \qquad (3.94)$$

It is now clear that the class of parametric block-strict-feedback nonlinear systems strictly contains the class of parametric strict-feedback nonlinear systems, since (3.70) can be obtained by setting $n_i = 1$, $\rho = n$, and $v = \beta(x)u$ in (3.91).

We now state and prove the generalization of Theorem 3.5 to block-strict-feedback systems of the form (3.91).

Theorem 3.6 (Parametric Block-Strict-Feedback Systems) *For the system (3.91), consider the adaptive controller*

$$v = \alpha_\rho(x_1, \ldots, x_\rho, \zeta_1, \ldots, \zeta_\rho, \vartheta_1, \ldots, \vartheta_\rho) \qquad (3.95)$$

$$\dot{\vartheta}_i = -\Gamma \left[\varphi_i^T - \sum_{j=1}^{i-1} \left(\frac{\partial \alpha_{i-1}}{\partial x_j} \varphi_j^T + \frac{\partial \alpha_{i-1}}{\partial \zeta_j} \Phi_i \right) \right]^T z_i, \quad i = 1, \ldots, \rho, \qquad (3.96)$$

where $\vartheta_i \in \mathbb{R}^p$ are multiple estimates of θ, $\Gamma = \Gamma^T > 0$ is the adaptation gain matrix, and the variables z_i and the stabilizing functions α_i, $i = 1, \ldots, \rho$, are defined by the following recursive expressions (with $c_i > 0$ being design constants, and $z_0 \equiv \alpha_0 \equiv 0$ used for notational convenience):

$$z_i = x_i - \alpha_{i-1}(x_1, \ldots, x_{i-1}, \zeta_1, \ldots, \zeta_{i-1}, \vartheta_1, \ldots, \vartheta_{i-1}) \tag{3.97}$$

$$\alpha_i = -c_i z_i - z_{i-1} - \left[\varphi_i^T - \sum_{j=1}^{i-1}\left(\frac{\partial \alpha_{i-1}}{\partial x_j}\varphi_j^T + \frac{\partial \alpha_{i-1}}{\partial \zeta_j}\Phi_j\right)\right]\vartheta_i + \sum_{j=1}^{i-1}\left\{\frac{\partial \alpha_{i-1}}{\partial x_j}x_{j+1}\right.$$

$$\left. + \frac{\partial \alpha_{i-1}}{\partial \zeta_j}\Phi_{j,0} - \frac{\partial \alpha_{i-1}}{\partial \vartheta_j}\Gamma\left[\varphi_j^T - \sum_{k=1}^{j-1}\left(\frac{\partial \alpha_{j-1}}{\partial x_k}\varphi_k^T + \frac{\partial \alpha_{j-1}}{\partial \zeta_k}\Phi_k\right)\right]^T z_j\right\}. \tag{3.98}$$

This overparametrized adaptive controller guarantees global boundedness of $x_1, \ldots, x_\rho, \zeta_1, \ldots, \zeta_\rho, \vartheta_1, \ldots, \vartheta_\rho$ and regulation of $y = x_1$.

Proof. As one would expect from the similarities between the systems (3.91) and (3.70), the expressions (3.95)–(3.98) are similar to (3.71)–(3.74). Using the same arguments as in the proof of Theorem 3.5, we write the derivative of the error variable z_i, $i = 1, \ldots, \rho$, as

$$\dot{z}_i = \dot{x}_i - \dot{\alpha}_{i-1}(x_1, \ldots, x_{i-1}, \zeta_1, \ldots, \zeta_{i-1}, \vartheta_1, \ldots, \vartheta_{i-1})$$

$$= x_{i+1} + \varphi_i^T \theta - \sum_{j=1}^{i-1}\frac{\partial \alpha_{i-1}}{\partial x_j}\underbrace{(x_{j+1} + \varphi_j^T \theta)}_{\dot{x}_j} - \sum_{j=1}^{i-1}\frac{\partial \alpha_{i-1}}{\partial \zeta_j}\underbrace{(\Phi_{j,0} + \Phi_j \theta)}_{\dot{\zeta}_j}$$

$$+ \sum_{j=1}^{i-1}\frac{\partial \alpha_{i-1}}{\partial \vartheta_j}\underbrace{\Gamma\left[\varphi_j^T - \sum_{k=1}^{j-1}\left(\frac{\partial \alpha_{j-1}}{\partial x_k}\varphi_k^T + \frac{\partial \alpha_{j-1}}{\partial \zeta_k}\Phi_k\right)\right]^T z_j}_{-\dot{\vartheta}_j}$$

$$= \alpha_i + z_{i+1} + \left[\varphi_i^T - \sum_{j=1}^{i-1}\left(\frac{\partial \alpha_{i-1}}{\partial x_j}\varphi_j^T + \frac{\partial \alpha_{i-1}}{\partial \zeta_j}\Phi_j\right)\right]\theta - \sum_{j=1}^{i-1}\left\{\frac{\partial \alpha_{i-1}}{\partial x_j}x_{j+1}\right.$$

$$\left. + \frac{\partial \alpha_{i-1}}{\partial \zeta_j}\Phi_{j,0} - \frac{\partial \alpha_{i-1}}{\partial \vartheta_j}\Gamma\left[\varphi_j^T - \sum_{k=1}^{j-1}\left(\frac{\partial \alpha_{j-1}}{\partial x_k}\varphi_k^T + \frac{\partial \alpha_{j-1}}{\partial \zeta_k}\Phi_k\right)\right]^T z_j\right\}$$

$$= -c_i z_i - z_{i-1} + z_{i+1} + \left[\varphi_i^T - \sum_{j=1}^{i-1}\left(\frac{\partial \alpha_{i-1}}{\partial x_j}\varphi_j^T + \frac{\partial \alpha_{i-1}}{\partial \zeta_j}\Phi_j\right)\right](\theta - \vartheta_i). \tag{3.99}$$

3.3 RECURSIVE DESIGN PROCEDURES

The closed-loop error system can thus be expressed in the matrix form

$$\frac{d}{dt}\begin{bmatrix} z_1 \\ z_2 \\ \vdots \\ z_{\rho-1} \\ z_\rho \end{bmatrix} = \begin{bmatrix} -c_1 & 1 & 0 & \cdots & 0 \\ -1 & -c_2 & 1 & \ddots & \vdots \\ 0 & \ddots & \ddots & \ddots & 0 \\ \vdots & \ddots & -1 & -c_{\rho-1} & 1 \\ 0 & \cdots & 0 & -1 & -c_\rho \end{bmatrix} \begin{bmatrix} z_1 \\ z_2 \\ \vdots \\ z_{\rho-1} \\ z_\rho \end{bmatrix}$$

$$+ \begin{bmatrix} \bar{w}_1 & 0 & \cdots & 0 \\ 0 & \bar{w}_2 & \ddots & \vdots \\ \vdots & \ddots & \ddots & 0 \\ 0 & \cdots & 0 & \bar{w}_\rho \end{bmatrix}^\mathrm{T} \begin{bmatrix} \theta - \vartheta_1 \\ \theta - \vartheta_2 \\ \vdots \\ \theta - \vartheta_\rho \end{bmatrix} \quad (3.100)$$

$$\frac{d}{dt}\begin{bmatrix} \vartheta_1 \\ \vartheta_2 \\ \vdots \\ \vartheta_\rho \end{bmatrix} = \begin{bmatrix} \Gamma & 0 & \cdots & 0 \\ 0 & \Gamma & \ddots & \vdots \\ \vdots & \ddots & \ddots & 0 \\ 0 & \cdots & 0 & \Gamma \end{bmatrix} \begin{bmatrix} \bar{w}_1 & 0 & \cdots & 0 \\ 0 & \bar{w}_2 & \ddots & \vdots \\ \vdots & \ddots & \ddots & 0 \\ 0 & \cdots & 0 & \bar{w}_\rho \end{bmatrix} \begin{bmatrix} z_1 \\ z_2 \\ \vdots \\ z_\rho \end{bmatrix},$$

with the notation

$$\bar{w}_1 = \varphi_1, \quad \bar{w}_i = \varphi_i - \sum_{j=1}^{i-1}\left[\frac{\partial \alpha_{i-1}}{\partial x_j}\varphi_j + \left(\frac{\partial \alpha_{i-1}}{\partial \zeta_j}\Phi_j\right)^\mathrm{T}\right], \quad i=2,\ldots,\rho. \quad (3.101)$$

The structural properties of this error system are once again apparent. The derivative of the nonnegative function

$$V_n(z_1,\ldots,z_\rho,\vartheta_1,\ldots,\vartheta_\rho) = \frac{1}{2}\sum_{i=1}^{\rho}\left[z_i^2 + (\theta-\vartheta_i)^\mathrm{T}\Gamma^{-1}(\theta-\vartheta_i)\right] \quad (3.102)$$

along the solutions of (3.100) is

$$\dot{V}_n = -\sum_{i=1}^{\rho} c_i z_i^2. \quad (3.103)$$

The LaSalle-Yoshizawa theorem (Theorem 2.1) now guarantees the global uniform boundedness of $z(t)$, $\vartheta_1(t),\ldots,\vartheta_\rho(t)$, as well as the regulation of $z(t)$. Since $z_1 = x_1$, we see that $y = x_1$ is also bounded and regulated. The boundedness of $x_2,\ldots,x_\rho$, $\zeta_1,\ldots,\zeta_\rho$ and of the transformed control variable v is then established via an induction argument for $i = 2,\ldots,\rho+1$ (with $x_{\rho+1} \equiv v$): If $x_1,\ldots,x_{i-1}$ and $\zeta_1,\ldots,\zeta_{i-2}$ are bounded, the ISS property of the ζ_{i-1}-subsystem guarantees that ζ_{i-1} is bounded. Hence, $\alpha_{i-1}(x_1,\ldots,x_{i-1},\zeta_1,\ldots,\zeta_{i-1},\vartheta_1,\ldots,\vartheta_{i-1})$ is bounded, which implies that $x_i = z_i + \alpha_{i-1}$ is also bounded. □

We should note that the adaptive controller (3.95)–(3.96), when applied to the system (3.87) using the expressions (3.89) and (3.90), guarantees the global uniform boundedness of χ and u, as well as the regulation of $y = h_1(\chi_1)$. This follows from the fact that $\frac{\partial h_j}{\partial \chi_j} g_j \neq 0$, which guarantees that the transformation relating $(y_1, \ldots, y_n, \zeta_1, \ldots, \zeta_n)$ to $(x_1, \ldots, x_n, \zeta_1, \ldots, \zeta_n)$ is a global diffeomorphism, and the feedback transformation (3.90) from u to v is nonsingular.

3.4 Extended Matching Design

The increase in the number of parameter estimates caused by overparametrization can be an undesirable feature, since it rapidly increases the dynamic order of the resulting adaptive controller. In Chapter 4, the overparametrization will be eliminated by the tuning functions method. As preliminary to this development, we now show how the overparametrization can be avoided in the case of extended matching, that is, when the uncertain parameters are only one integrator away from the control.

3.4.1 Reducing the overparametrization

We consider again the nonlinear system (3.29),

$$\dot{x}_1 = x_2 + \theta \varphi(x_1)$$
$$\dot{x}_2 = u,$$

and modify its two-step design.

Step 1. With $z_1 = x_1$ and x_2 viewed as the virtual control in the $\dot{z}_1$-equation, we define the first stabilizing function α_1 as in (3.34):

$$\alpha_1 = -c_1 z_1 - \hat{\theta}\varphi. \tag{3.104}$$

Comparing (3.104) with (3.34), we see that the parameter estimate $\hat{\vartheta}_1$ has been replaced by the parameter estimate $\hat{\theta}$. The difference in notation indicates that in this design procedure only one estimate $\hat{\theta}$ of the unknown parameter will be used.

The first Lyapunov function is now chosen as

$$V_1(z_1, \hat{\theta}) = \frac{1}{2} z_1^2 + \frac{1}{2\gamma} \tilde{\theta}^2, \tag{3.105}$$

where $\tilde{\theta} = \theta - \hat{\theta}$ is the parameter error, and $\gamma > 0$ is the adaptation gain. With $z_2 = x_2 - \alpha_1$, the derivative of V_1 is

$$\dot{V}_1 = z_1 z_2 - c_1 z_1^2 + \tilde{\theta}\left(\varphi z_1 - \frac{1}{\gamma}\dot{\hat{\theta}}\right). \tag{3.106}$$

3.4 EXTENDED MATCHING DESIGN

We postpone the choice of update law for $\hat{\theta}$ until the next step. The first error subsystem becomes

$$\dot{z}_1 = -c_1 z_1 + z_2 + \tilde{\theta}\varphi. \tag{3.107}$$

Step 2. The derivative of $z_2 = x_2 - \alpha_1$ is

$$\begin{aligned}
\dot{z}_2 &= u - \frac{\partial \alpha_1}{\partial x_1}(x_2 + \theta\varphi) - \frac{\partial \alpha_1}{\partial \hat{\theta}}\dot{\hat{\theta}} \\
&= u - \frac{\partial \alpha_1}{\partial x_1}x_2 - \hat{\theta}\frac{\partial \alpha_1}{\partial x_1}\varphi - \tilde{\theta}\frac{\partial \alpha_1}{\partial x_1}\varphi - \frac{\partial \alpha_1}{\partial \hat{\theta}}\dot{\hat{\theta}}.
\end{aligned} \tag{3.108}$$

To design the control u, we consider the augmented Lyapunov function

$$V_2 = V_1 + \frac{1}{2}z_2^2 = \frac{1}{2}z_1^2 + \frac{1}{2}z_2^2 + \frac{1}{2\gamma}\tilde{\theta}^2. \tag{3.109}$$

The only difference between (3.109) and (3.42) is the absence of the new parameter error $(\theta - \vartheta_2)$ in (3.109). In view of (3.106) and (3.108), the derivative of V_2 is

$$\begin{aligned}
\dot{V}_2 &= z_1 z_2 - c_1 z_1^2 + \tilde{\theta}\left(\varphi z_1 - \frac{1}{\gamma}\dot{\hat{\theta}}\right) \\
&\quad + z_2\left[u - \frac{\partial \alpha_1}{\partial x_1}x_2 - \hat{\theta}\frac{\partial \alpha_1}{\partial x_1}\varphi - \tilde{\theta}\frac{\partial \alpha_1}{\partial x_1}\varphi - \frac{\partial \alpha_1}{\partial \hat{\theta}}\dot{\hat{\theta}}\right] \\
&= -c_1 z_1^2 + \tilde{\theta}\left[\varphi z_1 - z_2 \frac{\partial \alpha_1}{\partial x_1}\varphi - \frac{1}{\gamma}\dot{\hat{\theta}}\right] \\
&\quad + z_2\left[z_1 + u - \frac{\partial \alpha_1}{\partial x_1}x_2 - \hat{\theta}\frac{\partial \alpha_1}{\partial x_1}\varphi - \frac{\partial \alpha_1}{\partial \hat{\theta}}\dot{\hat{\theta}}\right].
\end{aligned} \tag{3.110}$$

In the last equation, all the terms containing $\tilde{\theta}$ have been grouped together. To eliminate them, the update law is chosen as

$$\dot{\hat{\theta}} = \gamma\left(\varphi z_1 - \frac{\partial \alpha_1}{\partial x_1}\varphi z_2\right). \tag{3.111}$$

Then, the last bracketed term in (3.110) will be rendered equal to $-c_2 z_2^2$ with the control

$$u = -z_1 - c_2 z_2 + \frac{\partial \alpha_1}{\partial x_1}x_2 + \hat{\theta}\frac{\partial \alpha_1}{\partial x_1}\varphi + \frac{\partial \alpha_1}{\partial \hat{\theta}}\dot{\hat{\theta}}, \tag{3.112}$$

where for $\dot{\hat{\theta}}$ we use the analytical expression of the update law (3.111). Substituting the expressions (3.111) and (3.112) into (3.110) we obtain

$$\dot{V}_2 = -c_1 z_1^2 - c_2 z_2^2 \leq 0, \tag{3.113}$$

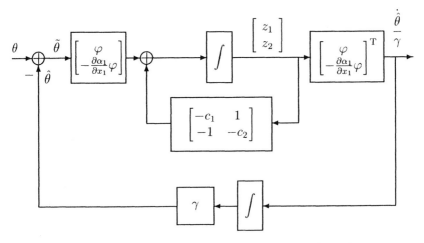

Figure 3.6: The closed-loop adaptive system (3.114).

and the error system becomes (see block diagram in Figure 3.6)

$$\frac{d}{dt}\begin{bmatrix} z_1 \\ z_2 \end{bmatrix} = \begin{bmatrix} -c_1 & 1 \\ -1 & -c_2 \end{bmatrix}\begin{bmatrix} z_1 \\ z_2 \end{bmatrix} + \begin{bmatrix} \varphi \\ -\frac{\partial \alpha_1}{\partial x_1}\varphi \end{bmatrix}\tilde{\theta}$$
$$\dot{\hat{\theta}} = \gamma \begin{bmatrix} \varphi & -\frac{\partial \alpha_1}{\partial x_1}\varphi \end{bmatrix}\begin{bmatrix} z_1 \\ z_2 \end{bmatrix}.$$
(3.114)

Comparing (3.114) with (3.47), we see that the system matrix in (3.114) has preserved the important structural properties it had in (3.47): Its diagonal terms are negative and its off-diagonal terms are skew-symmetric. Furthermore, we see that, as in (3.47), the matrix that multiplies the parameter error $\tilde{\theta}$ in the $\dot{z}$-equation is used (in its transposed form) in the update law for the parameter estimates. It is also instructive to compare the expressions for the parameter update laws in (3.114) and (3.47): Even though the update law for $\hat{\theta}$ appears in the form of the sum of the update laws for ϑ_1 and ϑ_2, the expressions (3.34) and (3.104) for α_1 depend on different parameter estimates (ϑ_1 and $\hat{\theta}$, respectively), and thus z_2 and the partial derivative $\frac{\partial \alpha_1}{\partial x_1}$ will have different values in (3.44) and (3.111).

Due to the structure of the error system (3.114), its stability and convergence properties are derived in a manner almost identical to those of (3.47) and are therefore omitted here.

In the extended matching case we avoided the overparametrization by postponing the choice of the update law until the second step. When $\dot{\hat{\theta}}$ appeared in the second step, it was replaced by its known analytical expression. Beyond the extended matching case we need more than two steps, so that $\ddot{\hat{\theta}}$ and higher derivatives of $\hat{\theta}$ appear. Instead of the simple idea of postponing the choice

3.4 EXTENDED MATCHING DESIGN

of the update law, we need the more intricate *tuning functions method*, to be developed in Chapter 4.

3.4.2 Example: biochemical process

Extended matching design is also applicable to pure-feedback systems, introduced in Section 2.3.2, provided that the unknown parameters appear linearly. While the general case of these *parametric pure-feedback systems* is presented in Section 4.5.3, the extended matching design will be illustrated on a simplified model of a biotechnological process which goes as far back as Monod [135]. In spite of its simplicity and somewhat unrealistic assumptions, this example is representative of several successful applications of adaptive nonlinear control to more complex processes described by Bastin [7]. In a model of a fed-batch process, S is the concentration of the growth limiting substrate, X is the concentration of the growing microbial population, k is the yield constant, D is the dilution rate, and the control u is the substrate feed rate. In a batch process, that is, when both $D = 0$ and $u = 0$, the rate of microbial growth $\dot{X}$ is modeled as $\dot{X} = \mu(S)X$, where $\mu(S)$ is the "specific growth rate." The nonlinear function $\mu(S)$ is usually poorly known, and for our illustrative purpose we parametrize it using unknown parameters:

$$\mu(S) = \varphi_0(S) + \theta_1\varphi_1(S) + \theta_2\varphi_2(S). \tag{3.115}$$

Note that X, S, and $\mu(S)$ are nonnegative quantities. With this parametrization the fed-batch process operating at constant temperature is modeled by the following two mass-balance equations:

$$\dot{X} = [\varphi_0(S) + \theta_1\varphi_1(S) + \theta_2\varphi_2(S)] X - DX \tag{3.116a}$$
$$\dot{S} = -k[\varphi_0(S) + \theta_1\varphi_1(S) + \theta_2\varphi_2(S)] X - DS + u. \tag{3.116b}$$

The control objective is regulation of X to the set point X_r. To further simplify the system (3.116), we use the change of coordinates $x_1 = \ln X$, $x_2 = S$, which is well-defined and invertible since $X > 0$. Then (3.116) becomes

$$\dot{x}_1 = \varphi_0(x_2) + \theta_1\varphi_1(x_2) + \theta_2\varphi_2(x_2) - D \tag{3.117a}$$
$$\dot{x}_2 = -k[\varphi_0(x_2) + \theta_1\varphi_1(x_2) + \theta_2\varphi_2(x_2)] e^{x_1} - Dx_2 + u. \tag{3.117b}$$

This system is clearly not in the parametric strict-feedback form (3.70), since the nonlinearities in (3.117a) depend on the second state variable x_2. However, we can still apply the design procedure illustrated in Section 3.4.1 to this system. In Section 2.3 we saw that our recursive design procedures can be applied not only to strict-feedback systems (2.165), but also to pure-feedback systems (2.180), whose nonlinearities are allowed to depend on one more state variable. The price to be paid is that the stability properties are no longer

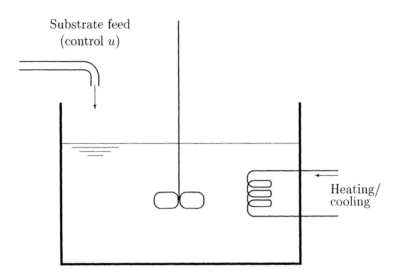

Figure 3.7: Fed-batch stirred tank reactor.

global, but regional: They are guaranteed only for a compact set of initial conditions. The same is possible for our adaptive designs. For the system (3.117), our design proceeds by choosing $\varphi_0(x_2)$ as the virtual control variable in (3.117a) and designing for it the stabilizing function

$$\alpha_1 = -c_1 z_1 - \hat{\theta}_1 \varphi_1 - \hat{\theta}_2 \varphi_2 + D, \qquad (3.118)$$

where $z_1 = x_1 - \ln X_r$. With $z_2 = \varphi_0(x_2) - \alpha_1$, the error system becomes

$$\dot{z}_1 = -c_1 z_1 + z_2 + (\theta - \hat{\theta}_1)\varphi_1 + (\theta - \hat{\theta}_2)\varphi_2 \qquad (3.119)$$

$$\dot{z}_2 = \left(\frac{\partial \varphi_0}{\partial x_2} + \hat{\theta}_1 \frac{\partial \varphi_1}{\partial x_2} + \hat{\theta}_2 \frac{\partial \varphi_2}{\partial x_2}\right)\left\{-k[\varphi_0 + \theta_1 \varphi_1 + \theta_2 \varphi_2]e^{x_1} - Dx_2 + u\right\}$$
$$+ c_1\left[-c_1 z_1 + z_2 + (\theta - \hat{\theta}_1)\varphi_1 + (\theta - \hat{\theta}_2)\varphi_2\right] + \dot{\hat{\theta}}_1 \varphi_1 + \dot{\hat{\theta}}_2 \varphi_2. \qquad (3.120)$$

Following the development of Section 3.4.1, we choose the update law

$$\dot{\hat{\theta}} = \Gamma \begin{bmatrix} \varphi_1 \\ \varphi_2 \end{bmatrix} \left\{ z_1 + \left[c_1 - k\left(\frac{\partial \varphi_0}{\partial x_2} + \hat{\theta}_1 \frac{\partial \varphi_1}{\partial x_2} + \hat{\theta}_2 \frac{\partial \varphi_2}{\partial x_2}\right)\right] z_2 \right\}, \qquad (3.121)$$

where $\theta^T = [\theta_1, \theta_2]$. The corresponding control law

$$u = \frac{1}{\left(\frac{\partial \varphi_0}{\partial x_2} + \hat{\theta}_1 \frac{\partial \varphi_1}{\partial x_2} + \hat{\theta}_2 \frac{\partial \varphi_2}{\partial x_2}\right)} \left\{ -c_2 z_2 - z_1 - c_1\left[-c_1 z_1 + z_2 - \hat{\theta}_1 \varphi_1 - \hat{\theta}_2 \varphi_2\right] \right.$$
$$\left. - [\varphi_1, \varphi_2]\Gamma \begin{bmatrix} \varphi_1 \\ \varphi_2 \end{bmatrix} \left\{ z_1 + \left[c_1 - k\left(\frac{\partial \varphi_0}{\partial x_2} + \hat{\theta}_1 \frac{\partial \varphi_1}{\partial x_2} + \hat{\theta}_2 \frac{\partial \varphi_2}{\partial x_2}\right)\right] z_2 \right\} \right\}$$
$$+ k\left[\varphi_0 + \hat{\theta}_1 \varphi_1 + \hat{\theta}_2 \varphi_2\right] e^{x_1} + Dx_2, \qquad (3.122)$$

3.4 EXTENDED MATCHING DESIGN

is feasibile only in the region in which $\frac{\partial\varphi_0}{\partial x_2} + \hat{\theta}_1 \frac{\partial\varphi_1}{\partial x_2} + \hat{\theta}_2 \frac{\partial\varphi_2}{\partial x_2} \neq 0$. With these choices, the derivative of the Lyapunov function

$$V = \frac{1}{2}z_1^2 + \frac{1}{2}z_2^2 + \frac{1}{2}(\theta - \hat{\theta})^{\mathrm{T}}\Gamma^{-1}(\theta - \hat{\theta}) \qquad (3.123)$$

is nonpositive:

$$\dot{V} = -c_1 z_1^2 - c_2 z_2^2. \qquad (3.124)$$

As we will see in Section 4.5.3, stability is guaranteed for all initial conditions inside the largest level set of the Lyapunov function 3.123 contained in the feasibility region.

3.4.3 Transient performance improvement

The nonlinear damping with κ-terms introduced in Section 2.5 can easily be incorporated into the adaptive design procedures we have discussed so far. The resulting adaptive controllers guarantee boundedness even when the adaptation is switched off, and their transient performance can be improved in a systematic way through *trajectory initialization* and the choice of design parameters.

To illustrate the design with κ-terms and the process of trajectory initialization, we consider again the system (3.29) with the output $y = x_1$:

$$\begin{aligned}
\dot{x}_1 &= x_2 + \theta\varphi(x_1) \\
\dot{x}_2 &= u \\
y &= x_1.
\end{aligned} \qquad (3.125)$$

The control objective is to asymptotically track a reference output $y_{\mathrm{r}}(t)$ with the output y of the system (3.125). We assume that not only y_{r}, but also its first two derivatives $\dot{y}_{\mathrm{r}}$, $\ddot{y}_{\mathrm{r}}$ are known and uniformly bounded, and, in addition, $\ddot{y}_{\mathrm{r}}$ is piecewise continuous.

Step 1. The first error variable is now the *tracking error*

$$z_1 = y - y_{\mathrm{r}} = x_1 - y_{\mathrm{r}}, \qquad (3.126)$$

whose derivative is

$$\dot{z}_1 = x_2 + \theta^{\mathrm{T}}\varphi_1(x_1) - \dot{y}_{\mathrm{r}}. \qquad (3.127)$$

Viewing x_2 as the virtual control we define the stabilizing function

$$\alpha_1 = -c_1 z_1 - \kappa_1 z_1 \varphi^2 - \hat{\theta}\varphi + \dot{y}_{\mathrm{r}}. \qquad (3.128)$$

Comparing (3.128) with (3.104) we note two new terms in (3.128). The term $\dot{y}_{\mathrm{r}}$, which is intended to cancel the corresponding term in (3.127), is due to the tracking objective. The nonlinear damping term $-\kappa_1 z_1 \varphi^2$ is motivated by Lemma 2.26. It contains the square of the term (φ) which multiplies the parametric uncertainty in the error equation obtained by substituting $z_2 = x_2 - \alpha_1$ and (3.128) into (3.127):

$$\dot{z}_1 = -c_1 z_1 - \kappa_1 z_1 \varphi^2 + z_2 + \tilde{\theta}\varphi. \qquad (3.129)$$

The derivative of the Lyapunov function $V_1 = \frac{1}{2}z_1^2 + \frac{1}{2\gamma}\tilde{\theta}^2$ becomes

$$\dot{V}_1 = z_1 z_2 - c_1 z_1^2 - \kappa_1 z_1^2 \varphi^2 + \tilde{\theta}\left(\varphi z_1 - \frac{1}{\gamma}\dot{\hat{\theta}}\right). \qquad (3.130)$$

Step 2. As in (3.108), the derivative of $z_2 = x_2 - \alpha_1$ is

$$\begin{aligned}\dot{z}_2 &= u - \frac{\partial \alpha_1}{\partial x_1}(x_2 + \theta\varphi) - \frac{\partial \alpha_1}{\partial y_r}\dot{y}_r - \frac{\partial \alpha_1}{\partial \dot{y}_r}\ddot{y}_r - \frac{\partial \alpha_1}{\partial \hat{\theta}}\dot{\hat{\theta}} \\ &= u - \frac{\partial \alpha_1}{\partial x_1}x_2 - \hat{\theta}\frac{\partial \alpha_1}{\partial x_1}\varphi - \tilde{\theta}\frac{\partial \alpha_1}{\partial x_1}\varphi - \frac{\partial \alpha_1}{\partial y_r}\dot{y}_r - \ddot{y}_r - \frac{\partial \alpha_1}{\partial \hat{\theta}}\dot{\hat{\theta}}, \quad (3.131)\end{aligned}$$

where in the last equality we have used the identity $\frac{\partial \alpha_1}{\partial \dot{y}_r} = 1$. Using (3.130) and (3.131), the derivative of the Lyapunov function

$$V_2 = V_1 + \frac{1}{2}z_2^2 = \frac{1}{2}z_1^2 + \frac{1}{2}z_2^2 + \frac{1}{2\gamma}\tilde{\theta}^2 \qquad (3.132)$$

is expressed as

$$\begin{aligned}\dot{V}_2 &= z_1 z_2 - c_1 z_1^2 - \kappa_1 z_1^2 \varphi^2 + \tilde{\theta}\left(\varphi z_1 - \frac{1}{\gamma}\dot{\hat{\theta}}\right) \\ &\quad + z_2\left[u - \frac{\partial \alpha_1}{\partial x_1}x_2 - \hat{\theta}\frac{\partial \alpha_1}{\partial x_1}\varphi - \tilde{\theta}\frac{\partial \alpha_1}{\partial x_1}\varphi - \frac{\partial \alpha_1}{\partial y_r}\dot{y}_r - \ddot{y}_r - \frac{\partial \alpha_1}{\partial \hat{\theta}}\dot{\hat{\theta}}\right] \\ &= -c_1 z_1^2 - \kappa_1 z_1^2 \varphi^2 + \tilde{\theta}\left[\varphi z_1 - z_2\frac{\partial \alpha_1}{\partial x_1}\varphi - \frac{1}{\gamma}\dot{\hat{\theta}}\right] \\ &\quad + z_2\left[z_1 + u - \frac{\partial \alpha_1}{\partial x_1}x_2 - \hat{\theta}\frac{\partial \alpha_1}{\partial x_1}\varphi - \frac{\partial \alpha_1}{\partial y_r}\dot{y}_r - \ddot{y}_r - \frac{\partial \alpha_1}{\partial \hat{\theta}}\dot{\hat{\theta}}\right]. \quad (3.133)\end{aligned}$$

As in (3.110), all the terms containing $\tilde{\theta}$ have been grouped together. To eliminate them, the update law is chosen as

$$\dot{\hat{\theta}} = \gamma\left(\varphi z_1 - \frac{\partial \alpha_1}{\partial x_1}\varphi z_2\right). \qquad (3.134)$$

The control law is now chosen to render the last bracketed term in (3.133) equal to $-c_2 z_2^2 - \kappa_2 z_2^2 \left(\frac{\partial \alpha_1}{\partial x_1}\varphi\right)^2$, instead of just equal to $-c_2 z_2^2$ as in (3.112):

$$u = -z_1 - c_2 z_2 - \kappa_2 z_2 \left(\frac{\partial \alpha_1}{\partial x_1}\varphi\right)^2 + \frac{\partial \alpha_1}{\partial x_1}x_2 + \hat{\theta}\frac{\partial \alpha_1}{\partial x_1}\varphi + \frac{\partial \alpha_1}{\partial y_r}\dot{y}_r + \ddot{y}_r + \frac{\partial \alpha_1}{\partial \hat{\theta}}\dot{\hat{\theta}}. \quad (3.135)$$

3.4 EXTENDED MATCHING DESIGN

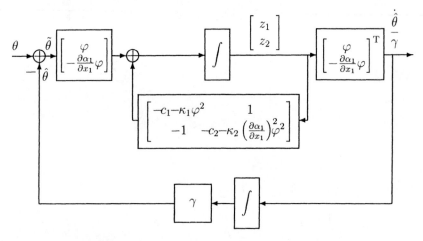

Figure 3.8: The closed-loop adaptive system (3.137).

To implement this control law, we will replace $\dot{\hat{\theta}}$ with the analytical expression of the update law (3.134).

Substituting the expressions (3.134) and (3.135) into (3.133) we obtain

$$\dot{V}_2 = -c_1 z_1^2 - \kappa_1 z_1^2 \varphi^2 - c_2 z_2^2 - \kappa_2 z_2^2 \left(\frac{\partial \alpha_1}{\partial x_1} \varphi\right)^2 \leq 0, \qquad (3.136)$$

while the complete error system becomes (see block diagram in Figure 3.8)

$$\frac{d}{dt}\begin{bmatrix} z_1 \\ z_2 \end{bmatrix} = \begin{bmatrix} -c_1 - \kappa_1 \varphi^2 & 1 \\ -1 & -c_2 - \kappa_2 \left(\frac{\partial \alpha_1}{\partial x_1} \varphi\right)^2 \end{bmatrix} \begin{bmatrix} z_1 \\ z_2 \end{bmatrix} + \begin{bmatrix} \varphi \\ -\frac{\partial \alpha_1}{\partial x_1} \varphi \end{bmatrix} \tilde{\theta}$$

$$\dot{\hat{\theta}} = \gamma \begin{bmatrix} \varphi & -\frac{\partial \alpha_1}{\partial x_1} \varphi \end{bmatrix} \begin{bmatrix} z_1 \\ z_2 \end{bmatrix}. \qquad (3.137)$$

Comparing (3.137) with (3.114), we see that the system matrix in (3.137) is not constant: Its diagonal terms have been "fortified" with additional nonlinear damping terms. These terms contain the squares of the elements of the vector that multiplies the parameter error $\tilde{\theta}$.

Let us now study the properties of the error system (3.137):

Global stability and asymptotic tracking. Using (3.132) and (3.136) we conclude that the $(z, \tilde{\theta})$-system has a globally uniformly stable equilibrium at the origin, and

$$\lim_{t \to \infty} z(t) = 0. \qquad (3.138)$$

In particular, this implies that the state of the system (3.125) is globally uniformly bounded (since $y_r, \dot{y}_r,$ and $\ddot{y}_r$ are bounded) and that the tracking error $z_1 = y - y_r$ converges to zero asymptotically.

Boundedness without adaptation. It is also straightforward to see that the designed controller guarantees global uniform boundedness even when the adaptation is turned off, that is, even with $\gamma = 0$. In that case, the closed-loop system (3.137) becomes

$$\frac{d}{dt}\begin{bmatrix} z_1 \\ z_2 \end{bmatrix} = \begin{bmatrix} -c_1 - \kappa_1\varphi^2 & 1 \\ -1 & -c_2 - \kappa_2\left(\frac{\partial\alpha_1}{\partial x_1}\varphi\right)^2 \end{bmatrix}\begin{bmatrix} z_1 \\ z_2 \end{bmatrix} + \begin{bmatrix} \varphi \\ -\frac{\partial\alpha_1}{\partial x_1}\varphi \end{bmatrix}\tilde{\theta}. \tag{3.139}$$

A candidate Lyapunov function for this system is given by

$$V(z) = \frac{1}{2}|z|^2 = \frac{1}{2}\left(z_1^2 + z_2^2\right). \tag{3.140}$$

Its derivative along the solutions of (3.139) satisfies

$$\begin{aligned}
\dot{V}_{(3.139)} &= -c_1 z_1^2 - c_2 z_2^2 - \kappa_1 z_1^2 \varphi^2 - \kappa_2 z_2^2 \left(\frac{\partial\alpha_1}{\partial x_1}\varphi\right)^2 + z_1\varphi\tilde{\theta} - z_2\frac{\partial\alpha_1}{\partial x_1}\varphi\tilde{\theta} \\
&\leq -c_1 z_1^2 - c_2 z_2^2 - \kappa_1\left(z_1\varphi - \frac{\tilde{\theta}}{2\kappa_1}\right)^2 + \frac{\tilde{\theta}^2}{4\kappa_1} - \kappa_2\left(z_2\frac{\partial\alpha_1}{\partial x_1}\varphi + \frac{\tilde{\theta}}{2\kappa_2}\right)^2 + \frac{\tilde{\theta}^2}{4\kappa_2} \\
&\leq -c_1 z_1^2 - c_2 z_2^2 + \frac{\tilde{\theta}^2}{4\kappa_1} + \frac{\tilde{\theta}^2}{4\kappa_2} \\
&\leq -c_0|z|^2 + \frac{\tilde{\theta}^2}{4\kappa_0},
\end{aligned} \tag{3.141}$$

where the constants c_0 and κ_0 are defined as

$$c_0 = \min\{c_1, c_2\}, \quad \frac{1}{\kappa_0} = \frac{1}{\kappa_1} + \frac{1}{\kappa_2}. \tag{3.142}$$

It is clear from (3.141) that, for any positive values of c_0 and κ_0, the state of the error system (and hence the state of the plant) is uniformly bounded, since $\dot{V} < 0$ whenever $|z|^2 > |\tilde{\theta}|^2/4\kappa_0 c_0$, where $\tilde{\theta} = \theta - \hat{\theta}(0)$ is constant since adaptation is turned off.

Transient performance improvement with trajectory initialization. Let us now investigate the transient performance of the adaptive closed-loop system (3.137). The derivative of the nonnegative function $V(z)$ defined in (3.140) along the solutions of (3.137) satisfies the same inequality as in (3.141):

$$\frac{d}{dt}\left(\frac{1}{2}|z|^2\right) \leq -c_0|z|^2 + \frac{\tilde{\theta}^2}{4\kappa_0}. \tag{3.143}$$

Since the boundedness of $\tilde{\theta}$ has already been established from (3.132) and (3.136), we can strengthen the inequality in (3.143) by replacing $\tilde{\theta}^2$ with its

3.4 EXTENDED MATCHING DESIGN

bound $\|\tilde{\theta}\|_\infty^2$. This bound is estimated from (3.132) using the fact that V_2 is nonincreasing:

$$\frac{1}{2\gamma}|\tilde{\theta}(t)|^2 \leq \frac{1}{2}|z(t)|^2 + \frac{1}{2\gamma}\tilde{\theta}(t)^2 = V_2(t)$$

$$\leq V_2(0) = \frac{1}{2}|z(0)|^2 + \frac{1}{2\gamma}\tilde{\theta}(0)^2. \tag{3.144}$$

This implies

$$\|\tilde{\theta}\|_\infty^2 \leq \gamma|z(0)|^2 + \tilde{\theta}(0)^2. \tag{3.145}$$

Combining (3.143) and (3.145) we obtain

$$\frac{d}{dt}\left(|z|^2\right) \leq -2c_0|z|^2 + \frac{1}{2\kappa_0}\left[\gamma|z(0)|^2 + \tilde{\theta}(0)^2\right]. \tag{3.146}$$

Multiplying both sides of (3.146) by $e^{2c_0 t}$ and integrating over the interval $[0, t]$ results in

$$|z(t)|^2 \leq |z(0)|^2 e^{-2c_0 t} + \frac{1}{4\kappa_0 c_0}\left[\gamma|z(0)|^2 + \tilde{\theta}(0)^2\right]. \tag{3.147}$$

The bound (3.147) suggests that the transient behavior of the error system can be influenced through the choice of design constants c_0, κ_0 and γ. What is not clear, however, is that an increase of $\kappa_0 c_0$ alone may not reduce the maximum value of $|z(t)|$ and will certainly not reduce the computable $\mathcal{L}_\infty$-bound of z. In fact, it may even *increase* this bound by increasing the initial value $|z(0)|$. To clarify this point, let us recall the definitions of z_1 and z_2:

$$z_1 = x_1 - y_r$$
$$z_2 = x_2 - \alpha_1 = x_2 + c_1 z_1 + \kappa_1 z_1 \varphi^2 + \hat{\theta}\varphi - \dot{y}_r.$$

Suppose now that $z_1(0)$ is different than zero. In that case, an increase of c_1 and κ_1 may increase the value of $z_2(0)$ and thus also the value of $|z(0)|$. Moreover, this increase may more than offset the decreasing effect of the term $1/4\kappa_0 c_0$ in (3.147), since $|z(0)|^2$ will increase in proportion to c_1^2 and κ_1^2.

It would seem that the dependence of $z(0)$ on the design constants $c_1, c_2, \kappa_1, \kappa_2$ eliminates any possibility of systematically improving the transient performance of the error system through the choice of c_0 and κ_0. Fortunately, it is not so. The remedy for this problem is to use *trajectory initialization* to render $z(0) = 0$ *independently* of the choice of these design constants. The initialization procedure, presented for the general case in Section 4.3.2, is straightforward and is dictated by the definitions of the z-variables:

- Starting with z_1, set $z_1(0) = 0$ by choosing

$$y_r(0) = x_1(0). \tag{3.148}$$

- Since $z_1(0) = 0$, (3.128) shows that

$$\alpha_1(0) = \dot{y}_r(0) - \hat{\theta}(0)\varphi(0), \qquad (3.149)$$

where we use the notation $\varphi(0) = \varphi(x_1(0))$. From (3.149) it is clear that we can set $z_2(0) = 0$ with the choice

$$\dot{y}_r(0) = x_2(0) + \hat{\theta}(0)\varphi(0). \qquad (3.150)$$

With the trajectory initialization defined by (3.148) and (3.150), we have set $z(0) = 0$. In the case of model reference control, this is achieved by adjusting the initial conditions of the reference model. If, on the other hand, the reference trajectory is given as a precomputed function of time, then it can be initialized through the addition of exponentially decaying terms which define the *reference transients*.

We note that (3.148) and (3.150) are independent of the design constants $c_1, c_2, \kappa_1, \kappa_2$. This means that different choices of c_0 and κ_0 will still result in $z(0) = 0$ with the same values of $y_r(0)$ and $\dot{y}_r(0)$. Returning to (3.147), we substitute $z(0) = 0$ to obtain

$$|z(t)|^2 \leq \frac{1}{4\kappa_0 c_0} \tilde{\theta}(0)^2, \qquad (3.151)$$

which implies

$$\|z\|_\infty \leq \frac{1}{2\sqrt{\kappa_0 c_0}} |\tilde{\theta}(0)|. \qquad (3.152)$$

Hence, the $\mathcal{L}_\infty$-bound on the transient performance of the error system is directly proportional to the initial parametric uncertainty and can be reduced arbitrarily by increasing the values of c_0 and κ_0. In particular, this implies that the transients of the tracking error $z_1 = y - y_r$ are directly influenced by the design constants c_i and κ_i. This possibility of arbitrary reduction may seem peculiar, since it can be achieved for all initial conditions. We must remember, however, that this error is defined with respect to the reference signals which have in turn been initialized to set $z(0) = 0$. Hence, the effect of the plant initial conditions has been "absorbed" into the reference transients.

To provide some further insight into the process of trajectory initialization, let us return to the Lyapunov function (3.132). When $z(0) = 0$, the initial value of this function is reduced to the initial value of the parametric uncertainty. If we interpret the value of this function as a distance between the actual system trajectory and the reference trajectory, we see that *trajectory initialization places the initial point of the reference trajectory as close as possible to the initial point of the system trajectory*. If the parametric uncertainty were zero, trajectory initialization would have placed the reference output and its derivatives at the true values of the plant output and its derivatives. This is easily seen if $\hat{\theta}(0)$ is replaced by θ in (3.148) and (3.150):

$$y_r(0) = x_1(0) = y(0)$$
$$\dot{y}_r(0) = x_2(0) + \theta\varphi(0) = \dot{y}(0).$$

However, since the parameter θ is unknown, trajectory initialization placed only the reference output at the true value of the plant output, while its derivatives were placed at the *estimated* values of the plant output derivatives. Through this process, the initial value of the Lyapunov function is

$$V_2(0) = \frac{1}{2\gamma}\tilde{\theta}(0)^2, \qquad (3.153)$$

which, in the presence of parametric uncertainty, is its smallest possible value.

Notes and References

Adaptive backstepping (Kanellakopoulos, Kokotović, and Morse [69]), first presented in a Grainger lecture [87], was a culmination of an intensive effort of several groups of authors. The path to adaptive backstepping was not as direct as it may appear from this chapter. It led through the matched case in Taylor, Kokotović, Marino, and Kanellakopoulos [186], and then the case of extended matching in Kanellakopoulos, Kokotović, and Marino [65], and Bastin and Campion [8]. Even though nonadaptive backstepping was available from Saberi, Kokotović, and Sussmann [163], the steps beyond the extended matching case were delayed and alternative approaches were explored. The focus was on estimation-based designs summarized in Praly, Bastin, Pomet, and Jiang [157].

The class of parametric strict-feedback systems was characterized via coordinate-free geometric conditions by Kanellakopoulos *et al.* [63, 69], and the class of parametric pure-feedback systems by Akhrif and Blankenship [1]. The multi-input design for parametric pure-feedback systems was presented in [87].

Teel [188] increased the feasibility region for parametric pure-feedback systems by casting the scheme [69] in an observer-based setting. Several extensions of [69] were proposed by Seto, Annaswamy, and Baillieul [3, 167].

Chapter 4
Tuning Functions Design

The adaptive backstepping solution to the problem of nonlinear stabilization and tracking in the presence of unknown parameters is a starting point for more elaborate adaptive designs which lead to new properties of the designed controller and the resulting feedback system. One of the improvements to be achieved with the tuning functions design in this chapter is the reduction of the dynamic order of the adaptive controller to its minimum: The number of parameter estimates is equal to the number of unknown parameters. This minimum-order design is advantageous not only for implementation, but also because it guarantees the strongest achievable stability and convergence properties.

In the tuning functions procedure the parameter update law is designed recursively. At each consecutive step we design a tuning function as a potential update law. In contrast to adaptive backstepping in Chapter 3, these intermediate update laws are not implemented. Instead, the controller uses them to compensate for the effect of parameter estimation transients. Only the final tuning function is used as the parameter update law.

We start this chapter with Section 4.1, which introduces a general framework for Lyapunov-based adaptive design via *adaptive control Lyapunov functions (aclf)*. We depart from the certainty equivalence principle and approach the problem of adaptive stabilization of the original nonlinear system as a problem of nonadaptive stabilization of a modified system. In this setting, the tuning functions design is a method for recursively generating aclf's.

The design procedure is presented in Sections 4.2 and 4.3, which are independent from Section 4.1 and can be read first. In Section 4.4 we derive transient performance bounds on the error state of the adaptive system. An essential part of the technique for improving the transients is the trajectory initialization presented in Section 4.3.2. Several extensions of the design are presented in Section 4.5. In Section 4.6 the tuning functions design is applied to suppress the wing rock instability in aircraft flying at high angle-of-attack.

4.1 Adaptive Control Lyapunov Functions

The basic idea of the Lyapunov approach to adaptive control is to design a control law and a parameter update law to guarantee that the derivative of a suitable Lyapunov function is nonpositive. We are therefore sent to search for a triple: Lyapunov function, control law, and update law. For a class of nonlinear systems called parametric-strict-feedback systems we will be able to make this search systematic.

To begin with, let us investigate the possibility of adaptive design for the system

$$\dot{x} = f(x) + F(x)\theta + g(x)u, \quad x \in \mathbb{R}^n, \ u \in \mathbb{R}, \tag{4.1}$$

where $\theta \in \mathbb{R}^p$ is a vector of unknown constant parameters, and $f(x)$, $F(x)$ and $g(x)$ are smooth. For simplicity let $f(0) = 0$, $F(0) = 0$, so that $x = 0$ is an equilibrium of the uncontrolled plant.

4.1.1 Departure from certainty equivalence

Much of the traditional adaptive control employs some form of "certainty equivalence" thinking. Following this path one first performs a design for the case when the exact value of θ is known. Suppose that this nontrivial task is completed and that its result is a feedback control $u = \alpha_c(x, \theta)$ which stabilizes the equilibrium $x = 0$ with respect to a known Lyapunov function $V_c(x, \theta)$. The subscript 'c' stands for "certainty equivalence." We know that $V_c(x, \theta)$ is positive definite and radially unbounded in x for all θ, and that there exists a function $W(x, \theta)$, which is also positive definite in x for all θ, such that[1]

$$\frac{\partial V_c}{\partial x}[f(x) + F(x)\theta + g(x)\alpha_c(x, \theta)] \leq -W(x, \theta). \tag{4.2}$$

How can we exploit the knowledge of $\alpha_c(x, \theta)$ and $V_c(x, \theta)$ for adaptive design when θ is not known? The certainty equivalence idea is to replace θ by an estimate $\hat{\theta}(t)$ obtained from a parameter update law

$$\dot{\hat{\theta}} = \Gamma \tau(x, \hat{\theta}), \tag{4.3}$$

where the adaptation gain matrix Γ is positive definite. We want to select u and τ to guarantee that the derivative of a Lyapunov function is nonpositive. For the system (4.1), (4.3), a Lyapunov function candidate is

$$V(x, \hat{\theta}) = V_c(x, \hat{\theta}) + \frac{1}{2}\tilde{\theta}^{\mathrm{T}}\Gamma^{-1}\tilde{\theta}, \tag{4.4}$$

---

[1] Throughout the chapter, we will drop the arguments in $\frac{\partial V(x,\theta)}{\partial x}$ and $\frac{\partial V(x,\theta)}{\partial \theta}$, and write shortly $\frac{\partial V}{\partial x}$ and $\frac{\partial V}{\partial \theta}$. However, we will keep the arguments in $f(x)$, $F(x)$, $g(x)$, and $\alpha(x, \theta)$.

4.1 ADAPTIVE CONTROL LYAPUNOV FUNCTIONS

where the "certainty equivalence" form of V_c is augmented by a term quadratic in the parameter estimation error

$$\tilde{\theta} = \theta - \hat{\theta}. \tag{4.5}$$

Upon the substitution of $F(x)\theta = F(x)\hat{\theta} + F(x)\tilde{\theta}$, the derivative of $V(x,\hat{\theta})$ along the solutions of (4.1), (4.3) is

$$\dot{V} = \frac{\partial V_c}{\partial x}\left(f(x) + F(x)\hat{\theta} + g(x)u\right) + \frac{\partial V_c}{\partial \hat{\theta}}\Gamma\tau + \tilde{\theta}^{\mathrm{T}}\left(\frac{\partial V_c}{\partial x}F(x)\right)^{\mathrm{T}} - \tilde{\theta}^{\mathrm{T}}\tau. \tag{4.6}$$

To eliminate the indefinite dependence of $\dot{V}$ on the unknown parameter error $\tilde{\theta}$, we select τ to cancel the last two terms in (4.6):

$$\tau(x,\hat{\theta}) = \left(\frac{\partial V_c}{\partial x}F(x)\right)^{\mathrm{T}}. \tag{4.7}$$

With this choice of τ, the expression (4.6) is reduced to

$$\dot{V} = \frac{\partial V_c}{\partial x}\left(f(x) + F(x)\hat{\theta} + g(x)u\right) + \frac{\partial V_c}{\partial \hat{\theta}}\Gamma\left(\frac{\partial V_c}{\partial x}F(x)\right)^{\mathrm{T}}. \tag{4.8}$$

Our next task is to select a control law $u = \alpha(x,\hat{\theta})$ to make $\dot{V}$ nonpositive. The "certainty equivalence" control $u = \alpha_c(x,\hat{\theta})$ fails to achieve this because then (4.2) and (4.8) yield

$$\dot{V} \leq -W(x,\hat{\theta}) + \frac{\partial V_c}{\partial \hat{\theta}}\Gamma\left(\frac{\partial V_c}{\partial x}F(x)\right)^{\mathrm{T}}. \tag{4.9}$$

Clearly, $\dot{V}$ is not nonpositive because a sign-indefinite term is added to $-W(x,\hat{\theta})$. In search of a better control law $\alpha(x,\hat{\theta})$, we augment $\alpha_c(x,\hat{\theta})$ by $\alpha_\tau(x,\hat{\theta})$,

$$\alpha(x,\hat{\theta}) = \alpha_c(x,\hat{\theta}) + \alpha_\tau(x,\hat{\theta}). \tag{4.10}$$

The substitution of (4.10) into (4.8) shows that the desired nonpositivity $\dot{V} \leq -W(x,\hat{\theta})$ will be achieved if α_τ can be found to satisfy

$$\frac{\partial V_c}{\partial x}g(x)\alpha_\tau(x,\hat{\theta}) + \frac{\partial V_c}{\partial \hat{\theta}}\Gamma\left(\frac{\partial V_c}{\partial x}F(x)\right)^{\mathrm{T}} = 0. \tag{4.11}$$

This condition for α_τ demonstrates the difficulty of adaptive designs for a general nonlinear system (4.1). It is easy to see that α_τ satisfying (4.11) is unlikely to exist: The scalar quantity $\frac{\partial V_c}{\partial x}g(x)$ may be zero at a set of points. Still, the condition (4.11) is of interest because of an important special case, which will be the starting point of our recursive design. This special case is the "extended matching" studied in Section 3.4. In this case, a smooth vector-valued function $\varphi : \mathbb{R}^{n+p} \to \mathbb{R}^p$ is known such that $\frac{\partial V_c}{\partial \hat{\theta}}$ can be factored as follows:

$$\frac{\partial V_c}{\partial \hat{\theta}} = \frac{\partial V_c}{\partial x} g(x) \varphi(x, \hat{\theta})^\mathrm{T}. \tag{4.12}$$

Then, irrespective of the zeros of $\frac{\partial V_c}{\partial x} g(x)$, an α_τ which satisfies (4.11) is

$$\alpha_\tau(x, \hat{\theta}) = -\varphi(x, \hat{\theta})^\mathrm{T} \Gamma \left(\frac{\partial V_c}{\partial x} F(x) \right)^\mathrm{T} = -\varphi(x, \hat{\theta})^\mathrm{T} \Gamma \tau(x, \hat{\theta}). \tag{4.13}$$

We observe that, in addition to its "certainty equivalence" part α_c, the adaptive control law α contains a part α_τ which is proportional to τ, that is, to $\dot{\hat{\theta}}$ (see (4.3), (4.10), and (4.13)). In this way the adaptive control law takes into account the parameter estimation transients. When the parameter estimate is constant, the control law reduces to the "certainty equivalence" control. Let us examine an example of a system for which (4.12) is satisfied.

Example 4.1 Consider the problem of designing an adaptive controller for the system

$$\begin{aligned} \dot{x}_1 &= x_2 + \varphi(x_1)^\mathrm{T} \theta \\ \dot{x}_2 &= u, \end{aligned} \tag{4.14}$$

where $\theta = [\theta_1, \theta_2]^\mathrm{T}$ is an unknown constant parameter vector, and the vector-valued function $\varphi(x_1) = [\varphi_1(x_1), \varphi_2(x_1)]^\mathrm{T}$ is known and smooth. We dealt with this system in Section 3.4.1. If the parameter θ were known, backstepping would result in the θ-dependent change of coordinates

$$\begin{aligned} z_1 &= x_1 \\ z_2 &= x_2 + \varphi(x_1)^\mathrm{T} \theta + c_1 x_1, \end{aligned} \tag{4.15}$$

and the control law

$$u = \alpha_c(x, \theta) = -z_1 - c_2 z_2 - \left(\frac{\partial \varphi^\mathrm{T}}{\partial x_1} \theta + c_1 \right) \left(x_2 + \varphi(x_1)^\mathrm{T} \theta \right) \tag{4.16}$$

with $c_1, c_2 > 0$, which results in the closed-loop system

$$\dot{z} = Az, \qquad A = \begin{bmatrix} -c_1 & 1 \\ -1 & -c_2 \end{bmatrix}. \tag{4.17}$$

Due to the structure of A, an appropriate Lyapunov function is

$$V_c(x, \theta) = \frac{1}{2} z(x, \theta)^\mathrm{T} z(x, \theta), \tag{4.18}$$

Observing from (4.1) and (4.14) that

$$f(x) = \begin{bmatrix} x_2 \\ 0 \end{bmatrix}, \quad F(x) = \begin{bmatrix} \varphi(x_1)^\mathrm{T} \\ 0 \end{bmatrix}, \quad g(x) = \begin{bmatrix} 0 \\ 1 \end{bmatrix} \tag{4.19}$$

4.1 ADAPTIVE CONTROL LYAPUNOV FUNCTIONS

and evaluating

$$\frac{\partial V_c}{\partial x} = z^{\mathrm{T}}\begin{bmatrix} 1 & 0 \\ \frac{\partial \varphi^{\mathrm{T}}}{\partial x_1}\theta + c_1 & 1 \end{bmatrix}, \qquad (4.20)$$

with (4.19), (4.20), and (4.16), it is easy to show that

$$\frac{\partial V_c}{\partial x}[f(x) + F(x)\theta + g(x)\alpha_c(x,\theta)] = -c_1 z_1^2 - c_2 z_2^2. \qquad (4.21)$$

Let us now evaluate the partial derivatives appearing in (4.11):

$$\frac{\partial V_c}{\partial \theta} = z^{\mathrm{T}} e_2 \varphi^{\mathrm{T}} = z_2 \varphi^{\mathrm{T}} \qquad (4.22)$$

$$\frac{\partial V_c}{\partial x} g = z^{\mathrm{T}} e_2 = z_2, \qquad (4.23)$$

where (4.23) is immediate from (4.20) and (4.19). A comparison of (4.22) and (4.23) reveals that $\frac{\partial V_c}{\partial \theta} = \frac{\partial V_c}{\partial x} g\varphi^{\mathrm{T}}$, so that α_τ is given by (4.13):

$$\alpha_\tau(x,\hat{\theta}) = -\varphi^{\mathrm{T}}\Gamma\left(\frac{\partial V_c}{\partial x}F(x)\right)^{\mathrm{T}} = -\varphi^{\mathrm{T}}\Gamma\varphi\left[1, \frac{\partial \varphi^{\mathrm{T}}}{\partial x_1}\hat{\theta} + c_1\right]z. \qquad (4.24)$$

Taking for simplicity $\Gamma = I$, the resulting adaptive control law is

$$\begin{aligned} u = \alpha(x,\hat{\theta}) &= -z_1 - c_2 z_2 - \left(c_1 + \frac{\partial \varphi^{\mathrm{T}}}{\partial x_1}\hat{\theta}\right)\left(x_2 + \varphi(x_1)^{\mathrm{T}}\hat{\theta}\right) \\ &\quad -\varphi^{\mathrm{T}}\varphi\left[1, \frac{\partial \varphi^{\mathrm{T}}}{\partial x_1}\hat{\theta} + c_1\right]z, \end{aligned} \qquad (4.25)$$

and the corresponding parameter update law (4.7) is

$$\dot{\hat{\theta}} = \tau(x,\hat{\theta}) = \left(\frac{\partial V_c}{\partial x}F(x)\right)^{\mathrm{T}} = \varphi\left[1, \frac{\partial \varphi^{\mathrm{T}}}{\partial x_1}\hat{\theta} + c_1\right]z. \qquad (4.26)$$

Note that in (4.25) and (4.26) we use $z(x,\hat{\theta})$ instead of $z(x,\theta)$. With the choice of α and τ given by (4.25) and (4.26), the derivative $\dot{V}$ of the Lyapunov function $V(x,\hat{\theta}) = \frac{1}{2}z(x,\hat{\theta})^{\mathrm{T}} z(x,\hat{\theta}) + \frac{1}{2}\tilde{\theta}^{\mathrm{T}}\tilde{\theta}$ is guaranteed to be nonpositive: $\dot{V} = -c_1 z_1^2 - c_2 z_2^2$. This assures that both x and $\hat{\theta}$ are bounded. A standard argument using the LaSalle-Yoshizawa theorem (Theorem 2.1) proves that also $x(t) \to 0$. ◊

In the above example the desired factorization (4.12) of $\frac{\partial V_c}{\partial \theta}$ is a consequence of a particular feature of the system (4.14). The unknown parameter appears in the first, while the control appears only in the second equation. It is not hard to see that the same factorization (4.12) would be possible for a higher-order plant, provided that *the unknown parameter is separated from the control input by at most one integrator.* So the factorization (4.12) is not a fortuitous event, but a structural property. For systems with this "extended matching" property, the above simple adaptive design is feasible. However, most systems fail to possess the "extended matching" property.

A benchmark example is the third-order system

$$\begin{aligned} \dot{x}_1 &= x_2 + \varphi(x_1)^T \theta \\ \dot{x}_2 &= x_3 \\ \dot{x}_3 &= u \end{aligned} \qquad (4.27)$$

which has the form of (4.14) augmented by an integrator. In this system, θ and u are separated by two integrators and we are unable to find α_τ which satisfies (4.11). We will solve this problem with a recursive design which will circumvent the obstacle posed by the restrictive condition (4.11).

4.1.2 Certainty equivalence for a modified system

Condition (4.11) was dictated by our choice of the Lyapunov function $V_c(x, \hat{\theta})$ as the "certainty equivalence" form of $V_c(x, \theta)$. The only good thing we know about $V_c(x, \hat{\theta})$ is that it works when the factorization (4.12) is possible. Otherwise we do not know how to remove the indefinite term preventing the nonpositivity of $\dot{V}$ in (4.9). Having recognized that a cause of our difficulties is $V_c(x, \theta)$, we now embark on a search for Lyapunov functions more suitable for adaptive control. The key idea is to counteract the effect of $\dot{\hat{\theta}}$ and thus prevent the parameter estimate transients from destroying the nonpositivity of the Lyapunov derivative.

We say that the system

$$\dot{x} = f(x) + F(x)\theta + g(x)u \qquad (4.28)$$

is **globally adaptively stabilizable** if there exist a function $\alpha(x, \hat{\theta})$ smooth on $(\mathbb{R}^n \setminus \{0\}) \times \mathbb{R}^p$ with $\alpha(0, \hat{\theta}) \equiv 0$, a smooth function $\tau(x, \hat{\theta})$, and a positive definite symmetric $p \times p$ matrix Γ, such that the dynamic controller

$$u = \alpha(x, \hat{\theta}) \qquad (4.29)$$
$$\dot{\hat{\theta}} = \Gamma\tau(x, \hat{\theta}), \qquad (4.30)$$

guarantees that the solution $(x(t), \hat{\theta}(t))$ is globally bounded, and $x(t) \to 0$ as $t \to \infty$, for all $\theta \in \mathbb{R}^p$.

Our approach is to replace the problem of adaptive stabilization of the original system (4.28) by a problem of nonadaptive stabilization of a modified system.

4.1 ADAPTIVE CONTROL LYAPUNOV FUNCTIONS

Definition 4.2 *A smooth function $V_a : \mathbb{R}^n \times \mathbb{R}^p \to \mathbb{R}_+$, positive definite and radially unbounded in x for each θ, is called an **adaptive control Lyapunov function (aclf)** for (4.28) if there exists a positive definite symmetric matrix $\Gamma \in \mathbb{R}^{p \times p}$ such that for each $\theta \in \mathbb{R}^p$, $V_a(x, \theta)$ is a **clf** for the modified system*

$$\dot{x} = f(x) + F(x)\left(\theta + \Gamma\left(\frac{\partial V_a}{\partial \theta}\right)^{\mathrm{T}}\right) + g(x)u, \qquad (4.31)$$

that is, V_a satisfies

$$\inf_{u \in \mathbb{R}}\left\{\frac{\partial V_a}{\partial x}\left[f(x) + F(x)\left(\theta + \Gamma\left(\frac{\partial V_a}{\partial \theta}\right)^{\mathrm{T}}\right) + g(x)u\right]\right\} < 0. \qquad (4.32)$$

We now show how to design an adaptive controller (4.29)–(4.30) when an aclf is known.

Theorem 4.3 *The following two statements are equivalent:*

1. *There exists a triple (α, V_a, Γ) such that $\alpha(x, \theta)$ globally asymptotically stabilizes (4.31) at $x = 0$ for each $\theta \in \mathbb{R}^p$ with respect to the Lyapunov function $V_a(x, \theta)$.*

2. *There exists an aclf $V_a(x, \theta)$ for (4.28).*

Moreover, if an aclf $V_a(x, \theta)$ exists, then (4.28) is globally adaptively stabilizable.

Proof. ($1 \Rightarrow 2$) Obvious because 1 implies that there exists a continuous function $W : \mathbb{R}^n \times \mathbb{R}^p \to \mathbb{R}_+$, positive definite in x for each θ, such that

$$\frac{\partial V_a}{\partial x}\left[f(x) + F(x)\left(\theta + \Gamma\left(\frac{\partial V_a}{\partial \theta}\right)^{\mathrm{T}}\right) + g(x)\alpha(x, \theta)\right] \leq -W(x, \theta). \qquad (4.33)$$

Thus $V_a(x, \theta)$ is a clf for (4.31) for each $\theta \in \mathbb{R}^p$, and therefore it is an aclf for (4.28).

($2 \Rightarrow 1$) The proof of this part is based on Sontag's constructive proof [171] of Artstein's theorem [4]. We assume that V_a is an aclf for (4.28), that is, a clf for (4.31). Sontag's formula (2.19) applied to (4.31) gives a control law smooth on $(\mathbb{R}^n \setminus \{0\}) \times \mathbb{R}^p$:

$$\alpha(x, \theta) = \begin{cases} -\dfrac{\frac{\partial V_a}{\partial x}\tilde{f} + \sqrt{\left(\frac{\partial V_a}{\partial x}\tilde{f}\right)^2 + \left(\frac{\partial V_a}{\partial x}g\right)^4}}{\frac{\partial V_a}{\partial x}g} & , \quad \frac{\partial V_a}{\partial x}g(x, \theta) \neq 0 \\ 0 & , \quad \frac{\partial V_a}{\partial x}g(x, \theta) = 0, \end{cases} \qquad (4.34)$$

where
$$\tilde{f}(x,\theta) = f(x) + F(x)\left(\theta + \Gamma\left(\frac{\partial V_a}{\partial \theta}\right)^{\mathrm{T}}\right). \tag{4.35}$$

With the choice (4.34), inequality (4.33) is satisfied with the continuous function
$$W(x,\theta) = \sqrt{\left(\frac{\partial V_a}{\partial x}\tilde{f}(x,\theta)\right)^2 + \left(\frac{\partial V_a}{\partial x}g(x,\theta)\right)^4}, \tag{4.36}$$

which is positive definite in x for each θ, because (4.32) implies that $\frac{\partial V_a}{\partial x}\tilde{f}(x,\theta) < 0$ whenever $\frac{\partial V_a}{\partial x}g(x,\theta) = 0$ and $x \neq 0$. We note that the control law $\alpha(x,\theta)$ will be continuous at $x = 0$ if and only if the aclf V_a satisfies the following property, called the *small control property* [171]: For each $\theta \in \mathbb{R}^p$ and for any $\varepsilon > 0$ there is a $\delta > 0$ such that, if $x \neq 0$ satisfies $|x| \leq \delta$, then there is some u with $|u| \leq \varepsilon$ such that

$$\frac{\partial V_a}{\partial x}\left[f(x) + F(x)\left(\theta + \Gamma\left(\frac{\partial V_a}{\partial \theta}\right)^{\mathrm{T}}\right) + g(x)u\right] < 0. \tag{4.37}$$

Assuming the existence of an aclf we now show that (4.28) is globally adaptively stabilizable. Since $(2 \Rightarrow 1)$, there exists a triple (α, V_a, Γ) and a function W such that (4.33) is satisfied, that is,

$$\frac{\partial V_a}{\partial x}[f(x) + F(x)\theta + g(x)\alpha(x,\theta)] + \frac{\partial V_a}{\partial \theta}\Gamma\left(\frac{\partial V_a}{\partial x}F(x)\right)^{\mathrm{T}} \leq -W(x,\theta). \tag{4.38}$$

Consider the Lyapunov function candidate
$$V(x,\hat{\theta}) = V_a(x,\hat{\theta}) + \frac{1}{2}(\theta - \hat{\theta})^{\mathrm{T}}\Gamma^{-1}(\theta - \hat{\theta}). \tag{4.39}$$

With the help of (4.38), the derivative of V along the solutions of (4.28), (4.29), (4.30), is

$$\begin{aligned}
\dot{V} &= \frac{\partial V_a}{\partial x}\left[f + F\theta + g\alpha(x,\hat{\theta})\right] + \frac{\partial V_a}{\partial \hat{\theta}}\Gamma\tau(x,\hat{\theta}) - \tilde{\theta}^{\mathrm{T}}\tau(x,\hat{\theta}) \\
&= \frac{\partial V_a}{\partial x}\left[f + F\hat{\theta} + g\alpha(x,\hat{\theta})\right] + \frac{\partial V_a}{\partial \hat{\theta}}\Gamma\tau(x,\hat{\theta}) + \frac{\partial V_a}{\partial x}F\tilde{\theta} - \tilde{\theta}^{\mathrm{T}}\tau(x,\hat{\theta}) \\
&\leq -W(x,\hat{\theta}) - \frac{\partial V_a}{\partial \hat{\theta}}\Gamma\left(\frac{\partial V_a}{\partial x}F\right)^{\mathrm{T}} + \frac{\partial V_a}{\partial \hat{\theta}}\Gamma\tau(x,\hat{\theta}) \\
&\quad + \tilde{\theta}^{\mathrm{T}}\left(\frac{\partial V_a}{\partial x}F\right)^{\mathrm{T}} - \tilde{\theta}^{\mathrm{T}}\tau(x,\hat{\theta}).
\end{aligned} \tag{4.40}$$

Choosing
$$\tau(x,\hat{\theta}) = \left(\frac{\partial V_a}{\partial x}(x,\hat{\theta})F(x)\right)^{\mathrm{T}}, \tag{4.41}$$

4.1 ADAPTIVE CONTROL LYAPUNOV FUNCTIONS

we get
$$\dot{V} \leq -W(x,\hat{\theta}), \quad \forall \theta \in \mathbb{R}^p. \tag{4.42}$$

Thus, the equilibrium $x = 0, \hat{\theta} = \theta$ of (4.28), (4.29), (4.30) is globally stable, and by the LaSalle-Yoshizawa theorem (Theorem 2.1), $x(t) \to 0$, that is, (4.28) is globally adaptively stabilizable. □

The adaptive controller constructed in the proof of Theorem 4.3 consists of a control law $u = \alpha(x, \hat{\theta})$ given by (4.34), and an update law $\dot{\hat{\theta}} = \Gamma\tau(x,\hat{\theta})$ with (4.41).

It is of interest to interpret this controller as a certainty equivalence controller. The control law $\alpha(x, \theta)$ given by (4.34) is stabilizing for the modified system (4.31) but may not be stabilizing for the original system (4.28). However, as the proof of Theorem 4.3 shows, its certainty equivalence form $\alpha(x, \hat{\theta})$ is an adaptive globally stabilizing control law for the original system (4.28). Hence, if a certainty equivalence approach is to be applied to a nonlinear system, the system is to be modified to require a control law which anticipates the parameter estimation transients. In the proof of Theorem 4.3, this is achieved by incorporating the *tuning function* τ in the control law α. Indeed, the formula (4.34) for α depends on τ via

$$\frac{\partial V_a}{\partial x}\tilde{f}(x,\theta) = \frac{\partial V_a}{\partial x}f + \tau(x,\theta)^{\mathrm{T}}\left(\theta + \Gamma\left(\frac{\partial V_a}{\partial \theta}\right)^{\mathrm{T}}\right), \tag{4.43}$$

which is obtained by combining (4.35) and (4.41). Using (4.41) to rewrite the inequality (4.38) as

$$\frac{\partial V_a}{\partial x}[f(x) + F(x)\theta + g(x)\alpha(x,\theta)] + \frac{\partial V_a}{\partial \theta}\Gamma\tau(x,\theta) \leq -W(x,\theta), \tag{4.44}$$

it is not difficult to see that the control law (4.34) containing (4.43) prevents τ from destroying the nonpositivity of the Lyapunov derivative.

Remark 4.4 A relevant question remains unanswered: If there exists an aclf for (4.28), is this system globally asymptotically stabilizable for *each* θ (and vice versa)? In other words, does the existence of a pair (α, V_a) satisfying (4.33) for some $\Gamma > 0$ imply the existence of a pair (α^0, V_a^0) satisfying (4.33) for $\Gamma = 0$ (and vice versa)? Adaptive Lyapunov designs available in the literature [59, 65, 69, 94, 156, 157, 186] are all for systems which are not only globally adaptively stabilizable, but also globally asymptotically stabilizable for each θ. ◇

As is always the case in adaptive control, in the proof of Theorem 4.3 we used a Lyapunov function $V(x,\hat{\theta})$ given by (4.39), which is quadratic in the parameter error $\theta - \hat{\theta}$. The quadratic form is suggested by the linear dependence of (4.28) on θ, and the fact that θ cannot be used for feedback. We will now show that the quadratic form of (4.39) is both necessary and sufficient for the existence of an aclf.

We say that system (4.28) is **globally adaptively quadratically stabilizable** if it is *globally adaptively stabilizable* and, in addition, there exist a smooth function $V_a(x,\theta)$ positive definite and radially unbounded in x for each θ, and a continuous function $W(x,\theta)$ positive definite in x for each θ, such that for all $(x(0), \hat\theta(0)) \in \mathbb{R}^{n+p}$ and all $\theta \in \mathbb{R}^p$, the derivative of (4.39) along the solutions of (4.28), (4.29), (4.30) is given by (4.42).

Corollary 4.5 *The system (4.28) is globally adaptively quadratically stabilizable if and only if there exists an* aclf $V_a(x,\theta)$.

Proof. The 'if' part is contained in the proof of Theorem 4.3 where the Lyapunov function $V(x,\hat\theta)$ is in the form (4.39). To prove the 'only if' part, we start by assuming global adaptive quadratic stabilizability of (4.28), and first show that $\tau(x,\hat\theta)$ must be given by (4.41). The derivative of V along the solutions of (4.28), (4.29), (4.30), given by (4.40), is rewritten as

$$\dot V = \frac{\partial V_a}{\partial x}\left[f + F\hat\theta + g\alpha(x,\hat\theta)\right] + \frac{\partial V_a}{\partial \hat\theta}\Gamma\tau(x,\hat\theta) - \hat\theta^T\left(\left(\frac{\partial V_a}{\partial x}F\right)^T - \tau\right)$$
$$+ \theta^T\left(\left(\frac{\partial V_a}{\partial x}F\right)^T - \tau\right). \qquad (4.45)$$

This expression has to be nonpositive to satisfy (4.42). Since it is affine in θ, it can be nonpositive for all $(x,\hat\theta) \in \mathbb{R}^{n+p}$ and all $\theta \in \mathbb{R}^p$ only if the last term is zero, that is, only if τ is defined as in (4.41). Then, it is straightforward to verify that

$$\frac{\partial V_a}{\partial x}\left[f(x) + F(x)\left(\hat\theta + \Gamma\left(\frac{\partial V_a}{\partial \hat\theta}\right)^T\right) + g(x)\alpha(x,\hat\theta)\right]$$
$$= \dot V + \left(\tilde\theta^T - \frac{\partial V_a}{\partial \hat\theta}\Gamma\right)\left(\tau - \left(\frac{\partial V_a}{\partial x}F\right)^T\right)$$
$$\leq -W(x,\hat\theta) \qquad (4.46)$$

for all $(x,\hat\theta) \in \mathbb{R}^{n+p}$. By $(1 \Rightarrow 2)$ in Theorem 4.3, $V_a(x,\theta)$ is an aclf for (4.28). □

The above analysis applies also to the case where the unknown parameters enter the control vector field:

$$\dot x = f(x) + F(x)\theta + [g(x) + G(x)\theta]u. \qquad (4.47)$$

4.1 ADAPTIVE CONTROL LYAPUNOV FUNCTIONS

In this case the existence of an aclf V_a is equivalent to the existence of a clf for the system

$$\dot{x} = f(x) + F(x)\left(\theta + \Gamma\left(\frac{\partial V_a}{\partial \hat{\theta}}\right)^{\mathrm{T}}\right) + \left[g(x) + G(x)\left(\theta + \Gamma\left(\frac{\partial V_a}{\partial \hat{\theta}}\right)^{\mathrm{T}}\right)\right]u. \tag{4.48}$$

The extension to the multi-input case is also straightforward.

It is of interest to examine the input-output properties of the system resulting from the application of the adaptive control law $\alpha(x,\hat{\theta})$ to the plant (4.1):

$$\dot{x} = f(x) + F(x)\hat{\theta} + g(x)\alpha(x,\hat{\theta}) + F(x)\tilde{\theta}. \tag{4.49}$$

In early Lyapunov designs for linear systems of relative degree one, an important property was the strict positive realness of the transfer function between the parameter error and the output error [142]. For an analogous passivity property of the nonlinear system (4.49), let us consider that its input is $\tilde{\theta}$.

Corollary 4.6 (Passivity) *Suppose a function $V_a(x,\theta)$ is known to be an aclf with an associated control law $\alpha(x,\theta)$. Then the system*

$$\begin{aligned}\dot{x} &= f(x) + F(x)\hat{\theta} + g(x)\alpha(x,\hat{\theta}) + F(x)\tilde{\theta} \\ \tau &= \left(\frac{\partial V_a}{\partial x}(x,\hat{\theta})F(x)\right)^{\mathrm{T}}\end{aligned} \tag{4.50}$$

with $\tilde{\theta}$ as the input and τ as the output is strictly passive.

Proof. Along the solutions of (4.49) we have

$$\begin{aligned}\dot{V}_a &= \frac{\partial V_a}{\partial x}\left[f + F\hat{\theta} + g\alpha(x,\hat{\theta})\right] + \frac{\partial V_a}{\partial \hat{\theta}}\Gamma\tau(x,\hat{\theta}) + \frac{\partial V_a}{\partial x}F\tilde{\theta} \\ &\leq -W(x,\hat{\theta}) + \tau(x,\hat{\theta})^{\mathrm{T}}\tilde{\theta},\end{aligned} \tag{4.51}$$

which, upon integration, yields

$$\int_0^t \tau^{\mathrm{T}}(x(s),\hat{\theta}(s))\tilde{\theta}(s)ds \geq V_a(x(t),\hat{\theta}(t)) - V_a(x(0),\hat{\theta}(0)) + \int_0^t W(x(s),\hat{\theta}(s))ds. \tag{4.52}$$

Using $V_a(x,\hat{\theta})$ as a storage function and $W(x,\hat{\theta})$ as a dissipation rate, and noting that V_a and W are positive definite in x for each $\hat{\theta}$, the inequality (4.52) establishes strict passivity by Definition D.2. □

Hence, our closed-loop adaptive system represents a negative feedback connection of the strictly passive system (4.50) and the integrator

$$-\dot{\tilde{\theta}} = \frac{\Gamma}{s}\tau, \tag{4.53}$$

which is passive (positive real) because

$$\frac{d}{dt}\left(\frac{1}{2}\tilde{\theta}^T\Gamma^{-1}\tilde{\theta}\right) = -\tilde{\theta}^T\tau \qquad (4.54)$$

implies that

$$\int_0^t (-\tilde{\theta})^T\tau ds = \frac{1}{2}\tilde{\theta}(t)^T\Gamma^{-1}\tilde{\theta}(t) - \frac{1}{2}\tilde{\theta}(0)^T\Gamma^{-1}\tilde{\theta}(0). \qquad (4.55)$$

For such a feedback connection, Theorem D.4 establishes that the equilibrium $x = 0, \tilde{\theta} = 0$ is globally stable, and $x(t) \to 0$ as $t \to 0$. Thus, the problem of adaptive stabilization can be approached as the problem of finding an output τ with respect to which it is possible to achieve strict passivity from $\tilde{\theta}$ as the input.

4.1.3 Adaptive backstepping via aclf

With Theorem 4.3, the problem of adaptive stabilization is reduced to the problem of finding an aclf. We now address the problem of systematic construction of an aclf. Our aim is a recursive approach because we already know how to find aclf's for systems with the extended matching property, and expect to recursively enlarge this initial class of systems with repeated use of backstepping. So, we assume that an aclf is known for an initial system, and construct a new aclf for the initial system augmented by an integrator.

Lemma 4.7 *If the system*

$$\dot{x} = f(x) + F(x)\theta + g(x)u, \qquad (4.56)$$

is globally adaptively quadratically stabilizable with $\alpha \in C^1$, then the augmented system

$$\begin{aligned} \dot{x} &= f(x) + F(x)\theta + g(x)\xi \\ \dot{\xi} &= u, \end{aligned} \qquad (4.57)$$

is also globally adaptively quadratically stabilizable.

Proof. Since system (4.56) is globally adaptively stabilizable, then by Corollary 4.5 there exists an aclf $V_a(x,\theta)$, and by Theorem 4.3 it satisfies (4.33) with a control law $u = \alpha(x,\theta)$. We will now show that

$$V_1(x,\xi,\theta) = V_a(x,\theta) + \frac{1}{2}(\xi - \alpha(x,\theta))^2 \qquad (4.58)$$

is an aclf for the augmented system (4.57) by showing that it satisfies

$$\frac{\partial V_1}{\partial(x,\xi)}\begin{bmatrix} f + F\left(\theta + \Gamma\left(\frac{\partial V_1}{\partial \theta}\right)^T\right) + g\xi \\ \alpha_1(x,\xi,\theta) \end{bmatrix} \leq -W - (\xi - \alpha)^2 \qquad (4.59)$$

4.1 ADAPTIVE CONTROL LYAPUNOV FUNCTIONS

with the control law

$$\begin{aligned} u = \alpha_1(x,\xi,\theta) &= -\frac{\partial V_a}{\partial x}g - (\xi - \alpha) + \frac{\partial \alpha}{\partial x}(f + F\theta + g\xi) \\ &\quad + \frac{\partial \alpha}{\partial \theta}\Gamma\left(\frac{\partial V_1}{\partial x}F\right)^T + \frac{\partial V_a}{\partial \theta}\Gamma\left(\frac{\partial \alpha}{\partial x}F\right)^T. \end{aligned} \quad (4.60)$$

Let us start by introducing for brevity $z = \xi - \alpha(x,\theta)$. With (4.58) we compute

$$\begin{aligned} \frac{\partial V_1}{\partial(x,\xi)}\begin{bmatrix} f + F\theta + g\xi \\ \alpha_1(x,\xi,\theta) \end{bmatrix} \\ = \frac{\partial V_1}{\partial x}(f + F\theta + g\xi) + \frac{\partial V_1}{\partial \xi}\alpha_1(x,\xi,\theta) \\ = \left(\frac{\partial V_a}{\partial x} - z\frac{\partial \alpha}{\partial x}\right)(f + F\theta + g\xi) + z\alpha_1 \\ = \frac{\partial V_a}{\partial x}(f + F\theta + g\alpha) + \frac{\partial V_a}{\partial x}gz - z\frac{\partial \alpha}{\partial x}(f + F\theta + g\xi) + z\alpha_1 \\ = \frac{\partial V_a}{\partial x}(f + F\theta + g\alpha) + z\left(\alpha_1 + \frac{\partial V_a}{\partial x}g - \frac{\partial \alpha}{\partial x}(f + F\theta + g\xi)\right). \quad (4.61) \end{aligned}$$

On the other hand, in view of (4.58), we have

$$\begin{aligned} \frac{\partial V_1}{\partial(x,\xi)}\begin{bmatrix} F\Gamma\left(\frac{\partial V_1}{\partial \theta}\right)^T \\ 0 \end{bmatrix} &= \frac{\partial V_1}{\partial x}F\Gamma\left(\frac{\partial V_1}{\partial \theta}\right)^T \\ &= \left(\frac{\partial V_a}{\partial x} - z\frac{\partial \alpha}{\partial x}\right)F\Gamma\left(\frac{\partial V_a}{\partial \theta} - z\frac{\partial \alpha}{\partial \theta}\right)^T \\ &= \frac{\partial V_a}{\partial x}F\Gamma\left(\frac{\partial V_a}{\partial \theta}\right)^T \\ &\quad -z\left(\frac{\partial \alpha}{\partial \theta}\Gamma\left(\frac{\partial V_1}{\partial x}F\right)^T + \frac{\partial V_a}{\partial \theta}\Gamma\left(\frac{\partial \alpha}{\partial x}F\right)^T\right). \quad (4.62) \end{aligned}$$

Adding (4.61) and (4.62), with (4.33) and (4.60) we get

$$\begin{aligned} \frac{\partial V_1}{\partial(x,\xi)}&\begin{bmatrix} f + F\left(\theta + \Gamma\left(\frac{\partial V_1}{\partial \theta}\right)^T\right) + g\xi \\ \alpha_1(x,\xi,\theta) \end{bmatrix} \\ &= \frac{\partial V_a}{\partial x}(f + F\theta + g\alpha) + \frac{\partial V_a}{\partial x}F\Gamma\left(\frac{\partial V_a}{\partial \theta}\right)^T \\ &\quad + z\left(\alpha_1 + \frac{\partial V_a}{\partial x}g - \frac{\partial \alpha}{\partial x}(f + F\theta + g\xi)\right. \\ &\quad\left. -\frac{\partial \alpha}{\partial \theta}\Gamma\left(\frac{\partial V_1}{\partial x}F\right)^T - \frac{\partial V_a}{\partial \theta}\Gamma\left(\frac{\partial \alpha}{\partial x}F\right)^T\right) \\ &\leq -W(x,\theta) - z^2. \quad (4.63) \end{aligned}$$

This proves by Theorem 4.3 that $V_1(x, \xi, \theta)$ is an aclf for system (4.57), and by Corollary 4.5 this system is globally adaptively quadratically stabilizable. □

The new tuning function for system (4.57) is determined by the new aclf V_1 and given by

$$\tau_1(x, \xi, \theta) = \left(\frac{\partial V_1}{\partial(x, \xi)} \begin{bmatrix} F \\ 0 \end{bmatrix}\right)^T = \left(\frac{\partial V_1}{\partial x} F\right)^T = \left[\left(\frac{\partial V_a}{\partial x} - (\xi - \alpha)\frac{\partial \alpha}{\partial x}\right) F\right]^T$$

$$= \tau(x, \theta) - \left(\frac{\partial \alpha}{\partial x} F\right)^T (\xi - \alpha). \tag{4.64}$$

We note that the new tuning function τ_1 is obtained by augmenting the initial tuning function τ with the term $-\left(\frac{\partial \alpha}{\partial x} F\right)^T (\xi - \alpha)$ which accounts for the fact that the aclf V_a is augmented by $\frac{1}{2}(\xi - \alpha(x, \theta))^2$.

The form of the control law $\alpha_1(x, \xi, \theta)$ in (4.60) is of particular interest. It consists of two parts, $\alpha_1 = \alpha_{1,c} + \alpha_{1,\tau}$. The first part,

$$\alpha_{1,c}(x, \xi, \theta) = -\frac{\partial V_a}{\partial x} g - (\xi - \alpha) + \frac{\partial \alpha}{\partial x}(f + F\theta + g\xi), \tag{4.65}$$

would become the "certainty equivalence" control law for the augmented system (4.57) if we were to set $\Gamma = 0$.[2] The second part consists of two terms,

$$\alpha_{1,\tau}(x, \xi, \theta) = \frac{\partial \alpha}{\partial \theta} \Gamma \left(\frac{\partial V_1}{\partial x} F\right)^T + \frac{\partial V_a}{\partial \theta} \Gamma \left(\frac{\partial \alpha}{\partial x} F\right)^T. \tag{4.66}$$

Their role is to produce $\frac{\partial V_a}{\partial x} F \Gamma \left(\frac{\partial V_a}{\partial \theta}\right)^T$ in the aclf inequality (4.59). Observe that the first term in (4.66) incorporates $\tau_1 = \left(\frac{\partial V_1}{\partial x} F\right)^T$.

The control law $\alpha_1(x, \xi, \theta)$ in (4.60) is only one out of many possible control laws. Once we have shown that V_1 given by (4.58) is an aclf for (4.57), we can use, for example, the C^0 control law α_1 given by Sontag's formula (4.34) with $\frac{\partial V_1}{\partial(x,\xi)} g_1 = z$ and

$$\frac{\partial V_1}{\partial(x,\xi)} \tilde{f}_1(x, \xi, \theta) = \frac{\partial V_1}{\partial(x,\xi)} \left[\begin{matrix} f + F\left(\theta + \Gamma\left(\frac{\partial V_1}{\partial \theta}\right)^T\right) + g\xi \\ 0 \end{matrix} \right]$$

$$= \frac{\partial V_1}{\partial x}(f + g\xi) + \tau_1(x, \xi, \theta)^T \left(\theta + \Gamma\left(\frac{\partial V_1}{\partial \theta}\right)^T\right). \tag{4.67}$$

It can be shown that the following function, used as a clf in [158], is a more general aclf than (4.58):

$$V_1(x, \xi, \theta) = V_a(x, \theta) + \int_0^{\xi - \alpha(x,\theta)} \eta(s) ds, \tag{4.68}$$

---

[2] Note, however, that $\alpha_{1,c}$ is *not obtained* by setting $\Gamma = 0$ in α_1 since $\alpha(x, \theta)$ and $V_a(x, \theta)$ are also functions of Γ.

4.1 ADAPTIVE CONTROL LYAPUNOV FUNCTIONS

where η is a C^0 function such that $s\eta(s) > 0$ whenever $s \neq 0$, $\eta'(0) > 0$, and $\eta \notin \mathcal{L}^1((-\infty, 0]) \cup \mathcal{L}^1([0, +\infty))$.

The following example illustrates the use of Lemma 4.7.

Example 4.8 Let us consider the system

$$\begin{aligned} \dot{x}_1 &= x_2 + \varphi(x_1)^T \theta \\ \dot{x}_2 &= x_3 \\ \dot{x}_3 &= u. \end{aligned} \quad (4.69)$$

We will treat the state x_3 as an integrator added to the (x_1, x_2)-subsystem from Example 4.1. In that example, we have already designed an adaptive control law for the system

$$\begin{aligned} \dot{x}_1 &= x_2 + \varphi(x_1)^T \theta \\ \dot{x}_2 &= x_3, \end{aligned} \quad (4.70)$$

considering x_3 as a control input. With (4.18), (4.19), (4.20), (4.22), it can be shown that

$$\frac{\partial V_c}{\partial x}\left[f(x) + F(x)\left(\theta + \left(\frac{\partial V_c}{\partial \theta}\right)^T\right) + g(x)\alpha(x, \theta)\right] = -c_1 z_1^2 - c_2 z_2^2, \quad (4.71)$$

which means that $V_a(x_1, x_2, \theta) = V_c(x_1, x_2, \theta) = \frac{1}{2}(z_1^2 + z_2^2)$ is an aclf for the system (4.70) considering x_3 as a control input. Therefore, Lemma 4.7 is directly applicable. We define $z = x_3 - \alpha(x, \theta)$. By Lemma 4.7, the function

$$V_1(x, \theta) = \frac{1}{2}\left(z_1^2 + z_2^2 + z_3^2\right) \quad (4.72)$$

is an aclf for the system (4.69). With (4.60) and (4.64) we obtain

$$\begin{aligned} \alpha_1(x, \theta) &= -z_2 - c_3 z_3 - \frac{\partial \alpha}{\partial(x_1, x_2)}\begin{bmatrix} x_2 + \varphi^T \theta \\ x_3 \end{bmatrix} \\ &\quad + \frac{\partial \alpha}{\partial \theta}\tau_1 + z_2 \varphi^T \frac{\partial \alpha}{\partial x_1}\varphi \end{aligned} \quad (4.73)$$

$$\tau_1(x, \theta) = \tau - \frac{\partial \alpha}{\partial x_1}\varphi z_3. \quad (4.74)$$

With the following adaptive control law and the parameter update law:

$$u = \alpha_1(x, \hat{\theta}) \quad (4.75)$$

$$\dot{\hat{\theta}} = \tau_1(x, \hat{\theta}), \quad (4.76)$$

it is straightforward to verify that the closed-loop adaptive system is

$$\begin{bmatrix} \dot{z}_1 \\ \dot{z}_2 \\ \dot{z}_3 \end{bmatrix} = \begin{bmatrix} -c_1 & 1 & 0 \\ -1 & -c_2 & 1 - \frac{\partial \alpha}{\partial x_1}|\varphi|^2 \\ 0 & -1 + \frac{\partial \alpha}{\partial x_1}|\varphi|^2 & -c_3 \end{bmatrix} \begin{bmatrix} z_1 \\ z_2 \\ z_3 \end{bmatrix}$$

$$+ \begin{bmatrix} 1 \\ \frac{\partial \varphi^T}{\partial x_1}\hat{\theta} + c_1 \\ -\frac{\partial \alpha}{\partial x_1} \end{bmatrix} \varphi^T \tilde{\theta} \qquad (4.77)$$

$$\dot{\hat{\theta}} = \varphi \begin{bmatrix} 1, & \frac{\partial \varphi^T}{\partial x_1}\hat{\theta} + c_1, & -\frac{\partial \alpha}{\partial x_1} \end{bmatrix} \begin{bmatrix} z_1 \\ z_2 \\ z_3 \end{bmatrix}, \qquad (4.78)$$

where z_1, z_2, z_3 are used with $\hat{\theta}$ as an argument. The global stability of this system is established using the Lyapunov function $V(x,\hat{\theta}) = V_1(x,\hat{\theta}) + \frac{1}{2}\tilde{\theta}^T\tilde{\theta}$.
◇

While in Lemma 4.7 the initial system is augmented only by an integrator, a minor modification is sufficient to obtain an analogous result for the more general system

$$\begin{aligned} \dot{x} &= f(x) + F(x)\theta + g(x)\xi \\ \dot{\xi} &= u + F_1(x,\xi)\theta. \end{aligned} \qquad (4.79)$$

Corollary 4.9 *The function $V_1(x,\xi,\theta)$ defined in (4.58) is an aclf for the system (4.79) with the control law and the tuning function given as*

$$\alpha_1(x,\xi,\theta) = \alpha_1(x,\xi,\theta)_{(4.60)} - F_1(x,\xi)\left(\theta + \Gamma\left(\frac{\partial V_a}{\partial \theta}\right)^T\right) \qquad (4.80)$$

$$\tau_1(x,\xi,\theta) = \tau_1(x,\xi,\theta)_{(4.64)} + (\xi - \alpha)F_1(x,\xi)^T. \qquad (4.81)$$

A repeated application of Corollary 4.9 will further extend the class of nonlinear systems for this type of adaptive design. With the knowledge of V_a, τ, and α for the system (4.79), it is not hard to see that by applying Corollary 4.9 twice we can find V_2, τ_2, and α_2 for the system

$$\begin{aligned} \dot{x} &= f(x) + F(x)\theta + g(x)\xi_1 \\ \dot{\xi}_1 &= \xi_2 + F_1(x,\xi_1)\theta \\ \dot{\xi}_2 &= u + F_2(x,\xi_1,\xi_2)\theta. \end{aligned} \qquad (4.82)$$

4.2 SET-POINT REGULATION

In fact, it is clear that an n-fold application of Corollary 4.9 will provide us with V_n, τ_n, and α_n for the system

$$
\begin{aligned}
\dot{x} &= f(x) + F(x)\theta + g(x)\xi_1 \\
\dot{\xi}_1 &= \xi_2 + F_1(x, \xi_1)\theta \\
&\vdots \\
\dot{\xi}_{n-1} &= \xi_n + F_{n-1}(x, \xi_1, \ldots, \xi_{n-1})\theta \\
\dot{\xi}_n &= u + F_n(x, \xi_1, \ldots, \xi_n)\theta .
\end{aligned}
\tag{4.83}
$$

We will now develop a detailed design procedure for such systems.

4.2 Set-Point Regulation

With repeated use of Corollary 4.9, we can design an adaptive controller to globally stabilize a desired equilibrium x^e of the *parametric strict-feedback system* (3.70):

$$
\begin{aligned}
\dot{x}_1 &= x_2 + \varphi_1(x_1)^T\theta \\
\dot{x}_2 &= x_3 + \varphi_2(x_1, x_2)^T\theta \\
&\vdots \\
\dot{x}_{n-1} &= x_n + \varphi_{n-1}(x_1, \ldots, x_{n-1})^T\theta \\
\dot{x}_n &= \beta(x)u + \varphi_n(x)^T\theta
\end{aligned}
\tag{4.84}
$$

where $\theta \in \mathbb{R}^p$ is a vector of unknown constant parameters, β and

$$
F = [\varphi_1, \cdots, \varphi_n]
\tag{4.85}
$$

are smooth nonlinear functions taking arguments in $\mathbb{R}^n$, and $\beta(x) \neq 0$, $\forall x \in \mathbb{R}^n$.

In this section we develop a procedure for adaptive regulation of the output $y = x_1$ to a given set-point y_s. With a constant control u^e, the first $n-1$ equilibrium equations of $\dot{x}^e = 0$ in (4.84) can be successively solved for $x_2^e, \ldots, x_n^e$ as functions of x_1^e and θ:

$$
\begin{aligned}
x_2^e &= -\varphi_1(x_1^e)^T\theta \\
x_3^e &= -\varphi_2(x_1^e, x_2^e)^T\theta \\
&\vdots \\
x_n^e &= -\varphi_{n-1}(x_1^e, \ldots, x_{n-1}^e)^T\theta
\end{aligned}
\tag{4.86}
$$

Then the nth equation $\dot{x}_n^e = 0$ yields a relationship between x_1^e, u^e, and θ. When θ is known, then $\dot{x}_n^e = 0$ can be solved for u^e needed to keep x_1^e at a desired set-point $x_1^e = y_s$. The corresponding values $x_2^e, \ldots, x_n^e$ will be dictated by (4.86). Therefore, for each value of θ and a prescribed y_s, the equilibrium x^e

and the corresponding control value u^e are uniquely defined. In the special case where $\varphi_1(0) = \cdots = \varphi_{n-1}(0) = 0$, the choice $y_s = 0$ results in the equilibrium being $x^e = 0$ for all values of θ.

Our problem now is to globally stabilize this equilibrium when θ is unknown and also to achieve set-point regulation: $x(t) \to x^e$ as $t \to \infty$.

Comparing the systems (4.84) and (4.79), we observe that if x_3 were the control variable, then Corollary 4.9 would provide the desired adaptive control for the subsystem made of the first two equations of (4.84). Therefore, we can initiate our recursive design procedure by augmenting this subsystem by the third equation, as in (4.82). For convenience, we will do this in a self-contained fashion, independent of Section 4.1. An additional feature of the procedure in this section is a set of error coordinates in which the stability properties of the resulting closed-loop adaptive system are clearly displayed without an explicit use of the aclf concept.

4.2.1 Design procedure

We will start by adaptively stabilizing the first equation of (4.84) considering x_2 to be its control. At each subsequent step we will augment the designed subsystem by one equation. At the ith step, an ith-order subsystem is stabilized with respect to a Lyapunov function V_i by the design of a *stabilizing function* α_i and a *tuning function* τ_i. The update law for the parameter estimate $\hat{\theta}(t)$ and the adaptive feedback control u are designed at the final step. The third step is crucial for understanding the general design procedure.

Step 1. Introducing the first two error variables

$$z_1 = x_1 - y_s \tag{4.87}$$
$$z_2 = x_2 - \alpha_1, \tag{4.88}$$

we rewrite $\dot{x}_1 = x_2 + \varphi_1(x_1)^T \theta$, the first equation of (4.84), as

$$\dot{z}_1 = z_2 + \alpha_1 + w_1(x_1)^T \theta, \tag{4.89}$$

where, for uniformity with subsequent steps, we have defined the first regressor vector as

$$w_1(x_1) = \varphi_1(x_1). \tag{4.90}$$

Our task in this step is to stabilize (4.89) with respect to the Lyapunov function

$$V_1 = \frac{1}{2} z_1^2 + \frac{1}{2} \tilde{\theta}^T \Gamma^{-1} \tilde{\theta}, \tag{4.91}$$

whose derivative along the solutions of (4.89) is

$$\dot{V}_1 = z_1(z_2 + \alpha_1 + w_1^T \theta) - \tilde{\theta}^T \Gamma^{-1} (\dot{\hat{\theta}} - \Gamma w_1 z_1). \tag{4.92}$$

4.2 SET-POINT REGULATION

We can eliminate $\tilde{\theta}$ from $\dot{V}_1$ with the update law $\dot{\hat{\theta}} = \Gamma \tau_1$, where

$$\tau_1(x_1) = w_1(x_1)z_1. \tag{4.93}$$

If x_2 were our actual control, we would let $z_2 \equiv 0$, that is, $x_2 \equiv \alpha_1$. Then, to make $\dot{V}_1 = -c_1 z_1^2$, we would choose[3]

$$\alpha_1(x_1, \hat{\theta}) = -c_1 z_1 - w_1(x_1)^T \hat{\theta}. \tag{4.94}$$

Since x_2 is not our control, we have $z_2 \not\equiv 0$, and we do *not* use $\dot{\hat{\theta}} = \Gamma \tau_1$ as an update law. Instead, we retain τ_1 as our first *tuning function* and tolerate the presence of $\tilde{\theta}$ in $\dot{V}_1$:

$$\dot{V}_1 = -c_1 z_1^2 + z_1 z_2 + \tilde{\theta}^T (\Gamma^{-1} \dot{\hat{\theta}} - \tau_1). \tag{4.95}$$

The second term $z_1 z_2$ in $\dot{V}_1$ will be cancelled at the next step. With $\alpha_1(x_1, \hat{\theta})$ as in (4.94), the z_1-system becomes

$$\dot{z}_1 = -c_1 z_1 + z_2 + w_1(x_1)^T \tilde{\theta}. \tag{4.96}$$

Step 2. We now consider that x_3 is the control variable in the second equation of (4.84). Introducing

$$z_3 = x_3 - \alpha_2, \tag{4.97}$$

we rewrite $\dot{x}_2 = x_3 + \varphi_2(x_1, x_2)^T \theta$ as

$$\dot{z}_2 = z_3 + \alpha_2 - \frac{\partial \alpha_1}{\partial x_1} x_2 + w_2(x_1, x_2, \hat{\theta})^T \theta - \frac{\partial \alpha_1}{\partial \hat{\theta}} \dot{\hat{\theta}}, \tag{4.98}$$

where the second regressor vector w_2 is defined as

$$w_2(x_1, x_2, \hat{\theta}) = \varphi_2 - \frac{\partial \alpha_1}{\partial x_1} \varphi_1. \tag{4.99}$$

Our task in this step is to stabilize the (z_1, z_2)-system (4.96), (4.98) with respect to

$$V_2 = V_1 + \frac{1}{2} z_2^2, \tag{4.100}$$

whose derivative along the solutions of (4.96), (4.98) is

$$\begin{aligned}\dot{V}_2 &= -c_1 z_1^2 + z_2 \left[z_1 + z_3 + \alpha_2 - \frac{\partial \alpha_1}{\partial x_1} x_2 + w_2^T \hat{\theta} - \frac{\partial \alpha_1}{\partial \hat{\theta}} \dot{\hat{\theta}} \right] \\ &+ \tilde{\theta}^T \left(\tau_1 + w_2 z_2 - \Gamma^{-1} \dot{\hat{\theta}} \right).\end{aligned} \tag{4.101}$$

---

[3] The character of the control law α_1 is "certainty equivalence" because the aclf $\frac{1}{2} z_1^2 = \frac{1}{2}(x_1 - y_s)^2$, with respect to which the design is performed, is independent of θ, so it is also a clf.

We can eliminate $\tilde{\theta}$ from $\dot{V}_2$ with the update law $\dot{\hat{\theta}} = \Gamma\tau_2$, where

$$\tau_2(x_1, x_2, \hat{\theta}) = \tau_1 + w_2 z_2 = [w_1,\ w_2] \begin{bmatrix} z_1 \\ z_2 \end{bmatrix}. \quad (4.102)$$

If x_3 were our actual control and, hence, $z_3 \equiv 0$, we would achieve $\dot{V}_2 = -c_1 z_1^2 - c_2 z_2^2$ by designing α_2 to make the bracketed term multiplying z_2 in (4.101) equal to $-c_2 z_2$, namely[4]

$$\alpha_2(x_1, x_2, \hat{\theta}) = -z_1 - c_2 z_2 + \frac{\partial \alpha_1}{\partial x_1} x_2 - w_2^T \hat{\theta} + \frac{\partial \alpha_1}{\partial \hat{\theta}} \Gamma\tau_2. \quad (4.103)$$

We retain τ_2 as our second tuning function in the term $\Gamma\tau_2$ which replaces $\dot{\hat{\theta}}$ in (4.103). However, we do not use $\dot{\hat{\theta}} = \Gamma\tau_2$ as an update law, so that the resulting $\dot{V}_2$ is

$$\dot{V}_2 = -c_1 z_1^2 - c_2 z_2^2 + z_2 z_3 + z_2 \frac{\partial \alpha_1}{\partial \hat{\theta}}(\Gamma\tau_2 - \dot{\hat{\theta}}) + \tilde{\theta}^T(\tau_2 - \Gamma^{-1}\dot{\hat{\theta}}). \quad (4.104)$$

The first two terms in $\dot{V}_2$ are negative definite, the third term will be cancelled at the next step, while the discrepancy between $\Gamma\tau_2$ and $\dot{\hat{\theta}}$ in the last two terms remains. By substituting (4.103) into (4.98), the (z_1, z_2)-subsystem becomes

$$\begin{bmatrix} \dot{z}_1 \\ \dot{z}_2 \end{bmatrix} = \begin{bmatrix} -c_1 & 1 \\ -1 & -c_2 \end{bmatrix} \begin{bmatrix} z_1 \\ z_2 \end{bmatrix} + \begin{bmatrix} w_1^T \\ w_2^T \end{bmatrix} \tilde{\theta} + \begin{bmatrix} 0 \\ z_3 + \frac{\partial \alpha_1}{\partial \hat{\theta}}(\Gamma\tau_2 - \dot{\hat{\theta}}) \end{bmatrix}. \quad (4.105)$$

Step 3. Proceeding to the third equation in (4.84) we introduce

$$z_4 = x_4 - \alpha_3 \quad (4.106)$$

and rewrite $\dot{x}_3 = x_4 + \varphi_3(x_1, x_2, x_3)^T \theta$ as

$$\dot{z}_3 = z_4 + \alpha_3 - \frac{\partial \alpha_2}{\partial x_1} x_2 - \frac{\partial \alpha_2}{\partial x_2} x_3 + w_3(x_1, x_2, x_3, \hat{\theta})^T \theta - \frac{\partial \alpha_2}{\partial \hat{\theta}} \dot{\hat{\theta}}, \quad (4.107)$$

where the third regressor vector w_3 is defined as

$$w_3(x_1, x_2, x_3, \hat{\theta}) = \varphi_3 - \frac{\partial \alpha_2}{\partial x_1} \varphi_1 - \frac{\partial \alpha_2}{\partial x_2} \varphi_2. \quad (4.108)$$

---

[4]While α_1 was a "certainty equivalence" control, the control law α_2 is not of the "certainty equivalence" type because it is designed with respect to the θ-dependent aclf $\frac{1}{2}z_1^2 + \frac{1}{2}z_2^2$. The term $\frac{\partial \alpha_1}{\partial \hat{\theta}}\Gamma\tau_2$ in α_2 corresponds to the first term in the second row of (4.60) in Lemma 4.7. In α_2, there is no equivalent of the second term in the second row of (4.60) because the aclf $\frac{1}{2}z_1^2$ is independent of θ.

4.2 SET-POINT REGULATION

Our task is to stabilize the (z_1, z_2, z_3)-system with respect to

$$V_3 = V_2 + \frac{1}{2}z_3^2, \tag{4.109}$$

whose derivative along (4.105) and (4.107) is

$$\begin{aligned}\dot{V}_3 &= -c_1 z_1^2 - c_2 z_2^2 + z_2 \frac{\partial \alpha_1}{\partial \hat{\theta}}(\Gamma \tau_2 - \dot{\hat{\theta}}) \\ &+ z_3 \left[z_2 + z_4 + \alpha_3 - \frac{\partial \alpha_2}{\partial x_1} x_2 - \frac{\partial \alpha_2}{\partial x_2} x_3 + w_3^T \hat{\theta} - \frac{\partial \alpha_2}{\partial \hat{\theta}} \dot{\hat{\theta}} \right] \\ &+ \tilde{\theta}^T \left(\tau_2 + w_3 z_3 - \Gamma^{-1} \dot{\hat{\theta}} \right). \end{aligned} \tag{4.110}$$

We can eliminate $\tilde{\theta}$ from $\dot{V}_3$ with the update law $\dot{\hat{\theta}} = \Gamma \tau_3$, where τ_3 is our tuning function

$$\begin{aligned}\tau_3(x_1, x_2, x_3, \hat{\theta}) &= \tau_2 + w_3 z_3 \\ &= [w_1, \ w_2, \ w_3] \begin{bmatrix} z_1 \\ z_2 \\ z_3 \end{bmatrix}. \end{aligned} \tag{4.111}$$

If x_3 were our actual control, we would have $z_4 \equiv 0$ and achieve $\dot{V}_3 = -c_1 z_1^2 - c_2 z_2^2 - c_3 z_3^2$ by designing α_3 to make the bracketed term multiplying z_3 equal to $-c_3 z_3$, namely

$$\begin{aligned}\alpha_3(x_1, x_2, x_3, \hat{\theta}) &= -z_2 - c_3 z_3 + \frac{\partial \alpha_2}{\partial x_1} x_2 + \frac{\partial \alpha_2}{\partial x_2} x_3 - w_3^T \hat{\theta} \\ &+ \frac{\partial \alpha_2}{\partial \hat{\theta}} \Gamma \tau_3 + \nu_3, \end{aligned} \tag{4.112}$$

where ν_3 is a correction term yet to be chosen.[5] Substituting (4.112) into (4.110), and noting that

$$\begin{aligned}\dot{\hat{\theta}} - \Gamma \tau_2 &= \dot{\hat{\theta}} - \Gamma \tau_3 + \Gamma \tau_3 - \Gamma \tau_2 \\ &= \dot{\hat{\theta}} - \Gamma \tau_3 + \Gamma w_3 z_3, \end{aligned} \tag{4.113}$$

(4.110) is rewritten as

$$\begin{aligned}\dot{V}_3 &= -c_1 z_1^2 - c_2 z_2^2 + z_3 \left(\nu_3 - \frac{\partial \alpha_1}{\partial \hat{\theta}} \Gamma w_3 z_2 \right) \\ &+ z_3 z_4 + \left(z_2 \frac{\partial \alpha_1}{\partial \hat{\theta}} + z_3 \frac{\partial \alpha_2}{\partial \hat{\theta}} \right)(\Gamma \tau_3 - \dot{\hat{\theta}}) + \tilde{\theta}^T (\tau_3 - \Gamma^{-1} \dot{\hat{\theta}}), \end{aligned} \tag{4.114}$$

---

[5]It is of interest to compare this control law with the control law (4.60) in Lemma 4.7: The last two terms in (4.112) are analogous to the last two terms in (4.60).

and the (z_1, z_2, z_3)-subsystem becomes

$$\begin{bmatrix} \dot{z}_1 \\ \dot{z}_2 \\ \dot{z}_3 \end{bmatrix} = \begin{bmatrix} -c_1 & 1 & 0 \\ -1 & -c_2 & 1 \\ 0 & -1 & -c_3 \end{bmatrix} \begin{bmatrix} z_1 \\ z_2 \\ z_3 \end{bmatrix} + \begin{bmatrix} w_1^\mathrm{T} \\ w_2^\mathrm{T} \\ w_3^\mathrm{T} \end{bmatrix} \tilde{\theta}$$

$$+ \begin{bmatrix} 0 \\ -\frac{\partial \alpha_1}{\partial \hat{\theta}} \Gamma w_3 z_3 \\ \nu_3 \end{bmatrix} + \begin{bmatrix} 0 \\ \frac{\partial \alpha_1}{\partial \hat{\theta}} (\Gamma \tau_3 - \dot{\hat{\theta}}) \\ z_4 + \frac{\partial \alpha_2}{\partial \hat{\theta}} (\Gamma \tau_3 - \dot{\hat{\theta}}) \end{bmatrix}. \quad (4.115)$$

If x_4 were our control, we would have $z_4 = 0$, and with the update law $\dot{\hat{\theta}} = \Gamma \tau_3$, the last vector in (4.115) would be zero. However, the potentially destabilizing term $-\frac{\partial \alpha_1}{\partial \hat{\theta}} \Gamma w_3 z_3$ would still remain. This unmatched term must be accommodated by a choice of the correction term ν_3. From (4.114), the choice of ν_3 is immediate:

$$\nu_3(x_1, x_2, x_3, \hat{\theta}) = \frac{\partial \alpha_1}{\partial \hat{\theta}} \Gamma w_3 z_2. \quad (4.116)$$

We again postpone the decision about $\dot{\hat{\theta}}$ and do not use $\dot{\hat{\theta}} = \Gamma \tau_3$ as an update law. The resulting $\dot{V}_3$ is

$$\dot{V}_3 = -c_1 z_1^2 - c_2 z_2^2 - c_3 z_3^2 + z_3 z_4$$
$$+ \left(z_2 \frac{\partial \alpha_1}{\partial \hat{\theta}} + z_3 \frac{\partial \alpha_2}{\partial \hat{\theta}} \right) (\Gamma \tau_3 - \dot{\hat{\theta}}) + \tilde{\theta}^\mathrm{T} (\tau_3 - \Gamma^{-1} \dot{\hat{\theta}}) \quad (4.117)$$

and the (z_1, z_2, z_3)-subsystem becomes

$$\begin{bmatrix} \dot{z}_1 \\ \dot{z}_2 \\ \dot{z}_3 \end{bmatrix} = \begin{bmatrix} -c_1 & 1 & 0 \\ -1 & -c_2 & 1 - \frac{\partial \alpha_1}{\partial \hat{\theta}} \Gamma w_3 \\ 0 & -1 + \frac{\partial \alpha_1}{\partial \hat{\theta}} \Gamma w_3 & -c_3 \end{bmatrix} \begin{bmatrix} z_1 \\ z_2 \\ z_3 \end{bmatrix} + \begin{bmatrix} w_1^\mathrm{T} \\ w_2^\mathrm{T} \\ w_3^\mathrm{T} \end{bmatrix} \tilde{\theta}$$

$$+ \begin{bmatrix} 0 \\ 0 \\ z_4 \end{bmatrix} + \begin{bmatrix} 0 \\ \frac{\partial \alpha_1}{\partial \hat{\theta}} \\ \frac{\partial \alpha_2}{\partial \hat{\theta}} \end{bmatrix} (\Gamma \tau_3 - \dot{\hat{\theta}}). \quad (4.118)$$

The 'system matrix' in (4.118) has a significant property: the skew symmetry of the nonlinear term $\frac{\partial \alpha_1}{\partial \hat{\theta}} \Gamma w_3$ achieved by the choice of ν_3 in (4.116). This term is analogous to the second term in (4.66) and the skew symmetry is crucial for stabilization.

Step i. Introducing

$$z_{i+1} = x_{i+1} - \alpha_i, \quad (4.119)$$

we rewrite $\dot{x}_i = x_{i+1} + \varphi_i(x_1, \ldots, x_i)^\mathrm{T} \theta$ as

$$\dot{z}_i = z_{i+1} + \alpha_i - \sum_{k=1}^{i-1} \frac{\partial \alpha_{i-1}}{\partial x_k} x_{k+1} + w_i(x_1, \ldots, x_i, \hat{\theta})^\mathrm{T} \theta - \frac{\partial \alpha_{i-1}}{\partial \hat{\theta}} \dot{\hat{\theta}}, \quad (4.120)$$

4.2 SET-POINT REGULATION

where the ith regressor vector is defined as

$$w_i(x_1,\ldots,x_i,\hat{\theta}) = \varphi_i - \sum_{k=1}^{i-1} \frac{\partial \alpha_{i-1}}{\partial x_k}\varphi_k. \tag{4.121}$$

Our objective is to stabilize the $(z_1,\ldots,z_i)$-system with respect to

$$V_i = V_{i-1} + \frac{1}{2}z_i^2, \tag{4.122}$$

whose derivative is

$$\begin{aligned}\dot{V}_i &= -\sum_{k=1}^{i-1} c_k z_k^2 + \left(\sum_{k=1}^{i-2} z_{k+1}\frac{\partial \alpha_k}{\partial \hat{\theta}}\right)(\Gamma\tau_{i-1} - \dot{\hat{\theta}}) \\ &\quad + z_i\left[z_{i-1} + z_{i+1} + \alpha_i - \sum_{k=1}^{i-1}\frac{\partial \alpha_{i-1}}{\partial x_k}x_{k+1} + w_i^{\mathrm{T}}\hat{\theta} - \frac{\partial \alpha_{i-1}}{\partial \hat{\theta}}\dot{\hat{\theta}}\right] \\ &\quad + \tilde{\theta}^{\mathrm{T}}\left(\tau_{i-1} + w_i z_i - \Gamma^{-1}\dot{\hat{\theta}}\right).\end{aligned} \tag{4.123}$$

We can eliminate $\tilde{\theta}$ from $\dot{V}_i$ with the update law $\dot{\hat{\theta}} = \Gamma\tau_i$, where

$$\begin{aligned}\tau_i(x_1,\ldots,x_i,\hat{\theta}) &= \tau_{i-1} + z_i w_i \\ &= [w_1,\ldots,w_i]\begin{bmatrix}z_1\\\vdots\\z_i\end{bmatrix}.\end{aligned} \tag{4.124}$$

Then, in the absence of z_{i+1}, we would achieve $\dot{V}_i = -\sum_{k=1}^{i} c_k z_k^2$, by designing α_i to make the bracketed term multiplying z_i equal to $-c_i z_i$, namely

$$\begin{aligned}\alpha_i(x_1,\ldots,x_i,\hat{\theta}) &= -z_{i-1} - c_i z_i + \sum_{k=1}^{i-1}\frac{\partial \alpha_{i-1}}{\partial x_k}x_{k+1} - w_i^{\mathrm{T}}\hat{\theta} \\ &\quad + \frac{\partial \alpha_{i-1}}{\partial \hat{\theta}}\Gamma\tau_i + \nu_i,\end{aligned} \tag{4.125}$$

where ν_i is a correction term yet to be chosen. Noting that

$$\begin{aligned}\dot{\hat{\theta}} - \Gamma\tau_{i-1} &= \dot{\hat{\theta}} - \Gamma\tau_i + \Gamma\tau_i - \Gamma\tau_{i-1} \\ &= \dot{\hat{\theta}} - \Gamma\tau_i + \Gamma w_i z_i,\end{aligned} \tag{4.126}$$

we rewrite $\dot{V}_i$ as

$$\begin{aligned}\dot{V}_i &= -\sum_{k=1}^{i-1} c_k z_k^2 + z_i\left[z_{i+1} + \nu_i - \frac{\partial \alpha_{i-1}}{\partial \hat{\theta}}(\Gamma\tau_i - \dot{\hat{\theta}})\right] \\ &\quad + \left(\sum_{k=1}^{i-2} z_{k+1}\frac{\partial \alpha_k}{\partial \hat{\theta}}\right)(\Gamma\tau_{i-1} - \dot{\hat{\theta}}) + \tilde{\theta}^{\mathrm{T}}(\tau_i - \Gamma^{-1}\dot{\hat{\theta}})\end{aligned}$$

$$= -\sum_{k=1}^{i-1} c_k z_k^2 + z_i \left[z_{i+1} + \nu_i - \sum_{k=1}^{i-2} z_{k+1} \frac{\partial \alpha_k}{\partial \hat{\theta}} \Gamma w_i \right]$$
$$+ \left(\sum_{k=1}^{i-1} z_{k+1} \frac{\partial \alpha_k}{\partial \hat{\theta}} \right) (\Gamma \tau_i - \dot{\hat{\theta}}) - \tilde{\theta}^{\mathrm{T}} (\tau_i - \Gamma^{-1} \dot{\hat{\theta}}), \quad (4.127)$$

and represent the $(z_1, \ldots, z_i)$-subsystem as

$$\begin{bmatrix} \dot{z}_1 \\ \vdots \\ \dot{z}_i \end{bmatrix} = \begin{bmatrix} -c_1 & 1 & 0 & \cdots & 0 & 0 \\ -1 & -c_2 & 1+\sigma_{23} & \cdots & \sigma_{2,i-1} & 0 \\ 0 & -1-\sigma_{23} & \ddots & \ddots & \vdots & \vdots \\ \vdots & \vdots & \ddots & \ddots & 1+\sigma_{i-2,i-1} & 0 \\ 0 & -\sigma_{2,i-1} & \cdots & -1-\sigma_{i-2,i-1} & -c_{i-1} & 1 \\ 0 & 0 & \cdots & 0 & -1 & -c_i \end{bmatrix} \begin{bmatrix} z_1 \\ \vdots \\ z_i \end{bmatrix}$$
$$+ \begin{bmatrix} w_1^{\mathrm{T}} \\ \vdots \\ w_i^{\mathrm{T}} \end{bmatrix} \tilde{\theta} + \begin{bmatrix} 0 \\ \sigma_{2,i} z_i \\ \vdots \\ \sigma_{i-1,i} z_i \\ \nu_i \end{bmatrix} + \begin{bmatrix} 0 \\ \vdots \\ 0 \\ z_{i+1} \end{bmatrix} + \begin{bmatrix} 0 \\ \frac{\partial \alpha_1}{\partial \hat{\theta}} \\ \vdots \\ \frac{\partial \alpha_{i-1}}{\partial \hat{\theta}} \end{bmatrix} (\Gamma \tau_i - \dot{\hat{\theta}}), \quad (4.128)$$

where

$$\sigma_{jk}(x, \hat{\theta}) = -\frac{\partial \alpha_{j-1}}{\partial \hat{\theta}} \Gamma w_k. \quad (4.129)$$

Now the correction term is chosen as

$$\nu_i(x_1, \ldots, x_i, \hat{\theta}) = \sum_{k=1}^{i-2} z_{k+1} \frac{\partial \alpha_k}{\partial \hat{\theta}} \Gamma w_i = -\sum_{k=2}^{i-1} \sigma_{k,i} z_k. \quad (4.130)$$

Because we do not use $\dot{\hat{\theta}} = \Gamma \tau_i$ as an update law, the resulting $\dot{V}_i$ is

$$\dot{V}_i = -\sum_{k=1}^{i} c_k z_k^2 + z_i z_{i+1} + \left(\sum_{k=1}^{i-1} z_{k+1} \frac{\partial \alpha_k}{\partial \hat{\theta}} \right) (\Gamma \tau_i - \dot{\hat{\theta}}) + \tilde{\theta}^{\mathrm{T}} (\tau_i - \Gamma^{-1} \dot{\hat{\theta}}), \quad (4.131)$$

and the $(z_1, \ldots, z_i)$-subsystem becomes

$$\begin{bmatrix} \dot{z}_1 \\ \vdots \\ \dot{z}_i \end{bmatrix} = \begin{bmatrix} -c_1 & 1 & 0 & \cdots & 0 \\ -1 & -c_2 & 1+\sigma_{23} & \cdots & \sigma_{2i} \\ 0 & -1-\sigma_{23} & \ddots & \ddots & \vdots \\ \vdots & \vdots & \ddots & \ddots & 1+\sigma_{i-1,i} \\ 0 & -\sigma_{2i} & \cdots & -1-\sigma_{i-1,i} & -c_i \end{bmatrix} \begin{bmatrix} z_1 \\ \vdots \\ z_i \end{bmatrix}$$
$$+ \begin{bmatrix} w_1^{\mathrm{T}} \\ \vdots \\ w_i^{\mathrm{T}} \end{bmatrix} \tilde{\theta} + \begin{bmatrix} 0 \\ \vdots \\ 0 \\ z_{i+1} \end{bmatrix} + \begin{bmatrix} 0 \\ \frac{\partial \alpha_1}{\partial \hat{\theta}} \\ \vdots \\ \frac{\partial \alpha_{i-1}}{\partial \hat{\theta}} \end{bmatrix} (\Gamma \tau_i - \dot{\hat{\theta}}). \quad (4.132)$$

4.2 SET-POINT REGULATION

Step n. At the final step, we introduce

$$z_n = x_n - \alpha_{n-1} \tag{4.133}$$

and rewrite the last equation $\dot{x}_n = \beta(x)u + \varphi_n(x)^T\theta$ as

$$\dot{z}_n = \beta u + \varphi_n^T \theta - \sum_{k=1}^{n-1} \frac{\partial \alpha_{n-1}}{\partial x_k}(x_{k+1} + \varphi_k^T \theta) - \frac{\partial \alpha_{n-1}}{\partial \hat{\theta}}\dot{\hat{\theta}}, \tag{4.134}$$

where the last regressor vector is defined as

$$w_n(x, \hat{\theta}) = \varphi_n - \sum_{k=1}^{n-1} \frac{\partial \alpha_{n-1}}{\partial x_k}\varphi_k. \tag{4.135}$$

In this equation, the actual control input is at our disposal. We are finally in the position to design our actual update law $\dot{\hat{\theta}} = \Gamma \tau_n$ and feedback control u to stabilize the full z-system with respect to

$$\begin{aligned} V_n &= V_{n-1} + \frac{1}{2}z_n^2 \\ &= \frac{1}{2}z^T z + \frac{1}{2}\tilde{\theta}^T \Gamma^{-1}\tilde{\theta}. \end{aligned} \tag{4.136}$$

Our goal is to make $\dot{V}_n$ nonpositive:

$$\begin{aligned} \dot{V}_n &= -\sum_{k=1}^{n-1} c_k z_k^2 + \left(\sum_{k=1}^{n-2} z_{k+1}\frac{\partial \alpha_k}{\partial \hat{\theta}}\right)(\Gamma\tau_{n-1} - \dot{\hat{\theta}}) \\ &\quad + z_n\left[z_{n-1} + \beta u - \sum_{k=1}^{n-1}\frac{\partial \alpha_{n-1}}{\partial x_k}x_{k+1} + w_n^T\hat{\theta} - \frac{\partial \alpha_{n-1}}{\partial \hat{\theta}}\dot{\hat{\theta}}\right] \\ &\quad + \tilde{\theta}^T\left(\tau_{n-1} + w_n z_n - \Gamma^{-1}\dot{\hat{\theta}}\right). \end{aligned} \tag{4.137}$$

To eliminate $\tilde{\theta}$ from $\dot{V}_n$ we choose the update law

$$\begin{aligned} \dot{\hat{\theta}} &= \Gamma\tau_n(z,\hat{\theta}) = \Gamma\tau_{n-1} + \Gamma w_n z_n \\ &= \Gamma W(z,\hat{\theta})z, \end{aligned} \tag{4.138}$$

where the regressor matrix W is composed of the regressor vectors $w_1, \ldots, w_n$:

$$W(z,\hat{\theta}) = [w_1, \ldots, w_n]. \tag{4.139}$$

We choose the control u to make the bracketed term multiplying z_n equal to $-c_n z_n$:

$$u = \frac{1}{\beta}\left(-z_{n-1} - c_n z_n + \sum_{k=1}^{n-1}\frac{\partial \alpha_{n-1}}{\partial x_k}x_{k+1} - w_n^T \hat{\theta} + \frac{\partial \alpha_{n-1}}{\partial \hat{\theta}}\Gamma\tau_n + \nu_n\right), \tag{4.140}$$

where ν_n is a correction term yet to be chosen. With (4.140), $\dot{V}_n$ becomes

$$\dot{V}_n = -\sum_{k=1}^{n-1} c_k z_k^2 + \left(\sum_{k=1}^{n-2} z_{k+1}\frac{\partial \alpha_k}{\partial \hat{\theta}}\right)(\Gamma\tau_{n-1} - \dot{\hat{\theta}}) + z_n \nu_n. \tag{4.141}$$

Then, noting that

$$\dot{\hat{\theta}} - \Gamma\tau_{n-1} = \Gamma\tau_n - \Gamma\tau_{n-1} = \Gamma w_n z_n, \tag{4.142}$$

we rewrite $\dot{V}_n$ as

$$\dot{V}_n = -\sum_{k=1}^{n-1} c_k z_k^2 + z_n\left(\nu_n - \sum_{k=1}^{n-2} z_{k+1}\frac{\partial \alpha_k}{\partial \hat{\theta}}\Gamma w_n\right). \tag{4.143}$$

Now the correction term ν_n is chosen as

$$\nu_n(x,\hat{\theta}) = \sum_{k=1}^{n-2} z_{k+1}\frac{\partial \alpha_k}{\partial \hat{\theta}}\Gamma w_n = -\sum_{k=2}^{n-1} \sigma_{k,n} z_k. \tag{4.144}$$

We have thus reached our goal:

$$\dot{V}_n = -\sum_{k=1}^{n} c_k z_k^2. \tag{4.145}$$

The overall closed-loop system is

$$\dot{z} = A_z(z,\hat{\theta})z + W(z,\hat{\theta})^{\mathrm{T}}\tilde{\theta} \tag{4.146}$$

$$\dot{\tilde{\theta}} = \Gamma W(z,\hat{\theta})z, \tag{4.147}$$

where

$$A_z(z,\hat{\theta}) = \begin{bmatrix} -c_1 & 1 & 0 & \cdots & 0 \\ -1 & -c_2 & 1+\sigma_{23} & \cdots & \sigma_{2n} \\ 0 & -1-\sigma_{23} & \ddots & \ddots & \vdots \\ \vdots & \vdots & \ddots & \ddots & 1+\sigma_{n-1,n} \\ 0 & -\sigma_{2n} & \cdots & -1-\sigma_{n-1,n} & -c_n \end{bmatrix}. \tag{4.148}$$

The system (4.146) will be referred to as the *error system*. It is important to note that a major portion of the design effort was invested into achieving

$$A_z(z,\hat{\theta}) + A_z(z,\hat{\theta})^{\mathrm{T}} = -2\begin{bmatrix} c_1 & & \\ & \ddots & \\ & & c_n \end{bmatrix}, \quad \forall (z,\hat{\theta}) \in \mathrm{I\!R}^{n+p} \tag{4.149}$$

which yields (4.145) with the simple quadratic Lyapunov function (4.136). We observe that, as desired, the system (4.146)–(4.147) has an equilibrium at $(z,\tilde{\theta}) = (0,0)$. The stability properties of this equilibrium will be established in Section 4.2.2.

4.2 SET-POINT REGULATION

Remark 4.10 Skew symmetry is not the only tool we can use for achieving nonpositivity of $\dot{V}_n$. For instance, instead of (4.130) we can use

$$\nu_i(x_1, \ldots, x_i, \hat{\theta}) = -z_i \sum_{k=2}^{i-1} \frac{n-k}{2c_k} \sigma_{k,i}^2, \tag{4.150}$$

which results in

$$A_z(z, \hat{\theta}) = \begin{bmatrix} -c_1 & 1 & 0 & \cdots & 0 \\ -1 & -c_2 & 1 + \sigma_{23} & \cdots & \sigma_{2n} \\ 0 & -1 & -c_3 - \frac{n-2}{2c_2}\sigma_{23}^2 & \ddots & \vdots \\ \vdots & \ddots & \ddots & \ddots & 1 + \sigma_{n-1,n} \\ 0 & \cdots & 0 & -1 & -c_n - \sum_{k=2}^{n-1} \frac{n-k}{2c_k}\sigma_{k,n}^2 \end{bmatrix}. \tag{4.151}$$

It can be shown that with (4.150) we obtain

$$A_z(z, \hat{\theta}) + A_z(z, \hat{\theta})^{\mathrm{T}} \leq -2 \begin{bmatrix} c_1 & & & \\ & \frac{1}{2}c_2 & & \\ & & \ddots & \\ & & & \frac{1}{2}c_n \end{bmatrix}, \quad \forall (z, \hat{\theta}) \in \mathrm{I\!R}^{n+p} \tag{4.152}$$

which yields

$$\dot{V}_n \leq -\frac{1}{2} \sum_{k=1}^{n} c_k z_k^2. \tag{4.153}$$

We can carry this idea a step further. By replacing the stabilizing function (4.125) by

$$\alpha_i(x_1, \ldots, x_i, \hat{\theta}) = -z_{i-1} - c_i z_i + \sum_{k=1}^{i-1} \frac{\partial \alpha_{i-1}}{\partial x_k} x_{k+1} - w_i^{\mathrm{T}} \hat{\theta}$$
$$- \sum_{k=2}^{i} (n-k+1) \left(\frac{\sigma_{k,i}^2}{c_k} + \frac{\sigma_{i,k-1}^2}{c_{k-1}} \right) z_i, \tag{4.154}$$

we arrive at the error system (4.146) with

$$A_z(z, \hat{\theta}) = \begin{bmatrix} -c_1 & 1 & & \\ -1 & \ddots & & \\ & \ddots & 1 & \\ & & -1 & -c_n \end{bmatrix} + \begin{bmatrix} 0 & \cdots & 0 \\ \sigma_{2,1} & \cdots & \sigma_{2,n} \\ \vdots & & \vdots \\ \sigma_{n,1} & \cdots & \sigma_{n,n} \end{bmatrix}$$
$$+ \begin{bmatrix} 0 & & \\ & \ddots & \\ & & -\sum_{k=2}^{i}(n-k+1)\left(\frac{\sigma_{k,i}^2}{c_k} + \frac{\sigma_{i,k-1}^2}{c_{k-1}}\right) \\ & & & \ddots \end{bmatrix}. \tag{4.155}$$

It can be shown that A_z satisfies

$$A_z(z,\hat{\theta}) + A_z(z,\hat{\theta})^{\mathrm{T}} \leq -2 \begin{bmatrix} \frac{3}{4}c_1 & & & \\ & \frac{1}{2}c_2 & & \\ & & \ddots & \\ & & & \frac{1}{2}c_n \end{bmatrix}, \quad \forall (z,\hat{\theta}) \in \mathbb{R}^{n+p} \quad (4.156)$$

which yields (4.153). Instead of cancelling the tuning function τ_i, the stabilizing function α_i in (4.154) dominates it. We consider (4.130) preferable over (4.150) and (4.154) because the latter two lead to a much faster growth of nonlinearities in the control law. ◇

Example 4.11 In applications of the tuning functions procedure we do not need to repeat the Lyapunov argument. All we need for a specific design are the final analytical expressions provided by the procedure. Let us now illustrate this by designing an adaptive controller for the benchmark system from Example 4.8:

$$\begin{aligned} \dot{x}_1 &= x_2 + \varphi(x_1)^{\mathrm{T}}\theta \\ \dot{x}_2 &= x_3 \\ \dot{x}_3 &= u \,. \end{aligned} \quad (4.157)$$

The design objective is the regulation of the output $y = x_1$ to the set-point y_s. The first three expressions provided by the procedure are the definitions (4.87), (4.88), and (4.97) of the error variables

$$\begin{aligned} z_1 &= x_1 - y_s \\ z_2 &= x_2 - \alpha_1(x_1, \hat{\theta}) \\ z_3 &= x_3 - \alpha_2(x_1, x_2, \hat{\theta}), \end{aligned} \quad (4.158)$$

where α_1 and α_2 are the stabilizing functions given by (4.94) and (4.103):

$$\begin{aligned} \alpha_1 &= -c_1 z_1 - \varphi^{\mathrm{T}}\hat{\theta} \\ \alpha_2 &= -c_2 z_2 - z_1 + \frac{\partial \alpha_1}{\partial x_1}(x_2 + \varphi^{\mathrm{T}}\hat{\theta}) + \frac{\partial \alpha_1}{\partial \hat{\theta}}\tau_2\,. \end{aligned} \quad (4.159)$$

The tuning functions, determined from (4.93), (4.102), and (4.111), are

$$\begin{aligned} \tau_1 &= z_1 \varphi \\ \tau_2 &= \tau_1 - z_2 \frac{\partial \alpha_1}{\partial x_1}\varphi \\ \tau_3 &= \tau_2 - z_3 \frac{\partial \alpha_2}{\partial x_1}\varphi\,. \end{aligned} \quad (4.160)$$

With the above expressions and the choice $\Gamma = I$, the parameter update law and the feedback control are obtained from (4.138) and (4.112), respectively. They are

$$\dot{\hat{\theta}} = \tau_3 = z_1 \varphi - z_2 \frac{\partial \alpha_1}{\partial x_1}\varphi - z_3 \frac{\partial \alpha_2}{\partial x_1}\varphi \quad (4.161)$$

$$u = -c_3 z_3 - z_2 + \frac{\partial \alpha_2}{\partial x_1}(x_2 + \varphi^{\mathrm{T}}\hat{\theta}) + \frac{\partial \alpha_2}{\partial x_2}x_3 + \frac{\partial \alpha_2}{\partial \hat{\theta}}\tau_3 - z_2 \frac{\partial \alpha_1}{\partial \hat{\theta}}\frac{\partial \alpha_2}{\partial x_1}\varphi\,. \quad (4.162)$$

4.2 SET-POINT REGULATION

This completes the design of the adaptive controller for (4.157). In the $(z,\tilde{\theta})$-coordinates the designed system is

$$\dot{z} = \begin{bmatrix} -c_1 & 1 & 0 \\ -1 & -c_2 & 1 - \frac{\partial \alpha_2}{\partial x_1}|\varphi|^2 \\ 0 & -1 + \frac{\partial \alpha_2}{\partial x_1}|\varphi|^2 & -c_3 \end{bmatrix} z + \begin{bmatrix} 1 \\ -\frac{\partial \alpha_1}{\partial x_1} \\ -\frac{\partial \alpha_2}{\partial x_1} \end{bmatrix} \varphi^\mathrm{T}\tilde{\theta} \quad (4.163)$$

$$\dot{\tilde{\theta}} = -\varphi\left[1, -\frac{\partial \alpha_1}{\partial x_1}, -\frac{\partial \alpha_2}{\partial x_1}\right] z. \quad (4.164)$$

It is of interest to relate the stabilizing functions α_1 and α_2 and the control law u to the material from Section 4.1. The stabilizing function α_1 has a "certainty equivalence" form. The stabilizing function α_2 has the term $\frac{\partial \alpha_1}{\partial \hat{\theta}} \tau_2$ which accounts for parameter estimation transients, while the rest of it is in the "certainty equivalence" form. The control law u departs from the "certainty equivalence" form in its last two terms whose role is the same as that of (4.66). The last term in u is particularly important. Since $\frac{\partial \alpha_1}{\partial \hat{\theta}} = -\varphi^\mathrm{T}$, this term contributes with $+\frac{\partial \alpha_2}{\partial x_1}|\varphi|^2$ in the 'system matrix' in (4.163) and achieves the skew symmetry, which is crucial for stability. $\diamond$

4.2.2 Stability and convergence

To investigate stability properties of the closed-loop adaptive system (4.146)–(4.147), we express φ_i, α_i, τ_i, and w_i in the z-coordinates. Then, by Theorem A.5, the global stability of the equilibrium $(z,\tilde{\theta}) = 0$ follows from the fact that the derivative $\dot{V}_n$ of V_n along the solutions of (4.146)-(4.147) is given by (4.145).

From LaSalle's Invariance Theorem (Theorem 2.2), it further follows that the $((n+p)$-dimensional) state $(z(t),\tilde{\theta}(t))$ converges to the largest invariant set M of (4.146)–(4.147) contained in $E = \{(z,\tilde{\theta}) \in \mathbb{R}^{n+p} \mid z = 0\}$, that is, in the set where $\dot{V}_n = 0$. This means, in particular, that $z(t) \to 0$ as $t \to 0$.

We now set out to determine M. On this invariant set, we have $z \equiv 0$ and $\dot{z} \equiv 0$. Setting $z = 0$, $\dot{z} = 0$ in (4.146) we obtain $\dot{\hat{\theta}} = 0$ and

$$W(z,\hat{\theta})^\mathrm{T}(\theta - \hat{\theta}) = 0, \quad \forall (z,\hat{\theta}) \in M. \quad (4.165)$$

From (4.121) and (4.139) it is easily seen that

$$W(z,\hat{\theta})^\mathrm{T} = \begin{bmatrix} 1 & 0 & \cdots & 0 \\ -\frac{\partial \alpha_1}{\partial x_1} & 1 & \ddots & \vdots \\ \vdots & \ddots & \ddots & 0 \\ -\frac{\partial \alpha_{n-1}}{\partial x_1} & \cdots & -\frac{\partial \alpha_{n-1}}{\partial x_{n-1}} & 1 \end{bmatrix} F(x)^\mathrm{T} \triangleq N(z,\hat{\theta})F(x)^\mathrm{T}. \quad (4.166)$$

Since $N(z,\hat{\theta})$ is obviously nonsingular for all $(z,\hat{\theta}) \in M$, then (4.165) and (4.166) imply

$$F(x)^\mathrm{T}(\theta - \hat{\theta}) = 0 \quad \text{on } M. \quad (4.167)$$

Now we show that $x = x^e$ on M. Since $z_1 = x_1 - y_s$ then $x_1 = y_s = x_1^e$ on M. In view of (4.167), we get

$$(\theta - \hat{\theta})^T \varphi_1(x_1^e) = 0 \text{ on } M. \tag{4.168}$$

Recall from (4.94) that $\alpha_1 = -c_1 z_1 - \hat{\theta}^T \varphi_1$. Therefore, on M we have $\alpha_1 = -\hat{\theta}^T \varphi_1(x_1^e) = -\theta^T \varphi_1(x_1^e)$. Combining this with $z_2 = 0 = x_2 - \alpha_1$ and (4.86), we get $x_2 = x_2^e$ on M. Using (4.167), we obtain

$$(\theta - \hat{\theta})^T \varphi_2(x_1^e, x_2^e) = 0 \text{ on } M. \tag{4.169}$$

Continuing in the same fashion, we prove that $x_i = x_i^e$ and $(\theta - \hat{\theta})^T \varphi_i(x_1^e, \ldots, x_i^e) = 0$ on M, $i = 1, \ldots, n$.

Thus, the largest invariant set M in E is

$$\begin{aligned}M &= \left\{ (z, \tilde{\theta}) \in \mathbb{R}^{n+p} \mid z = 0,\ F_e^T \tilde{\theta} = 0 \right\} \\ &= \left\{ (x, \hat{\theta}) \in \mathbb{R}^{n+p} \mid x = x^e,\ F_e^T \hat{\theta} = F_e^T \theta \right\},\end{aligned} \tag{4.170}$$

where $F_e = F(x^e)$. The two equivalent expressions for M and the convergence of $(z(t), \tilde{\theta}(t))$ to M prove that $x(t) \to x^e$ as $t \to \infty$.

An important property of M is its dimension, $p - \text{rank}\{F_e\}$. When $\text{rank}\{F_e\} = p$, then $\dim M = 0$, that is, M becomes the equilibrium point $x = x^e$, $\hat{\theta} = \theta$. This means that the parameter estimates converge to their true values, so that the equilibrium $x = x^e$, $\hat{\theta} = \theta$ is globally *asymptotically* stable.

The above facts prove the following result:

Theorem 4.12 *The closed-loop adaptive system consisting of the plant (4.84), the controller (4.140), and the update law (4.138) has a globally stable equilibrium* $(x, \hat{\theta}) = (x^e, \theta)$. *Furthermore, its state* $\left(x(t), \hat{\theta}(t)\right)$ *converges to the* $(p - \text{rank}\{F_e\})$-*dimensional equilibrium manifold* M *given by (4.170), which means, in particular, that*

$$\lim_{t \to \infty} x(t) = x^e. \tag{4.171}$$

If $y_s = 0$ *and* $F(0) = 0$, *then* $\lim_{t \to \infty} x(t) = 0$. *The equilibrium* $x = x^e, \hat{\theta} = \theta$ *is globally asymptotically stable if and only if* $\text{rank}\{F_e\} = p$.

As the dimension of M reduces, the stability properties of the adaptive system improve. The most desirable case is when M is an equilibrium point, in which case this equilibrium is globally asymptotically stable, and the parameter estimates converge to the actual parameter values. Global asymptotic stability can be achieved with as many as $p = n$ unknown parameters. This is among the main advantages of eliminating overparametrization.

We now discuss the basic stability properties established in Theorem 4.12 on a simple example.

4.2 SET-POINT REGULATION

Example 4.13 We consider the second order system with an unknown parameter vector $\theta \in \mathbb{R}^p$:

$$\begin{aligned} \dot{x}_1 &= x_2 + \varphi_1(x_1)^\mathrm{T}\theta \\ \dot{x}_2 &= u + \varphi_2(x)^\mathrm{T}\theta. \end{aligned} \quad (4.172)$$

The control objective is to regulate x to zero ($x_1^e = 0$). We define the error variables

$$\begin{aligned} z_1 &= x_1 \\ z_2 &= x_2 - \alpha_1(x_1, \hat{\theta}). \end{aligned} \quad (4.173)$$

The controller is designed applying (4.94) and (4.103) as

$$\begin{aligned} \alpha_1 &= -c_1 z_1 - \varphi_1(x_1)^\mathrm{T}\hat{\theta} \\ \tfrac{\partial \alpha_1}{\partial x_1} &= -c_1 - \tfrac{\partial \varphi_1(x_1)^\mathrm{T}}{\partial x_1}\hat{\theta}, \qquad \tfrac{\partial \alpha_1}{\partial \hat{\theta}} = -\varphi_1(x_1)^\mathrm{T} \\ u &= -z_1 - c_2 z_2 + \tfrac{\partial \alpha_1}{\partial x_1} x_2 - \left(\varphi_2(x)^\mathrm{T} - \tfrac{\partial \alpha_1}{\partial x_1}\varphi_1(x_1)^\mathrm{T}\right)\hat{\theta} + \tfrac{\partial \alpha_1}{\partial \hat{\theta}}\dot{\hat{\theta}}, \end{aligned} \quad (4.174)$$

while the parameter update law is

$$\dot{\hat{\theta}} = \Gamma \left[\varphi_1, \ \varphi_2 - \tfrac{\partial \alpha_1}{\partial x_1}\varphi_1\right] z. \quad (4.175)$$

The resulting error system is

$$\dot{z} = \begin{bmatrix} -c_1 & 1 \\ -1 & -c_2 \end{bmatrix} z + \begin{bmatrix} \varphi_1^\mathrm{T} \\ \varphi_2^\mathrm{T} - \tfrac{\partial \alpha_1}{\partial x_1}\varphi_1^\mathrm{T} \end{bmatrix} \tilde{\theta}. \quad (4.176)$$

Now we illustrate and discuss the stability properties established by Theorem 4.12. From (4.172) we see that $x_1^e = 0$, $x_2^e = -\varphi_1(0)^\mathrm{T}\theta$. By Theorem 4.12, the point

$$\begin{bmatrix} x_1 \\ x_2 \\ \hat{\theta} \end{bmatrix} = \begin{bmatrix} 0 \\ -\varphi_1(0)^\mathrm{T}\theta \\ \theta \end{bmatrix} \quad (4.177)$$

is a globally stable equilibrium, and the state of the closed-loop system converges to the equilibrium manifold

$$M = \left\{ (x, \hat{\theta}) \in \mathbb{R}^{2+p} \ \bigg| \ \begin{bmatrix} x_1 \\ x_2 \end{bmatrix} = \begin{bmatrix} 0 \\ -\varphi_1(0)^\mathrm{T}\hat{\theta} \end{bmatrix}, \ \begin{bmatrix} \varphi_1(0)^\mathrm{T} \\ \varphi_2(0, -\varphi_1(0)^\mathrm{T}\hat{\theta})^\mathrm{T} \end{bmatrix}(\theta - \hat{\theta}) = 0 \right\}. \quad (4.178)$$

A basic question that one would ask is: What type of figure in $\mathbb{R}^{2+p}$ is M? Our further discussion will, without loss of generality, be limited to $p \leq 2$.

In the simplest case where $\dim \theta = p = 1$, the following two possibilities exist:

- If both $\varphi_1(0) = 0$ and $\varphi_2(0,0) = 0$, then the manifold M is the subspace $x = 0$ in $\mathbb{R}^3$, that is, M is the $\hat{\theta}$-axis.

- If either $\varphi_1(0) \neq 0$ or $\varphi_2(0, -\varphi_1(0)\theta) \neq 0$, then the manifold M is the single point $x_1 = 0$, $x_2 = -\varphi_1(0)\theta$, $\hat{\theta} = \theta$. This point is an equilibrium which is not only globally stable, but also globally *asymptotically* stable.

Next, we analyze the case $p = 2$.

- Suppose $\begin{bmatrix} \varphi_1(x_1)^{\mathrm{T}} \\ \varphi_2(x_1)^{\mathrm{T}} \end{bmatrix} = \begin{bmatrix} x_1^2 & e^{x_1} \\ \cos x_1 & 0 \end{bmatrix}$. Since $\begin{bmatrix} \varphi_1(0)^{\mathrm{T}} \\ \varphi_2(0)^{\mathrm{T}} \end{bmatrix} = \begin{bmatrix} 0 & 1 \\ 1 & 0 \end{bmatrix}$ has full rank, the manifold M is the single point $x_1 = 0$, $x_2 = -\theta_2$, $\hat{\theta}_1 = \theta_1$, $\hat{\theta}_2 = \theta_2$, which is a globally *asymptotically* stable equilibrium.

- Suppose $\begin{bmatrix} \varphi_1(x_1)^{\mathrm{T}} \\ \varphi_2(x_1)^{\mathrm{T}} \end{bmatrix} = \begin{bmatrix} -\cos x_1 & e^{x_1} \\ \sin x_1 & 0 \end{bmatrix}$. Since $\begin{bmatrix} \varphi_1(0)^{\mathrm{T}} \\ \varphi_2(0)^{\mathrm{T}} \end{bmatrix} = \begin{bmatrix} -1 & 1 \\ 0 & 0 \end{bmatrix}$, the manifold M is the linear variety $x_1 = 0$, $x_2 = \theta_1 - \theta_2$, $\hat{\theta}_2 - \hat{\theta}_1 = \theta_2 - \theta_1$. Neither of the parameter estimates is guaranteed to converge to the actual parameter value, but they are jointly converging to the line $\hat{\theta}_2 = \hat{\theta}_1 + \theta_2 - \theta_1$ in the plane $x_1 = 0$, $x_2 = \theta_1 - \theta_2$.

- Suppose $\begin{bmatrix} \varphi_1(x_1)^{\mathrm{T}} \\ \varphi_2(x_1)^{\mathrm{T}} \end{bmatrix} = \begin{bmatrix} x_1^2 & e^{x_1} - 1 \\ \sin x_1 & 0 \end{bmatrix}$. Since $\begin{bmatrix} \varphi_1(0)^{\mathrm{T}} \\ \varphi_2(0)^{\mathrm{T}} \end{bmatrix} = \begin{bmatrix} 0 & 0 \\ 0 & 0 \end{bmatrix}$, the manifold M is the plane (linear variety) $x = 0$. This is the case of the weakest convergence properties because one cannot guarantee that the parameter estimates converge to any submanifold in the plane M. ◇

4.2.3 Passivity

The closed-loop adaptive system (4.146)–(4.147) has an important passivity property. Rewriting this system in the following $(z, \tilde{\theta})$-form

$$\dot{z} = A_z(z, \hat{\theta})z + W(z, \hat{\theta})^{\mathrm{T}} \tilde{\theta} \qquad (4.179)$$

$$\dot{\tilde{\theta}} = -\Gamma W(z, \hat{\theta})z, \qquad (4.180)$$

we see that due to the structure of $A_z(z, \hat{\theta})$ in (4.148), along the solutions of (4.179) we have

$$\frac{d}{dt}\left(\frac{1}{2}z^{\mathrm{T}}z\right) = -\sum_{i=1}^{n} c_i z_i^2 + z^{\mathrm{T}} W^{\mathrm{T}} \tilde{\theta}$$

$$= -\sum_{i=1}^{n} c_i z_i^2 + \tau_n^{\mathrm{T}} \tilde{\theta}. \qquad (4.181)$$

4.2 SET-POINT REGULATION

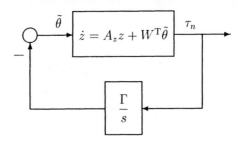

Figure 4.1: Negative feedback connection of the strictly passive z-system and the passive system $\frac{\Gamma}{s}$.

By integrating (4.181) over $[0, t]$, we get

$$\int_0^t \tau_n(s)^T \tilde{\theta}(s) ds = \frac{1}{2} z(t)^T z(t) - \frac{1}{2} z(0)^T z(0) + \int_0^t \sum_{i=1}^n c_i z_i(s)^2 ds. \quad (4.182)$$

By Definition D.2, (4.182) implies that the system

$$\begin{aligned} \dot{z} &= A_z(z, \hat{\theta})z + W(z, \hat{\theta})^T \tilde{\theta} \\ \tau_n &= W(z, \hat{\theta})z \end{aligned} \quad (4.183)$$

is strictly passive with $\tilde{\theta}$ as its input, τ_n as its output, $V(z) = \frac{1}{2} z^T z$ as the storage function, and $\psi(z) = \sum_{i=1}^n c_i z_i^2$ as the dissipation rate. On the other hand, the integrator system

$$-\dot{\tilde{\theta}} = \frac{\Gamma}{s} \tau_n \quad (4.184)$$

is passive from τ_n to $-\dot{\tilde{\theta}}$. Hence, the closed-loop adaptive system represents a negative feedback connection of the strictly passive system (4.183) with the passive system (4.184), as shown in Figure 4.1.

By Theorem D.4, the equilibrium $(z, \tilde{\theta}) = (0, 0)$ of the feedback system in Figure 4.1 is globally stable, and $z(t) \to 0$ as $t \to 0$.

An interpretation of the tuning functions design procedure is that at each step i, a tuning function τ_i is chosen as an output of the $(z_1, \ldots, z_i)$-subsystem, and a control law α_i is designed to make this subsystem strictly passive. A schematic representation of the recursive procedure is given in Figure 4.2. The design is completed at the nth step, where the last tuning function τ_n is used to close the adaptive feedback loop via the passive parameter update law $\dot{\hat{\theta}} = \Gamma \tau_n$.

As an illustration of the use of passivity at each step of the design procedure Figure 4.3 shows the adaptive feedback connection at step $i = 2$. The property that the output matrix is the transpose of the input matrix, a distinctive feature of passive systems, is clearly displayed for the upper system.

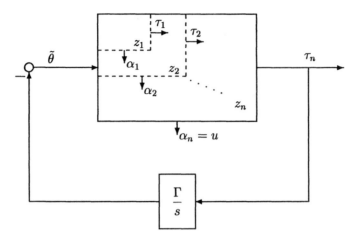

Figure 4.2: A schematic representation of the design procedure.

4.3 Tracking

The set-point regulation design is readily extended to the task of tracking.

The control objective is to force the output $y = x_1$ of the system

$$\begin{aligned}
\dot{x}_1 &= x_2 + \varphi_1(x_1)^{\mathrm{T}}\theta \\
\dot{x}_2 &= x_3 + \varphi_2(x_1, x_2)^{\mathrm{T}}\theta \\
&\vdots \\
\dot{x}_{n-1} &= x_n + \varphi_{n-1}(x_1, \ldots, x_{n-1})^{\mathrm{T}}\theta \\
\dot{x}_n &= \beta(x)u + \varphi_n(x)^{\mathrm{T}}\theta
\end{aligned} \quad (4.185)$$

to asymptotically track the reference output $y_r(t)$ whose first n derivatives are assumed to be known, bounded, and piecewise continuous.

An alternative control objective would be to asymptotically track the output of a known asymptotically stable linear reference model

$$y_r = G_m(s)r(s) = \frac{k_m}{s^n + m_{n-1}z^{n-1} + \cdots + m_0}r(s), \quad (4.186)$$

where $s^n + m_{n-1}z^{n-1} + \cdots + m_0$ is Hurwitz, $k_m > 0$, and $r(t)$ is bounded and piecewise continuous. A realization of (4.186) which is of particular interest is

$$\begin{aligned}
\dot{x}_m &= \begin{bmatrix} 0 & & \\ \vdots & I_{n-1} & \\ 0 & & \\ -m_0 & \cdots & -m_{n-1} \end{bmatrix} x_m + \begin{bmatrix} 0 \\ \vdots \\ 0 \\ k_m \end{bmatrix} r \\
y_r &= x_{m,1}
\end{aligned} \quad (4.187)$$

because, in this case, the derivatives of y_r are available as the states of the reference model: $y_r^{(i)} = x_{m,i+1}$, $i = 0, \ldots, n-1$.

4.3 TRACKING

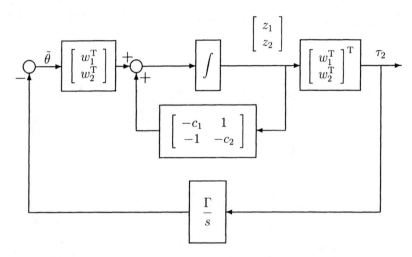

Figure 4.3: The feedback connection of the strictly passive (z_1, z_2)-system with a passive update law.

4.3.1 Design procedure

The design for tracking is only a minor modification of the set-point design procedure. As before, the first z-variable is the tracking error, $z_1 = x_1 - y_r$. However, because the reference signal $y_r(t)$ is not constant, its derivative $y_r^{(i-1)}(t)$ appears in the definition of the i-th error state z_i, $i = 1, \ldots, n$. The only change this creates in the design is the addition of the sum $\sum_{k=1}^{i-1} \frac{\partial \alpha_{i-1}}{\partial y_r^{(k-1)}} y_r^{(k)}$ in the definition of α_i.

As we showed in Section 3.4.3, *nonlinear damping* can be used to guarantee global boundedness in the absence of adaptation, as well as to enhance performance. Therefore, the general design also incorporates the nonlinear damping terms

$$- \kappa_i |w_i|^2 z_i, \quad \kappa_i \geq 0 \qquad (4.188)$$

in the definition of α_i's.

It is sufficient that we now only give a complete set of recursive expressions for the stabilizing functions α_i and the tuning functions τ_i leading to the final adaptive control law u and the final update law for $\hat{\theta}$. These expressions, organized in Table 4.1, give a succinct summary of the tuning functions design.[6]

It can be checked that the resulting error system has the following form similar to the set-point regulation case:

$$\dot{z} = A_z(z, \hat{\theta}, t) z + W(z, \hat{\theta}, t)^T \tilde{\theta}, \quad z \in \mathbb{R}^n \qquad (4.195)$$

---

[6]For notational convenience we define $z_0 \triangleq 0$, $\alpha_0 \triangleq 0$, $\tau_0 \triangleq 0$.

Table 4.1: Tuning Functions Design for Tracking

$$z_i = x_i - y_r^{(i-1)} - \alpha_{i-1} \quad (4.189)$$

$$\alpha_i(\bar{x}_i, \hat{\theta}, \bar{y}_r^{(i-1)}) = -z_{i-1} - c_i z_i - w_i^T \hat{\theta} + \sum_{k=1}^{i-1} \left(\frac{\partial \alpha_{i-1}}{\partial x_k} x_{k+1} + \frac{\partial \alpha_{i-1}}{\partial y_r^{(k-1)}} y_r^{(k)} \right)$$

$$-\kappa_i |w_i|^2 z_i + \frac{\partial \alpha_{i-1}}{\partial \hat{\theta}} \Gamma \tau_i + \sum_{k=2}^{i-1} \frac{\partial \alpha_{k-1}}{\partial \hat{\theta}} \Gamma w_i z_k \quad (4.190)$$

$$\tau_i(\bar{x}_i, \hat{\theta}, \bar{y}_r^{(i-1)}) = \tau_{i-1} + w_i z_i \quad (4.191)$$

$$w_i(\bar{x}_i, \hat{\theta}, \bar{y}_r^{(i-2)}) = \varphi_i - \sum_{k=1}^{i-1} \frac{\partial \alpha_{i-1}}{\partial x_k} \varphi_k \quad (4.192)$$

$$i = 1, \ldots, n$$

$$\bar{x}_i = (x_1, \ldots, x_i), \qquad \bar{y}_r^{(i)} = (y_r, \dot{y}_r, \ldots, y_r^{(i)})$$

Adaptive control law:

$$u = \frac{1}{\beta(x)} \left[\alpha_n(x, \hat{\theta}, \bar{y}_r^{(n-1)}) + y_r^{(n)} \right] \quad (4.193)$$

Parameter update law:

$$\dot{\hat{\theta}} = \Gamma \tau_n(x, \hat{\theta}, \bar{y}_r^{(n-1)}) = \Gamma W z \quad (4.194)$$

where, with the definition $\sigma_{ik} = -\frac{\partial \alpha_{i-1}}{\partial \hat{\theta}} \Gamma w_k$, the matrix $A_z(z, \hat{\theta}, t)$ has the form of (4.148) with the addition of the nonlinear damping terms $-\kappa_i |w_i|^2 z_i$:

$$A_z = \begin{bmatrix} -c_1 - \kappa_1 |w_1|^2 & 1 & 0 & \cdots & 0 \\ -1 & -c_2 - \kappa_2 |w_2|^2 & 1 + \sigma_{23} & \cdots & \sigma_{2n} \\ 0 & -1 - \sigma_{23} & \ddots & \ddots & \vdots \\ \vdots & \vdots & \ddots & \ddots & 1 + \sigma_{n-1,n} \\ 0 & -\sigma_{2n} & \cdots & -1 - \sigma_{n-1,n} & -c_n - \kappa_n |w_n|^2 \end{bmatrix}$$

(4.196)

and $W(z, \hat{\theta}, t)$ has the same form as in (4.139). Although the functions σ_{ik} and w_i may appear to be the same as in the set-point regulation case, this is not so, because now they include $y_r^{(i)}(t)$ through the partial derivatives of α_{i-1}, which is reflected in the dependence of $A_z(z, \hat{\theta}, t)$ and $W(z, \hat{\theta}, t)$ on t.

4.3 TRACKING

The change of coordinates (4.189)–(4.192), which we compactly write as

$$z = \Psi(x, \hat{\theta}, t), \tag{4.197}$$

is smooth in x and $\hat{\theta}$ and bounded in t. Note also that the inverse transformation

$$x = \Phi(z, \hat{\theta}, t) \tag{4.198}$$

is smooth in z and $\hat{\theta}$ and bounded in t.

Theorem 4.14 *The closed-loop adaptive system consisting of the plant (4.185), the controller (4.193), and the update law (4.194). has a globally uniformly stable equilibrium at $(z, \tilde{\theta}) = 0$, and $\lim_{t\to\infty} z(t) = 0$, which means, in particular, that global asymptotic tracking is achieved:*

$$\lim_{t\to\infty} [y(t) - y_r(t)] = 0. \tag{4.199}$$

Moreover, if $\lim_{t\to\infty} y_r^{(i)}(t) = 0$, $i = 0, \ldots, n-1$, and $F(0) = 0$, then $\lim_{t\to\infty} x(t) = 0$.

Proof. Denote $c_0 = \min_{1 \le i \le n} c_i$. The derivative of the Lyapunov function

$$V_n = \frac{1}{2} z^T z + \frac{1}{2} \tilde{\theta}^T \Gamma^{-1} \tilde{\theta} \tag{4.200}$$

along the solutions of (4.195) and (4.194) is

$$\dot{V}_n = -\sum_{k=1}^n c_k z_k^2 - \sum_{i=1}^n \kappa_i |w_i|^2 z_i^2 \le -c_0 |z|^2, \tag{4.201}$$

which proves that the equilibrium $(z, \tilde{\theta}) = 0$ is globally uniformly stable. From the LaSalle-Yoshizawa theorem (Theorem 2.1), it further follows that, as $t \to \infty$, all the solutions converge to the manifold $z = 0$. From the definitions in (4.189)–(4.192) we conclude that, if $\lim_{t\to\infty} y_r^{(i)}(t) = 0$, $i = 0, \ldots, n-1$, and $F(0) = 0$, then $x(t) \to 0$ as $t \to \infty$. □

The proof of Theorem 4.14 reveals the stabilization mechanism employed in the tuning functions design. The update law is chosen so as to make the derivative of the Lyapunov function nonpositive. The update law is fast because it does not use any form of normalization common in traditional certainty equivalence adaptive control. The speed of adaptation is dictated by the speed of the nonlinear behavior captured by the Lyapunov function. The tuning functions controller incorporates the knowledge of the update law and eliminates the disturbing effect of the parameter estimation transients on the error system. The controller and the update law designs are interlaced.

The next theorem shows that in the absence of adaptation the nonlinear damping terms guarantee boundedness. In addition, global asymptotic stability is achieved for sufficiently small parameter error $\tilde{\theta}$.

Theorem 4.15 (Boundedness and Stability Without Adaptation)
Consider the closed-loop adaptive system consisting of the plant (4.185), the controller (4.193), and the update law (4.194), with $\Gamma = 0$ and $\kappa_i > 0$, $i = 1, \ldots, n$. All the solutions are globally uniformly bounded. Furthermore, if $F(0) = 0$ and $y_{\mathrm{r}}(t) \equiv 0$, then there exist $R_\theta > 0$ such that for each constant $\hat{\theta} \in \mathbb{R}^p$, $|\theta - \hat{\theta}| \leq R_\theta$, the equilibrium $x = 0$ is globally asymptotically stable.

Proof. Since $\Gamma = 0$, the parameter estimate $\hat{\theta}$ is constant. Denote $\kappa_0 = \left(\sum_{i=1}^{n} \frac{1}{\kappa_i}\right)^{-1}$. For the nonadaptive system (4.195) we have

$$\begin{aligned}
\frac{d}{dt}\left(\frac{1}{2}|z|^2\right) &= -\sum_{i=1}^{n} c_i z_i^2 - \sum_{i=1}^{n} \kappa_i |w_i|^2 z_i^2 + \sum_{i=1}^{n} z_i w_i^{\mathrm{T}} \tilde{\theta} \\
&\leq -c_0 |z|^2 - \sum_{i=1}^{n} \kappa_i \left|w_i z_i - \frac{1}{2\kappa_i}\tilde{\theta}\right|^2 + \left(\sum_{i=1}^{n} \frac{1}{4\kappa_i}\right)|\tilde{\theta}|^2 \\
&\leq -c_0 |z|^2 + \frac{1}{4\kappa_0}|\tilde{\theta}|^2 .
\end{aligned} \qquad (4.202)$$

From Lemma C.5*(i)*, taking $v = z^2$ and $\rho = \frac{1}{\sqrt{\kappa_0}}|\tilde{\theta}|$, it follows that

$$|z(t)| \leq |z(0)| e^{-c_0 t} + \frac{1}{2\sqrt{c_0 \kappa_0}}|\tilde{\theta}| . \qquad (4.203)$$

This proves that the solution $z(t)$ is globally uniformly bounded and, since $x = \Phi(z, \hat{\theta}, t)$ is smooth in z and $\hat{\theta}$ and bounded in t, this also proves that $x(t)$ is globally uniformly bounded.

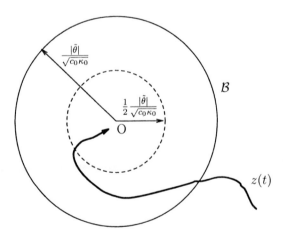

Figure 4.4: The solution enters the larger ball (solid) in finite time. Then, the equilibrium $z = 0$, which is exponentially stable for sufficiently small $\tilde{\theta}$, takes over and attracts the solution.

4.3 Tracking

Now, suppose that $F(0) = 0$ and $y_r(t) \equiv 0$. Since $x = 0$ whenever $z = 0$, in view of (4.166), we have $W(0, \hat{\theta}, t) \equiv 0$. Thus $z = 0$ is an equilibrium of (4.195). To prove asymptotic stability of $z = 0$, we only need to consider the case $\tilde{\theta} \neq 0$. The situation is depicted in Figure 4.4. From (4.203) we know that $z(t)$ converges exponentially to the ball of radius $\frac{1}{2}\frac{|\tilde{\theta}|}{\sqrt{c_0 \kappa_0}}$ around $z = 0$. This means that $z(t)$ enters a larger ball, say the ball $\mathcal{B}$ of radius $\frac{|\tilde{\theta}|}{\sqrt{c_0 \kappa_0}}$, in finite time $t \leq T = \max\left\{0, \frac{1}{c_0} \ln \frac{2\sqrt{c_0 \kappa_0}|z(0)|}{|\tilde{\theta}|}\right\}$. Note that before the solution enters $\mathcal{B}$ it decays exponentially:

$$|z(t)| \leq 2|z(0)|e^{-c_0 t}, \qquad t \leq T. \tag{4.204}$$

We now examine the trajectories inside $\mathcal{B}$. Because $W(z, \hat{\theta}, t)$ is locally Lipschitz and vanishes at $z = 0$, there exists a finite positive number L such that for all $z \in \mathcal{B}$ we have $|W(z, \hat{\theta}, t)| \leq L|z|$, and therefore

$$\begin{aligned}\frac{d}{dt}\left(\frac{1}{2}|z|^2\right) &\leq -c_0|z|^2 + z^T W(z, \hat{\theta}, t)^T \tilde{\theta} \\ &\leq -\left(c_0 - L|\tilde{\theta}|\right)|z|^2, \qquad t \geq T.\end{aligned} \tag{4.205}$$

The equilibrium $z = 0$ is (locally) exponentially stable provided $|\tilde{\theta}| < \frac{c_0}{L}$. For $z(0) \in \mathcal{B}$ we have $T = 0$ and

$$|z(t)| \leq |z(0)|e^{-(c_0 - L|\tilde{\theta}|)t}. \tag{4.206}$$

When $z(0) \notin \mathcal{B}$ then $T > 0$ and, for $t \geq T$ we have

$$\begin{aligned}|z(t)| &\leq \frac{|\tilde{\theta}|}{\sqrt{c_0 \kappa_0}} e^{-(c_0 - L|\tilde{\theta}|)(t-T)} \\ &= \frac{|\tilde{\theta}|}{\sqrt{c_0 \kappa_0}} e^{\frac{c_0 - L|\tilde{\theta}|}{c_0} \ln \frac{2\sqrt{c_0 \kappa_0}|z(0)|}{|\tilde{\theta}|}} e^{-(c_0 - L|\tilde{\theta}|)t} \\ &= \frac{|\tilde{\theta}|}{\sqrt{c_0 \kappa_0}} \left(\frac{2\sqrt{c_0 \kappa_0}|z(0)|}{|\tilde{\theta}|}\right)^{\frac{c_0 - L|\tilde{\theta}|}{c_0}} e^{-(c_0 - L|\tilde{\theta}|)t}.\end{aligned} \tag{4.207}$$

Since $z(0) \notin \mathcal{B}$, it follows that $\frac{2\sqrt{c_0 \kappa_0}|z(0)|}{|\tilde{\theta}|} > 1$, which because of $0 < \frac{c_0 - L|\tilde{\theta}|}{c_0} < 1$ yields

$$\begin{aligned}|z(t)| &\leq \frac{|\tilde{\theta}|}{\sqrt{c_0 \kappa_0}} \frac{2\sqrt{c_0 \kappa_0}|z(0)|}{|\tilde{\theta}|} e^{-(c_0 - L|\tilde{\theta}|)t} \\ &= 2|z(0)|e^{-(c_0 - L|\tilde{\theta}|)t}.\end{aligned} \tag{4.208}$$

Combining this with (4.204), we get

$$\begin{aligned}|z(t)| &\leq \begin{cases} 2|z(0)|e^{-c_0 t} & t \leq T \\ 2|z(0)|e^{-(c_0 - L|\tilde{\theta}|)t} & t \geq T \end{cases} \\ &\leq 2|z(0)|e^{-(c_0 - L|\tilde{\theta}|)t}, \qquad \forall t \geq 0.\end{aligned} \tag{4.209}$$

This proves that for each $\hat{\theta} \in \mathbb{R}^p$ such that

$$|\theta - \hat{\theta}| < \frac{c_0}{L} \triangleq R_\theta, \qquad (4.210)$$

the equilibrium $z = 0$ is globally exponentially stable, so the equilibrium $x = 0$ is globally asymptotically stable. □

4.3.2 Trajectory initialization

One of the goals of adaptation is to reduce uncertainty, that is, to make $\tilde{\theta}(t)$ smaller. From the proof of Theorem 4.14 we know that $\tilde{\theta}(t)^T \Gamma^{-1} \tilde{\theta}(t) \leq 2V_n(t) \leq 2V_n(0)$, that is,

$$\tilde{\theta}(t)^T \Gamma^{-1} \tilde{\theta}(t) \leq z(0)^T z(0) + \tilde{\theta}(0)^T \Gamma^{-1} \tilde{\theta}(0). \qquad (4.211)$$

This bound shows that a possibility for reducing $\tilde{\theta}(t)$ lies in $z(0)$. We will now explain how $z(0)$ can be set to zero by an appropriate initialization of the reference trajectory. As we shall see in the next section, an even more important benefit of setting $z(0) = 0$ is in reducing a bound on $z(t)$, that is, in improving the adaptive system's transient performance.

Let us consider the change of coordinates (4.189) which defines z in terms of y_r and its derivatives:

$$\begin{aligned}
z_1 &= x_1 - y_r \\
z_2 &= x_2 - \dot{y}_r - \alpha_1(x_1, \hat{\theta}, y_r) \\
z_3 &= x_3 - \ddot{y}_r - \alpha_2(x_1, x_2, \hat{\theta}, y_r, \dot{y}_r) \\
&\vdots \\
z_n &= x_n - y_r^{(n-1)} - \alpha_{n-1}\left(x_1, \ldots, x_{n-1}, \hat{\theta}, y_r, \ldots, y_r^{(n-2)}\right).
\end{aligned} \qquad (4.212)$$

Clearly, in order to set $z(0) = 0$, the reference trajectory $y_r(t)$ has to satisfy the initial conditions

$$\begin{aligned}
y_r(0) &= x_1(0) \\
\dot{y}_r(0) &= x_2(0) - \alpha_1(x_1(0), \hat{\theta}(0), y_r(0)) \\
\ddot{y}_r(0) &= x_3(0) - \alpha_2(x_1(0), x_2(0), \hat{\theta}(0), y_r(0), \dot{y}_r(0)) \\
&\vdots \\
y_r^{(n-1)}(0) &= x_n(0) - \alpha_{n-1}\left(x_1(0), \ldots, x_{n-1}(0), \hat{\theta}(0), y_r(0), \ldots, y_r^{(n-2)}(0)\right).
\end{aligned} \qquad (4.213)$$

It is true that $y_r(0), \dot{y}_r(0), \ldots, y_r^{(n-1)}(0)$ are prescribed by $y_r(t)$ and appear not to be free for us to chose. However, it is also true that the tracking objective $y(t) = x_1(t) = y_r(t)$ cannot be achieved instantaneously, but only *asymptotically* as $t \to \infty$. We are therefore allowed to modify a prescribed

4.3 TRACKING

reference signal $y_{r0}(t)$ by an additional signal $\delta_r(t)$ which vanishes as $t \to \infty$. So our system will be designed to track

$$y_r(t) = y_{r0}(t) + \delta_r(t), \qquad (4.214)$$

and the asymptotic tracking of $y_{r0}(t)$ will still be achieved because $\delta_r(t) \to 0$. By selecting $\delta_r(0), \dot{\delta}_r(0), \ldots, \delta_r^{(n-1)}(0)$, we are free to choose $y_r(0), \dot{y}_r(0), \ldots, y_r^{(n-1)}(0)$ as desired,

$$\begin{aligned}
\delta_r(0) &= y_r(0) - y_{r0}(0) \\
\dot{\delta}_r(0) &= \dot{y}_r(0) - \dot{y}_{r0}(0) \\
&\vdots \\
\delta_r^{(n-1)}(0) &= y_r^{(n-1)}(0) - y_{r0}^{(n-1)}(0).
\end{aligned} \qquad (4.215)$$

The signal $\delta_r(t)$ can, for example, be generated as the first state $\delta_r = \delta_1$ of the exponentially stable linear system in phase-canonic form,

$$\dot{\delta} = \begin{bmatrix} 0 & & & \\ \vdots & & I_{n-1} & \\ 0 & & & \\ -m_0 & \cdots & & -m_{n-1} \end{bmatrix} \delta \qquad (4.216)$$

with initial conditions $\delta_i(0) = \delta_r^{(i-1)}(0)$. Alternatively, if the control objective is to track the output of a reference model as in (4.187), then the trajectory initialization amounts to setting the reference model initial conditions at

$$\begin{aligned}
x_{m,1}(0) &= x_1(0) \\
x_{m,2}(0) &= x_2(0) - \alpha_1(x_1(0), \hat{\theta}(0), x_{m,1}(0)) \\
x_{m,3}(0) &= x_3(0) - \alpha_2(x_1(0), x_2(0), \hat{\theta}(0), x_{m,1}(0), x_{m,2}(0)) \\
&\vdots \\
x_{m,n}(0) &= x_n(0) - \alpha_{n-1}(x_1(0), \ldots, x_{n-1}(0), \hat{\theta}(0), x_{m,1}(0), \ldots, x_{m,n-1}(0)).
\end{aligned} \qquad (4.217)$$

Let us gain more insight into the initialization algorithm (4.213). Since $y = x_1$, from the first equation in (4.213), we see that the initial value of the reference output is placed at the initial value of the system output,

$$y_r(0) = y(0). \qquad (4.218)$$

Next, we evaluate the derivatives of y from the plant model (4.185):

$$\begin{aligned}
y &= x_1 \triangleq h_0(x_1) \\
y^{(i)} &= \sum_{k=1}^{i} \frac{\partial h_{i-1}}{\partial x_k}(\bar{x}_i, \theta) \left(x_{k+1} + \varphi_k(\bar{x}_k)^T \theta\right) \triangleq h_i(\bar{x}_{i+1}, \theta), \\
& \qquad\qquad i = 1, \ldots, n-1.
\end{aligned} \qquad (4.219)$$

For $i = 1$, we have
$$\dot{y} = x_2 + \varphi_1(x_1)^T \theta = h_1(\bar{x}_2, \theta). \tag{4.220}$$
On the other hand, from (4.213), in view of (4.190), we have
$$\begin{aligned}\dot{y}_r(0) &= x_2(0) - \alpha_1(x_1(0), \hat{\theta}(0), y_r(0)) \\ &= x_2(0) + c_1 z_1(0) + \varphi_1(x_1(0))^T \hat{\theta}(0) \\ &= x_2(0) + \varphi_1(x_1(0))^T \hat{\theta}(0) \\ &= h_1(\bar{x}_2(0), \hat{\theta}(0)), \end{aligned} \tag{4.221}$$
which means that $\dot{y}_r(0)$ is placed at an *estimated* value of $\dot{y}(0)$. For $i = 2$, we have
$$\begin{aligned}\ddot{y} &= \frac{\partial h_1}{\partial x_1}(x_2 + \varphi_1^T \theta) + \frac{\partial h_1}{\partial x_2}(x_3 + \varphi_2^T \theta) \\ &= \frac{\partial \varphi_1}{\partial x_1}(x_2 + \varphi_1^T \theta) + x_3 + \varphi_2^T \theta \\ &= h_2(\bar{x}_3, \theta). \end{aligned} \tag{4.222}$$
From (4.213), in view of (4.190) for $i = 2$, we obtain
$$\begin{aligned}\ddot{y}_r(0) &= x_3(0) - \alpha_3(x_1(0), x_2(0), \hat{\theta}(0), y_r(0), \dot{y}_r(0)) \\ &= x_3(0) + z_1(0) + c_2 z_2(0) - \frac{\partial \alpha_1}{\partial x_1} x_2(0) - \frac{\partial \alpha_1}{\partial y_r} \dot{y}_r(0) \\ &\quad + \varphi_2(\bar{x}_2(0))^T \hat{\theta}(0) - \frac{\partial \alpha_1}{\partial x_1} \varphi_1(x_1(0))^T \hat{\theta}(0) \\ &\quad - \frac{\partial \alpha_1}{\partial \hat{\theta}} \Gamma \tau_1(x_1(0), y_r(0)). \end{aligned} \tag{4.223}$$
Since $\frac{\partial \alpha_1}{\partial x_1} = -c_1 - \frac{\partial \varphi_1^T \theta}{\partial x_1}$, $\frac{\partial \alpha_1}{\partial y_r} = c_1$, and $\tau_1(x_1(0), x_1(0)) = 0$, we have
$$\begin{aligned}\ddot{y}_r(0) &= x_3(0) + \varphi_2(\bar{x}_2(0))^T \hat{\theta}(0) + \frac{\partial \varphi_1^T \hat{\theta}}{\partial x_1}\left(x_2(0) + \varphi_1(x_1(0))^T \hat{\theta}(0)\right) \\ &= h_2(\bar{x}_3(0), \hat{\theta}(0)), \end{aligned} \tag{4.224}$$
which means that $\ddot{y}_r(0)$ is placed at an *estimated* value of $\ddot{y}(0)$.

By continuing in the same fashion, we reach the following conclusion.

Lemma 4.16 *For the coordinate change (4.189)–(4.192), the following equivalence holds:*
$$z = 0 \iff y_r^{(i)} = h_i(\bar{x}_{i+1}, \hat{\theta}), \quad i = 0, \ldots, n-1. \tag{4.225}$$

Remark 4.17 The importance of this conclusion is twofold. First, the trajectory initialization can be performed using $y_r^{(i)}(0) = h_i(\bar{x}_{i+1}(0), \hat{\theta}(0))$ instead of (4.213). Second, and much more important, the trajectory initial conditions do not depend on the design parameters c_i, κ_i, and Γ but only on the initial state $x(0)$ and the initial parameter estimate $\hat{\theta}(0)$. ◇

4.4 Transient Performance

In this section we will derive $\mathcal{L}_2$ and $\mathcal{L}_\infty$ transient performance bounds for the error state z, which include a bound for the tracking error $z_1 = y - y_r$.

We will make use of the constants $c_0 = \min_{1 \leq i \leq n} c_i$ and $\kappa_0 = \left(\sum_{i=1}^n \frac{1}{\kappa_i}\right)^{-1}$. For simplicity, we let $\Gamma = \gamma I$.

4.4.1 $\mathcal{L}_2$ performance

We first derive a bound on the $\mathcal{L}_2$ norm of the state of the error system.

Theorem 4.18 *In the adaptive system (4.185), (4.194), (4.193), the following inequality holds*

$$\|z\|_2 \leq \frac{|\tilde{\theta}(0)|}{\sqrt{2c_0\gamma}} + \frac{1}{\sqrt{2c_0}}|z(0)|. \qquad (4.226)$$

Proof. From (4.201), we have

$$\dot{V}_n \leq -c_0|z|^2 \leq 0. \qquad (4.227)$$

Since $V_n = \frac{1}{2}|z|^2 + \frac{1}{2\gamma}|\tilde{\theta}|^2$ is nonincreasing and bounded from below by zero, it has a limit as $t \to \infty$, so

$$\begin{aligned}
\|z\|_2^2 &= \int_0^\infty |z(\tau)|^2 d\tau \leq -\frac{1}{c_0}\int_0^\infty \dot{V}_n(\tau) d\tau \\
&= \frac{1}{c_0}V_n(0) - \frac{1}{c_0}V_n(\infty) \leq \frac{1}{c_0}V_n(0) \\
&\leq \frac{1}{c_0}\left(\frac{1}{2}|z(0)|^2 + \frac{1}{2\gamma}|\tilde{\theta}(0)|^2\right),
\end{aligned} \qquad (4.228)$$

which implies (4.226). $\square$

It may appear that the bound (4.226) can be reduced by increasing c_0 alone. This is not so. Although the term $\frac{|\tilde{\theta}(0)|}{\sqrt{2c_0\gamma}}$ is reduced by increasing c_0, the initial condition $z(0)$ may, in fact, increase by increasing c_0 ($|z_i(0)|$ may grow as c_0^{i-1}). Fortunately, we can circumvent this difficulty using trajectory initialization. By applying the initialization algorithm (4.213), we set $z(0) = 0$ and obtain

$$\|z\|_2 \leq \frac{|\tilde{\theta}(0)|}{\sqrt{2c_0\gamma}}. \qquad (4.229)$$

At the same time, as we noted in Remark 4.17, the trajectory initial conditions are independent of c_i. Therefore, the $\mathcal{L}_2$ transient performance of the z-system can be systematically improved by increasing c_0.

Another way of improving the $\mathcal{L}_2$ transient performance suggested by the bound (4.229) is by increasing the adaptation gain γ. The tuning functions scheme is unique in this respect because it directly compensates for the time-varying effect of parameter estimation. As we shall see, the bounds for estimation-based schemes are increasing functions of γ for large values of γ.

4.4.2 $\mathcal{L}_\infty$ performance

Now we turn our attention to $\mathcal{L}_\infty$ performance. For the adaptive system (4.195) with (4.196), the following bound is obtained from (4.227) and (4.200):

$$|z(t)| \leq \frac{1}{\sqrt{\gamma}}|\tilde{\theta}(0)| + |z(0)|. \tag{4.230}$$

Although this bound suggests that performance improvement is possible through increasing γ and initializing $z(0) = 0$, it does not capture the effect of the design coefficients c_i on the system transients.

In the general tuning functions procedure summarized in Table 4.1 we introduced the nonlinear damping terms $-\kappa_i|w_i|^2 z_i$ for achieving boundedness. With these terms, we are also able to derive the following $\mathcal{L}_\infty$ bound.[7]

Theorem 4.19 *In the adaptive system (4.185), (4.193), (4.194), with $\kappa_0 > 0$, the following inequality holds*

$$|z(t)| \leq \frac{1}{2\sqrt{c_0 \kappa_0}}\left(|\tilde{\theta}(0)|^2 + \gamma|z(0)|^2\right)^{\frac{1}{2}} + |z(0)|e^{-c_0 t}. \tag{4.231}$$

Proof. For the adaptive system (4.195) inequality (4.202) gives

$$\frac{d}{dt}\left(\frac{1}{2}|z|^2\right) \leq -c_0|z|^2 + \frac{1}{4\kappa_0}|\tilde{\theta}|^2. \tag{4.232}$$

From Lemma C.5(i), taking $v = z^2$ and $\rho = \frac{1}{\sqrt{\kappa_0}}|\tilde{\theta}|$, it follows that

$$|z(t)|^2 \leq |z(0)|^2 e^{-2c_0 t} + \frac{1}{4c_0\kappa_0}\|\tilde{\theta}\|_\infty^2. \tag{4.233}$$

A bound on $\|\tilde{\theta}\|_\infty$ is obtained from (4.211) as $\|\tilde{\theta}\|_\infty^2 \leq |\tilde{\theta}(0)|^2 + \gamma|z(0)|^2$. By substituting the latter expression into (4.233), we complete the proof. □

Remark 4.20 *An elegant way of removing the term $\gamma|z(0)|^2$ from the bound (4.231) without trajectory initialization, is using the "observer"*

$$\dot{\hat{z}} = A_z(z, \hat{\theta}, t)\hat{z}, \qquad \hat{z}(0) = z(0). \tag{4.234}$$

---

[7] As we shall see later, a similar bound can be derived for *linear systems* even without these terms.

4.4 TRANSIENT PERFORMANCE

We introduce the observer error

$$\tilde{z} = z - \hat{z}, \tag{4.235}$$

and instead of (4.194) we employ the update law driven by the observer error:

$$\dot{\hat{\theta}} = \Gamma W(z, \hat{\theta}, t)\tilde{z}. \tag{4.236}$$

Combining (4.234) and (4.235) with (4.195) we obtain the observer error system

$$\dot{\tilde{z}} = A_z(z, \hat{\theta}, t)\tilde{z} + W(z, \hat{\theta}, t)^{\mathrm{T}}\tilde{\theta}. \tag{4.237}$$

Now, along the solutions of (4.236)–(4.237) we have

$$\frac{d}{dt}\left(\frac{1}{2}|\tilde{z}|^2 + \frac{1}{2\gamma}|\tilde{\theta}|^2\right) = -\sum_{i=1}^{n} c_i \tilde{z}_i^2 \leq 0, \tag{4.238}$$

which implies that

$$\|\tilde{\theta}\|_\infty^2 \leq |\tilde{\theta}(0)|^2 + \gamma|\tilde{z}(0)|^2 = |\tilde{\theta}(0)|^2, \tag{4.239}$$

because $\tilde{z}(0) = z(0) - \hat{z}(0) = 0$. By substituting (4.239) into (4.233) we get

$$|z(t)| \leq \frac{1}{2\sqrt{c_0 \kappa_0}}|\tilde{\theta}(0)| + |z(0)|e^{-c_0 t} \tag{4.240}$$

instead of (4.231). Using this method, the $\mathcal{L}_2$ bound (4.226) and the $\mathcal{L}_\infty$ bound (4.230) remain unchanged. ◇

With our standard initialization $z(0) = 0$, combining the bound (4.231) with the bound (4.230), we get

$$|z(t)| \leq \min\left\{\frac{1}{2\sqrt{c_0 \kappa_0}}, \frac{1}{\sqrt{\gamma}}\right\}|\tilde{\theta}(0)|. \tag{4.241}$$

It follows that the $\mathcal{L}_\infty$ bound for the tuning functions scheme can be systematically reduced by increasing any one of the design parameters: c_0, κ_0, or γ. The tuning functions scheme is unique in its capability to use the adaptation gain γ for improvement of transient performance.

In Theorem 4.15 we proved that nonlinear damping guarantees boundedness without adaptation for any (unknown) amount of parametric uncertainty. For the nonadaptive system we also established an $\mathcal{L}_\infty$ performance bound (4.203), which with $z(0) = 0$ becomes

$$|z(t)| \leq \frac{1}{2\sqrt{c_0 \kappa_0}}|\tilde{\theta}|, \tag{4.242}$$

where $\tilde{\theta}$ is a constant parameter error. Let us suppose that the parameter estimate initial condition $\hat{\theta}(0)$ in the adaptive design is the same as the constant parameter estimate $\hat{\theta}$ in the nonadaptive design. (This assumption is reasonable because the a priori knowledge is the same.) Comparing the adaptive bound (4.241) with the nonadaptive bound (4.242) we come to an easy but fundamental conclusion: With adaptation gain $\gamma > 4c_0\kappa_0$, the adaptive bound is lower than the nonadaptive bound. This indicates that the adaptive controller may achieve improvement of performance over the nonadaptive controller for the same values of coefficients c_i and κ_i. As we shall see, this is a unique property of the tuning functions scheme, not shared by the modular schemes. We will return to this important issue and discuss it in much more detail when we deal with adaptive control of linear systems. Here we just summarize our conclusion in the form of the following corollary.

Corollary 4.21 *For $\gamma > 4c_0\kappa_0$, the adaptive bound (4.241) is lower than the nonadaptive bound (4.242).*

4.5 Extensions

For the sake of clarity, the adaptive design in this chapter was presented for the class of parametric strict-feedback systems. We now give three extensions of the tuning functions design. We first consider strict-feedback systems with unknown virtual control coefficients. Second, we explain how the design is modified for parametric block-strict-feedback systems. Third, we extend the design to parametric pure-feedback systems.

4.5.1 Unknown virtual control coefficients

We consider systems of the form

$$\begin{aligned}
\dot{x}_i &= b_i x_{i+1} + \varphi_i(x_1, \ldots, x_i)^\mathrm{T} \theta, & i = 1, \ldots, n-1 \\
\dot{x}_n &= b_n \beta(x) u + \varphi_n(x_1, \ldots, x_n)^\mathrm{T} \theta,
\end{aligned} \qquad (4.243)$$

where, in addition to the unknown vector θ, the constant coefficients b_i are also unknown. We refer to the coefficients b_i as the 'virtual control coefficients'. The occurrence of the unknown b_i-coefficients is frequent in applications ranging from electric motors and robotic manipulators to flight dynamics.

Assumption 4.22 *The signs of b_i, $i = 1, \ldots, n$, are known.*

We consider two special cases of (4.243). The extension to the general case is straightforward but tedious.

4.5 EXTENSIONS

The first special case is when the only unknown virtual control coefficient is the 'high-frequency gain' b_n:

$$\begin{aligned} \dot{x}_i &= x_{i+1} + \varphi_i(x_1, \ldots, x_i)^T \theta, & i &= 1, \ldots, n-1 \\ \dot{x}_n &= b_n \beta(x) u + \varphi_n(x_1, \ldots, x_n)^T \theta. \end{aligned} \qquad (4.244)$$

For this case the modification of the tuning functions design is simple. In Table 4.1 we only need to change the control law (4.193):

$$u = \frac{\hat{\varrho}}{\beta(x)} \left[\alpha_n(x, \hat{\theta}, \bar{y}_r^{(n-1)}) + y_r^{(n)} \right], \qquad (4.245)$$

where $\hat{\varrho}$ is the estimate of $\varrho = 1/b_n$ computed as

$$\dot{\hat{\varrho}} = -\gamma \operatorname{sgn}(b_n) \left(\alpha_n + y_r^{(n)} \right) z_n, \qquad \gamma > 0. \qquad (4.246)$$

We are only using the knowledge of the *sign* of the unknown parameter b_n. In this simple case it is not necessary to estimate b_n itself. It can be checked that the resulting error system has the form (4.195) with an additional term due to $\tilde{\varrho} = \varrho - \hat{\varrho}$:

$$\dot{z} = A_z(z, \hat{\theta}, t) z + W(z, \hat{\theta}, t)^T \tilde{\theta} + \begin{bmatrix} 0 \\ \vdots \\ 0 \\ -b_n \left(\alpha_n + y_r^{(n)} \right) \end{bmatrix} \tilde{\varrho}. \qquad (4.247)$$

Consider the Lyapunov function

$$V = \frac{1}{2} z^T z + \frac{1}{2} \tilde{\theta}^T \Gamma^{-1} \tilde{\theta} + \frac{|b_n|}{2\gamma} \tilde{\varrho}^2. \qquad (4.248)$$

Its derivative along the solutions of (4.247) and (4.246) is

$$\dot{V} \leq -c_0 |z|^2, \qquad (4.249)$$

which leads us to the same conclusion as in Theorem 4.14: All the states are bounded and asymptotic tracking is achieved.

Now we move on to a more difficult case:

$$\begin{aligned} \dot{x}_i &= x_{i+1} + \varphi_i(x_1, \ldots, x_i)^T \theta, & i &= 1, \ldots, m-1, m+1, \ldots, n-1 \\ \dot{x}_m &= b_m x_{m+1} + \varphi_m(x_1, \ldots, x_m)^T \theta, \\ \dot{x}_n &= \beta(x) u + \varphi_n(x_1, \ldots, x_n)^T \theta, \end{aligned}$$
$$(4.250)$$

where b_m, $m < n$, is the only unknown coefficient. From step m on, the design procedure for this case differs considerably from the procedure in Table 4.1.

We now need $\hat{b}_m$ and $\hat{\varrho}$, the estimates of b_m and $\varrho = 1/b_m$. The estimate $\hat{\varrho}$ is introduced to avoid the division by $\hat{b}_m(t)$ which can occasionally take value zero. The complete design procedure is given by the following expressions (with $z_0 = 0$, $\alpha_0 = 0$, $\tau_0 = 0$):

Coordinate transformation:

$$z_i = x_i - y_r^{(i-1)} - \alpha_{i-1}, \quad i = 1, \ldots, m \quad (4.251)$$
$$z_j = x_j - \hat{\varrho} y_r^{(j-1)} - \alpha_{j-1}, \quad j = m+1, \ldots, n \quad (4.252)$$

Regressor:

$$w_i = \varphi_i - \sum_{k=1}^{i-1} \frac{\partial \alpha_{i-1}}{\partial x_k} \varphi_k, \quad i = 1, \ldots, n \quad (4.253)$$

Tuning functions for $\hat{\theta}$:

$$\tau_i = \tau_{i-1} + w_i z_i, \quad i = 1, \ldots, n \quad (4.254)$$

Tuning functions for $\hat{b}_m$:

$$\pi_m = z_{m+1} z_m \quad (4.255)$$
$$\pi_j = \pi_{j-1} - \frac{\partial \alpha_m}{\partial x_m} x_{m+1} z_j, \quad j = m+1, \ldots, n \quad (4.256)$$

Stabilizing functions:

$$\alpha_i(\bar{x}_i, \hat{\theta}, \bar{y}_r^{(i-1)}) = \bar{\alpha}_i, \quad i = 1, \ldots, m-1 \quad (4.257)$$
$$\alpha_m(\bar{x}_m, \hat{\theta}, \bar{y}_r^{(m-1)}, \hat{\varrho}) = \hat{\varrho} \bar{\alpha}_m \quad (4.258)$$
$$\alpha_j(\bar{x}_j, \hat{\theta}, \bar{y}_r^{(j-1)}, \hat{b}_m, \hat{\varrho}) = \bar{\alpha}_j \quad j = m+1, \ldots, n \quad (4.259)$$

$$\bar{\alpha}_i = -z_{i-1} - c_i z_i - w_i^T \hat{\theta} + \sum_{k=1}^{i-1} \left(\frac{\partial \alpha_{i-1}}{\partial x_k} x_{k+1} + \frac{\partial \alpha_{i-1}}{\partial y_r^{(k-1)}} y_r^{(k)} \right)$$
$$+ \frac{\partial \alpha_{i-1}}{\partial \hat{\theta}} \Gamma \tau_i + \sum_{k=2}^{i-1} \frac{\partial \alpha_{k-1}}{\partial \hat{\theta}} \Gamma w_i z_k, \quad i = 1, \ldots, m \quad (4.260)$$

$$\bar{\alpha}_{m+1} = -\hat{b}_m z_m - c_{m+1} z_{m+1} - w_{m+1}^T \hat{\theta}$$
$$+ \sum_{k=1}^{m-1} \frac{\partial \alpha_m}{\partial x_k} x_{k+1} + \hat{b}_m \frac{\partial \alpha_m}{\partial x_m} x_{m+1} + \sum_{k=1}^{m} \frac{\partial \alpha_m}{\partial y_r^{(k-1)}} y_r^{(k)}$$
$$+ \frac{\partial \alpha_m}{\partial \hat{\theta}} \Gamma \tau_{m+1} + \left(y_r^{(m)} + \frac{\partial \alpha_m}{\partial \hat{\varrho}} \right) \dot{\hat{\varrho}} + \sum_{k=2}^{m} \frac{\partial \alpha_{k-1}}{\partial \hat{\theta}} \Gamma w_{m+1} z_k \quad (4.261)$$

4.5 Extensions

$$\bar{\alpha}_j = -z_{j-1} - c_j z_j - w_j^T \hat{\theta}$$
$$+ \sum_{\substack{k=1 \\ k \neq m}}^{j-1} \frac{\partial \alpha_{j-1}}{\partial x_k} x_{k+1} + \hat{b}_m \frac{\partial \alpha_{j-1}}{\partial x_m} x_{m+1} + \sum_{k=1}^{j-1} \frac{\partial \alpha_{j-1}}{\partial y_r^{(k-1)}} y_r^{(k)}$$
$$+ \frac{\partial \alpha_{j-1}}{\partial \hat{\theta}} \Gamma \tau_j + \frac{\partial \alpha_{j-1}}{\partial \hat{b}_m} \gamma \pi_j + \left(y_r^{(j-1)} + \frac{\partial \alpha_{j-1}}{\partial \hat{\varrho}} \right) \dot{\hat{\varrho}}$$
$$+ \sum_{k=2}^{j-1} \frac{\partial \alpha_{k-1}}{\partial \hat{\theta}} \Gamma w_j z_k - \sum_{k=m+1}^{j-1} \frac{\partial \alpha_{k-1}}{\partial \hat{b}_m} \gamma \frac{\partial \alpha_{j-1}}{\partial x_m} x_{m+1} z_k$$
$$j = m+2, \ldots, n \quad (4.262)$$

Adaptive control law:
$$u = \frac{1}{\hat{\beta}(x)} \left[\alpha_n + \hat{\varrho} y_r^{(n)} \right] \quad (4.263)$$

Parameter update laws:
$$\dot{\hat{\theta}} = \Gamma \tau_n = \Gamma W z \quad (4.264)$$
$$\dot{\hat{b}}_m = \gamma \pi_n = \gamma \left[z_{m+1} z_m - \sum_{j=m+1}^{n} \frac{\partial \alpha_{j-1}}{\partial x_m} x_{m+1} z_j \right] \quad (4.265)$$
$$\dot{\hat{\varrho}} = -\gamma \text{sgn}(b_m) \left(y_r^{(m)} + \bar{\alpha}_m \right) z_m \quad (4.266)$$

Lengthy but straightforward calculations show that the design procedure (4.251)–(4.266) results in the closed-loop system

$$\dot{z}_i = -c_i z_i - \sum_{k=2}^{i-1} \sigma_{ki} z_k - z_{i-1} + z_{i+1} + \sum_{k=i+1}^{n} \sigma_{ik} z_k + w_i^T \tilde{\theta},$$
$$i = 1, \ldots, m-1 \quad (4.267)$$
$$\dot{z}_m = -c_m z_m - \sum_{k=2}^{m-1} \sigma_{km} z_k - z_{m-1} + \hat{b}_m z_{m+1} + \sum_{k=m+1}^{n} \sigma_{mk} z_k$$
$$+ w_m^T \tilde{\theta} - b_m \left(y_r^{(m)} + \bar{\alpha}_m \right) \tilde{\varrho} + z_{m+1} \tilde{b}_m \quad (4.268)$$
$$\dot{z}_{m+1} = -c_{m+1} z_{m+1} - \sum_{k=2}^{m} \sigma_{k,m+1} z_k - b_m z_m + z_{m+2} + \sum_{k=m+2}^{n} \sigma_{m+1,k} z_k$$
$$+ w_{m+1}^T \tilde{\theta} - \frac{\partial \alpha_m}{\partial x_m} x_{m+1} \tilde{b}_m \quad (4.269)$$
$$\dot{z}_j = -c_j z_j - \sum_{k=2}^{j-1} \sigma_{kj} z_k - z_{j-1} + z_{j+1} + \sum_{k=j+1}^{n} \sigma_{jk} z_k$$
$$+ w_j^T \tilde{\theta} - \frac{\partial \alpha_{j-1}}{\partial x_m} x_{m+1} \tilde{b}_m, \quad i = m+2, \ldots, n, \quad (4.270)$$

where σ_{ik} is defined for $k = i+1, \ldots, n$ as

$$\sigma_{ik} = \begin{cases} 0, & i = 1 \\ -\dfrac{\partial \alpha_{i-1}}{\partial \hat{\theta}} \Gamma w_k, & i = 2, \ldots, m+1 \\ -\dfrac{\partial \alpha_{i-1}}{\partial \hat{\theta}} \Gamma w_k + \dfrac{\partial \alpha_{i-1}}{\partial \hat{b}_m} \gamma \dfrac{\partial \alpha_{k-1}}{\partial x_m} x_{m+1}, & i = m+2, \ldots, n-1 \end{cases} \quad (4.271)$$

A Lyapunov function for this system is

$$V = \frac{1}{2} z^T z + \frac{1}{2} \tilde{\theta}^T \Gamma^{-1} \tilde{\theta} + \frac{1}{2\gamma} \tilde{b}_m^2 + \frac{|b_m|}{2\gamma} \tilde{\varrho}^2. \quad (4.272)$$

Its derivative along the solutions of (4.264)–(4.266) and (4.267)–(4.270),

$$\dot{V} = -\sum_{k=1}^{n} c_k z_k^2, \quad (4.273)$$

leads us to the same conclusion as in Theorem 4.14: All the states are bounded and asymptotic tracking is achieved.

For clarity, we have not included in the stabilizing functions the nonlinear damping terms which guarantee boundedness in the absence of adaptation. To achieve this property, we augment the functions (4.260), (4.261), and (4.262), respectively, by the nonlinear damping terms

$$-\kappa_i |w_i|^2 z_i$$

$$-\kappa_{m+1} \left[|w_{m+1}|^2 + \left(z_m - \frac{\partial \alpha_m}{\partial x_m} x_{m+1} \right)^2 \right] z_{m+1}$$

$$-\kappa_j \left[|w_j|^2 + \left(\frac{\partial \alpha_{j-1}}{\partial x_m} x_{m+1} \right)^2 \right] z_j,$$

and we augment the stabilizing function (4.258) by the term $-\kappa_m \operatorname{sgn}(b_m) \left(y_r^{(m)} + \bar{\alpha}_m \right)^2 z_m$,

$$\alpha_m = \hat{\varrho} \bar{\alpha}_m - \kappa_m \operatorname{sgn}(b_m) \left(y_r^{(m)} + \bar{\alpha}_m \right)^2 z_m. \quad (4.274)$$

It can be verified that the resulting closed-loop system is

$$\begin{aligned} \dot{z}_i &= -c_i z_i - \sum_{k=2}^{i-1} \sigma_{ki} z_k - z_{i-1} + z_{i+1} + \sum_{k=i+1}^{n} \sigma_{ik} z_k \\ &\quad - \kappa_i |w_i|^2 z_i + w_i^T \tilde{\theta}, \qquad i = 1, \ldots, m-1 \end{aligned} \quad (4.275)$$

4.5 EXTENSIONS

$$\dot{z}_m = -c_m z_m - \sum_{k=2}^{m-1} \sigma_{km} z_k - z_{m-1} + b_m z_{m+1} + \sum_{k=m+1}^{n} \sigma_{mk} z_k$$
$$-\kappa_m |w_m|^2 z_m - \kappa_m |b_m| \left(y_{\mathrm{r}}^{(m)} + \bar{\alpha}_m \right)^2 z_m$$
$$+ w_m^{\mathrm{T}} \tilde{\theta} - b_m \left(y_{\mathrm{r}}^{(m)} + \bar{\alpha}_m \right) \tilde{\varrho} \qquad (4.276)$$

$$\dot{z}_{m+1} = -c_{m+1} z_{m+1} - \sum_{k=2}^{m} \sigma_{k,m+1} z_k - b_m z_m + z_{m+2} + \sum_{k=m+2}^{n} \sigma_{m+1,k} z_k$$
$$-\kappa_{m+1} |w_{m+1}|^2 z_{m+1} - \kappa_{m+1} \left(z_m - \frac{\partial \alpha_m}{\partial x_m} x_{m+1} \right)^2 z_{m+1}$$
$$+ w_{m+1}^{\mathrm{T}} \tilde{\theta} + \left(z_m - \frac{\partial \alpha_m}{\partial x_m} x_{m+1} \right) \tilde{b}_m \qquad (4.277)$$

$$\dot{z}_j = -c_j z_j - \sum_{k=2}^{j-1} \sigma_{kj} z_k - z_{j-1} + z_{j+1} + \sum_{k=j+1}^{n} \sigma_{jk} z_k$$
$$-\kappa_j |w_j|^2 z_j - \kappa_j \left(\frac{\partial \alpha_{j-1}}{\partial x_m} x_{m+1} \right)^2 z_j$$
$$+ w_j^{\mathrm{T}} \tilde{\theta} - \frac{\partial \alpha_{j-1}}{\partial x_m} x_{m+1} \tilde{b}_m, \qquad i = m+2, \ldots, n \qquad (4.278)$$

and then show that the following inequality is satisfied by the solutions of (4.275)–(4.278):

$$\frac{d}{dt} \left(\frac{1}{2} |z|^2 \right) \leq -c_0 |z|^2 + \frac{1}{4\kappa_0} \left(|\tilde{\theta}|^2 + \tilde{b}_m^2 + |b_m| \tilde{\varrho}^2 \right), \qquad (4.279)$$

which proves that all the signals are bounded even when the adaptation is turned off.

4.5.2 Block-strict-feedback systems

We consider systems of the form

$$\begin{aligned}
\dot{x}_i &= x_{i+1} + \varphi_i(x_1, \ldots, x_i, \bar{\zeta}_1, \ldots, \bar{\zeta}_i)^{\mathrm{T}} \theta, & i = 1, \ldots, \rho-1 \\
\dot{x}_\rho &= \beta(x, \zeta) u + \varphi_\rho(x, \zeta)^{\mathrm{T}} \theta & (4.280) \\
\dot{\zeta}_i &= \Phi_{i,0}(\bar{x}_i, \bar{\zeta}_i) + \Phi_i(\bar{x}_i, \bar{\zeta}_i)^{\mathrm{T}} \theta, & i = 1, \ldots, \rho
\end{aligned}$$

with the following notation: $\bar{x}_i = [x_1, \ldots, x_i]^{\mathrm{T}}$, $\bar{\zeta}_i = \left[\zeta_1^{\mathrm{T}}, \ldots, \zeta_i^{\mathrm{T}} \right]^{\mathrm{T}}$, $x = \bar{x}_\rho$, and $\zeta = \bar{\zeta}_\rho$.

Assumption 4.23 *Each ζ_i-subsystem of (4.280) has a bounded-input bounded-state (BIBS) property with respect to $(\bar{x}_i, \bar{\zeta}_{i-1})$ as the input.*

A nonlinear system has a BIBS property if for any initial condition and for any uniformly bounded continuous input the solution exists and is uniformly bounded.[8]

For this class of systems it is quite simple to modify the procedure in Table 4.1. Because of the dependence of φ_i on $\bar{\zeta}_i$, the stabilizing function (4.190) is augmented by the term $\sum_{k=1}^{i-1} \frac{\partial \alpha_{i-1}}{\partial \zeta_k} \Phi_{k,0}$, namely,

$$\alpha_i(\bar{x}_i, \hat{\theta}, \bar{y}_r^{(i-1)}, \bar{\zeta}_i) = -z_{i-1} - c_i z_i - w_i^T \hat{\theta}$$
$$+ \sum_{k=1}^{i-1} \left(\frac{\partial \alpha_{i-1}}{\partial x_k} x_{k+1} + \frac{\partial \alpha_{i-1}}{\partial y_r^{(k-1)}} y_r^{(k)} + \frac{\partial \alpha_{i-1}}{\partial \zeta_k} \Phi_{k,0} \right)$$
$$- \kappa_i |w_i|^2 z_i + \frac{\partial \alpha_{i-1}}{\partial \hat{\theta}} \Gamma \tau_i + \sum_{k=2}^{i-1} \frac{\partial \alpha_{k-1}}{\partial \hat{\theta}} \Gamma w_i z_k , \quad (4.281)$$

and the regressor w_i is augmented by $-\sum_{k=1}^{i-1} \Phi_i \left(\frac{\partial \alpha_{i-1}}{\partial \zeta_k} \right)^T$,

$$w_i(\bar{x}_i, \hat{\theta}, \bar{y}_r^{(i-2)}, \bar{\zeta}_i) = \varphi_i - \sum_{k=1}^{i-1} \frac{\partial \alpha_{i-1}}{\partial x_k} \varphi_k - \sum_{k=1}^{i-1} \Phi_i \left(\frac{\partial \alpha_{i-1}}{\partial \zeta_k} \right)^T . \quad (4.282)$$

Note that the change in w_i in (4.192) also has an impact on the tuning function (4.191). The control law and the parameter update law are as in (4.193) and (4.194), respectively.

It can be checked that the resulting error system has the same form (4.195), (4.196) as the basic tuning functions design. Since the parameter update law (4.194) is also the same, the Lyapunov function

$$V_\rho = \frac{1}{2} z^T z + \frac{1}{2} \tilde{\theta}^T \Gamma^{-1} \tilde{\theta} \quad (4.283)$$

has the same derivative:

$$\dot{V}_\rho \leq -c_0 |z|^2 , \quad (4.284)$$

which proves that $(z, \tilde{\theta}) = 0$ is globally uniformly stable. The boundedness of z_1 implies that x_1 is bounded. Then, by Assumption 4.23, ζ_1 is bounded. This along with the boundedness of z_2 establishes the boundedness of x_2. Then, by Assumption 4.23, ζ_2 is bounded. Continuing in the same fashion, we arrive at the conclusion that x, ζ, and $\hat{\theta}$ are all bounded. As a continuous function of its arguments, the control u is also bounded. The asymptotic tracking property is established as in the proof of Theorem 4.14.

---

[8]Input-to-state stability implies BIBS, but the converse is not true. This can be seen, from the system $\dot{x} = -\left(1 - \sin^2 u\right) x$ for the particular input $u = \pi/2$, see [114, Remark 3.1.5].

4.5 EXTENSIONS

4.5.3 Pure-feedback systems

Strict-feedback systems have a 'triangular' form. We now consider a broader class of *parametric pure-feedback* systems which include some upper triangular entries[9]

$$\begin{aligned}\dot{x}_i &= x_{i+1} + \varphi_i(x_1, \ldots, x_{i+1})^T \theta, \qquad i = 1, \ldots, n-1 \\ \dot{x}_n &= \left(\beta_0(x) + \beta(x)^T \theta\right) u + \varphi_0(x) + \varphi_n(x)^T \theta,\end{aligned} \qquad (4.285)$$

where

$$\varphi_0(0) = 0, \quad \varphi_1(0) = \cdots = \varphi_n(0) = 0, \qquad \beta_0(0) \neq 0, \qquad (4.286)$$

and $\varphi_0, \varphi_1, \ldots, \varphi_n$, as well as β_0, β, are smooth functions in B_x, a neighborhood of the origin $x = 0$.

In this section we consider only the case of regulation. In general, stability is achieved only in a well-defined *region* around the origin.

The tuning functions design for parametric pure-feedback systems (4.285) is summarized in Table 4.2. The differences from the design in Table 4.1 are:

- α_i, τ_i, w_i depend on $\bar{x}_{i+1}$ rather than $\bar{x}_i$. This is because φ_i depends on x_{i+1}, so the sums $\sum \frac{\partial \alpha_{i-1}}{\partial x_k} x_{k+1}$ in (4.290) and $-\sum \frac{\partial \alpha_{i-1}}{\partial x_k} \varphi_k$ in (4.292) run to i rather than to $i-1$.

- The update law (4.294) contains the term $\left(1 - \frac{\partial \alpha_{n-1}}{\partial x_n}\right) \beta \alpha z_n$ which is due to the term $\beta^T \theta$ multiplying u in (4.285).

- A factor $\left(1 - \frac{\partial \alpha_{n-1}}{\partial x_n}\right)$ appears in the denominator of the control law in (4.293). In addition, instead of $\beta_0 + \beta^T \hat{\theta}$, the denominator of the control law has a factor $\beta_0 + \beta^T \left(\hat{\theta} - \sum_{k=2}^n \Gamma \left(\frac{\partial \alpha_{k-1}}{\partial \hat{\theta}}\right)^T z_k\right)$.

One can verify that with these modifications, the closed-loop system is

$$\dot{z} = A_z z + \left[W + \beta e_n^T \left(1 - \frac{\partial \alpha_{n-1}}{\partial x_n}\right) \alpha\right]^T \tilde{\theta}, \qquad (4.287)$$

where

$$A_z(z, \hat{\theta}) = \begin{bmatrix} -c_1 & 1 & 0 & \cdots & 0 \\ -1 & -c_2 & 1+\sigma_{23} & \cdots & \sigma_{2n} \\ 0 & -1-\sigma_{23} & \ddots & \ddots & \vdots \\ \vdots & \vdots & \ddots & \ddots & 1+\sigma_{n-1,n} \\ 0 & -\sigma_{2n} & \cdots & -1-\sigma_{n-1,n} & -c_n \end{bmatrix} \qquad (4.288)$$

---

[9] A coordinate-free characterization of these systems is given in Appendix G, Theorem G.9.

Table 4.2: Tuning Functions Design for Pure-Feedback Systems

$$z_i = x_i - \alpha_{i-1} \tag{4.289}$$

$$\alpha_i(\bar{x}_{i+1}, \hat{\theta}) = -z_{i-1} - c_i z_i - w_i^T \hat{\theta} + \sum_{k=1}^{i} \frac{\partial \alpha_{i-1}}{\partial x_k} x_{k+1}$$

$$+ \frac{\partial \alpha_{i-1}}{\partial \hat{\theta}} \Gamma \tau_i + \sum_{k=2}^{i-1} \frac{\partial \alpha_{k-1}}{\partial \hat{\theta}} \Gamma w_i z_k \tag{4.290}$$

$$\tau_i(\bar{x}_{i+1}, \hat{\theta}) = \tau_{i-1} + w_i z_i \tag{4.291}$$

$$w_i(\bar{x}_{i+1}, \hat{\theta}) = \varphi_i - \sum_{k=1}^{i} \frac{\partial \alpha_{i-1}}{\partial x_k} \varphi_k \tag{4.292}$$

$$i = 1, \ldots, n$$

$$\bar{x}_i = (x_1, \ldots, x_i), \qquad \bar{y}_r^{(i)} = (y_r, \dot{y}_r, \ldots, y_r^{(i)})$$

Adaptive control law:

$$u = \alpha(x, \hat{\theta}) = \frac{1}{\left(1 - \frac{\partial \alpha_{n-1}}{\partial x_n}\right) \left[\beta_0 + \beta^T \left(\hat{\theta} - \sum_{k=2}^{n} \Gamma \left(\frac{\partial \alpha_{k-1}}{\partial \hat{\theta}}\right)^T z_k\right)\right]} \times$$

$$\left[\alpha_n - \left(1 - \frac{\partial \alpha_{n-1}}{\partial x_n}\right) \varphi_0\right] \tag{4.293}$$

Parameter update law:

$$\dot{\hat{\theta}} = \Gamma \left[\tau_n + \left(1 - \frac{\partial \alpha_{n-1}}{\partial x_n}\right) \beta \alpha z_n\right] = \Gamma \left[W + \beta e_n^T \left(1 - \frac{\partial \alpha_{n-1}}{\partial x_n}\right) \alpha\right] z \tag{4.294}$$

and σ_{ik} is defined for $k = i+1, \ldots, n$ as

$$\sigma_{ik} = \begin{cases} 0, & i = 1 \\ -\dfrac{\partial \alpha_{i-1}}{\partial \hat{\theta}} \Gamma w_k, & i = 2, \ldots, n-1, \ k \neq n \\ -\dfrac{\partial \alpha_{i-1}}{\partial \hat{\theta}} \Gamma \left[w_n + \left(1 - \dfrac{\partial \alpha_{n-1}}{\partial x_n}\right) \beta \alpha \right], & k = n. \end{cases} \tag{4.295}$$

4.5 EXTENSIONS

In deriving (4.287), (4.288), (4.295) it is helpful to rewrite the control law (4.293) as

$$u = \alpha(x, \hat{\theta}) = \frac{1}{\left(1 - \frac{\partial \alpha_{n-1}}{\partial x_n}\right)\left(\beta_0 + \beta^T \hat{\theta}\right)} \left\{ -z_{n-1} - c_n z_n - \left(1 - \frac{\partial \alpha_{n-1}}{\partial x_n}\right)\varphi_0 \right.$$
$$- w_i^T \hat{\theta} + \sum_{k=1}^{n-1} \frac{\partial \alpha_{n-1}}{\partial x_k} x_{k+1} + \frac{\partial \alpha_{n-1}}{\partial \hat{\theta}} \dot{\hat{\theta}}$$
$$\left. + \sum_{k=2}^{n-1} \frac{\partial \alpha_{k-1}}{\partial \hat{\theta}} \Gamma \left[w_n + \left(1 - \frac{\partial \alpha_{n-1}}{\partial x_n}\right) \beta \alpha(x, \hat{\theta}) \right] z_k \right\}. \quad (4.296)$$

Equation (4.296), which is implicit in $u = \alpha(x, \hat{\theta})$, is more convenient for analysis, whereas (4.293) is used for implementation.

The design procedure from Table 4.2 is *feasible* only if the denominator in the control law (4.293) never becomes zero, namely, only if

$$\left| 1 - \frac{\partial \alpha_{n-1}}{\partial x_n} \right| > 0 \quad (4.297)$$

$$\left| \beta_0 + \beta^T \left(\hat{\theta} - \sum_{k=2}^{n} \Gamma \left(\frac{\partial \alpha_{k-1}}{\partial \hat{\theta}} \right)^T z_k \right) \right| > 0. \quad (4.298)$$

Let us first analyze condition (4.297). From (4.290) we compute

$$\frac{\partial \alpha_{n-1}}{\partial x_n} = \frac{\partial}{\partial x_n} \left\{ -z_{n-2} - c_{n-1} z_{n-1} - w_{n-1}^T \hat{\theta} + \sum_{k=1}^{n-1} \frac{\partial \alpha_{n-2}}{\partial x_k} x_{k+1} \right.$$
$$\left. + \frac{\partial \alpha_{n-2}}{\partial \hat{\theta}} \Gamma \tau_{n-1} + \sum_{k=2}^{n-2} \frac{\partial \alpha_{k-1}}{\partial \hat{\theta}} \Gamma w_{n-1} z_k \right\}$$
$$= \frac{\partial}{\partial x_n} \left\{ -w_{n-1}^T \hat{\theta} + \frac{\partial \alpha_{n-2}}{\partial x_{n-1}} x_n + \frac{\partial \alpha_{n-2}}{\partial \hat{\theta}} \Gamma \tau_{n-1} + \sum_{k=2}^{n-2} \frac{\partial \alpha_{k-1}}{\partial \hat{\theta}} \Gamma w_{n-1} z_k \right\}$$
$$= \frac{\partial}{\partial x_n} \left\{ -w_{n-1}^T \left(\hat{\theta} - \sum_{k=2}^{n-1} \Gamma \left(\frac{\partial \alpha_{k-1}}{\partial \hat{\theta}} \right)^T z_k \right) + \frac{\partial \alpha_{n-2}}{\partial x_{n-1}} x_n \right\}$$
$$= \frac{\partial}{\partial x_n} \left\{ -\left(1 - \frac{\partial \alpha_{n-2}}{\partial x_{n-1}}\right) \varphi_{n-1}^T \left(\hat{\theta} - \sum_{k=2}^{n-1} \Gamma \left(\frac{\partial \alpha_{k-1}}{\partial \hat{\theta}} \right)^T z_k \right) \right.$$
$$\left. + \frac{\partial \alpha_{n-2}}{\partial x_{n-1}} x_n \right\}$$
$$= -\left(1 - \frac{\partial \alpha_{n-2}}{\partial x_{n-1}}\right) \frac{\partial \varphi_{n-1}^T}{\partial x_n} \left(\hat{\theta} - \sum_{k=2}^{n-1} \Gamma \left(\frac{\partial \alpha_{k-1}}{\partial \hat{\theta}} \right)^T z_k \right) + \frac{\partial \alpha_{n-2}}{\partial x_{n-1}}. \quad (4.299)$$

Thus, condition (4.297) becomes

$$\left|\left(1 - \frac{\partial \alpha_{n-2}}{\partial x_{n-1}}\right)\left[1 + \frac{\partial \varphi_{n-1}^T}{\partial x_n}\left(\hat{\theta} - \sum_{k=2}^{n-1} \Gamma \left(\frac{\partial \alpha_{k-1}}{\partial \hat{\theta}}\right)^T z_k\right)\right]\right| > 0. \qquad (4.300)$$

By repeating the computations as in (4.299), condition (4.297) is brought to the form

$$\left|\prod_{i=1}^{n-1}\left[1 + \frac{\partial \varphi_i^T}{\partial x_{i+1}}\left(\hat{\theta} - \sum_{k=2}^{i} \Gamma \left(\frac{\partial \alpha_{k-1}}{\partial \hat{\theta}}\right)^T z_k\right)\right]\right| > 0. \qquad (4.301)$$

With (4.301) and (4.298), we have the following proposition.

Proposition 4.24 *Let $B_x \subset \mathbb{R}^n$ and $B_\theta \subset \mathbb{R}^p$ be open sets such that $(x, \hat{\theta}) = (0, \theta) \in B_x \times B_\theta$, and for all $(x, \hat{\theta}) \in B_x \times B_\theta$,*

$$\left|1 + \frac{\partial \varphi_i^T}{\partial x_{i+1}}\left(\hat{\theta} - \sum_{k=2}^{i} \Gamma \left(\frac{\partial \alpha_{k-1}}{\partial \hat{\theta}}\right)^T z_k\right)\right| > 0, \qquad i = 1, \ldots, n-1 \qquad (4.302)$$

$$\left|\beta_0 + \beta^T \left(\hat{\theta} - \sum_{k=2}^{n} \Gamma \left(\frac{\partial \alpha_{k-1}}{\partial \hat{\theta}}\right)^T z_k\right)\right| > 0. \qquad (4.303)$$

Then $\mathcal{F} = B_x \times B_\theta$ is a subset of the set in which the design procedure of Table 4.2 is feasible.

Let us now examine the change of coordinates in Table 4.2. For all $i = 1, \ldots, n$ we have

$$\begin{aligned}
z_i &= x_i + z_{i-2} + c_{i-1} z_{i-1} + w_{i-1}^T \hat{\theta} - \sum_{k=1}^{i-1} \frac{\partial \alpha_{i-2}}{\partial x_k} x_{k+1} \\
&\quad - \frac{\partial \alpha_{i-2}}{\partial \hat{\theta}} \Gamma \tau_{i-1} - \sum_{k=2}^{i-2} \frac{\partial \alpha_{k-1}}{\partial \hat{\theta}} \Gamma w_{i-1} z_k \\
&= \left(1 - \frac{\partial \alpha_{i-2}}{\partial x_{i-1}}\right)\left[x_i + \varphi_{i-1}^T \left(\hat{\theta} - \sum_{k=2}^{i-1} \Gamma \left(\frac{\partial \alpha_{k-1}}{\partial \hat{\theta}}\right)^T z_k\right)\right] \\
&\quad + z_{i-2} + c_{i-1} z_{i-1} - \sum_{k=1}^{i-2} \frac{\partial \alpha_{i-2}}{\partial x_k} x_{k+1} \\
&\quad - \sum_{k=1}^{i-2} \frac{\partial \alpha_{i-2}}{\partial x_k} \varphi_k^T \left(\hat{\theta} - \sum_{k=2}^{i-1} \Gamma \left(\frac{\partial \alpha_{k-1}}{\partial \hat{\theta}}\right)^T z_k\right) - \frac{\partial \alpha_{i-2}}{\partial \hat{\theta}} \Gamma \tau_{i-2}. \qquad (4.304)
\end{aligned}$$

We showed above that

$$1 - \frac{\partial \alpha_{i-2}}{\partial x_{i-1}} = \prod_{j=1}^{i-2}\left[1 + \frac{\partial \varphi_j^T}{\partial x_{j+1}}\left(\hat{\theta} - \sum_{k=2}^{j} \Gamma \left(\frac{\partial \alpha_{k-1}}{\partial \hat{\theta}}\right)^T z_k\right)\right]. \qquad (4.305)$$

4.5 EXTENSIONS

By substituting (4.305) into (4.304), and inductively applying the implicit function theorem, we conclude that the change of coordinates

$$(x, \hat{\theta}) \mapsto (z, \tilde{\theta}) \tag{4.306}$$

is one-to-one, onto, smooth, and has a smooth inverse on $\mathcal{F}$. Moreover, since $\varphi_0(0) = 0$, $\varphi_1(0) = \cdots = \varphi_n(0) = 0$, by examining (4.304), one can show inductively that

$$x = 0 \iff z = 0. \tag{4.307}$$

Because of (4.307) it is not hard to show that a feasibility condition, equivalent to Proposition 4.24, is that there exists a set $\mathcal{F}' = B'_x \times B'_\theta$ containing $(x, \hat{\theta}) = (0, \theta)$ such that for all $(x, \hat{\theta}) \in \mathcal{F}'$,

$$\left| 1 + \frac{\partial \varphi_i(x)^{\mathrm{T}}}{\partial x_{i+1}} \hat{\theta} \right| > 0, \quad i = 1, \ldots, n-1 \tag{4.308}$$

$$\left| \beta_0(x) + \beta(x)^{\mathrm{T}} \hat{\theta} \right| > 0. \tag{4.309}$$

It is then possible to find a subset of $\mathcal{F}'$ which is another estimate of the feasibility region.

In general, the feasibility region is not global. However, this is not due to the adaptive scheme. Even when the parameters θ are known, the feedback linearization of the system (4.285) can only be guaranteed for $\theta \in B_\theta \subset \mathbb{R}^p$, an open set such that for all $(x, \theta) \in B_x \times B_\theta$,

$$\left| 1 + \frac{\partial \varphi_i(x)^{\mathrm{T}}}{\partial x_{i+1}} \theta \right| > 0, \quad i = 1, \ldots, n-1 \tag{4.310}$$

$$\left| \beta_0(x) + \beta(x)^{\mathrm{T}} \theta \right| > 0. \tag{4.311}$$

Let us now return to the closed-loop adaptive system (4.294), (4.287),

$$\begin{aligned} \dot{z} &= A_z z + \left[W + \beta e_n^{\mathrm{T}} \left(1 - \frac{\partial \alpha_{n-1}}{\partial x_n} \right) \alpha \right]^{\mathrm{T}} \tilde{\theta} \\ \dot{\tilde{\theta}} &= -\Gamma \left[W + \beta e_n^{\mathrm{T}} \left(1 - \frac{\partial \alpha_{n-1}}{\partial x_n} \right) \alpha \right] z . \end{aligned} \tag{4.312}$$

The derivative of the Lyapunov function

$$V = \frac{1}{2} z^{\mathrm{T}} z + \frac{1}{2} \tilde{\theta}^{\mathrm{T}} \Gamma^{-1} \tilde{\theta} \tag{4.313}$$

along the solutions of (4.312) and (4.288) is

$$\dot{V} = -\sum_{i=1}^{n} c_i z_i^2 . \tag{4.314}$$

This proves that the equilibrium $z = 0, \tilde{\theta} = 0$ is stable. Since we showed above that the coordinate change (4.306) is a diffeomorphism which preserves the origin, we conclude that the equilibrium $x = 0, \hat{\theta} = \theta$ is also stable.

We now give an estimate $\Omega \subset \mathcal{F}$ of the region of attraction of this equilibrium. Let $\Omega(c)$ be the invariant set defined by $V(x, \hat{\theta}) < c$, and let c^* be the largest constant c such that $\Omega(c) \subset \mathcal{F}$. Then, an estimate Ω of the region of attraction is

$$\Omega = \Omega(c^*) = \left\{ (x, \hat{\theta}) \mid V(x, \hat{\theta}) < c^* \right\}, \qquad c^* = \arg\sup_{\Omega(c) \subset \mathcal{F}} \{c\}. \qquad (4.315)$$

Finally, by applying [81, Theorem 4.8] (a local version of the LaSalle-Yoshizawa theorem (Theorem A.8)) to (4.314), it follows that $z(t) \to 0$ as $t \to \infty$. In view of (4.307), this means that $x(t) \to 0$.

Theorem 4.25 *Suppose that system (4.285) satisfies Proposition 4.24. Then the closed-loop adaptive system consisting of the plant (4.285), the control law (4.293), and the update law (4.294) has a stable equilibrium $x = 0, \hat{\theta} = \theta$, and its region of attraction includes the set Ω defined in (4.315). Furthermore, for all $(x(0), \hat{\theta}(0)) \in \Omega$, we have*

$$\lim_{t \to \infty} x(t) = 0. \qquad (4.316)$$

4.6 Example: Aircraft Wing Rock

Wing rock is a limit cycling oscillation in the roll angle ϕ and the roll rate $\dot{\phi}$ which can occur in high-performance aircraft with slender forebodies when flying in high angle-of-attack. Conventional methods of eliminating wing rock include a redesign of the airframe configuration and limiting of the angle-of-attack. These methods may reduce maneuverability of the aircraft. An effective method for suppressing wing rock without degrading maneuverability is using feedback control.

Several one-degree-of-freedom models of wing rock have been proposed in Nguyen, Whipple, and Brandon [143], Hsu and Lan [47], and Elzebda, Nayfeh, and Mook [32]. They are all nonlinear and contain parameters θ_i which depend on the angle-of-attack, dynamic pressure, wing reference area, wing span, roll moment of inertia, and flight velocity. We now present an adaptive controller of Monahemi, Barlow, and Krstić [133, 134] which allows these parameters to be unknown and eliminates the wing rock phenomenon by achieving global stabilization. The model we consider here,

$$\ddot{\phi} = \theta_1 + \theta_2 \phi + \theta_3 \dot{\phi} + \theta_4 |\phi| \dot{\phi} + \theta_5 |\dot{\phi}| \dot{\phi}, \qquad (4.317)$$

is based on wind tunnel tests [143] at NASA Langley Research Center. In these tests, physical scaled wings were mounted on an apparatus which allows free

4.6 Example: Aircraft Wing Rock

rotation about the roll axis. These model wings are aerodynamically similar to the wings of an F-18 HARV aircraft. There are no control surfaces in the model (4.317). With ailerons modeled as first-order actuator dynamics, the state-space form of the wing-rock model is

$$\begin{aligned} \dot{\phi} &= p \\ \dot{p} &= \theta_1 + \theta_2\phi + \theta_3 p + \theta_4|\phi|p + \theta_5|p|p + b\delta_A \\ \tau\dot{\delta}_A &= -\delta_A + u, \end{aligned} \quad (4.318)$$

where δ_A is the aileron deflection angle, u is the control input, τ is the aileron time constant, and b is an unknown constant parameter. Denoting $\varphi(\phi, p) = [1, \phi, p, |\phi|p, |p|p]^T$ and $\theta = [\theta_1, \theta_2, \theta_3, \theta_4, \theta_5]^T$, we rewrite (4.318) as

$$\begin{aligned} \dot{\phi} &= p \\ \dot{p} &= b\delta_A + \varphi(\phi, p)^T\theta \\ \dot{\delta}_A &= \tfrac{1}{\tau}u - \tfrac{1}{\tau}\delta_A. \end{aligned} \quad (4.319)$$

This model is in the parametric strict-feedback form with unknown virtual control coefficient b, so we apply the design from Section 4.5.1.

Our control objective is to asymptotically track a given reference $\phi_r(t)$ with the roll angle ϕ. We use the error variables

$$\begin{aligned} z_1 &= \phi - \phi_r \\ z_2 &= p - \dot{\phi}_r - \alpha_1(\phi, \phi_r) \\ z_3 &= \delta_A - \hat{\varrho}\ddot{\phi}_r - \alpha_2(\phi, p, \phi_r, \dot{\phi}_r, \hat{\theta}, \hat{\varrho}), \end{aligned} \quad (4.320)$$

and derive the stabilizing functions

$$\begin{aligned} \alpha_1 &= -c_1 z_1 \\ \alpha_2 &= \hat{\varrho}\bar{\alpha}_2, \quad \bar{\alpha}_2 = -z_1 - c_2 z_2 - \varphi^T\hat{\theta} - c_1(p - \dot{\phi}_r). \end{aligned} \quad (4.321)$$

The design procedure from Section 4.5.1 results in an adaptive controller consisting of the control law

$$\begin{aligned} u = \tau\Bigg[&\frac{1}{\tau}\delta_A - \hat{b}z_2 - c_3 z_3 + \frac{\partial\alpha_2}{\partial\phi}p + \frac{\partial\alpha_2}{\partial p}\left(\hat{b}\delta_A + \varphi^T\hat{\theta}\right) \\ &+ \frac{\partial\alpha_2}{\partial\phi_r}\dot{\phi}_r + \frac{\partial\alpha_2}{\partial\dot{\phi}_r}\ddot{\phi}_r + \hat{\varrho}\phi_r^{(3)} + \frac{\partial\alpha_2}{\partial\hat{\theta}}\dot{\hat{\theta}} + \left(\ddot{\phi}_r + \bar{\alpha}_2\right)\dot{\hat{\varrho}}\Bigg] \end{aligned} \quad (4.322)$$

and the update laws

$$\dot{\hat{\theta}} = \Gamma\varphi\left(z_2 - \frac{\partial\alpha_2}{\partial p}z_3\right) \quad (4.323)$$

$$\dot{\hat{b}} = \gamma\left(z_2 - \frac{\partial\alpha_2}{\partial p}\delta_A\right)z_3 \quad (4.324)$$

$$\dot{\hat{\varrho}} = -\gamma\operatorname{sgn}(b)\left(\ddot{\phi}_r + \bar{\alpha}_2\right)z_2. \quad (4.325)$$

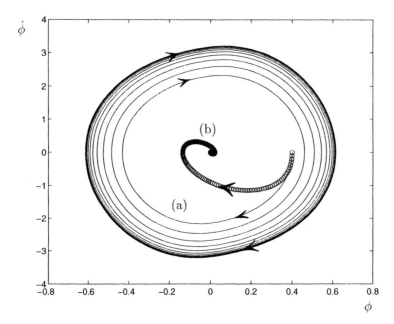

Figure 4.5: (a) Uncontrolled wing rock (—). (b) Suppression of wing rock by adaptive nonlinear control (⊙⊙⊙).

For the resulting error system

$$\begin{bmatrix} \dot{z}_1 \\ \dot{z}_2 \\ \dot{z}_3 \end{bmatrix} = \begin{bmatrix} -c_1 & 1 & 0 \\ -1 & -c_2 & b \\ 0 & -b & -c_3 \end{bmatrix} \begin{bmatrix} z_1 \\ z_2 \\ z_3 \end{bmatrix} + \begin{bmatrix} 0 \\ 1 \\ -\frac{\partial \alpha_2}{\partial p} \end{bmatrix} \varphi^{\mathrm{T}} \tilde{\theta}$$
$$+ \begin{bmatrix} 0 \\ 0 \\ z_2 - \frac{\partial \alpha_2}{\partial p} \delta_{\mathrm{A}} \end{bmatrix} \tilde{b} + \begin{bmatrix} 0 \\ -b \left(\ddot{\phi}_{\mathrm{r}} + \bar{\alpha}_2 \right) \\ 0 \end{bmatrix} \tilde{\varrho}, \quad (4.326)$$

the equilibrium $z = 0, \tilde{\theta} = 0, \tilde{b} = 0, \tilde{\varrho} = 0$ is globally stable and, moreover, $z(t) \to 0$. This means that the wing rock phenomenon is eliminated and the tracking objective $\phi(t) - \phi_{\mathrm{r}}(t) \to 0$ is achieved.

This is illustrated in Figure 4.5. Without feedback control the response to an initial condition $\phi(0) = 0.4, p(0) = \delta_{\mathrm{A}}(0) = 0$ is the trajectory (a) which represents limit cycling oscillations typical for wing rock. They are obtained for the model (4.317) with wind tunnel data provided in [143] at angle-of-attack $\alpha = 30°$: $\theta_1 = 0$, $\theta_2 = -26.67$, $\theta_3 = 0.76485$, $\theta_4 = -2.9225$, $\theta_5 = 0$. The amplitude of the wing rock oscillations is about $35°$ and the frequency is about 1Hz.

The trajectory (b) in Figure 4.5 shows the wing rock suppressing effect of the adaptive controller (4.322)–(4.325) acting through the aileron with $b = 1.5$

and $\tau = 1/15$. Parameter estimates are initialized as $\hat{\theta}(0) = 1.35\theta$, $\hat{b}(0) = 1.35b$, and $\hat{\varrho}(0) = 1/\hat{b}(0)$. The controller coefficients are $c_1 = c_2 = c_3 = 5$ and the adaptation gains are $\Gamma = 0.02I$, $\gamma = 0.02$. For softer regulation to the origin, we have employed an exponentially decaying reference trajectory $\phi_r(t)$ governed by the equation $(s+10)(s^2 + 4s + 24.25)\phi_r(s) = 0$.

Notes and References

The adaptive backstepping design of Kanellakopoulos, Kokotović, and Morse [69] required multiple estimates of the same parameter. This overparametrization is impractical for high-order multi-parameter systems. Jiang and Praly [59] were quick to notice that the extended matching idea can be employed to reduce the number of estimates by half. Still a significant overparametrization remained until Krstić, Kanellakopoulos, and Kokotović [94] introduced the tuning functions method which completely eliminated overparametrization. The removal of overparametrization strengthened the stability and convergence properties of the resulting adaptive system. By studying global invariant manifolds it was further shown in Krstić [93] that, except for a measure zero set of initial conditions in the state space of the complete adaptive system, the parameter estimates converge to values with which a nonadaptive controller would guarantee (at least local) asymptotic stability.

Trajectory initialization was employed in Krstić, Kanellakopoulos, and Kokotović [95] and in Kanellakopoulos, Krstić, and Kokotović [79].

Krstić and Kokotović [100] developed the adaptive control Lyapunov function framework which is behind the tuning functions method. A Lyapunov-based design by Praly [156] offered a solution alternative to [94] for strict feedback systems with nonlinearities of polynomial growth.

Jiang and Pomet [58] extended the tuning functions design to a class of nonholonomic systems, while Jain and Khorrami [55] employed it in a decentralized design. Polycarpou and Ioannou [151] and Yao and Tomizuka [199] presented robust extensions of the tuning functions design. Freeman and Kokotović [40] used the tuning functions technique in a design of partial-state feedback controllers for nonlinear systems linear in the unmeasured states.

The designs in this chapter involve expressions with partial derivatives which are difficult to derive for systems of high order. Ghanadan [42] developed symbolic software for mechanizing the derivations of tuning functions controllers and update laws.

Chapter 5

Modular Design with Passive Identifiers

The tuning functions approach developed in the preceding chapter has two distinguishing features: First, a single Lyapunov function encompasses the complete state $(z, \tilde{\theta})$ of the closed-loop system, and second, the dynamic order of the resulting controller is as low as the number of unknown parameters. However, the tuning functions controller is somewhat complicated because it cancels the effects of $\dot{\hat{\theta}}$ in the z-system. Another drawback of this approach is that the choice of a parameter update law is limited to a Lyapunov-type algorithm.

In the traditional adaptive linear control the restriction to the Lyapunov update law is removed by estimation-based designs which achieve a *modularity* of the controller-identifier pair: Any stabilizing controller can be combined with any identifier. The controller module is capable of stabilizing the plant when all the parameters are known. This is its *certainty equivalence* property. The identifier module, in turn, guarantees certain boundedness properties independently of the controller module. The modularity of the estimation-based designs makes them much more versatile than the Lyapunov-based designs.

In this chapter we begin the development of a modular approach to adaptive nonlinear control. We first show that a major obstacle to earlier attempts to apply estimation-based designs to nonlinear systems was the weakness of their certainty equivalence controllers. Such controllers cannot achieve any controller-identifier separation without severely restricting the system nonlinearities.

To overcome the weakness of certainty equivalence controllers, we develop a new controller with strong parametric robustness properties. In the presence of unknown parameters this nonlinear controller achieves boundedness without adaptation. It guarantees boundedness not only in the presence of constant parameter errors, but also in the presence of time-varying parameter estimates. The strong controller is suitable for modular designs of adaptive control schemes for nonlinear systems. It can be combined with any standard identifier.

In this chapter we develop two *passive identifiers*. They employ 'observers' with passive error systems and unnormalized gradient-type update laws. The first identifier is based on the open-loop plant model, while the second is based on the closed-loop model.

We derive bounds for transient performance of the new adaptive schemes and compare them with the tuning functions design.

In addition to strong controllers, we also design controllers with a small gain property, analogous to those in adaptive linear control. These weaker controllers reduce some of the nonlinear complexity of the strong controllers, but their performance is not as high.

This chapter also serves as an introduction to the next chapter where we develop a fully modular design with standard identifiers, including those with least-squares update laws.

Section 5.2 introduces a general framework for modular adaptive design. The problem of adaptive stabilization is approached as a problem of input-to-state stabilization (ISS) with respect to the parameter estimation error considered as a disturbance input. This section introduces *ISS-control Lyapunov functions (ISS-clf)* which are the main tool in constructing the control law in the modular design.

5.1 Weakness of Certainty Equivalence

A major obstacle to the certainty equivalence design for nonlinear systems is a fundamental difference between the instability phenomena in linear and nonlinear systems. The states of an unstable linear system remain bounded over any finite time interval, so that there is enough time for the identifier to "catch up" and generate stabilizing values of parameter estimates. The situation is dramatically different in systems with nonlinearities whose growth is faster than linear (x^2, $x_1 x_2$, e^x, etc.). Even a small parameter estimation error may drive the state to infinity in finite time. For this reason, only a few nonlinear estimation-based results go beyond the linear growth constraints [152, 153, 154, 157]. In [157], nonlinear certainty equivalence designs are characterized by relationships between nonlinear growth constraints and controller stabilizing properties. In the absence of matching conditions, all these designs involve growth restrictions.

Let us explain why the certainty equivalence approach fails to achieve global stability for systems whose nonlinearities are not linearly bounded. Consider the scalar system

$$\dot{x} = u + \varphi(x)^{\mathrm{T}}\theta, \qquad (5.1)$$

5.1 WEAKNESS OF CERTAINTY EQUIVALENCE

where $\varphi(x)$ is a vector of smooth nonlinear functions and θ is a vector of unknown parameters. For this system, an obvious certainty equivalence controller is

$$u = -x - \varphi(x)^\mathrm{T}\hat{\theta}, \tag{5.2}$$

where $\hat{\theta}$ is the estimate of θ. With this controller, the resulting certainty equivalence feedback system is

$$\dot{x} = -x + \varphi(x)^\mathrm{T}\tilde{\theta}. \tag{5.3}$$

In Section 3.1 we have shown that the update law $\dot{\hat{\theta}} = \varphi(x)x$ globally stabilizes this system at the equilibrium $(x, \tilde{\theta}) = 0$. This was achieved by a Lyapunov design with $V = x^2 + \tilde{\theta}^\mathrm{T}\tilde{\theta}$. We now want to employ a standard identifier to provide the estimate $\hat{\theta}$ for the certainty equivalence controller (5.2). This means that we want to connect two independently designed modules, the controller (5.2) and a standard identifier. From among the identifiers found in the adaptive control literature [51, 165] we will take the following identifier with a normalized gradient update law:

$$\dot{\chi} = -\chi + \varphi(x), \qquad \chi(0) = 0 \tag{5.4}$$
$$\dot{\xi} = -\xi - \varphi(x)^\mathrm{T}\hat{\theta}, \qquad \xi(0) = x(0) \tag{5.5}$$
$$\dot{\hat{\theta}} = \frac{\chi}{1+\chi^\mathrm{T}\chi}\left(x - \chi^\mathrm{T}\hat{\theta} - \xi\right). \tag{5.6}$$

(This identifier was used in the early adaptive designs for nonlinear systems [141, 166].) To understand this identifier, observe that if the estimate $\hat{\theta}$ were correct, $\hat{\theta} = \theta$, then $x = \chi^\mathrm{T}\hat{\theta} + \xi$. This can be verified by direct substitution. When $\hat{\theta} \neq \theta$, then $x - \chi^\mathrm{T}\hat{\theta} - \xi$ represents the state estimation error which is used to drive the parameter update law (5.6). The update is in the direction of the regressor vector χ, but its rate is 'normalized' by the factor $(1+\chi^\mathrm{T}\chi)^{-1}$. This identifier is convenient for our illustration, because its stability properties are well known [51, 165]. It is easy to check that $x - \chi^\mathrm{T}\hat{\theta} - \xi = \chi^\mathrm{T}\tilde{\theta}$ so that (5.6) is rewritten as

$$\dot{\tilde{\theta}} = -\frac{\chi\chi^\mathrm{T}}{1+\chi^\mathrm{T}\chi}\tilde{\theta}. \tag{5.7}$$

Using the inequality $\tilde{\theta}^\mathrm{T}\chi\chi^\mathrm{T}\tilde{\theta} \leq \tilde{\theta}^\mathrm{T}\chi^\mathrm{T}\chi\tilde{\theta}$, we see that

$$\begin{aligned}\frac{d}{dt}\left(\frac{1}{2}\tilde{\theta}^\mathrm{T}\tilde{\theta}\right) &= -\tilde{\theta}^\mathrm{T}\frac{\chi\chi^\mathrm{T}}{1+\chi^\mathrm{T}\chi}\tilde{\theta} \geq -\tilde{\theta}^\mathrm{T}\frac{\chi^\mathrm{T}\chi}{1+\chi^\mathrm{T}\chi}\tilde{\theta}\\ &\geq -\tilde{\theta}^\mathrm{T}\tilde{\theta},\end{aligned} \tag{5.8}$$

which implies that

$$|\tilde{\theta}(t)| \geq |\tilde{\theta}(0)|\mathrm{e}^{-t}. \tag{5.9}$$

Thus the convergence of this common identifier can be at best exponential. Let us now connect this identifier with the certainty equivalence controller (5.2). To be specific, we let θ and φ be scalars:

$$\varphi(x) = x^2, \qquad \tilde{\theta}(t) = \tilde{\theta}(0)e^{-t}. \tag{5.10}$$

Then the certainty equivalence feedback system (5.3) is described by

$$\dot{x} = -x + x^2 \tilde{\theta}(0)e^{-t}. \tag{5.11}$$

The analytical solution to this equation was obtained in (2.243)–(2.246):

$$x(t) = x(0) \frac{2}{x(0)\tilde{\theta}(0)e^{-t} + (2 - x(0)\tilde{\theta}(0))e^t}. \tag{5.12}$$

It is not hard to see that whenever

$$x(0)\tilde{\theta}(0) > 2, \tag{5.13}$$

the solution (5.12) escapes to infinity in finite time, that is

$$x(t) \to \infty \quad \text{as} \quad t \to \frac{1}{2} \ln \frac{x(0)\tilde{\theta}(0)}{x(0)\tilde{\theta}(0) - 2}. \tag{5.14}$$

This catastrophic instability is due to $\varphi(x) = x^2$. If, instead, we had a nonlinearity $\varphi(x)$ with a linearly bounded growth, $|\varphi(x)| \leq k|x|$, the above instability would not have occurred.

A far-reaching conclusion to be drawn from the above example is that a modular design, with a certainty equivalence controller and a standard identifier as its modules, is not applicable to systems with nonlinearities whose growth is higher than linear. The mechanism of instability is clear: The identifier, whose speed of convergence is at best exponential, is not fast enough to cope with potentially explosive instabilities of nonlinear systems.

This simple example shows that to achieve stability we need either a much faster identifier, as in our tuning functions design, or a stronger controller with a bigger stability margin for disturbances such as $\tilde{\theta}$ in (5.3).

We consider the parameter estimation error and its derivative as two independent disturbance inputs and design stronger controllers that guarantee boundedness of all the states of the closed-loop system whenever

1. the parameter estimation error $\tilde{\theta} = \theta - \hat{\theta}$ is bounded, and

2. its derivative $\dot{\tilde{\theta}} = -\dot{\hat{\theta}}$ is *either* bounded or square-integrable.

These new controllers create a possibility for a complete identifier-controller modularity. Such complete controller-identifier separation has been an unachieved goal of adaptive control, even for linear systems.

5.2 ISS-Control Lyapunov Functions

In Section 4.1 we introduced a framework for Lyapunov design of adaptive stabilizers where the central role was played by adaptive control Lyapunov functions (aclf's). In this section we develop a framework for modular design of adaptive nonlinear controllers.

Let us consider the system

$$\dot{x} = f(x) + F(x)\theta + g(x)u \tag{5.15}$$

where $f(x), F(x)$, and $g(x)$ are smooth, and $f(0) = 0$, $F(0) = 0$.

We say that system (5.15) is **globally adaptively stabilizable**[1] if there exist a function $\alpha(x, \hat{\theta})$ continuous on $(\mathbb{R}^n \setminus \{0\}) \times \mathbb{R}^p$ with $\alpha(0, \hat{\theta}) \equiv 0$, continuous functions $\tau(x, \hat{\theta}, \eta)$ and $H(x, \hat{\theta}, \eta)$, and a positive definite symmetric $p \times p$ matrix Γ, such that the dynamic controller

$$u = \alpha(x, \hat{\theta}) \tag{5.16}$$
$$\dot{\hat{\theta}} = \Gamma\tau(x, \hat{\theta}, \eta) \tag{5.17}$$
$$\dot{\eta} = H(x, \hat{\theta}, \eta) \tag{5.18}$$

guarantees that the solution $(x(t), \hat{\theta}(t), \eta(t))$ is globally bounded, and $x(t) \to 0$ as $t \to \infty$, for all $\theta \in \mathbb{R}^p$.

We refer to (5.17)–(5.18) as an identifier, with (5.17) being its update law. We start by rewriting (5.15) with (5.16) as

$$\dot{x} = f(x) + F(x)\hat{\theta} + g(x)\alpha(x, \hat{\theta}) + F(x)\tilde{\theta}. \tag{5.19}$$

Suppose we know how to find a control law $\alpha(x, \hat{\theta})$ that stabilizes this system with $\tilde{\theta} = 0$. As we saw in Section 5.1, in the presence of the disturbance input $\tilde{\theta}$, this control law, in general, does not preserve stability even if $\tilde{\theta}$ is bounded and exponentially decaying. To preserve stability, we need a stronger controller. Since the standard parameter estimators guarantee that $\hat{\theta}$ is bounded, we are interested in designing controllers which can guarantee input-to-state stability of (5.19) with respect to $\tilde{\theta}$ as input, in the sense of Definition C.1. However, the time-varying character of the parameter estimate $\hat{\theta}(t)$ forces us to consider $\dot{\hat{\theta}}(t)$ as another disturbance input, even though it is not explicitly present in (5.19).

Our goal is to find a control law $\alpha(x, \hat{\theta})$ continuous on $(\mathbb{R}^n \setminus \{0\}) \times \mathbb{R}^p$ with $\alpha(0, \hat{\theta}) \equiv 0$, such that the following *input-to-state stability* (ISS) property is satisfied:

$$|x(t)| \leq \beta(|x(0)|, t) + \gamma\left(\sup_{0 \leq \tau \leq t}\left\|\begin{bmatrix} \tilde{\theta}(\tau) \\ \dot{\hat{\theta}}(\tau) \end{bmatrix}\right\|\right), \tag{5.20}$$

---

[1] This definition is more general than the one from Section 4.1.2 to suit the modular approach.

where β is a class $\mathcal{KL}$ function and γ is a class $\mathcal{K}$ function (see Appendix C for details). Lin, Sontag, and Wang [115, Theorem 3] recently proved that a necessary and sufficient condition for (5.20) is the existence of an *ISS-Lyapunov function*. (The proof of sufficiency is given in Theorem C.2.) We say that a smooth function $V : \mathbb{R}^n \times \mathbb{R}^p \to \mathbb{R}_+$, positive definite and radially unbounded in x for each $\hat{\theta}$ is an ISS-Lyapunov function for (5.15) if there exists a class $\mathcal{K}_\infty$ function ρ such that the following implication holds for all $x \neq 0$ and all $\hat{\theta}, \tilde{\theta}, \dot{\hat{\theta}} \in \mathbb{R}^p$:

$$|x| \geq \rho\left(\left\|\begin{bmatrix} \tilde{\theta} \\ \dot{\hat{\theta}} \end{bmatrix}\right\|\right)$$
$$\Downarrow \qquad (5.21)$$
$$\frac{\partial V}{\partial x}\left[f(x) + F(x)\hat{\theta} + g(x)\alpha(x,\hat{\theta})\right] + \frac{\partial V}{\partial x}F(x)\tilde{\theta} + \frac{\partial V}{\partial \hat{\theta}}\dot{\hat{\theta}} < 0.$$

If there exists such a triple (α, V, ρ) we say that system (5.15) is input-to-state stabilizable with respect to $\left(\tilde{\theta}, \dot{\hat{\theta}}\right)$.

We have thus chosen to study the problem of modular adaptive stabilization as a problem of input-to-state stabilization. Similar to Section 4.1 where we cast Lyapunov adaptive stabilization in the framework of control Lyapunov functions, in this section we develop a clf framework for modular adaptive stabilization.

Definition 5.1 *A smooth function $V : \mathbb{R}^n \times \mathbb{R}^p \to \mathbb{R}_+$, positive definite and radially unbounded in x for each $\hat{\theta}$, is called an **ISS-control Lyapunov function (ISS-clf)** for (5.15) if there exists a class $\mathcal{K}_\infty$ function ρ such that the following implication holds for all $x \neq 0$ and all $\hat{\theta}, \tilde{\theta}, \dot{\hat{\theta}} \in \mathbb{R}^p$:*

$$|x| \geq \rho\left(\left\|\begin{bmatrix} \tilde{\theta} \\ \dot{\hat{\theta}} \end{bmatrix}\right\|\right)$$
$$\Downarrow \qquad (5.22)$$
$$\inf_{u \in \mathbb{R}} \left\{\frac{\partial V}{\partial x}\left[f(x) + F(x)\hat{\theta} + g(x)u\right] + \frac{\partial V}{\partial x}F(x)\tilde{\theta} + \frac{\partial V}{\partial \hat{\theta}}\dot{\hat{\theta}}\right\} < 0.$$

We now show that the existence of an ISS-clf is a necessary and sufficient condition for input-to-state stabilizability. The proof of sufficiency is constructive—we design a control law starting from a given ISS-clf.

Lemma 5.2 (Input-to-State Stabilization) *System (5.15) is input-to-state stabilizable with respect to $\left(\tilde{\theta}, \dot{\hat{\theta}}\right)$ if and only if there exists an ISS-clf.*

5.2 ISS-CONTROL LYAPUNOV FUNCTIONS

Proof. The 'only if' part is obvious because (5.21) implies that there exists a particular control law $u = \alpha(x, \hat{\theta})$ which satisfies (5.22). We will now prove the 'if' part by showing that the following control law achieves input-to-state stabilization:

$$\alpha(x, \hat{\theta}) = \begin{cases} -\dfrac{\omega + \sqrt{\omega^2 + \left(\frac{\partial V}{\partial x}g\right)^4}}{\frac{\partial V}{\partial x}g}, & \frac{\partial V}{\partial x}g(x, \hat{\theta}) \neq 0 \\ 0, & \frac{\partial V}{\partial x}g(x, \hat{\theta}) = 0, \end{cases} \quad (5.23)$$

where

$$\omega(x, \hat{\theta}) = \frac{\partial V}{\partial x}\left[f(x) + F(x)\hat{\theta}\right] + \left|\left[\frac{\partial V}{\partial x}F, \frac{\partial V}{\partial \hat{\theta}}\right]^T\right| \rho^{-1}(|x|). \quad (5.24)$$

We first show that (5.23) is continuous on $(\mathbb{R}^n \setminus \{0\}) \times \mathbb{R}^p$. In [171], Sontag proved that the function (5.23) is smooth provided that its arguments ω and $\frac{\partial V}{\partial x}g$ are such that

$$\frac{\partial V}{\partial x}g = 0 \Rightarrow \omega < 0. \quad (5.25)$$

We show that V being an ISS-clf implies that (5.25) is satisfied. By Definition 5.1, if $x \neq 0$ and $\frac{\partial V}{\partial x}g = 0$, then

$$|x| \geq \rho\left(\left|\begin{bmatrix} \tilde{\theta} \\ \dot{\hat{\theta}} \end{bmatrix}\right|\right)$$

$$\Downarrow \quad (5.26)$$

$$\frac{\partial V}{\partial x}\left[f(x) + F(x)\hat{\theta}\right] + \frac{\partial V}{\partial x}F(x)\tilde{\theta} + \frac{\partial V}{\partial \hat{\theta}}\dot{\hat{\theta}} < 0.$$

Let us consider the particular input

$$\begin{bmatrix} \tilde{\theta} \\ \dot{\hat{\theta}} \end{bmatrix} = \frac{\left[\frac{\partial V}{\partial x}F, \frac{\partial V}{\partial \hat{\theta}}\right]^T}{\left|\left[\frac{\partial V}{\partial x}F, \frac{\partial V}{\partial \hat{\theta}}\right]^T\right|} \rho^{-1}(|x|). \quad (5.27)$$

This input satisfies the upper part of implication (5.26):

$$\rho^{-1}\left(\left|\begin{bmatrix} \tilde{\theta} \\ \dot{\hat{\theta}} \end{bmatrix}\right|\right) = |x|. \quad (5.28)$$

Therefore, substituting (5.27) into the lower part of (5.26), we conclude that, if $x \neq 0$ and $\frac{\partial V}{\partial x}g = 0$, then

$$\frac{\partial V}{\partial x}\left[f(x) + F(x)\hat{\theta}\right] + \left|\left[\frac{\partial V}{\partial x}F, \frac{\partial V}{\partial \hat{\theta}}\right]^T\right| \rho^{-1}(|x|) < 0, \quad (5.29)$$

that is, (5.25) is satisfied for $x \neq 0$. Therefore, (5.23) is a smooth function of ω and $\frac{\partial V}{\partial x}g$ whenever $x \neq 0$. Since $\omega(x,\hat{\theta})$ is continuous and $\frac{\partial V}{\partial x}g(x,\hat{\theta})$ is smooth, the control law (5.23) is continuous for $x \neq 0$. We note that the control law $\alpha(x,\hat{\theta})$ given by (5.23) is also continuous at $x = 0$ if and only if the ISS-clf V satisfies the following *small control property* [171]: For each $\hat{\theta} \in \mathbb{R}^p$ and for any $\varepsilon > 0$ there is a $\delta > 0$ such that, if $x \neq 0$ satisfies $\rho\left(\left\|\begin{bmatrix}\tilde{\theta}\\\dot{\hat{\theta}}\end{bmatrix}\right\|\right) \leq |x| \leq \delta$, then there is some u with $|u| \leq \varepsilon$ such that

$$\frac{\partial V}{\partial x}\left[f(x) + F(x)\hat{\theta} + g(x)u\right] + \frac{\partial V}{\partial x}F(x)\tilde{\theta} + \frac{\partial V}{\partial \hat{\theta}}\dot{\hat{\theta}} < 0. \tag{5.30}$$

Now we show that the control law (5.23) achieves input-to-state stabilization. Along the solutions of (5.19) and (5.23), the derivative of V is

$$\begin{aligned}\dot{V} &= -\left\|\begin{bmatrix}\frac{\partial V}{\partial x}F, & \frac{\partial V}{\partial \hat{\theta}}\end{bmatrix}^T\right\|\rho^{-1}(|x|) + \frac{\partial V}{\partial x}F(x)\tilde{\theta} + \frac{\partial V}{\partial \hat{\theta}}\dot{\hat{\theta}} \\ &\quad -\sqrt{\left(\frac{\partial V}{\partial x}\left[f(x) + F(x)\hat{\theta}\right] + \left\|\begin{bmatrix}\frac{\partial V}{\partial x}F, & \frac{\partial V}{\partial \hat{\theta}}\end{bmatrix}^T\right\|\rho^{-1}(|x|)\right)^2 + \left(\frac{\partial V_a}{\partial x}g\right)^4} \\ &\leq -\left\|\begin{bmatrix}\frac{\partial V}{\partial x}F, & \frac{\partial V}{\partial \hat{\theta}}\end{bmatrix}^T\right\|\left(\rho^{-1}(|x|) - \left\|\begin{bmatrix}\tilde{\theta}\\\dot{\hat{\theta}}\end{bmatrix}\right\|\right) \\ &\quad -\sqrt{\left(\frac{\partial V}{\partial x}\left[f(x) + F(x)\hat{\theta}\right] + \left\|\begin{bmatrix}\frac{\partial V}{\partial x}F, & \frac{\partial V}{\partial \hat{\theta}}\end{bmatrix}^T\right\|\rho^{-1}(|x|)\right)^2 + \left(\frac{\partial V_a}{\partial x}g\right)^4}.\end{aligned} \tag{5.31}$$

In view of (5.29) this proves that $\dot{V} < 0$, $\forall x \neq 0$, whenever $|x| \geq \rho\left(\left\|\begin{bmatrix}\tilde{\theta}\\\dot{\hat{\theta}}\end{bmatrix}\right\|\right)$, that is, V is an ISS-Lyapunov function, which by Theorem C.2 establishes that (5.15) is input-to-state stable with respect to $\left(\tilde{\theta}, \dot{\hat{\theta}}\right)$. □

This lemma shows that, if an ISS-clf is available, it is easy to design a control law which guarantees boundedness of the state x whenever $\tilde{\theta}$ and $\dot{\hat{\theta}}$ are bounded. Therefore, we need identifiers which can guarantee that $\tilde{\theta}$ and $\dot{\hat{\theta}}$ are bounded. As we shall see in Chapter 6, the swapping identifiers guarantee that both $\tilde{\theta}$ and $\dot{\hat{\theta}}$ are bounded. The passive identifiers that we present in this chapter can guarantee only the boundedness of $\tilde{\theta}$. Fortunately, even though they cannot guarantee that $\dot{\hat{\theta}}$ is bounded, they can guarantee that it is square-integrable. The boundedness of $\tilde{\theta}$ and the square-integrability of $\dot{\hat{\theta}}$ will be enough to establish the boundedness of x because we will design controllers

5.2 ISS-Control Lyapunov Functions

which guarantee the following linear-like relationship:

$$\dot{V} \leq -\lambda V + \lambda \left(|\tilde{\theta}|^2 + |\dot{\hat{\theta}}|^2 \right), \qquad \lambda > 0. \tag{5.32}$$

With inequality (5.32) we will show that x is bounded whenever $\tilde{\theta}$ is bounded and $\dot{\hat{\theta}}$ is square-integrable.

In contrast to the Lyapunov design, where it is not clear if the global asymptotic stabilizability for each θ is a necessary condition for the existence of an aclf (see Remark 4.4), the global asymptotic stabilizability for each θ *is* a necessary condition for the existence of an ISS-clf. This becomes obvious by setting $\hat{\theta}(t) \equiv \theta$, which implies $\tilde{\theta}(t) = \dot{\hat{\theta}}(t) \equiv 0$, into (5.20).

Next we give sufficient conditions under which $x(t) \to 0$ as $t \to \infty$. Due to the ISS property (5.20), one sufficient condition is that both $\tilde{\theta}$ and $\dot{\hat{\theta}}$ tend to zero. However, in general, identifiers cannot guarantee that $\tilde{\theta}$ goes to zero, so the next lemma gives a less demanding condition. Both the passive (see Remark 5.15) and the swapping (see Remark 6.8) identifiers will be able to guarantee that these conditions are satisfied.

Lemma 5.3 (Regulation) *Suppose the control law $u = \alpha(x, \hat{\theta})$ guarantees that system (5.15) is ISS with respect to $\left(\tilde{\theta}, \dot{\hat{\theta}}\right)$. If $x(t)$ is bounded, and $F(x(t))\tilde{\theta}(t)$ and $\dot{\hat{\theta}}(t)$ converge to zero as $t \to \infty$, then $\lim_{t\to\infty} x(t) = 0$.*

Proof. Since the system

$$\dot{x} = f(x) + F(x)\theta + g(x)\alpha(x, \hat{\theta}) \tag{5.33}$$

is ISS with respect to $\left(\tilde{\theta}, \dot{\hat{\theta}}\right)$, then the same system with $\tilde{\theta} = \dot{\hat{\theta}} = 0$, namely the system

$$\dot{x} = f(x) + F(x)\hat{\theta} + g(x)\alpha(x, \hat{\theta}) \tag{5.34}$$

is globally asymptotically stable. Therefore by [174, Theorem 2] there exist $\beta \in \mathcal{KL}, \gamma \in \mathcal{K}$, and a continuous function $\sigma : \mathbb{R}_+ \to \mathbb{R}_+$, $\sigma(s) > 0$ for $s > 0$ such that for each continuous and bounded input $w(t) \stackrel{\triangle}{=} \begin{bmatrix} F(x(t))\tilde{\theta}(t) \\ \dot{\hat{\theta}}(t) \end{bmatrix}$, for each $x(t_0) \in \mathbb{R}^n$, and for all $t \geq t_0 \geq 0$, the following implication holds:

$$\sup_{t_0 \leq \tau \leq t} |w(\tau)| \leq \sigma(|x(t_0)|)$$

$$\Downarrow \tag{5.35}$$

$$|x(t)| \leq \beta(|x(t_0)|, t - t_0) + \gamma \left(\sup_{t_0 \leq \tau \leq t} |w(\tau)| \right).$$

Let M be such that $|x(t)| \leq M$ for all $t \geq 0$. Let $\varepsilon = \min\{\sigma(r) \mid r \leq M\} > 0$, and let $T \geq 0$ be such that $|w(t)| \leq \varepsilon$ for all $t \geq T$. Then from (5.35) we obtain

$$|x(t)| \leq \beta(|x(t_0)|, t - t_0) + \gamma \left(\sup_{t_0 \leq \tau \leq t} |w(\tau)| \right) \quad (5.36)$$

for all $t \geq t_0 \geq T$. Thus, from time T onward, the system satisfies the ISS inequality (5.36). To complete the proof, we have to show that the ISS property implies that $x(t) \to 0$ as $t \to \infty$. Our computations follow those in the proof of [173, Proposition 7.2]. First, we note that there exists a monotonically decreasing to zero function η continuous on $[T, \infty)$ such that

$$|w(t)| \leq \eta(t - t_0), \qquad \forall t \geq t_0 \geq T. \quad (5.37)$$

Then we have

$$\begin{aligned}
|x(t)| &\leq \beta\left(\left|x\left(\frac{t+T}{2}\right)\right|, \frac{t-T}{2}\right) + \gamma\left(\sup_{\frac{t+T}{2} \leq \tau \leq t} |w(\tau)|\right) \\
&\leq \beta\left(\beta\left(|x(T)|, \frac{t-T}{2}\right) + \gamma\left(\sup_{T \leq \tau \leq \frac{t+T}{2}} |w(\tau)|\right), \frac{t-T}{2}\right) \\
&\quad + \gamma\left(\sup_{\frac{t+T}{2} \leq \tau \leq t} \eta(\tau - T)\right).
\end{aligned} \quad (5.38)$$

Noting that for any class $\mathcal{K}$ function δ, $\delta(a+b) \leq \delta(2a) + \delta(2b)$ for any nonnegative a and b, we proceed from (5.38) with

$$\begin{aligned}
|x(t)| &\leq \beta\left(2\beta\left(|x(T)|, \frac{t-T}{2}\right), \frac{t-T}{2}\right) + \beta\left(2\gamma\left(\sup_{T \leq \tau \leq \frac{t+T}{2}} |w(\tau)|\right), \frac{t-T}{2}\right) \\
&\quad + \gamma\left(\eta\left(\frac{t-T}{2}\right)\right) \\
&\leq \beta\left(2\beta\left(|x(T)|, \frac{t-T}{2}\right), \frac{t-T}{2}\right) \\
&\quad + \beta\left(2\gamma\left(\sup_{T \leq \tau \leq \frac{t+T}{2}} \eta(\tau - T)\right), \frac{t-T}{2}\right) + \gamma\left(\eta\left(\frac{t-T}{2}\right)\right) \\
&\leq \beta\left(2\beta\left(M, \frac{t-T}{2}\right), \frac{t-T}{2}\right) \\
&\quad + \beta\left(2\gamma(\eta(0)), \frac{t-T}{2}\right) + \gamma\left(\eta\left(\frac{t-T}{2}\right)\right),
\end{aligned} \quad (5.39)$$

which converges to zero as $t \to \infty$. □

As we shall see later in this chapter, as well as in Chapter 6, both the passive and the swapping identifiers guarantee that $F(x(t))\tilde{\theta}(t)$ and $\dot{\tilde{\theta}}(t)$ converge to zero whenever $x(t)$ and $u(t)$ are bounded.

5.2 ISS-CONTROL LYAPUNOV FUNCTIONS

The two lemmas in this section outline a framework for modular adaptive design. Lemma 5.2 shows how to design a control law once an ISS-clf is known. While Lemma 5.2 gives sufficient conditions that the identifier has to satisfy to guarantee boundedness of the plant state x, Lemma 5.3 gives sufficient identifier conditions for regulation of x. Identifiers satisfying these conditions will be designed in this chapter and Chapter 6.

Therefore the main task is to find an ISS-clf for a given system. For the simple scalar system

$$\dot{x} = f(x) + F(x)\theta + g(x)u, \quad x \in \mathbb{R}, \tag{5.40}$$

where $g(x) \not\equiv 0$, a valid ISS-clf is $V(x) = x^2$. This is easy to see because the control law

$$u = \alpha(x, \hat{\theta}) = \frac{1}{g(x)}\left(-f(x) - F(x) - x - |F(x)|x\right) \tag{5.41}$$

yields

$$\frac{\partial V}{\partial x}\left[f(x) + F(x)\hat{\theta} + g(x)\alpha(x,\hat{\theta})\right] + \frac{\partial V}{\partial x}F(x)\tilde{\theta} + \frac{\partial V}{\partial \hat{\theta}} \cdot 0$$
$$= -2x^2 - 2|F(x)|x^2 + 2xF(x)\tilde{\theta}$$
$$\leq -2x^2 - 2|xF(x)|(|x| - |\tilde{\theta}|)$$
$$\leq -2x^2 \quad \text{whenever} \quad |x| \geq |\tilde{\theta}|, \tag{5.42}$$

which implies that system (5.40)–(5.41) is ISS with respect to $\left(\tilde{\theta}, \dot{\hat{\theta}}\right)$. (To make the control law $\alpha(x, \hat{\theta})$ smooth, we replace $|F(x)|$ in (5.41) by $\sqrt{F(x)F(x)^{\mathrm{T}} + 1}$.) Since we know how to design ISS-clf's for scalar systems, our approach is recursive: We assume that an ISS-clf is known for an initial system, and construct a new ISS-clf for the initial system augmented by an integrator using backstepping.

Lemma 5.4 (ISS-Backstepping) *If the system*

$$\dot{x} = f(x) + F(x)\theta + g(x)u \tag{5.43}$$

is input-to-state stabilizable with respect to $\left(\tilde{\theta}, \dot{\hat{\theta}}\right)$ *using* $\alpha \in C^1$, *and* $\hat{\theta}$ *is bounded, then the augmented system*

$$\begin{aligned} \dot{x} &= f(x) + F(x)\theta + g(x)\xi \\ \dot{\xi} &= u \end{aligned} \tag{5.44}$$

is also input-to-state stabilizable with respect to $\left(\tilde{\theta}, \dot{\hat{\theta}}\right)$.

Proof. Since (5.43) is input-to-state stabilizable with respect to $\left(\tilde{\theta}, \dot{\hat{\theta}}\right)$, there exists a triple (α, V, ρ) and a class $\mathcal{K}$ function μ such that

$$|x| \geq \rho\left(\left\|\begin{bmatrix}\tilde{\theta}\\\dot{\hat{\theta}}\end{bmatrix}\right\|\right)$$
$$\Downarrow$$
$$\frac{\partial V}{\partial x}\left[f(x) + F(x)\hat{\theta} + g(x)\alpha(x,\hat{\theta})\right] + \frac{\partial V}{\partial x}F(x)\tilde{\theta} + \frac{\partial V}{\partial \hat{\theta}}\dot{\hat{\theta}} \leq -\mu(|x|).$$
(5.45)

In fact, without loss of generality we assume that μ is class $\mathcal{K}_\infty$. It was shown in [173] that if μ is only in class $\mathcal{K}$, the given Lyapunov function V can be modified so that the new μ be in class $\mathcal{K}_\infty$. For $\mu \in \mathcal{K}_\infty$, it was shown in [177] that (5.45) is equivalent to the following 'dissipation' type of characterization:

$$\frac{\partial V}{\partial x}\left[f(x) + F(x)\hat{\theta} + g(x)\alpha(x,\hat{\theta})\right] + \frac{\partial V}{\partial x}F(x)\tilde{\theta} + \frac{\partial V}{\partial \hat{\theta}}\dot{\hat{\theta}} \leq -\mu(|x|) + \pi\left(\left\|\begin{bmatrix}\tilde{\theta}\\\dot{\hat{\theta}}\end{bmatrix}\right\|\right),$$
(5.46)

where π is a class $\mathcal{K}$ function. Since the proof of the affine case considered here is simple, we give it for completeness. It is clear that (5.46) implies (5.45). To see the converse, one only needs to consider the case $|x| \leq \rho\left(\left\|\begin{bmatrix}\tilde{\theta}\\\dot{\hat{\theta}}\end{bmatrix}\right\|\right)$. Since $\hat{\theta}$ is bounded, with Young's inequality one obtains

$$\frac{\partial V}{\partial x}\left[f(x) + F(x)\hat{\theta} + g(x)\alpha(x,\hat{\theta})\right] + \frac{\partial V}{\partial x}F(x)\tilde{\theta} + \frac{\partial V}{\partial \hat{\theta}}\dot{\hat{\theta}} + \mu(|x|)$$
$$\leq \frac{\partial V}{\partial x}\left[f(x) + F(x)\hat{\theta} + g(x)\alpha(x,\hat{\theta})\right] + \mu(|x|)$$
$$+ \frac{1}{4}\left\|\left[\frac{\partial V}{\partial x}F(x), \frac{\partial V}{\partial \hat{\theta}}\right]^T\right\|^2 + \left\|\begin{bmatrix}\tilde{\theta}\\\dot{\hat{\theta}}\end{bmatrix}\right\|^2$$
$$\leq \bar{\mu}(|x|) + \left\|\begin{bmatrix}\tilde{\theta}\\\dot{\hat{\theta}}\end{bmatrix}\right\|^2 \leq \bar{\mu} \circ \rho\left(\left\|\begin{bmatrix}\tilde{\theta}\\\dot{\hat{\theta}}\end{bmatrix}\right\|\right) + \left\|\begin{bmatrix}\tilde{\theta}\\\dot{\hat{\theta}}\end{bmatrix}\right\|^2$$
$$\stackrel{\Delta}{=} \pi\left(\left\|\begin{bmatrix}\tilde{\theta}\\\dot{\hat{\theta}}\end{bmatrix}\right\|\right),$$
(5.47)

where $\bar{\mu}$ is a class $\mathcal{K}_\infty$ function. This completes the proof of (5.46). We will now use (5.46) to show that

$$V_1(x, \xi, \hat{\theta}) = V(x, \hat{\theta}) + \frac{1}{2}(\xi - \alpha(x, \hat{\theta}))^2$$
(5.48)

5.2 ISS-CONTROL LYAPUNOV FUNCTIONS

is an ISS-clf for (5.44). We do this by showing that the control law

$$u = \alpha_1(x,\xi,\hat\theta) = -\frac{\partial V}{\partial x}g - (\xi - \alpha) + \frac{\partial \alpha}{\partial x}\left(f + F\hat\theta + g\xi\right)$$
$$- \left\|\left[\frac{\partial \alpha}{\partial x}F(x), \frac{\partial \alpha}{\partial \hat\theta}\right]^{\mathrm{T}}\right\|^2 (\xi - \alpha) \tag{5.49}$$

achieves input-to-state stabilization of (5.44). Towards this end, consider

$$\dot V_1 = \frac{\partial V}{\partial x}\left[f(x) + F(x)\hat\theta + g(x)\alpha(x,\hat\theta)\right] + \frac{\partial V}{\partial x}F(x)\tilde\theta + \frac{\partial V}{\partial \hat\theta}\dot{\hat\theta} + \frac{\partial V}{\partial x}g(\xi - \alpha)$$
$$+ (\xi - \alpha)\left(u - \frac{\partial \alpha}{\partial x}(f + F\theta + g\xi) - \frac{\partial \alpha}{\partial \hat\theta}\dot{\hat\theta}\right)$$
$$\leq -\mu(|x|) + \pi\left(\left\|\begin{bmatrix}\tilde\theta\\ \dot{\hat\theta}\end{bmatrix}\right\|\right)$$
$$+ (\xi - \alpha)\left(u + \frac{\partial V}{\partial x}g - \frac{\partial \alpha}{\partial x}(f + F\hat\theta + g\xi)\right)$$
$$- (\xi - \alpha)\frac{\partial \alpha}{\partial x}F\tilde\theta - (\xi - \alpha)\frac{\partial \alpha}{\partial \hat\theta}\dot{\hat\theta}$$
$$\leq -\mu(|x|) + \pi\left(\left\|\begin{bmatrix}\tilde\theta\\ \dot{\hat\theta}\end{bmatrix}\right\|\right) - (\xi - \alpha)^2$$
$$- \left\|\left[\frac{\partial \alpha}{\partial x}F, \frac{\partial \alpha}{\partial \hat\theta}\right]^{\mathrm{T}}\right\|^2 (\xi - \alpha)^2 - (\xi - \alpha)\left[\frac{\partial \alpha}{\partial x}F, \frac{\partial \alpha}{\partial \hat\theta}\right]^{\mathrm{T}}\begin{bmatrix}\tilde\theta\\ \dot{\hat\theta}\end{bmatrix}$$
$$\leq -\mu(|x|) + \pi\left(\left\|\begin{bmatrix}\tilde\theta\\ \dot{\hat\theta}\end{bmatrix}\right\|\right) - (\xi - \alpha)^2 + \frac{1}{4}\left\|\begin{bmatrix}\tilde\theta\\ \dot{\hat\theta}\end{bmatrix}\right\|^2. \tag{5.50}$$

Denoting $\pi_1(r) = \pi(r) + \frac{1}{4}r^2$ and picking a class $\mathcal{K}_\infty$ function $\mu_1(r) \leq \min\{\mu(r), r^2\}$, because of the boundedness of $\hat\theta$, we get

$$\dot V_1 \leq -\mu_1\left(\left\|\begin{bmatrix}x\\ \xi - \alpha(x,\xi,\hat\theta)\end{bmatrix}\right\|\right) + \pi_1\left(\left\|\begin{bmatrix}\tilde\theta\\ \dot{\hat\theta}\end{bmatrix}\right\|\right)$$
$$\leq -\mu_2\left(\left\|\begin{bmatrix}x\\ \xi\end{bmatrix}\right\|\right) + \pi_1\left(\left\|\begin{bmatrix}\tilde\theta\\ \dot{\hat\theta}\end{bmatrix}\right\|\right), \tag{5.51}$$

where μ_2 is a class $\mathcal{K}_\infty$ function. Thus, V_1 is an ISS-Lyapunov function. By applying Theorem C.2 to (5.51) with $\rho = \mu_2^{-1} \circ 2\pi_1$, we prove that system (5.44) with control law (5.49) is ISS with respect to $\left(\tilde\theta,\dot{\hat\theta}\right)$. □

The control law $\alpha_1(x,\xi,\theta)$ in (5.49) is only one out of many possible control laws. Once we have shown that V_1 given by (5.48) is an ISS-clf for (5.44) (with

$\rho = \mu_2^{-1} \circ 2\pi_1)$, we can use, for example, the C^0 control law α_1 given by the formula (5.23).

While in Lemma 5.4 the initial system is augmented only by an integrator, a minor modification is sufficient to obtain an analogous result for the more general system

$$\begin{aligned} \dot{x} &= f(x) + F(x)\theta + g(x)\xi \\ \dot{\xi} &= u + F_1(x,\xi)\theta. \end{aligned} \quad (5.52)$$

Corollary 5.5 *The function $V_1(x,\xi,\hat{\theta})$ defined in (5.48) is an ISS-clf for system (5.52) with the control law*

$$\alpha_1(x,\xi,\hat{\theta}) = \alpha_1(x,\xi,\hat{\theta})_{(5.49)} - F_1(x,\xi)\hat{\theta} - |F_1(x,\xi)|^2(\xi - \alpha(x,\hat{\theta})). \quad (5.53)$$

An n-fold application of Corollary 4.9 will provide us with V_n and α_n for the system

$$\begin{aligned} \dot{x} &= f(x) + F(x)\theta + g(x)\xi_1 \\ \dot{\xi}_1 &= \xi_2 + F_1(x,\xi_1)\theta \\ &\vdots \\ \dot{\xi}_{n-1} &= \xi_n + F_{n-1}(x,\xi_1,\ldots,\xi_{n-1})\theta \\ \dot{\xi}_n &= u + F_n(x,\xi_1,\ldots,\xi_n)\theta. \end{aligned} \quad (5.54)$$

We will now develop a detailed design procedure for such systems.

5.3 ISS-Controller Design

Our goal is to develop a modular adaptive design for nonlinear systems in the *parametric strict-feedback* form

$$\begin{aligned} \dot{x}_1 &= x_2 + \varphi_1(x_1)^{\mathrm{T}}\theta \\ \dot{x}_2 &= x_3 + \varphi_2(x_1,x_2)^{\mathrm{T}}\theta \\ &\vdots \\ \dot{x}_{n-1} &= x_n + \varphi_{n-1}(x_1,\ldots,x_{n-1})^{\mathrm{T}}\theta \\ \dot{x}_n &= \beta(x)u + \varphi_n(x)^{\mathrm{T}}\theta, \end{aligned} \quad (5.55)$$

where $\theta \in \mathbb{R}^p$ is a vector of unknown constant parameters, β and $F = [\varphi_1,\cdots,\varphi_n]$ are smooth nonlinear functions in $\mathbb{R}^n$, and $\beta(x) \neq 0$, $\forall x \in \mathbb{R}^n$.

The control objective is to force the output $y = x_1$ of the system (5.55) to asymptotically track the reference output y_r whose first n derivatives are known, bounded and piecewise continuous.

As we announced in Section 5.2, our modular adaptive design for (5.55) will place the burden of achieving boundedness on the controller module. We

5.3 ISS-Controller Design

require that the controller guarantees that the state x be bounded whenever the parameter error $\tilde{\theta} = \theta - \hat{\theta}$ and its derivative $\dot{\tilde{\theta}} = -\dot{\hat{\theta}}$ are bounded.

The next lemma, which builds upon Lemma 2.26, is the main tool for the controller module design.

Lemma 5.6 (Nonlinear Damping) *Assume that for the system*

$$\dot{x} = f(x,t) + g(x,t)\left[u + p(x,t)^T d(t)\right], \qquad x \in \mathbb{R}^n, u \in \mathbb{R} \qquad (5.56)$$

a feedback control $u = \alpha(x,t)$ guarantees

$$\frac{\partial V}{\partial x}[f(x,t) + g(x,t)\alpha(x,t)] + \frac{\partial V}{\partial t} \leq -U(x,t), \qquad \forall x \in \mathbb{R}^n, \forall t \geq 0 \quad (5.57)$$

where the functions V, U are positive definite and radially unbounded in x and V is decrescent, f, g, p, α are continuously differentiable in x and piecewise continuous and bounded in t, and d is piecewise continuous. Then the feedback control

$$u = \alpha(x,t) - \lambda|p(x,t)|^2 \frac{\partial V}{\partial x} g(x,t), \qquad \lambda > 0 \qquad (5.58)$$

guarantees that:

1. *If either of the conditions*

   (a) $d \in \mathcal{L}_\infty$ *or*

   (b) $d \in \mathcal{L}_2$ *and* $U(x,t) \geq cV(x,t), c > 0$

   is satisfied, then $x \in \mathcal{L}_\infty$.

2. *If $d \in \mathcal{L}_2 \cap \mathcal{L}_\infty$ and $U(x,t) \geq c|x|^2$, then $\lim_{t\to\infty} x(t) = 0$.*

Proof. 1. Due to (5.57), the derivative of V along (5.56), (5.58) is

$$\begin{aligned}
\dot{V} &= \frac{\partial V}{\partial x}\left[f + g\alpha + g\left(-\lambda p^T p \frac{\partial V}{\partial x} g + p^T d\right)\right] + \frac{\partial V}{\partial t} \\
&\leq -U - \lambda\left|p\frac{\partial V}{\partial x}g - \frac{1}{2\lambda}d\right|^2 + \frac{1}{4\lambda}|d|^2 \\
&\leq -U + \frac{1}{4\lambda}|d|^2. \qquad (5.59)
\end{aligned}$$

(a) Since U is positive definite and radially unbounded, there exists a class $\mathcal{K}_\infty$ function γ such that $U(x,t) \geq \gamma(|x|)$, and therefore

$$\dot{V} \leq -\gamma(|x|) + \frac{1}{4\lambda}|d|^2. \qquad (5.60)$$

It follows that if $|x| \geq \gamma^{-1}\left(\frac{1}{2\lambda}|d|^2\right)$, then $\dot{V} \leq -\frac{1}{2}\gamma(|x|)$. By Theorem C.2, system (5.56) with control (5.58) is ISS with respect to d. Hence, if $d \in \mathcal{L}_\infty$, then $x \in \mathcal{L}_\infty$. (b) In the case when $U(x,t) \geq cV(x,t)$, from (5.59) we get

$$\dot{V} \leq -cV + \frac{1}{4\lambda}|d|^2. \tag{5.61}$$

With $d \in \mathcal{L}_2$, by Lemma B.5, $x \in \mathcal{L}_\infty$.

2. Integrating (5.59) over $[0,t]$, we obtain

$$c\int_0^t |x(\tau)|^2 d\tau \leq \int_0^t U(x(\tau),\tau)d\tau \leq \frac{1}{4\lambda}\int_0^t |d(\tau)|^2 d\tau + V(0) - V(t)$$
$$\leq \frac{1}{4\lambda}\|d\|_2^2 + V(0), \tag{5.62}$$

which implies that $x \in \mathcal{L}_2$. By part 1 of this lemma, $x \in \mathcal{L}_\infty$, and therefore $u \in \mathcal{L}_\infty$. Hence $\dot{x} \in \mathcal{L}_\infty$. By Barbalat's lemma (Corollary A.7), $x(t) \to 0$ as $t \to \infty$. □

Our primary interest is in part 1 of Lemma 5.6, which states that x is bounded if d is either bounded or square-integrable. In our design, $\tilde{\theta}$ and $\dot{\hat{\theta}}$ play the role of d. The identifiers that we shall use will guarantee that $\tilde{\theta}$ is bounded, and $\dot{\hat{\theta}}$ is either bounded or square-integrable. Part 2 of Lemma 5.6 is also useful. It will help us to achieve tracking because our identifiers will guarantee that $\dot{\hat{\theta}}$ is square-integrable.

Before we develop a general design of a control law for the parametric strict-feedback systems (5.55) we illustrate the use of Lemma 5.6 on a second-order example.

Example 5.7 Let us consider the system

$$\begin{aligned} \dot{x}_1 &= x_2 + \varphi(x_1)^T \theta \\ \dot{x}_2 &= u. \end{aligned} \tag{5.63}$$

Viewing x_2 as a control input, we first design a control law $\alpha_1(x_1, \hat{\theta})$ to guarantee that the state x_1 in $\dot{x}_1 = x_2 + \varphi(x_1)^T \theta$ is bounded whenever $\hat{\theta}$ is bounded. Following Lemma 5.6, we design

$$\alpha_1(x_1, \hat{\theta}) = -c_1 x_1 - \varphi(x_1)^T \hat{\theta} - \kappa_1 |\varphi(x_1)|^2 x_1, \quad c_1, \kappa_1 > 0. \tag{5.64}$$

Then we define the error variable $z_2 = x_2 - \alpha_1(x_1, \hat{\theta})$, and for uniformity denote $z_1 = x_1$. The first equation is now

$$\dot{z}_1 = -c_1 z_1 - \kappa_1 |\varphi|^2 z_1 + \varphi^T \tilde{\theta} + z_2. \tag{5.65}$$

If z_2 were zero, the Lyapunov function $V_1 = \frac{1}{2} z_1^2$ would have the derivative

$$\dot{V}_1 = -c_1 z_1^2 - \kappa_1 |\varphi|^2 z_1^2 + z_1 \varphi^T \tilde{\theta} \leq -c_1 z_1^2 + \frac{1}{4\kappa_1}|\tilde{\theta}|^2, \tag{5.66}$$

5.3 ISS-CONTROLLER DESIGN

which would mean that z_1 is bounded whenever $\tilde{\theta}$ is bounded. With $z_2 \neq 0$ this is no longer clear because then

$$\dot{V}_1 \leq -c_1 z_1^2 + \frac{1}{4\kappa_1}|\tilde{\theta}|^2 + z_1 z_2. \tag{5.67}$$

The second equation in (5.63) yields

$$\dot{z}_2 = \dot{x}_2 - \dot{\alpha}_1 = u - \frac{\partial \alpha_1}{\partial x_1}\left(x_2 + \varphi^T \theta\right) - \frac{\partial \alpha_1}{\partial \hat{\theta}}\dot{\hat{\theta}}. \tag{5.68}$$

The derivative of the Lyapunov function $V_2 = V_1 + \frac{1}{2}z_2^2 = \frac{1}{2}|z|^2$ is

$$\begin{aligned}
\dot{V}_2 &\leq -c_1 z_1^2 + \frac{1}{4\kappa_1}|\tilde{\theta}|^2 + z_1 z_2 + z_2\left[u - \frac{\partial \alpha_1}{\partial x_1}\left(x_2 + \varphi^T \theta\right) - \frac{\partial \alpha_1}{\partial \hat{\theta}}\dot{\hat{\theta}}\right] \\
&\leq -c_1 z_1^2 + \frac{1}{4\kappa_1}|\tilde{\theta}|^2 + z_2\left[u + z_1 - \frac{\partial \alpha_1}{\partial x_1}\left(x_2 + \varphi^T \hat{\theta}\right)\right] \\
&\quad - z_2\left(\frac{\partial \alpha_1}{\partial x_1}\varphi^T \tilde{\theta} + \frac{\partial \alpha_1}{\partial \hat{\theta}}\dot{\hat{\theta}}\right).
\end{aligned} \tag{5.69}$$

Our design is now led by Lemma 5.6. To make the bracketed term equal to $-c_2 z_2 - \kappa_2 \left|\frac{\partial \alpha_1}{\partial x_1}\varphi\right|^2 z_2 - g_2 \left|\frac{\partial \alpha_1}{\partial \hat{\theta}}^T\right|^2 z_2$, we design the control law

$$u = -z_1 - c_2 z_2 - \kappa_2 \left|\frac{\partial \alpha_1}{\partial x_1}\varphi\right|^2 z_2 - g_2 \left|\frac{\partial \alpha_1}{\partial \hat{\theta}}^T\right|^2 z_2 + \frac{\partial \alpha_1}{\partial x_1}\left(x_2 + \varphi^T \hat{\theta}\right), \tag{5.70}$$

where $c_2, \kappa_2, g_2 > 0$. Thus, using completion of the squares as in (5.59), we get

$$\dot{V}_2 \leq -c_1 z_1^2 - c_2 z_2^2 + \left(\frac{1}{4\kappa_1} + \frac{1}{4\kappa_2}\right)|\tilde{\theta}|^2 + \frac{1}{4g_2}|\dot{\hat{\theta}}|^2, \tag{5.71}$$

which means that the state of the error system

$$\dot{z} = \begin{bmatrix} -c_1 - \kappa_1|\varphi|^2 & 1 \\ -1 & -c_2 - \kappa_2\left|\frac{\partial \alpha_1}{\partial x_1}\varphi\right|^2 - g_2\left|\frac{\partial \alpha_1}{\partial \hat{\theta}}^T\right|^2 \end{bmatrix} z$$
$$+ \begin{bmatrix} \varphi^T \\ -\frac{\partial \alpha_1}{\partial x_1}\varphi^T \end{bmatrix} \tilde{\theta} + \begin{bmatrix} 0 \\ -\frac{\partial \alpha_1}{\partial \hat{\theta}} \end{bmatrix} \dot{\hat{\theta}} \tag{5.72}$$

is bounded whenever the disturbance inputs $\tilde{\theta}$ and $\dot{\hat{\theta}}$ are bounded. $\diamond$

We now consider the parametric strict-feedback systems (5.55). The recursive design procedure is given in Table 5.1.[2]

---

[2] For notational convenience we define $z_0 \triangleq 0$, $\alpha_0 \triangleq 0$.

Table 5.1: ISS-Controller

$$z_i = x_i - y_r^{(i-1)} - \alpha_{i-1} \tag{5.73}$$

$$\alpha_i(\bar{x}_i, \hat{\theta}, \bar{y}_r^{(i-1)}) = -z_{i-1} - c_i z_i - w_i^T \hat{\theta} + \sum_{k=1}^{i-1}\left(\frac{\partial \alpha_{i-1}}{\partial x_k} x_{k+1} + \frac{\partial \alpha_{i-1}}{\partial y_r^{(k-1)}} y_r^{(k)}\right)$$
$$- s_i z_i \tag{5.74}$$

$$w_i(\bar{x}_i, \hat{\theta}, \bar{y}_r^{(i-2)}) = \varphi_i - \sum_{k=1}^{i-1} \frac{\partial \alpha_{i-1}}{\partial x_k} \varphi_k \tag{5.75}$$

$$s_i(\bar{x}_i, \hat{\theta}, \bar{y}_r^{(i-2)}) = \kappa_i |w_i|^2 + g_i \left|\frac{\partial \alpha_{i-1}}{\partial \hat{\theta}}^T\right|^2 \tag{5.76}$$

$$i = 1, \ldots, n$$

$$\bar{x}_i = (x_1, \ldots, x_i), \qquad \bar{y}_r^{(i)} = (y_r, \dot{y}_r, \ldots, y_r^{(i)})$$

Adaptive control law:

$$u = \frac{1}{\beta(x)}\left[\alpha_n(x, \hat{\theta}, \bar{y}_r^{(n-1)}) + y_r^{(n)}\right] \tag{5.77}$$

By comparing the expression for the stabilizing function (5.74) in the modular design with the expression (4.190) for the tuning functions design we see that the difference is in the second lines. While the stabilization in the tuning functions design is achieved using the terms $\frac{\partial \alpha_{i-1}}{\partial \hat{\theta}} \Gamma \tau_i + \sum_{k=2}^{i-1} \frac{\partial \alpha_{k-1}}{\partial \hat{\theta}} \Gamma w_i z_k$, the stabilization in the modular design is achieved with the nonlinear damping term $-s_i z_i$.

Claim. The closed-loop system obtained by applying the design procedure (5.73)–(5.77) to system (5.55) is

$$\dot{z} = A_z(z, \hat{\theta}, t)z + W(z, \hat{\theta}, t)^T \tilde{\theta} + Q(z, \hat{\theta}, t)^T \dot{\hat{\theta}}, \qquad z \in \mathbb{R}^n, \tag{5.78}$$

where A_z, W, and Q are matrix-valued functions of z, $\hat{\theta}$, and t:

$$A_z(z, \hat{\theta}, t) = \begin{bmatrix} -c_1 - s_1 & 1 & 0 & \cdots & 0 \\ -1 & -c_2 - s_2 & 1 & \ddots & \vdots \\ 0 & -1 & \ddots & \ddots & 0 \\ \vdots & & \ddots & \ddots & 1 \\ 0 & \cdots & 0 & -1 & -c_n - s_n \end{bmatrix}$$

5.3 ISS-Controller Design

$$W(z,\hat{\theta},t)^{\mathrm{T}} = \begin{bmatrix} w_1^{\mathrm{T}} \\ w_2^{\mathrm{T}} \\ \vdots \\ w_n^{\mathrm{T}} \end{bmatrix} \in \mathbb{R}^{n \times p}, \quad Q(z,\hat{\theta},t)^{\mathrm{T}} = \begin{bmatrix} 0 \\ -\frac{\partial \alpha_1}{\partial \hat{\theta}} \\ \vdots \\ -\frac{\partial \alpha_{n-1}}{\partial \hat{\theta}} \end{bmatrix} \in \mathbb{R}^{n \times p}. \quad (5.79)$$

Proof. For $i = 1$ we have

$$\begin{aligned} \dot{z}_1 &= \dot{x}_1 - \dot{y}_\mathrm{r} \\ &= x_2 + \varphi_1^{\mathrm{T}}\theta - \dot{y}_\mathrm{r} \\ &= z_2 + \alpha_1 + w_1^{\mathrm{T}}\theta \\ &= -(c_1 + s_1)z_1 + z_2 + w_1^{\mathrm{T}}\tilde{\theta}. \end{aligned} \quad (5.80)$$

For $i = 2, \ldots, n-1$, since α_{i-1} is a function of only the variables $x_1, \ldots, x_{i-1}$, $\hat{\theta}$, $y_\mathrm{r}, \ldots, y_\mathrm{r}^{(i-2)}$, we have

$$\begin{aligned} \dot{z}_i &= \dot{x}_i - \dot{y}_\mathrm{r}^{(i-1)} - \dot{\alpha}_{i-1} \\ &= x_{i+1} + \varphi_i^{\mathrm{T}}\theta - y_\mathrm{r}^{(i)} - \sum_{k=1}^{i-1} \frac{\partial \alpha_{i-1}}{\partial x_k}\left(x_{k+1} + \varphi_k^{\mathrm{T}}\theta\right) \\ &\quad - \frac{\partial \alpha_{i-1}}{\partial \hat{\theta}}\dot{\hat{\theta}} - \sum_{k=1}^{i-1} \frac{\partial \alpha_{i-1}}{\partial y_\mathrm{r}^{(k-1)}} y_\mathrm{r}^{(k)} \\ &= z_{i+1} + \alpha_i + w_i^{\mathrm{T}}\theta - \sum_{k=1}^{i-1}\left(\frac{\partial \alpha_{i-1}}{\partial x_k}x_{k+1} + \frac{\partial \alpha_{i-1}}{\partial y_\mathrm{r}^{(k-1)}} y_\mathrm{r}^{(k)}\right) - \frac{\partial \alpha_{i-1}}{\partial \hat{\theta}}\dot{\hat{\theta}} \\ &= -z_{i-1} - (c_i + s_i)z_i + z_{i+1} + w_i^{\mathrm{T}}\tilde{\theta} - \frac{\partial \alpha_{i-1}}{\partial \hat{\theta}}\dot{\hat{\theta}}. \end{aligned} \quad (5.81)$$

For $i = n$ we have

$$\begin{aligned} \dot{z}_n &= \dot{x}_n - \dot{y}_\mathrm{r}^{(n-1)} - \dot{\alpha}_{n-1} \\ &= \beta u + y_\mathrm{r}^{(n)} + w_n^{\mathrm{T}}\theta \\ &\quad - \sum_{k=1}^{n-1}\left(\frac{\partial \alpha_{n-1}}{\partial x_k}x_{k+1} + \frac{\partial \alpha_{n-1}}{\partial y_\mathrm{r}^{(k-1)}} y_\mathrm{r}^{(k)}\right) - \frac{\partial \alpha_{n-1}}{\partial \hat{\theta}}\dot{\hat{\theta}} \\ &= -z_{n-1} - (c_n + s_n)z_n + z_{n+1} + w_n^{\mathrm{T}}\tilde{\theta} - \frac{\partial \alpha_{n-1}}{\partial \hat{\theta}}\dot{\hat{\theta}}. \end{aligned} \quad (5.82)$$

Writing (5.80)–(5.82) in vector form yields (5.78)–(5.79). □

System (5.78) will be referred to as the *error system*. Note that the first component of its state, $z_1 = x_1 - y_\mathrm{r} = y - y_\mathrm{r}$, represents the tracking error.

The change of coordinates (5.73)–(5.75), which we compactly write as

$$z = \Psi(x,\hat{\theta},t), \quad (5.83)$$

is smooth in x and $\hat{\theta}$ and is bounded in t. Note also that the inverse transformation

$$x = \Phi(z, \hat{\theta}, t) \tag{5.84}$$

is smooth in z and $\hat{\theta}$ and is bounded in t.

Except for the term $Q(z,\hat{\theta},t)^{\mathrm{T}}\dot{\hat{\theta}}$, the error system (5.78) is similar to the error system (4.195) in which the term $Q(z,\hat{\theta},t)^{\mathrm{T}}\dot{\hat{\theta}}$ was accounted for by using *tuning functions*. Here we let both $\tilde{\theta}$ and $\dot{\hat{\theta}}$ appear as disturbance inputs. In the modular design their boundedness will be guaranteed by parameter identifiers.

We now establish the basic input-to-state properties of the error system (5.76), (5.78), (5.79), making use of the following constants:

$$c_0 = \min_{1 \le i \le n} c_i, \quad \kappa_0 = \left(\sum_{i=1}^{n} \frac{1}{\kappa_i}\right)^{-1}, \quad g_0 = \left(\sum_{i=1}^{n} \frac{1}{g_i}\right)^{-1}. \tag{5.85}$$

Lemma 5.8 *In the error system (5.78), (5.79), (5.76), the following input-to-state properties hold:*

(i) If $\tilde{\theta}, \dot{\hat{\theta}} \in \mathcal{L}_\infty[0, t_f)$, then $z, x \in \mathcal{L}_\infty[0, t_f)$, and

$$|z(t)| \le \frac{1}{2\sqrt{c_0}}\left(\frac{1}{\kappa_0}\|\tilde{\theta}\|_\infty^2 + \frac{1}{g_0}\|\dot{\hat{\theta}}\|_\infty^2\right)^{\frac{1}{2}} + |z(0)|e^{-c_0 t}. \tag{5.86}$$

(ii) If $\tilde{\theta} \in \mathcal{L}_\infty[0, t_f)$ and $\dot{\hat{\theta}} \in \mathcal{L}_2[0, t_f)$, then $z, x \in \mathcal{L}_\infty[0, t_f)$, and

$$|z(t)| \le \left(\frac{1}{4c_0\kappa_0}\|\tilde{\theta}\|_\infty^2 + \frac{1}{2g_0}\|\dot{\hat{\theta}}\|_2^2\right)^{\frac{1}{2}} + |z(0)|e^{-c_0 t}. \tag{5.87}$$

Proof. Differentiating $\frac{1}{2}|z|^2$ along the solutions of (5.78) we compute

$$\begin{aligned}\frac{d}{dt}\left(\frac{1}{2}|z|^2\right) &= -\sum_{i=1}^{n} c_i z_i^2 - \sum_{i=1}^{n}\left(\kappa_i|w_i|^2 + g_i\left|\frac{\partial \alpha_{i-1}}{\partial \hat{\theta}}^{\mathrm{T}}\right|^2\right)z_i^2 \\ &\quad + \sum_{i=1}^{n} z_i\left(w_i^{\mathrm{T}}\tilde{\theta} - \frac{\partial \alpha_{i-1}}{\partial \hat{\theta}}\dot{\hat{\theta}}\right) \\ &\le -c_0|z|^2 - \sum_{i=1}^{n}\kappa_i\left|w_i z_i - \frac{1}{2\kappa_i}\tilde{\theta}\right|^2 - \sum_{i=1}^{n} g_i\left|\frac{\partial \alpha_{i-1}}{\partial \hat{\theta}}^{\mathrm{T}} z_i + \frac{1}{2g_i}\dot{\hat{\theta}}\right|^2 \\ &\quad + \left(\sum_{i=1}^{n}\frac{1}{4\kappa_i}\right)|\tilde{\theta}|^2 + \left(\sum_{i=1}^{n}\frac{1}{4g_i}\right)|\dot{\hat{\theta}}|^2 \end{aligned} \tag{5.88}$$

and arrive at

$$\frac{d}{dt}\left(\frac{1}{2}|z|^2\right) \le -c_0|z|^2 + \frac{1}{4}\left(\frac{1}{\kappa_0}|\tilde{\theta}|^2 + \frac{1}{g_0}|\dot{\hat{\theta}}|^2\right). \tag{5.89}$$

5.3 ISS-Controller Design

(i) From Lemma C.5*(i)*, taking $v = z^2$ and $\rho = \left(\frac{1}{\kappa_0}|\tilde{\theta}|^2 + \frac{1}{g_0}|\dot{\hat{\theta}}|^2\right)^{1/2}$, it follows that

$$|z(t)|^2 \leq |z(0)|^2 e^{-2c_0 t} + \frac{1}{4c_0}\left(\frac{1}{\kappa_0}\|\tilde{\theta}\|_\infty^2 + \frac{1}{g_0}\|\dot{\hat{\theta}}\|_\infty^2\right), \qquad (5.90)$$

which proves $z \in \mathcal{L}_\infty[0, t_f)$ and (5.86), and by (5.84), $x \in \mathcal{L}_\infty[0, t_f)$.

(ii) By Lemma C.5, taking $\rho_1 = \tilde{\theta}$ and $\rho_2 = \dot{\hat{\theta}}$, from (5.89) we get

$$|z(t)|^2 \leq |z(0)|^2 e^{-2c_0 t} + \frac{1}{4c_0\kappa_0}\|\tilde{\theta}\|_\infty^2 + \frac{1}{2g_0}\|\dot{\hat{\theta}}\|_2^2, \qquad (5.91)$$

which proves $z \in \mathcal{L}_\infty[0, t_f)$ and (5.87), and by (5.84), $x \in \mathcal{L}_\infty[0, t_f)$. □

As we can see from Lemma 5.8, with nonlinear damping we achieve not only input-to-state stability (5.86) in the sense of Definition C.1, but also the input-to-state property (5.87). With respect to $\dot{\hat{\theta}}$, this property can be understood as "$\mathcal{L}_2$-input → $\mathcal{L}_\infty$-output" stability, but it can also be seen as ISS with $\|\dot{\hat{\theta}}\|_2$ considered as input. While this property is not important in swapping-based schemes, it is crucial in passive schemes where boundedness of $\dot{\hat{\theta}}$ can be independently guaranteed by the identifier only in the $\mathcal{L}_2$ sense.

The quadratic form of the nonlinear damping functions is only one out of many possible forms. Any power greater than one would yield an ISS property, but the proof with quadratic nonlinear damping is by far the simplest.

A consequence of Lemma 5.8 is that, even when the adaptation is switched off, that is, when the parameter estimate $\hat{\theta}$ is constant ($\dot{\hat{\theta}} = 0$) and the only disturbance input is $\tilde{\theta}$, the state z of the error system (5.78), (5.79), (5.76) remains bounded and converges exponentially to a positively invariant compact set. (Note that the terms $-g_i \left|\frac{\partial \alpha_{i-1}}{\partial \hat{\theta}}\right|^2 z_i$ are not needed when $\dot{\hat{\theta}} = 0$.) Moreover, when the adaptation is switched off, this boundedness result holds even when the unknown parameter is time-varying.

Corollary 5.9 (Boundedness Without Adaptation) *If $\theta : \mathbb{R}_+ \to \mathbb{R}^p$ is piecewise continuous and bounded and $\hat{\theta}$ is constant, then $z, x \in \mathcal{L}_\infty$, and*

$$|z(t)| \leq \frac{1}{2\sqrt{c_0\kappa_0}} \sup_{\tau \geq 0} |\theta(\tau) - \hat{\theta}| + |z(0)|e^{-c_0 t}. \qquad (5.92)$$

Proof. Since $\dot{\hat{\theta}}(t) \equiv 0$, (5.89) holds with $\tilde{\theta}(t) = \theta(t) - \hat{\theta}$. □

Thus, the controller module alone guarantees boundedness, and the task of adaptation is to achieve tracking.

In fact, a stronger result given by Theorem 4.15 for the tuning functions scheme also holds: if adaptation is disconnected and the parameter error is sufficiently small, not only will the global boundedness be achieved, but also the global asymptotic stability.

5.4 Observers for Strict Passivity

Having designed a controller module which achieves input-to-state stability with respect to $\tilde{\theta}$ and $\dot{\hat{\theta}}$, we turn our attention to the design of an identifier module which guarantees the boundedness of $\tilde{\theta}$ and $\dot{\hat{\theta}}$. This chapter deals only with those modular schemes which use passive identifiers. Swapping schemes will be the subject of the next chapter.

In order to design identifiers which guarantee boundedness of $\tilde{\theta}$, let us consider the negative feedback connection in Figure 5.1. It consists of a transfer matrix $\dfrac{\Gamma}{s}$, $\Gamma = \Gamma^T > 0$, which is passive, and a nonlinear dynamical system Σ whose input is the parameter error $\tilde{\theta}$. If we can design the system Σ and select an output τ so that Σ is strictly passive from $\tilde{\theta}$ to τ, then by Theorem D.4 the equilibrium at the origin of the interconnected system in Figure 5.1 is globally uniformly stable (and, in addition, the state of system Σ converges to zero). Thus, in order to guarantee the boundedness of $\tilde{\theta}$, it suffices to design a strictly passive system Σ and let $\dot{\tilde{\theta}} = -\dot{\hat{\theta}} = -\Gamma\tau$.

Now we present the design of observers whose error systems are strictly passive from $\tilde{\theta}$ as the input, to a judiciously selected output τ. These error systems will play the role of Σ in our identifier design.

The parametric z-model

Let us first discuss the parametric model

$$\dot{z} = A_z(z,\hat{\theta},t)z + W(z,\hat{\theta},t)^T\tilde{\theta} + Q(z,\hat{\theta},t)^T\dot{\hat{\theta}}. \tag{5.93}$$

If the term $Q(z,\hat{\theta},t)^T\dot{\hat{\theta}}$ were not present, we would have strict passivity from the input $\tilde{\theta}$ to the output $W(z,\hat{\theta},t)z$. To see this, let us consider the system

$$\dot{\tilde{z}} = A_z(z,\hat{\theta},t)\tilde{z} + W(z,\hat{\theta},t)^T\tilde{\theta}, \tag{5.94}$$

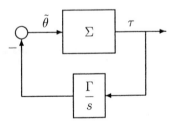

Figure 5.1: Negative feedback interconnection of the passive system Γ/s, $\Gamma = \Gamma^T > 0$, with a dynamic nonlinear system Σ. We have to design the system Σ and select an output τ such that Σ is strictly passive with input $\tilde{\theta}$.

5.4 Observers for Strict Passivity

which is a copy of (5.93) without $Q(z,\hat{\theta},t)^T\dot{\hat{\theta}}$. In view of (5.79), system (5.94) satisfies

$$\frac{d}{dt}\left(\frac{1}{2}|\tilde{z}|^2\right) \leq -c|\tilde{z}|^2 + \tilde{z}^T W(z,\hat{\theta},t)^T\tilde{\theta}. \tag{5.95}$$

Integrating over $[0,t]$ we obtain

$$\int_0^t (W\tilde{z})^T\tilde{\theta}d\tau \geq \frac{1}{2}|\tilde{z}(t)|^2 - \frac{1}{2}|\tilde{z}(0)|^2 + c\int_0^t |\tilde{z}(\tau)|^2 d\tau, \tag{5.96}$$

which by Definition D.2 proves that (5.94) possesses a strict passivity property from the input $\tilde{\theta}$ to the output $W(z,\hat{\theta},t)\tilde{z}$, or in other words the nonlinear operator

$$\Sigma_z : \tilde{\theta} \mapsto W(z,\hat{\theta},t)\tilde{z} \tag{5.97}$$

is strictly passive. To eliminate the term $Q(z,\hat{\theta},t)^T\dot{\hat{\theta}}$ from (5.93), we introduce the observer

$$\dot{\hat{z}} = A_z(z,\hat{\theta},t)\hat{z} + Q(z,\hat{\theta},t)^T\dot{\hat{\theta}} \tag{5.98}$$

and define the observer error

$$\tilde{z} = z - \hat{z}. \tag{5.99}$$

It is readily verified that $\tilde{z}$ is governed by (5.94). Hence, with the addition of an observer, we have generated a strictly passive error system whose state is available.

The parametric x-model

Our goal is to design a parameter identifier for nonlinear systems in the parametric strict-feedback form (5.55):

$$\begin{aligned}\dot{x}_i &= x_{i+1} + \varphi_i(x_1,\ldots,x_i)^T\theta, \quad 1 \leq i \leq n-1 \\ \dot{x}_n &= \beta(x)u + \varphi_n(x)^T\theta,\end{aligned} \tag{5.100}$$

where $\theta \in \mathbb{R}^p$ is the vector of unknown constant parameters, and the complete state x is assumed to be available for measurement. System (5.100) is a special case of the general affine parametric model:

$$\dot{x} = f(x,u) + F(x,u)^T\theta, \qquad x \in \mathbb{R}^n, \tag{5.101}$$

where vector f and the "regressor" matrix F are defined by[3]

$$f(x,u) = \begin{bmatrix} x_2 \\ \vdots \\ x_n \\ \beta_0(x)u \end{bmatrix}, \quad F(x,u)^T = \begin{bmatrix} \varphi_1(x_1)^T \\ \vdots \\ \varphi_{n-1}(x_1,\ldots,x_{n-1})^T \\ \varphi_n(x)^T \end{bmatrix}. \tag{5.102}$$

---

[3]Even though F in (5.102) does not depend on u, we allow this dependence in (5.101) because our identifier design will be applicable to general linearly parametrized nonlinear systems.

It was easy to achieve strict passivity with the parametric z-model because the undriven system was exponentially stable. How can we bring the parametric x-model (5.101) into a form similar to (5.94)? First, we need the presence of $\tilde{\theta}$ instead of θ; second, we need an exponentially stable homogeneous part; and, third, we must remove $f(x, u)$. Namely, we would prefer to have the model

$$\dot{\tilde{x}} = A(x, t)\tilde{x} + F(x, u)^\mathrm{T}\tilde{\theta}, \tag{5.103}$$

whose homogeneous part is exponentially stable:

$$PA(x, t) + A(x, t)^\mathrm{T} P \leq -I, \qquad P = P^\mathrm{T} > 0. \tag{5.104}$$

To obtain (5.103), we introduce the observer

$$\dot{\hat{x}} = A(x, t)(\hat{x} - x) + f(x, u) + F(x, u)^\mathrm{T}\hat{\theta}. \tag{5.105}$$

Its error state

$$\tilde{x} = x - \hat{x} \tag{5.106}$$

is governed by the error system (5.103). This error system satisfies

$$\frac{d}{dt}\left(|\tilde{x}|_P^2\right) \leq -|\tilde{x}|^2 + 2\tilde{x}^\mathrm{T} PF(x, u)^\mathrm{T}\tilde{\theta}, \tag{5.107}$$

which upon integration turns into

$$2\int_0^t (FP\tilde{x})^\mathrm{T}\tilde{\theta}\,d\tau \geq |\tilde{x}(t)|_P^2 - |\tilde{x}(0)|_P^2 + \int_0^t |\tilde{x}(\tau)|^2\,d\tau. \tag{5.108}$$

By Definition D.2, this establishes the strict passivity from the input $\tilde{\theta}$ to the output $F(x, u)P\tilde{x}$, that is, the strict passivity of the nonlinear operator

$$\Sigma_x : \tilde{\theta} \mapsto F(x, u)P\tilde{x}. \tag{5.109}$$

It is important to note that with passivity we can only claim the boundedness of $\tilde{\theta}$. The boundedness of $\hat{\theta}$ is yet to be dealt with. It turns out that with passive identifiers we can achieve boundedness of $\hat{\theta}$ in the $\mathcal{L}_2$ sense but not in the $\mathcal{L}_\infty$ sense, which means that we will depend on part (ii) of Lemma 5.8.

We present two passive schemes: the z-passive scheme and the x-passive scheme. The z-passive identifier is based on the parametric z-model, whereas the x-passive identifier is based on the parametric x-model.

5.5 z-Passive Scheme

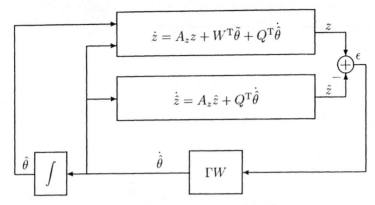

Figure 5.2: The z-passive identifier.

5.5 z-Passive Scheme

We consider the parametric z-model

$$\dot{z} = A_z(z,\hat{\theta},t)z + W(z,\hat{\theta},t)^T\tilde{\theta} + Q(z,\hat{\theta},t)^T\dot{\hat{\theta}} \qquad (5.110)$$

and the observer

$$\dot{\hat{z}} = A_z(z,\hat{\theta},t)\hat{z} + Q(z,\hat{\theta},t)^T\dot{\hat{\theta}}, \qquad (5.111)$$

which is a copy of the system (5.110) with the term $W(z,\hat{\theta},t)^T\tilde{\theta}$ omitted. The observer error

$$\epsilon = z - \hat{z} \qquad (5.112)$$

is governed by an equation driven by $\tilde{\theta}$:

$$\dot{\epsilon} = A_z(z,\hat{\theta},t)\epsilon + W(z,\hat{\theta},t)^T\tilde{\theta}. \qquad (5.113)$$

As we have explained in Section 5.4, the observer error system possesses a strict passivity property from the input $\tilde{\theta}$ to the output $W(z,\hat{\theta},t)\epsilon$, that is, the operator Σ_z defined in (5.97) is strictly passive. Therefore, we choose $\dot{\hat{\theta}} = -\Gamma\Sigma_z\{\tilde{\theta}\}$, that is,

$$\dot{\hat{\theta}} = \Gamma W(z,\hat{\theta},t)\epsilon, \qquad \Gamma = \Gamma^T > 0. \qquad (5.114)$$

The basic properties of the z-passive identifier (Figure 5.2) are as follows.

Lemma 5.10 *Let the maximum interval of existence of solutions of (5.110), (5.113) and (5.114) be $[0,t_f)$. Then the following identifier properties hold:*

$$\begin{align}
(i) \quad & \tilde{\theta} \in \mathcal{L}_\infty[0,t_f) & (5.115)\\
(ii) \quad & \epsilon \in \mathcal{L}_2[0,t_f) \cap \mathcal{L}_\infty[0,t_f) & (5.116)\\
(iii) \quad & \dot{\hat{\theta}} \in \mathcal{L}_2[0,t_f). & (5.117)
\end{align}$$

Proof. Let us introduce a Lyapunov-like function

$$V = \frac{1}{2}|\tilde{\theta}|^2_{\Gamma^{-1}} + \frac{1}{2}|\epsilon|^2 . \tag{5.118}$$

Its derivative along the solutions of (5.113)–(5.114) is

$$\begin{aligned}
\dot{V} &= -\tilde{\theta}^T \Gamma^{-1} \dot{\hat{\theta}} + \frac{1}{2}\epsilon^T \left(A_z + A_z^T\right)\epsilon + \epsilon^T W^T \tilde{\theta} \\
&= -c|\epsilon|^2 - \sum_{i=1}^{n} \left(\kappa_i |w_i|^2 + g_i \left|\frac{\partial \alpha_{i-1}}{\partial \hat{\theta}}^T\right|^2\right)\epsilon_i^2 + \tilde{\theta}^T \Gamma^{-1}\left(\Gamma W \epsilon - \dot{\hat{\theta}}\right) \\
&\leq -c|\epsilon|^2 - \sum_{i=1}^{n} \kappa_i |w_i|^2 \epsilon_i^2 .
\end{aligned} \tag{5.119}$$

The nonpositivity of $\dot{V}$ proves that $\tilde{\theta}$ and ϵ are in $\mathcal{L}_\infty[0, t_f)$. Integrating (5.119) we get

$$c\int_0^t |\epsilon|^2 d\tau \leq -\int_0^t \dot{V} d\tau \leq V(0) - V(t) \leq V(0) < \infty \tag{5.120}$$

which proves that $\epsilon \in \mathcal{L}_2[0, t_f)$. To prove (iii), let us consider

$$\begin{aligned}
|\dot{\hat{\theta}}|^2 &= \epsilon^T W^T \Gamma^2 W \epsilon \leq \bar{\lambda}(\Gamma)^2 \epsilon^T W^T W \epsilon \\
&= \bar{\lambda}(\Gamma)^2 \left|\sum_{i=1}^{n} w_i \epsilon_i\right|^2 \\
&\leq \bar{\lambda}(\Gamma)^2 n \sum_{i=1}^{n} |w_i|^2 \epsilon_i^2 \\
&\leq \frac{\bar{\lambda}(\Gamma)^2 n}{\kappa_m} \sum_{i=1}^{n} \kappa_i |w_i|^2 \epsilon_i^2 ,
\end{aligned} \tag{5.121}$$

where $\kappa_m \triangleq \min_{1 \leq i \leq n} \kappa_i$. Substituting (5.121) into (5.119) we get

$$\dot{V} \leq -c|\epsilon|^2 - \frac{\kappa_m}{\bar{\lambda}(\Gamma)^2 n}|\dot{\hat{\theta}}|^2 , \tag{5.122}$$

which upon integration yields

$$\int_0^t |\dot{\hat{\theta}}|^2 d\tau \leq \frac{\bar{\lambda}(\Gamma)^2 n}{\kappa_m} V(0) < \infty , \tag{5.123}$$

and hence $\dot{\hat{\theta}} \in \mathcal{L}_2[0, t_f)$. □

The most important fact in this lemma is the $\mathcal{L}_2$-property of $\dot{\hat{\theta}}$ achieved with the nonlinear damping terms $\kappa_1|w_1|^2, \ldots, \kappa_n|w_n|^2$.

5.5 z-PASSIVE SCHEME

The properties established for the identifier hold only as long as the solution to the plant differential equation exists. Because the right-hand side of this equation is locally Lipschitz in the state variables and piecewise continuous in time, the existence of solutions over an open interval $[0, t_f)$ is assured (see, e.g., [81, Theorem 2.2]). Lemma 5.10 establishes that even if the plant state escapes to infinity as $t \to t_f$, the solutions of the identifier equation is *uniformly* bounded by a constant independent of t_f. The same boundedness property of all the signals on $[0, t_f)$ can then be deduced with Lemma 5.8. The independence of the bounds of t_f implies that $t_f = \infty$. All our proofs of boundedness for modular schemes will follow this pattern because of the lack of Lyapunov functions encompassing both the states of the identifier and those of the plant.

The only exception is the z-passive scheme where we can construct a single Lyapunov function. For this scheme we establish a stronger stability property than for other modular schemes: In addition to global uniform boundedness and asymptotic tracking, we prove global uniform stability of the reference trajectory.

Theorem 5.11 ($\hat{z}$-Passive) *The closed-loop adaptive system consisting of the plant (5.55), controller (5.77), observer (5.111), and update law (5.114) has a globally uniformly stable equilibrium at the origin $z = 0, \epsilon = 0, \tilde{\theta} = 0$, and $\lim_{t\to\infty} z(t) = \lim_{t\to\infty} \epsilon(t) = 0$. This means, in particular, that global asymptotic tracking is achieved:*

$$\lim_{t\to\infty} [y(t) - y_{\mathrm{r}}(t)] = 0. \tag{5.124}$$

Moreover, if $\lim_{t\to\infty} y_{\mathrm{r}}^{(i)}(t) = 0$, $i = 0, \ldots, n-1$, and $F(0) = 0$, then $\lim_{t\to\infty} x(t) = 0$.

Proof. We make use of a constant $\mu > 0$ to be chosen later. Along the solutions of (5.111), (5.113), (5.114), we have

$$\begin{aligned}
\frac{d}{dt}\left(\frac{\mu}{2}|\hat{z}|^2 + \frac{1}{2}|\epsilon|^2 + \frac{1}{2}|\tilde{\theta}|^2_{\Gamma^{-1}}\right) &= -\mu \sum_{i=1}^{n}\left(c_i + \kappa_i|w_i|^2 + g_i\left|\frac{\partial \alpha_{i-1}}{\partial \hat{\theta}}^{\mathrm{T}}\right|^2\right)\hat{z}_i^2 \\
&\quad -\mu \sum_{i=1}^{n} \hat{z}_i \frac{\partial \alpha_{i-1}}{\partial \hat{\theta}}\dot{\hat{\theta}} \\
&\quad -\sum_{i=1}^{n}\left(c_i + \kappa_i|w_i|^2 + g_i\left|\frac{\partial \alpha_{i-1}}{\partial \hat{\theta}}^{\mathrm{T}}\right|^2\right)\epsilon_i^2 \\
&\quad +\epsilon^{\mathrm{T}}W^{\mathrm{T}}\tilde{\theta} - \tilde{\theta}^{\mathrm{T}}\Gamma^{-1}\dot{\hat{\theta}} \\
&\leq -\mu c_0|\hat{z}|^2 - \mu\sum_{i=1}^n g_i\left|\frac{\partial \alpha_{i-1}}{\partial \hat{\theta}}^{\mathrm{T}}\hat{z}_i + \frac{1}{2g_i}\dot{\hat{\theta}}\right|^2 \\
&\quad +\frac{\mu}{4g_0}|\dot{\hat{\theta}}|^2 - c_0|\epsilon|^2 - \kappa_m\sum_{i=1}^n |w_i|^2\epsilon_i^2. \tag{5.125}
\end{aligned}$$

Substituting (5.121), we get

$$\frac{d}{dt}\left(\frac{\mu}{2}|\hat{z}|^2 + \frac{1}{2}|\epsilon|^2 + \frac{1}{2}|\tilde{\theta}|^2_{\Gamma^{-1}}\right) \leq -\mu c_0|\hat{z}|^2 - c_0|\epsilon|^2$$
$$- \left(\kappa_m - \mu\frac{\bar{\lambda}(\Gamma)^2 n}{4g_0}\right)\sum_{i=1}^{n}|w_i|^2\epsilon_i^2. \quad (5.126)$$

Choosing $\mu < \dfrac{4g_0\kappa_m}{n\bar{\lambda}(\Gamma)^2}$ we get

$$\frac{d}{dt}\left(\frac{\mu}{2}|\hat{z}|^2 + \frac{1}{2}|\epsilon|^2 + \frac{1}{2}|\tilde{\theta}|^2_{\Gamma^{-1}}\right) \leq -\mu c_0|\hat{z}|^2 - c_0|\epsilon|^2, \quad (5.127)$$

which proves that the equilibrium $(\hat{z}, \epsilon, \tilde{\theta}) = 0$ is globally uniformly stable. From the LaSalle-Yoshizawa theorem (Theorem 2.1), it further follows that all the solutions converge to the manifold $\hat{z} = \epsilon = 0$ as $t \to \infty$. Since $z = \hat{z} - \epsilon$, the last two conclusions imply that the equilibrium $(z, \epsilon, \tilde{\theta}) = 0$ is globally uniformly stable, and all solutions converge to the manifold $z = \epsilon = 0$ as $t \to \infty$.

From the definitions in (5.73)–(5.75) we conclude that if $\lim_{t\to\infty} y_r^{(i)}(t) = 0$, $i = 0, \ldots, n-1$, and $F(0) = 0$, then $x(t) \to 0$ as $t \to \infty$. $\square$

Note from (5.125) and (5.126) that the nonlinear damping terms $\kappa_i|w_i|^2$ in the matrix A_z of the observer error equation (5.113) are crucial for counteracting the destabilizing effects of $\dot{\hat{\theta}}$.

5.6 x-Passive Scheme

We consider the parametric x-model

$$\dot{x} = f(x, u) + F(x, u)^{\mathrm{T}}\theta, \quad (5.128)$$

which encompasses the class of parametric-strict feedback plants (5.100). Following the passivity motivation from Section 5.4, we implement the observer

$$\dot{\hat{x}} = A(x, t)(\hat{x} - x) + f(x, u) + F(x, u)^{\mathrm{T}}\hat{\theta}, \quad (5.129)$$

where $A(x, t)$ is exponentially stable (uniformly in x and t): $PA(x, t) + A(x, t)^{\mathrm{T}}P \leq -I$, $P = P^{\mathrm{T}} > 0$. The observer error

$$\epsilon = x - \hat{x} \quad (5.130)$$

is governed by

$$\dot{\epsilon} = A(x, t)\epsilon + F(x, u)^{\mathrm{T}}\tilde{\theta}, \quad (5.131)$$

5.6 x-Passive Scheme

which is a system shown to be strictly passive from the input $\tilde{\theta}$ to the output $F(x,u)P\epsilon$ in Section 5.4. Therefore, recalling the definition of the strictly passive nonlinear operator Σ_x from (5.109), we choose the update law as $\dot{\hat{\theta}} = -\Gamma\Sigma_x\{\tilde{\theta}\}$, that is,

$$\dot{\hat{\theta}} = \Gamma F(x,u)P\epsilon, \qquad \Gamma = \Gamma^T > 0. \tag{5.132}$$

When $A(x,t)$ is a constant Hurwitz matrix, (5.129), (5.132) is a passive identifier with standard properties: $\tilde{\theta}, \epsilon$ are bounded and ϵ is square-integrable. Nevertheless, these properties are not enough for our purpose because we need also $\dot{\hat{\theta}}$ to be either bounded or square-integrable. It is not known if the boundedness of $\dot{\hat{\theta}} = \Gamma F(x,u)P\epsilon$ can be guaranteed irrespectively of the boundedness of $F(x,u)$. However, even when $F(x,u)$ is growing unbounded, we can guarantee that the signal $\dot{\hat{\theta}} = \Gamma F(x,u)P\epsilon$ is square-integrable if we choose

$$A(x,t) = A_0 - \lambda F(x,u)^T F(x,u) P, \tag{5.133}$$

where $\lambda > 0$ and A_0 is an arbitrary constant matrix such that

$$PA_0 + A_0^T P = -I, \quad P = P^T > 0. \tag{5.134}$$

Thus, our strengthened observer error system becomes

$$\dot{\epsilon} = \left(A_0 - \lambda F(x,u)^T F(x,u) P\right)\epsilon + F(x,u)^T \tilde{\theta}, \tag{5.135}$$

while the update law is

$$\dot{\hat{\theta}} = \Gamma F(x,u) P\epsilon. \tag{5.136}$$

This sets the stage for the following lemma.

Lemma 5.12 *Suppose the solutions of (5.128), (5.135), and (5.136) are defined on $[0, t_f)$. Then the following identifier properties hold:*

$$(i) \qquad \tilde{\theta} \in \mathcal{L}_\infty[0, t_f) \tag{5.137}$$
$$(ii) \qquad \epsilon \in \mathcal{L}_2[0, t_f) \cap \mathcal{L}_\infty[0, t_f) \tag{5.138}$$
$$(iii) \qquad \dot{\hat{\theta}} \in \mathcal{L}_2[0, t_f). \tag{5.139}$$

Proof. Let us introduce a Lyapunov-like function

$$V = |\tilde{\theta}|^2_{\Gamma^{-1}} + |\epsilon|^2_P \tag{5.140}$$

whose derivative along the solutions of (5.135), (5.136) is

$$\begin{aligned}\dot{V} &= -2\tilde{\theta}^T \Gamma^{-1} \dot{\hat{\theta}} + \epsilon^T\left(PA_0 + A_0^T P\right)\epsilon - 2\lambda\epsilon^T PF^T FP\epsilon + 2\epsilon^T PF^T\tilde{\theta} \\ &= -|\epsilon|^2 - 2\lambda|FP\epsilon|^2. \end{aligned} \tag{5.141}$$

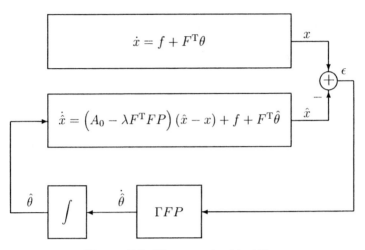

Figure 5.3: The x-passive identifier.

The nonpositivity of $\dot{V}$ proves that $\tilde{\theta}$ and ϵ are in $\mathcal{L}_\infty[0, t_f)$. Integrating (5.141) we get

$$\int_0^t |\epsilon|^2 d\tau \leq -\int_0^t \dot{V} d\tau \leq V(0) - V(t) \leq V(0) < \infty, \qquad (5.142)$$

which proves that $\epsilon \in \mathcal{L}_2[0, t_f)$. As for (iii), noting that

$$|\dot{\hat{\theta}}|^2 = \epsilon^\mathrm{T} P F^\mathrm{T} \Gamma^2 F P \epsilon \leq \bar{\lambda}(\Gamma)^2 |FP\epsilon|^2 \qquad (5.143)$$

and substituting into (5.141), we get

$$\dot{V} \leq -|\epsilon|^2 - \frac{\lambda}{\bar{\lambda}(\Gamma)^2} |\dot{\hat{\theta}}|^2. \qquad (5.144)$$

Upon integration we arrive at

$$\int_0^t |\dot{\hat{\theta}}|^2 d\tau \leq \frac{\bar{\lambda}(\Gamma)^2}{\lambda} V(0) < \infty, \qquad (5.145)$$

and hence $\dot{\hat{\theta}} \in \mathcal{L}_2[0, t_f)$. □

We reiterate that all the $\mathcal{L}_\infty$ and $\mathcal{L}_2$ bounds that we have established on $[0, t_f)$ are independent of t_f.

Theorem 5.13 (x-Passive) *All the signals in the closed-loop adaptive system consisting of the plant (5.55), controller (5.77), observer (5.129), and the update law (5.136) are globally uniformly bounded, and $\lim_{t\to\infty} z(t) = \lim_{t\to\infty} \epsilon(t) = 0$. This means, in particular, that global asymptotic tracking is achieved:*

$$\lim_{t\to\infty} [y(t) - y_r(t)] = 0. \qquad (5.146)$$

Moreover, if $\lim_{t\to\infty} y_r^{(i)}(t) = 0$, $i = 0, \ldots, n-1$, and $F(0) = 0$, then $\lim_{t\to\infty} x(t) = 0$.

5.6 x-PASSIVE SCHEME

Proof. Due to the piecewise continuity of $y_r(t), \ldots, y_r^{(n)}(t)$ and the smoothness of the nonlinearities in (5.55), the solution of the closed-loop adaptive system exists and is unique. Let its maximum interval of existence be $[0, t_f)$.

From Lemma 5.12 we have $\tilde{\theta} \in \mathcal{L}_\infty[0, t_f)$ and $\dot{\hat{\theta}} \in \mathcal{L}_2[0, t_f)$, which in view of Lemma 5.8 implies that $z \in \mathcal{L}_\infty[0, t_f)$ and $x \in \mathcal{L}_\infty[0, t_f)$. Because by Lemma 5.12, $\epsilon \in \mathcal{L}_\infty[0, t_f)$, then by (5.130), $\hat{x} \in \mathcal{L}_\infty[0, t_f)$.

We have thus shown that all of the signals of the closed-loop adaptive system are bounded on $[0, t_f)$ by constants depending on the initial conditions, design coefficients, and the external signals $y_r(t), \ldots, y_r^{(n)}(t)$, but not on t_f. The independence of the bound of t_f proves that $t_f = \infty$ (if t_f were finite, then the solution would escape any compact set as $t \to t_f$, which would contradict the existence of a bound independent of t_f). Hence, all signals are globally uniformly bounded on $[0, \infty)$.

To prove convergence of z to zero, we recall first from Lemma 5.12 that $\epsilon, \dot{\hat{\theta}} \in \mathcal{L}_2$. Let us factor the regressor matrix W, using (5.79) and (5.75), as

$$W^T(z, \hat{\theta}, t) = \begin{bmatrix} 1 & 0 & \cdots & 0 \\ -\frac{\partial \alpha_1}{\partial x_1} & 1 & \ddots & \vdots \\ \vdots & \ddots & \ddots & 0 \\ -\frac{\partial \alpha_{n-1}}{\partial x_1} & \cdots & -\frac{\partial \alpha_{n-1}}{\partial x_{n-1}} & 1 \end{bmatrix} F^T(x) \triangleq N(z, \hat{\theta}, t) F^T(x). \tag{5.147}$$

Considering now the state

$$\zeta \triangleq z - N\epsilon, \tag{5.148}$$

consisting of the state of the error system z and the observer error ϵ, which are both driven by $\tilde{\theta}$, we obtain

$$\dot{\zeta} = A_z(z, \hat{\theta}, t)\zeta + \left[\dot{N} + A_z(z, \hat{\theta}, t)N - N\left(A_0 - \lambda F(x)^T F(x) P\right)\right]\epsilon$$
$$+ Q(z, \hat{\theta}, t)^T \dot{\hat{\theta}}. \tag{5.149}$$

What we have arrived at is a system which is not driven by $\tilde{\theta}$ but, instead, by the $\mathcal{L}_2$ signals ϵ and $\dot{\hat{\theta}}$. We now compute

$$\frac{d}{dt}\left(\frac{1}{2}|\zeta|^2\right) \leq -c_0|\zeta|^2 - \sum_{i=1}^n g_i \left|\frac{\partial \alpha_{i-1}}{\partial \hat{\theta}}\right|^T \zeta_i^2 + \sum_{i=1}^n \frac{\partial \alpha_{i-1}}{\partial \hat{\theta}} \dot{\hat{\theta}} \zeta_i$$
$$+ \zeta^T \left[\dot{N} + A_z N - N\left(A_0 - \lambda F^T F P\right)\right] \epsilon$$
$$\leq -\frac{c_0}{2}|\zeta|^2 + \frac{1}{4g_0}|\dot{\hat{\theta}}|^2$$
$$+ \frac{1}{2c_0}\left|\dot{N} + A_z N - N\left(A_0 - \lambda F^T F P\right)\right|_2^2 |\epsilon|^2. \tag{5.150}$$

Examining (5.74), we see that the terms $\frac{\partial \alpha_i}{\partial x_j}$ appearing in N in (5.147) are continuous functions of z and $\hat{\theta}$ and bounded functions of t. Hence N is bounded. Likewise, we can show that $\frac{\partial N}{\partial z}$, $\frac{\partial N}{\partial \hat{\theta}}$, $\frac{\partial N}{\partial t}$ are bounded. Since, in view of (5.78) and (5.136), $\dot{z}$ and $\dot{\hat{\theta}}$ are bounded,

$$\dot{N} = \frac{\partial N}{\partial z}\dot{z} + \frac{\partial N}{\partial \hat{\theta}}\dot{\hat{\theta}} + \frac{\partial N}{\partial t} \tag{5.151}$$

is also bounded. Thus $\dot{N} + A_z N - N\left(A_0 - \lambda F^{\mathrm{T}} F P\right)$ is bounded. By applying Lemma B.5 to (5.150) with $v = |\zeta|^2$, it follows from $\epsilon, \dot{\hat{\theta}} \in \mathcal{L}_2$ that $\zeta \in \mathcal{L}_2$. Therefore, by (5.148) and the boundedness of N, we conclude that $z \in \mathcal{L}_2$. To finish the proof of convergence of z and ϵ to zero by applying Barbalat's lemma, we recall that $z, \dot{z}, \epsilon \in \mathcal{L}_\infty$ and note that (5.135) implies $\dot{\epsilon} \in \mathcal{L}_\infty$. Therefore, by Barbalat's lemma (Corollary A.7) $z(t), \epsilon(t) \to 0$ as $t \to \infty$. □

Remark 5.14 Writing the z-system (5.78) in the form

$$\begin{aligned} \dot{z} &= A_z(z, \hat{\theta}, t)z - W(z, \hat{\theta}, t)^{\mathrm{T}}\tilde{\theta} + Q(z, \hat{\theta}, t)^{\mathrm{T}}\dot{\hat{\theta}} + W(z, \hat{\theta}, t)^{\mathrm{T}}\theta \\ &\triangleq f(z, \hat{\theta}, t) + F(z, \hat{\theta}, t)^{\mathrm{T}}\theta \end{aligned} \tag{5.152}$$

and comparing it with (5.128), we see that the parametric z-model could be considered to be a special case of the parametric x-model. One can also see that the z-passive identifier is a special case of the x-passive identifier. However, this special case is also the most important one. We have already seen this by going through the proofs of Theorems 5.11 and 5.13: The proof of convergence for the z-scheme is much simpler than the proof of convergence for the x-scheme. In the sequel, we will encounter more instances where the parametric z-model will be more convenient than the parametric x-model. ◇

Remark 5.15 An alternative to the proof that $z(t) \to 0$ in Theorem 5.12 is as follows. For convenience, we first copy (5.135):

$$\dot{\epsilon} = \left(A_0 - \lambda F(x)^{\mathrm{T}} F(x) P\right)\epsilon + F(x)^{\mathrm{T}}\tilde{\theta}. \tag{5.153}$$

Because of the boundedness of all the signals, $\dot{\epsilon}$ is bounded. Since $\epsilon \in \mathcal{L}_2$, by Barbalat's lemma (Corollary A.7), $\epsilon(t) \to 0$. By virtue of the smoothness of F (which implies that its partial derivatives are bounded for bounded values of their arguments) and boundedness of all the states, $\ddot{\epsilon}$ is also bounded, and therefore $\dot{\epsilon}$ is uniformly continuous. Since $\epsilon(t) \to 0$, then

$$\lim_{t \to \infty} \int_0^t \dot{\epsilon}(\tau)d\tau = \lim_{t \to \infty} \epsilon(t) - \epsilon(0) = -\epsilon(0) < \infty. \tag{5.154}$$

Then by Barbalat's lemma (Lemma A.6), $\dot{\epsilon}(t) \to 0$. From (5.153) we conclude $F(x(t))^{\mathrm{T}}\tilde{\theta}(t) \to 0$, and therefore, by (5.147), $W(z(t), \hat{\theta}(t), t)^{\mathrm{T}}\tilde{\theta}(t) \to 0$. Since $\dot{\hat{\theta}} \in \mathcal{L}_2$, using a Barbalat's lemma argument, we also have $\dot{\hat{\theta}}(t) \to 0$, so

5.6 x-PASSIVE SCHEME

$Q(z(t), \hat{\theta}(t), t)^{\mathrm{T}}\dot{\hat{\theta}}(t) \to 0$. (Note that while $Q^{\mathrm{T}}\dot{\hat{\theta}}$ is in $\mathcal{L}_2$, $W^{\mathrm{T}}\tilde{\theta}$ may not be.) Thus the input $W^{\mathrm{T}}\tilde{\theta} + Q^{\mathrm{T}}\dot{\hat{\theta}}$ to the error system

$$\dot{z} = A_z(z, \hat{\theta}, t)z + W(z, \hat{\theta}, t)^{\mathrm{T}}\tilde{\theta} + Q(z, \hat{\theta}, t)^{\mathrm{T}}\dot{\hat{\theta}} \tag{5.155}$$

converges to zero. In view of

$$\begin{aligned}
\frac{d}{dt}\left(\frac{1}{2}|z|^2\right) &\leq -c_0|z|^2 + z^{\mathrm{T}}\left(W^{\mathrm{T}}\tilde{\theta} + Q^{\mathrm{T}}\dot{\hat{\theta}}\right) \\
&\leq -\frac{c_0}{2}|z|^2 + \frac{1}{2c_0}\left|W^{\mathrm{T}}\tilde{\theta} + Q^{\mathrm{T}}\dot{\hat{\theta}}\right|^2,
\end{aligned} \tag{5.156}$$

by applying Lemma B.8 with $\rho = 0$, $\beta_1 = 0$, $\beta_2 = \frac{1}{c_0}\left|W^{\mathrm{T}}\tilde{\theta} + Q^{\mathrm{T}}\dot{\hat{\theta}}\right|^2$ (or [28, Theorem IV.1.9.(e)]), we arrive to the conclusion that $z(t) \to 0$. At this point we remind the reader of Lemma 5.3 where we proved that if $F(x(t))\tilde{\theta}(t)$ (as we just showed) and $\dot{\hat{\theta}}(t)$ converge to zero as $t \to \infty$, then $\lim_{t\to\infty} x(t) = 0$. ◇

Remark 5.16 Even though Theorem 5.12 states only global boundedness and tracking, and its proof used an input-to-state rather than a Lyapunov argument, we can prove that the equilibrium $z = 0$, $\epsilon = 0$, $\tilde{\theta} = 0$ is globally uniformly stable, as we did in Theorem 5.11 for the z-passive scheme and Theorem 4.14 for the tuning functions scheme. It is enough to note from (5.149), (5.151), (5.78), and (5.136) that

$$\begin{aligned}
\dot{\zeta} =\ & A_z(z, \hat{\theta}, t)\zeta \\
& + \left[\frac{\partial N(z, \hat{\theta}, t)}{\partial z}\left(A_z(z, \hat{\theta}, t)z + W(z, \hat{\theta}, t)^{\mathrm{T}}\tilde{\theta} + Q(z, \hat{\theta}, t)^{\mathrm{T}}\Gamma F(x)P\epsilon\right)\right. \\
& + \frac{\partial N(z, \hat{\theta}, t)}{\partial \hat{\theta}}\Gamma F(x)P\epsilon + \frac{\partial N(z, \hat{\theta}, t)}{\partial t} \\
& + A_z(z, \hat{\theta}, t)N(z, \hat{\theta}, t) - N(z, \hat{\theta}, t)\left(A_0 - \lambda F(x)^{\mathrm{T}}F(x)P\right) \\
& \left. + Q(z, \hat{\theta}, t)^{\mathrm{T}}\Gamma F(x)P\right]\epsilon,
\end{aligned} \tag{5.157}$$

which, in view of (5.148), can be expressed as

$$\dot{\zeta} = A_\zeta(\zeta, \epsilon, \tilde{\theta}, t)\zeta + W_\epsilon(\zeta, \epsilon, \tilde{\theta}, t)^{\mathrm{T}}\epsilon, \tag{5.158}$$

where W_ϵ is smooth in ζ, ϵ and $\tilde{\theta}$, and bounded in t, and the homogeneous part of the system (5.158) is exponentially stable. Starting from (5.158), one can prove that the equilibrium $\zeta = 0$, $\epsilon = 0$, $\tilde{\theta} = 0$ is globally uniformly stable by Definition A.4, which along with (5.148) implies that the equilibrium $z = 0$, $\epsilon = 0$, $\tilde{\theta} = 0$ is globally uniformly stable. The difference from Theorems 5.11 and 4.14 is that the stability proof is not based on a quadratic Lyapunov function encompassing the complete state of the closed-loop system. ◇

Remark 5.17 At the first glance, one may think that a passive identifier can be designed using only one of the state equations in (5.55), for example,

$$\dot{x}_1 = x_2 + \varphi_1(x_1)^{\mathrm{T}}\theta \tag{5.159}$$

(or a combination of several state equations). The observer

$$\dot{\hat{x}}_1 = -\left(c_1 + \kappa_1|\varphi_1|^2\right)(\hat{x}_1 - x_1) + x_2 + \varphi_1^{\mathrm{T}}\hat{\theta} \tag{5.160}$$

would yield the observer error system

$$\dot{e} = -\left(c_1 + \kappa_1|\varphi_1|^2\right)e + \varphi_1^{\mathrm{T}}\tilde{\theta}, \tag{5.161}$$

where $e = x_1 - \hat{x}_1$. The update law would be

$$\dot{\hat{\theta}} = \varphi_1 e. \tag{5.162}$$

A simple analysis with a Lyapunov function $V = e^2 + |\tilde{\theta}|^2$ would show that $\tilde{\theta}$ is bounded, $\dot{\hat{\theta}}$ is square-integrable, and e is both bounded and square-integrable. Therefore all the states are bounded. Unfortunately, we cannot prove that z converges to zero. If we apply the same approach as in the proof of Theorem 5.13, namely, if we define $\zeta = z_1 - e$ and subtract (5.161) from the z_1-equation

$$\dot{z}_1 = -\left(c_1 + \kappa_1|\varphi_1|^2\right)z_1 + z_2 + \varphi_1^{\mathrm{T}}\tilde{\theta}, \tag{5.163}$$

then we get

$$\dot{\zeta} = -\left(c_1 + \kappa_1|\varphi_1|^2\right)\zeta + z_2. \tag{5.164}$$

Since e can easily be shown to converge to zero, in order to show that z_1 does so too, it suffices to show that ζ converges to zero. However, z_2 in (5.164), which we only know is bounded, keeps us from concluding that ζ goes to zero. ◇

5.7 Transient Performance

Transient performance of our adaptive system will be estimated by $\mathcal{L}_2$ and $\mathcal{L}_\infty$ bounds for the error state z. Since $z_1 = y - y_r$, these bounds also bound the tracking error $y - y_r$.

We analyze first the z-passive scheme and then the x-passive scheme. To simplify the derivations, without loss of generality, we assume that $\hat{z}(0) = z(0)$ (in the z-passive scheme) and $\hat{x}(0) = x(0)$ (in the x-passive scheme), which implies $\epsilon(0) = 0$. Such observer initializations should be performed in practice to eliminate the disturbing effect of the initial observer error. For simplicity, we also let $\Gamma = \gamma I$.

5.7 Transient Performance

Theorem 5.18 (z-Passive) *In the adaptive system (5.55), (5.77), (5.111), (5.114), the following inequalities hold:*

$$(i) \quad \|z\|_2 \leq \frac{|\tilde{\theta}(0)|}{\sqrt{c_0\gamma}}\left(1 + \frac{n\gamma^2}{2g_0\kappa_m}\right)^{1/2} + \frac{1}{\sqrt{c_0}}|z(0)|, \qquad (5.165)$$

$$(ii) \quad |z(t)| \leq \frac{|\tilde{\theta}(0)|}{2\sqrt{c_0\kappa_0}}\left(1 + \frac{2n\gamma^2}{g_0\kappa_m}\right)^{1/2} + |z(0)|e^{-c_0 t}. \qquad (5.166)$$

Proof. *(i)* Since $\epsilon(0) = 0$, (5.119) implies that

$$\|\tilde{\theta}\|_\infty = |\tilde{\theta}(0)|. \qquad (5.167)$$

Substituted into (5.122), $\epsilon(0) = 0$ gives

$$\|\epsilon\|_2 \leq \frac{1}{\sqrt{2c_0\gamma}}|\tilde{\theta}(0)|. \qquad (5.168)$$

Now, for $\mu < \frac{4g_0\kappa_m}{n\gamma^2}$, by integrating (5.127) over $[0,\infty]$, we get

$$\mu\|\hat{z}\|_2^2 + \|\epsilon\|_2^2 \leq \frac{1}{c_0}\left(\frac{\mu|\hat{z}(0)|^2 + |\epsilon(0)|^2}{2} + \frac{1}{2\gamma}|\tilde{\theta}(0)|^2\right), \qquad (5.169)$$

and, since $\hat{z}(0) = z(0)$, then

$$\|\hat{z}\|_2^2 \leq \frac{1}{2c_0\gamma\mu}|\tilde{\theta}(0)|^2 + \frac{1}{2c_0}|z(0)|^2. \qquad (5.170)$$

Letting

$$\mu = \frac{2g_0\kappa_m}{n\gamma^2}, \qquad (5.171)$$

and adding (5.168) and (5.170) in

$$\|z\|_2^2 = \|\epsilon + \hat{z}\|_2^2 \leq 2\|\epsilon\|_2^2 + 2\|\hat{z}\|_2^2, \qquad (5.172)$$

we arrive at (5.165).

(ii) In a fashion similar to (5.126), we compute

$$\begin{aligned}\frac{d}{dt}\left(\frac{\mu|z|^2 + |\epsilon|^2}{2}\right) &\leq -c_0(\mu|z|^2 + |\epsilon|^2) - \mu\sum_{i=1}^n \kappa_i|w_i|^2 z_i^2 - \sum_{i=1}^n \kappa_i|w_i|^2\epsilon_i^2 \\ &\quad + \mu\sum_{i=1}^n z_i w_i^T\tilde{\theta} + \sum_{i=1}^n \epsilon_i w_i^T\tilde{\theta} \\ &\quad - \sum_{i=1}^n g_i\left|\frac{\partial\alpha_{i-1}}{\partial\hat{\theta}}\right|^{T^2} z_i^2 + \sum_{i=1}^n \frac{\partial\alpha_{i-1}}{\partial\hat{\theta}}\dot{\hat{\theta}} z_i\end{aligned}$$

$$\leq -c_0(\mu|z|^2 + |\epsilon|^2) - \frac{\kappa_m}{2}\sum_{i=1}^{n}|w_i|^2\epsilon_i^2$$
$$+\frac{\mu}{4\kappa_0}|\tilde{\theta}|^2 + \frac{1}{2\kappa_0}|\tilde{\theta}|^2 + \frac{\mu}{4g_0}|\dot{\tilde{\theta}}|^2$$
$$\leq -c_0(\mu|z|^2 + |\epsilon|^2) + \frac{\mu}{4\kappa_0}|\tilde{\theta}|^2 + \frac{1}{2\kappa_0}|\tilde{\theta}|^2$$
$$-\left(\frac{\kappa_m}{2} - \mu\frac{n\gamma^2}{4g_0}\right)\sum_{i=1}^{n}|w_i|^2\epsilon_i^2, \qquad (5.173)$$

where for the last inequality we have used (5.121). Setting

$$\mu = \frac{g_0\kappa_m}{n\gamma^2} \qquad (5.174)$$

in (5.173) and applying Lemma B.5 part *(ii)*, we get

$$\mu|z(t)|^2 + |\epsilon(t)|^2 \leq \left(\mu|z(0)|^2 + |\epsilon(0)|^2\right)e^{-2c_0 t} + \frac{\mu+2}{2\kappa_0}\int_0^t e^{-2c_0(t-\tau)}|\tilde{\theta}(\tau)|^2 d\tau \qquad (5.175)$$

which, by taking the supremum of $\tilde{\theta}(\tau)$, implies

$$|z(t)| \leq \frac{1}{2\sqrt{c_0\kappa_0}}\left(1 + \frac{2}{\mu}\right)^{1/2}\|\tilde{\theta}\|_\infty + |z(0)|e^{-c_0 t}. \qquad (5.176)$$

Substituting (5.167) and (5.174) into (5.176) results in (5.166).[4] □

When we perform the trajectory initialization introduced in Section 4.3.2, which sets $z(0) = 0$, the bounds (5.165) and (5.166) become

$$\|z\|_2 \leq \frac{1}{\sqrt{c_0\gamma}}\left(1 + \frac{n\gamma^2}{2g_0\kappa_m}\right)^{1/2}|\tilde{\theta}(0)|, \qquad (5.177)$$

$$|z(t)| \leq \frac{1}{2\sqrt{c_0\kappa_0}}\left(1 + \frac{2n\gamma^2}{g_0\kappa_m}\right)^{1/2}|\tilde{\theta}(0)|. \qquad (5.178)$$

These bounds depend only on the design parameters c_i, κ_i, g_i, and γ and the initial parameter error $\tilde{\theta}(0)$. It should be clear that, since the plant initial conditions and the reference model initial conditions are the same, the tracking error transient is caused only by the initial parametric uncertainty $\tilde{\theta}(0)$. While by increasing g_0 and κ_m we can reduce the bounds somewhat, by increasing c_0 we can make them arbitrarily small. By increasing κ_0 we can arbitrarily reduce the $\mathcal{L}_\infty$ bound but not the $\mathcal{L}_2$ bound. This general property of the κ-terms,

---

[4]Inequality (5.176) establishes an ISS property from $\tilde{\theta}$ to z. Note that in contrast to (5.86) and (5.87), $\dot{\tilde{\theta}}$ is absent from (5.176). It is also easy to see that $|\epsilon(t)| \leq \frac{1}{2\sqrt{c_0\kappa_0}}\|\tilde{\theta}\|_\infty$ describes an ISS property from $\tilde{\theta}$ to ϵ.

5.7 Transient Performance

of reducing the transient peaks while at the same time making the transient tails longer by slowing down the adaptation, will be illustrated by simulations. The dependence of the bounds on the adaptation gain γ is less clear. The $\mathcal{L}_\infty$ bound is an increasing function of γ, and the $\mathcal{L}_2$ bound is decreasing for small values of γ and increasing for large values of γ. With regard to γ, the $\mathcal{L}_\infty$ bound does not seem to be tight because, as we shall see, the simulations do not corroborate its increasing dependence on γ.

Let us now compare the bounds for the z-passive scheme with the bounds for the tuning functions scheme. Examining the expressions (5.177) and (4.229), we see that the $\mathcal{L}_2$ bound for the z-passive scheme has the term $\frac{n\gamma^2}{2g_0\kappa_m}$, which is not present in the $\mathcal{L}_2$ bound for the tuning functions scheme. Also, examining the expressions (5.178) and (4.241), we see that the $\mathcal{L}_\infty$ bound for the z-passive scheme has the term $\frac{2n\gamma^2}{g_0\kappa_0}$, which is not present in the $\mathcal{L}_\infty$ bound for the tuning functions scheme. These γ-dependent terms occur because in the modular approach $\hat{\theta}$ is not eliminated from the z-system, but only "damped." In other words, by ignoring the time-varying character of the parameter estimates, the modular approach results in performance inferior to that of the tuning functions design.

Now we give an $\mathcal{L}_\infty$ bound for the x-passive scheme, for a special choice of design parameters.

Theorem 5.19 (x-Passive) *In the adaptive system (5.55), (5.77), (5.129), (5.136) with the choice[5] $A_0 = -c_0$, $P = I$, $\lambda = \kappa_0$, the following inequality holds:*

$$|z(t)| \leq \frac{|\tilde{\theta}(0)|}{2\sqrt{c_0\kappa_0}} \left(1 + \frac{2\gamma^2}{g_0\kappa_0}\right)^{1/2} + |z(0)|e^{-c_0 t}. \qquad (5.179)$$

Proof. With this choice of design parameters, we have

$$\dot{\epsilon} = -c_0\epsilon - \kappa_0 F^T F\epsilon + F^T\tilde{\theta} \qquad (5.180)$$

$$\dot{\hat{\theta}} = \gamma F\epsilon. \qquad (5.181)$$

It follows that

$$\begin{aligned}\frac{d}{dt}\left(\frac{1}{2}|\epsilon|^2\right) &= -c_0|\epsilon|^2 - \kappa_0|F\epsilon|^2 + \tilde{\theta}^T F\epsilon \\ &\leq -c_0|\epsilon|^2 - \frac{\kappa_0}{2}|F\epsilon|^2 + \frac{1}{2\kappa_0}|\tilde{\theta}|^2 \\ &= -c_0|\epsilon|^2 - \frac{\kappa_0}{2\gamma^2}|\dot{\hat{\theta}}|^2 + \frac{1}{2\kappa_0}|\tilde{\theta}|^2. \end{aligned} \qquad (5.182)$$

---

[5] Note that this choice of A_0 and P gives $PA_0 + A_0^T P = -2c_0 I$, which differs from (5.134). The proof takes this into account.

Combining this with (5.89), we get

$$\frac{d}{dt}\left(\frac{\mu|z|^2+|\epsilon|^2}{2}\right) \le -c_0(\mu|z|^2+|\epsilon|^2) + \frac{\mu+2}{4\kappa_0}|\tilde{\theta}|^2$$

$$-\left(-\frac{\kappa_0}{2\gamma^2}-\frac{\mu}{4g_0}\right)|\dot{\hat{\theta}}|^2. \quad (5.183)$$

Setting

$$\mu = \frac{g_0\kappa_0}{\gamma^2} \quad (5.184)$$

in (5.183) and applying Lemma B.5 part (ii), we get inequality (5.175) from the proof of Theorem 5.18. By proceeding in the same fashion as there, we arrive at (5.19). □

Noting that $\kappa_m \le n\kappa_0$, the $\mathcal{L}_\infty$ bound (5.179) for the x-passive scheme becomes identical to the $\mathcal{L}_\infty$ bound (5.166) for the z-passive scheme. We can thus conclude that both schemes are equally capable of reducing the transient peaks.

However, an $\mathcal{L}_2$ bound similar to (5.165) is not available for the x-passive scheme. At this point we recall that the proof of convergence of z to zero was much more involved for the x-passive scheme than for the z-passive scheme. Thus, the parametric z-model is once again more convenient than the parametric x-model.

5.8 SG-Scheme (Weak Modularity)

So far in this chapter we have been using the ISS-controller from Table 5.1. This controller achieves a complete controller-identifier separation thanks to the strength of its nonlinear damping terms. Now we examine other possibilities for achieving boundedness and tracking.

In traditional adaptive linear control, the strong ISS controller-identifier separation does not exist. Instead, boundedness and tracking are achieved employing a small-gain (SG) stabilization mechanism. Mostly thanks to the fact that $\dot{\hat{\theta}}$ is $\mathcal{L}_2$-bounded, the loop gain in adaptive linear systems is asymptotically small, namely, a stabilizing small-gain property is achieved as $t \to \infty$. It is this feedback phenomenon that allows the use of the certainty equivalence principle and makes adaptive linear control possible without nonlinear damping. When the parameter estimates of certainty equivalence adaptive controllers are frozen, these controllers become linear. This is not the case with our ISS-controllers. Even when the plant is linear, the underlying nonadaptive controller is nonlinear because of the nonlinear damping terms. There are advantages and disadvantages of nonlinear damping. It improves the tracking performance but, since it acts as high gain for large signals, it may result

5.8 SG-SCHEME (WEAK MODULARITY)

in large control effort (and could potentially have an adverse effect on robustness). This prompts us to try to reduce the growth of nonlinearities in our controllers. The weaker controllers, which become linear when applied to linear systems, are also of interest because they will enable us to make a comparison between adaptive controllers designed using the linear certainty equivalence principle, and those designed using ISS.

5.8.1 Controller design

The only difference between the stronger ISS-controllers and the weaker SG-controllers is that the SG-controllers employ weaker nonlinear damping functions s_i. For example, for an uncertain term $\varphi(x_1)\theta$ in the first equation of the plant (5.55), the ISS and SG nonlinear damping functions are, respectively,

$$s_1^{\text{ISS}}(x_1) = \varphi(x_1)^2 \quad \text{and} \quad s_1^{\text{SG}}(x_1) = \left[\frac{\varphi(x_1) - \varphi(0)}{x_1}\right]^2 \triangleq \omega(x_1)^2. \quad (5.185)$$

In this way the growth of s_1^{SG} is reduced by a factor of x_1^2. In the process of backstepping, this reduction becomes more pronounced. However, the SG-controller can no longer guarantee the ISS property with respect to $\tilde{\theta}$ and $\hat{\theta}$. Instead, it relies on a small gain property achieved by adaptation.

To derive the nonlinear damping expressions for the SG-controller we rewrite the regressor vectors w_i as follows

$$w_i(\bar{z}_i, \hat{\theta}, t) = w_i(0, \hat{\theta}, t) + \omega_i(\bar{z}_i, \hat{\theta}, t)^{\text{T}} \bar{z}_i, \quad (5.186)$$

where $\omega_i : \mathbb{R}^i \times \mathbb{R}^p \times \mathbb{R}_+ \to \mathbb{R}^{i \times p}$ is a matrix-valued function smooth in the first two arguments and continuous and bounded in the third argument, and $\bar{z}_i \triangleq [z_1, \ldots, z_i]^{\text{T}}$. (With a slight abuse of notation relative to (5.75) we now express w_i as a function of $\bar{z}_i, \hat{\theta}, t$.) Thus we have

$$W(z, \hat{\theta}, t)^{\text{T}} = W(0, \hat{\theta}, t)^{\text{T}} + \begin{bmatrix} \bar{z}_1^{\text{T}} \omega_1 \\ \vdots \\ \bar{z}_n^{\text{T}} \omega_n \end{bmatrix}. \quad (5.187)$$

To reduce future notation, we denote $W_0 \triangleq W(0, \hat{\theta}, t)$. Likewise, we rewrite Q from (5.79):

$$Q(z, \hat{\theta}, t)^{\text{T}} = Q(0, \hat{\theta}, t)^{\text{T}} + \begin{bmatrix} 0 \\ \bar{z}_1^{\text{T}} \delta_2 \\ \vdots \\ \bar{z}_{n-1}^{\text{T}} \delta_n \end{bmatrix}, \quad (5.188)$$

where $\delta_i : \mathbb{R}^{i-1} \times \mathbb{R}^p \times \mathbb{R}_+ \to \mathbb{R}^{(i-1) \times p}$ are matrix-valued functions smooth in the first two arguments and continuous and bounded in the third argument. To reduce notation, we denote $Q_0 \triangleq Q(0, \hat{\theta}, t)$.

The SG-controller has the same form (5.77) as the ISS-controller, but its nonlinear damping functions are different:

$$s_i = \kappa_i |\omega_i|_\mathcal{F}^2 + g_i |\delta_i|_\mathcal{F}^2. \tag{5.189}$$

The form of the error system (5.78)–(5.79) is unchanged:

$$\dot{z} = A_z(z, \hat{\theta}, t) z + W(z, \hat{\theta}, t)^\mathrm{T} \tilde{\theta} + Q(z, \hat{\theta}, t)^\mathrm{T} \dot{\hat{\theta}} \tag{5.190}$$

where

$$A_z(z, \hat{\theta}, t) = \begin{bmatrix} -c_1 - s_1 & 1 & 0 & \cdots & 0 \\ -1 & -c_2 - s_2 & 1 & \ddots & \vdots \\ 0 & -1 & \ddots & \ddots & 0 \\ \vdots & \ddots & \ddots & \ddots & 1 \\ 0 & \cdots & 0 & -1 & -c_n - s_n \end{bmatrix}. \tag{5.191}$$

Underlying nonadaptive controller

Before we combine the SG-controller with an identifier, we explain the underlying nonadaptive controller. Substituting (5.187) and (5.188) into (5.190), we obtain

$$\dot{z} = A_z z + \begin{bmatrix} \bar{z}_1^\mathrm{T} \omega_1 \\ \vdots \\ \bar{z}_n^\mathrm{T} \omega_n \end{bmatrix} \tilde{\theta} + \begin{bmatrix} 0 \\ \bar{z}_1^\mathrm{T} \delta_2 \\ \vdots \\ \bar{z}_{n-1}^\mathrm{T} \delta_n \end{bmatrix} \dot{\hat{\theta}} + W_0^\mathrm{T} \tilde{\theta} + Q_0^\mathrm{T} \dot{\hat{\theta}}. \tag{5.192}$$

Then, along the solutions of (5.192) we compute

$$\frac{d}{dt}\left(\frac{1}{2}|z|^2\right) \leq -c_0|z|^2 - \sum_{i=1}^n \left(\kappa_i|\omega_i|_\mathcal{F}^2 + g_i|\delta_i|_\mathcal{F}^2\right) z_i^2$$

$$+ \sum_{i=1}^n z_i \left(\bar{z}_i^\mathrm{T} \omega_i \tilde{\theta} + \bar{z}_{i-1}^\mathrm{T} \delta_i \dot{\hat{\theta}}\right) + z^\mathrm{T}\left(W_0^\mathrm{T}\tilde{\theta} + Q_0^\mathrm{T}\dot{\hat{\theta}}\right)$$

$$\leq -\frac{c_0}{2}|z|^2 + \left(\frac{1}{4\kappa_0}|\tilde{\theta}|^2 + \frac{1}{4g_0}|\dot{\hat{\theta}}|^2\right)|z|^2 + \frac{1}{2c_0}\left|W_0^\mathrm{T}\tilde{\theta} + Q_0^\mathrm{T}\dot{\hat{\theta}}\right|^2. \tag{5.193}$$

Now let $\hat{\theta}(t)$ be constant and bounded. Then (5.193) becomes

$$\frac{d}{dt}\left(\frac{1}{2}|z|^2\right) \leq -\left(\frac{c_0}{2} - \frac{1}{4\kappa_0}|\tilde{\theta}|^2\right)|z|^2 + \frac{1}{2c_0}\left|W_0^\mathrm{T}\tilde{\theta}\right|^2. \tag{5.194}$$

Since W_0 is a continuous function of $\hat{\theta}$ bounded in t, (5.194) implies that z will be bounded provided that

$$2c_0\kappa_0 > |\tilde{\theta}|^2. \tag{5.195}$$

5.8 SG-SCHEME (WEAK MODULARITY)

In other words, boundedness of z can be assured in the absence of adaptation by using high gain. It is crucial that, in order to satisfy (5.195), the designer must know a bound on the parameter error $\tilde{\theta}$. This is a weaker result than the one given in Corollary 5.9, where no such a priori knowledge is required.

We will see next that with adaptation we circumvent both the requirement of a priori knowledge and the use of high gain.

5.8.2 Scheme with strengthened identifier

We now present an adaptive scheme which combines the SG-controller with a strengthened passive identifier designed from the parametric z-model.

To achieve some sort of small-gain property, we need to have $\dot{\hat{\theta}} \in \mathcal{L}_2$. However, if we employ the observer

$$\dot{\hat{z}} = A_z(z, \hat{\theta}, t)\hat{z} + Q(z, \hat{\theta}, t)^T \dot{\hat{\theta}}, \tag{5.196}$$

and the update law

$$\dot{\hat{\theta}} = \Gamma W(z, \hat{\theta}, t)\epsilon, \qquad \Gamma = \Gamma^T > 0 \tag{5.197}$$

where the observer error is given by

$$\epsilon = z - \hat{z}, \tag{5.198}$$

we can prove that $\tilde{\theta}$ is bounded and ϵ is bounded and square-integrable, but we cannot prove that $\dot{\hat{\theta}} \in \mathcal{L}_2$. The reason for this obstacle is that, because of the weaker nonlinear damping functions (5.189) in A_z, the stability of the observer error system

$$\dot{\epsilon} = A_z(z, \hat{\theta}, t)\epsilon + W(z, \hat{\theta}, t)^T \tilde{\theta} \tag{5.199}$$

is not as strong as before.

We strengthen the observer (5.196) by augmenting it with the term $W^T \Gamma W \epsilon$:

$$\begin{aligned}
\dot{\hat{z}} &= A_z(z, \hat{\theta}, t)\hat{z} + W(z, \hat{\theta}, t)^T \Gamma W(z, \hat{\theta}, t)\epsilon + Q(z, \hat{\theta}, t)^T \dot{\hat{\theta}} \\
&= A_z(z, \hat{\theta}, t)\hat{z} + \left(W(z, \hat{\theta}, t) + Q(z, \hat{\theta}, t)\right)^T \dot{\hat{\theta}},
\end{aligned} \tag{5.200}$$

and get the observer error system

$$\dot{\epsilon} = \left[A_z(z, \hat{\theta}, t) - W(z, \hat{\theta}, t)^T \Gamma W(z, \hat{\theta}, t)\right] \epsilon + W(z, \hat{\theta}, t)^T \tilde{\theta}. \tag{5.201}$$

We will show that this strengthening yields $\dot{\hat{\theta}} \in \mathcal{L}_2$. As we mentioned above, the boundedness of $\hat{\theta}$ and ϵ is guaranteed by the identifier (5.200), (5.197), so what remains to be proven is the boundedness of either z or $\hat{z}$. It turns out

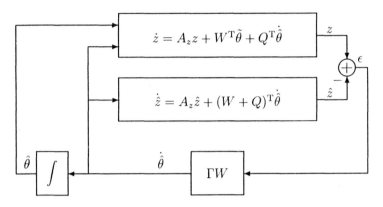

Figure 5.4: The strengthened z-passive identifier.

that the boundedness of $\hat{z}$ has to be proven first. It should be noted that the term $W^\mathrm{T}\Gamma W\epsilon = W^\mathrm{T}\dot{\tilde{\theta}}$, which enhances the stability of the ϵ-system, acts as a perturbation in the $\hat{z}$-system. We deal with this in the proof of the following theorem.

Theorem 5.20 (SG-Scheme) *All the signals in the closed-loop adaptive system consisting of the plant (5.55), the SG-controller (5.77) with (5.189), the observer (5.200), and the update law (5.197) are globally uniformly bounded, and $\lim_{t\to\infty} z(t) = \lim_{t\to\infty} \epsilon(t) = 0$. This means, in particular, that global asymptotic tracking is achieved:*

$$\lim_{t\to\infty}[y(t) - y_\mathrm{r}(t)] = 0. \tag{5.202}$$

Moreover, if $\lim_{t\to\infty} y_\mathrm{r}^{(i)}(t) = 0$, $i = 0, \ldots, n-1$, and $F(0) = 0$, then $\lim_{t\to\infty} x(t) = 0$.

Proof. The solution of the closed-loop adaptive system exists and is unique on its maximum interval of existence $[0, t_f)$. We first study the observer error system (5.201) and the update law (5.197). From

$$\frac{d}{dt}\left(\frac{1}{2}|\tilde{\theta}|^2_{\Gamma^{-1}} + \frac{1}{2}|\epsilon|^2\right) \leq -c_0|\epsilon|^2 - \epsilon^\mathrm{T} W^\mathrm{T}\Gamma W\epsilon$$

$$\leq -c_0|\epsilon|^2 - \frac{1}{\bar{\lambda}(\Gamma)}|\dot{\tilde{\theta}}|^2 \tag{5.203}$$

we conclude that $\hat{\theta}, \epsilon \in \mathcal{L}_\infty[0, t_f)$. By integrating (5.203) we get

$$\int_0^t |\epsilon(\tau)|^2 d\tau \leq \frac{1}{c_0}\left(\frac{1}{2}|\tilde{\theta}(0)|^2_{\Gamma^{-1}} + \frac{1}{2}|\epsilon(0)|^2\right) \tag{5.204}$$

$$\int_0^t |\dot{\tilde{\theta}}(\tau)|^2 d\tau \leq \bar{\lambda}(\Gamma)\left(\frac{1}{2}|\tilde{\theta}(0)|^2_{\Gamma^{-1}} + \frac{1}{2}|\epsilon(0)|^2\right), \tag{5.205}$$

5.8 SG-Scheme (Weak Modularity)

which implies that $\epsilon, \dot{\hat{\theta}} \in \mathcal{L}_2[0, t_f)$. In order to prove that $\hat{z} \in \mathcal{L}_\infty[0, t_f)$, we first rewrite (5.200) as

$$\dot{\hat{z}} = A_z \hat{z} + (W + Q)^T \dot{\hat{\theta}}$$

$$= A_z \hat{z} + (W_0 + Q_0)^T \dot{\hat{\theta}} + \begin{bmatrix} \bar{z}_1^T \omega_1 \\ \bar{z}_2^T \omega_2 + \bar{z}_1^T \delta_2 \\ \vdots \\ \bar{z}_n^T \omega_n + \bar{z}_{n-1}^T \delta_n \end{bmatrix} \dot{\hat{\theta}} \qquad (5.206)$$

and then compute

$$\frac{d}{dt}\left(\frac{1}{2}|\hat{z}|^2\right) \leq -c_0 |\hat{z}|^2 - \sum_{i=1}^n \left(\kappa_i |\omega_i|_\mathcal{F}^2 + g_i |\delta_i|_\mathcal{F}^2\right) \hat{z}_i^2$$

$$+ \hat{z}^T (W_0 + Q_0)^T \dot{\hat{\theta}} + \hat{z}^T \begin{bmatrix} \bar{z}_1^T \omega_1 \\ \bar{z}_2^T \omega_2 + \bar{z}_1^T \delta_2 \\ \vdots \\ \bar{z}_n^T \omega_n + \bar{z}_{n-1}^T \delta_n \end{bmatrix} \dot{\hat{\theta}}$$

$$\leq -c_0 |\hat{z}|^2 - \sum_{i=1}^n \left(\kappa_i |\omega_i|_\mathcal{F}^2 + g_i |\delta_i|_\mathcal{F}^2\right) \hat{z}_i^2$$

$$+ \frac{c_0}{2}|\hat{z}|^2 + \frac{1}{2c_0}|W_0 + Q_0|^2 |\dot{\hat{\theta}}|^2 + \sum_{i=1}^n \hat{z}_i \left(\bar{z}_i^T \omega_i + \bar{z}_{i-1}^T \delta_i\right) \dot{\hat{\theta}}$$

$$\leq -\frac{c_0}{2}|\hat{z}|^2 - \sum_{i=1}^n \left(\kappa_i |\omega_i|_\mathcal{F}^2 + g_i |\delta_i|_\mathcal{F}^2\right) \hat{z}_i^2 + \frac{1}{2c_0}|W_0 + Q_0|^2 |\dot{\hat{\theta}}|^2$$

$$+ \sum_{i=1}^n \left(\kappa_i |\omega_i|_\mathcal{F}^2 + g_i |\delta_i|_\mathcal{F}^2\right) \hat{z}_i^2 + \sum_{i=1}^n \left(\frac{1}{4\kappa_i}|\bar{z}_i|^2 + \frac{1}{4g_i}|\bar{z}_{i-1}|^2\right) |\dot{\hat{\theta}}|^2$$

$$\leq -\frac{c_0}{2}|\hat{z}|^2 + \frac{1}{2c_0}|W_0 + Q_0|^2 |\dot{\hat{\theta}}|^2$$

$$+ \left(\frac{1}{4\kappa_0} + \frac{1}{4g_0}\right)|z|^2 |\dot{\hat{\theta}}|^2. \qquad (5.207)$$

Since

$$|z|^2 \leq 2|\hat{z}|^2 + 2|\epsilon|^2, \qquad (5.208)$$

from (5.207) we obtain

$$\frac{d}{dt}\left(|\hat{z}|^2\right) \leq -c_0 |\hat{z}|^2 + \left(\frac{1}{\kappa_0} + \frac{1}{g_0}\right)|\dot{\hat{\theta}}|^2 |\hat{z}|^2$$

$$+ \left[\left(\frac{1}{\kappa_0} + \frac{1}{g_0}\right)\|\epsilon\|_\infty^2 + \frac{1}{c_0}\|W_0 + Q_0\|_\infty^2\right] |\dot{\hat{\theta}}|^2, \qquad (5.209)$$

where we recall that ϵ is bounded, and note that $\|W_0 + Q_0\|_\infty$ is finite because $W(0, \hat{\theta}, t)$ and $Q(0, \hat{\theta}, t)$ are continuous functions of $\hat{\theta}$ and bounded functions of t. Now we apply Lemma B.6 with $v = |\hat{z}|^2$ to conclude that $\hat{z} \in \mathcal{L}_\infty[0, t_f) \cap \mathcal{L}_2[0, t_f)$. Hence $z \in \mathcal{L}_\infty[0, t_f) \cap \mathcal{L}_2[0, t_f)$. By the same argument as in the proof of Theorem 5.13, $t_f = \infty$. The proof of convergence uses Barbalat's lemma and the fact that $\dot{z}, \dot{\epsilon} \in \mathcal{L}_\infty$. □

Remark 5.21 In contrast to the schemes with the ISS-controller, for the SG scheme we cannot derive transient performance bounds similar to those from Section 5.7. The $\mathcal{L}_\infty$ and $\mathcal{L}_2$ bounds that we get by applying Lemma B.6 to (5.209) involve $\|\dot{\tilde{\theta}}\|_2$, $\|\epsilon\|_\infty$ and $\|W_0 + Q_0\|_\infty$. While for the first two quantities we have bounds which depend only on $\epsilon(0), \tilde{\theta}(0)$ and Γ, the third quantity, $\|W_0 + Q_0\|_\infty$, depends on the reference signals and $\hat{\theta}$ in a complicated nonlinear fashion. ◇

The above SG-scheme employs a passive identifier designed from a parametric z-model. It does not seem possible to obtain an analogous result with the parametric x-model. This adds to the list of advantages of the parametric z-model.

The main advantage of the SG-controller over the ISS-controller is that it has slower growth of nonlinearities in the control law. The main contributors to the growth of nonlinearities are the nonlinear damping functions s_i. The strengthening nonlinear term $W^T \Gamma W \epsilon$ added in the observer (5.200) does not appear in the controller. The following example illustrates the differences in the growth of nonlinearities between the ISS and the SG controllers.

Example 5.22 Consider the relative-degree two plant

$$\begin{aligned} \dot{x}_1 &= x_2 + \theta \varphi(x_1), \quad \varphi(0) = 0 \\ \dot{x}_2 &= u. \end{aligned} \qquad (5.210)$$

We define the error variables

$$\begin{aligned} z_1 &= x_1 \\ z_2 &= x_2 - \alpha_1(x_1, \hat{\theta}). \end{aligned} \qquad (5.211)$$

The two-step ISS-controller design is

$$\begin{aligned} \alpha_1 &= -c_1 z_1 - \kappa_1 \varphi(x_1)^2 z_1 - \hat{\theta} \varphi(x_1) \\ \frac{\partial \alpha_1}{\partial x_1} &= -c_1 - \kappa_1 \varphi(x_1)^2 - 2\kappa_1 \varphi(x_1) \varphi'(x_1) x_1 - \hat{\theta} \varphi'(x_1) \\ u &= -z_1 - c_2 z_2 - \kappa_2 \left(\frac{\partial \alpha_1}{\partial x_1} \right)^2 \varphi(x_1)^2 z_2 - g_2 \varphi(x_1)^2 z_2 + \frac{\partial \alpha_1}{\partial x_1} \left(x_2 + \hat{\theta} \varphi(x_1) \right). \end{aligned} \qquad (5.212)$$

The two-step SG-controller design is

$$\begin{aligned} \alpha_1 &= -c_1 z_1 - \kappa_1 \left(\frac{\varphi(x_1)}{x_1} \right)^2 z_1 - \hat{\theta} \varphi(x_1) \\ \frac{\partial \alpha_1}{\partial x_1} &= -c_1 - \kappa_1 \left(\frac{\varphi(x_1)}{x_1} \right)^2 - 2\kappa_1 \frac{\varphi(x_1)}{x_1} \left(\varphi'(x_1) - \frac{\varphi(x_1)}{x_1} \right) - \hat{\theta} \varphi'(x_1) \\ u &= -z_1 - c_2 z_2 - \kappa_2 \left(\frac{\partial \alpha_1}{\partial x_1} \right)^2 \left(\frac{\varphi(x_1)}{x_1} \right)^2 z_2 - g_2 \varphi(x_1)^2 z_2 + \frac{\partial \alpha_1}{\partial x_1} \left(x_2 + \hat{\theta} \varphi(x_1) \right). \end{aligned} \qquad (5.213)$$

Let us now make a comparison of the nonlinear growths when the plant nonlinearity is

$$\varphi(x_1) = x_1^2. \tag{5.214}$$

First, we compare the stabilizing functions α_1. While the highest power in the ISS design is 3, in the SG design it is 2. The reduction of the nonlinearity growth with the SG-controller is more dramatic when we compare the final control laws u. While the highest power in the ISS-control law is 17, in the SG-control law it is only 7! This difference is even more pronounced for systems of higher relative degree where the recursive controller design consists of more steps. ◇

This disadvantage of the ISS-controller is more than compensated by a superior transient performance. We shall illustrate this in the next chapter by making a simulation comparison of two swapping-based schemes: one using the ISS-controller, and the other using the SG-controller.

5.9 Unknown Virtual Control Coefficients

As in Section 4.5.1, we consider systems of the form

$$\begin{aligned}
\dot{x}_i &= x_{i+1} + \varphi_i(x_1, \ldots, x_i)^T \theta, \quad i = 1, \ldots, m-1, m+1, \ldots, n-1 \\
\dot{x}_m &= b_m x_{m+1} + \varphi_m(x_1, \ldots, x_m)^T \theta, \\
\dot{x}_n &= \beta(x) u + \varphi_n(x_1, \ldots, x_n)^T \theta,
\end{aligned} \tag{5.215}$$

where, in addition to the unknown vector θ, the constant coefficients $b_m, m < n$, referred to as the "virtual control coefficient," is also unknown. The extension of the result of this section to more general systems of the form (4.243) is straightforward but tedious.

Assumption 5.23 *In addition to* $\operatorname{sgn} b_m$, *a positive constant* ς_m *is known such that* $|b_m| \geq \varsigma_m$.

5.9.1 Controller design

From step m on, the design procedure considerably differs from the procedure in Table 5.1. It is also quite different from the tuning functions design of Section 4.5.1. Instead of introducing an additional parameter $\hat{\varrho}$ to avoid a division by $\hat{b}_m$, in the modular design we do allow division by $\hat{b}_m$, because we employ parameter projection which keeps $\hat{b}_m(t)$ from becoming zero.

The complete controller design is given by the following expressions (with $z_0 \stackrel{\triangle}{=} 0$, $\alpha_0 \stackrel{\triangle}{=} 0$, $\tau_0 \stackrel{\triangle}{=} 0$):

Coordinate transformation:

$$z_i = x_i - y_r^{(i-1)} - \alpha_{i-1}, \quad i = 1,\ldots,m \quad (5.216)$$

$$z_j = x_j - \frac{1}{\hat{b}_m} y_r^{(j-1)} - \alpha_{j-1}, \quad j = m+1,\ldots,n \quad (5.217)$$

Regressor:

$$w_i = \varphi_i - \sum_{k=1}^{i-1} \frac{\partial \alpha_{i-1}}{\partial x_k} \varphi_k, \quad i = 1,\ldots,n \quad (5.218)$$

Stabilizing functions:

$$\alpha_i(\bar{x}_i, \hat{\theta}, \bar{y}_r^{(i-1)}) = -z_{i-1} - (c_i + s_i)z_i - w_i^T \hat{\theta}$$
$$+ \sum_{k=1}^{i-1} \left(\frac{\partial \alpha_{i-1}}{\partial x_k} x_{k+1} + \frac{\partial \alpha_{i-1}}{\partial y_r^{(k-1)}} y_r^{(k)} \right)$$
$$i = 1,\ldots,m-1 \quad (5.219)$$

$$\alpha_m(\bar{x}_m, \hat{\theta}, \bar{y}_r^{(m-1)}, \hat{b}_m) = \frac{1}{\hat{b}_m} \bar{\alpha}_m - \frac{\mathrm{sgn}\, b_m}{\varsigma_m}(c_m + s_m)z_m \quad (5.220)$$

$$\bar{\alpha}_m = -z_{m-1} - w_m^T \hat{\theta} + \sum_{k=1}^{m-1} \left[\frac{\partial \alpha_{m-1}}{\partial x_k} x_{k+1} + \frac{\partial \alpha_{m-1}}{\partial y_r^{(k-1)}} y_r^{(k)} \right]$$

$$\alpha_{m+1}(\bar{x}_{m+1}, \hat{\theta}, \bar{y}_r^{(m)}, \hat{b}_m) = -\hat{b}_m z_m - (c_{m+1} + s_{m+1})z_{m+1} - w_{m+1}^T \hat{\theta}$$
$$+ \sum_{k=1}^{m-1} \frac{\partial \alpha_m}{\partial x_k} x_{k+1} + \hat{b}_m \frac{\partial \alpha_m}{\partial x_m} x_{m+1}$$
$$+ \sum_{k=1}^{m} \frac{\partial \alpha_m}{\partial y_r^{(k-1)}} y_r^{(k)} \quad (5.221)$$

$$\alpha_j(\bar{x}_j, \hat{\theta}, \bar{y}_r^{(j-1)}, \hat{b}_m) = -z_{j-1} - (c_j + s_j)z_j - w_j^T \hat{\theta} + \sum_{\substack{k=1 \\ k \neq m}}^{j-1} \frac{\partial \alpha_{j-1}}{\partial x_k} x_{k+1}$$
$$+ \hat{b}_m \frac{\partial \alpha_{j-1}}{\partial x_m} x_{m+1} + \sum_{k=1}^{j-1} \frac{\partial \alpha_{j-1}}{\partial y_r^{(k-1)}} y_r^{(k)}$$
$$j = m+2,\ldots,n \quad (5.222)$$

Nonlinear damping functions:

$$s_i = \kappa_i |w_i|^2 + g_i \left| \frac{\partial \alpha_{i-1}}{\partial \hat{\theta}}^T \right|^2, \quad i = 1,\ldots,m-1 \quad (5.223)$$

$$s_m = \kappa_m \left(|w_m|^2 + \left(\frac{y_r^{(m)} + \bar{\alpha}_m}{\hat{b}_m} \right)^2 \right) + g_m \left| \frac{\partial \alpha_{m-1}}{\partial \hat{\theta}}^T \right|^2 \quad (5.224)$$

5.9 UNKNOWN VIRTUAL CONTROL COEFFICIENTS

$$s_{m+1} = \kappa_{m+1}\left(|w_{m+1}|^2 + \left(z_m - \frac{\partial \alpha_m}{\partial x_m}x_{m+1}\right)^2\right)$$
$$+ g_{m+1}\left(\left|\frac{\partial \alpha_m}{\partial \hat{\theta}}^{\mathrm{T}}\right|^2 + \left(\frac{\partial \alpha_m}{\partial \hat{b}_m} - \frac{y_{\mathrm{r}}^{(m)}}{\hat{b}_m^2}\right)^2\right) \quad (5.225)$$

$$s_j = \kappa_j\left(|w_j|^2 + \left(\frac{\partial \alpha_{j-1}}{\partial x_m}x_{m+1}\right)^2\right)$$
$$+ g_j\left(\left|\frac{\partial \alpha_{j-1}}{\partial \hat{\theta}}^{\mathrm{T}}\right|^2 + \left(\frac{\partial \alpha_{j-1}}{\partial \hat{b}_m} - \frac{y_{\mathrm{r}}^{(j-1)}}{\hat{b}_m^2}\right)^2\right)$$
$$j = m+2,\ldots,n \quad (5.226)$$

Adaptive control law:

$$u = \frac{1}{\beta(x)}\left[\alpha_n + \frac{1}{\hat{b}_m}y_{\mathrm{r}}^{(n)}\right] \quad (5.227)$$

Lengthy calculations show that the design procedure (5.216)–(5.227) results in the closed-loop system

$$\dot{z}_i = -(c_i + s_i)z_i - z_{i-1} + z_{i+1} + w_i^{\mathrm{T}}\tilde{\theta} - \frac{\partial \alpha_{i-1}}{\partial \hat{\theta}}\dot{\hat{\theta}},$$
$$i = 1,\ldots,m-1 \quad (5.228)$$

$$\dot{z}_m = -\frac{|b_m|}{\varsigma_m}(c_m + s_m)z_m - z_{m-1} + b_m z_{m+1}$$
$$+ w_m^{\mathrm{T}}\tilde{\theta} + \frac{1}{\hat{b}_m}\left(y_{\mathrm{r}}^{(m)} + \bar{\alpha}_m\right)\tilde{b}_m - \frac{\partial \alpha_{m-1}}{\partial \hat{\theta}}\dot{\hat{\theta}} \quad (5.229)$$

$$\dot{z}_{m+1} = -(c_{m+1} + s_{m+1})z_{m+1} - b_m z_m + z_{m+2}$$
$$+ w_{m+1}^{\mathrm{T}}\tilde{\theta} + \left(z_m - \frac{\partial \alpha_m}{\partial x_m}x_{m+1}\right)\tilde{b}_m$$
$$- \frac{\partial \alpha_m}{\partial \hat{\theta}}\dot{\hat{\theta}} - \left(\frac{\partial \alpha_m}{\partial \hat{b}_m} - \frac{y_{\mathrm{r}}^{(m)}}{\hat{b}_m^2}\right)\dot{\hat{b}}_m \quad (5.230)$$

$$\dot{z}_j = -(c_j + s_j)z_j - z_{j-1} + z_{j+1} + w_j^{\mathrm{T}}\tilde{\theta} - \frac{\partial \alpha_{j-1}}{\partial x_m}x_{m+1}\tilde{b}_m$$
$$- \frac{\partial \alpha_{j-1}}{\partial \hat{\theta}}\dot{\hat{\theta}} - \left(\frac{\partial \alpha_{j-1}}{\partial \hat{b}_m} - \frac{y_{\mathrm{r}}^{(j-1)}}{\hat{b}_m^2}\right)\dot{\hat{b}}_m, \quad i = m+2,\ldots,n. \quad (5.231)$$

Lemma 5.24 *In the error system (5.228)–(5.231), (5.223)–(5.226), the same input-to-state properties hold as in Lemma 5.8, with $(\tilde{\theta}, \tilde{b}_m)$ and $(\dot{\hat{\theta}}, \dot{\hat{b}}_m)$ as inputs.*

Proof. Noting that Assumption 5.23 implies that $-\frac{|b_m|}{\varsigma_m} \leq -1$, by completing squares as in the proof of Lemma 5.8, we can prove that

$$\frac{d}{dt}\left(\frac{1}{2}|z|^2\right) \leq -c_0|z|^2 + \frac{1}{4}\left[\frac{1}{\kappa_0}\left(|\tilde{\theta}|^2 + |\tilde{b}_m|^2\right) + \frac{1}{g_0}\left(|\dot{\hat{\theta}}|^2 + |\dot{\hat{b}}_m|^2\right)\right], \quad (5.232)$$

and the rest of the proof is identical as for Lemma 5.8. □

5.9.2 Passive adaptive scheme

Now we design an identifier which, in addition to the standard properties of passive identifiers, guarantees that $|\hat{b}_m(t)| \geq \varsigma_m$.

Let us start by writing system (5.215) in the following compact form:

$$\dot{x} = f(x,u) + F(x,u)^T \vartheta, \qquad \vartheta = \begin{bmatrix} b_m \\ \theta \end{bmatrix}. \quad (5.233)$$

We use our standard observer

$$\dot{\hat{x}} = \left(A_0 - \lambda F(x,u)^T F(x,u) P\right)(\hat{x} - x) + f(x,u) + F(x,u)^T \hat{\vartheta}, \quad (5.234)$$

where $\lambda > 0$ and A_0 is an arbitrary constant matrix such that $PA_0 + A_0^T P = -I$, $P = P^T > 0$. The estimation error

$$\epsilon = x - \hat{x} \quad (5.235)$$

is used to drive the update law for the estimate $\hat{\vartheta}$ of ϑ:

$$\dot{\hat{\vartheta}} = \operatorname*{Proj}_{\hat{b}_m}\{\Gamma F(x,u) P \epsilon\}, \qquad \hat{b}_m(0)\operatorname{sgn} b_m > \varsigma_m, \quad (5.236)$$

where the projection operator is employed to guarantee that $|\hat{b}_m(t)| \geq \varsigma_m > 0$, $\forall t \geq 0$. For a detailed treatment of parameter projection the reader is referred to Appendix E.

Lemma 5.25 *Suppose the solutions of (5.233), (5.234) and (5.236) are defined on $[0, t_f)$. Then the following identifier properties hold:*

$$\begin{align}
(o) \quad & |\hat{b}_m(t)| \geq \varsigma_m > 0, \; \forall t \in [0, t_f) & (5.237) \\
(i) \quad & \tilde{\vartheta} \in \mathcal{L}_\infty[0, t_f) & (5.238) \\
(ii) \quad & \epsilon \in \mathcal{L}_2[0, t_f) \cap \mathcal{L}_\infty[0, t_f) & (5.239) \\
(iii) \quad & \dot{\hat{\vartheta}} \in \mathcal{L}_2[0, t_f). & (5.240)
\end{align}$$

Proof. The proof of this lemma is only slightly different from the proof of Lemma 5.12. The differences are due to projection whose properties are summarized in Lemma E.1. First, from Lemma E.1 we conclude that $|\hat{b}_m(t)| \geq \varsigma_m > 0$, $\forall t \in [0, t_f)$. Equation (5.141) is modified using the fact from Lemma E.1 that $-\tilde{\vartheta}^T \Gamma^{-1} \dot{\hat{\vartheta}} = -\tilde{\vartheta}^T \Gamma^{-1} \operatorname{Proj}\{\Gamma F P \epsilon\} \leq -\tilde{\vartheta}^T \Gamma^{-1} \Gamma F P \epsilon$, to yield the same conclusion about boundedness of $\tilde{\vartheta}$. Finally, we note that Lemma E.1 implies that $|\dot{\hat{\vartheta}}|^2 \leq \frac{\bar{\lambda}(\Gamma)}{\underline{\lambda}(\Gamma)}|\Gamma F P \epsilon|^2 \leq \frac{\bar{\lambda}(\Gamma)^3}{\underline{\lambda}(\Gamma)}|F P \epsilon|^2$. This fact is now used as (5.143) in the proof of Lemma 5.12 to establish square-integrability of $\dot{\hat{\vartheta}}$. □

By combining Lemmas 5.24 and 5.25, we obtain the following result. Its proof is the same as the proof of Theorem 5.13.

Theorem 5.26 *All the boundedness and convergence properties from Theorem 5.13 hold also for the closed-loop adaptive system consisting of the plant (5.215), control law (5.227), observer (5.234), and the update law (5.236).*

Remark 5.27 The proof of Theorem 5.13 involves a discussion of the existence of solutions on $[0, t_f)$, which is a condition in Lemma 5.25. The projection operator we introduced in Appendix E is locally Lipschitz, as stated in Lemma E.1. Therefore, the conclusions about the existence and uniqueness of solutions in the proof of Theorem 5.13 remain unchanged. ◇

Notes and References

The foundations for estimation-based certainty equivalence adaptive control of linear systems were given by Egardt [31] and, in a more recent unified framework, by Goodwin and Mayne [43]. The estimation-based certainty equivalence design for nonlinear systems was summarized by Praly, Bastin, Pomet, and Jiang [157]. All of the results in [157] involved some form of either matching or growth conditions. The modular adaptive nonlinear design which removed these conditions was introduced by Krstić and Kokotović [98], while some of the passive designs in this chapter were presented in Krstić and Kokotović [97]. The usefulness of nonlinear damping for achieving boundedness without adaptation was stressed by Kanellakopoulos [64].

Passive identifiers (also known as observer-based identifiers and equation-error filtering identifiers) have earlier been used by Campion and Bastin [15], Teel, Kadiyala, Kokotović, and Sastry [190], and Praly, Bastin, Pomet, and Jiang [157]. Various restrictions were imposed in these early results: in [15] the system structure was restricted by extended matching conditions, in [190] the nonlinearities had to be linearly bounded, and in [157] the nonlinearities and the Lyapunov function had to satisfy nonlinear growth conditions. Ghanadan and Blankenship [41] obtained local results for approximately feedback linearizable systems.

The ISS-clf formalism proposed in Krstić and Kokotović [101] is suitable for modular adaptive design because it takes advantage of the fact that the system is affine both in the control input and in the disturbance input. An ISS-clf is a particular form of a *robust control Lyapunov function (rclf)* introduced by Freeman and Kokotović [39].

Chapter 6

Modular Design with Swapping Identifiers

In the preceding chapter we initiated a modular approach in which the controller and the identifier are designed separately. The strong ISS-controller allowed the use of any identifier which can independently guarantee that the parameter error and its derivative are bounded. However, we have not yet employed the standard gradient and least-squares update laws. The least-squares is of particular interest because it is considered to have better convergence properties and is the only estimation algorithm that guarantees that the parameter estimates always converge to constant values. Instead, in Chapter 5 we retained simple passive identifiers similar to those in Chapter 4.

In this chapter we fully exploit the modular approach and develop adaptive schemes with standard gradient and least-squares update laws. We accomplish this by extending the well-known swapping technique to nonlinear systems. With this technique we convert dynamic parametric models into static ones to which the standard update laws are applicable.

For the swapping modular schemes we derive transient performance bounds and compare them with the bounds obtained for the tuning functions and passive designs.

In addition to the strong ISS-controllers, we employ the SG-controller from Section 5.8.1 and design an adaptive scheme which achieves stabilization via a small gain property. We further enrich the choice of modular controllers by a weaker ISS-controller which expends less control effort for large signals.

6.1 ISS-Controller

As in Chapters 4 and 5, our main focus is on nonlinear systems in the *parametric strict-feedback* form

$$\begin{aligned}\dot{x}_{i-1} &= x_i + \varphi_{i-1}(x_1,\ldots,x_{i-1})^{\mathrm{T}}\theta, & i = 1,\ldots,n-1 \\ \dot{x}_n &= \beta(x)u + \varphi_n(x)^{\mathrm{T}}\theta,\end{aligned} \qquad (6.1)$$

Table 6.1: ISS-Controller

$$z_i = x_i - y_r^{(i-1)} - \alpha_{i-1} \qquad (6.2)$$

$$\alpha_i(\bar{x}_i, \hat{\theta}, \bar{y}_r^{(i-1)}) = -z_{i-1} - c_i z_i - w_i^T \hat{\theta} + \sum_{k=1}^{i-1} \left(\frac{\partial \alpha_{i-1}}{\partial x_k} x_{k+1} + \frac{\partial \alpha_{i-1}}{\partial y_r^{(k-1)}} y_r^{(k)} \right)$$
$$- s_i z_i \qquad (6.3)$$

$$w_i(\bar{x}_i, \hat{\theta}, \bar{y}_r^{(i-2)}) = \varphi_i - \sum_{k=1}^{i-1} \frac{\partial \alpha_{i-1}}{\partial x_k} \varphi_k \qquad (6.4)$$

$$s_i(\bar{x}_i, \hat{\theta}, \bar{y}_r^{(i-2)}) = \kappa_i |w_i|^2 + g_i \left| \frac{\partial \alpha_{i-1}}{\partial \hat{\theta}}^T \right|^2 \qquad (6.5)$$

$$i = 1, \ldots, n$$

$$\bar{x}_i = (x_1, \ldots, x_i), \qquad \bar{y}_r^{(i)} = (y_r, \dot{y}_r, \ldots, y_r^{(i)})$$

Adaptive control law:

$$u = \frac{1}{\beta(x)} \left[\alpha_n(x, \hat{\theta}, \bar{y}_r^{(n-1)}) + y_r^{(n)} \right] \qquad (6.6)$$

where $\theta \in \mathbb{R}^p$ is a vector of unknown constant parameters, β and $F = [\varphi_1, \cdots, \varphi_n]$ are smooth nonlinear functions in $\mathbb{R}^n$, and $\beta(x) \neq 0$, $\forall x \in \mathbb{R}^n$. The control objective is to force the output $y = x_1$ of the system (6.1) to asymptotically track the reference output y_r.

We briefly remind the reader of the ISS-controller designed in Section 5.3. We use the same controller in the swapping designs. The general design procedure is repeated in Table 6.1.

The closed-loop system obtained by applying the design procedure (6.2)–(6.6) to system (6.1) is

$$\dot{z} = A_z(z, \hat{\theta}, t) z + W(z, \hat{\theta}, t)^T \tilde{\theta} + Q(z, \hat{\theta}, t)^T \dot{\hat{\theta}}, \qquad (6.7)$$

where A_z, W, and Q are matrix-valued functions of z, $\hat{\theta}$ and t:

$$A_z(z, \hat{\theta}, t) = \begin{bmatrix} -c_1 - s_1 & 1 & 0 & \cdots & 0 \\ -1 & -c_2 - s_2 & 1 & \ddots & \vdots \\ 0 & -1 & \ddots & \ddots & 0 \\ \vdots & & \ddots & \ddots & 1 \\ 0 & \cdots & 0 & -1 & -c_n - s_n \end{bmatrix}$$

$$W(z,\hat{\theta},t)^{\mathrm{T}} = \begin{bmatrix} w_1^{\mathrm{T}} \\ w_2^{\mathrm{T}} \\ \vdots \\ w_n^{\mathrm{T}} \end{bmatrix}, \quad Q(z,\hat{\theta},t)^{\mathrm{T}} = \begin{bmatrix} 0 \\ -\frac{\partial \alpha_1}{\partial \hat{\theta}} \\ \vdots \\ -\frac{\partial \alpha_{n-1}}{\partial \hat{\theta}} \end{bmatrix}. \tag{6.8}$$

Input-to-state properties achieved by this controller were established in Lemma 5.8. As we shall see, the swapping identifiers will be able to independently guarantee that both $\tilde{\theta}$ and $\hat{\theta}$ are bounded. For this reason, in the swapping design only the property (5.86) is of interest:

$$|z(t)| \leq \frac{1}{2\sqrt{c_0}} \left(\frac{1}{\kappa_0} \|\tilde{\theta}\|_\infty^2 + \frac{1}{g_0} \|\dot{\hat{\theta}}\|_\infty^2 \right)^{\frac{1}{2}} + |z(0)| e^{-c_0 t}, \tag{6.9}$$

where $c_0 = \min_{1 \leq i \leq n} c_i$, $\kappa_0 = \left(\sum_{i=1}^n \frac{1}{\kappa_i} \right)^{-1}$, $g_0 = \left(\sum_{i=1}^n \frac{1}{g_i} \right)^{-1}$.

6.2 Swapping and Static Parametric Models

In Section 5.1 we considered the scalar system

$$\dot{x} = -x + \varphi(x)^{\mathrm{T}} \tilde{\theta} \tag{6.10}$$

and used the identifier

$$\dot{\chi} = -\chi + \varphi(x), \quad \chi(0) = 0 \tag{6.11}$$
$$\dot{\xi} = -\xi - \varphi(x)^{\mathrm{T}} \hat{\theta}, \quad \xi(0) = x(0) \tag{6.12}$$
$$\dot{\hat{\theta}} = \frac{\chi}{1 + \chi^{\mathrm{T}} \chi} \left(x - \chi^{\mathrm{T}} \hat{\theta} - \xi \right) \tag{6.13}$$

to illustrate the weakness of the certainty equivalence approach. This identifier was based on the swapping technique which is standard in adaptive linear control. In the swapping technique, the goal is to convert a dynamic parametric model into a static form, so that standard parameter estimation algorithms can be used.

To clarify this, let us consider the dynamic parametric model (6.10). Even though this model is linear in the parameter error $\tilde{\theta}$, standard parameter estimation algorithms are not applicable because $\dot{x}$ is not available for measurement. The swapping technique circumvents this obstacle by employing the filters (6.11) and (6.12) which, as is easily verified, convert (6.10) into the static parametric model

$$\epsilon \triangleq x - \chi^{\mathrm{T}} \hat{\theta} - \xi = \chi^{\mathrm{T}} \tilde{\theta}. \tag{6.14}$$

The term 'swapping' is descriptive of the fact that we exchange the order of the transfer function $\frac{1}{s+1}$ in (6.10) and the time-varying parameter error $\tilde{\theta}(t)$, that is, we replace x from

$$x = \frac{1}{s+1}\left[\varphi(x)^{\mathrm{T}}\tilde{\theta}\right] \tag{6.15}$$

by

$$\epsilon = \frac{1}{s+1}\left[\varphi(x)^{\mathrm{T}}\right]\tilde{\theta}. \tag{6.16}$$

With the help of the filter χ, the signal ϵ is available from the static expression (6.14). The effect of adding the filters to compensate for the time-varying nature of $\tilde{\theta}(t)$ is visible from $\psi \triangleq -\chi^{\mathrm{T}}\hat{\theta} - \xi$, which is governed by

$$\psi = \frac{1}{s+1}\left[\chi^{\mathrm{T}}\dot{\tilde{\theta}}\right]. \tag{6.17}$$

Details of the swapping technique and a nonlinear generalization of the swapping lemma [138] are given in Appendix F.

The swapping technique can also be interpreted by treating $\epsilon = \chi^{\mathrm{T}}\tilde{\theta}$ as the "prediction error." Consider the signal $\mathcal{Y} \triangleq x - \xi$ governed by

$$\mathcal{Y} = \frac{1}{s+1}\left[\varphi(x)^{\mathrm{T}}\theta\right] = \chi^{\mathrm{T}}\theta. \tag{6.18}$$

The signal $\hat{\mathcal{Y}} \triangleq \chi^{\mathrm{T}}\hat{\theta}$ is referred to as 'prediction' because the unknown parameter θ is replaced by its estimate $\hat{\theta}$. Thus we obtain a static expression

$$\epsilon \triangleq \mathcal{Y} - \hat{\mathcal{Y}} = \chi^{\mathrm{T}}\tilde{\theta}, \tag{6.19}$$

which is the same as (6.14).

The linear parametric model (6.14) is suitable for any standard parameter update law including the normalized gradient (6.13) or the least squares:

$$\dot{\hat{\theta}} = \Gamma\frac{\chi}{1+\chi^{\mathrm{T}}\Gamma\chi}\left(x - \chi^{\mathrm{T}}\hat{\theta} - \xi\right) \tag{6.20}$$

$$\dot{\Gamma} = -\Gamma\frac{\chi\chi^{\mathrm{T}}}{1+\chi^{\mathrm{T}}\Gamma\chi}\Gamma. \tag{6.21}$$

In the next two sections we present two swapping schemes, one using a z-swapping identifier derived from the parametric z-model

$$\dot{z} = A_z(z,\hat{\theta},t)z + W(z,\hat{\theta},t)^{\mathrm{T}}\tilde{\theta} + Q(z,\hat{\theta},t)^{\mathrm{T}}\dot{\hat{\theta}} \tag{6.22}$$

and the other using an x-swapping identifier derived from the parametric x-model

$$\dot{x} = f(x,u) + F(x,u)^{\mathrm{T}}\theta. \tag{6.23}$$

For each of the two we can employ any standard update law: gradient or least-squares, normalized or unnormalized.

6.3 z-Swapping Scheme

Let us consider the parametric z-model

$$\dot{z} = A_z(z,\hat{\theta},t)z + W(z,\hat{\theta},t)^{\mathrm{T}}\tilde{\theta} + Q(z,\hat{\theta},t)^{\mathrm{T}}\dot{\hat{\theta}}. \tag{6.24}$$

This model is in the form (F.1) of the Nonlinear Swapping Lemma (Lemma F.1) with $l(z,t) = 0$, $g(z,t) = h(z,t) = I_n$, $A(z,t) = A_z(z,\hat{\theta},t)$, and $e(t) \equiv 0$. By directly applying Lemma F.1, with the filters

$$\dot{\Omega}^{\mathrm{T}} = A_z(z,\hat{\theta},t)\Omega^{\mathrm{T}} + W(z,\hat{\theta},t)^{\mathrm{T}}, \qquad \Omega \in \mathbb{R}^{p\times n}, \tag{6.25}$$

$$\dot{\psi} = A_z(z,\hat{\theta},t)\psi - \Omega^{\mathrm{T}}\dot{\hat{\theta}} - Q(z,\hat{\theta},t)^{\mathrm{T}}\dot{\hat{\theta}}, \qquad \psi \in \mathbb{R}^n, \tag{6.26}$$

we get the static linear parametric model

$$z + \psi = \Omega^{\mathrm{T}}\tilde{\theta} + \tilde{\epsilon} \tag{6.27}$$

where z, ψ, and Ω are available and $\tilde{\epsilon}$ is an exponentially decaying signal governed by

$$\dot{\tilde{\epsilon}} = A_z(z,\hat{\theta},t)\tilde{\epsilon}, \qquad \tilde{\epsilon} \in \mathbb{R}^n. \tag{6.28}$$

On the left-hand side of (6.27) we observe that the state z of the closed-loop error system has been augmented by ψ to get a static error equation linear in the parameter error. To bring the *augmented error* $\epsilon \stackrel{\triangle}{=} z + \psi$ into the form where the swapping terms explicitly appear, we introduce

$$\dot{\Omega}_0 = A_z(z,\hat{\theta},t)\Omega_0 + W(z,\hat{\theta},t)^{\mathrm{T}}\hat{\theta} - Q(z,\hat{\theta},t)^{\mathrm{T}}\dot{\hat{\theta}}, \qquad \Omega_0 \in \mathbb{R}^n \tag{6.29}$$

and replace ψ by $\Omega_0 - \Omega^{\mathrm{T}}\hat{\theta}$, that is, we write the augmented error as

$$\epsilon = z + \Omega_0 - \Omega^{\mathrm{T}}\hat{\theta} = \Omega^{\mathrm{T}}\tilde{\theta} + \tilde{\epsilon}. \tag{6.30}$$

It is sometimes simpler to view the swapping identifier design as a method to generate a "prediction error." Therefore, we give a prediction error interpretation of (6.30). First we note that $\mathcal{Y} \stackrel{\triangle}{=} z + \Omega_0$ satisfies

$$\dot{\mathcal{Y}} = A_z(z,\hat{\theta},t)\mathcal{Y} + W(z,\hat{\theta},t)^{\mathrm{T}}\theta, \tag{6.31}$$

which implies that

$$\mathcal{Y} = \Omega^{\mathrm{T}}\theta + \tilde{\epsilon}. \tag{6.32}$$

Hence, we have a static model linear in θ, plus an exponentially decaying term. Since Ω is available, we define the "prediction" $\hat{\mathcal{Y}} = \Omega^{\mathrm{T}}\hat{\theta}$ and the "prediction error"

$$\epsilon = \mathcal{Y} - \hat{\mathcal{Y}} = \Omega^{\mathrm{T}}\tilde{\theta} + \tilde{\epsilon}. \tag{6.33}$$

We shall not dwell on the difference between the augmented error and the prediction error views of the swapping technique. We will adopt one or the other view depending on what is more appropriate for a particular parametric model. Instead of either augmented or prediction error, ϵ will be called *estimation error*.

Before we proceed to the selection of update laws, let us summarize the implemented filters and the estimation error:

$$\dot{\Omega}^T = A_z(z,\hat{\theta},t)\Omega^T + W(z,\hat{\theta},t)^T \tag{6.34}$$

$$\dot{\Omega}_0 = A_z(z,\hat{\theta},t)\Omega_0 + W(z,\hat{\theta},t)^T\hat{\theta} - Q(z,\hat{\theta},t)^T\dot{\hat{\theta}} \tag{6.35}$$

$$\epsilon = z + \Omega_0 - \Omega^T\hat{\theta}. \tag{6.36}$$

The choice of update laws is intimately connected with the fact that the estimation error

$$\epsilon = \Omega^T\tilde{\theta} + \tilde{\epsilon} \tag{6.37}$$

is linear in $\tilde{\theta}$. This allows us to use either the gradient or the least-squares update laws. On the other hand, the enhanced exponential stability of $A_z(z,\hat{\theta},t)$ allows the update laws to be either normalized or unnormalized. We unify the notation by denoting by Γ both the constant adaptation gain matrix in the gradient algorithm, and the time-varying covariance matrix in the least-squares algorithm.

The update law for $\hat{\theta}$ is either the gradient:

$$\dot{\hat{\theta}} = \Gamma\frac{\Omega\epsilon}{1+\nu|\Omega|_{\mathcal{F}}^2}, \qquad \Gamma = \Gamma^T > 0, \quad \nu \geq 0 \tag{6.38}$$

or the least squares:

$$\begin{aligned}\dot{\hat{\theta}} &= \Gamma\frac{\Omega\epsilon}{1+\nu\mathrm{tr}\{\Omega^T\Gamma\Omega\}} \\ \dot{\Gamma} &= -\Gamma\frac{\Omega\Omega^T}{1+\nu\mathrm{tr}\{\Omega^T\Gamma\Omega\}}\Gamma, \qquad \Gamma(0) = \Gamma(0)^T > 0, \quad \nu \geq 0,\end{aligned} \tag{6.39}$$

where by allowing $\nu = 0$, we encompass unnormalized update laws. The complete z-swapping identifier is shown in Figure 6.1.

With the regressor Ω being a matrix, our use of the Frobenius norm $|\Omega|_{\mathcal{F}}$ in the gradient update law and the "Γ-weighted Frobenius norm" $\mathrm{tr}\{\Omega^T\Gamma\Omega\}$ in the least-squares update law avoids unnecessary algebraic complications in the stability arguments that arise from applying the normalized gradient update law in the form $\dot{\hat{\theta}} = \Gamma\Omega\left(I_n + \nu\Omega^T\Omega\right)^{-1}\epsilon$ or the normalized least-squares with the covariance matrix computed from $\dot{\Gamma} = -\Gamma\Omega\left(I_n + \nu\Omega^T\Gamma\Omega\right)^{-1}\Omega^T\Gamma$. It is even more important that this eliminates the need for on-line matrix inversion.

6.3 z-SWAPPING SCHEME

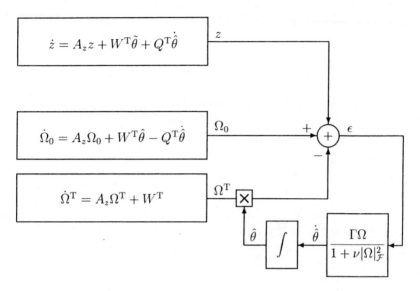

Figure 6.1: The z-swapping identifier.

The following lemma establishes properties of the identifier module which consists of the filters (6.34)–(6.35) and either the gradient (6.38) or the least-squares (6.39) update law. The properties hold for both unnormalized ($\nu = 0$) and normalized ($\nu > 0$) update laws. In contrast to standard identifiers with *unnormalized* update laws, which are able only to guarantee that $\tilde{\theta}$ is bounded and ϵ is square integrable, this identifier also guarantees that ϵ is bounded and $\dot{\hat{\theta}}$ is bounded and square integrable.

Lemma 6.1 *Let the maximal interval of existence of solutions of (6.24), (6.34)–(6.35) with either (6.38) or (6.39) be $[0, t_f)$. Then for $\nu \geq 0$, the following identifier properties hold:*

$$
\begin{align}
(i) \quad & \tilde{\theta} \in \mathcal{L}_\infty[0, t_f) & (6.40) \\
(ii) \quad & \epsilon \in \mathcal{L}_2[0, t_f) \cap \mathcal{L}_\infty[0, t_f) & (6.41) \\
(iii) \quad & \dot{\hat{\theta}} \in \mathcal{L}_2[0, t_f) \cap \mathcal{L}_\infty[0, t_f). & (6.42)
\end{align}
$$

Proof. First we prove that the filter state Ω is uniformly bounded on $[0, t_f)$, irrespectively of the boundedness of its input W. Then we prove the stated properties separately for the gradient (6.38) and for the least-squares (6.39) update laws.

Along the solutions of (6.34) we have

$$\frac{d}{dt}\left(\frac{1}{2}\Omega^T\Omega\right) = \frac{1}{2}\left(A_z\Omega^T\Omega + \Omega^T\Omega A_z^T\right) + \frac{1}{2}\left(W^T\Omega + \Omega^T W\right). \tag{6.43}$$

Using (6.8) and (6.5), the linearity of the trace operator, and the fact that $\text{tr}\{AB\} = \text{tr}\{BA\}$, we compute

$$\begin{aligned}
\frac{d}{dt}\text{tr}\left\{\frac{1}{2}\Omega^T\Omega\right\} &= \text{tr}\left\{\frac{1}{2}\Omega\left(A_z + A_z^T\right)\Omega^T\right\} + \text{tr}\{W^T\Omega\} \\
&\leq -c_0\,\text{tr}\{\Omega\Omega^T\} - \text{tr}\left\{\Omega\,\text{diag}\left(\kappa_1|w_1|^2,\ldots,\kappa_n|w_n|^2\right)\Omega^T\right\} \\
&\quad + \text{tr}\left\{\begin{bmatrix} w_1^T \\ \vdots \\ w_n^T \end{bmatrix}[\Omega_1 \cdots \Omega_n]\right\} \\
&= -c_0|\Omega|_{\mathcal{F}}^2 + \sum_{i=1}^n\left(-\kappa_i|w_i|^2|\Omega_i|^2 + w_i^T\Omega_i\right) \\
&\leq -c_0|\Omega|_{\mathcal{F}}^2 - \sum_{i=1}^n \kappa_i\left(|w_i|^2|\Omega_i|^2 - \frac{1}{2\kappa_i}\right)^2 + \sum_{i=1}^n \frac{1}{4\kappa_i} \quad (6.44)
\end{aligned}$$

and arrive at

$$\frac{d}{dt}\left(\frac{1}{2}|\Omega|_{\mathcal{F}}^2\right) \leq -c_0|\Omega|_{\mathcal{F}}^2 + \frac{1}{4\kappa_0}. \quad (6.45)$$

This proves that Ω is uniformly bounded on $[0,t_f)$ even if W is only in $\mathcal{L}_{\infty e}[0,t_f)$, that is, even if W is escaping to infinity as $t \to t_f$.

In view of (6.28) and (6.8), which give

$$\frac{d}{dt}\left(\frac{1}{2}|\tilde{\epsilon}|^2\right) = \frac{1}{2}\tilde{\epsilon}^T\left(A_z + A_z^T\right)\tilde{\epsilon} \leq -c_0|\tilde{\epsilon}|^2, \quad (6.46)$$

it is clear that $\tilde{\epsilon} \in \mathcal{L}_2[0,t_f) \cap \mathcal{L}_\infty[0,t_f)$.

Gradient update law (6.38). Let us consider the positive definite function

$$V = \frac{1}{2}|\tilde{\theta}|_{\Gamma^{-1}}^2 + \frac{1}{2c_0}|\tilde{\epsilon}|^2. \quad (6.47)$$

Along (6.38) and (6.28) the derivative of V is

$$\begin{aligned}
\dot{V} &\leq -\tilde{\theta}^T\Gamma^{-1}\dot{\tilde{\theta}} - |\tilde{\epsilon}|^2 = -\frac{\tilde{\theta}^T\Omega\epsilon}{1+\nu|\Omega|_{\mathcal{F}}^2} - |\tilde{\epsilon}|^2 \\
&= -\frac{|\epsilon|^2}{1+\nu|\Omega|_{\mathcal{F}}^2} + \frac{\epsilon^T}{1+\nu|\Omega|_{\mathcal{F}}^2}\tilde{\epsilon} - |\tilde{\epsilon}|^2 \\
&\leq -\frac{3}{4}\frac{|\epsilon|^2}{1+\nu|\Omega|_{\mathcal{F}}^2} - \frac{1}{4}\frac{|\epsilon|^2}{(1+\nu|\Omega|_{\mathcal{F}}^2)^2} + \frac{\epsilon^T}{1+\nu|\Omega|_{\mathcal{F}}^2}\tilde{\epsilon} - |\tilde{\epsilon}|^2 \\
&= -\frac{3}{4}\frac{|\epsilon|^2}{1+\nu|\Omega|_{\mathcal{F}}^2} - \left|\frac{1}{2}\frac{\epsilon}{1+\nu|\Omega|_{\mathcal{F}}^2} - \tilde{\epsilon}\right|^2 \quad (6.48)
\end{aligned}$$

and hence

$$\dot{V} \leq -\frac{3}{4}\frac{|\epsilon|^2}{1+\nu|\Omega|_{\mathcal{F}}^2}. \quad (6.49)$$

6.3 z-SWAPPING SCHEME

The nonpositivity of $\dot{V}$ proves that $\tilde{\theta} \in \mathcal{L}_\infty[0, t_f)$. Recalling that $\epsilon = \Omega^{\mathrm{T}}\tilde{\theta} + \tilde{\epsilon}$, from the boundedness of Ω it follows that $\epsilon \in \mathcal{L}_\infty[0, t_f)$. This in turn shows that $\dot{\hat{\theta}} = \Gamma \dfrac{\Omega \epsilon}{1 + \nu |\Omega|_{\mathcal{F}}^2} \in \mathcal{L}_\infty[0, t_f)$. Let us now prove the square-integrability statements. Integrating (6.49) we get

$$\frac{3}{4}\int_0^t \frac{|\epsilon|^2}{1 + \nu|\Omega|_{\mathcal{F}}^2}d\tau \leq -\int_0^t \dot{V}d\tau \leq V(0) - V(t) \leq V(0) < \infty, \qquad (6.50)$$

where we have used the fact that $V(t)$ is nonnegative. Thus

$$\frac{\epsilon}{\sqrt{1+\nu|\Omega|_{\mathcal{F}}^2}} \in \mathcal{L}_2[0, t_f). \qquad (6.51)$$

This, along with the boundedness of Ω, yields

$$\int_0^t |\epsilon|^2 d\tau \leq \left\|1 + \nu|\Omega|_{\mathcal{F}}^2\right\|_\infty \int_0^t \frac{|\epsilon|^2}{1+\nu|\Omega|_{\mathcal{F}}^2}d\tau < \infty, \qquad (6.52)$$

so $\epsilon \in \mathcal{L}_2[0, t_f)$. Finally

$$\int_0^t |\dot{\hat{\theta}}|^2 d\tau \leq \int_0^t \frac{\epsilon^{\mathrm{T}}\Omega^{\mathrm{T}}\Gamma^2\Omega\epsilon}{(1+\nu|\Omega|_{\mathcal{F}}^2)^2}d\tau \leq \bar{\lambda}(\Gamma)^2 \left\|\frac{|\Omega|_{\mathcal{F}}^2}{1+\nu|\Omega|_{\mathcal{F}}^2}\right\|_\infty \int_0^t \frac{|\epsilon|^2}{1+\nu|\Omega|_{\mathcal{F}}^2}d\tau < \infty, \qquad (6.53)$$

and we get $\dot{\hat{\theta}} \in \mathcal{L}_2[0, t_f)$.

Least-squares update law (6.39). First we establish properties of the covariance matrix $\Gamma(t)$. Since $\dot{\Gamma} \leq 0$, then $\Gamma(t) \leq \Gamma(0)$, which implies that Γ is bounded. From $\dfrac{d}{dt}\left(\Gamma^{-1}\right) = -\Gamma^{-1}\dot{\Gamma}\Gamma^{-1} = \dfrac{\Omega\Omega^{\mathrm{T}}}{1+\nu\mathrm{tr}\{\Omega^{\mathrm{T}}\Gamma\Omega\}} \geq 0$ and the fact that $\Gamma^{-1}(0) > 0$, we see that $\Gamma(t)^{-1}$ is positive definite for each t. Now we introduce the function

$$V = |\tilde{\theta}|^2_{\Gamma(t)^{-1}} + \frac{1}{2c_0}|\tilde{\epsilon}|^2, \qquad (6.54)$$

which is positive definite. Its derivative along the solutions of (6.39) and (6.28) is

$$\dot{V} \leq -\dot{\hat{\theta}}^{\mathrm{T}}\Gamma^{-1}\tilde{\theta} + \tilde{\theta}^{\mathrm{T}}\frac{d}{dt}\left(\Gamma^{-1}\tilde{\theta}\right) - |\tilde{\epsilon}|^2. \qquad (6.55)$$

Upon the examination of

$$\begin{aligned}\frac{d}{dt}\left(\Gamma^{-1}\tilde{\theta}\right) &= -\Gamma^{-1}\dot{\Gamma}\Gamma^{-1}\tilde{\theta} - \Gamma^{-1}\dot{\hat{\theta}} \\ &= \frac{\Omega\Omega^{\mathrm{T}}\tilde{\theta}}{1+\nu\mathrm{tr}\{\Omega^{\mathrm{T}}\Gamma\Omega\}} - \frac{\Omega\epsilon}{1+\nu\mathrm{tr}\{\Omega^{\mathrm{T}}\Gamma\Omega\}} \\ &= -\frac{\Omega\tilde{\epsilon}}{1+\nu\mathrm{tr}\{\Omega^{\mathrm{T}}\Gamma\Omega\}}\end{aligned} \qquad (6.56)$$

and its substitution in (6.55) we obtain

$$\begin{aligned}
\dot{V} &\leq -\frac{\epsilon^T \Omega^T \tilde{\theta}}{1+\nu\mathrm{tr}\{\Omega^T\Gamma\Omega\}} - \frac{\tilde{\epsilon}^T\Omega^T\tilde{\theta}}{1+\nu\mathrm{tr}\{\Omega^T\Gamma\Omega\}} - |\tilde{\epsilon}|^2 \\
&= -\frac{|\epsilon|^2}{1+\nu\mathrm{tr}\{\Omega^T\Gamma\Omega\}} + \frac{|\tilde{\epsilon}|^2}{1+\nu\mathrm{tr}\{\Omega^T\Gamma\Omega\}} - |\tilde{\epsilon}|^2 \\
&\leq -\frac{|\epsilon|^2}{1+\nu\mathrm{tr}\{\Omega^T\Gamma\Omega\}}.
\end{aligned} \qquad (6.57)$$

In view of the positive definiteness of $\Gamma(t)^{-1}$ this proves that $\tilde{\theta} \in \mathcal{L}_\infty[0, t_f)$. It also proves that $\dfrac{\epsilon}{\sqrt{1+\nu\mathrm{tr}\{\Omega^T\Gamma\Omega\}}} \in \mathcal{L}_2[0, t_f)$. Now using the boundedness of Γ and Ω, following the same line of argument as for the gradient update law, we prove that $\epsilon, \dot{\hat{\theta}} \in \mathcal{L}_2[0, t_f) \cap \mathcal{L}_\infty[0, t_f)$. □

The proofs for other swapping identifiers throughout this chapter have similar steps and arguments and will therefore be given in much less detail.

The fact that all the $\mathcal{L}_\infty$ and $\mathcal{L}_2$ bounds that we have established on $[0, t_f)$ are independent of t_f will be used for extending t_f to infinity in the stability proofs for the complete closed-loop adaptive systems.

As explained in [43], various modifications of the least-squares algorithm—covariance resetting, exponential data weighting, etc.—do not affect the properties established by Lemma 6.1. A priori knowledge of parameter bounds can also be included in the form of projection (see Appendix E).

It is possible to establish additional properties of the least-squares algorithm not stated in Lemma 6.1: $\hat{\theta}(t)$ converges to a constant, and $\dot{\hat{\theta}} \in \mathcal{L}_1$.

Remark 6.2 It is possible to note a very interesting connection between the z-passive identifier and the z-swapping identifier. Denoting $\hat{z} = -\psi$, (6.26) becomes

$$\dot{\hat{z}} = A_z(z, \hat{\theta}, t)\hat{z} + Q(z, \hat{\theta}, t)^T\hat{\theta} + \Omega^T\dot{\hat{\theta}}, \qquad (6.58)$$

so that the estimation error is

$$\epsilon = z - \hat{z} = \Omega^T\tilde{\theta} + \tilde{\epsilon}. \qquad (6.59)$$

Let us compare (6.58) with the observer from the z-passive identifier (5.111),

$$\dot{\hat{z}} = A_z(z, \hat{\theta}, t)\hat{z} + Q(z, \hat{\theta}, t)^T\hat{\theta}. \qquad (6.60)$$

The 'observer' in the z-swapping identifier is augmented by the term $\Omega^T\dot{\hat{\theta}}$ to achieve the static parametrization (6.59). It is important to understand that

6.3 z-SWAPPING SCHEME

this parametrization also leads to a strict passivity property that is different from the strict passivity property for the operator (5.97) in the z-passive scheme. Let us consider the operator

$$\Sigma_z : \tilde{\theta} \mapsto \Omega \epsilon, \tag{6.61}$$

with ϵ defined in (6.59) and Ω defined in (6.25). In view of (6.59) we have

$$\int_0^t (\Omega \epsilon)^T \tilde{\theta} d\tau = \int_0^t \epsilon^T \Omega^T \tilde{\theta} d\tau = \int_0^t \epsilon^T (\epsilon - \tilde{\epsilon}) d\tau = \int_0^t \left(\left| \epsilon - \frac{1}{2}\tilde{\epsilon} \right|^2 - \frac{1}{4}|\tilde{\epsilon}|^2 \right) d\tau$$

$$\geq -\frac{1}{4} \int_0^t |\tilde{\epsilon}|^2 d\tau. \tag{6.62}$$

On the other hand, integrating (6.46) we get

$$-\frac{1}{4} \int_0^t |\tilde{\epsilon}|^2 d\tau \geq \frac{1}{2c_0}|\tilde{\epsilon}(t)|^2 - \frac{1}{2c_0}|\tilde{\epsilon}(0)|^2 + \frac{3}{4} \int_0^t |\tilde{\epsilon}|^2 d\tau. \tag{6.63}$$

Substituting (6.63) in (6.62) we obtain

$$\int_0^t (\Omega \epsilon)^T \tilde{\theta} d\tau \geq \frac{1}{2c_0}|\tilde{\epsilon}(t)|^2 - \frac{1}{2c_0}|\tilde{\epsilon}(0)|^2 + \frac{3}{4} \int_0^t |\tilde{\epsilon}|^2 d\tau, \tag{6.64}$$

which by Definition D.2 proves that the operator (6.61) is strictly passive. This explains via passivity why (6.38) and (6.39) are valid update laws. The additional term $\Omega^T \dot{\tilde{\theta}}$ in (6.58) is the price paid in the swapping approach to obtain not only passivity but also a static parametrization (6.59). $\diamond$

Lemma 6.1 sets the stage for the following theorem that characterizes stability properties of the closed-loop adaptive system.

Theorem 6.3 (z-Swapping Scheme) *All the signals in the closed-loop adaptive system consisting of the plant (6.1), controller (6.6), filters (6.34), (6.35), and either the gradient (6.38) or the least-squares (6.39) update law are globally uniformly bounded, and $\lim_{t\to\infty} z(t) = \lim_{t\to\infty} \epsilon(t) = 0$. This means, in particular, that global asymptotic tracking is achieved:*

$$\lim_{t\to\infty} [y(t) - y_r(t)] = 0. \tag{6.65}$$

Moreover, if $\lim_{t\to\infty} y_r^{(i)}(t) = 0$, $i = 0, \ldots, n-1$, and $F(0) = 0$, then $\lim_{t\to\infty} x(t) = 0$.

Proof. Due to the piecewise continuity of $y_r(t), \ldots, y_r^{(n)}(t)$ and the smoothness of the nonlinearities in (6.1), the solution of the closed-loop adaptive system exists and is unique. Let its maximum interval of existence be $[0, t_f)$.

For both normalized and unnormalized update laws, from Lemma 6.1 we obtain $\tilde{\theta}, \dot{\tilde{\theta}}, \epsilon \in \mathcal{L}_\infty[0, t_f)$. Therefore, by Lemma 5.8, $z, x \in \mathcal{L}_\infty[0, t_f)$. In

Lemma 6.1 we also proved that $\Omega \in \mathcal{L}_\infty[0, t_f)$. Finally, by (6.36), $\Omega_0 \in \mathcal{L}_\infty[0, t_f)$.

We have thus shown that all of the signals of the closed-loop adaptive system are bounded on $[0, t_f)$ by constants depending only on the initial conditions, design gains, and the external signals $y_r(t), \ldots, y_r^{(n)}(t)$, but not on t_f. The independence of the bound of t_f proves that $t_f = \infty$ (if t_f were finite, then the solution would escape any compact set as $t \to t_f$, which would contradict the existence of a bound independent of t_f). Hence, all signals are globally uniformly bounded on $[0, \infty)$.

Now we set out to prove that $z \in \mathcal{L}_2$ and that $z(t) \to 0$ as $t \to \infty$. For both normalized and unnormalized update laws, from Lemma 6.1 we obtain $\dot{\hat{\theta}}, \epsilon \in \mathcal{L}_2$. Consequently $\Omega^T \tilde{\theta} \in \mathcal{L}_2$ because $\tilde{\epsilon} \in \mathcal{L}_2$. With $V = \frac{1}{2}|\zeta|^2$, all the conditions of Lemmas F.1 and F.4 are satisfied. Thus, by Lemma F.4, $z - \Omega^T \tilde{\theta} \in \mathcal{L}_2$. Hence $z \in \mathcal{L}_2$. To prove the convergence of z to zero, we note that (6.7) implies that $\dot{z} \in \mathcal{L}_\infty$. Therefore, by Barbalat's lemma (Corollary A.7) $z(t) \to 0$ as $t \to \infty$.

From the definitions in (6.2)–(6.4) we conclude that if $\lim_{t \to \infty} y_r^{(i)}(t) = 0$, $i = 0, \ldots, n-1$, and $F(0) = 0$, then $x(t) \to 0$ as $t \to \infty$. $\square$

The standard parameter estimators cannot guarantee that $\dot{\hat{\theta}}$ is bounded or square-integrable unless they use normalized update laws. For this reason, normalization is common in traditional adaptive linear control.[1] It is, therefore, significant that Theorem 6.3 holds also for unnormalized update laws. The proof of Lemma 6.1 reveals that normalization is not necessary because the nonlinear damping, which is built into the error system (6.7), guarantees that the filter state Ω is bounded even if its input, the regressor W, is growing unbounded. Therefore $\dot{\hat{\theta}}$ is bounded, which means that nonlinear damping acts as some form of normalization. Both the update law normalization and the nonlinear damping act to slow down the adaptation. The slow adaptation is a basic ingredient in the modular approach to adaptive control. The controller module tolerates the presence of the disturbance $\dot{\hat{\theta}}$ in the error system (6.7), but requires that this disturbance be bounded.

Controller with $g_i = 0$

In the swapping approach we can eliminate the nonlinear damping terms counteracting $\dot{\hat{\theta}}$, that is, we can set $g_i = 0$ in (6.5) without destroying the stability result of Theorem 6.3. To achieve this, we need an identifier which guarantees not only $\dot{\hat{\theta}} \in \mathcal{L}_\infty$, but also $Q^T \dot{\hat{\theta}} \in \mathcal{L}_\infty$.

---

[1] An exception is the unnormalized least-squares update law which can guarantee that $\dot{\hat{\theta}} \in \mathcal{L}_1$ [157, Lemma (264)].

6.3 z-SWAPPING SCHEME

Let us first discuss the construction of such an identifier. It uses the same filters (6.34) and (6.35) but with some modifications in the update laws.

The *gradient* update law (6.38) is modified as

$$\dot{\hat{\theta}} = \frac{\Gamma}{1+\nu'|Q|_{\mathcal{F}}} \frac{\Omega\epsilon}{1+\nu|\Omega|_{\mathcal{F}}^2}, \qquad \nu' > 0 \tag{6.66}$$

so that the following inequality is readily established

$$\frac{d}{dt}\left(\frac{1}{2}|\tilde{\theta}|_{\Gamma^{-1}}^2 + \frac{1}{2c_0}|\tilde{\epsilon}|^2\right) \leq -\frac{3}{4} \frac{|\epsilon|^2}{(1+\nu'|Q|_{\mathcal{F}})(1+\nu|\Omega|_{\mathcal{F}}^2)}. \tag{6.67}$$

This inequality shows that $\tilde{\theta} \in \mathcal{L}_\infty$. Since the boundedness of Ω established by (6.45) is unaffected, then by (6.37), $\epsilon \in \mathcal{L}_\infty$. Now the boundedness of Ω and ϵ, along with

$$|Q^{\mathrm{T}}\dot{\hat{\theta}}| \leq \frac{|Q|_{\mathcal{F}}}{1+\nu'|Q|_{\mathcal{F}}} \frac{|\Gamma\Omega\epsilon|}{1+\nu|\Omega|_{\mathcal{F}}^2} \leq \frac{1}{\nu'} \frac{|\Gamma\Omega\epsilon|}{1+\nu|\Omega|_{\mathcal{F}}^2}, \tag{6.68}$$

proves that $Q^{\mathrm{T}}\dot{\hat{\theta}} \in \mathcal{L}_\infty$. As for the square-integrability properties, by integrating (6.67) we get $\dfrac{\epsilon}{\sqrt{(1+\nu'|Q|_{\mathcal{F}})(1+\nu|\Omega|_{\mathcal{F}}^2)}} \in \mathcal{L}_2$. Because of the boundedness of Ω, this implies that $\dfrac{\epsilon}{\sqrt{1+\nu'|Q|_{\mathcal{F}}}} \in \mathcal{L}_2$. Finally,

$$|\dot{\hat{\theta}}| \leq \frac{\Gamma\Omega}{\sqrt{(1+\nu'|Q|_{\mathcal{F}})(1+\nu|\Omega|_{\mathcal{F}}^2)}} \frac{\epsilon}{\sqrt{1+\nu'|Q|_{\mathcal{F}}}}, \tag{6.69}$$

where the first factor is bounded and the second is square-integrable, proves that $\dot{\hat{\theta}} \in \mathcal{L}_2$.

Now, let us consider a modified *least-squares* update law:

$$\begin{aligned} \dot{\hat{\theta}} &= \frac{\Gamma}{1+\nu'|Q|_{\mathcal{F}}} \frac{\Omega\epsilon}{1+\nu\mathrm{tr}\{\Omega^{\mathrm{T}}\Gamma\Omega\}}, \qquad \nu' > 0 \\ \dot{\Gamma} &= -\frac{1}{1+\nu'|Q|_{\mathcal{F}}}\Gamma\frac{\Omega\Omega^{\mathrm{T}}}{1+\nu\mathrm{tr}\{\Omega^{\mathrm{T}}\Gamma\Omega\}}\Gamma. \end{aligned} \tag{6.70}$$

Like in the case of the original least-squares update law (6.39), we readily derive

$$\frac{d}{dt}\left(|\tilde{\theta}|_{\Gamma(t)^{-1}}^2 + \frac{1}{2c_0}|\tilde{\epsilon}|^2\right) \leq -\frac{|\epsilon|^2}{(1+\nu'|Q|_{\mathcal{F}})(1+\nu\mathrm{tr}\{\Omega^{\mathrm{T}}\Gamma\Omega\})}. \tag{6.71}$$

Following similar lines of argument as for the modified gradient update law, we establish that $\tilde{\theta}, \epsilon, Q^{\mathrm{T}}\dot{\hat{\theta}}$ are bounded and $\dfrac{\epsilon}{\sqrt{1+\nu'|Q|_{\mathcal{F}}}}, \dot{\hat{\theta}}$ are square-integrable.

To summarize, for both the modified gradient (6.66) and the modified least-squares (6.70) update laws, for $\nu \geq 0$ and $\nu' > 0$, the following properties hold:

$$(i) \quad \tilde{\theta} \in \mathcal{L}_\infty[0, t_f) \tag{6.72}$$

$$(ii) \quad \epsilon \in \mathcal{L}_\infty[0, t_f), \quad \frac{\epsilon}{\sqrt{1 + \nu'|Q|_\mathcal{F}}} \in \mathcal{L}_2[0, t_f) \tag{6.73}$$

$$(iii) \quad Q^\mathrm{T}\dot{\tilde{\theta}} \in \mathcal{L}_\infty[0, t_f), \quad \dot{\tilde{\theta}} \in \mathcal{L}_2[0, t_f). \tag{6.74}$$

It is important to note that the boundedness of $Q^\mathrm{T}\dot{\tilde{\theta}}$ can be achieved only with swapping identifiers. Passive identifiers do not seem to be able to guarantee this property independently of the boundedness of Q.

Since we have set $g_i = 0$, Lemma 5.8 is no longer applicable. However, in a fashion analogous to the proof of Lemma 5.8, we arrive at

$$|z(t)| \leq \frac{1}{\sqrt{2c_0}} \left(\frac{1}{\kappa_0} \|\tilde{\theta}\|_\infty^2 + \frac{2}{c_0} \|Q\dot{\tilde{\theta}}\|_\infty^2 \right)^{\frac{1}{2}} + |z(0)|e^{-c_0 t}, \tag{6.75}$$

which, in view of the boundedness properties (6.72) and (6.74), means that z and x are bounded. The boundedness of other signals follows readily, and the convergence of $z(t)$ to zero is established with the square-integrability properties (6.73) and (6.74).

We have thus proven the following result.

Theorem 6.4 *All the signals in the closed-loop adaptive system consisting of the plant (6.1), controller (6.6) with $g_i = 0$, $i = 1, \ldots, n$, filters (6.34), (6.35), and either the modified gradient (6.66) or the modified least-squares (6.70) update law, are globally uniformly bounded, and $\lim\limits_{t \to \infty} z(t) = 0$.*

6.4 x-Swapping Scheme

Let us consider the parametric x-model

$$\dot{x} = f(x, u) + F(x, u)^\mathrm{T} \theta, \tag{6.76}$$

which encompasses the class of parametric-strict feedback plants (6.1) with

$$f(x, u) = \begin{bmatrix} x_2 \\ \vdots \\ x_n \\ \beta_0(x)u \end{bmatrix}, \quad F(x, u)^\mathrm{T} = \begin{bmatrix} \varphi_1(x_1)^\mathrm{T} \\ \vdots \\ \varphi_{n-1}(x_1, \ldots, x_{n-1})^\mathrm{T} \\ \varphi_n(x)^\mathrm{T} \end{bmatrix}. \tag{6.77}$$

We introduce two filters

$$\dot{\Omega}_0 = A(x, t)(\Omega_0 + x) - f(x, u), \quad \Omega_0 \in \mathbb{R}^n \tag{6.78}$$

$$\dot{\Omega}^\mathrm{T} = A(x, t)\Omega^\mathrm{T} + F(x, u)^\mathrm{T}, \quad \Omega \in \mathbb{R}^{p \times n}, \tag{6.79}$$

6.4 x-SWAPPING SCHEME

where $A(x,t)$ is exponentially stable for each x continuous in t. Combining (6.76) and (6.78), we define $\mathcal{Y} = x + \Omega_0$ so that

$$\dot{\mathcal{Y}} = A(x,t)\mathcal{Y} + F(x,u)^\mathrm{T}\theta. \tag{6.80}$$

Since θ is constant, it follows that

$$\mathcal{Y} = \Omega^\mathrm{T}\theta + \tilde{\epsilon}, \tag{6.81}$$

where $\tilde{\epsilon} \triangleq x + \Omega_0 - \Omega^\mathrm{T}\theta$ is exponentially decaying because it is governed by

$$\dot{\tilde{\epsilon}} = A(x,t)\tilde{\epsilon}, \qquad \tilde{\epsilon} \in \mathbb{R}^n. \tag{6.82}$$

Now we introduce the "prediction" of $\mathcal{Y}$ as

$$\hat{\mathcal{Y}} = \Omega^\mathrm{T}\hat{\theta}. \tag{6.83}$$

The "prediction error" $\epsilon \triangleq \mathcal{Y} - \hat{\mathcal{Y}} = x + \Omega_0 - \Omega^\mathrm{T}\hat{\theta}$ can be written in the form

$$\epsilon = \Omega^\mathrm{T}\tilde{\theta} + \tilde{\epsilon}. \tag{6.84}$$

Like the z-swapping identifier, the x-swapping identifier should guarantee boundedness of $\tilde{\theta}, \epsilon, \hat{\theta}$, and square-integrability of $\epsilon, \dot{\hat{\theta}}$ with both normalized and unnormalized update laws. However, if $A(x,t)$ is constant and Hurwitz, the above properties would be achieved only by normalized update laws, whereas the unnormalized update laws would just guarantee that $\tilde{\theta}$ is bounded and ϵ is square-integrable [165]. To enforce the rest of the properties with unnormalized update laws, we need to guarantee the boundedness of Ω even when $F(x,u)$ is growing unbounded. We do so by strengthening the stability of $A(x,t)$, namely by choosing

$$A(x,t) = A_0 - \lambda F(x,u)^\mathrm{T} F(x,u) P \tag{6.85}$$

where $\lambda > 0$ and A_0 is an arbitrary constant matrix such that

$$PA_0 + A_0^\mathrm{T} P = -I, \quad P = P^\mathrm{T} > 0. \tag{6.86}$$

With this choice the filter and estimation error equations are

$$\dot{\Omega}^\mathrm{T} = \left(A_0 - \lambda F(x,u)^\mathrm{T} F(x,u) P\right) \Omega^\mathrm{T} + F(x,u)^\mathrm{T} \tag{6.87}$$

$$\dot{\Omega}_0 = \left(A_0 - \lambda F(x,u)^\mathrm{T} F(x,u) P\right) (\Omega_0 + x) - f(x,u) \tag{6.88}$$

$$\epsilon = x + \Omega_0 - \Omega^\mathrm{T}\hat{\theta}. \tag{6.89}$$

The update law for $\hat{\theta}$ is either the gradient:

$$\dot{\hat{\theta}} = \Gamma \frac{\Omega \epsilon}{1 + \nu |\Omega|_\mathcal{F}^2}, \qquad \Gamma = \Gamma^\mathrm{T} > 0, \quad \nu \geq 0 \tag{6.90}$$

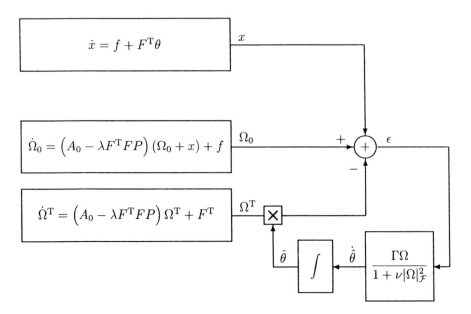

Figure 6.2: The x-swapping identifier.

or the least squares:

$$\begin{aligned}\dot{\hat{\theta}} &= \Gamma\frac{\Omega\epsilon}{1+\nu\operatorname{tr}\{\Omega^{\mathrm{T}}\Gamma\Omega\}} \\ \dot{\Gamma} &= -\Gamma\frac{\Omega\Omega^{\mathrm{T}}}{1+\nu\operatorname{tr}\{\Omega^{\mathrm{T}}\Gamma\Omega\}}\Gamma, \qquad \Gamma(0)=\Gamma(0)^{\mathrm{T}}>0, \quad \nu\geq 0.\end{aligned} \qquad(6.91)$$

Again, by allowing $\nu = 0$, we encompass unnormalized gradient and least-squares. The complete x-swapping identifier is shown in Figure 6.2.

Lemma 6.5 *Let the maximal interval of existence of solutions of (6.76), (6.87)–(6.88) with either (6.90) or (6.91) be $[0,t_f)$. Then for $\nu \geq 0$, the following identifier properties hold:*

$$\begin{aligned}&(i)&& \tilde{\theta} \in \mathcal{L}_\infty[0,t_f) &&(6.92)\\ &(ii)&& \epsilon \in \mathcal{L}_2[0,t_f)\cap \mathcal{L}_\infty[0,t_f) &&(6.93)\\ &(iii)&& \dot{\hat{\theta}} \in \mathcal{L}_2[0,t_f)\cap \mathcal{L}_\infty[0,t_f).&&(6.94)\end{aligned}$$

Proof (Outline). Along the solutions of (6.87) we have

$$\begin{aligned}\frac{d}{dt}\left(\Omega P\Omega^{\mathrm{T}}\right) &= \Omega\left(PA_0+A_0^{\mathrm{T}}P\right)\Omega^{\mathrm{T}} - 2\lambda\Omega PF^{\mathrm{T}}FP\Omega^{\mathrm{T}} + \Omega PF^{\mathrm{T}} + F P\Omega^{\mathrm{T}} \\ &= -\Omega\Omega^{\mathrm{T}} - 2\lambda\left(FP\Omega^{\mathrm{T}}-\frac{1}{2\lambda}I_p\right)^{\mathrm{T}}\left(FP\Omega^{\mathrm{T}}-\frac{1}{2\lambda}I_p\right)+\frac{1}{2\lambda}I_p.\end{aligned} \qquad(6.95)$$

6.4 x-SWAPPING SCHEME

Using the Frobenius norm we obtain

$$\frac{d}{dt}\operatorname{tr}\left\{\Omega P \Omega^{\mathrm{T}}\right\} = -|\Omega|_{\mathcal{F}}^2 - 2\lambda\left|F P \Omega^{\mathrm{T}} - \frac{1}{2\lambda}I_p\right|_{\mathcal{F}}^2 + \frac{1}{2\lambda}\operatorname{tr}\{I_p\}$$
$$\leq -|\Omega|_{\mathcal{F}}^2 + \frac{p}{2\lambda}. \tag{6.96}$$

In view of the fact that $\underline{\lambda}(P)|\Omega|_{\mathcal{F}}^2 \leq \operatorname{tr}\left\{\Omega P \Omega^{\mathrm{T}}\right\}$, (6.96) proves that $\Omega \in \mathcal{L}_\infty[0, t_f)$. From (6.82) and (6.85) it follows that

$$\frac{d}{dt}\left(|\tilde{\epsilon}|_P^2\right) \leq -|\tilde{\epsilon}|^2, \tag{6.97}$$

which implies $\tilde{\epsilon} \in \mathcal{L}_2[0, t_f) \cap \mathcal{L}_\infty[0, t_f)$.

Gradient update law (6.90). We consider the positive definite function

$$V = \frac{1}{2}|\tilde{\theta}|_{\Gamma^{-1}}^2 + |\tilde{\epsilon}|_P^2 \tag{6.98}$$

whose derivative is readily shown to satisfy

$$\dot{V} \leq -\frac{3}{4}\frac{|\epsilon|^2}{1 + \nu|\Omega|_{\mathcal{F}}^2}. \tag{6.99}$$

The nonpositivity of $\dot{V}$ proves that $\tilde{\theta} \in \mathcal{L}_\infty[0, t_f)$. Due to $\epsilon = \Omega^{\mathrm{T}}\tilde{\theta} + \tilde{\epsilon}$ and the boundedness of Ω it follows that $\epsilon \in \mathcal{L}_\infty[0, t_f)$, which, in turn proves that $\dot{\hat{\theta}} = \Gamma \frac{\Omega \epsilon}{1 + \nu|\Omega|_{\mathcal{F}}^2} \in \mathcal{L}_\infty[0, t_f)$. Integrating (6.99) we get $\frac{\epsilon}{\sqrt{1 + \nu|\Omega|_{\mathcal{F}}^2}} \in \mathcal{L}_2[0, t_f)$. Since Ω is bounded, then $\epsilon \in \mathcal{L}_2[0, t_f)$. The boundedness of Ω and the square-integrability of ϵ prove that $\dot{\hat{\theta}} = \Gamma \frac{\Omega \epsilon}{1 + \nu|\Omega|_{\mathcal{F}}^2} \in \mathcal{L}_2[0, t_f)$.

Least-squares update law (6.91). We consider the function

$$V = |\tilde{\theta}|_{\Gamma(t)^{-1}}^2 + |\tilde{\epsilon}|_P^2 \tag{6.100}$$

which is positive definite because $\Gamma(t)^{-1}$ is positive definite for each t. After routine calculations we get

$$\dot{V} \leq -\frac{|\epsilon|^2}{1 + \nu\operatorname{tr}\{\Omega^{\mathrm{T}}\Gamma\Omega\}}, \tag{6.101}$$

which, due to the positive definiteness of $\Gamma(t)^{-1}$, proves that $\tilde{\theta} \in \mathcal{L}_\infty[0, t_f)$. Integration of inequality (6.101) yields $\frac{\epsilon}{\sqrt{1 + \nu\operatorname{tr}\{\Omega^{\mathrm{T}}\Gamma\Omega\}}} \in \mathcal{L}_2[0, t_f)$. Now using the boundedness of Γ and Ω, following the same line of argument as for the gradient update law, we prove that $\epsilon, \dot{\hat{\theta}} \in \mathcal{L}_2[0, t_f) \cap \mathcal{L}_\infty[0, t_f)$. □

Our x-swapping identifier, unlike the standard linear parameter estimators, guarantees boundedness and square-integrability of $\dot{\hat{\theta}}$ even with unnormalized update laws. This is achieved by including nonlinear damping into filters (6.87)–(6.88) to slow down the adaptation.

Remark 6.6 As we noted in Remark 6.2 for the z-swapping scheme, there is a passivity interpretation of the x-swapping identifier. The signal $\hat{x} = \Omega_0 + \Omega^T \hat{\theta}$ is driven by the 'observer'

$$\dot{\hat{x}} = \left(A_0 - \lambda F(x,u)^T F(x,u) P \right) (\hat{x} - x) + f(x,u) + F(x,u)^T \hat{\theta} + \Omega^T \dot{\hat{\theta}}, \quad (6.102)$$

which differs from the observer (5.129) in the x-passive identifier in the additional term $\Omega^T \dot{\hat{\theta}}$. The filter employed in this term is the cost of making the operator $\Sigma_x : \tilde{\theta} \mapsto \Omega \epsilon$ strictly passive and achieving the static parametrization (6.84). $\diamond$

Lemma 6.5 sets the stage for the following result.

Theorem 6.7 (x-Swapping Scheme) *All the signals in the closed-loop adaptive system consisting of the plant (6.1), controller (6.6), filters (6.87), (6.88), and either the gradient (6.90) or the least-squares (6.91) update law are globally uniformly bounded, and $\lim_{t \to \infty} z(t) = \lim_{t \to \infty} \epsilon(t) = 0$. This means, in particular, that global asymptotic tracking is achieved:*

$$\lim_{t \to \infty} [y(t) - y_r(t)] = 0. \quad (6.103)$$

Moreover, if $\lim_{t \to \infty} y_r^{(i)}(t) = 0$, $i = 0, \ldots, n-1$, and $F(0) = 0$, then $\lim_{t \to \infty} x(t) = 0$.

Proof. As in the proof of Theorem 6.3, we show that $\hat{\theta}, \dot{\hat{\theta}}, z, x \in \mathcal{L}_\infty[0, t_f)$ and hence $u \in \mathcal{L}_\infty[0, t_f)$. From (6.87) and (6.88) it follows that Ω_0, Ω, and therefore ϵ are in $\mathcal{L}_\infty[0, t_f)$. By the same argument as in the proof of Theorem 6.3 we conclude that $t_f = \infty$.

Now we set out to prove that $z \in \mathcal{L}_2$. From Lemma 6.5, we have $\dot{\hat{\theta}}, \epsilon \in \mathcal{L}_2$. Consequently, $\Omega^T \tilde{\theta} \in \mathcal{L}_2$ because $\tilde{\epsilon} \in \mathcal{L}_2$. As in Theorem 6.3, we invoke Lemma F.4 to deduce that $z - \chi^T \tilde{\theta} \in \mathcal{L}_2$, where

$$\dot{\chi}^T = A_z(z, \hat{\theta}, t) \chi^T + W(z, \hat{\theta}, t)^T. \quad (6.104)$$

In order to show that $z \in \mathcal{L}_2$, we need to prove that $\Omega^T \tilde{\theta} \in \mathcal{L}_2$ implies $\chi^T \tilde{\theta} \in \mathcal{L}_2$, or, in the notation of Lemma F.7, that $T_{(A_0 - \lambda F^T F P)}[F^T] \tilde{\theta} \in \mathcal{L}_2$ implies $T_{A_z}[W^T] \tilde{\theta} \in \mathcal{L}_2$. To apply this lemma to our adaptive system we note from

6.4 x-SWAPPING SCHEME

(6.8) and (6.4) that

$$W^{\mathrm{T}}(z,\hat{\theta},t) = \begin{bmatrix} 1 & 0 & \cdots & 0 \\ -\frac{\partial \alpha_1}{\partial x_1} & 1 & \ddots & \vdots \\ \vdots & \ddots & \ddots & 0 \\ -\frac{\partial \alpha_{n-1}}{\partial x_1} & \cdots & -\frac{\partial \alpha_{n-1}}{\partial x_{n-1}} & 1 \end{bmatrix} F^{\mathrm{T}}(x) \triangleq N(z,\hat{\theta},t) F^{\mathrm{T}}(x). \tag{6.105}$$

Since $F(x(t))$ is continuous and bounded, and $N(z(t),\hat{\theta}(t),t)$ is bounded and has a bounded derivative (due to the smoothness of the components of $F(x)$), they satisfy the conditions of Lemma F.7. Then $\chi^{\mathrm{T}}\tilde{\theta} \in \mathcal{L}_2$ and hence $z \in \mathcal{L}_2$. The rest of the proof is the same as for Theorem 6.3. $\square$

Remark 6.8 An alternative proof of $z(t) \to 0$ which avoids Lemma F.7 is as follows. Combining (6.87), (6.88), and (6.89), we get

$$\dot{\epsilon} = \left(A_0 - \lambda F(x)^{\mathrm{T}} F(x) P\right)\epsilon + F(x)^{\mathrm{T}}\tilde{\theta} - \Omega^{\mathrm{T}}\dot{\hat{\theta}}. \tag{6.106}$$

Because of the boundedness of all the signals, $\dot{\epsilon}$ is bounded. Since $\epsilon \in \mathcal{L}_2$, by Barbalat's lemma (Corollary A.7), $\epsilon(t) \to 0$. By virtue of the smoothness of F (which implies that its partial derivatives are bounded for bounded values of their arguments) and boundedness of all the states, $\ddot{\epsilon}$ is also bounded, and therefore $\dot{\epsilon}$ is uniformly continuous. Since $\epsilon(t) \to 0$, then

$$\lim_{t\to\infty} \int_0^t \dot{\epsilon}(\tau)d\tau = \lim_{t\to\infty} \epsilon(t) - \epsilon(0) = -\epsilon(0) < \infty. \tag{6.107}$$

Then by Barbalat's lemma (Lemma A.6), $\dot{\epsilon}(t) \to 0$. Since $\dot{\hat{\theta}} \in \mathcal{L}_\infty \cap \mathcal{L}_2$ and $\ddot{\hat{\theta}} \in \mathcal{L}_\infty$, it follows that $\dot{\hat{\theta}}(t) \to 0$. From (6.106) we conclude $F(x(t))^{\mathrm{T}}\tilde{\theta}(t) \to 0$, and therefore, by (6.105), $W(z(t),\hat{\theta}(t),t)^{\mathrm{T}}\tilde{\theta}(t) \to 0$. Since $\dot{\hat{\theta}}(t) \to 0$, we have $Q(z(t),\hat{\theta}(t),t)^{\mathrm{T}}\dot{\hat{\theta}}(t) \to 0$. (Note that while $Q^{\mathrm{T}}\dot{\hat{\theta}}$ is in $\mathcal{L}_2$, $W^{\mathrm{T}}\tilde{\theta}$ may not be.) Thus the input $W^{\mathrm{T}}\tilde{\theta} + Q^{\mathrm{T}}\dot{\hat{\theta}}$ to the error system

$$\dot{z} = A_z(z,\hat{\theta},t)z + W(z,\hat{\theta},t)^{\mathrm{T}}\tilde{\theta} + Q(z,\hat{\theta},t)^{\mathrm{T}}\dot{\hat{\theta}} \tag{6.108}$$

converges to zero. By the same argument as in (5.156), followed by the application of Lemma B.8, we arrive at the conclusion that $z(t) \to 0$. $\diamond$

Remark 6.9 As in the z-swapping identifier, we can modify the gradient

$$\dot{\hat{\theta}} = \frac{\Gamma}{1+\nu'|Q|_{\mathcal{F}}} \frac{\Omega\epsilon}{1+\nu|\Omega|_{\mathcal{F}}^2}, \qquad \nu' > 0 \tag{6.109}$$

and the least-squares

$$\dot{\hat{\theta}} = \frac{\Gamma}{1+\nu'|Q|_{\mathcal{F}}} \frac{\Omega\epsilon}{1+\nu\text{tr}\{\Omega^T\Gamma\Omega\}}, \quad \nu' > 0$$

$$\dot{\Gamma} = -\frac{1}{1+\nu'|Q|_{\mathcal{F}}} \Gamma \frac{\Omega\Omega^T}{1+\nu\text{tr}\{\Omega^T\Gamma\Omega\}} \Gamma \tag{6.110}$$

update laws and obtain the following properties

$$(i) \quad \tilde{\theta} \in \mathcal{L}_\infty[0, t_f) \tag{6.111}$$

$$(ii) \quad \epsilon \in \mathcal{L}_\infty[0, t_f), \quad \frac{\epsilon}{\sqrt{1+\nu'|Q|_{\mathcal{F}}}} \in \mathcal{L}_2[0, t_f) \tag{6.112}$$

$$(iii) \quad Q^T\dot{\hat{\theta}} \in \mathcal{L}_\infty[0, t_f), \quad \dot{\hat{\theta}} \in \mathcal{L}_2[0, t_f). \tag{6.113}$$

This allows us to set $g_i = 0$ in the controller in Table 6.1 as we did in Theorem 6.4 with the z-swapping identifier. ◇

6.5 Transient Performance

In this section we will derive transient performance bounds for the error state z. Since $z_1 = y - y_r$, these bounds also bound the tracking error $y - y_r$.

We analyze first the z-swapping scheme and then the x-swapping scheme. To simplify the derivations without loss of generality, we assume that $\Omega(0) = 0$, as well as $\Omega_0(0) = -z(0)$ in the z-passive scheme, and $\Omega_0(0) = x(0)$ in the x-passive scheme. These filter initializations guarantee that $\tilde{\epsilon}(0) = 0$ and should always be performed to eliminate the disturbing effect of the initial estimation error.

We will derive performance bounds only for schemes using gradient update law. For simplicity we let $\Gamma = \gamma I$. We will then briefly explain how the bounds are modified for the least-squares update law.

We first give a lemma which establishes performance properties of the z-swapping identifier with the gradient update law.

Lemma 6.10 (z-Swapping Identifier with Gradient Update) *The identifier (6.34), (6.35), (6.38) guarantees*

$$(i) \quad \|\tilde{\theta}\|_\infty = |\tilde{\theta}(0)| \tag{6.114}$$

$$(ii) \quad \|\,|\Omega|_{\mathcal{F}}\|_\infty \leq \frac{1}{2\sqrt{c_0\kappa_0}} \tag{6.115}$$

$$(iii) \quad \|\dot{\hat{\theta}}\|_\infty \leq \gamma \min\left\{\frac{1}{\nu}, \frac{1}{4c_0\kappa_0}\right\} |\tilde{\theta}(0)| \tag{6.116}$$

$$(iv) \quad \|\dot{\hat{\theta}}\|_2 \leq \sqrt{\frac{\gamma}{2} \min\left\{\frac{1}{\nu}, \frac{1}{4c_0\kappa_0}\right\}} |\tilde{\theta}(0)| \tag{6.117}$$

$$(v) \quad \|\epsilon\|_2 \leq \frac{1}{\sqrt{2\gamma}} \left(1 + \frac{\nu}{4c_0\kappa_0}\right)^{1/2} |\tilde{\theta}(0)|. \tag{6.118}$$

6.5 Transient Performance

Proof. Since $\tilde{\epsilon}(0) = 0$, we have $\tilde{\epsilon}(t) \equiv 0$, so (6.37) becomes

$$\epsilon = \Omega^T \tilde{\theta}. \tag{6.119}$$

Thus, (6.47) and (6.49) imply (6.114).

From (6.45), by Lemma C.5 we conclude (6.115).

With (6.38) we write

$$|\dot{\tilde{\theta}}|^2 = \gamma^2 \frac{\epsilon^T \Omega^T \Omega \epsilon}{(1 + \nu |\Omega|_{\mathcal{F}}^2)^2} \leq \gamma^2 \frac{|\epsilon|^2 |\Omega|_{\mathcal{F}}^2}{(1 + \nu |\Omega|_{\mathcal{F}}^2)^2} \leq \frac{\gamma^2}{\nu} \frac{|\epsilon|^2}{1 + \nu |\Omega|_{\mathcal{F}}^2}. \tag{6.120}$$

By using (6.119) we get

$$|\dot{\tilde{\theta}}|^2 \leq \frac{\gamma^2}{\nu} \frac{\tilde{\theta}^T \Omega \Omega^T \tilde{\theta}}{1 + \nu |\Omega|_{\mathcal{F}}^2} \leq \frac{\gamma^2}{\nu} \frac{|\tilde{\theta}|^2 |\Omega|_{\mathcal{F}}^2}{1 + \nu |\Omega|_{\mathcal{F}}^2} \leq \left(\frac{\gamma}{\nu}\right)^2 |\tilde{\theta}|^2, \tag{6.121}$$

which, in view of (6.114), proves that

$$\|\dot{\tilde{\theta}}\|_\infty \leq \frac{\gamma}{\nu} |\tilde{\theta}(0)|. \tag{6.122}$$

By noting that $|\dot{\tilde{\theta}}| = \gamma \left|\frac{\Omega \epsilon}{1+\nu |\Omega|_{\mathcal{F}}^2}\right| \leq \gamma |\Omega \epsilon|$, with (6.119) we get $|\dot{\tilde{\theta}}| \leq \gamma |\Omega|_{\mathcal{F}}^2 |\tilde{\theta}|$, which in view of (6.115) and (6.114) yields

$$\|\dot{\tilde{\theta}}\|_\infty \leq \frac{\gamma}{4 c_0 \kappa_0} |\tilde{\theta}(0)|. \tag{6.123}$$

Combining (6.122) and (6.123), we arrive at (6.116).

Since $\tilde{\epsilon}(0) = 0$, the first row of (6.48) gives

$$\dot{V} \leq -\frac{|\epsilon|^2}{1 + \nu |\Omega|_{\mathcal{F}}^2}. \tag{6.124}$$

Integrating (6.124) over $[0, \infty)$ we obtain

$$\left\|\frac{\epsilon}{\sqrt{1 + \nu |\Omega|_{\mathcal{F}}^2}}\right\|_2 \leq \sqrt{V(0)} = \frac{1}{\sqrt{2\gamma}} |\tilde{\theta}(0)|. \tag{6.125}$$

The integration of (6.120) over $[0, \infty)$ and substitution of (6.125) yields

$$\|\dot{\tilde{\theta}}\|_2 \leq \sqrt{\frac{\gamma}{2\nu}} |\tilde{\theta}(0)|. \tag{6.126}$$

In view of (6.38) we write

$$|\dot{\tilde{\theta}}|^2 = \gamma^2 \frac{\epsilon^T \Omega^T \Omega \epsilon}{(1 + \nu |\Omega|_{\mathcal{F}}^2)^2} \leq \gamma^2 \frac{|\epsilon|^2 |\Omega|_{\mathcal{F}}^2}{(1 + \nu |\Omega|_{\mathcal{F}}^2)^2} \leq \gamma^2 |\Omega|_{\mathcal{F}}^2 \frac{|\epsilon|^2}{1 + \nu |\Omega|_{\mathcal{F}}^2}. \tag{6.127}$$

Integrating (6.127) over $[0, \infty)$ and substituting (6.115), we obtain

$$\|\dot{\hat{\theta}}\|_2 \leq \sqrt{\frac{\gamma}{2}} \frac{1}{2\sqrt{c_0 \kappa_0}} |\tilde{\theta}(0)|. \tag{6.128}$$

Combining (6.126) and (6.128) we get (6.117).

A bound on the $\mathcal{L}_2$ norm of ϵ is obtained using

$$\int_0^\infty |\epsilon(\tau)|^2 d\tau \leq \int_0^\infty \frac{|\epsilon(\tau)|^2}{1 + \nu|\Omega|_{\mathcal{F}}^2}(1 + \nu \|\|\Omega|_{\mathcal{F}}\|_\infty^2) d\tau$$

$$\leq (1 + \nu \|\|\Omega|_{\mathcal{F}}\|_\infty^2) \left\| \frac{\epsilon}{\sqrt{1 + \nu|\Omega|_{\mathcal{F}}^2}} \right\|_2^2. \tag{6.129}$$

By substituting (6.125) and (6.115) into (6.129) we prove (6.118). $\square$

Lemma 6.10 is now used to characterize the transient performance of the closed-loop adaptive system.

Theorem 6.11 (z-Swapping, Gradient Update) *In the adaptive system (6.1), (6.6), (6.34), (6.35), (6.38), the following inequalities hold:*

(i) $\quad |z(t)| \leq \dfrac{|\tilde{\theta}(0)|}{2\sqrt{c_0}} \left(\dfrac{1}{\kappa_0} + \dfrac{\gamma^2}{g_0} \min\left\{\dfrac{1}{\nu^2}, \dfrac{1}{(4c_0\kappa_0)^2}\right\} \right)^{1/2} + |z(0)|e^{-c_0 t}$ (6.130)

(ii) $\quad \|z\|_2 \leq \dfrac{|\tilde{\theta}(0)|}{\sqrt{2c_0}} \left[\dfrac{\nu}{2\kappa_0\gamma} + \gamma\left(\dfrac{1}{g_0} + \dfrac{1}{2c_0^2\kappa_0}\right) \min\left\{\dfrac{1}{\nu}, \dfrac{1}{4c_0\kappa_0}\right\} \right]^{1/2}$

$$+ \frac{|\tilde{\theta}(0)|}{\sqrt{\gamma}} + \frac{1}{\sqrt{c_0}}|z(0)|. \tag{6.131}$$

Proof. By direct substitution of (6.114) and (6.116) into (6.9) we obtain (6.130).

We now set out to derive (6.131). We will calculate the $\mathcal{L}_2$ norm bound for z as $\|z\|_2 \leq \|\epsilon\|_2 + \|\psi\|_2$, where

$$\psi \triangleq \Omega_0 - \Omega^T \hat{\theta}. \tag{6.132}$$

A bound on $\|\epsilon\|_2$ is given by (6.118). To obtain a bound on $\|\psi\|_2$, we examine

$$\dot{\psi} = A_z(z, \hat{\theta}, t)\psi - Q(z, \hat{\theta}, t)^T \hat{\theta} - \Omega^T \dot{\hat{\theta}}. \tag{6.133}$$

By using (6.8) and repeating the completion of squares with g-terms as in (5.88), we derive

$$\frac{d}{dt}\left(\frac{1}{2}|\psi|^2\right) \leq -c_0|\psi|^2 + \frac{1}{4g_0}|\dot{\hat{\theta}}|^2 - \psi^T \Omega^T \dot{\hat{\theta}}$$

$$\leq -\frac{c_0}{2}|\psi|^2 - \frac{c_0}{2}\left|\psi - \frac{1}{c_0}\Omega^T \dot{\hat{\theta}}\right|^2 + \frac{1}{4g_0}|\dot{\hat{\theta}}|^2 + \frac{1}{2c_0}|\Omega^T \dot{\hat{\theta}}|^2$$

$$\leq -\frac{c_0}{2}|\psi|^2 + \left(\frac{1}{4g_0} + \frac{|\Omega|_{\mathcal{F}}^2}{2c_0}\right)|\dot{\hat{\theta}}|^2, \tag{6.134}$$

6.5 TRANSIENT PERFORMANCE

which with (6.115) gives

$$\frac{d}{dt}\left(|\psi|^2\right) \leq -c_0|\psi|^2 + \left(\frac{1}{2g_0} + \frac{1}{4c_0^2\kappa_0}\right)|\dot{\tilde{\theta}}|^2. \tag{6.135}$$

By applying Lemma B.5.(ii) to (6.135), we arrive at

$$\|\psi\|_2 \leq \frac{1}{\sqrt{c_0}}\left[|\psi(0)| + \left(\frac{1}{2g_0} + \frac{1}{4c_0^2\kappa_0}\right)^{1/2}\|\dot{\tilde{\theta}}\|_2\right]. \tag{6.136}$$

Since we use initial conditions $\Omega(0) = 0$, $\Omega_0(0) = -z(0)$, which imply $\psi(0) = 0$, inequality (6.136) becomes

$$\|\psi\|_2 \leq \frac{1}{\sqrt{c_0}}\left[|z(0)| + \left(\frac{1}{2g_0} + \frac{1}{4c_0^2\kappa_0}\right)^{1/2}\|\dot{\tilde{\theta}}\|_2\right]. \tag{6.137}$$

By substituting (6.117) into (6.137) we get

$$\|\psi\|_2 \leq \frac{1}{\sqrt{c_0}}\left[|z(0)| + \left(\frac{1}{2g_0} + \frac{1}{4c_0^2\kappa_0}\right)^{1/2}\sqrt{\frac{\gamma}{2}\min\left\{\frac{1}{\nu},\frac{1}{4c_0\kappa_0}\right\}}|\tilde{\theta}(0)|\right]. \tag{6.138}$$

Combining this with (6.118), and rearranging the terms, we obtain (6.131). □

When we perform the standard initialization $z(0) = 0$ introduced in Section 4.3.2, the bounds (6.130) and (6.131) become

$$|z(t)| \leq \frac{1}{2\sqrt{c_0\kappa_0}}\left(1 + \frac{\gamma^2}{16c_0^2\kappa_0 g_0}\right)^{1/2}|\tilde{\theta}(0)| \tag{6.139}$$

$$\|z\|_2 \leq \left\{\frac{1}{2\sqrt{c_0\kappa_0}}\left[\frac{\nu}{\gamma} + \frac{\gamma}{2c_0}\left(\frac{1}{g_0} + \frac{1}{2c_0^2\kappa_0}\right)\right]^{1/2} + \frac{1}{\sqrt{\gamma}}\right\}|\tilde{\theta}(0)|. \tag{6.140}$$

These bounds depend only on the design parameters $c_i, \kappa_i, g_i, \gamma$, and ν, and the initial parameter error $\tilde{\theta}(0)$. By increasing c_0 and κ_0 we can make the $\mathcal{L}_\infty$ bound (6.139) arbitrarily small. These gains alone are not enough to arbitrarily reduce the $\mathcal{L}_2$ bound because of the last term in (6.140). However, this bound can be systematically reduced, for example, by increasing γ simultaneously with c_0 or κ_0. As we shall see, the increasing dependence of the bounds on the adaptation gain γ is not corroborated by simulations, which indicates that the dependence on γ may not be very tight.

Let us now compare the bounds for the z-swapping scheme with the bounds for the tuning functions scheme and the z-passive scheme.

Examining the $\mathcal{L}_\infty$ bounds (4.241), (5.178) and (6.139), we note that the bounds for the modular schemes (z-passive and z-swapping) are higher than the bounds for the tuning functions scheme. The additional γ-dependent terms in the modular schemes occur due to the fact that $\dot{\hat{\theta}}$ is not eliminated from the z-system, but only "damped." Because of these terms, the result from Corollary 4.21 does not hold for the modular schemes. Thus, we arrive at an important conclusion: The capability to outperform the nonadaptive design is a unique feature of the tuning functions adaptive design. We will illustrate this point in Example 6.15.

Examining the $\mathcal{L}_2$ bounds (4.229), (5.177), and (6.140), we again see that the bound for the tuning functions approach is lower.

In Examples 6.14 and 6.15 we illustrate the performance bounds with simulations.

Remark 6.12 For $\nu = 0$ the bounds from Lemma 6.10, as well as those from Theorem 6.11, are readily modified to hold for the least-squares update law (6.39). While the bounds (6.114), (6.115), and (6.116) remain the same, the bounds (6.117) and (6.118) increase by a factor of $\sqrt{2}$. $\diamond$

Finally we give an $\mathcal{L}_\infty$ bound for the x-swapping scheme, for a special choice of design parameters.

Theorem 6.13 (x-Swapping, Gradient Update) *In the adaptive system (6.1), (6.6), (6.87), (6.88), (6.90) with the choice[2] $A_0 = -c_0$, $P = I$, $\lambda = p\kappa_0$, the following inequality holds:*

$$|z(t)| \leq \frac{|\tilde{\theta}(0)|}{2\sqrt{c_0}} \left(\frac{1}{\kappa_0} + \frac{\gamma^2}{g_0} \min\left\{ \frac{1}{\nu^2}, \frac{1}{(4c_0\kappa_0)^2} \right\} \right)^{1/2} + |z(0)|e^{-c_0 t}. \qquad (6.141)$$

Proof. With this choice of design parameters, we have

$$\dot{\Omega}^{\mathrm{T}} = -c_0 \Omega^{\mathrm{T}} - p\kappa_0 F^{\mathrm{T}} F \Omega + F^{\mathrm{T}} \qquad (6.142)$$

$$\dot{\hat{\theta}} = \gamma \frac{\Omega \epsilon}{1 + \nu |\Omega|_\mathcal{F}^2}. \qquad (6.143)$$

From (6.99), we obtain the same bound on $\|\tilde{\theta}\|_\infty$ as in (6.114) for the z-swapping scheme. In view of (6.96), we obtain the same bound on $\| |\Omega|_\mathcal{F} \|_\infty$ as in (6.115) for the z-swapping scheme. By repeating the third part of the proof of Lemma 6.10, we get the same bound on $\|\dot{\hat{\theta}}\|_\infty$ as in (6.116). It then follows that the resulting bound (6.141) is the same as (6.130). $\square$

Since the bounds (6.141) and (6.130) are the same, we conclude that the z-swapping and the x-swapping schemes are equally capable of reducing transient peaks. However, an $\mathcal{L}_2$ bound similar to (6.131) is not available for the x-swapping scheme.

---

[2]Note that this choice of A_0 and P gives $PA_0 + A_0^{\mathrm{T}} P = -2c_0 I$, which differs from (6.86). The proof takes this into account.

6.5.1 Simulation examples

We now present simulation results for the swapping and the tuning functions schemes. They will illustrate the derived performance properties and serve for a qualitative comparison between the tuning functions and the modular approach.

The first example illustrates the performance bounds established in Theorem 6.11.

Example 6.14 (Performance of the Swapping Design) We consider the second-order plant

$$\begin{aligned} \dot{x}_1 &= x_2 + \varphi(x_1)\theta, \quad \varphi(x_1) = x_1^2 \\ \dot{x}_2 &= u. \end{aligned} \qquad (6.144)$$

The ISS-controller for this system is

$$\begin{aligned} \alpha_1 &= -c_1 z_1 - \kappa_1 \varphi^2 z_1 - \hat{\theta}\varphi \\ u &= -z_1 - c_2 z_2 - \kappa_2 \left(\frac{\partial \alpha_1}{\partial x_1}\right)^2 \varphi^2 z_2 - g_2 \left(\frac{\partial \alpha_1}{\partial \hat{\theta}}\right)^2 z_2 + \frac{\partial \alpha_1}{\partial x_1}(x_2 + \hat{\theta}\varphi). \end{aligned} \qquad (6.145)$$

It results in the closed-loop system

$$\dot{z} = \begin{bmatrix} -c_1 - \kappa_1 \varphi^2 & 1 \\ -1 & -c_2 - \kappa_2 \left(\frac{\partial \alpha_1}{\partial x_1}\varphi\right)^2 - g_2 \left(\frac{\partial \alpha_1}{\partial \hat{\theta}}\right)^2 \end{bmatrix} z$$

$$+ \begin{bmatrix} \varphi \\ -\frac{\partial \alpha_1}{\partial x_1}\varphi \end{bmatrix} \tilde{\theta} + \begin{bmatrix} 0 \\ -\frac{\partial \alpha_1}{\partial \hat{\theta}} \end{bmatrix} \dot{\hat{\theta}}. \qquad (6.146)$$

The *z-swapping identifier* uses the filters

$$\dot{\Omega}^T = \begin{bmatrix} -c_1 - \kappa_1 \varphi^2 & 1 \\ -1 & -c_2 - \kappa_2 \left(\frac{\partial \alpha_1}{\partial x_1}\varphi\right)^2 - g_2 \left(\frac{\partial \alpha_1}{\partial \hat{\theta}}\right)^2 \end{bmatrix} \Omega^T$$

$$+ \begin{bmatrix} \varphi \\ -\frac{\partial \alpha_1}{\partial x_1}\varphi \end{bmatrix} \qquad (6.147)$$

$$\dot{\Omega}_0 = \begin{bmatrix} -c_1 - \kappa_1 \varphi^2 & 1 \\ -1 & -c_2 - \kappa_2 \left(\frac{\partial \alpha_1}{\partial x_1}\varphi\right)^2 - g_2 \left(\frac{\partial \alpha_1}{\partial \hat{\theta}}\right)^2 \end{bmatrix} \Omega_0$$

$$+ \begin{bmatrix} \varphi \\ -\frac{\partial \alpha_1}{\partial x_1}\varphi \end{bmatrix} \hat{\theta} - \begin{bmatrix} 0 \\ -\frac{\partial \alpha_1}{\partial \hat{\theta}} \end{bmatrix} \dot{\hat{\theta}}. \qquad (6.148)$$

The gradient update law is

$$\dot{\hat{\theta}} = \gamma \frac{\Omega \epsilon}{1 + \nu |\Omega|^2}, \qquad (6.149)$$

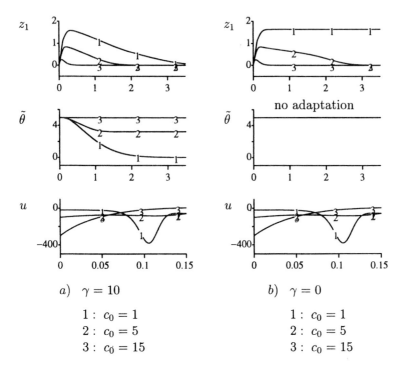

Figure 6.3: Dependence of the transients on c_0 with $2\kappa_0 = g_0 = 1$. (Note an expanded time scale for control u.)

where the estimation error is implemented as

$$\epsilon = z + \Omega_0 - \Omega^T \hat{\theta}. \tag{6.150}$$

Simulations were carried out with nominal values $c_1 = c_2 = c_0 = \kappa_1 = \kappa_2 = 2\kappa_0 = g_2 = g_0 = 1$, $\gamma = 10$, $\nu = 1$, $\theta = 5$, and $\hat{\theta}(0) = 0$, which were judged to give representative responses. All simulations are with following initial conditions: $x(0) = -\Omega_0(0) = [0,\ 10]^T$, $\Omega(0) = 0$ (to set $\tilde{\epsilon}(0) = 0$).

Figure 6.3a illustrates Theorem 6.11. The design parameter c_0 can be used for systematically improving the transient performance. Up to a certain point the error transients and the control effort in Fig. 6.3a are simultaneously decreasing as c_0 increases. Beyond that point the control effort starts increasing. The control u is given in an expanded time scale in order to clearly display the main qualitative differences among the three cases.

Figure 6.3b illustrates Corollary 5.9. When adaptation is switched off, the states are uniformly bounded and converge to (or remain inside) a compact residual set. Corollary 5.9 does not describe the behavior inside the residual set, which may contain multiple equilibria, limit cycles, and so on. For this example (but not in general), there is an asymptotically stable equilibrium at

6.5 TRANSIENT PERFORMANCE

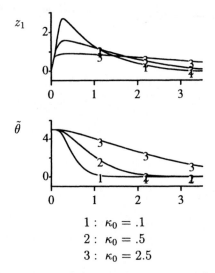

1: $\kappa_0 = .1$
2: $\kappa_0 = .5$
3: $\kappa_0 = 2.5$

Figure 6.4: Dependence of the transients on κ_0 with $c_0 = g_0 = 1$, $\gamma = 10$.

the origin for any value of the parameter error. The origin is asymptotically stable because $\varphi(x_1) = x_1^2$ is quadratic (around the origin), but it is not globally stable for nonzero values of the parameter error. For large values of the parameter error, this equilibrium has a region of attraction which is strictly inside the residual set, as shown in Figure 6.3b, curve 1. For small values of the parameter error, according to Theorem 4.15, global asymptotic stability is achieved.

Figure 6.4 shows the influence of κ_0 on transients. According to Theorem 6.11, the peak values can be decreased by increasing κ_0, which is confirmed by the plot. An important observation is that the κ-terms slow down the adaptation and make the transients longer. The effect of the g-terms was shown to be significant only for very small c_0 and κ_0 or for very large γ.

Figure 6.5 demonstrates the influence of the adaptation gain γ on transients. Due to the slow initial adaptation, which should be attributed not only to the normalized gradient update law but also to the fact that the regressor is filtered, there is a clear separation of action between the nonadaptive controller, which at the beginning brings the state z quickly to the residual set, and the adaptive controller, which takes over to drive the state to the origin. The property that the $\mathcal{L}_\infty$ bounds are increasing functions of γ, to be expected from Theorem 6.11, was exhibited in simulations only with extremely high values of γ. This indicates that some of the bounds derived are not very tight over the entire range of design parameter values.

In our simulations we used the plant initial condition $x(0) = [0, 10]^T$ and hence $z(0) = [0, 10]^T$, which is independent of the design gains c_0, κ_0, g_0. This is why the peak of z_1 decreases monotonically as any of these gains increases.

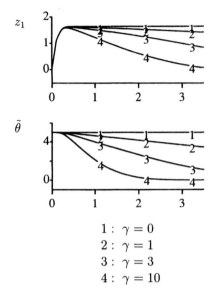

Figure 6.5: Dependence of the transients on γ with $c_0 = g_0 = 2\kappa_0 = 1$

If, instead, we used $x_1(0) \neq 0$, then, according to Section 4.3.2, we would have added an appropriately initialized reference trajectory to set $z(0) = 0$. In this way, bad transients would be eliminated by following a less aggressive path to the origin.

In the plant (6.144) the unknown parameter appears only in the first equation. This opens a possibility for a reduction in the dynamic order of the identifier. Instead of the z-swapping identifier we can use an x-*swapping identifier*

$$
\begin{align}
\dot{\Omega} &= -(c + \kappa\varphi^2)\Omega + \varphi, & \Omega &\in \mathbb{R} & (6.151) \\
\dot{\Omega}_0 &= -(c + \kappa\varphi^2)(\Omega_0 + x_1) - x_2, & \Omega_0 &\in \mathbb{R} & (6.152) \\
\dot{\hat{\theta}} &= \gamma\frac{\Omega\epsilon}{1 + \nu\Omega^2} & & & (6.153) \\
\epsilon &= x_1 + \Omega_0 - \Omega\hat{\theta}, & & & (6.154)
\end{align}
$$

which is uncertainty-specific and avoids unnecessary filtering. In simulations, the only difference between the z-swapping and the x-swapping approach was in the value of γ needed to achieve the same speed of adaptation—higher value was needed in the z-swapping case. Since the responses were similar we show them only for the z-swapping scheme.

Simulations were also carried out for the z-passive scheme and were qualitatively the same as for the z-swapping scheme and are therefore omitted.

6.5 TRANSIENT PERFORMANCE

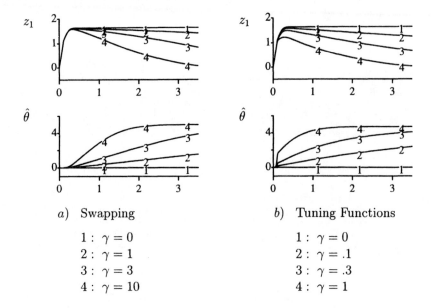

Figure 6.6: Comparison between the swapping and the tuning functions designs with $c_0 = g_0 = 2\kappa_0 = 1$

The next example gives a simulation comparison between the z-swapping scheme and the tuning functions scheme.

Example 6.15 (Swapping vs. Tuning Functions) The tuning functions controller designed for system (6.144) is

$$\begin{aligned} \alpha_1 &= -c_1 z_1 - \kappa_1 \varphi^2 z_1 - \hat{\theta}\varphi \\ u &= -z_1 - c_2 z_2 - \kappa_2 \left(\frac{\partial \alpha_1}{\partial x_1}\right)^2 \varphi^2 z_2 + \frac{\partial \alpha_1}{\partial \hat{\theta}}\dot{\hat{\theta}} + \frac{\partial \alpha_1}{\partial x_1}(x_2 + \hat{\theta}\varphi), \end{aligned} \quad (6.155)$$

and the parameter estimator is

$$\dot{\hat{\theta}} = \gamma \left[\varphi, -\frac{\partial \alpha_1}{\partial x_1}\varphi\right] z. \quad (6.156)$$

The resulting error system is

$$\dot{z} = \begin{bmatrix} -c_1 - \kappa_1 \varphi^2 & 1 \\ -1 & -c_2 - \kappa_2 \left(\frac{\partial \alpha_1}{\partial x_1}\right)^2 \varphi^2 \end{bmatrix} z + \begin{bmatrix} \varphi \\ -\frac{\partial \alpha_1}{\partial x_1}\varphi \end{bmatrix} \tilde{\theta}. \quad (6.157)$$

Simulations were carried out for the tuning functions scheme using the same design parameters and initial conditions as for the z-swapping scheme in Example 6.14. A comparison is given in Figure 6.6. The initial adaptation with the tuning functions scheme is rapid because it has no filters or normalization

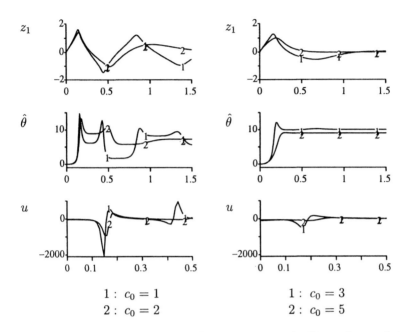

Figure 6.7: Transients with tuning functions approach. Dependence of the transients on $c_1 = c_2 = c_0$ for $\gamma = 1$. (Note a different time scale for control u.)

to slow it down, as the swapping identifier does. Figure 6.6 corroborates the difference between the $\mathcal{L}_\infty$ bounds (4.241) and (6.139). The increasing of the adaptation gain γ reduces the $\mathcal{L}_\infty$ bound in the tuning functions design but not in the swapping design. This confirms our earlier conclusion that the tuning functions design is unique in its capability to improve the performance of the underlying nonadaptive design.

The importance of fast adaptation in the tuning functions design is even more pronounced when we set the nonlinear damping terms to zero, $\kappa_0 = 0$. Simulation results that illustrate this point are shown in Fig. 6.7. The rapid changes of $\hat{\theta}$ when the values of $c_1 = c_2 = c_0$ are small reflect the fact that a fast identifier is indispensable for catching up with fast nonlinear instability. The identifier is as fast as necessary to keep the derivative of the complete Lyapunov function $V_2 = \frac{1}{2}(z_1^2 + z_2^2 + \tilde{\theta}^2/\gamma)$ nonpositive. It is important to understand that the responses in Figure 6.7 are neither the best nor very typical. They only show that when all other means of achieving stability and performance are absent (high gain, for instance), the fast adaptation, an inherent part of the tuning functions design, comes to rescue the system from instability.

While nonlinear damping is not necessary for stabilization in the tuning functions design, it is helpful for transient performance improvement. However, it is important to keep in mind that the drawback of nonlinear damping is the rapid growth of the controller nonlinearity. For example, while the tuning functions controller with $\kappa_0 = 0$ has the highest power of a nonlinear term equal to 7, with $\kappa_0 > 0$ the highest power is 17. Therefore, the tuning functions scheme without nonlinear damping has a big advantage in this design aspect over the modular schemes which require nonlinear damping.

While very small values of the nonlinear damping coefficients κ_i and g_i will result in poor performance, very large values will result in excessive control effort (and could potentially have an adverse effect on robustness). Fortunately, there will always be a range of values of κ_i and g_i that will result in good performance without large control effort. $\diamond$

6.6 SG-Scheme

The SG-controller introduced in Section 5.8.1 achieves a weak form of modularity using weaker nonlinear damping terms. This controller was combined in Section 5.8.2 with a z-passive identifier to design an SG adaptive scheme which guarantees stability due to a small-gain property.

In this section we follow the same idea and design an SG-scheme with a z-swapping identifier. This scheme is of interest because in the case of linear plants it is a certainty equivalence scheme and serves for a comparison of the ISS modular design with traditional certainty equivalence designs. The main benefit of this scheme is a reduction of the growth of nonlinearities in the control law relative to the ISS-controller.

We first briefly review the controller design from Section 5.8.1. The only modification of the ISS-controller in Table 6.1 is in the nonlinear damping terms (6.5). Following the notation of Section 5.8.1, we first rewrite the matrices W and Q from (6.7) as

$$W(z,\hat{\theta},t)^{\mathrm{T}} = W(0,\hat{\theta},t)^{\mathrm{T}} + \begin{bmatrix} \bar{z}_1^{\mathrm{T}}\omega_1 \\ \vdots \\ \bar{z}_n^{\mathrm{T}}\omega_n \end{bmatrix} \qquad (6.158)$$

$$Q(z,\hat{\theta},t)^{\mathrm{T}} = Q(0,\hat{\theta},t)^{\mathrm{T}} + \begin{bmatrix} 0 \\ \bar{z}_1^{\mathrm{T}}\delta_2 \\ \vdots \\ \bar{z}_{n-1}^{\mathrm{T}}\delta_n \end{bmatrix} \qquad (6.159)$$

and, to reduce notation, denote $W_0 \triangleq W(0,\hat{\theta},t)$ and $Q_0 \triangleq Q(0,\hat{\theta},t)$. The modified nonlinear damping functions are defined as

$$s_i = \kappa_i |\omega_i|_{\mathcal{F}}^2 + g_i |\delta_i|_{\mathcal{F}}^2 . \qquad (6.160)$$

Although the nonlinear damping functions (6.160) are different from (6.5), the form of the error system (6.7)–(6.8) is unchanged:

$$\dot{z} = A_z(z,\hat{\theta},t)z + W(z,\hat{\theta},t)^{\mathrm{T}}\tilde{\theta} + Q(z,\hat{\theta},t)^{\mathrm{T}}\dot{\hat{\theta}} \qquad (6.161)$$

where

$$A_z(z,\hat{\theta},t) = \begin{bmatrix} -c_1-s_1 & 1 & 0 & \cdots & 0 \\ -1 & -c_2-s_2 & 1 & \ddots & \vdots \\ 0 & -1 & \ddots & \ddots & 0 \\ \vdots & \ddots & \ddots & \ddots & 1 \\ 0 & \cdots & 0 & -1 & -c_n-s_n \end{bmatrix}. \quad (6.162)$$

As is usually the case in linear estimation-based (certainty equivalence) design, we use an identifier with a *normalized* update law. We employ the z-swapping identifier consisting of the filters

$$\dot{\Omega}^{\mathrm{T}} = A_z(z,\hat{\theta},t)\Omega^{\mathrm{T}} + W(z,\hat{\theta},t)^{\mathrm{T}} \quad (6.163)$$

$$\dot{\Omega}_0 = A_z(z,\hat{\theta},t)\Omega_0 + W(z,\hat{\theta},t)^{\mathrm{T}}\hat{\theta} - Q(z,\hat{\theta},t)^{\mathrm{T}}\dot{\hat{\theta}}, \quad (6.164)$$

the estimation error

$$\epsilon = z + \Omega_0 - \Omega^{\mathrm{T}}\hat{\theta}, \quad (6.165)$$

and either the normalized gradient,

$$\dot{\hat{\theta}} = \Gamma \frac{\Omega\epsilon}{1+\nu|\Omega|_{\mathcal{F}}^2}, \qquad \Gamma = \Gamma^{\mathrm{T}} > 0, \quad \nu > 0 \quad (6.166)$$

or the normalized least-squares update law

$$\begin{aligned} \dot{\hat{\theta}} &= \Gamma \frac{\Omega\epsilon}{1+\nu|\Omega|_{\mathcal{F}}^2} \\ \dot{\Gamma} &= -\Gamma \frac{\Omega\Omega^{\mathrm{T}}}{1+\nu|\Omega|_{\mathcal{F}}^2}\Gamma, \qquad \Gamma(0)=\Gamma(0)^{\mathrm{T}} > 0, \quad \nu > 0. \end{aligned} \quad (6.167)$$

This identifier is different from the original z-swapping identifier (6.34), (6.35), (6.38), (6.39), because of the difference in the nonlinear damping functions (6.160). The new nonlinear damping functions s_i appearing in the matrix (6.162) are no longer strong enough to guarantee the boundedness of Ω (which implied parts (ii) and (iii) in Lemma 6.1). Therefore we cannot use Lemma 6.1. Fortunately, if we use the normalization, $\nu > 0$, the identifier can guarantee certain boundedness properties sufficient for global stability. These properties are summarized in the following standard lemma.

Lemma 6.16 *Let the maximal interval of existence of solutions of (6.161), (6.163)–(6.164) with either (6.166) or (6.167) be $[0,t_f)$. Then the following identifier properties hold:*

$$(i) \qquad \tilde{\theta} \in \mathcal{L}_\infty[0,t_f) \quad (6.168)$$

$$(ii) \qquad \frac{\epsilon}{\sqrt{1+\nu|\Omega|_{\mathcal{F}}^2}} \in \mathcal{L}_2[0,t_f) \cap \mathcal{L}_\infty[0,t_f) \quad (6.169)$$

$$(iii) \qquad \dot{\hat{\theta}} \in \mathcal{L}_2[0,t_f) \cap \mathcal{L}_\infty[0,t_f). \quad (6.170)$$

6.6 SG-SCHEME

Proof (Outline). The proof of boundedness of $\tilde{\theta}$ and square-integrability of $\frac{\epsilon}{\sqrt{1+\nu|\Omega|_{\mathcal{F}}^2}}$ is the same as in Lemma 6.1. To establish the boundedness of $\frac{\epsilon}{\sqrt{1+\nu|\Omega|_{\mathcal{F}}^2}}$, we first recall that

$$\epsilon = \Omega^T \tilde{\theta} + \tilde{\epsilon}, \tag{6.171}$$

where $\tilde{\epsilon}$ is a bounded, exponentially decaying signal. Therefore,

$$\begin{aligned}\frac{|\epsilon|}{\sqrt{1+\nu|\Omega|_{\mathcal{F}}^2}} &= \frac{|\Omega^T\tilde{\theta}+\tilde{\epsilon}|}{\sqrt{1+\nu|\Omega|_{\mathcal{F}}^2}} \leq \frac{|\Omega|_2|\tilde{\theta}|+|\tilde{\epsilon}|}{\sqrt{1+\nu|\Omega|_{\mathcal{F}}^2}} \leq \frac{|\Omega|_{\mathcal{F}}}{\sqrt{1+\nu|\Omega|_{\mathcal{F}}^2}}|\tilde{\theta}| + \frac{|\tilde{\epsilon}|}{\sqrt{1+\nu|\Omega|_{\mathcal{F}}^2}} \\ &\leq \frac{1}{\sqrt{\nu}}|\tilde{\theta}| + |\tilde{\epsilon}| < \infty.\end{aligned} \tag{6.172}$$

Then the boundedness of $\dot{\tilde{\theta}}$ follows readily:

$$\begin{aligned}|\dot{\tilde{\theta}}| &= \frac{|\Gamma\Omega\epsilon|}{1+\nu|\Omega|_{\mathcal{F}}^2} \leq \bar{\lambda}(\Gamma)\frac{|\Omega|}{\sqrt{1+\nu|\Omega|_{\mathcal{F}}^2}}\frac{|\epsilon|}{\sqrt{1+\nu|\Omega|_{\mathcal{F}}^2}} \\ &\leq \bar{\lambda}(\Gamma)\frac{1}{\sqrt{\nu}}\frac{|\epsilon|}{\sqrt{1+\nu|\Omega|_{\mathcal{F}}^2}}.\end{aligned} \tag{6.173}$$

From (6.173) it also follows that $\dot{\tilde{\theta}}$ is square integrable. $\square$

The next theorem establishes the stability properties of the SG-scheme.

Theorem 6.17 (SG-Scheme) *All the signals in the closed-loop adaptive system consisting of the plant (6.1), the SG-controller (6.6) with (6.160), the filters (6.163), (6.164), and either the normalized gradient (6.166) or the normalized least-squares (6.167) update law are globally uniformly bounded, and $\lim_{t\to\infty} z(t) = \lim_{t\to\infty} \epsilon(t) = 0$. This means, in particular, that global asymptotic tracking is achieved:*

$$\lim_{t\to\infty}[y(t) - y_{\mathrm{r}}(t)] = 0. \tag{6.174}$$

Moreover, if $\lim_{t\to\infty} y_{\mathrm{r}}^{(i)}(t) = 0$, $i = 0,\ldots,n-1$, and $F(0) = 0$, then $\lim_{t\to\infty} x(t) = 0$.

Proof. Using (6.158) we write (6.163) as

$$\dot{\Omega}^T = A_z \Omega^T + W_0^T + \begin{bmatrix} \bar{z}_1^T \omega_1 \\ \vdots \\ \bar{z}_n^T \omega_n \end{bmatrix}. \tag{6.175}$$

In a fashion similar to (6.44) we compute

$$\begin{aligned}\frac{d}{dt}\left(\frac{1}{2}|\Omega|_{\mathcal{F}}^2\right) &\leq -c_0|\Omega|_{\mathcal{F}}^2 - \sum_{i=1}^n \kappa_i|\omega_i|_{\mathcal{F}}^2|\Omega_i|^2 + \sum_{i=1}^n \bar{z}_i^{\mathrm{T}}\omega_i\Omega_i + \mathrm{tr}\left\{W_0^{\mathrm{T}}\Omega\right\} \\ &\leq -\frac{c_0}{2}|\Omega|_{\mathcal{F}}^2 + \sum_{i=1}^n \frac{1}{4\kappa_i}|\bar{z}_i|^2 + \frac{1}{2c_0}|W_0|_{\mathcal{F}}^2 \\ &\leq -\frac{c_0}{2}|\Omega|_{\mathcal{F}}^2 + \frac{1}{4\kappa_0}|z|^2 + \frac{1}{2c_0}|W_0|_{\mathcal{F}}^2 .\end{aligned} \qquad (6.176)$$

On the other hand, using (6.159) we write (6.133) as

$$\dot{\psi} = A_z \psi - (\Omega + Q_0)^{\mathrm{T}}\dot{\hat{\theta}} - \begin{bmatrix} 0 \\ \delta_2^{\mathrm{T}}\bar{z}_1 \\ \vdots \\ \delta_n^{\mathrm{T}}\bar{z}_{n-1} \end{bmatrix}\dot{\hat{\theta}} \qquad (6.177)$$

. and compute

$$\begin{aligned}\frac{d}{dt}\left(\frac{1}{2}|\psi|^2\right) &\leq -c_0|\psi|^2 - \sum_{i=1}^n g_i|\delta_i|_{\mathcal{F}}^2\psi_i^2 - \psi^{\mathrm{T}}(\Omega+Q_0)^{\mathrm{T}}\dot{\hat{\theta}} - \sum_{i=1}^n \bar{z}_{i-1}^{\mathrm{T}}\delta_i\psi_i\dot{\hat{\theta}} \\ &\leq -\frac{c_0}{2}|\psi|^2 + \frac{1}{2c_0}|(\Omega+Q_0)^{\mathrm{T}}\dot{\hat{\theta}}|^2 + \sum_{i=1}^n \frac{1}{4g_i}|\bar{z}_{i-1}|^2|\dot{\hat{\theta}}|^2 \\ &\leq -\frac{c_0}{2}|\psi|^2 + \frac{1}{c_0}|\Omega|_{\mathcal{F}}^2|\dot{\hat{\theta}}|^2 + \frac{1}{4g_0}|z|^2|\dot{\hat{\theta}}|^2 + \frac{1}{c_0}|Q_0|_{\mathcal{F}}^2|\dot{\hat{\theta}}|^2 .\end{aligned} \qquad (6.178)$$

The system (6.176), (6.178) is summarized as

$$\frac{d}{dt}\left(|\Omega|_{\mathcal{F}}^2\right) \leq -c_0|\Omega|_{\mathcal{F}}^2 + \frac{1}{2\kappa_0}|z|^2 + \frac{1}{c_0}|W_0|_{\mathcal{F}}^2 \qquad (6.179)$$

$$\frac{d}{dt}\left(|\psi|^2\right) \leq -c_0|\psi|^2 + \frac{2}{c_0}|\dot{\hat{\theta}}|^2|\Omega|_{\mathcal{F}}^2 + \frac{1}{2g_0}|\dot{\hat{\theta}}|^2|z|^2 + \frac{2}{c_0}|Q_0|_{\mathcal{F}}^2|\dot{\hat{\theta}}|^2 . \qquad (6.180)$$

From Lemma 6.16 we have $\hat{\theta} \in \mathcal{L}_\infty[0, t_f)$. Because the functions W_0 and Q_0 are smooth in $\hat{\theta}$ and bounded in t, then $|W_0|_{\mathcal{F}}^2 < k$ and $|Q_0|_{\mathcal{F}}^2 < k$, where k denotes a generic positive finite constant. From Lemma 6.16 we also have $\dot{\hat{\theta}}, \dfrac{\epsilon}{\sqrt{1+\nu|\Omega|_{\mathcal{F}}^2}} \in \mathcal{L}_2[0,t_f) \cap \mathcal{L}_\infty[0,t_f)$. Let us denote by l_1 a generic function in $\mathcal{L}_1[0,t_f) \cap \mathcal{L}_\infty[0,t_f)$. (Note that this notation allows us to say $kl_1 \leq l_1$ for any finite k.) Hence

$$|\dot{\hat{\theta}}|^2 + \frac{|\epsilon|^2}{1+\nu|\Omega|_{\mathcal{F}}^2} \leq l_1 . \qquad (6.181)$$

Since

$$z = \epsilon - \psi , \qquad (6.182)$$

6.6 SG-SCHEME

we have

$$|z|^2 \leq 2\frac{|\epsilon|^2}{1+\nu|\Omega|_{\mathcal{F}}^2}(1+\nu|\Omega|_{\mathcal{F}}^2) + 2|\psi|^2 \leq 2l_1(1+\nu|\Omega|_{\mathcal{F}}^2) + 2|\psi|^2. \quad (6.183)$$

Thus with (6.182) and (6.183), inequalities (6.179)–(6.180) become

$$\frac{d}{dt}\left(|\Omega|_{\mathcal{F}}^2\right) \leq -(c_0-l_1)|\Omega|_{\mathcal{F}}^2 + \frac{1}{\kappa_0}|\psi|^2 + k \quad (6.184)$$

$$\frac{d}{dt}\left(|\psi|^2\right) \leq -(c_0-l_1)|\psi|^2 + l_1|\Omega|_{\mathcal{F}}^2 + k. \quad (6.185)$$

These two differential inequalities are interconnected and form a loop with small gain because $|\Omega|_{\mathcal{F}}^2$ appears multiplied by l_1 in (6.185). To finish the proof we define the "superstate"

$$X \triangleq |\Omega|_{\mathcal{F}}^2 + \frac{2}{\kappa_0 c_0}|\psi|^2, \quad (6.186)$$

differentiate it, and substitute (6.184)–(6.185):

$$\dot{X} \leq -c_0|\Omega|_{\mathcal{F}}^2 - \frac{2}{\kappa_0}|\psi|^2 + l_1\left(|\Omega|_{\mathcal{F}}^2 + \frac{2}{\kappa_0 c_0}|\psi|^2 + \frac{2}{\kappa_0 c_0}|\Omega|_{\mathcal{F}}^2\right) + \frac{1}{\kappa_0}|\psi|^2 + k$$

$$\leq -c_0|\Omega|_{\mathcal{F}}^2 - \frac{1}{\kappa_0}|\psi|^2 + l_1 X + k; \quad (6.187)$$

thus we get

$$\dot{X} \leq -\left(\frac{c_0}{2} - l_1\right)X + k. \quad (6.188)$$

By applying Lemma B.6 we conclude that X is bounded on $[0, t_f)$, so Ω and ψ are also bounded. In view of (6.171), ϵ is bounded, which, along with the boundedness of ψ, proves that z is bounded. Hence all the signals are bounded, and thus $t_f = \infty$. The rest of the proof is the same as for Theorem 6.3 and again uses Lemma F.4 for proving convergence. □

Performance bounds comparable to those in Section 6.5 for the ISS design are not available for the SG design.

By examining the design procedure in Table 6.1 with the modification (6.160), one can see that if the plant (6.1) is linear, then the SG-controller is nonlinear in $\hat{\theta}$ but linear in x. (To arrive at this conclusion, by carefully examining the nonlinear damping terms one should observe that they are independent of x.) Because of the linear dependence of the control law on x, the SG design for linear systems should have qualitatively the same behavior as the traditional certainty equivalence adaptive linear controllers. Therefore, the SG design will serve for comparison of the strong ISS design with traditional adaptive control.

In the next example we compare the SG and ISS designs for a linear plant.

Example 6.18 (ISS vs. SG) Let us consider system (6.144) with $\varphi(x_1) = x_1$. For this linear system we make a comparison between the ISS and SG designs. The only difference is that the terms $\kappa_1 \varphi^2$, $\kappa_2 \left(\frac{\partial \alpha_1}{\partial x_1}\right)^2 \varphi^2$, $g_2 \left(\frac{\partial \alpha_1}{\partial \hat{\theta}}\right)^2$ in the ISS design (6.145)–(6.148) are, respectively, replaced by κ_1, $\kappa_2 \left(\frac{\partial \alpha_1}{\partial x_1}\right)^2$, g_2 in the SG design. The same design coefficients and initial conditions are used as in Example 6.14, except for $\theta = 3$. The adaptation gains, $\gamma = 5$ for the ISS design, and $\gamma = 1.5$ for the SG design, are chosen so that the rate of parameter convergence is the same for both designs. The control law of the SG design is linear in x and nonlinear in $\hat{\theta}$.

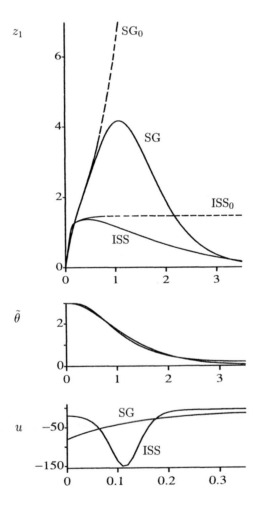

Figure 6.8: ISS design vs. SG design. The dashed lines show $z_1(t)$ when $\gamma = 0$. (Note the expanded time scale for the control u.)

Figure 6.8 shows the difference in performance between the two designs. The ISS design achieves a better attenuation of the z_1-transient but with a larger control effort. Hence, the ISS and SG designs offer a clear trade-off between performance improvement and control effort.

Let us now focus attention the dashed responses in Figure 6.8, marked ISS$_0$ and SG$_0$. They illustrate the underlying nonadaptive behavior ($\gamma = 0$) of the two controllers. According to Corollary 5.9, as shown by curve ISS$_0$ in Figure 6.8, the ISS design results in global boundedness (although the origin is unstable). In contrast, the SG design results in instability, as shown by curve SG$_0$ in Figure 6.8. The instability occurs because the constant parameter estimate has a destabilizing value. The closed-loop system with the SG controller is linear, and the instability is exponential.

The trade-offs between the ISS and SG designs revealed by this example also apply qualitatively if the SG design is replaced by any of the traditional certainty equivalence adaptive linear designs. ◇

6.7 Schemes with Weak ISS-Controller

The SG-controller reduces the growth of nonlinearities in the control law. The weak ISS-controller designed in this section achieves a more significant reduction of the nonlinearity growth. However, in contrast to the SG-controller, the weak ISS-controller is nonlinear in the plant state even when the plant is linear.

We start with a motivating example.

Example 6.19 Let us consider the scalar system

$$\dot{x} = u + x^4\theta. \tag{6.189}$$

The (strong) ISS-controller for this system is

$$\begin{aligned} u &= -x - x^4\hat{\theta} - \left(x^4\right)^2 x \\ &= -x - x^4\hat{\theta} - x^9, \end{aligned} \tag{6.190}$$

We now show how the desired ISS property is achieved with the following weaker controller:

$$\begin{aligned} u &= -x - x^4\hat{\theta} - \left|x^4\right|x \\ &= -x - x^4\hat{\theta} - x^5. \end{aligned} \tag{6.191}$$

The closed-loop system with this controller is

$$\dot{x} = -x - x^5 + x^4\tilde{\theta}. \tag{6.192}$$

Along the solutions of (6.192) we have

$$\frac{d}{dt}\left(\frac{x^2}{2}\right) = -x^2 - x^6 + x^5\tilde{\theta}$$
$$\leq -x^2 - |x|^5\left(|x| - |\tilde{\theta}|\right). \tag{6.193}$$

By Theorem C.2, it follows that input-to-state stability is achieved with respect to $\tilde{\theta}$ as input:

$$|x(t)| \leq |x(0)|e^{-t} + \|\tilde{\theta}\|_\infty. \tag{6.194}$$

Other ISS bounds are also possible.[3] The ISS bound (6.194) is somewhat higher than the one that would follow from the (strong) ISS control law (6.190) substituted in (6.189):

$$|x(t)| \leq |x(0)|e^{-t} + \frac{1}{2}\|\tilde{\theta}\|_\infty. \tag{6.196}$$

The ISS property (6.194) allows us to combine the weak ISS-controller with any identifier that independently guarantees boundedness of $\tilde{\theta}$. The weak ISS controller (6.191) whose highest-order term is x^5 is expected to use less control effort than the ISS-controller (6.190) with x^9. It is also important to note that the weak ISS-controller is 'less nonlinear' than the SG-controller where the highest-order term is x^7:

$$u = -x - x^4\hat{\theta} - \left(x^3\right)^2 x$$
$$= -x - x^4\hat{\theta} - x^7. \tag{6.197}$$

This reduction with the weak ISS-controller is even more pronounced when the plant nonlinearity is of higher order, say x^n. While the highest-order terms in the strong ISS-controller and the SG-controller are x^{2n+1} and x^{2n-1}, respectively, in the weak ISS-controller it is x^{n+1}. ◊

Example 6.19 suggests that the nonlinear damping functions (6.5) should be modified to contain the norms, rather than the squares of the norms, of the columns of W and Q, namely,

$$s_i = \kappa_i|w_i| + g_i\left|\frac{\partial \alpha_{i-1}}{\partial \hat{\theta}}^{\mathrm{T}}\right|. \tag{6.198}$$

---

[3]Consider, for example, the expression $-|x|^5\left(|x| - |\tilde{\theta}|\right)$. When $|x| \geq |\tilde{\theta}|$, this expression is nonpositive. When $|x| \leq |\tilde{\theta}|$ we have $-|x|^5\left(|x| - |\tilde{\theta}|\right) \leq |x|^5|\tilde{\theta}| \leq \tilde{\theta}^6$, which, substituted into (6.193), gives $\frac{d}{dt}\left(\frac{x^2}{2}\right) \leq -x^2 + \tilde{\theta}^6$. The ISS bound has a cubic gain with respect to $\tilde{\theta}$:

$$|x(t)| \leq |x(0)|e^{-t} + \|\tilde{\theta}\|_\infty^3, \tag{6.195}$$

which is tighter than (6.194) for small $\tilde{\theta}$.

6.7 Schemes with Weak ISS-Controller

Table 6.2: Weak ISS-Controller

$$z_i = x_i - y_r^{(i-1)} - \alpha_{i-1} \tag{6.200}$$

$$\alpha_i(\bar{x}_i, \hat{\theta}, \bar{y}_r^{(i-1)}) = -(c_i + s_i)z_i - w_i^{\mathrm{T}}\hat{\theta} + \sum_{k=1}^{i-1}\left(\frac{\partial \alpha_{i-1}}{\partial x_k}x_{k+1} + \frac{\partial \alpha_{i-1}}{\partial y_r^{(k-1)}}y_r^{(k)}\right) \tag{6.201}$$

$$w_i(\bar{x}_i, \hat{\theta}, \bar{y}_r^{(i-2)}) = \varphi_i - \sum_{k=1}^{i-1}\frac{\partial \alpha_{i-1}}{\partial x_k}\varphi_k \tag{6.202}$$

$$s_i(\bar{x}_i, \hat{\theta}, \bar{y}_r^{(i-2)}) = \kappa_i\sqrt{|w_i|^2 + 1} + g_i\sqrt{\left|\frac{\partial \alpha_{i-1}}{\partial \hat{\theta}}^{\mathrm{T}}\right|^2 + 1} \tag{6.203}$$

$$i = 1, \ldots, n$$

$$\bar{x}_i = (x_1, \ldots, x_i), \qquad \bar{y}_r^{(i)} = (y_r, \dot{y}_r, \ldots, y_r^{(i)})$$

Adaptive control law:

$$u = \frac{1}{\beta(x)}\left[\alpha_n(x, \hat{\theta}, \bar{y}_r^{(n-1)}) + y_r^{(n)}\right] \tag{6.204}$$

A slight problem with this form of nonlinear damping is that the function $|\cdot|$ is not differentiable, which means that it would pose a problem in the process of backstepping. Consequently, we replace (6.198) with

$$s_i = \kappa_i\sqrt{|w_i|^2 + 1} + g_i\sqrt{\left|\frac{\partial \alpha_{i-1}}{\partial \hat{\theta}}^{\mathrm{T}}\right|^2 + 1}, \tag{6.199}$$

which is a smooth function that overbounds (6.198). To simplify the analysis we make an additional modification to the stabilizing functions (6.3) — we eliminate the term $-z_{i-1}$. The weak ISS-controller is summarized in Table 6.2.

The closed-loop error system obtained by applying the design procedure (6.200)–(6.204) to system (6.1) is

$$\dot{z} = A_z(z, \hat{\theta}, t)z + W(z, \hat{\theta}, t)^{\mathrm{T}}\tilde{\theta} + Q(z, \hat{\theta}, t)^{\mathrm{T}}\dot{\hat{\theta}}, \tag{6.205}$$

where W and Q are as in (6.8), and A_z is without the -1's below the diagonal:

$$A_z(z, \hat{\theta}, t) = \begin{bmatrix} -c_1 - s_1 & 1 & & \\ & \ddots & \ddots & \\ & & \ddots & 1 \\ & & & -c_n - s_n \end{bmatrix} \tag{6.206}$$

Before we proceed to the identifier design, we establish the ISS property of the error system (6.205).

Lemma 6.20 *The error system (6.205) is ISS with respect to $\left(\tilde{\theta}, \dot{\hat{\theta}}\right)$.*

Proof. Along the solutions of (6.205), for $i = 1, \ldots, n-1$, we have

$$\begin{aligned}\frac{d}{dt}\left(\frac{z_i^2}{2}\right) &= -c_i z_i^2 - \kappa_i \sqrt{|w_i|^2 + 1}\, z_i^2 - g_i \sqrt{\left|\frac{\partial \alpha_{i-1}}{\partial \hat{\theta}}\right|^2 + 1}\, z_i^2 \\ &\quad + z_i z_{i+1} + z_i w_i^T \tilde{\theta} - z_i \frac{\partial \alpha_{i-1}}{\partial \hat{\theta}} \dot{\hat{\theta}} \\ &\leq -\frac{c_i}{2} z_i^2 - \frac{c_i}{2}|z_i|\left(|z_i| - \frac{2}{c_i}|z_{i+1}|\right) \\ &\quad - \kappa_i |z_i w_i|\left(|z_i| - \frac{1}{\kappa_i}|\tilde{\theta}|\right) - g_i \left|z_i \frac{\partial \alpha_{i-1}}{\partial \hat{\theta}}^T\right|\left(|z_i| - \frac{1}{g_i}|\dot{\hat{\theta}}|\right).\end{aligned} \quad (6.207)$$

Thus we have the following implication:

$$|z_i| \geq \left(\frac{2}{c_i}|z_{i+1}| + \frac{1}{\kappa_i}|\tilde{\theta}| + \frac{1}{g_i}|\dot{\hat{\theta}}|\right) \quad \Rightarrow \quad \frac{d}{dt}\left(z_i^2\right) \leq -c_i z_i^2. \quad (6.208)$$

By Theorem C.2, for all $0 \leq s \leq t$,

$$|z_i(t)| \leq |z_i(s)| e^{-c_i t/2} + \sup_{s \leq \tau \leq t}\left\{\frac{2}{c_i}|z_{i+1}(\tau)| + \frac{1}{\kappa_i}|\tilde{\theta}(\tau)| + \frac{1}{g_i}|\dot{\hat{\theta}}(\tau)|\right\}, \quad (6.209)$$

that is, each z_i-equation is ISS with respect to $\left(z_{i+1}, \tilde{\theta}, \dot{\hat{\theta}}\right)$. The z_n-equation, however, has only $\left(\tilde{\theta}, \dot{\hat{\theta}}\right)$ as input. When the same arguments as (6.207)–(6.209) are applied to the z_n-equation, we get

$$|z_n(t)| \leq |z_n(s)| e^{-c_n t} + \sup_{s \leq \tau \leq t}\left\{\frac{1}{\kappa_n}|\tilde{\theta}(\tau)| + \frac{1}{g_n}|\dot{\hat{\theta}}(\tau)|\right\}. \quad (6.210)$$

Inequalities (6.209) and (6.210) define a set of cascaded ISS inequalities. By repeatedly applying Lemma C.4 we show that there exist positive real numbers β_1, ρ_1, and σ independent of initial conditions such that

$$|z(t)| \leq \beta_1 |z(0)| e^{-\sigma t} + \rho_1 \sup_{s \leq \tau \leq t}\left\{|\tilde{\theta}(\tau)| + |\dot{\hat{\theta}}(\tau)|\right\}, \quad (6.211)$$

which completes the proof. □

Lemma 6.20 implies that we can use any identifier that guarantees boundedness of $\tilde{\theta}$ and $\dot{\hat{\theta}}$. This means, in particular, that the result on boundedness

6.7 SCHEMES WITH WEAK ISS-CONTROLLER

without adaptation given in Corollary 5.9 for the ISS-controller also holds for the weak ISS-controller.

We now look for identifiers which guarantee the boundedness of $\tilde{\theta}$ and $\dot{\hat{\theta}}$.

We already know from Lemma 6.5 that these properties are guaranteed by the *x-swapping identifier* (6.87)–(6.91) (see Section 6.4 for details).

The *z-swapping identifier*, as presented in Section 6.3, is not directly applicable because the matrix A_z in (6.206), obtained with the weak ISS-controller, is different from the matrix (6.8) resulting from the strong ISS-controller. The z-swapping identifier that we use here has the filters and the update laws of the same form as in Section 6.3, but with matrix A_z defined in (6.206):

$$\dot{\Omega}^T = A_z(z,\hat{\theta},t)\Omega^T + W(z,\hat{\theta},t)^T \tag{6.212}$$

$$\dot{\Omega}_0 = A_z(z,\hat{\theta},t)\Omega_0 + W(z,\hat{\theta},t)^T\hat{\theta} - Q(z,\hat{\theta},t)^T\dot{\hat{\theta}} \tag{6.213}$$

$$\epsilon = z + \Omega_0 - \Omega^T\hat{\theta}, \tag{6.214}$$

with either the gradient

$$\dot{\hat{\theta}} = \Gamma\frac{\Omega\epsilon}{1+\nu|\Omega|_{\mathcal{F}}^2}, \qquad \Gamma = \Gamma^T > 0, \quad \nu \geq 0 \tag{6.215}$$

or the least squares

$$\dot{\hat{\theta}} = \Gamma\frac{\Omega\epsilon}{1+\nu\mathrm{tr}\{\Omega^T\Gamma\Omega\}}$$
$$\dot{\Gamma} = -\Gamma\frac{\Omega\Omega^T}{1+\nu\mathrm{tr}\{\Omega^T\Gamma\Omega\}}\Gamma, \qquad \Gamma(0) = \Gamma(0)^T > 0, \quad \nu \geq 0. \tag{6.216}$$

The differences between this z-swapping identifier and the one from Section 6.3 are emphasized in the proof of the following lemma.

Lemma 6.21 *Let the maximal interval of existence of solutions of (6.205), (6.206), (6.212)–(6.213) with either (6.215) or (6.216) be $[0,t_f)$. Then for $\nu \geq 0$, the following identifier properties hold:*

$$(i) \qquad \tilde{\theta} \in \mathcal{L}_\infty[0,t_f) \tag{6.217}$$

$$(ii) \qquad \epsilon \in \mathcal{L}_2[0,t_f) \cap \mathcal{L}_\infty[0,t_f) \tag{6.218}$$

$$(iii) \qquad \dot{\hat{\theta}} \in \mathcal{L}_2[0,t_f) \cap \mathcal{L}_\infty[0,t_f). \tag{6.219}$$

Proof (Outline). The main difference from the proof of Lemma 6.1 is in establishing the boundedness of Ω. In view of (6.212) with (6.206), we have

$$\dot{\Omega}_i = -(c_i + s_i)\Omega_i + \Omega_{i+1} + w_i, \qquad i = 1,\ldots,n-1 \tag{6.220}$$

$$\dot{\Omega}_n = -(c_n + s_n)\Omega_n + w_n. \tag{6.221}$$

With (6.203) we get

$$\frac{d}{dt}\left(\frac{|\Omega_i|^2}{2}\right) \leq -\frac{c_i}{2}|\Omega_i|^2 - \frac{c_i}{2}|\Omega_i|\left(|\Omega_i| - \frac{2}{c_i}|\Omega_{i+1}|\right)$$
$$-\kappa_i|w_i||\Omega_i|\left(|\Omega_i| - \frac{1}{\kappa_i}\right), \quad i=1,\ldots,n-1 \quad (6.222)$$

$$\frac{d}{dt}\left(\frac{|\Omega_n|^2}{2}\right) \leq -c_n|\Omega_n|^2 - \kappa_i|w_n||\Omega_n|\left(|\Omega_n| - \frac{1}{\kappa_n}\right). \quad (6.223)$$

From (6.223) it follows that Ω_n is bounded, which, by (6.222), implies that Ω_{n-1} is bounded. Continuing in the same fashion, (6.222) proves that Ω is bounded. Let us now define the positive definite diagonal matrix

$$P = \text{diag}\{p_1,\ldots,p_n\}, \quad p_n = 1, \quad p_i = p_{i+1}\frac{c_i c_{i+1}}{4}. \quad (6.224)$$

It can be shown that A_z defined in (6.206) and P defined in (6.224) satisfy

$$PA_z + A_z^T P \leq -\text{diag}\{p_1(c_1 + 2s_1),\ldots,p_n(c_n + 2s_n)\}, \quad (6.225)$$

which implies

$$\frac{1}{2}(PA_z + A_z^T P) \leq -\frac{c_0}{2}\min_{1\leq i\leq n}\{p_i\}I$$
$$\leq -\frac{c_0}{2}\min\left\{1,\left(\frac{c_0}{2}\right)^{2n-2}\right\}I$$
$$\leq -\min\left\{1,\left(\frac{c_0}{2}\right)^{2n-1}\right\}I$$
$$\stackrel{\triangle}{=} -m_0 I. \quad (6.226)$$

For system

$$\dot{\tilde{\epsilon}} = A_z(z,\hat{\theta},t)\tilde{\epsilon} \quad (6.227)$$

by virtue of (6.226) it follows that

$$\frac{d}{dt}\left(\frac{1}{2}|\tilde{\epsilon}|_P^2\right) \leq -m_0|\tilde{\epsilon}|^2. \quad (6.228)$$

The proof of the lemma is completed with the same arguments as the proof of Lemma 6.1, with the Lyapunov function $V = \frac{1}{2}|\tilde{\theta}|_{\Gamma^{-1}}^2 + \frac{1}{2m_0}|\tilde{\epsilon}|^2$ for the gradient update law and the Lyapunov function $V = |\tilde{\theta}|_{\Gamma(t)^{-1}}^2 + \frac{1}{2m_0}|\tilde{\epsilon}|^2$ for the least-squares update law. □

Having established the properties of the x-swapping and the z-swapping identifiers, we are ready to conclude the stability properties of the closed-loop adaptive system.

6.8 UNKNOWN VIRTUAL CONTROL COEFFICIENTS

Theorem 6.22 (Schemes with Weak ISS-Controller) *All the signals in the closed-loop adaptive system consisting of the plant (6.1), the weak ISS-controller (6.204), and either the x-swapping or the z-swapping identifier are globally uniformly bounded, and* $\lim_{t\to\infty} z(t) = \lim_{t\to\infty} \epsilon(t) = 0$. *This means, in particular, that global asymptotic tracking is achieved:*

$$\lim_{t\to\infty} [y(t) - y_{\mathrm{r}}(t)] = 0. \qquad (6.229)$$

Moreover, if $\lim_{t\to\infty} y_{\mathrm{r}}^{(i)}(t) = 0$, $i = 0, \ldots, n-1$, *and* $F(0) = 0$, *then* $\lim_{t\to\infty} x(t) = 0$.

Proof. With Lemma 6.20 and either Lemma 6.5 or Lemma 6.21, the boundedness of all signals is established as in Theorems 6.3 and 6.7. The convergence properties are established as in Theorem 6.3 (for the scheme with the z-swapping identifier) and either Theorem 6.7 or Remark 6.8 (for the scheme with the x-swapping identifier). □

Remark 6.23 As in Theorem 6.4 and Remark 6.9, we can normalize the update law with $1 + \nu'|Q|_{\mathcal{F}}$, which guarantees that $Q^{\mathrm{T}}\dot{\hat{\theta}} \in \mathcal{L}_{\infty}$ and allows us to set $g_i = 0$. ◇

6.8 Unknown Virtual Control Coefficients

As in Sections 4.5.1 and 6.8, we consider systems of the form

$$\begin{aligned}
\dot{x}_i &= x_{i+1} + \varphi_i(x_1, \ldots, x_i)^{\mathrm{T}}\theta, \quad i = 1, \ldots, m-1, m+1, \ldots, n-1 \\
\dot{x}_m &= b_m x_{m+1} + \varphi_m(x_1, \ldots, x_m)^{\mathrm{T}}\theta \\
\dot{x}_n &= \beta(x)u + \varphi_n(x_1, \ldots, x_n)^{\mathrm{T}}\theta,
\end{aligned} \qquad (6.230)$$

where b_m is unknown, but, as stated in Assumption 5.23, in addition to sgn b_m, a positive constant ς_m is known such that $|b_m| \geq \varsigma_m$.

We present swapping adaptive schemes with the controller (5.216)–(5.227) from Section 5.9.1, which, by Lemma 5.24, guarantees input-to-state properties with respect to the inputs $(\tilde{\theta}, \tilde{b}_m)$ and $(\dot{\hat{\theta}}, \dot{\hat{b}}_m)$. The swapping identifiers use parameter projection to keep $\hat{b}_m(t)$ from becoming zero.

The x-swapping scheme

Let us start by writing system (6.230) in the following compact form:

$$\dot{x} = f(x, u) + F(x, u)^{\mathrm{T}}\vartheta, \qquad \vartheta = \begin{bmatrix} b_m \\ \theta \end{bmatrix}. \qquad (6.231)$$

We use our standard filters

$$\dot{\Omega}^T = \left(A_0 - \lambda F(x,u)^T F(x,u) P\right)\Omega^T + F(x,u)^T \qquad (6.232)$$
$$\dot{\Omega}_0 = \left(A_0 - \lambda F(x,u)^T F(x,u) P\right)(\Omega_0 - x) - f(x,u) \qquad (6.233)$$
$$\epsilon = x + \Omega_0 - \Omega^T \hat{\vartheta} \qquad (6.234)$$

and either the gradient,

$$\dot{\hat{\vartheta}} = \underset{\hat{b}_m}{\text{Proj}}\left\{\Gamma\frac{\Omega\epsilon}{1+\nu|\Omega|_{\mathcal{F}}^2}\right\}, \qquad \begin{array}{l}\hat{b}_m(0)\,\text{sgn}\,b_m > \varsigma_m \\ \Gamma = \Gamma^T > 0,\ \nu \geq 0\end{array} \qquad (6.235)$$

or the least squares update law

$$\dot{\hat{\vartheta}} = \underset{\hat{b}_m}{\text{Proj}}\left\{\Gamma\frac{\Omega\epsilon}{1+\nu\text{tr}\{\Omega^T\Gamma\Omega\}}\right\}, \qquad \hat{b}_m(0)\,\text{sgn}\,b_m > \varsigma_m$$
$$\dot{\Gamma} = -\Gamma\frac{\Omega\Omega^T}{1+\nu\text{tr}\{\Omega^T\Gamma\Omega\}}\Gamma, \qquad \Gamma(0) = \Gamma(0)^T > 0,\ \nu \geq 0, \qquad (6.236)$$

where the projection operator is employed to guarantee that $|\hat{b}_m(t)| \geq \varsigma_m > 0$, $\forall t \geq 0$. For a detailed treatment of parameter projection the reader is referred to Appendix E.

Lemma 6.24 *Let the maximal interval of existence of solutions of (6.231), (6.232)–(6.234) with either (6.235) or (6.236) be $[0, t_f)$. Then for $\nu \geq 0$, the following identifier properties hold:*

$$(o) \quad |\hat{b}_m(t)| \geq \varsigma_m > 0,\ \forall t \in [0, t_f) \qquad (6.237)$$
$$(i) \quad \tilde{\theta} \in \mathcal{L}_\infty[0, t_f) \qquad (6.238)$$
$$(ii) \quad \epsilon \in \mathcal{L}_2[0, t_f) \cap \mathcal{L}_\infty[0, t_f) \qquad (6.239)$$
$$(iii) \quad \dot{\hat{\theta}} \in \mathcal{L}_2[0, t_f) \cap \mathcal{L}_\infty[0, t_f). \qquad (6.240)$$

Proof. The proof of this lemma is only slightly different from the proof of Lemma 6.5. The differences are due to projection. First, from Lemma E.1 we conclude that $|\hat{b}_m(t)| \geq \varsigma_m > 0$, $\forall t \in [0, t_f)$. To establish (6.99) in presence of projection, we use the fact from Lemma E.1 that $-\tilde{\vartheta}^T\Gamma^{-1}\dot{\hat{\vartheta}} = -\tilde{\vartheta}^T\Gamma^{-1}\text{Proj}\left\{\Gamma\frac{\Omega\epsilon}{1+\nu|\Omega|_{\mathcal{F}}^2}\right\} \leq -\tilde{\vartheta}^T\Gamma^{-1}\Gamma\frac{\Omega\epsilon}{1+\nu|\Omega|_{\mathcal{F}}^2}$. Finally, we note that Lemma E.1 implies that $|\dot{\hat{\vartheta}}| \leq \sqrt{\frac{\bar{\lambda}(\Gamma)}{\underline{\lambda}(\Gamma)}}\left|\Gamma\frac{\Omega\epsilon}{1+\nu|\Omega|_{\mathcal{F}}^2}\right|$. This fact is now used as in the proof of Lemma 6.5 to establish boundedness and square-integrability of $\dot{\hat{\vartheta}}$. □

By combining Lemmas 5.24 and 6.24, we obtain the following result. Its proof is the same as the proof of Theorem 6.7.

6.8 UNKNOWN VIRTUAL CONTROL COEFFICIENTS

Theorem 6.25 *All the boundedness and convergence properties from Theorem 6.7 hold also for the closed-loop adaptive system consisting of the plant (6.230), control law (5.227), filters (6.232), (6.233), and either the gradient (6.235) or the least-squares (6.236) update law.*

The z-swapping scheme

Consider the closed-loop system (5.228)–(5.231). Our z-swapping identifier is not directly applicable to this parametric model because the terms $-\frac{|b_m|}{\varsigma_m}(c_m + s_m)z_m$ and $b_m z_{m+1}$ in (5.229), as well as the term $-b_m z_m$ in (5.230), depend on the unknown parameter b_m. However, writing b_m in (5.229) and (5.230) as $\hat{b}_m + \tilde{b}_m$, in view of (5.220) we obtain

$$\dot{z}_i = -(c_i + s_i)z_i - z_{i-1} + z_{i+1} + w_i^T \tilde{\theta} - \frac{\partial \alpha_{i-1}}{\partial \hat{\theta}}\dot{\hat{\theta}},$$
$$i = 1,\ldots,m-1 \quad (6.241)$$

$$\dot{z}_m = -\frac{\hat{b}_m \text{sgn} b_m}{\varsigma_m}(c_m + s_m)z_m - z_{m-1} + \hat{b}_m z_{m+1}$$
$$+ w_m^T \tilde{\theta} + x_{m+1}\tilde{b}_m - \frac{\partial \alpha_{m-1}}{\partial \hat{\theta}}\dot{\hat{\theta}} \quad (6.242)$$

$$\dot{z}_{m+1} = -(c_{m+1} + s_{m+1})z_{m+1} - \hat{b}_m z_m + z_{m+2} + w_{m+1}^T \tilde{\theta} - \frac{\partial \alpha_m}{\partial x_m}x_{m+1}\tilde{b}_m$$
$$- \frac{\partial \alpha_m}{\partial \hat{\theta}}\dot{\hat{\theta}} - \left(\frac{\partial \alpha_m}{\partial \hat{b}_m} - \frac{y_r^{(m)}}{\hat{b}_m^2}\right)\dot{\hat{b}}_m \quad (6.243)$$

$$\dot{z}_j = -(c_j + s_j)z_j - z_{j-1} + z_{j+1} + w_j^T \tilde{\theta} - \frac{\partial \alpha_{j-1}}{\partial x_m}x_{m+1}\tilde{b}_m$$
$$- \frac{\partial \alpha_{j-1}}{\partial \hat{\theta}}\dot{\hat{\theta}} - \left(\frac{\partial \alpha_{j-1}}{\partial \hat{b}_m} - \frac{y_r^{(j-1)}}{\hat{b}_m^2}\right)\dot{\hat{b}}_m, \quad i = m+2,\ldots,n. \quad (6.244)$$

Let us rewrite the system (6.241)–(6.244) in the following compact form

$$\dot{z} = A_z(z,\hat{\vartheta},t)z + W_\vartheta(z,\hat{\vartheta},t)^T\tilde{\vartheta} + Q_\vartheta(z,\hat{\vartheta},t)^T\dot{\hat{\vartheta}}, \qquad \vartheta = \begin{bmatrix} b_m \\ \theta \end{bmatrix}, \quad (6.245)$$

where

$$A_z = \begin{bmatrix} -c_1 - s_1 & 1 & & & & \\ -1 & \ddots & \ddots & & & \\ & \ddots & -\frac{\hat{b}_m \text{sgn}\, b_m}{\varsigma_m}(c_m + s_m) & \hat{b}_m & & \\ & & -\hat{b}_m & \ddots & \ddots & \\ & & & \ddots & \ddots & 1 \\ & & & & -1 & -c_n - s_n \end{bmatrix}. \quad (6.246)$$

Since we will use projection to guarantee that $|\hat{b}_m(t)| \geq \varsigma_m$ and $\operatorname{sgn} \hat{b}_m(t) = \operatorname{sgn} b_m$, which implies $\hat{b}_m \operatorname{sgn} b_m \geq \varsigma_m$, the matrix A_z will satisfy

$$A_z(z,\hat{\vartheta},t) + A_z(z,\hat{\vartheta},t)^T \leq -2c_0 I, \qquad \forall z, \hat{\vartheta}, t. \tag{6.247}$$

With the parametric z-model (6.245), the form of the filters is the same as in Section 6.3:

$$\dot{\Omega}^T = A_z(z,\hat{\vartheta},t)\Omega^T + W_\vartheta(z,\hat{\vartheta},t)^T \tag{6.248}$$

$$\dot{\Omega}_0 = A_z(z,\hat{\vartheta},t)\Omega_0 + W_\vartheta(z,\hat{\vartheta},t)^T\hat{\vartheta} - Q_\vartheta(z,\hat{\vartheta},t)^T\dot{\hat{\vartheta}} \tag{6.249}$$

$$\epsilon = z + \Omega_0 - \Omega^T\hat{\vartheta}, \tag{6.250}$$

with either the gradient

$$\dot{\hat{\vartheta}} = \operatorname*{Proj}_{\hat{b}_m}\left\{\Gamma\frac{\Omega\epsilon}{1+\nu|\Omega|_{\mathcal{F}}^2}\right\}, \qquad \hat{b}_m(0)\operatorname{sgn}b_m > \varsigma_m, \quad \Gamma = \Gamma^T > 0 \\ \nu > 0 \tag{6.251}$$

or the least squares update law

$$\dot{\hat{\vartheta}} = \operatorname*{Proj}_{\hat{b}_m}\left\{\Gamma\frac{\Omega\epsilon}{1+\nu|\Omega|_{\mathcal{F}}^2}\right\}, \qquad \hat{b}_m(0)\operatorname{sgn}b_m > \varsigma_m$$

$$\dot{\Gamma} = -\Gamma\frac{\Omega\Omega^T}{1+\nu|\Omega|_{\mathcal{F}}^2}\Gamma, \qquad \Gamma(0) = \Gamma(0)^T > 0, \quad \nu > 0, \tag{6.252}$$

where the projection operator is employed to guarantee that $|\hat{b}_m(t)| \geq \varsigma_m > 0$, $\forall t \geq 0$.

The reader should note in (6.251) and (6.252) that $\nu > 0$, which means that we allow only normalized update laws. This is in contrast to the z-swapping identifier from Section 6.3, where the normalization is not necessary because the nonlinear damping substitutes the normalization in the task of slowing down the adaptation. From the proof of Lemma 6.1 we recall that the nonlinear damping slows down the adaptation by guaranteeing the boundedness of Ω for any input W. It is important to understand why we have to use normalization here. To this end, let us note from (6.241)–(6.244) that the regressor W_ϑ is given by

$$W_\vartheta(z,\hat{\vartheta},t)^T = \begin{bmatrix} 0 & , & w_i^T \\ x_{m+1} & , & w_m^T \\ -\dfrac{\partial\alpha_m}{\partial x_m}x_{m+1} & , & w_{m+1}^T \\ -\dfrac{\partial\alpha_{j-1}}{\partial x_m}x_{m+1} & , & w_j^T \end{bmatrix} \begin{matrix} i=1,\ldots,m-1 \\ \\ \\ j=m+2,\ldots,n \end{matrix} \tag{6.253}$$

6.8 UNKNOWN VIRTUAL CONTROL COEFFICIENTS

We recall from (5.223)–(5.226) that the relevant terms in the nonlinear damping functions are

$$s_i = \kappa_i |w_i|^2 + \cdots \qquad (6.254)$$

$$s_m = \kappa_m \left(\left(\frac{y_r^{(m)} + \bar{\alpha}_m}{\hat{b}_m} \right)^2 + |w_m|^2 \right) + \cdots \qquad (6.255)$$

$$s_{m+1} = \kappa_{m+1} \left(\left(z_m - \frac{\partial \alpha_m}{\partial x_m} x_{m+1} \right)^2 + |w_{m+1}|^2 \right) + \cdots \qquad (6.256)$$

$$s_j = \kappa_j \left(\left(\frac{\partial \alpha_{j-1}}{\partial x_m} x_{m+1} \right)^2 + |w_j|^2 \right) + \cdots . \qquad (6.257)$$

By comparing the nonlinear damping terms (6.255) and (6.256) with the mth and the $(m+1)$st rows of (6.253), we see that the the nonlinear damping terms cannot be expected to guarantee the boundedness of Ω when W_ϑ grows unbounded because they no longer correspond to the columns of the regressor W_ϑ. Therefore, we have to use normalization. This is the price paid for replacing the original nonimplementable parametric model (5.228)–(5.231) by the parametric model (6.241)–(6.244).

Lemma 6.26 *Let the maximal interval of existence of solutions of (6.245), (6.248)–(6.249) with either (6.251) or (6.252) be $[0, t_f)$. Then the following identifier properties hold:*

$$(o) \quad |\hat{b}_m(t)| \geq \varsigma_m > 0, \ \forall t \in [0, t_f) \qquad (6.258)$$

$$(i) \quad \tilde{\vartheta} \in \mathcal{L}_\infty[0, t_f) \qquad (6.259)$$

$$(ii) \quad \frac{\epsilon}{\sqrt{1 + \nu |\Omega|_{\mathcal{F}}^2}} \in \mathcal{L}_2[0, t_f) \cap \mathcal{L}_\infty[0, t_f) \qquad (6.260)$$

$$(iii) \quad \dot{\hat{\vartheta}} \in \mathcal{L}_2[0, t_f) \cap \mathcal{L}_\infty[0, t_f). \qquad (6.261)$$

Proof. Because of the update law normalization, the proof of this lemma is similar to the proof of Lemma 6.16, which, in turn, connects to the proof of Lemma 6.1. We emphasize only the differences which are due to projection. From Lemma E.1 we conclude that $|\hat{b}_m(t)| \geq \varsigma_m > 0$, $\forall t \in [0, t_f)$. Therefore, inequality (6.247) holds, and so does (6.46). To establish (6.49) in presence of projection, we modify (6.48) with the following fact implied by Lemma E.1: $-\tilde{\vartheta}^T \Gamma^{-1} \dot{\hat{\vartheta}} = -\tilde{\vartheta}^T \Gamma^{-1} \operatorname{Proj} \left\{ \Gamma \frac{\Omega \epsilon}{1 + \nu |\Omega|_{\mathcal{F}}^2} \right\} \leq -\tilde{\vartheta}^T \Gamma^{-1} \Gamma \frac{\Omega \epsilon}{1 + \nu |\Omega|_{\mathcal{F}}^2}$. This establishes the boundedness of $\tilde{\vartheta}$. The proof of $\frac{\epsilon}{\sqrt{1 + \nu |\Omega|_{\mathcal{F}}^2}} \in \mathcal{L}_2[0, t_f)$ is as in

Lemma 6.1, while the proof of $\dfrac{\epsilon}{\sqrt{1+\nu|\Omega|_{\mathcal{F}}^2}} \in \mathcal{L}_\infty[0,t_f)$ is as in Lemma 6.16.
With Lemma E.1 we have

$$\begin{aligned}
|\dot{\hat{\vartheta}}|^2 &\leq \frac{1}{\underline{\lambda}(\Gamma^{-1})}|\dot{\hat{\vartheta}}|_{\Gamma^{-1}}^2 \leq \frac{1}{\underline{\lambda}(\Gamma^{-1})} \frac{\epsilon^T \Omega^T \Gamma \Omega \epsilon}{(1+\nu|\Omega|_{\mathcal{F}}^2)^2} \leq \frac{\bar{\lambda}(\Gamma)}{\underline{\lambda}(\Gamma^{-1})} \frac{|\Omega|_2^2 \epsilon^T \epsilon}{(1+\nu|\Omega|_{\mathcal{F}}^2)^2} \\
&\leq \bar{\lambda}(\Gamma)^2 \frac{|\Omega|_{\mathcal{F}}^2}{1+\nu|\Omega|_{\mathcal{F}}^2} \frac{|\epsilon|^2}{1+\nu|\Omega|_{\mathcal{F}}^2} \leq \frac{\bar{\lambda}(\Gamma)^2}{\nu} \frac{|\epsilon|^2}{1+\nu|\Omega|_{\mathcal{F}}^2}\,,
\end{aligned} \qquad (6.262)$$

which proves that $\dot{\hat{\vartheta}} \in \mathcal{L}_2[0,t_f) \cap \mathcal{L}_\infty[0,t_f)$. □

Theorem 6.27 *All the boundedness and convergence properties from Theorem 6.3 hold also for the closed-loop adaptive system consisting of the plant (6.230), control law (5.227), filters (6.248), (6.249), and either the gradient (6.251) or the least-squares (6.252) update law.*

Proof. As we commented in Remark 5.27, because of the local Lipschitzness of the projection operator, the argument about the existence of solutions remains as in the proof of Theorem 6.3. The difference between this proof and that of Theorem 6.3 is that the identifier here does not independently guarantee the boundedness of Ω. By combining Lemmas 5.24 and 6.26, we conclude that z is bounded. This implies the boundedness of W_ϑ and Q_ϑ, so Ω and Ω_0 are also bounded. To prove convergence, we point out that Lemma 6.26 establishes that $\dfrac{\epsilon}{\sqrt{1+\nu|\Omega|_{\mathcal{F}}^2}} \in \mathcal{L}_2$, so, due to the boundedness of Ω, we have $\epsilon \in \mathcal{L}_2$. The proof of the convergence properties is finished as in Theorem 6.3. □

Notes and References

Many early approaches to adaptive control of nonlinear systems were estimation-based and employed swapping identifiers: Sastry and Isidori [166], Nam and Arapostathis [141], Middleton and Goodwin [130], Bastin and Campion [8], Teel, Kadiyala, Kokotović, and Sastry [190], Pomet and Praly [152, 153, 154], Kanellakopoulos, Kokotović, and Middleton [67, 68], and Praly, Bastin, Pomet, and Jiang [157]. Various restrictions were imposed in these results: [8, 130] contained structural conditions, [141, 166, 190] imposed linear growth conditions, [67, 68] had both structural and growth conditions. Only in [152, 153, 154, 157] did the growth conditions have a less demanding nonlinear form. Both the structural and the growth conditions were removed in the modular design with swapping identifiers developed by Krstić and Kokotović [98].

Part II

Output Feedback

Chapter 7
Output-Feedback Design Tools

All the results in Part I have been obtained under the assumption that the full state of the system is measured. We now remove this assumption and consider more realistic problems where only a part of the state or just the scalar plant output is available for measurement. In the case of linear systems, the *separation principle* allows output-feedback problems to be solved by combining state-feedback controllers with state observers. However, the separation principle does not hold for nonlinear systems.

In this chapter we introduce backstepping procedures for output-feedback and partial-state-feedback designs for nonlinear systems, whose nonlinearities depend only on the measured signals. For these systems, we build exponentially convergent nonlinear observers and replace the unmeasured states by their estimates. We show that global results can be obtained if nonlinear damping is used to counteract the destabilizing effect of the observer errors. As we did in Chapters 2 and 3, we start with systems whose parameters are known, and then present adaptive schemes for systems with unknown constant parameters. Such adaptive schemes are further developed by the tuning functions design in Chapter 8 and the modular design in Chapter 9.

7.1 Observer Backstepping

7.1.1 Unmeasured states

To illustrate the difficulties that arise when some of the states of the system are not measured, as well as the means to overcome these difficulties, we use two simple examples.

State feedback revisited. First, let us consider the system

$$\dot{x} = -x + x^4 + x^2\xi \tag{7.1a}$$
$$\dot{\xi} = -k\xi + u, \tag{7.1b}$$

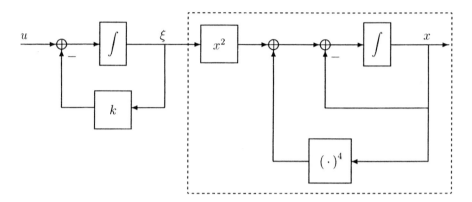

Figure 7.1: The block diagram of system (7.1).

with $k > 0$. Its block diagram is given in Figure 7.1. When both states x and ξ are measured, this system is stabilizable via backstepping. Using ξ as the virtual control in (7.1a), an obvious choice of stabilizing function is $\alpha_1(x) = -x^2$. Indeed, if $\xi = -x^2$, then (7.1a) is reduced to $\dot{x} = -x$. Then, with the error variable $z = \xi - \alpha_1(x) = \xi + x^2$, the system (7.1) becomes

$$\dot{x} = -x + x^2 z \tag{7.2a}$$
$$\dot{z} = -k\xi + u + 2x\left(-x + x^4 + x^2\xi\right) = -k\xi + u + 2x\left(-x + x^2 z\right). \tag{7.2b}$$

The derivative of $V(x, \xi) = \tfrac{1}{2}(x^2 + z^2)$ along the solutions of (7.2) is

$$\dot{V} = -x^2 + z\left[x^3 - k\xi + u + 2x\left(-x + x^2 z\right)\right]. \tag{7.3}$$

Hence, the choice of control

$$u = -cz - x^3 + k\xi - 2x\left(-x + x^2 z\right), \tag{7.4}$$

with $c > 0$ a design constant, yields $\dot{V} = -x^2 - cz^2$ and renders $x = 0, \xi = 0$ the GAS equilibrium of the closed-loop system:

$$\dot{x} = -x + x^2 z \tag{7.5a}$$
$$\dot{z} = -cz - x^3. \tag{7.5b}$$

State estimates as virtual controls. Suppose now that in (7.1) the state ξ is not measured. This implies that $z = \xi + x^2$ is not available for feedback, and hence the control (7.4) can not be implemented. We see that the unmeasured state ξ can not be chosen as the virtual control, and has to be replaced by a measured variable. Fortunately, (7.1b) suggests that the state ξ can be estimated by $\hat{\xi}$, where

$$\dot{\hat{\xi}} = -k\hat{\xi} + u. \tag{7.6}$$

7.1 OBSERVER BACKSTEPPING

Indeed, subtracting (7.6) from (7.1b) shows that the state estimation error $\tilde{\xi} = \xi - \hat{\xi}$ converges exponentially to zero:

$$\dot{\tilde{\xi}} = -k\tilde{\xi} \Rightarrow \tilde{\xi}(t) = \tilde{\xi}(0)e^{-kt}. \tag{7.7}$$

To utilize this estimate, we replace ξ by $\hat{\xi} + \tilde{\xi}$ in (7.1a):

$$\dot{x} = -x + x^4 + x^2\hat{\xi} + x^2\tilde{\xi}. \tag{7.8}$$

In this equation, the only possible choice of virtual control is $\hat{\xi}$, which means that in the backstepping procedure we must replace (7.1b) with (7.6). In essence, we start with the system composed of (7.1) and the observer (7.6), shown in Figure 7.2, and manipulate it into the form shown in Figure 7.3:

$$\dot{x} = -x + x^4 + x^2\hat{\xi} + x^2\tilde{\xi} \tag{7.9a}$$

$$\dot{\hat{\xi}} = -k\hat{\xi} + u \tag{7.9b}$$

$$\dot{\tilde{\xi}} = -k\tilde{\xi}. \tag{7.9c}$$

The next step is to design a control law for (7.9). Since $\tilde{\xi}$ is exponentially convergent, one might be tempted to ignore its effect on the rest of the system, as is often done in linear control design. This leads to the error variable

$$z = \hat{\xi} - \alpha_1(x) = \hat{\xi} + x^2, \tag{7.10}$$

and the control (7.4), which now yields the closed-loop system

$$\dot{x} = -x + x^2 z + x^2\tilde{\xi} \tag{7.11a}$$

$$\dot{z} = -cz - x^3 + 2x^3\tilde{\xi} \tag{7.11b}$$

$$\dot{\tilde{\xi}} = -k\tilde{\xi}. \tag{7.11c}$$

The reader familiar with Section 2.5 of Chapter 2 will recognize the flaw of this "certainty-equivalence" design, namely the presence of the terms $x^2\tilde{\xi}$ in (7.11a) and $2x^3\tilde{\xi}$ in (7.11b). As we saw in Section 2.5, even an exponentially decaying disturbance like $\tilde{\xi}$ can destabilize a nonlinear system and lead to finite escape time from certain initial conditions. In (2.242)–(2.246) we examined the system

$$\dot{x} = -x + x^2\tilde{\xi}, \quad \tilde{\xi} = \tilde{\xi}(0)e^{-kt},$$

which is identical to (7.11a) when $z \equiv 0$. There we showed that this system has the solution

$$x(t) = \frac{x(0)(1+k)}{[1 + k - \tilde{\xi}(0)x(0)]e^t + \tilde{\xi}(0)x(0)e^{-kt}},$$

which escapes to infinity in finite time

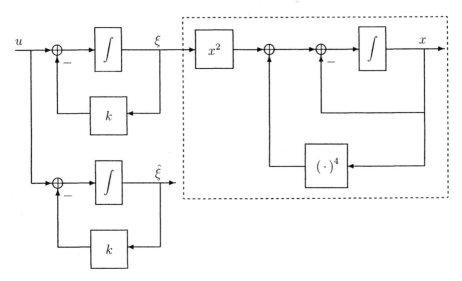

Figure 7.2: Adding the observer (7.6) to the system (7.1).

$$t_f = \frac{1}{1+k} \ln \left\{ \frac{\tilde{\xi}(0)x(0)}{\tilde{\xi}(0)x(0) - (1+k)} \right\}$$

from all initial conditions for which $\tilde{\xi}(0)x(0) > 1 + k$. This implies that irrespective of how fast $\tilde{\xi}(t)$ converges to zero or how small its initial condition $\tilde{\xi}(0)$ is, there always exist initial conditions $x(0)$ from which the system escapes to infinity in finite time.

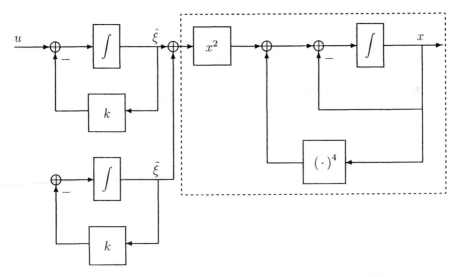

Figure 7.3: Replacing ξ by $\hat{\xi}$ as the virtual control in (7.9).

7.1 OBSERVER BACKSTEPPING

Nonlinear damping. The remedy for this problem is, again, nonlinear damping. Starting with (7.9a) and using $\hat{\xi}$ as the virtual control, we modify the stabilizing function $\alpha_1(x)$ by adding to it a nonlinear damping term $-s(x)x$:

$$\alpha_1(x) = -x^2 - s(x)x. \tag{7.12}$$

Following the development in Section 2.5, equations (2.248)–(2.252), we design $s(x)$ using the function $V(x) = \frac{1}{2}x^2$, whose derivative is

$$\dot{V} = -x^2 + x^3 z - x^2 s(x) + x(x^2)\tilde{\xi}. \tag{7.13}$$

The choice of nonlinear damping

$$s(x) = d_1(x^2)^2, \quad d_1 > 0 \tag{7.14}$$

yields the stabilizing function

$$\alpha_1(x) = -x^2 - d_1 x^5, \tag{7.15}$$

the error variable

$$z = \hat{\xi} - \alpha_1(x) = \hat{\xi} + x^2 + d_1 x^5, \tag{7.16}$$

and the closed-loop expression for (7.9a)

$$\dot{x} = -x - d_1 x^5 + x^2 z + x^2 \tilde{\xi}. \tag{7.17}$$

The derivative of V becomes

$$\begin{aligned}
\dot{V} &= -x^2 - d_1 x^6 + x^3 z + x^3 \tilde{\xi} \\
&= -x^2 + x^3 z - d_1 \left(x^3 - \frac{1}{2 d_1}\tilde{\xi} \right)^2 + \frac{1}{4 d_1}\tilde{\xi}^2 \\
&\leq -x^2 + x^3 z + \frac{1}{4 d_1}\tilde{\xi}^2 .
\end{aligned} \tag{7.18}$$

The last inequality implies that if $z \equiv 0$, x will remain bounded if $\tilde{\xi}$ is bounded. Here, however, we can achieve more than that by exploiting the fact that $\tilde{\xi}(t)$ is the error of an exponentially converging observer. To this end, we augment the function $V(x)$ with a quadratic term in $\tilde{\xi}$:

$$V_1(x, \tilde{\xi}) = V(x) + \frac{1}{2 d_1 k}\tilde{\xi}^2 = \frac{1}{2}x^2 + \frac{1}{2 d_1 k}\tilde{\xi}^2. \tag{7.19}$$

Using (7.18), we see that the derivative of V_1 satisfies

$$\begin{aligned}
\dot{V}_1 &= \dot{V} - \frac{1}{d_1}\tilde{\xi}^2 \\
&\leq -x^2 + x^3 z + \frac{1}{4 d_1}\tilde{\xi}^2 - \frac{1}{d_1}\tilde{\xi}^2 \\
&= -x^2 + x^3 z - \frac{3}{4 d_1}\tilde{\xi}^2 .
\end{aligned} \tag{7.20}$$

Hence, if $z \equiv 0$ the control (7.15) would render $(0,0)$ the GAS equilibrium of the $(x, \tilde{\xi})$ system.

The derivative of z is now expressed as

$$\begin{aligned}\dot{z} &= -k\hat{\xi} + u - \frac{\partial \alpha_1}{\partial x}(x)\left(-x + x^4 + x^2\xi\right) \\ &= -k\hat{\xi} + u - \frac{\partial \alpha_1}{\partial x}(x)\left(-x + x^4 + x^2\hat{\xi}\right) - \frac{\partial \alpha_1}{\partial x}(x)x^2\tilde{\xi}. \end{aligned} \quad (7.21)$$

In this equation, the state estimation error appears again, so its effect will have to be compensated by another nonlinear damping term. This is reflected in the corresponding Lyapunov function, which is augmented not only by a z^2-term, but also by an additional $\tilde{\xi}^2$-term:

$$V_2(x, z, \tilde{\xi}) = V_1(x, \tilde{\xi}) + \frac{1}{2}z^2 + \frac{1}{2d_2 k}\tilde{\xi}^2 = \frac{1}{2}\left[x^2 + z^2 + \left(\frac{1}{d_1 k} + \frac{1}{d_2 k}\right)\tilde{\xi}^2\right]. \quad (7.22)$$

Its derivative is

$$\begin{aligned}\dot{V}_2 &\leq -x^2 + x^3 z - \frac{3}{4d_1}\tilde{\xi}^2 - \frac{1}{d_2}\tilde{\xi}^2 \\ &\quad + z\left[-k\hat{\xi} + u - \frac{\partial \alpha_1}{\partial x}\left(-x + x^4 + x^2\hat{\xi}\right) - \frac{\partial \alpha_1}{\partial x}x^2\tilde{\xi}\right] \\ &= -x^2 - \frac{3}{4d_1}\tilde{\xi}^2 - \frac{1}{d_2}\tilde{\xi}^2 \\ &\quad + z\left[x^3 - k\hat{\xi} + u - \frac{\partial \alpha_1}{\partial x}\left(-x + x^4 + x^2\hat{\xi}\right)\right] - z\frac{\partial \alpha_1}{\partial x}x^2\tilde{\xi}. \quad (7.23)\end{aligned}$$

The choice of control

$$u = -cz - d_2 z\left(\frac{\partial \alpha_1}{\partial x}(x)x^2\right)^2 - x^3 + k\hat{\xi} + \frac{\partial \alpha_1}{\partial x}(x)\left(-x + x^4 + x^2\hat{\xi}\right) \quad (7.24)$$

yields

$$\begin{aligned}\dot{V}_2 &\leq -x^2 - cz^2 - \frac{3}{4d_1}\tilde{\xi}^2 - d_2 z^2\left(\frac{\partial \alpha_1}{\partial x}x^2\right)^2 - z\frac{\partial \alpha_1}{\partial x}x^2\tilde{\xi} - \frac{1}{d_2}\tilde{\xi}^2 \\ &= -x^2 - cz^2 - \frac{3}{4d_1}\tilde{\xi}^2 - d_2\left(z\frac{\partial \alpha_1}{\partial x}x^2 + \frac{1}{2d_2}\tilde{\xi}\right)^2 - \frac{3}{4d_2}\tilde{\xi}^2 \\ &\leq -x^2 - cz^2 - \frac{3}{4}\left(\frac{1}{d_1} + \frac{1}{d_2}\right)\tilde{\xi}^2. \quad (7.25)\end{aligned}$$

Hence, the origin is the GAS equilibrium of the resulting closed-loop system:

$$\dot{x} = -x - d_1 x^5 + x^2 z + x^2\tilde{\xi} \quad (7.26a)$$

$$\dot{z} = -cz - x^3 - d_2 z\left(\frac{\partial \alpha_1}{\partial x}x^2\right)^2 - \frac{\partial \alpha_1}{\partial x}x^2\tilde{\xi} \quad (7.26b)$$

$$\dot{\tilde{\xi}} = -k\tilde{\xi}. \quad (7.26c)$$

7.1.2 Output-feedback systems

The above example illustrated the design tool of *observer backstepping*: First a nonlinear observer is designed which provides exponentially convergent estimates of the unmeasured states. Then, backstepping is applied to a new system, in which the equations of the unmeasured states have been replaced by the corresponding equations of their estimates from the observer. At each step of the procedure, observation errors are treated as disturbances and accounted for using nonlinear damping.

Observer backstepping can be used to construct systematic design procedures applicable to nonlinear systems for which exponential observers are available. One such class consists of *output-feedback systems*, whose output y is the only measured signal. These systems can be transformed into the *output-feedback form*, in which nonlinearities depend only on the output y: [1]

$$\begin{aligned}
\dot{x}_1 &= x_2 + \varphi_1(y) \\
\dot{x}_2 &= x_3 + \varphi_2(y) \\
&\vdots \\
\dot{x}_{\rho-1} &= x_\rho + \varphi_{\rho-1}(y) \\
\dot{x}_\rho &= x_{\rho+1} + \varphi_\rho(y) + b_m \beta(y) u \\
&\vdots \\
\dot{x}_{n-1} &= x_n + \varphi_{n-1}(y) + b_1 \beta(y) u \\
\dot{x}_n &= \varphi_n(y) + b_0 \beta(y) u \\
y &= x_1.
\end{aligned} \qquad (7.27)$$

We assume that (7.27) is minimum phase, that is, $b_m s^m + \cdots + b_1 s + b_0$ is a Hurwitz polynomial, and $\beta(y) \neq 0 \ \forall y \in \mathbb{R}$. The system (7.27) has relative degree $\rho = n - m$, and its m-dimensional zero dynamics are linear:

$$\dot{\zeta} = A_{zd} \zeta + \bar{\varphi}(y). \qquad (7.28)$$

The eigenvalues of the $m \times m$ matrix A_{zd} are the roots of the Hurwitz polynomial $b_m s^m + \cdots + b_1 s + b_0$, and the elements of $\bar{\varphi}(y)$ are linear combinations $\varphi_1(y), \ldots, \varphi_n(y)$.

We first derive an observer for the system (7.27), rewritten as

$$\begin{aligned}
\dot{x} &= Ax + \varphi(y) + b\beta(y)u \\
y &= c^T x
\end{aligned} \qquad (7.29)$$

---

[1] The coordinate-free characterization of these systems in terms of differential geometric conditions is given in Appendix G, Corollaries G.6 and G.7.

$$A = \begin{bmatrix} 0 & & \\ \vdots & I & \\ 0 & \cdots & 0 \end{bmatrix}, \quad b = \begin{bmatrix} 0 \\ \vdots \\ 0 \\ b_m \\ \vdots \\ b_0 \end{bmatrix}, \quad c = \begin{bmatrix} 1 \\ 0 \\ \vdots \\ 0 \end{bmatrix}, \quad \varphi(y) = \begin{bmatrix} \varphi_1(y) \\ \vdots \\ \varphi_n(y) \end{bmatrix}. \quad (7.30)$$

An exponential observer for (7.29) is

$$\begin{aligned} \dot{\hat{x}} &= A\hat{x} + k(y - \hat{y}) + \varphi(y) + b\beta(y)u \\ \hat{y} &= c^T \hat{x}, \end{aligned} \quad (7.31)$$

where K_0 is chosen so that $A_0 = A - kc^T$ is Hurwitz. Subtracting (7.31) from (7.29) shows that the observation error $\tilde{x} = x - \hat{x}$ exponentially decays:

$$\dot{\tilde{x}} = A_0 \tilde{x}. \quad (7.32)$$

Using the observer (7.31) and Lemmas 2.8 and 2.26, we now design a feedback controller to force the output y of (7.27) to track a reference signal $y_r(t)$.

Theorem 7.1 (Output-Feedback Systems) *For the nonlinear system (7.27), assume that $b_m s^m + \cdots + b_1 s + b_0$ is a Hurwitz polynomial, and that y_r, $\dot{y}_r, \ldots, y_r^{(\rho)}$ are known and bounded on $[0, \infty)$ and $y_r^{(\rho)}(t)$ is piecewise continuous. Then, there exists a feedback control which guarantees global boundedness of $x(t)$ and $\hat{x}(t)$ and regulation of the tracking error:*

$$\lim_{t \to \infty} [y(t) - y_r(t)] = 0. \quad (7.33)$$

One choice for this control is

$$u = \frac{1}{b_m \beta(y)} \left[\alpha_\rho - \hat{x}_{\rho+1} - y_r^{(\rho)} \right], \quad (7.34)$$

with $z_i, \alpha_i, \; i = 1, \ldots, \rho$ defined by the following recursive expressions ($c_i > 0$, $d_i > 0, \; i = 1, \ldots, \rho$):

$$z_1 = y - y_r \quad (7.35)$$
$$z_i = \hat{x}_i - \alpha_{i-1}(y, \hat{x}_1, \ldots, \hat{x}_{i-1}, y_r, \ldots, y_r^{(i-2)}) - y_r^{(i-1)}, \quad i = 2, \ldots, \rho \quad (7.36)$$
$$\alpha_1 = -c_1 z_1 - d_1 z_1 - \varphi_1(y) \quad (7.37)$$
$$\alpha_i = -c_i z_i - z_{i-1} - d_i \left(\frac{\partial \alpha_{i-1}}{\partial y} \right)^2 z_i - k_i(y - \hat{x}_1) - \varphi_i(y)$$
$$+ \frac{\partial \alpha_{i-1}}{\partial y} [\hat{x}_2 + \varphi_1(y)] + \sum_{j=1}^{i-1} \frac{\partial \alpha_{i-1}}{\partial \hat{x}_j} [\hat{x}_{j+1} + k_j(y - \hat{x}_1) + \varphi_j(y)]$$
$$+ \sum_{j=1}^{i-2} \frac{\partial \alpha_{i-1}}{\partial y_r^{(j)}} y_r^{(j+1)}, \quad i = 2, \ldots, \rho. \quad (7.38)$$

7.1 OBSERVER BACKSTEPPING

Proof. From (7.27), (7.31), and the definitions (7.34)–(7.38), we can express the derivatives of the error variables $z_1, \ldots, z_\rho$ as follows:

$$\begin{aligned}
\dot{z}_1 &= \dot{y} - \dot{y}_r = x_2 + \varphi_1(y) - \dot{y}_r = \hat{x}_2 + \tilde{x}_2 + \varphi_1(y) - \dot{y}_r \\
&= z_2 + \alpha_1 + \varphi_1(y) + \tilde{x}_2 = -c_1 z_1 + z_2 - d_1 z_1 + \tilde{x}_2
\end{aligned} \quad (7.39)$$

$$\begin{aligned}
\dot{z}_i &= \hat{x}_{i+1} + k_i(y - \hat{x}_1) + \varphi_i(y) - \frac{\partial \alpha_{i-1}}{\partial y}[\hat{x}_2 + \varphi_1(y) + \tilde{x}_2] \\
&\quad - \sum_{j=1}^{i-1} \frac{\partial \alpha_{i-1}}{\partial \hat{x}_j}[\hat{x}_{j+1} + k_j(y - \hat{x}_1) + \varphi_j(y)] - \sum_{j=1}^{i-2} \frac{\partial \alpha_{i-1}}{\partial y_r^{(j)}} y_r^{(j+1)} - y_r^{(i)} \\
&= -c_i z_i - z_{i-1} + z_{i+1} - d_i \left(\frac{\partial \alpha_{i-1}}{\partial y}\right)^2 z_i - \frac{\partial \alpha_{i-1}}{\partial y}\tilde{x}_2, \quad i = 2, \ldots, \rho - 1,
\end{aligned} \quad (7.40)$$

$$\begin{aligned}
\dot{z}_\rho &= \hat{x}_{\rho+1} + b_m\beta(y)u + k_\rho(y - \hat{x}_1) + \varphi_\rho(y) - \frac{\partial \alpha_{\rho-1}}{\partial y}[\hat{x}_2 + \varphi_1(y) + \tilde{x}_2] \\
&\quad - \sum_{j=1}^{\rho-1} \frac{\partial \alpha_{\rho-1}}{\partial \hat{x}_j}[\hat{x}_{j+1} + k_j(y - \hat{x}_1) + \varphi_j(y)] - \sum_{j=1}^{\rho-2} \frac{\partial \alpha_{\rho-1}}{\partial y_r^{(j)}} y_r^{(j+1)} - y_r^{(\rho)} \\
&= b_m\beta(y)u + \hat{x}_{\rho+1} - \alpha_\rho - c_\rho z_\rho - z_{\rho-1} - d_\rho\left(\frac{\partial \alpha_{\rho-1}}{\partial y}\right)^2 z_\rho - \frac{\partial \alpha_{\rho-1}}{\partial y}\tilde{x}_2 - y_r^{(\rho)} \\
&= -c_\rho z_\rho - z_{\rho-1} - d_\rho\left(\frac{\partial \alpha_{\rho-1}}{\partial y}\right)^2 z_\rho - \frac{\partial \alpha_{\rho-1}}{\partial y}\tilde{x}_2.
\end{aligned} \quad (7.41)$$

The resulting error system is

$$\begin{aligned}
\dot{z}_1 &= -c_1 z_1 + z_2 - d_1 z_1 + \tilde{x}_2 \\
&\vdots \\
\dot{z}_i &= -c_i z_i - z_{i-1} + z_{i+1} - d_i\left(\frac{\partial \alpha_{i-1}}{\partial y}\right)^2 z_i - \frac{\partial \alpha_{i-1}}{\partial y}\tilde{x}_2 \\
&\vdots \\
\dot{z}_\rho &= -c_\rho z_\rho - z_{\rho-1} - d_\rho\left(\frac{\partial \alpha_{\rho-1}}{\partial y}\right)^2 z_\rho - \frac{\partial \alpha_{\rho-1}}{\partial y}\tilde{x}_2 \\
\dot{\tilde{x}} &= A_0\tilde{x}.
\end{aligned} \quad (7.42)$$

Due to the piecewise continuity of $y_r^{(\rho)}(t)$ and the smoothness of the nonlinearities, the solution of the closed-loop system (7.42) exists. Let its maximum interval of existence be $[0, t_f)$. On this interval, the nonnegative function V_ρ defined by

$$V_\rho(z, \tilde{x}) = \sum_{j=1}^{\rho}\left[\frac{1}{2}z_j^2 + \frac{1}{d_j}\tilde{x}^T P_0 \tilde{x}\right], \quad (7.43)$$

where P_0 is the positive definite symmetric solution of the Lyapunov equation $P_0 A_0 + A_0^T P_0 = -I$, is nonincreasing, since its derivative along the solutions of (7.42) satisfies

$$\dot{V}_\rho \leq -\sum_{j=1}^{\rho}\left[c_j z_j^2 + \frac{3}{4d_j}|\tilde{x}|^2\right] \leq 0. \tag{7.44}$$

Thus, $z_1, \ldots, z_\rho$ are bounded on $[0, t_f)$ by some constants depending only on the initial conditions of (7.27) and (7.31). The boundedness of all other signals on $[0, t_f)$ is established as follows. Since z_1 and y_r are bounded, y is bounded. The boundedness of $\tilde{x}$ and $\hat{x}_1 = y - \tilde{x}_1$ imply that $\hat{x}_1$ is bounded. Since z_2 is bounded, $\hat{x}_2$ is bounded. In the same manner, it can be shown that $\hat{x}_1, \ldots, \hat{x}_\rho$ are bounded. Hence, $x_1, \ldots, x_\rho$ are bounded.

To prove the boundedness of $x_{\rho+1}, \ldots, x_n$, we note that the boundedness of y implies the boundedness of ζ from (7.28). Since $[x_1, \ldots, x_\rho, \zeta^T]^T = Tx$, where T is a nonsingular matrix [164, Theorem 2.1], we conclude that x is bounded. Finally, the feedback control u defined in (7.34) is bounded because $b_m\beta(y)$ is bounded away from zero.

We have thus shown that the state of the closed-loop system is bounded on its maximal interval of existence $[0, t_f)$. Hence, $t_f = \infty$.

The convergence of the tracking error to zero now follows from (7.43) and (7.44) and the LaSalle-Yoshizawa theorem (Theorem 2.1). □

7.2 MIMO Design: Induction Motor

Heretofore we have not dealt with multi-input systems in much detail, because the backstepping procedure is essentially the same, provided that the system is in an appropriate form. When it comes to backstepping with two or more controls, there are additional degrees of freedom which designers can skillfully employ. This additional flexibility is particularly welcome when some states are not measured but instead estimated through an observer. Rather than develop a general theory, we will illustrate such a MIMO design on an induction motor control problem.

The induction motor is indispensable because of its ruggedness and low cost, but, until recently, its control applications have been restricted by the difficulties in controlling its torque and speed. Advances in power electronics and the field orientation design methodology promise that induction motors will replace less reliable dc motors in industrial drives and servo systems.

With full-state feedback, field orientation can emulate the dc machine dynamics. Since full-state feedback requires expensive and unreliable sensors for rotor flux, flux observers are used in place of sensors.

In this design we combine nonlinear damping and observer backstepping with a simple flux observer. Under the assumption that the motor parameters are known, our design achieves stability with a guaranteed region of attraction. Simulation results indicate that this observer-based design recovers the performance of the full-state design.

7.2 MIMO Design: Induction Motor

The well-known two-phase equivalent model of the induction motor is

$$\begin{aligned}
\frac{d\omega}{dt} &= \frac{3}{2}\frac{n_p M}{JL_r}(\psi_{ra}i_{sb} - \psi_{rb}i_{sa}) - \frac{T_L}{J} \\
\frac{di_{sa}}{dt} &= \frac{MR_r}{\sigma L_r^2}\psi_{ra} + \frac{n_p M}{\sigma L_r}\omega\psi_{rb} - \gamma i_{sa} + \frac{1}{\sigma}u_{sa} \\
\frac{di_{sb}}{dt} &= -\frac{n_p M}{\sigma L_r}\omega\psi_{ra} + \frac{MR_r}{\sigma L_r^2}\psi_{rb} - \gamma i_{sb} + \frac{1}{\sigma}u_{sb} \qquad (7.45) \\
\frac{d\psi_{ra}}{dt} &= -\frac{R_r}{L_r}\psi_{ra} - n_p\omega\psi_{rb} + \frac{R_r}{L_r}Mi_{sa} \\
\frac{d\psi_{rb}}{dt} &= n_p\omega\psi_{ra} - \frac{R_r}{L_r}\psi_{rb} + \frac{R_r}{L_r}Mi_{sb},
\end{aligned}$$

where ω, i, ψ, u_s are the angular speed, current, flux linkage and stator voltage input to the machine; R, L, M, J, T_L, and n_p are the resistance, inductance, mutual inductance, rotor inertia, load torque, and number of pole pairs; the subscripts s and r stand for stator and rotor; (a, b) denote the components of a vector with respect to a fixed reference frame; and $\sigma = L_s - \frac{M^2}{L_r}$, $\gamma = \frac{M^2 R_r + L_r^2 R_s}{\sigma L_r^2}$.

We assume that the torque load is a known function of the rotor speed, $T_L(\omega) = k_0 + k_1\omega + k_2\omega^2$, which is realistic for loads such as pumps and compressors. Denoting

$$\begin{array}{llll}
x_1 = \omega & u_1 = u_{sa} & a_1 = \frac{3}{2}\frac{n_p M}{JL_r} & a_6 = \frac{n_p M}{\sigma L_r} \\
x_2 = i_{sa} & u_2 = u_{sb} & a_2 = \frac{k_0}{J} & a_7 = \gamma \\
x_3 = i_{sb} & b = \frac{1}{\sigma} & a_3 = \frac{k_1}{J} & a_8 = \frac{R_r}{L_r} \qquad (7.46) \\
x_4 = \psi_{ra} & & a_4 = \frac{k_2}{J} & a_9 = n_p \\
x_5 = \psi_{rb} & & a_5 = \frac{MR_r}{\sigma L_r^2} & a_{10} = \frac{R_r M}{L_r}
\end{array}$$

we rewrite (7.45) as

$$\begin{aligned}
\dot{x}_1 &= a_1(x_3 x_4 - x_2 x_5) - a_2 - a_3 x_1 - a_4 x_1^2 \\
\dot{x}_2 &= a_5 x_4 + a_6 x_1 x_5 - a_7 x_2 + bu_1 \\
\dot{x}_3 &= -a_6 x_1 x_4 + a_5 x_5 - a_7 x_3 + bu_2 \qquad (7.47) \\
\dot{x}_4 &= -a_8 x_4 - a_9 x_1 x_5 + a_{10} x_2 \\
\dot{x}_5 &= a_9 x_1 x_4 - a_8 x_5 + a_{10} x_3.
\end{aligned}$$

The control objective for $y_1 = x_1$ is to asymptotically track a given speed reference $\omega_{\text{ref}}(t)$ while keeping all the signals in the closed-loop system bounded. It is also required that $y_2 = \sqrt{x_4^2 + x_5^2} = \sqrt{\psi_{ra}^2 + \psi_{rb}^2}$ track a reference $\psi_{\text{ref}}(t)$ to prevent excessive flux magnitudes or demagnetization. The first two derivatives of $\omega_{\text{ref}}(t)$ and of $\psi_{\text{ref}}(t)$ are assumed to be known and bounded.

A simple flux observer is constructed as a copy of the last two-state equations:

$$\begin{aligned} \dot{\hat{x}}_4 &= -a_8\hat{x}_4 - a_9 x_1 \hat{x}_5 + a_{10} x_2 \\ \dot{\hat{x}}_5 &= a_9 x_1 \hat{x}_4 - a_8 \hat{x}_5 + a_{10} x_3 \, . \end{aligned} \qquad (7.48)$$

The resulting error system is ($\tilde{x}_4 = x_4 - \hat{x}_4, \tilde{x}_5 = x_5 - \hat{x}_5$)

$$\begin{aligned} \dot{\tilde{x}}_4 &= -a_8\tilde{x}_4 - a_9 x_1 \tilde{x}_5 \\ \dot{\tilde{x}}_5 &= -a_8\tilde{x}_5 + a_9 x_1 \tilde{x}_4 \, . \end{aligned} \qquad (7.49)$$

Since $a_8 > 0$, this second-order system is globally exponentially stable. This is established via the Lyapunov function

$$V_{\text{obs}}(\tilde{x}) = \frac{1}{2}(\tilde{x}_4^2 + \tilde{x}_5^2), \qquad (7.50)$$

whose derivative along the solutions of (7.49) is

$$\begin{aligned} \dot{V}_{\text{obs}} &= -a_8\tilde{x}_4^2 - a_9 x_1 \tilde{x}_5 \tilde{x}_4 - a_8\tilde{x}_5^2 + a_9 x_1 \tilde{x}_4 \tilde{x}_5 \\ &= -a_8(\tilde{x}_4^2 + \tilde{x}_5^2) \\ &= -2a_8 V_{\text{obs}} \le 0 \, . \end{aligned} \qquad (7.51)$$

With an exponentially convergent observer available, we proceed to our backstepping design.

Step 1. Let $c_1, c_2, c_3, c_4, d_1, d_2$, and d_4 be positive design constants. We first consider the speed tracking objective. We define $z_1 = x_1 - \omega_{\text{ref}}$ and write $\dot{z}_1$ as

$$\dot{z}_1 = a_1(x_3 x_4 - x_2 x_5) - a_2 - a_3 x_1 - a_4 x_1^2 - \dot{\omega}_{\text{ref}} \, . \qquad (7.52)$$

To initiate backstepping, we need to choose our first virtual control. The form of (7.52) suggests that instead of a single state variable we choose $a_1(x_3 x_4 - x_2 x_5)$. For an electric machine expert, this choice of virtual control is natural, because it represents the torque produced by the rotating magnetic field. If x_2 and x_4 were measured, the first stabilizing function would be

$$\bar{\alpha}_1 = -c_1 z_1 + a_2 + a_3 x_1 + a_4 x_1^2 + \dot{\omega}_{\text{ref}} \, . \qquad (7.53)$$

Since x_4 and x_5 are not measured but only estimated from (7.48), we apply observer backstepping: We rewrite (7.52) as

$$\dot{z}_1 = a_1(x_3\hat{x}_4 - x_2\hat{x}_5) + a_1(x_3\tilde{x}_4 - x_2\tilde{x}_5) - a_2 - a_3 x_1 - a_4 x_1^2 - \dot{\omega}_{\text{ref}} \qquad (7.54)$$

and choose the *estimated torque* $a_1(x_3\hat{x}_4 - x_2\hat{x}_5)$ as our virtual control. Viewing $\tilde{x}_4$ and $\tilde{x}_5$ in (7.54) as unknown disturbances, we apply the Nonlinear Damping Lemma (Lemma 2.26) to design the stabilizing function

$$\alpha_1 = -c_1 z_1 + a_2 + a_3 x_1 + a_4 x_1^2 + \dot{\omega}_{\text{ref}} - d_1 a_1^2 (x_2^2 + x_3^2) z_1 \, . \qquad (7.55)$$

7.2 MIMO Design: Induction Motor

Step 2. Defining the error variable

$$z_2 = a_1(x_3\hat{x}_4 - x_2\hat{x}_5) - \alpha_1, \tag{7.56}$$

we augment (7.54) with the $\dot{z}_2$-equation

$$\begin{aligned}
\dot{z}_2 &= (a_1\hat{x}_4 + 2d_1a_1^2 x_3 z_1)(-a_6 x_1 x_4 + a_5 x_5 - a_7 x_3 + bu_2) \\
&+ (-a_1\hat{x}_5 + 2d_1 a_1^2 x_2 z_1)(a_5 x_4 + a_6 x_1 x_5 - a_7 x_2 + bu_1) \\
&+ a_1 x_3(-a_8\hat{x}_4 - a_9 x_1 \hat{x}_5 + a_{10} x_2) - a_1 x_2(a_9 x_1 \hat{x}_4 - a_8 \hat{x}_5 + a_{10} x_3) \\
&+ [c_1 - a_3 - 2a_4 x_1 + d_1 a_1^2 (x_2^2 + x_3^2)][a_1(x_3\hat{x}_4 - x_2\hat{x}_5) + a_1(x_3\tilde{x}_4 - x_2\tilde{x}_5) \\
&- a_2 - a_3 x_1 - a_4 x_1^2 - \dot{\omega}_{\text{ref}}] - \ddot{\omega}_{\text{ref}} - (a_3 + 2a_4 x_1)\dot{\omega}_{\text{ref}}. \tag{7.57}
\end{aligned}$$

The resulting system is then rewritten as

$$\begin{aligned}
\dot{z}_1 &= -c_1 z_1 + a_1(x_3\tilde{x}_4 - x_2\tilde{x}_5) - d_1 a_1^2(x_2^2 + x_3^2)z_1 + z_2 \\
\dot{z}_2 &= (a_1\hat{x}_4 + 2d_1 a_1^2 x_3 z_1)bu_2 + (-a_1\hat{x}_5 + 2d_1 a_1^2 x_2 z_1)bu_1 \\
&+ (a_1\hat{x}_4 + 2d_1 a_1^2 x_3 z_1)(-a_6 x_1 \hat{x}_4 + a_5\hat{x}_5 - a_7 x_3) \\
&+ (-a_1\hat{x}_5 + 2d_1 a_1^2 x_2 z_1)(a_5\hat{x}_4 + a_6 x_1 \hat{x}_5 - a_7 x_2) \tag{7.58} \\
&+ a_1 x_3(-a_8\hat{x}_4 - a_9 x_1 \hat{x}_5 + a_{10} x_2) - a_1 x_2(a_9 x_1 \hat{x}_4 - a_8 \hat{x}_5 + a_{10} x_3) \\
&+ [c_1 - a_3 - 2a_4 x_1 + d_1 a_1^2(x_2^2 + x_3^2)][-c_1 z_1 - d_1 a_1^2(x_2^2 + x_3^2)z_1 + z_2] \\
&- \ddot{\omega}_{\text{ref}} - (a_3 + 2a_4 x_1)\dot{\omega}_{\text{ref}} \\
&+ \{(-a_1\hat{x}_5 + 2d_1 a_1^2 x_2 z_1)a_5 - (a_1\hat{x}_4 + 2d_1 a_1^2 x_3 z_1)a_6 x_1 \\
&+ [c_1 - a_3 - 2a_4 x_1 + d_1 a_1^2(x_2^2 + x_3^2)]a_1 x_3\}\tilde{x}_4 + \{(a_1\hat{x}_4 + 2d_1 a_1^2 x_3 z_1)a_5 \\
&+ (-a_1\hat{x}_5 + 2d_1 a_1^2 x_2 z_1)a_6 x_1 - [c_1 - a_3 - 2a_4 x_1 + d_1 a_1^2(x_2^2 + x_3^2)]a_1 x_2\}\tilde{x}_5 \\
&\stackrel{\triangle}{=} \alpha_{21} bu_1 + \alpha_{22} bu_2 + \psi_2 + \omega_{24}\tilde{x}_4 + \omega_{25}\tilde{x}_5.
\end{aligned}$$

Lemma 2.28 results in the choice of feedback

$$\alpha_{21} bu_1 + \alpha_{22} bu_2 = -c_2 z_2 - z_1 - \psi_2 - d_2(\omega_{24}^2 + \omega_{25}^2)z_2. \tag{7.59}$$

Of course, (7.59) is only one equation to be satisfied by the two controls u_1 and u_2. This leaves us with an additional degree of freedom which we will use for flux tracking in the following steps.

Step 3. We now turn our attention to the flux tracking objective. The flux error is $x_4^2 + x_5^2 - \psi_{\text{ref}}$. However, since x_4 and x_5 are not measured, we replace them by their estimates and define the error variable $z_3 = \hat{x}_4^2 + \hat{x}_5^2 - \psi_{\text{ref}}$. This error variable represents the *estimated flux error*, and its derivative contains no unknown terms such as $\tilde{x}_4$ or $\tilde{x}_5$:

$$\begin{aligned}
\dot{z}_3 &= 2(-a_8\hat{x}_4^2 - a_9 x_1 \hat{x}_5 \hat{x}_4 + a_{10} x_2 \hat{x}_4 + a_9 x_1 \hat{x}_4 \hat{x}_5 - a_8 \hat{x}_5^2 + a_{10} x_3 \hat{x}_5) - \dot{\psi}_{\text{ref}} \\
&= -2a_8(\hat{x}_4^2 + \hat{x}_5^2) + 2a_{10}(x_2 \hat{x}_4 + x_3 \hat{x}_5) - \dot{\psi}_{\text{ref}}. \tag{7.60}
\end{aligned}$$

Once again, we have to choose a virtual control. Since differentiation of x_2 and x_3 produces the actual control variables u_1 and u_2, we choose the term $2a_{10}(x_2\hat{x}_4 + x_3\hat{x}_5)$ as the virtual control, and we design for it the stabilizing function

$$\alpha_3 = -c_3 z_3 + 2a_8(\hat{x}_4^2 + \hat{x}_5^2) + \dot{\psi}_{\text{ref}}. \tag{7.61}$$

Step 4. Applying Lemma 2.8, we define the error variable

$$z_4 = 2a_{10}(x_2\hat{x}_4 + x_3\hat{x}_5) - \alpha_3 \tag{7.62}$$

and augment (7.60) with

$$\begin{aligned}
\dot{z}_4 =\ & 2a_{10}\hat{x}_4(a_5 x_4 + a_6 x_1 x_5 - a_7 x_2 + bu_1) \\
& +2a_{10}\hat{x}_5(-a_6 x_1 x_4 + a_5 x_5 - a_7 x_3 + bu_2) \\
& +2(a_{10}x_2 - 2a_8\hat{x}_4)(-a_8\hat{x}_4 - a_9 x_1\hat{x}_5 + a_{10}x_2) \\
& +2(a_{10}x_3 - 2a_8\hat{x}_5)(a_9 x_1\hat{x}_4 - a_8\hat{x}_5 + a_{10}x_3) \\
& +c_3[-2a_8(\hat{x}_4^2 + \hat{x}_5^2) + 2a_{10}(x_2\hat{x}_4 + x_3\hat{x}_5) - \dot{\psi}_{\text{ref}}] - \ddot{\psi}_{\text{ref}}.
\end{aligned} \tag{7.63}$$

The resulting system is then rewritten as

$$\begin{aligned}
\dot{z}_3 =\ & -c_3 z_3 + z_4 \\
\dot{z}_4 =\ & 2a_{10}\hat{x}_4 bu_1 + 2a_{10}\hat{x}_5 bu_2 + 2a_{10}\hat{x}_4(a_5\hat{x}_4 + a_6 x_1\hat{x}_5 - a_7 x_2) \\
& +2a_{10}\hat{x}_5(-a_6 x_1\hat{x}_4 + a_5\hat{x}_5 - a_7 x_3) \\
& +2(a_{10}x_2 - a_8\hat{x}_4)(-a_8\hat{x}_4 - a_9 x_1\hat{x}_5 + a_{10}x_2) \\
& +2(a_{10}x_3 - a_8\hat{x}_5)(a_9 x_1\hat{x}_4 - a_8\hat{x}_5 + a_{10}x_3) + c_3(-c_3 z_3 + z_4) - \ddot{\psi}_{\text{ref}} \\
& +2a_{10}(a_5\hat{x}_4 - a_6 x_1\hat{x}_5)\tilde{x}_4 + 2a_{10}(a_6 x_1\hat{x}_4 + a_5\hat{x}_5)\tilde{x}_5 \\
\stackrel{\triangle}{=}\ & \alpha_{41} bu_1 + \alpha_{42} bu_2 + \psi_4 + \omega_{44}\tilde{x}_4 + \omega_{45}\tilde{x}_5.
\end{aligned} \tag{7.64}$$

From Lemmas 2.8 and 2.26, the choice of feedback is

$$\alpha_{41} bu_1 + \alpha_{42} bu_2 = -c_4 z_4 - z_3 - \psi_4 - d_4(\omega_{44}^2 + \omega_{45}^2) z_4. \tag{7.65}$$

This is the second equation to be satisfied by the two controls u_1 and u_2. Combining (7.59) and (7.65) we obtain the following expression for u_1 and u_2:

$$\begin{bmatrix} u_1 \\ u_2 \end{bmatrix} = \frac{1}{b}\begin{bmatrix} \alpha_{21} & \alpha_{22} \\ \alpha_{41} & \alpha_{42} \end{bmatrix}^{-1} \begin{bmatrix} -c_2 z_2 - \psi_2 - d_2(\omega_{24}^2 + \omega_{25}^2) z_2 - z_1 \\ -c_4 z_4 - \psi_4 - d_4(\omega_{44}^2 + \omega_{45}^2) z_4 - z_3 \end{bmatrix}. \tag{7.66}$$

This concludes our design procedure, whose schematic representation is given in Figure 7.4.

7.2 MIMO Design: Induction Motor

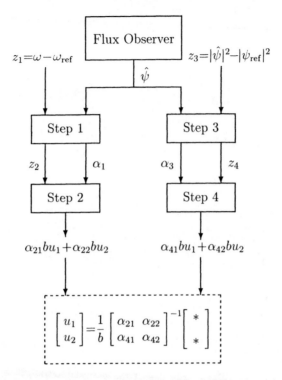

Figure 7.4: MIMO backstepping design for an induction motor with a flux observer.

Feasibility and Stability. From (7.66) we see that our control law is well defined only in the region where the matrix $D = \begin{bmatrix} \alpha_{21} & \alpha_{22} \\ \alpha_{41} & \alpha_{42} \end{bmatrix}$ is invertible. The determinant of this matrix is given by

$$\begin{aligned}
\det(D) &= \alpha_{21}\alpha_{42} - \alpha_{22}\alpha_{41} \\
&= (-a_1\hat{x}_5 + 2d_1 a_1^2 x_2 z_1) 2a_{10}\hat{x}_5 - (a_1\hat{x}_4 + 2d_1 a_1^2 x_3 z_1) 2a_{10}\hat{x}_4 \\
&= -2a_1 a_{10}(\hat{x}_4^2 + \hat{x}_5^2) + 4d_1 a_1^2 a_{10} z_1 (x_2\hat{x}_5 - x_3\hat{x}_4) \\
&= -2a_1 a_{10} z_3 - 2a_1 a_{10} \psi_{\text{ref}}(t) + 4d_1 a_1^2 a_{10} z_1 (x_2\hat{x}_5 - x_3\hat{x}_4). \quad (7.67)
\end{aligned}$$

Hence, D is not globally invertible, and (7.66) does not achieve the speed and flux tracking objectives globally. Nevertheless, under the natural assumption that the flux reference does not dictate demagnetization, $\psi_{\text{ref}}(t) \geq \delta > 0$ for all $t \geq 0$, we can guarantee that (7.66) achieves the control objectives from a set of initial conditions that can be determined a priori.

To see this, we first combine (7.49), (7.58), (7.64), and (7.66) to compose

the system
$$\begin{aligned}
\dot{z}_1 &= -c_1 z_1 + a_1(x_3 \tilde{x}_4 - x_2 \tilde{x}_5) - d_1 a_1^2(x_2^2 + x_3^2)z_1 + z_2 \\
\dot{z}_2 &= -c_2 z_2 + \omega_{24}\tilde{x}_4 + \omega_{25}\tilde{x}_5 - d_2(\omega_{24}^2 + \omega_{25}^2)z_2 - z_1 \\
\dot{z}_3 &= -c_3 z_3 + z_4 \\
\dot{z}_4 &= -c_4 z_4 + \omega_{44}\tilde{x}_4 + \omega_{45}\tilde{x}_5 - d_4(\omega_{44}^2 + \omega_{45}^2)z_4 - z_3 \\
\dot{\tilde{x}}_4 &= -a_8 \tilde{x}_4 - 2x_1 \tilde{x}_5 \\
\dot{\tilde{x}}_5 &= -a_8 \tilde{x}_5 + 2x_1 \tilde{x}_4 \, .
\end{aligned} \qquad (7.68)$$

This system has an equilibrium at $z_1 = z_2 = z_3 = z_4 = \tilde{x}_4 = \tilde{x}_5 = 0$. Furthermore, the derivative of the function

$$V = \frac{1}{2}(z_1^2 + z_2^2 + z_3^2 + z_4^2) + \frac{1}{2a_8}\left(\frac{1}{d_1} + \frac{1}{d_2} + \frac{1}{d_4}\right)(\tilde{x}_4^2 + \tilde{x}_5^2) \qquad (7.69)$$

along the solutions of (7.68) is nonpositive:

$$\begin{aligned}
\dot{V} &= -c_1 z_1^2 - c_2 z_2^2 - c_3 z_3^2 - c_4 z_4^2 - \frac{3}{4}\left(\frac{1}{d_1} + \frac{1}{d_2} + \frac{1}{d_4}\right)(\tilde{x}_4^2 + \tilde{x}_5^2) \\
&\quad - d_1\left(\frac{1}{2d_1}\tilde{x}_4 - z_1 a_1 x_3\right)^2 - d_1\left(\frac{1}{2d_1}\tilde{x}_5 - z_1 a_1 x_2\right)^2 \\
&\quad - d_2\left(\frac{1}{2d_2}\tilde{x}_4 - z_2 \omega_{24}\right)^2 - d_2\left(\frac{1}{2d_2}\tilde{x}_5 - z_2 \omega_{25}\right)^2 \\
&\quad - d_4\left(\frac{1}{2d_4}\tilde{x}_4 - z_4 \omega_{44}\right)^2 - d_4\left(\frac{1}{2d_4}\tilde{x}_5 - z_4 \omega_{45}\right)^2 \\
&\leq -c_1 z_1^2 - c_2 z_2^2 - c_3 z_3^2 - c_4 z_4^2 - \frac{3}{4}\left(\frac{1}{d_1} + \frac{1}{d_2} + \frac{1}{d_4}\right)(\tilde{x}_4^2 + \tilde{x}_5^2). \quad (7.70)
\end{aligned}$$

Now if $\psi_{\text{ref}}(t) \geq \delta > 0$, we see from (7.67) that $\det(D) \leq 2a_1 a_{10}\delta < 0$ whenever $z_1 = z_3 = 0$. From this fact, the definitions of z_1, z_2, z_3, z_4, and the Implicit Function Theorem (see, e.g., [81]), we conclude that there exist an open set $\mathcal{F}_0$ containing the equilibrium $z_1 = z_2 = z_3 = z_4 = \tilde{x}_4 = \tilde{x}_5 = 0$ and a function $\phi(z_1, z_2, z_3, z_4, \tilde{x}_4, \tilde{x}_5, \omega_{\text{ref}}(t), \dot{\omega}_{\text{ref}}(t), \psi_{\text{ref}}(t), \dot{\psi}_{\text{ref}}(t))$, which is continuous on $\mathcal{F}_0$, such that on $\mathcal{F}_0$ we have

$$\det(D) \leq -2a_1 a_{10} z_3 - 2a_{10} a_{10}\psi_{\text{ref}}(t) + 4d_1 a_1^2 a_{10} z_1 \phi. \qquad (7.71)$$

Then, we define $\mathcal{F}$ as the largest subset of $\mathcal{F}_0$ on which

$$\sup_{t \geq 0}\left\{-2a_1 a_{10} z_3 - 2a_1 a_{10}\psi_{\text{ref}}(t) + 4d_1 a_1^2 a_{10} z_1 \phi\right\} < 0, \qquad (7.72)$$

where in (7.72) we treat $z_1, z_2, z_3, z_4, \tilde{x}_4$, and $\tilde{x}_5$ as independent variables and not as functions of time. Since $\omega_{\text{ref}}(t), \psi_{\text{ref}}(t)$ and their first and second derivatives are bounded, the supremum in (7.72) exists.

From (7.69), (7.70), and the definition of $\mathcal{F}$, we see that any set of the form

$$\mathcal{U}(c) = \{(z_1, z_2, z_3, z_4, \tilde{x}_4, \tilde{x}_5) \,|\, V < c\}, \qquad (7.73)$$

where c is such that $\mathcal{U}(c) \subset \mathcal{F}$, has the following properties:

7.3 ADAPTIVE OBSERVER BACKSTEPPING

(i) $\mathcal{U}(c)$ is an invariant set of the closed-loop system, and

(ii) $\det(D) < 0$ on $\mathcal{U}(c)$.

From these properties, the definitions of z_1, z_2, z_3, z_4, u_1, and u_2, and the boundedness of $\omega_{\text{ref}}(t)$, $\psi_{\text{ref}}(t)$ and their first and second derivatives, we conclude that for any initial conditions such that $(z_1, z_2, z_3, z_4, \tilde{x}_4, \tilde{x}_5) \in \mathcal{U}(c^*)$, $c^* = \sup_{\mathcal{U}(c) \subset \mathcal{F}} \{c\}$, all of the signals in the closed-loop system are bounded (note that the boundedness of $\hat{x}_4^2 + \hat{x}_5^2 = z_3 + \psi_{\text{ref}}$ implies the boundedness of both $\hat{x}_4$ and $\hat{x}_5$).

Furthermore, from (7.69)–(7.70) and Theorem 2.1 we see that $z_1, z_2, z_3, z_4, \tilde{x}_4, \tilde{x}_5 \to 0$ as $t \to \infty$. In particular, this implies that $z_1 = \omega - \omega_{\text{ref}} \to 0$, $z_3 = \hat{x}_4^2 + \hat{x}_5^2 - \psi_{\text{ref}} \to 0$ as $t \to \infty$. Since $\tilde{x}_4 = x_4 - \hat{x}_4 \to 0$, $\tilde{x}_4 = x_5 - \hat{x}_5 \to 0$ as $t \to \infty$, and $\hat{x}_4, \tilde{x}_4, \hat{x}_5, \tilde{x}_5$ are bounded, we conclude that $x_4^2 + x_5^2 = \psi_{\text{ra}}^2 + \psi_{\text{rb}}^2 - \psi_{\text{ref}} \to 0$ as $t \to \infty$. Hence, the speed and flux tracking objectives are achieved for any initial conditions such that $(z_1, z_2, z_3, z_4, \tilde{x}_4, \tilde{x}_5) \in \mathcal{U}(c^*)$.

Simulation Results. We simulated the controller designed in this section with the flux observer and compared its performance with the performance obtained when flux measurements are used. In simulations the flux reference steps to its rated value at $t = 0$, the speed reference steps from standstill to the rated speed at $t = 0.5$ sec, and finally the load torque steps from no load to 40 Nm (roughly 60% of the rated value) at $t = 1.0$ sec.

Figure 7.5 shows the speed tracking for the controller with flux measurement (a) and with the flux observer (b). It is clear that the observer-based controller recovers the performance of the full-state-feedback controller. The beneficial effect of nonlinear damping is clear from Figure 7.5(c), which shows the performance of the controller without nonlinear damping, that is, with $d_1 = d_2 = d_4 = 0$. The system now shows high sensitivity to flux error, which results in deterioration of transient performance.

7.3 Adaptive Observer Backstepping

7.3.1 An introductory example

Our adaptive design for output-feedback systems will be first illustrated on a plant with one unknown parameter:

$$\begin{aligned} \dot{x}_1 &= x_2 + \theta \varphi_1(y) \\ \dot{x}_2 &= x_3 + \theta \varphi_2(y) + u \\ \dot{x}_3 &= u \\ y &= x_1. \end{aligned} \qquad (7.74)$$

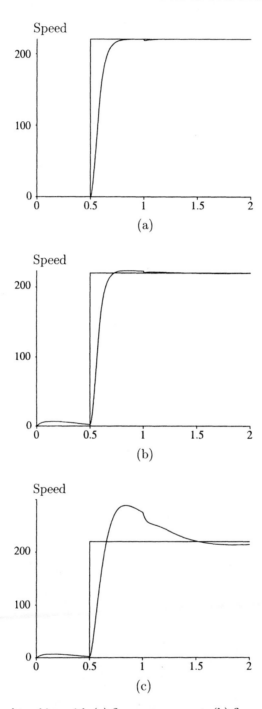

Figure 7.5: Speed tracking with (a) flux measurement, (b) flux observer and nonlinear damping, (c) flux observer without nonlinear damping.

7.3 ADAPTIVE OBSERVER BACKSTEPPING

We assume that only the output $y = x_1$ is measured, that is, the states x_2 and x_3 are not available for feedback. We want to design an adaptive nonlinear output-feedback controller that guarantees asymptotic tracking of the reference signal $y_r(t)$ by the output y while keeping all the states of the closed-loop system bounded.

We will attempt to reconstruct the full state of the system through the use of two filters, one for the part of the plant that does not contain θ and one for its unknown part. Denoting the states of the two filters by ξ_0 and ξ_1, our virtual estimate is $\xi_0 + \theta \xi_1$, which depends on the unknown parameter θ. The state is reconstructed as

$$x = \xi_0 + \theta \xi_1 + \varepsilon. \tag{7.75}$$

This expression is good enough for our backstepping design, provided that the error ε tends to zero asymptotically. To ensure this, we construct the two filters as follows:

$$\begin{aligned} \dot{\xi}_{01} &= k_1(y - \xi_{01}) + \xi_{02} & \dot{\xi}_{11} &= -k_1 \xi_{11} + \xi_{12} + \varphi_1(y) \\ \dot{\xi}_{02} &= k_2(y - \xi_{01}) + \xi_{03} + u & \dot{\xi}_{12} &= -k_2 \xi_{11} + \xi_{13} + \varphi_2(y) \\ \dot{\xi}_{03} &= k_3(y - \xi_{01}) + u & \dot{\xi}_{13} &= -k_3 \xi_{11}. \end{aligned} \tag{7.76}$$

The gain vector $k = [k_1, \ k_2, \ k_3]^{\mathrm{T}}$ is chosen so that the matrix

$$A_0 = \begin{bmatrix} -k_1 & 1 & 0 \\ -k_2 & 0 & 1 \\ -k_3 & 0 & 0 \end{bmatrix} \tag{7.77}$$

is Hurwitz. We then rewrite the plant and the filters ξ_0 and ξ_1 as follows:

$$\dot{x} = A_0 x + ky + bu + \theta \varphi(y), \quad b = \begin{bmatrix} 0 \\ 1 \\ 1 \end{bmatrix}, \quad \varphi(y) = \begin{bmatrix} \varphi_1(y) \\ \varphi_2(y) \\ 0 \end{bmatrix}. \tag{7.78}$$

The filters ξ_0 and ξ_1 are rewritten in the more transparent form:

$$\dot{\xi}_0 = A_0 \xi_0 + ky + bu, \quad \dot{\xi}_1 = A_0 \xi_1 + \varphi(y). \tag{7.79}$$

Combining these equations, we derive the equation for the error $\varepsilon = x - \xi_0 - \theta \xi_1$:

$$\begin{aligned} \dot{\varepsilon} &= \dot{x} - \dot{\xi}_0 - \theta \dot{\xi}_1 \\ &= A_0 x + ky + bu + \theta \varphi(y) - A_0 \xi_0 - ky - bu - A_0 \theta \xi_1 - \theta \varphi(y) \\ &= A_0(x - \xi_0 - \theta \xi_1) \\ &= A_0 \varepsilon. \end{aligned} \tag{7.80}$$

Thus ε converges to zero exponentially since A_0 is Hurwitz.

We are now ready to design our adaptive output-feedback controller.

Step 1. The control objective is to track the reference signal $y_r(t)$ with the output y, so the first error variable is the tracking error:

$$z_1 = y - y_r. \tag{7.81}$$

The derivative of z_1 is

$$\dot{z}_1 = \dot{y} - \dot{y}_r = x_2 + \theta \varphi_1(y) - \dot{y}_r. \tag{7.82}$$

Since x_2 is not measured, it cannot be our virtual control. We replace it with the sum of its "virtual estimate" and the corresponding error:

$$x_2 = \xi_{02} + \theta \xi_{12} + \varepsilon_2. \tag{7.83}$$

Substituting into (7.82) we obtain

$$\dot{z}_1 = \xi_{02} + \theta \underbrace{[\varphi_1(y) + \xi_{12}]}_{\omega} - \dot{y}_r + \varepsilon_2. \tag{7.84}$$

Now we must choose one of the known variables appearing in the above equation as the virtual control. Looking at the filter equations, we see that only the equation for ξ_{02} contains the control u, so it is the only candidate. We introduce the second error variable as

$$z_2 = \xi_{02} - \alpha_1 - \dot{y}_r \tag{7.85}$$

and substitute it into (7.84):

$$\dot{z}_1 = z_2 + \alpha_1 + \theta \omega + \varepsilon_2. \tag{7.86}$$

Denoting the first estimate of θ as ϑ_1, the first stabilizing function α_1 is chosen to be

$$\alpha_1 = -c_1 z_1 - d_1 z_1 - \vartheta_1 \omega. \tag{7.87}$$

With this choice, the $\dot{z}_1$-equation becomes

$$\dot{z}_1 = -c_1 z_1 + z_2 - d_1 z_1 + (\theta - \vartheta_1)\omega + \varepsilon_2. \tag{7.88}$$

Our first Lyapunov function is

$$V_1 = \frac{1}{2} z_1^2 + \frac{1}{2\gamma}(\theta - \vartheta_1)^2 + \frac{1}{d_1} \varepsilon^T P_0 \varepsilon, \tag{7.89}$$

where $\gamma > 0$ is the adaptation gain, and $P_0 = P_0^T > 0$ satisfies $P_0 A_0 + A_0^T P_0 =$

7.3 Adaptive Observer Backstepping

$-I$. The derivative of V_1 is

$$\begin{aligned}\dot{V}_1 &= z_1\dot{z}_1 - \frac{1}{\gamma}(\theta-\vartheta_1)\dot{\vartheta}_1 + \frac{1}{d_1}\frac{d}{dt}\left(\varepsilon^{\mathrm{T}} P_0 \varepsilon\right) \\ &= z_1 z_2 - c_1 z_1^2 - d_1 z_1^2 + (\theta-\vartheta_1)\left(z_1\omega - \frac{1}{\gamma}\dot{\vartheta}_1\right) + z_1\varepsilon_2 - \frac{1}{d_1}\varepsilon^{\mathrm{T}}\varepsilon \\ &= z_1 z_2 - c_1 z_1^2 + (\theta-\vartheta_1)\left(z_1\omega - \frac{1}{\gamma}\dot{\vartheta}_1\right) \\ &\quad - d_1\left[z_1 - \frac{1}{2d_1}\varepsilon_2\right]^2 + \frac{1}{4d_1}\varepsilon_2^2 - \frac{1}{d_1}\varepsilon^{\mathrm{T}}\varepsilon \\ &\le z_1 z_2 - c_1 z_1^2 + (\theta-\vartheta_1)\left(z_1\omega - \frac{1}{\gamma}\dot{\vartheta}_1\right) - \frac{3}{4d_1}\varepsilon^{\mathrm{T}}\varepsilon. \end{aligned} \quad (7.90)$$

The $(\theta-\vartheta_1)$-term is now eliminated from (7.90) by the choice of update law

$$\dot{\vartheta}_1 = \gamma z_1 \omega. \quad (7.91)$$

Step 2. The derivative of z_2 is expressed as

$$\begin{aligned}\dot{z}_2 &= \dot{\xi}_{02} - \dot{\alpha}_1 - \ddot{y}_r \\ &= u + k_2(y-\xi_{01}) + \xi_{03} - \frac{\partial\alpha_1}{\partial y}\dot{y} \\ &\quad - \frac{\partial\alpha_1}{\partial\xi_{12}}\underbrace{(-k_2\xi_{11}+\xi_{13}+\varphi_2(y))}_{\dot{\xi}_{12}} - \frac{\partial\alpha_1}{\partial y_r}\dot{y}_r - \frac{\partial\alpha_1}{\partial\vartheta_1}\dot{\vartheta}_1 - \ddot{y}_r \\ &= u + k_2(y-\xi_{01}) + \xi_{03} - \frac{\partial\alpha_1}{\partial y}\underbrace{(\xi_{02}+\theta\omega+\varepsilon_2)}_{\dot{y}} \\ &\quad - \frac{\partial\alpha_1}{\partial\xi_{12}}(-k_2\xi_{11}+\xi_{13}+\varphi_2(y)) - \frac{\partial\alpha_1}{\partial y_r}\dot{y}_r - \frac{\partial\alpha_1}{\partial\vartheta_1}\underbrace{\gamma\omega z_1}_{\dot{\vartheta}_1} - \ddot{y}_r \\ &= u + k_2(y-\xi_{01}) + \xi_{03} - \frac{\partial\alpha_1}{\partial y}\xi_{02} - \frac{\partial\alpha_1}{\partial\xi_{12}}(-k_2\xi_{11}+\xi_{13}+\varphi_2(y)) \\ &\quad - \frac{\partial\alpha_1}{\partial y_r}\dot{y}_r - \frac{\partial\alpha_1}{\partial\vartheta_1}\gamma\omega z_1 - \frac{\partial\alpha_1}{\partial y}\theta\omega - \frac{\partial\alpha_1}{\partial y}\varepsilon_2 - \ddot{y}_r. \end{aligned} \quad (7.92)$$

Since the unknown parameter θ appears in (7.92), we employ a new estimate ϑ_2. To counteract the disturbance ε_2, which is now multiplied by the nonlinear term $\frac{\partial\alpha_1}{\partial y}$, we employ nonlinear damping. The control is thus chosen as

$$\begin{aligned}u &= -c_2 z_2 - z_1 - d_2\left(\frac{\partial\alpha_1}{\partial y}\right)^2 z_2 - k_2(y-\xi_{01}) - \xi_{03} + \frac{\partial\alpha_1}{\partial y}(\xi_{02}+\vartheta_2\omega) \\ &\quad + \frac{\partial\alpha_1}{\partial\xi_{12}}(-k_2\xi_{11}+\xi_{13}+\varphi_2(y)) + \frac{\partial\alpha_1}{\partial y_r}\dot{y}_r + \frac{\partial\alpha_1}{\partial\vartheta_1}\gamma\omega z_1 + \ddot{y}_r, \end{aligned} \quad (7.93)$$

which results in

$$\dot{z}_2 = -c_2 z_2 - z_1 - \frac{\partial \alpha_1}{\partial y} \omega (\theta - \vartheta_2) - d_2 \left(\frac{\partial \alpha_1}{\partial y}\right)^2 z_2 - \frac{\partial \alpha_1}{\partial y} \varepsilon_2 . \qquad (7.94)$$

Using (7.90), (7.91) and (7.94), the derivative of the Lyapunov function

$$V_2 = V_1 + \frac{1}{2} z_2^2 + \frac{1}{2\gamma}(\theta - \vartheta_2)^2 + \frac{1}{d_2} \varepsilon^T P_0 \varepsilon \qquad (7.95)$$

is computed as

$$\begin{aligned}
\dot{V}_2 &= \dot{V}_1 + z_2 \dot{z}_2 - \frac{1}{\gamma} \dot{\vartheta}_2 (\theta - \vartheta_2) - \frac{1}{d_2} \varepsilon^T \varepsilon \\
&\leq z_1 z_2 - c_1 z_1^2 - \frac{3}{4 d_1} \varepsilon^T \varepsilon - c_2 z_2^2 - z_1 z_2 \\
&\quad - \left(\frac{\partial \alpha_1}{\partial y} \omega z_2 + \frac{1}{\gamma} \dot{\vartheta}_2\right)(\theta - \vartheta_2) - d_2 \left(\frac{\partial \alpha_1}{\partial y}\right)^2 z_2^2 - \frac{\partial \alpha_1}{\partial y} z_2 \varepsilon_2 - \frac{1}{d_2} \varepsilon^T \varepsilon \\
&\leq -c_1 z_1^2 - c_2 z_2^2 - \left(\frac{3}{4 d_1} + \frac{3}{4 d_2}\right) \varepsilon^T \varepsilon - \left(\frac{\partial \alpha_1}{\partial y} \omega z_2 + \frac{1}{\gamma} \dot{\vartheta}_2\right)(\theta - \vartheta_2) . \quad (7.96)
\end{aligned}$$

The $(\theta - \vartheta_2)$-term is then eliminated with the update law

$$\dot{\vartheta}_2 = -\gamma \frac{\partial \alpha_1}{\partial y} \omega z_2 , \qquad (7.97)$$

which yields

$$\dot{V}_2 \leq -c_1 z_1^2 - c_2 z_2^2 - \frac{3}{4}\left(\frac{1}{d_1} + \frac{1}{d_2}\right) \varepsilon^T \varepsilon . \qquad (7.98)$$

With (7.95) and (7.98), the LaSalle-Yoshizawa theorem (Theorem 2.1) guarantees the boundedness of $z_1, z_2, \vartheta_1, \vartheta_2, \varepsilon$ and the convergence of z_1, z_2, ε to zero. In particular, this implies that the tracking error $z_1 = y - y_r$ converges to zero and that $y = x_1 = z_1 + y_r$ is bounded. From the definition of ξ_1 in (7.76) and the boundedness of y we conclude that ξ_1 is also bounded. To prove the boundedness of ξ_0, x_2 and x_3, we first recall the definition $\varepsilon = x - \xi_0 - \theta \xi_1$, which implies that $\xi_{01} = x_1 - \varepsilon_1 - \theta \xi_1 = y - \varepsilon_1 - \theta \xi_1$ is bounded. Then we note that $\omega = \varphi_1(y) + \xi_{12}$ is bounded. Recalling (7.87), we see that α_1 is also bounded. Combining (7.85) with the boundedness of $z_2, \alpha_1, \dot{y}_r$ shows that ξ_{02} and $x_2 = \varepsilon_2 + \xi_{02} + \theta \xi_{12}$ are bounded. To show the boundedness of x_3, we define the variable $\zeta = x_3 - x_2 + y$ as suggested by (7.28). Its derivative is

$$\begin{aligned}
\dot{\zeta} &= \dot{x}_3 - \dot{x}_2 + \dot{x}_1 \\
&= u - x_3 - \theta \varphi_2(y) - u + x_2 + \theta \varphi_1(y) \\
&= -\zeta + y + \theta[\varphi_1(y) - \varphi_2(y)] .
\end{aligned} \qquad (7.99)$$

7.3 ADAPTIVE OBSERVER BACKSTEPPING

Since y is bounded, ζ is also bounded. Hence, $x_3 = \zeta + x_2 - y$ and $\xi_{03} = x_3 - \varepsilon_3 - \theta\xi_{13}$ are also bounded. Finally, from (7.93) we see that the control u is bounded.

The effect of the nonlinear damping terms is illustrated by the matrix form of the error system:

$$\begin{bmatrix} \dot{z}_1 \\ \dot{z}_2 \end{bmatrix} = \begin{bmatrix} -c_1 - d_1 & 1 \\ -1 & -c_2 - d_2\left(\frac{\partial \alpha_1}{\partial y}\right)^2 \end{bmatrix} \begin{bmatrix} z_1 \\ z_2 \end{bmatrix} + \begin{bmatrix} \omega & 0 \\ 0 & -\frac{\partial \alpha_1}{\partial y}\omega \end{bmatrix} \begin{bmatrix} \tilde{\vartheta}_1 \\ \tilde{\vartheta}_2 \end{bmatrix} + \begin{bmatrix} 1 \\ -\frac{\partial \alpha_1}{\partial y} \end{bmatrix} \varepsilon_2$$

$$\dot{\varepsilon} = A_0 \varepsilon \qquad (7.100)$$

$$\begin{bmatrix} \dot{\tilde{\vartheta}}_1 \\ \dot{\tilde{\vartheta}}_2 \end{bmatrix} = -\gamma \begin{bmatrix} \omega & 0 \\ 0 & -\frac{\partial \alpha_1}{\partial y}\omega \end{bmatrix} \begin{bmatrix} z_1 \\ z_2 \end{bmatrix}, \quad \begin{bmatrix} \tilde{\vartheta}_1 \\ \tilde{\vartheta}_2 \end{bmatrix} \triangleq \begin{bmatrix} \theta - \vartheta_1 \\ \theta - \vartheta_2 \end{bmatrix}.$$

We see that the nonlinear damping terms strengthen the negativity of the diagonal entries by including the squares of the factors multiplying the estimation error ε_2.

7.3.2 Parametric output-feedback systems

The above procedure is now generalized to the class of *parametric output-feedback systems*

$$\begin{aligned}
\dot{x}_1 &= x_2 + \varphi_{0,1}(y) + \sum_{j=1}^{p} \theta_j \varphi_{j,1}(y) \\
\dot{x}_2 &= x_3 + \varphi_{0,2}(y) + \sum_{j=1}^{p} \theta_j \varphi_{j,2}(y) \\
&\vdots \\
\dot{x}_{\rho-1} &= x_\rho + \varphi_{0,\rho-1}(y) + \sum_{j=1}^{p} \theta_j \varphi_{j,\rho-1}(y) \qquad (7.101) \\
\dot{x}_\rho &= x_{\rho+1} + \varphi_{0,\rho}(y) + \sum_{j=1}^{p} \theta_j \varphi_{j,\rho}(y) + b_m \beta(y) u \\
&\vdots \\
\dot{x}_n &= \varphi_{0,n}(y) + \sum_{j=1}^{p} \theta_j \varphi_{j,n}(y) + b_0 \beta(y) u \\
y &= x_1,
\end{aligned}$$

where $\theta_1, \ldots, \theta_p$ and $b_0, \ldots, b_m$ are unknown constant parameters.[2]

We make the following assumptions about the system (7.101):

---

[2] Coordinate-free characterizations of output-feedback systems are given in Appendix G, Corollaries G.6 and G.7.

Assumption 7.2 *The sign of b_m is known.*

Assumption 7.3 *The polynomial $B(s) = b_m s^m + \cdots + b_1 s + b_0$ is Hurwitz.*

Assumption 7.4 *$\beta(y) \neq 0 \ \forall y \in \mathbb{R}$.*

Assumption 7.5 *The reference signal $y_r(t)$ and its first ρ derivatives are known and bounded, and, in addition, $y_r^{(\rho)}(t)$ is piecewise continuous.*

Assuming that *only the output y is measured*, the control objective is to track the given reference signal $y_r(t)$ with the output y of the system (7.101), while keeping all of the signals in the closed-loop system globally bounded.

The first step in our design procedure is the choice of filters which will provide "virtual estimates" of the unmeasured state variables $x_2, \ldots, x_n$. As in the preceding subsection, we rewrite (7.101) in the form

$$\begin{aligned} \dot{x} &= Ax + \varphi_0(y) + \sum_{j=1}^{p} \theta_j \varphi_j(y) + b\beta(y)u \\ y &= c^{\mathrm{T}} x \end{aligned} \tag{7.102}$$

$$A = \begin{bmatrix} 0 & & \\ \vdots & I_{(n-1)\times(n-1)} \\ 0 & \cdots & 0 \end{bmatrix}, \quad b = \begin{bmatrix} 0_{(\rho-1)\times 1} \\ b_m \\ \vdots \\ b_0 \end{bmatrix}, \quad c = \begin{bmatrix} 1 \\ 0_{(n-1)\times 1} \end{bmatrix} \tag{7.103}$$

$$\varphi_j(y) = (\varphi_{j,1}(y), \ldots, \varphi_{j,n}(y))^{\mathrm{T}}, \quad 0 \leq j \leq p. \tag{7.104}$$

We then choose a gain vector k such that $A_0 = A - kc^{\mathrm{T}}$ is Hurwitz, and define the filters

$$\begin{aligned} \dot{\xi}_0 &= A_0 \xi_0 + ky + \varphi_0(y) \\ \dot{\xi}_j &= A_0 \xi_j + \varphi_j(y), \quad 1 \leq j \leq p \\ \dot{v}_j &= A_0 v_j + e_{n-j} \beta(y)u, \quad 0 \leq j \leq m, \end{aligned} \tag{7.105}$$

where e_i is the ith coordinate vector in $\mathbb{R}^n$. Note that the signals $v_0, \ldots, v_m$ can be obtained from the single filter

$$\dot{\lambda} = A_0 \lambda + e_n \beta(y) u \tag{7.106}$$

through the algebraic expressions

$$v_j = (A_0)^j \lambda, \ j = 0, \ldots, m. \tag{7.107}$$

From (7.102) and (7.105) it follows that

$$\dot{\varepsilon} = A_0 \varepsilon, \quad \varepsilon \stackrel{\triangle}{=} x - \left(\xi_0 + \sum_{j=1}^{p} \theta_j \xi_j + \sum_{j=0}^{m} b_j v_j \right), \tag{7.108}$$

7.3 Adaptive Observer Backstepping

which implies that ε converges exponentially to zero. Therefore, the virtual estimate of x is $\xi_0 + \sum_{j=1}^{p} \theta_j \xi_j + \sum_{j=0}^{m} b_j v_j$. In particular, the derivative of y can be expressed as

$$\dot{y} = \xi_{0,2} + \varphi_{0,1}(y) + \sum_{j=1}^{p} \theta_j \left[\varphi_{j,1}(y) + \xi_{j,2} \right] + \sum_{j=0}^{m} b_j v_{j,2} + \varepsilon_2. \tag{7.109}$$

The design procedure for the system (7.101) starts with this equation and backsteps on the variables of the filter v_m defined in (7.105).

Theorem 7.6 (Parametric Output-Feedback Systems) *Consider the nonlinear system (7.101) with Assumptions 7.2–7.5 and the filters (7.105). The adaptive controller*

$$u = \frac{1}{\beta(y)} \left[\alpha_\rho - v_{m,\rho+1} + \vartheta_{1,1} y_r^{(\rho)} \right] \tag{7.110}$$

$$\dot{\vartheta}_1 = \operatorname{sgn}(b_m) \Gamma \left[\omega_1(y, \bar{\xi}^{(2)}, \bar{v}^{(2)}, \dot{y}_r^{(1)}) - \dot{y}_r e_1 \right] z_1 \tag{7.111}$$

$$\dot{\vartheta}_2 = \Gamma \left[\omega_2(y, \bar{\xi}^{(2)}, \bar{v}^{(2)}, \bar{\vartheta}^{(2)}, \bar{y}_r^{(1)}) + z_1 e_{p+m+1} \right] z_2 \tag{7.112}$$

$$\dot{\vartheta}_i = \Gamma \omega_i(y, \bar{\xi}^{(i)}, \bar{v}^{(i)}, \bar{\vartheta}^{(i-1)}, \bar{y}_r^{(i-1)}) z_i, \quad i = 2, \ldots, \rho \tag{7.113}$$

with e_i the ith coordinate vector in $\mathbb{R}^{p+m+1}$ and $z_i, \omega_i, \alpha_i, i = 1, \ldots, \rho$ defined by the following recursive expressions ($c_i > 0, d_i > 0, i = 1, \ldots, \rho, \Gamma > 0$)

$$z_1 = y - y_r \tag{7.114}$$

$$z_i = v_{m,i} - \alpha_{i-1}(y, \bar{\xi}^{(i)}, \bar{v}^{(i)}, \bar{\vartheta}^{(i)}, \bar{y}_r^{(i-1)}) - \vartheta_{1,1} y_r^{(i)}, \quad i = 2, \ldots, \rho \tag{7.115}$$

$$\alpha_1 = -\vartheta_1^T \omega_1 \tag{7.116}$$

$$\alpha_2 = -c_2 z_2 - \vartheta_{2,p+m+1} z_1 - d_2 \left(\frac{\partial \alpha_1}{\partial y} \right)^2 z_2 + \frac{\partial \alpha_1}{\partial y} \left[\xi_{0,2} + \varphi_{0,1}(y) \right] - \vartheta_2^T \omega_2$$

$$+ k_2 v_{m,1} + \frac{\partial \alpha_1}{\partial \xi_0} \left[A_0 \xi_0 + ky + \varphi_0(y) \right] + \sum_{j=1}^{p} \frac{\partial \alpha_1}{\partial \xi_j} \left[A_0 \xi_j + \varphi_j(y) \right]$$

$$+ \sum_{j=0}^{m} \frac{\partial \alpha_1}{\partial v_j} A_0 v_j + \left(\frac{\partial \alpha_1}{\partial \vartheta_1} + \dot{y}_r e_1^T \right) \operatorname{sgn}(b_m) \Gamma \left[\omega_1 - \dot{y}_r e_1 \right] z_1 + \frac{\partial \alpha_1}{\partial y_r} \dot{y}_r \tag{7.117}$$

$$\alpha_i = -c_i z_i - z_{i-1} - d_i \left(\frac{\partial \alpha_{i-1}}{\partial y} \right)^2 z_i + \frac{\partial \alpha_{i-1}}{\partial y} \left[\xi_{0,2} + \varphi_{0,1}(y) \right] - \vartheta_i^T \omega_i$$

$$+ k_i v_{m,1} + \frac{\partial \alpha_{i-1}}{\partial \xi_0} \left[A_0 \xi_0 + ky + \varphi_0(y) \right] + \sum_{j=1}^{p} \frac{\partial \alpha_{i-1}}{\partial \xi_j} \left[A_0 \xi_j + \varphi_j(y) \right]$$

$$+ \sum_{j=0}^{m} \frac{\partial \alpha_{i-1}}{\partial v_j} A_0 v_j + \left(\frac{\partial \alpha_{i-1}}{\partial \vartheta_1} + y_r^{(i-1)} e_1^T \right) \operatorname{sgn}(b_m) \Gamma \left[\omega_1 - \dot{y}_r e_1 \right] z_1$$

$$+ \frac{\partial \alpha_{i-1}}{\partial \vartheta_2} \Gamma \left[\omega_2 + z_1 e_{p+m+1} \right] z_2 + \sum_{j=3}^{i-1} \frac{\partial \alpha_{i-1}}{\partial \vartheta_j} \Gamma \omega_j z_j + \sum_{j=1}^{i-2} \frac{\partial \alpha_{i-1}}{\partial y_r^{(j)}} y_r^{(j+1)}$$

$$i = 3, \ldots, \rho \tag{7.118}$$

$$\omega_1^{\text{T}} = [c_1 z_1 + d_1 z_1 + \xi_{0,2} + \varphi_{0,1}, \varphi_{1,1} + \xi_{1,2}, \ldots, \varphi_{p,1} + \xi_{p,2}, v_{0,2}, \ldots, v_{m-1,2}] \tag{7.119}$$

$$\omega_i^{\text{T}} = -\frac{\partial \alpha_{i-1}}{\partial y} [\varphi_{1,1} + \xi_{1,2}, \ldots, \varphi_{p,1} + \xi_{p,2}, v_{0,2}, \ldots, v_{m-1,2}, v_{m,2}],$$
$$\qquad\qquad\qquad i = 2, \ldots, \rho \tag{7.120}$$

$$\bar{\xi}^{(i)} = [\xi_{0,1}, \ldots, \xi_{0,i}, \ldots, \xi_{p,1}, \ldots, \xi_{p,i}], \quad i = 1, \ldots, \rho-1 \tag{7.121}$$

$$\bar{v}^{(i)} = [v_{0,1}, \ldots, v_{0,i}, \ldots, v_{m-1,1}, \ldots, v_{m-1,i}, v_{m,1}, \ldots, v_{m,i-1}],$$
$$\qquad\qquad\qquad i = 1, \ldots, \rho \tag{7.122}$$

$$\bar{\vartheta}^{(i)} = [\vartheta_1^{\text{T}}, \ldots, \vartheta_i^{\text{T}}], \quad i = 1, \ldots, \rho \tag{7.123}$$

$$\bar{y}_r^{(i)} = [y_r, \dot{y}_r, \ldots, y_r^{(i)}], \quad i = 1, \ldots, \rho \tag{7.124}$$

guarantees global boundedness of $x(t), \xi_0(t) \ldots, \xi_p(t),$ *and* $v_0(t), \ldots, v_m(t)$ *and regulation of the tracking error:*

$$\lim_{t \to \infty} [y(t) - y_r(t)] = 0. \tag{7.125}$$

Proof. *Design procedure.* Using the definition (7.114) of the output error $z_1 = y - y_r$, we write $\dot{z}_1$ as

$$\dot{z}_1 = x_2 + \varphi_{0,1}(y) + \sum_{j=1}^{p} \theta_j \varphi_{j,1}(y) - \dot{y}_r. \tag{7.126}$$

Since x_2 is not measured, we use (7.109) to rewrite (7.126) as

$$\dot{z}_1 = \xi_{0,2} + \varphi_{0,1}(y) + \sum_{j=1}^{p} \theta_j [\varphi_{j,1}(y) + \xi_{j,2}] + \sum_{j=0}^{m} b_j v_{j,2} - \dot{y}_r + \varepsilon_2. \tag{7.127}$$

The choice of virtual control in (7.127) is $v_{m,2}$, because (7.105) reveals that the control u appears in the ρth derivative of $v_{m,2}$, sooner than for any of the other variables in (7.127). If $v_{m,2}$ were the control and the parameters $\theta_1, \ldots, \theta_p, b_m, \ldots, b_0$ were known, then our choice of control law would be

$$v_{m,2} = -\frac{1}{b_m} [c_1 z_1 + d_1 z_1 + \xi_{0,2} + \varphi_{0,1} - \dot{y}_r] - \sum_{j=1}^{p} \frac{\theta_j}{b_m} [\varphi_{j,1} + \xi_{j,2}] - \sum_{j=0}^{m-1} \frac{b_j}{b_m} v_{j,2}. \tag{7.128}$$

To deal with the presence of the unknown coefficient b_m in front of $v_{m,2}$, we add and subtract the right-hand side of (7.128) to the right-hand side of (7.127),

$$\begin{aligned}
\dot{z}_1 &= -c_1 z_1 - d_1 z_1 + \varepsilon_2 + b_m \Bigg\{ \frac{1}{b_m} [c_1 z_1 + d_1 z_1 + \xi_{0,2} + \varphi_{0,1} - \dot{y}_r] \\
&\quad + \sum_{j=1}^{p} \frac{\theta_j}{b_m} [\varphi_{j,1} + \xi_{j,2}] + \sum_{j=0}^{m-1} \frac{b_j}{b_m} v_{j,2} + v_{m,2} \Bigg\} \\
&= -c_1 z_1 - d_1 z_1 + \varepsilon_2 + b_m \left\{ v_{m,2} + \bar{\theta}_0^{\text{T}} \omega_1 - \frac{1}{b_m} \dot{y}_r \right\}, \tag{7.129}
\end{aligned}$$

7.3 ADAPTIVE OBSERVER BACKSTEPPING

where we have used the definitions (7.119)–(7.124) and

$$\bar{\theta}_0^{\mathrm{T}} = \left[\frac{1}{b_m}, \frac{\theta_1}{b_m}, \ldots, \frac{\theta_p}{b_m}, \frac{b_0}{b_m}, \ldots, \frac{b_{m-1}}{b_m}\right]. \tag{7.130}$$

Denoting the estimate of $\bar{\theta}_0$ by ϑ_1, we obtain the stabilizing function (7.116): $\alpha_1 = -\vartheta_1 \omega_1$. Using $z_2 = v_{m,2} - \alpha_1 - \vartheta_{1,1} \dot{y}_r$ as in (7.115), we rewrite (7.129) as

$$\dot{z}_1 = -c_1 z_1 - d_1 z_1 + \varepsilon_2 + b_m z_2 + b_m(\bar{\theta}_0 - \vartheta_1)^{\mathrm{T}} \omega_1 - b_m\left(\frac{1}{b_m} - \vartheta_{1,1}\right)\dot{y}_r. \tag{7.131}$$

Exploiting the fact that the sign of b_m is known, we use the function

$$V_1 = \frac{1}{2}z_1^2 + \frac{|b_m|}{2}(\bar{\theta}_0 - \vartheta_1)^{\mathrm{T}} \Gamma^{-1}(\bar{\theta}_0 - \vartheta_1) + \frac{1}{d_1}\varepsilon^{\mathrm{T}} P_0 \varepsilon, \tag{7.132}$$

where $P_0 A_0 + A_0^{\mathrm{T}} P_0 = -I$. The update law (7.111) is then chosen to eliminate the $(\bar{\theta}_0 - \vartheta_1)$-term from the derivative of V_1:

$$\dot{V}_1 \leq -c_1 z_1^2 - \frac{3}{4d_1}\varepsilon^{\mathrm{T}} \varepsilon + b_m z_1 z_2. \tag{7.133}$$

Now the derivative of $z_2 = v_{m,2} - \alpha_1 - \vartheta_{1,1}\dot{y}_r$ is

$$\dot{z}_2 = v_{m,3} - k_2 v_{m,1} - \frac{\partial \alpha_1}{\partial y}[\xi_{0,2} + \varphi_{0,1}(y) + \varepsilon_2] + \bar{\theta}^{\mathrm{T}} \omega_2$$
$$- \frac{\partial \alpha_1}{\partial \xi_0}[A_0 \xi_0 + ky + \varphi_0(y)] - \sum_{j=1}^{p} \frac{\partial \alpha_1}{\partial \xi_j}[A_0 \xi_j + \varphi_j(y)] - \sum_{j=0}^{m} \frac{\partial \alpha_1}{\partial v_j} A_0 v_j$$
$$- \left(\frac{\partial \alpha_1}{\partial \vartheta_1} + \dot{y}_r e_1^{\mathrm{T}}\right)\mathrm{sgn}(b_m)\Gamma[\omega_1 - \dot{y}_r e_1]z_1 - \frac{\partial \alpha_1}{\partial y_r}\dot{y}_r - \vartheta_{1,1}\ddot{y}_r, \tag{7.134}$$

where for the last equality we used the definition (7.120) for ω_2 and

$$\bar{\theta}^{\mathrm{T}} = [\theta_1, \ldots, \theta_p, b_0, \ldots, b_m]. \tag{7.135}$$

Using the estimate ϑ_2 of $\bar{\theta}$, the stabilizing function α_2 defined in (7.117), and the definition $z_3 = v_{m,3} - \alpha_2 - \vartheta_{1,1}\ddot{y}_r$ from (7.115), we rewrite (7.134) as

$$\dot{z}_2 = -c_2 z_2 - b_m z_1 + z_3 - d_2\left(\frac{\partial \alpha_1}{\partial y}\right)^2 z_2 - \frac{\partial \alpha_1}{\partial y}\varepsilon_2 + (\bar{\theta} - \vartheta_2)^{\mathrm{T}}[\omega_2 + z_1 e_{p+m+1}]. \tag{7.136}$$

The derivative of the nonnegative function

$$V_2 = V_1 + \frac{1}{2}z_2^2 + \frac{1}{2}(\bar{\theta} - \vartheta_2)^{\mathrm{T}}\Gamma^{-1}(\bar{\theta} - \vartheta_2) + \frac{1}{d_2}\varepsilon^{\mathrm{T}} P_0 \varepsilon \tag{7.137}$$

is computed using (7.133), (7.136), and the update law (7.112):

$$\dot{V}_2 \leq -c_1 z_1^2 - c_2 z_2^2 - \frac{3}{4}\left(\frac{1}{d_1}+\frac{1}{d_2}\right)\varepsilon^T\varepsilon + z_2 z_3. \tag{7.138}$$

From here on the design proceeds along the lines of the example of Section 7.3.1. At the final step, the derivative of the nonnegative function V_ρ,

$$V_\rho = \sum_{i=1}^{\rho}\left[\frac{1}{2}z_i^2 + \frac{1}{d_i}\varepsilon^T P_0 \varepsilon\right] + \frac{|b_m|}{2}(\bar{\theta}_0 - \vartheta_1)^T \Gamma^{-1}(\bar{\theta}_0 - \vartheta_1)$$
$$+ \sum_{i=2}^{\rho}\frac{1}{2}(\bar{\theta}-\vartheta_i)^T \Gamma^{-1}(\bar{\theta}-\vartheta_i), \tag{7.139}$$

is rendered nonpositive:

$$\dot{V}_\rho \leq -\sum_{i=1}^{\rho}\left[c_i z_i^2 + \frac{3}{4d_i}\varepsilon^T\varepsilon\right]. \tag{7.140}$$

Stability and convergence. Due to the piecewise continuity of $y_r^{(\rho)}(t)$ and the smoothness of the nonlinearities in (7.101), the solution of the closed-loop adaptive system exists. Let its maximum interval of existence be $[0, t_f)$. On this interval, the nonnegative function V_ρ is nonincreasing because of (7.140). Thus, $z_1, \ldots, z_\rho$, $\tilde{\vartheta}_1, \ldots, \tilde{\vartheta}_\rho$, and hence $\vartheta_1, \ldots \vartheta_\rho$, are bounded on $[0, t_f)$ by constants depending only on the initial conditions of the adaptive system. In addition, (7.108) shows that ε is bounded.

The boundedness of all other signals on $[0, t_f)$ is established as follows. Since z_1 and y_r are bounded, it follows that y is bounded. This implies that $\beta(y)$ is bounded away from zero and, from (7.105), that $\xi_0, \ldots, \xi_p$ are bounded. From (7.105) we also see that

$$v_{m-j,i} = \left[e_i^T (sI - A_0)^{-1} e_{\rho+j}\right]\beta(y)u, \quad 0 \leq j \leq m, \tag{7.141}$$

where e_i is the ith coordinate vector in $\mathbb{R}^n$. Then, we express (7.101) in the differential equation form ($D = d/dt$)

$$D^n y = \sum_{i=1}^{n} D^{n-i}\left[\varphi_{0,i}(y) + \sum_{j=1}^{p}\theta_j \varphi_{j,i}(y)\right] + \sum_{i=0}^{m} b_i D^i \left[\beta(y)u\right]. \tag{7.142}$$

Since y is bounded and, by Assumption 7.3, the polynomial $B(s) = b_m s^m + \cdots + b_1 s + b_0$ is Hurwitz, we conclude from (7.142) that $H_\rho(s)[\beta(y)u]$ is bounded, where $H_i(s)$ denotes any exponentially stable transfer function of relative degree greater than or equal to i. By (7.141), this in turn implies that the vectors $\bar{v}_{m-j}^{(j+1)}$, $0 \leq j \leq m$, are bounded, where

$$\bar{v}_j^{(i)} = [v_{j,1}, \ldots, v_{j,i}]. \tag{7.143}$$

7.3 ADAPTIVE OBSERVER BACKSTEPPING

In particular, by (7.122), this implies that $\bar{v}^{(2)}$ is bounded. Combining (7.119) and (7.116) we conclude that ω_1 and α_1 are bounded, which in turn implies that $v_{m,2} = z_2 + \alpha_1 + \vartheta_{1,1}\dot{y}_r$ is bounded. Hence, by (7.141), $H_{\rho-1}(s)[\beta(y)u]$, and thus $\bar{v}_{m-j}^{(j+2)}$, $0 \leq j \leq m$, are bounded. This in turn implies that ω_2, α_2 and $v_{m,3}$ are bounded. Continuing in the same fashion, we use (7.115), and (7.141) to show that $H_i(s)[\beta(y)u]$, $\rho - 2 \geq i \geq 1$, are bounded, which implies that v is bounded. Since $\beta(y)$ is bounded away from zero, we conclude from (7.110) that u is bounded. Furthermore, from (7.108) we see that x is bounded.

We have thus shown that all of the signals in the closed-loop adaptive system are bounded on $[0, t_f)$ by constants depending only on initial conditions. Hence, $t_f = \infty$. The convergence of the tracking error to zero can now be deduced from the LaSalle-Yoshizawa theorem (Theorem 2.1), since $z_1, \ldots, z_\rho$ and ε converge to zero as $t \to \infty$. □

7.3.3 Example: single-link flexible robot

As an example, we consider a single-link robotic manipulator coupled to a dc motor with a nonrigid joint. When the joint is modeled as a linear torsional spring, the dynamic equations of the system are

$$J_1\ddot{q}_1 + F_1\dot{q}_1 + K\left(q_1 - \frac{q_2}{N}\right) + mgd\cos q_1 = 0$$
$$J_2\ddot{q}_2 + F_2\dot{q}_2 - \frac{K}{N}\left(q_1 - \frac{q_2}{N}\right) = K_t i \qquad (7.144)$$
$$L\dot{i} + Ri + K_b\dot{q}_2 = u,$$

where q_1 and q_2 are the angular positions of the link and the motor shaft, i is the armature current, and u is the armature voltage. The inertias J_1, J_2, the viscous friction constants F_1, F_2, the spring constant K, the torque constant K_t, the back-emf constant K_b, the armature resistance R and inductance L, the link mass M, the position of the link's center of gravity d, the gear ratio N and the acceleration of gravity g can all be unknown.

We assume that only the link position q_1 is measured so that $n = \rho = 5$ ($m = 0$). Let us examine whether our design procedure is applicable to this system. We first try the natural choice of state variables $\zeta_1 = q_1$, $\zeta_2 = \dot{q}_1$, $\zeta_3 = q_2$, $\zeta_4 = \dot{q}_2$, $\zeta_5 = i$. The dynamic equations (7.144) become

$$\dot{\zeta}_1 = \zeta_2$$
$$\dot{\zeta}_2 = -\frac{mgd}{J_1}\cos y - \frac{F_1}{J_1}\zeta_2 - \frac{K}{J_1}\left(\zeta_1 - \frac{\zeta_3}{N}\right)$$
$$\dot{\zeta}_3 = \zeta_4 \qquad (7.145)$$
$$\dot{\zeta}_4 = \frac{K}{J_2 N}\left(\zeta_1 - \frac{\zeta_3}{N}\right) - \frac{F_2}{J_2}\zeta_4 + \frac{K_t}{J_2}\zeta_5$$

$$\dot{\zeta}_5 = -\frac{R}{L}\zeta_5 - \frac{K_b}{L}\zeta_4 + \frac{1}{L}u$$
$$y = \zeta_1.$$

Clearly, (7.145) is not in the output-feedback form (7.101). However, there exists a different choice of coordinates which brings (7.144) into that form. To show this, we derive an input-output description of (7.145). Differentiating y twice, we obtain $\zeta_2 = Dy$ ($D = \frac{d}{dt}$ is the differentiation operator) and

$$D^2 y = -\frac{mgd}{J_1}\cos y - \frac{F_1}{J_1} Dy - \frac{K}{J_1}\left(y - \frac{\zeta_3}{N}\right), \qquad (7.146)$$

which implies that

$$\zeta_3 = \frac{J_1 N}{K}\left(D^2 y + \frac{mgd}{J_1}\cos y + \frac{F_1}{J_1} Dy + \frac{K}{J_1} y\right) \qquad (7.147)$$

$$\zeta_4 = D\zeta_3 = \frac{J_1 N}{K}\left(D^3 y + \frac{mgd}{J_1} D\cos y + \frac{F_1}{J_1} D^2 y + \frac{K}{J_1} Dy\right). \qquad (7.148)$$

Differentiating (7.148) and substituting ζ_3 and ζ_4 from (7.147) and (7.148), we obtain

$$\zeta_5 = \frac{J_1 J_2 N}{K_t K}\left[D^4 y + \left(\frac{F_1}{J_1} + \frac{F_2}{J_2}\right) D^3 y + \left(\frac{K}{J_1} + \frac{K}{J_2 N^2} + \frac{F_1 F_2}{J_1 J_2}\right) D^2 y\right.$$
$$+ \frac{mgd}{J_1} D^2 \cos y + \left(\frac{F_1 K}{J_1 J_2 N^2} + \frac{F_2 K}{J_1 J_2}\right) Dy$$
$$\left. + \frac{mgd F_2}{J_1 J_2} D \cos y + \frac{mgd K}{J_1 J_2 N^2} \cos y\right]. \qquad (7.149)$$

Finally, differentiating (7.149) and substituting ζ_4 and ζ_5 from (7.147) and (7.148), we arrive at the input-output description of (7.144):

$$D^5 y = \frac{K_t K}{J_1 J_2 N L} u - \left(\frac{R}{L} + \frac{F_1}{J_1} + \frac{F_2}{J_2}\right) D^4 y - \frac{mgd}{J_1} D^3 \cos y$$
$$- \left[\frac{R}{L}\left(\frac{F_1}{J_1} + \frac{F_2}{J_2}\right) + \frac{K_b K_t}{L J_2} + \left(\frac{K}{J_1} + \frac{K}{J_2 N^2} + \frac{F_1 F_2}{J_1 J_2}\right)\right] D^3 y$$
$$- \left[\frac{R}{L}\left(\frac{K}{J_1} + \frac{K}{J_2 N^2} + \frac{F_1 F_2}{J_1 J_2}\right) + \frac{F_1 K}{J_1 J_2 N^2} + \frac{F_2 K}{J_1 J_2} + \frac{K_b}{L}\frac{F_1 K_t}{J_1 J_2}\right] D^2 y$$
$$- \left(\frac{R}{L} + \frac{F_2}{J_2}\right)\frac{mgd}{J_1} D^2 \cos y - \left[\frac{R}{L}\left(\frac{F_1 K}{J_1 J_2 N^2} + \frac{F_2 K}{J_1 J_2}\right) + \frac{K_b}{L}\frac{K_t K}{J_1 J_2}\right] Dy$$
$$- \left(\frac{K}{N^2} + \frac{R F_2}{L} + \frac{K_b K_t}{L}\right)\frac{mgd}{J_1 J_2} D \cos y - \frac{R}{L}\frac{mgd K}{J_1 J_2 N^2} \cos y. \qquad (7.150)$$

Using (7.150), it is tedious but straightforward to find a choice of state variables which bring (7.144) into the form (7.101),

$$\begin{aligned}
\dot{x}_1 &= x_2 + \theta_1 y \\
\dot{x}_2 &= x_3 + \theta_2 y + \theta_3 \cos y \\
\dot{x}_3 &= x_4 + \theta_4 y + \theta_5 \cos y \\
\dot{x}_4 &= x_5 + \theta_6 y + \theta_7 \cos y \\
\dot{x}_5 &= \theta_8 \cos y + b_0 u \\
y &= x_1,
\end{aligned} \quad (7.151)$$

where the unknown parameters $\theta_1, \ldots, \theta_8, b_0$ are defined as

$$\begin{aligned}
\theta_1 &= -\left(\frac{R}{L} + \frac{F_1}{J_1} + \frac{F_2}{J_2}\right) \\
\theta_2 &= -\left[\frac{R}{L}\left(\frac{F_1}{J_1} + \frac{F_2}{J_2}\right) + \frac{K_b}{L}\frac{K_t}{J_2} + \left(\frac{K}{J_1} + \frac{K}{J_2 N^2} + \frac{F_1 F_2}{J_1 J_2}\right)\right] \\
\theta_3 &= -\frac{mgd}{J_1} \\
\theta_4 &= -\left[\frac{R}{L}\left(\frac{K}{J_1} + \frac{K}{J_2 N^2} + \frac{F_1 F_2}{J_1 J_2}\right) + \frac{F_1 K}{J_1 J_2 N^2} + \frac{F_2 K}{J_1 J_2} + \frac{K_b}{L}\frac{F_1 K_t}{J_1 J_2}\right] \\
\theta_5 &= -\left(\frac{R}{L} + \frac{F_2}{J_2}\right)\frac{mgd}{J_1} \\
\theta_6 &= -\left[\frac{R}{L}\left(\frac{F_1 K}{J_1 J_2 N^2} + \frac{F_2 K}{J_1 J_2}\right) + \frac{K_b}{L}\frac{K_t K}{J_1 J_2}\right] \\
\theta_7 &= -\left(\frac{K}{N^2} + \frac{RF_2}{L} + \frac{K_b K_t}{L}\right)\frac{mgd}{J_1 J_2} \\
\theta_8 &= -\frac{R}{L}\frac{mgdK}{J_1 J_2 N^2} \\
b_0 &= \frac{K_t K}{J_1 J_2 NL} > 0.
\end{aligned} \quad (7.152)$$

Hence, the design procedure of Theorem 7.6 is applicable to (7.151) and yields an adaptive controller that achieves bounded asymptotic position tracking from all initial conditions and for all positive values of the constants K, K_t, K_b, J_1, J_2, R, and L.

7.4 Extensions

7.4.1 Interlaced controller-observer design

In the observer-backstepping procedures presented in this chapter the observer design is independent of the controller. We first design a state observer (or filters when uncertainties are present) and then use nonlinear damping in the controller design to counteract the effect of observation errors ($\tilde{x}$ or ε). In particular, the quadratic nonlinear damping terms dominate the cross-terms containing observation errors in the derivative of the Lyapunov function, thereby rendering it nonpositive.

In this subsection we present an alternative approach, which introduces *correction terms* in the observer in order to cancel (instead of dominating) the cross-terms in the Lyapunov derivative. As a result, the design of the observer is *interlaced* with that of the controller. Inspired by the tuning functions of Chapter 4, this approach introduces *interlacing functions* at each intermediate step of the design procedure. These functions are analogs of the tuning functions for state estimation. At the final step, the last interlacing function becomes the correction term for the observer.

As an example, consider the system

$$\begin{aligned}\dot{x}_1 &= x_1^3 + (1+x_1^2)x_2 \\ \dot{x}_2 &= -x_2 + u,\end{aligned} \qquad (7.153)$$

where x_1 is measured but x_2 is not. The observer for x_2 includes the correction term χ yet to be designed:

$$\dot{\hat{x}}_2 = -\hat{x}_2 + u + \chi. \qquad (7.154)$$

If we were to choose $\chi \equiv 0$, we would need nonlinear damping terms to guarantee stability. Here we avoid those terms by designing χ to cancel all indefinite terms in the derivative of the Lyapunov function. The equation for the observer error is

$$\dot{\tilde{x}}_2 = -\tilde{x}_2 - \chi. \qquad (7.155)$$

Instead of applying backstepping to the system (7.153), we apply it to the system

$$\begin{aligned}\dot{x}_1 &= x_1^3 + (1+x_1^2)\hat{x}_2 + (1+x_1^2)\tilde{x}_2 & (7.156\text{a}) \\ \dot{\hat{x}}_2 &= -\hat{x}_2 + u + \chi. & (7.156\text{b})\end{aligned}$$

This form suggests how to employ the tuning functions technique. While the state estimation error $\tilde{x}_2$, which appears linearly in (7.156a), plays the same role as the parameter estimation error $\tilde{\theta}$ in the tuning functions design, the observer correction term $\chi = -\dot{\tilde{x}}_2 - \tilde{x}_2$ should be viewed as $\dot{\hat{\theta}} = -\dot{\tilde{\theta}} = \Gamma\tau_\rho$.

Using $z = \hat{x}_2 + x_1$ and (7.155), we compute the derivative of $V_1 = \frac{1}{2}x_1^2 + \frac{1}{2d_0}\tilde{x}_2^2$:

$$\begin{aligned}\dot{V}_1 &= x_1(1+x_1^2)z + x_1\left[-(1+x_1^2)x_1 + x_1^3\right] - \frac{1}{d_0}\tilde{x}_2^2 + \tilde{x}_2\left[x_1(1+x_1^2) - \frac{1}{d_0}\chi\right] \\ &= x_1(1+x_1^2)z - x_1^2 - \frac{1}{d_0}\tilde{x}_2^2 + \tilde{x}_2\left[x_1(1+x_1^2) - \frac{1}{d_0}\chi\right]. \quad (7.157)\end{aligned}$$

7.4 EXTENSIONS

If this were the last step of the design procedure, we would eliminate $\tilde{x}_2$ from (7.157) with the choice $\chi = d_0 \iota_1$, where

$$\iota_1 = x_1(1 + x_1^2). \tag{7.158}$$

Since we have one more design step to go, we do *not* use $\chi = d_0\iota_1$ as our correction term. Instead, we retain ι_1 as our first *interlacing function* and rewrite (7.157) as

$$\dot{V}_1 = -x_1^2 + x_1(1+x_1^2)z - \frac{1}{d_0}\tilde{x}_2^2 + \tilde{x}_2\left[\iota_1 - \frac{1}{d_0}\chi\right]. \tag{7.159}$$

The derivative of $z = \hat{x}_2 + x_1$ is

$$\dot{z} = -\hat{x}_2 + u + \chi + \left[x_1^3 + (1+x_1^2)\hat{x}_2\right] + (1+x_1^2)\tilde{x}_2. \tag{7.160}$$

Since the actual control u appears in (7.160), we design u and the *actual correcting term* χ to stabilize the $(x_1, z, \tilde{x}_2)$-system with respect to the augmented Lyapunov function $V_2 = V_1 + \frac{1}{2}z^2$. Using (7.159) and (7.160), the derivative of V_2 is computed as

$$\begin{aligned}\dot{V}_2 &= -x_1^2 - \frac{1}{d_0}\tilde{x}_2^2 + z\left\{x_1(1+x_1^2) - \hat{x}_2 + u + \kappa + \left[x_1^3 + (1+x_1^2)\hat{x}_2\right]\right\} \\ &\quad + \tilde{x}_2\left[\zeta_1 + z(1+x_1^2) - \frac{1}{d_0}\kappa\right]. \end{aligned} \tag{7.161}$$

To eliminate $\tilde{x}_2$ from (7.161), the correction term χ is chosen as

$$\chi = d_0\iota_2 = d_0\iota_1 + d_0 z(1+x_1^2) = d_0(2x_1 + \hat{x}_2)(1+x_1^2). \tag{7.162}$$

Then, the choice

$$u = -z - x_1(1+x_1^2) + \hat{x}_2 - \zeta_2 - \left[x_1^3 + (1+x_1^2)\hat{x}_2\right] \tag{7.163}$$

yields $\dot{V}_2 = -x_1^2 - z^2 - \frac{1}{d_0}\tilde{x}_2^2$, which implies that the $(x_1, z, \tilde{x}_2)$-system (or, equivalently, the $(x_1, x_2, \hat{x}_2)$-system) has the origin as a globally exponentially stable equilibrium.

This interlaced design can be generalized to output-feedback nonlinear systems of the form (7.27) as follows:

Corollary 7.7 (Output-Feedback Systems) *For the nonlinear system (7.27), assume that $b_m s^m + \cdots + b_1 s + b_0$ is a Hurwitz polynomial, and that $y_r, \dot{y}_r, \ldots, y_r^{(\rho)}$ are known and bounded on $[0, \infty)$ and $y_r^{(\rho)}(t)$ is piecewise continuous. Then, the control*

$$u = \frac{1}{b_m \beta(y)}\left[\alpha_\rho - \hat{x}_{\rho+1} - y_r^{(\rho)}\right], \tag{7.164}$$

and the observer
$$\dot{\hat{x}} = A\hat{x} + k(y - \hat{y}) + \varphi(y) + b\,\beta(y)u + D\iota_\rho \qquad (7.165)$$
$$\hat{y} = c^T\hat{x},$$

with $z_i, \alpha_i, \iota_i, i = 1, \ldots, \rho$ defined by the following recursive expressions ($P_0 = P_0^T > 0$, $P_0 A_0 + A_0^T P_0 = -2I$, $d_0 > 0$, $c_i > 0$, $i = 1, \ldots, \rho$)

$$z_1 = y - y_r \qquad (7.166)$$
$$z_i = \hat{x}_i - \alpha_{i-1}(y, \hat{x}_1, \ldots, \hat{x}_{i-1}, y_r, \ldots, y_r^{(i-2)}) - y_r^{(i-1)}, \quad i = 2, \ldots, \rho \qquad (7.167)$$
$$\alpha_1 = -c_1 z_1 - \varphi_1(y) \qquad (7.168)$$
$$\alpha_i = -c_i z_i - z_{i-1} - k_i(y - \hat{x}_1) - \varphi_i(y) - e_i^T d_0 \iota_i + \frac{\partial \alpha_{i-1}}{\partial y}[\hat{x}_2 + \varphi_1(y)]$$
$$+ \sum_{j=1}^{i-1} \frac{\partial \alpha_{i-1}}{\partial \hat{x}_j}\left[\hat{x}_{j+1} + k_j(y - \hat{x}_1) + \varphi_j(y) + e_j^T d_0 \iota_i\right] + \sum_{j=1}^{i-2} \frac{\partial \alpha_{i-1}}{\partial y_r^{(j)}} y_r^{(j+1)}$$
$$+ d_0 P_0^{-1} e_2 \sum_{k=2}^{i-1}\left(e_k - \sum_{j=1}^{k-1} \frac{\partial \alpha_{k-1}}{\partial \hat{x}_j} e_j\right)^T \frac{\partial \alpha_{i-1}}{\partial y} z_k, \quad i = 2, \ldots, \rho \qquad (7.169)$$
$$\iota_1 = P_0^{-1} e_2 z_1 \qquad (7.170)$$
$$\iota_i = \iota_{i-1} - P_0^{-1} e_2 \frac{\partial \alpha_{i-1}}{\partial y} z_i, \quad i = 2, \ldots, \rho \qquad (7.171)$$

guarantees global boundedness of $x(t)$, $\hat{x}(t)$, and regulation of the tracking error:
$$\lim_{t \to \infty}[y(t) - y_r(t)] = 0. \qquad (7.172)$$

Proof. From (7.27), (7.165), and the definitions (7.164)–(7.171), we can express the derivatives of the error variables $z_1, \ldots, z_\rho$ as follows:

$$\dot{z}_1 = \dot{y} - \dot{y}_r = x_2 + \varphi_1(y) - \dot{y}_r = \hat{x}_2 + \tilde{x}_2 + \varphi_1(y) - \dot{y}_r$$
$$= z_2 + \alpha_1 + \varphi_1(y) + \tilde{x}_2 = -c_1 z_1 + z_2 + \tilde{x}_2 \qquad (7.173)$$
$$\dot{z}_i = \hat{x}_{i+1} + k_i(y - \hat{x}_1) + e_i^T d_0 \iota_\rho + \varphi_i(y) - \frac{\partial \alpha_{i-1}}{\partial y}[\hat{x}_2 + \varphi_1(y) + \tilde{x}_2]$$
$$- \sum_{j=1}^{i-1} \frac{\partial \alpha_{i-1}}{\partial \hat{x}_j}\left[\hat{x}_{j+1} + k_j(y - \hat{x}_1) + \varphi_j(y) + e_j^T d_0 \iota_\rho\right]$$
$$- \sum_{j=1}^{i-2} \frac{\partial \alpha_{i-1}}{\partial y_r^{(j)}} y_r^{(j+1)} - y_r^{(i)}$$
$$= -c_i z_i - z_{i-1} + z_{i+1} - \frac{\partial \alpha_{i-1}}{\partial y}\tilde{x}_2 + d_0\left(e_i - \sum_{j=1}^{i-1} \frac{\partial \alpha_{i-1}}{\partial \hat{x}_j} e_j\right)^T (\iota_\rho - \iota_i)$$
$$+ d_0 P_0^{-1} e_2 \sum_{k=2}^{i-1}\left(e_k - \sum_{j=1}^{k-1} \frac{\partial \alpha_{k-1}}{\partial \hat{x}_j} e_j\right)^T \frac{\partial \alpha_{i-1}}{\partial y} z_k, \quad i = 2, \ldots, \rho \qquad (7.174)$$

7.4 EXTENSIONS

$$\begin{aligned}
\dot{z}_\rho &= \hat{x}_{\rho+1} + b_m\beta(y)u + k_\rho(y - \hat{x}_1) + \varphi_\rho(y) + e_\rho^T d_0 \iota_\rho \\
&\quad - \frac{\partial \alpha_{\rho-1}}{\partial y}[\hat{x}_2 + \varphi_1(y) + \tilde{x}_2] - \sum_{j=1}^{\rho-2} \frac{\partial \alpha_{\rho-1}}{\partial y_r^{(j)}} y_r^{(j+1)} - y_r^{(\rho)} \\
&\quad - \sum_{j=1}^{\rho-1} \frac{\partial \alpha_{\rho-1}}{\partial \hat{x}_j}\left[\hat{x}_{j+1} + k_j(y - \hat{x}_1) + \varphi_j(y) + e_j^T d_0 \iota_p\right] \\
&= b_m \beta(y) u + \hat{x}_{\rho+1} - \alpha_\rho - c_\rho z_\rho - z_{\rho-1} - \frac{\partial \alpha_{\rho-1}}{\partial y}\tilde{x}_2 - y_r^{(\rho)} \\
&\quad + d_0 P_0^{-1} e_2 \sum_{k=2}^{\rho-1}\left(e_k - \sum_{j=1}^{k-1} \frac{\partial \alpha_{k-1}}{\partial \hat{x}_j}e_j\right)^T \frac{\partial \alpha_{\rho-1}}{\partial y} z_k \\
&= -c_\rho z_\rho - z_{\rho-1} - \frac{\partial \alpha_{\rho-1}}{\partial y}\tilde{x}_2 + d_0 P_0^{-1} e_2 \sum_{k=2}^{\rho-1}\left(e_k - \sum_{j=1}^{k-1} \frac{\partial \alpha_{k-1}}{\partial \hat{x}_j}e_j\right)^T \frac{\partial \alpha_{\rho-1}}{\partial y} z_k.
\end{aligned} \tag{7.175}$$

The resulting error system is

$$\dot{z}_1 = -c_1 z_1 + z_2 + \tilde{x}_2$$
$$\vdots$$
$$\begin{aligned}
\dot{z}_i &= -c_i z_i - z_{i-1} + z_{i+1} - \frac{\partial \alpha_{i-1}}{\partial y}\tilde{x}_2 \\
&\quad - d_0 P_0^{-1} e_2 \left(e_i - \sum_{j=1}^{i-1} \frac{\partial \alpha_{i-1}}{\partial \hat{x}_j} e_j\right)^T \sum_{\ell=i+1}^{\rho} \frac{\partial \alpha_{\ell-1}}{\partial y} z_\ell \\
&\quad + d_0 P_0^{-1} e_2 \sum_{k=2}^{i-1}\left(e_k - \sum_{j=1}^{k-1} \frac{\partial \alpha_{k-1}}{\partial \hat{x}_j} e_j\right)^T \frac{\partial \alpha_{i-1}}{\partial y} z_k
\end{aligned} \tag{7.176}$$
$$\vdots$$
$$\dot{z}_\rho = -c_\rho z_\rho - z_{\rho-1} - \frac{\partial \alpha_{\rho-1}}{\partial y}\tilde{x}_2 + d_0 P_0^{-1} e_2 \sum_{k=2}^{\rho-1}\left(e_k - \sum_{j=1}^{k-1}\frac{\partial \alpha_{k-1}}{\partial \hat{x}_j} e_j\right)^T \frac{\partial \alpha_{\rho-1}}{\partial y} z_k$$
$$\dot{\tilde{x}} = A_0 \tilde{x} + d_0 P_0^{-1} e_2 \left[z_1 - \sum_{j=2}^{\rho} \frac{\partial \alpha_{j-1}}{\partial y} z_j\right].$$

For this error system, consider the Lyapunov function

$$V(z, \tilde{x}) = \frac{1}{2}\sum_{j=1}^{\rho} z_j^2 + \frac{1}{2d_0}\tilde{x}^T P_0 \tilde{x}. \tag{7.177}$$

Its derivative along the solutions of (7.176) is nonpositive:

$$\begin{aligned}
\dot{V} &= -\sum_{j=1}^{\rho} c_j z_j^2 - \frac{1}{d_0}|\tilde{x}|^2 + \left[z_1 - \sum_{j=2}^{\rho} \frac{\partial \alpha_{j-1}}{\partial y} z_j\right] \tilde{x}_2 \\
&\quad - \frac{1}{d_0} \tilde{x}^{\mathrm{T}} P_0 d_0 P_0^{-1} e_2 \left[z_1 - \sum_{j=2}^{\rho} \frac{\partial \alpha_{j-1}}{\partial y} z_j\right] \\
&\quad - \sum_{i=2}^{\rho-1} d_0 P_0^{-1} e_2 \left(e_i - \sum_{j=1}^{i-1} \frac{\partial \alpha_{i-1}}{\partial \hat{x}_j} e_j\right)^{\mathrm{T}} \sum_{\ell=i+1}^{\rho} \frac{\partial \alpha_{\ell-1}}{\partial y} z_\ell \\
&\quad + \sum_{i=3}^{\rho} d_0 P_0^{-1} e_2 \sum_{k=2}^{i-1} \left(e_k - \sum_{j=1}^{k-1} \frac{\partial \alpha_{k-1}}{\partial \hat{x}_j} e_j\right)^{\mathrm{T}} \frac{\partial \alpha_{i-1}}{\partial y} z_k \\
&= -\sum_{j=1}^{\rho} c_j z_j^2 - \frac{1}{d_0}|\tilde{x}|^2 - d_0 P_0^{-1} e_2 \sum_{\ell=3}^{\rho} \sum_{i=2}^{\ell-1} \left(e_i - \sum_{j=1}^{i-1} \frac{\partial \alpha_{i-1}}{\partial \hat{x}_j} e_j\right)^{\mathrm{T}} \frac{\partial \alpha_{\ell-1}}{\partial y} z_\ell \\
&\quad + d_0 P_0^{-1} e_2 \sum_{i=3}^{\rho} \sum_{k=2}^{i-1} \left(e_k - \sum_{j=1}^{k-1} \frac{\partial \alpha_{k-1}}{\partial \hat{x}_j} e_j\right)^{\mathrm{T}} \frac{\partial \alpha_{i-1}}{\partial y} z_k \\
&= -\sum_{j=1}^{\rho} c_j z_j^2 - \frac{1}{d_0}|\tilde{x}|^2 .
\end{aligned} \qquad (7.178)$$

Thus, $z_1, \ldots, z_\rho, \tilde{x}$ are bounded and converge to zero. The remainder of the proof is similar to that of Theorem 7.1. □

7.4.2 Design with partial-state feedback

All the nonadaptive and adaptive schemes we have presented so far, as well as those to be presented in the remainder of the book, assume either that the full state is measured, or that only one scalar output is available for feedback. In this section we present a natural generalization of these results which bridges the gap between full-state and single-output feedback. Combining the design tools we have developed for these two cases, we can easily incorporate information available from additional sensors and accommodate systems whose nonlinearities depend on measured state variables only. Such designs are applicable to systems in *partial-state-feedback form* which have k groups of *measured* state variables, denoted by $x_1, \ldots, x_{m_1}$ and $x_{n_i+1}, \ldots, x_{m_{i+1}}$, $i = 1, \ldots, k-1$, and k groups of *unmeasured* variables, denoted by $x_{m_i+1}, \ldots, x_{n_i}$, $i = 1, \ldots k$, where $n_k = n$ and $m_k \leq \rho$. For notational convenience, we adopt the following definitions of the vectors of measured variables for $1 \leq i \leq k$:

$$x^{m_i} = (x_1, \ldots, x_{m_1}, x_{n_1+1}, \ldots, x_{m_2}, \ldots, x_{n_{i-1}+1}, \ldots, x_{m_i}), \quad x^m = x^{m_k}. \qquad (7.179)$$

As we can see, the nonlinearities in (7.180) are strict-feedback nonlinearities, since they depend only on measured state variables which are fed back: In the

7.4 EXTENSIONS

$\dot{x}_i$-equation, the nonlinearities depend only on measured state variables up to x_i. An example of such a system is[3]

$$\dot{x}_1 = x_2 + \varphi_{0,1}(x_1) + \sum_{j=1}^{p} \theta_j \varphi_{j,1}(x_1)$$

$$\dot{x}_2 = x_3 + \varphi_{0,2}(x_1, x_2) + \sum_{j=1}^{p} \theta_j \varphi_{j,2}(x_1, x_2)$$

$$\vdots$$

$$\dot{x}_{m_1} = x_{m_1+1} + \varphi_{0,m_1}(x^{m_1}) + \sum_{j=1}^{p} \theta_j \varphi_{j,m_1}(x^{m_1})$$

$$\vdots$$

$$\dot{x}_{n_1} = x_{n_1+1} + \varphi_{0,n_1}(x^{m_1}) + \sum_{j=1}^{p} \theta_j \varphi_{j,n_1}(x^{m_1})$$

$$\dot{x}_{n_1+1} = x_{n_1+2} + \varphi_{0,n_1+1}(x^{m_1}, x_{n_1+1}) + \sum_{j=1}^{p} \theta_j \varphi_{j,n_1+1}(x^{m_1}, x_{n_1+1})$$

$$\vdots$$

$$\dot{x}_{m_2} = x_{m_2+1} + \varphi_{0,m_2}(x^{m_2}) + \sum_{j=1}^{p} \theta_j \varphi_{j,m_2}(x^{m_2}) \qquad (7.180)$$

$$\vdots$$

$$\dot{x}_{n_2} = x_{n_2+1} + \varphi_{0,n_2}(x^{m_2}) + \sum_{j=1}^{p} \theta_j \varphi_{j,n_2}(x^{m_2})$$

$$\vdots$$

$$\dot{x}_{m_k} = x_{m_k+1} + \varphi_{0,m_k}(x^m) + \sum_{j=1}^{p} \theta_j \varphi_{j,m_k}(x^m)$$

$$\vdots$$

$$\dot{x}_{\rho-1} = x_\rho + \varphi_{0,\rho-1}(x^m) + \sum_{j=1}^{p} \theta_j \varphi_{j,\rho-1}(x^m)$$

$$\dot{x}_\rho = x_{\rho+1} + \varphi_{0,\rho}(x^m) + \sum_{j=1}^{p} \theta_j \varphi_{j,\rho}(x^m) + b_{n-\rho}\beta(x^m)u$$

$$\vdots$$

$$\dot{x}_n = \varphi_{0,n}(x^m) + \sum_{j=1}^{p} \theta_j \varphi_{j,n}(x^m) + b_0 \beta(x^m) u$$

$$y = x_1.$$

---

[3]Differential geometric characterizations of partial-state-feedback systems are given in Appendix G, Theorems G.2 and G.5.

In (7.180), $x \in \mathbb{R}^n$ is the state, $u \in \mathbb{R}$ is the input, $y \in \mathbb{R}$ is the output, $\varphi_{j,i}$, $0 \leq j \leq p$, $1 \leq i \leq n$, and β are smooth nonlinear functions, and $\theta = [\theta_1, \ldots, \theta_p]^T \in \mathbb{R}^p$ and $b = [b_{n-\rho}, \ldots, b_0]^T \in \mathbb{R}^{n-\rho+1}$ are vectors of unknown constant parameters.

Under Assumptions 7.2, 7.3, and 7.4 (with $y \in \mathbb{R}$ replaced by $x^m \in \mathbb{R}^{\bar{m}}$, $\bar{m} = m_1 + \sum_{i=2}^{k}(m_i - n_{i-1}))$, we can design adaptive controllers which guarantee global stability and tracking of reference signals satisfying Assumption 7.5. The design procedure starts with the choice of filters which will provide "virtual estimates" of the unmeasured state variables. In (7.180) there are k groups of such variables: $x_{m_i+1}, \ldots, x_{n_i}$, $1 \leq i \leq k$, with $m_k \leq \rho$ and $n_k = n$. Following the development of Section 7.3, for each of the first $k-1$ such groups we consider the subsystem

$$\begin{aligned}
\dot{x}_{m_i} &= x_{m_i+1} + \varphi_{0,m_i}(x^{m_i}) + \sum_{j=1}^{p} \theta_j \varphi_{j,m_i}(x^{m_i}) \\
&\vdots \\
\dot{x}_{n_i} &= x_{n_i+1} + \varphi_{0,n_i}(x^{m_i}) + \sum_{j=1}^{p} \theta_j \varphi_{j,n_i}(x^{m_i})
\end{aligned} \quad (7.181)$$

where x_{m_i} and x_{n_i+1} are measured. The subsystem (7.181) can be rewritten in the form

$$\begin{aligned}
\dot{x}^{n_i} &= A^i x^{n_i} + \varphi_0^{n_i}(x^{m_i}) + \sum_{j=1}^{p} \theta_j \varphi_j^{n_i}(x^{m_i}) + b^i x_{n_i+1} \\
x_{m_i} &= c^{iT} x^{n_i},
\end{aligned} \quad (7.182)$$

where, denoting $k_i = n_i - m_i$, we have for $1 \leq i \leq k$

$$x^{n_i} = (x_{m_i}, x_{m_i+1}, \ldots, x_{n_i})^T \quad (7.183)$$

$$A^i = \begin{bmatrix} 0 & & \\ \vdots & I_{k_i \times k_i} & \\ 0 & \ldots & 0 \end{bmatrix}, \quad b^i = \begin{bmatrix} 0_{k_i \times 1} \\ 1 \end{bmatrix}, \quad c^i = \begin{bmatrix} 1 \\ 0_{k_i \times 1} \end{bmatrix} \quad (7.184)$$

$$\varphi_j^{n_i}(x^{m_i}) = (\varphi_{j,m_i}(x^{m_i}), \ldots, \varphi_{j,n_i}(x^{m_i}))^T, \quad 0 \leq j \leq p. \quad (7.185)$$

For each of the $k-1$ subsystems of the form (7.182), we choose a gain vector K^i such that $A_0^i = A^i - K^i c^{iT}$ is a Hurwitz matrix and we define the filters

$$\begin{aligned}
\dot{\xi}_0^i &= A_0^i \xi_0^i + K^i x_{m_i} + \varphi_0^{n_i}(x^{m_i}) + b^i x_{n_i+1} \\
\dot{\xi}_j^i &= A_0^i \xi_j^i + \varphi_j^{n_i}(x^{m_i}), \quad 1 \leq j \leq p.
\end{aligned} \quad (7.186)$$

7.4 EXTENSIONS

It is then straightforward to verify that

$$\dot{\varepsilon}_i = A_0^i \varepsilon_i, \quad \varepsilon_i \triangleq x^{n_i} - \left(\xi_0^i + \sum_{j=1}^{p} \theta_j \xi_j^i \right). \tag{7.187}$$

Since A_0^i is Hurwitz, (7.187) implies that the signals ε_i converge exponentially to zero. Furthermore, we note that the derivative of x_{m_i} can be expressed as

$$\dot{x}_{m_i} = \xi_{0,2}^i + \varphi_{0,m_i}(x^{m_i}) + \sum_{j=1}^{p} \theta_j \left[\varphi_{j,m_i}(x^{m_i}) + \xi_{j,2}^i \right] + \varepsilon_{i,2}, \quad 1 \leq i \leq k-1. \tag{7.188}$$

For the last group of unmeasured variables we consider the subsystem (recall that $m_k \leq \rho, n_k = n, x^{m_k} = x^m$)

$$\dot{x}_{m_k} = x_{m_k+1} + \varphi_{0,m_k}(x^m) + \sum_{j=1}^{p} \theta_j \varphi_{j,m_k}(x^m)$$

$$\vdots$$

$$\dot{x}_{\rho-1} = x_\rho + \varphi_{0,\rho-1}(x^m) + \sum_{j=1}^{p} \theta_j \varphi_{j,\rho-1}(x^m)$$

$$\dot{x}_\rho = x_{\rho+1} + \varphi_{0,\rho}(x^m) + \sum_{j=1}^{p} \theta_j \varphi_{j,\rho}(x^m) + b_{n-\rho} \beta(x^m) u \tag{7.189}$$

$$\vdots$$

$$\dot{x}_n = \varphi_{0,n}(x^m) + \sum_{j=1}^{p} \theta_j \varphi_{j,n}(x^m) + b_0 \beta(x^m) u,$$

which can be rewritten in the form

$$\dot{x}^{n_k} = A^k x^{n_k} + \varphi_0^{n_k}(x^m) + \sum_{j=1}^{p} \theta_j \varphi_j^{n_k}(x^m) + b^k \beta(x^m) u$$

$$x_{m_k} = c^{kT} x^{n_k}, \tag{7.190}$$

where

$$b^k = [\, 0_{1 \times (\rho - m_k)}, \, b_{n-\rho}, \ldots, b_0 \,]^T. \tag{7.191}$$

Again, we choose a gain vector K^k such that $A_0^k = A^k - K^k c^{kT}$ is Hurwitz, and we define the filters

$$\begin{aligned}
\dot{\xi}_0^k &= A_0^k \xi_0^k + K^k x_{m_k} + \varphi_0^{n_k}(x^m) \\
\dot{\xi}_j^k &= A_0^k \xi_0^k + \varphi_j^{n_k}(x^m), \quad 1 \leq j \leq p \\
\dot{\lambda} &= A_0^k \lambda + c^k \beta(x^m) u, \quad c^k = [0, \ldots, 0, 1]^T.
\end{aligned} \tag{7.192}$$

We also define the signals

$$v_j = (A_0^k)^j \lambda, \quad j = 0, \ldots, n - \rho. \tag{7.193}$$

It is then straightforward to verify that these signals satisfy the differential equations

$$\dot{v}_j = A_0^k v_j + e_{n-m_k+1-j}\beta(x^m)u, \qquad 0 \leq j \leq n-\rho, \tag{7.194}$$

where e_i is the ith coordinate vector in $\mathbb{R}^{n-m_k+1}$. From (7.190), (7.192) and (7.194) it follows that

$$\dot{\varepsilon}_k = A_0^k \varepsilon_k, \quad \varepsilon_k \triangleq x^{n_k} - \left(\xi_0^k + \sum_{j=1}^p \theta_j \xi_j^k + \sum_{j=0}^{n-\rho} b_j v_j\right), \tag{7.195}$$

which implies that ε_k converges exponentially to zero. We also note that the derivative of x_{m_k} can be expressed as

$$\dot{x}_{m_k} = \xi_{0,2}^k + \varphi_{0,m_k}(x^m) + \sum_{j=1}^p \theta_j \left[\varphi_{j,m_k}(x^m) + \xi_{j,2}^k\right] + \sum_{j=0}^{n-\rho} b_j v_{j,2} + \varepsilon_{k,2}. \tag{7.196}$$

Once these filters have been designed, the design proceeds as follows: Starting with $z_1 = x_1 - y_r$, any of the full-state-feedback design procedures can be applied to the first group of measured states $x_1, \ldots, x_{m_1-1}$. For each of the subsequent measured subsystems of (7.180), additional nonlinear damping terms are needed, since unmeasured states appear in the corresponding $\dot{z}_i$-equations and have to be replaced by their "virtual estimates." As for the unmeasured subsystems of (7.180), following Section 7.3.2, their equations are replaced by the corresponding equations from the filters (7.186), and then any of the design procedures presented in this or the following chapters can be applied.

Notes and References

Even before the development of adaptive backstepping with full-state feedback, the more challenging output-feedback problem was addressed by Kanellakopoulos, Kokotović, and Middleton under restrictive structural and growth conditions on the nonlinearities via an extension of adaptive linear techniques [67, 68]. Subsequently, the growth restrictions were removed by Kanellakopoulos, Kokotović, and Morse [70, 71], applying an early adaptive scheme for linear systems by Feuer and Morse [35]. Still, the output nonlinearities were not allowed to precede the control input. Adaptive backstepping [69] created the possibility for removal of this structural restriction.

Combining adaptive backstepping with their adaptive observers [121] and output-feedback linearization with filtered transformations [120], Marino and Tomei [122, 123] developed a new adaptive scheme for systems in the output feedback form. This is still the largest class of nonlinear systems for which global boundedness and tracking of arbitrary trajectories can be achieved using output feedback. A more general class of systems was considered in [124], but only for the set-point regulation problem.

The design summarized in Theorem 7.6 is an alternative approach for the same class of systems due to Kanellakopoulos, Kokotović and Morse [72]. This design employs the filters (7.105) patterned after Kreisselmeier [91]. Both of these output-feedback schemes, [122, 123] and [72], inherited overparametrization from [69].

Partial-state-feedback designs were first developed by Kanellakopoulos, Kokotović, Marino, and Tomei [66] using adaptive backstepping and filtered transformations. The design we presented in Section 7.4.2 was developed by Kanellakopoulos, Kokotović, and Morse [74].

The idea of interlacing the design of the observer with that of the controller was developed by Kanellakopoulos, Krstić, and Kokotović [78] and was subsequently used for the design of passive adaptive systems by Kanellakopoulos [64], as well as for induction motor control by Kanellakopoulos, Krein, and Disilvestro [77, 75].

An output-feedback result for systems with uncertain parameters (of known bound) entering nonlinearly was developed by Marino and Tomei [124] for the case of set point regulation. Praly and Jiang [159] solved the stabilization problem for a class of systems broader than the output-feedback form, where the nonlinearities also depend on the unmeasured state of nonlinear ISS inverse dynamics. Teel and Praly [191] extended the result of [159] for the case of uncertain nonlinearities.

Induction motors served as testing ground for multivariable nonlinear designs. The physically motivated field-orientation design by Blaschke [9] was followed by feedback linearization designs of Krzeminski [106] and Marino, Peresada, and Valigi [119]. While these schemes assume flux measurement, most implemented controllers use flux observers such as those proposed by Verghese and Sanders [194]. The scheme presented in Section 7.2 was developed by Kanellakopoulos, Krein, and Disilvestro [76] and subsequently improved by Hu, Dawson, and Qian [46], who obtained global stability and tracking results through a modification of the flux reference signal. A different, physically motivated approach, was recently developed by Espinosa and Ortega [33].

Chapter 8

Tuning Functions Designs

In Chapter 7 we presented an output-feedback design for systems in the output-feedback form, and started the development of adaptive schemes by designing output-feedback adaptive backstepping controllers with overparametrization. In this and the next chapter we develop output-feedback adaptive nonlinear controllers without overparametrization. This chapter extends the tuning functions design of Chapter 4 to the output-feedback case. The next chapter extends the modular designs of Chapters 5 and 6 to the output-feedback case. We present designs that use different filter structures and identifiers. This chapter can be read immediately after Chapter 4.

As in Chapter 7 we consider systems in the *output-feedback form*:

$$\dot{x}_1 = x_2 + \varphi_{0,1}(y) + \sum_{j=1}^{q} a_j \varphi_{j,1}(y)$$

$$\vdots$$

$$\dot{x}_{\rho-1} = x_\rho + \varphi_{0,\rho-1}(y) + \sum_{j=1}^{q} a_j \varphi_{j,\rho-1}(y)$$

$$\dot{x}_\rho = x_{\rho+1} + \varphi_{0,\rho}(y) + \sum_{j=1}^{q} a_j \varphi_{j,\rho}(y) + b_m \sigma(y) u \qquad (8.1)$$

$$\vdots$$

$$\dot{x}_{n-1} = x_n + \varphi_{0,n-1}(y) + \sum_{j=1}^{q} a_j \varphi_{j,n-1}(y) + b_1 \sigma(y) u$$

$$\dot{x}_n = \varphi_{0,n}(y) + \sum_{j=1}^{q} a_j \varphi_{j,n}(y) + b_0 \sigma(y) u$$

$$y = x_1,$$

where $x \in \mathbb{R}^n$ is the state, $u \in \mathbb{R}$ is the input, $y \in \mathbb{R}$ is the output, $\varphi_{j,i}$, $0 \le j \le q$, $1 \le i \le n$, and σ are smooth nonlinear functions, and

$$a = [a_1, \ldots, a_q]^T \in \mathbb{R}^q, \quad b = [b_m, \ldots, b_0]^T \in \mathbb{R}^{m+1} \qquad (8.2)$$

are vectors of unknown constant parameters. Only the output y is available for measurement. We rewrite (8.1) as

$$\dot{x} = Ax + \phi(y) + \Phi(y)a + \begin{bmatrix} 0 \\ b \end{bmatrix} \sigma(y)u, \qquad x \in \mathbb{R}^n \qquad (8.3)$$
$$y = e_1^T x,$$

where

$$A = \begin{bmatrix} 0 & & \\ \vdots & & I_{n-1} \\ 0 & \cdots & 0 \end{bmatrix} \qquad (8.4)$$

$$\phi(y) = \begin{bmatrix} \varphi_{0,1}(y) \\ \vdots \\ \varphi_{0,n}(y) \end{bmatrix}, \qquad \Phi(y) = \begin{bmatrix} \varphi_{1,1}(y) & \cdots & \varphi_{q,1}(y) \\ \vdots & & \vdots \\ \varphi_{1,n}(y) & \cdots & \varphi_{q,n}(y) \end{bmatrix}. \qquad (8.5)$$

The control objective is to track a given reference signal $y_r(t)$ with the output y, while keeping all the signals in the closed-loop system globally bounded.

We make the following assumptions about the system (8.1):

Assumption 8.1 *The sign of b_m is known.*

Assumption 8.2 *The polynomial $B(s) = b_m s^m + \cdots + b_1 s + b_0$ is known to be Hurwitz.*

Assumption 8.3 $\sigma(y) \neq 0 \ \forall y \in \mathbb{R}$.

The class of systems (8.1) is restrictive because the nonlinearities depend only on the output. However, this restriction, is not imposed because of adaptation. It is needed even when the parameters are known. It was shown in [127] that the system

$$\begin{aligned} \dot{x}_1 &= x_2 \\ \dot{x}_2 &= x_2^n + u \\ y &= x_1 \end{aligned} \qquad (8.6)$$

cannot be globally stabilized by dynamic output feedback for $n > 2$. In other words, the class of systems which are globally stabilizable by output feedback is not much broader than the class of output feedback systems (8.1).

In the absence of full state measurement we need filters, both in the design of an output-feedback control law and in the design of an identifier. We present designs which employ two different sets of filters:

- *K-filters*, originally proposed for adaptive observer design by Kreisselmeier [91].

- *MT-filters*, developed by Marino and Tomei [121, 123].

8.1 Design with K-Filters

We focus on the designs with K-filters, and briefly present the designs with MT-filters for completeness. There are considerable differences between the two families of designs, and we will comment on their relative merits.

In this chapter we develop output feedback designs using the tuning functions approach.

8.1 Design with K-Filters

8.1.1 Filters and observer

We start by rewriting (8.3) as

$$\begin{aligned} \dot{x} &= Ax + \phi(y) + F(y,u)^{\mathrm{T}}\theta \\ y &= e_1^{\mathrm{T}} x, \end{aligned} \tag{8.7}$$

where the $p = q + m + 1$-dimensional parameter vector θ is defined by

$$\theta = \begin{bmatrix} b \\ a \end{bmatrix}, \tag{8.8}$$

and

$$F(y,u)^{\mathrm{T}} = \left[\begin{bmatrix} 0_{(\rho-1)\times(m+1)} \\ I_{m+1} \end{bmatrix} \sigma(y)u, \ \Phi(y) \right]. \tag{8.9}$$

If θ were known, we would design an observer

$$\dot{\hat{x}} = A_0 \hat{x} + ky + \phi(y) + F(y,u)^{\mathrm{T}}\theta, \tag{8.10}$$

with the vector $k = [k_1, \ldots, k_n]^{\mathrm{T}}$ chosen so that the matrix

$$A_0 = A - k e_1^{\mathrm{T}} \tag{8.11}$$

is Hurwitz, that is,

$$PA_0 + A_0^{\mathrm{T}} P = -I, \quad P = P^{\mathrm{T}} > 0. \tag{8.12}$$

Then, the observer error $\tilde{x} = x - \hat{x}$ would be governed by the exponentially stable system

$$\dot{\tilde{x}} = A_0 \tilde{x}. \tag{8.13}$$

Since θ is not known, the observer (8.10) is not implementable but it provides motivation for the subsequent development. We define the state estimate

$$\hat{x} = \xi + \Omega^{\mathrm{T}} \theta, \tag{8.14}$$

which employs the filters

$$\begin{aligned} \dot{\xi} &= A_0 \xi + ky + \phi(y) \\ \dot{\Omega}^{\mathrm{T}} &= A_0 \Omega^{\mathrm{T}} + F(y,u)^{\mathrm{T}}. \end{aligned} \tag{8.15} \tag{8.16}$$

The state estimation error
$$\varepsilon = x - \hat{x} \tag{8.17}$$
is readily shown to satisfy
$$\dot{\varepsilon} = A_0 \varepsilon. \tag{8.18}$$
The nonminimal observer (8.14), (8.15), and (8.16) is still nonimplementable because it depends on θ. However, it has a key property not present in (8.10): The state x satisfies a static relationship with θ; that is,
$$x = \xi + \Omega^T \theta + \varepsilon. \tag{8.19}$$
This is easily verified by substituting (8.14) into (8.17).

Remark 8.4 The certainty equivalence counterpart of the estimate (8.14) is
$$\bar{x} = \xi + \Omega^T \hat{\theta}. \tag{8.20}$$
It can alternatively be generated via
$$\dot{\bar{x}} = A_0 \bar{x} + ky + \phi(y) + F(y,u)^T \hat{\theta} + \Omega^T \dot{\hat{\theta}}. \tag{8.21}$$
This observer is not a certainty equivalence version of (8.10) because of the term $\Omega^T \dot{\hat{\theta}}$. ◇

To reduce the dynamic order of the Ω-filter (8.16), we exploit the structure of $F(y,u)$. Denote the first $m+1$ columns of Ω^T by $v_m, \ldots, v_1, v_0$. Due to the special dependence of $F(y,u)$ on $\sigma(y)u$, the vectors $v_m, \ldots, v_1, v_0$ satisfy the equations
$$\dot{v}_j = A_0 v_j + e_{n-j}\, \sigma(y)u, \qquad j = 0, \ldots, m. \tag{8.22}$$
It easy to show that
$$A_0^j e_n = e_{n-j}, \qquad j = 0, \ldots, n-1. \tag{8.23}$$
Therefore, the vectors v_j are generated by only one input filter
$$\dot{\lambda} = A_0 \lambda + e_n \sigma u, \tag{8.24}$$
with the algebraic expressions
$$v_j = A_0^j \lambda, \qquad j = 0, \ldots, m. \tag{8.25}$$
While we always implement the filter (8.24), for analysis we use the equations (8.22). Now, in view of (8.9), (8.16), and (8.22), Ω is obtained as
$$\Omega^T = [v_m, \ldots, v_1, v_0, \Xi] \tag{8.26}$$

8.1 Design with K-Filters

Table 8.1: K-Filters

K-filters:	
$\dot{\xi} = A_0\xi + ky + \phi(y)$	(8.28)
$\dot{\Xi} = A_0\Xi + \Phi(y)$	(8.29)
$\dot{\lambda} = A_0\lambda + e_n\sigma(y)u$	(8.30)
$v_j = A_0^j\lambda, \quad j = 0,\ldots,m$	(8.31)
$\Omega^{\mathrm{T}} = [v_m,\ldots,v_1,v_0,\Xi]$	(8.32)

where the matrix Ξ is generated by

$$\dot{\Xi} = A_0\Xi + \Phi(y), \qquad \Xi \in \mathbb{R}^{n\times q}. \tag{8.27}$$

The implemented filters are summarized in Table 8.1. The total dynamic order of the K-filters is $n(q+2)$. As explained in [63], a further reduction is possible by using the reduced-order observer technique, so that the total filter dynamic order becomes $(n-1)(q+2)$.

To prepare for the backstepping procedure in the next subsection, we consider the equation for the output $y = x_1$ rewritten from (8.1):

$$\dot{y} = x_2 + \varphi_{0,1} + \Phi_{(1)}a. \tag{8.33}$$

We need to replace the unavailable state x_2 by available filter signals. From (8.19) we have

$$\begin{aligned}
x_2 &= \xi_2 + \Omega_{(2)}^{\mathrm{T}}\theta + \varepsilon_2 \\
&= \xi_2 + [v_{m,2}, v_{m-1,2},\ldots,v_{0,2}, \Xi_{(2)}]\theta + \varepsilon_2 \quad &(8.34) \\
&= b_m v_{m,2} + \xi_2 + [0, v_{m-1,2},\ldots,v_{0,2}, \Xi_{(2)}]\theta + \varepsilon_2. \quad &(8.35)
\end{aligned}$$

Substituting first (8.34) and then (8.35) into (8.33), we obtain the following two important expressions for $\dot{y}$:

$$\begin{aligned}
\dot{y} &= \omega_0 + \omega^{\mathrm{T}}\theta + \varepsilon_2 &(8.36) \\
&= b_m v_{m,2} + \omega_0 + \bar{\omega}^{\mathrm{T}}\theta + \varepsilon_2, &(8.37)
\end{aligned}$$

where the 'regressor' ω and the 'truncated regressor' $\bar{\omega}$ are defined as

$$\begin{aligned}
\omega &= [v_{m,2}, v_{m-1,2},\ldots,v_{0,2},\ \Phi_{(1)} + \Xi_{(2)}]^{\mathrm{T}} &(8.38) \\
\bar{\omega} &= [0, v_{m-1,2},\ldots,v_{0,2},\ \Phi_{(1)} + \Xi_{(2)}]^{\mathrm{T}}, &(8.39)
\end{aligned}$$

and
$$\omega_0 = \varphi_{0,1} + \xi_2. \tag{8.40}$$

8.1.2 Adaptive controller design

The tuning functions design with K-filters has many similarities with the state-feedback design presented in detail in Section 4.2.1. It also uses the same technique for dealing with unknown high frequency gain b_m as in Section 4.5.1. We present the design for the case $\rho > 1$.

The first obstacle to backstepping with output feedback is that the state x_2 is not measured. For this reason, (8.37) is written in a form which suggests that the filter signal $v_{m,2}$ be used for backstepping. Indeed, by comparing (8.1) with
$$\dot{v}_m = A_0 v_m + e_\rho \sigma(y) u \tag{8.41}$$
(the latter is obtained from (8.22) for $j = m$), we see that both x_2 and $v_{m,2}$ are separated from the control u by $\rho - 1$ integrators. A closer examination of the filters in Table 8.1 reveals that more integrators stand in the way of any other variable in (8.37). Therefore, the system to which we will apply backstepping is
$$\dot{y} = b_m v_{m,2} + \omega_0 + \bar{\omega}^T \theta + \varepsilon_2 \tag{8.42}$$
$$\dot{v}_{m,i} = v_{m,i+1} - k_i v_{m,1}, \qquad i = 2, \ldots, \rho - 1 \tag{8.43}$$
$$\dot{v}_{m,\rho} = \sigma(y) u + v_{m,\rho+1} - k_\rho v_{m,1}. \tag{8.44}$$

Our analysis will show that once we stabilize the system (8.42)–(8.44), all the closed-loop signals remain bounded.

The second output-feedback difficulty is the presence of the disturbance ε_2 in (8.42). In the process of backstepping, ε_2 will be multiplied by nonlinear terms, which can destabilize the system even though ε_2 is bounded and exponentially decaying. To prevent this, we will include nonlinear damping terms in our stabilizing functions.

To avoid repetition, the backstepping design will not be presented in a step-by-step fashion as in Section 4.2.1. We present only the first step in detail and then proceed to the *complete error system*, for which we choose the stabilizing functions and the tuning functions to achieve a desired skew-symmetric form.

For system (8.42)–(8.44) we define a change of coordinates
$$z_1 = y - y_r \tag{8.45}$$
$$z_i = v_{m,i} - \hat{\varrho} y_r^{(i-1)} - \alpha_{i-1}, \qquad i = 2, \ldots, \rho, \tag{8.46}$$
where $\hat{\varrho}$ is an estimate of $\varrho = 1/b_m$.

Step 1. The equation for the tracking error z_1 obtained from (8.45) and (8.42) is
$$\dot{z}_1 = b_m v_{m,2} + \omega_0 + \bar{\omega}^T \theta + \varepsilon_2 - \dot{y}_r. \tag{8.47}$$

8.1 Design with K-Filters

By substituting (8.46) for $i = 2$ into (8.47), and scaling the first stabilizing function α_1 as
$$\alpha_1 = \hat{\varrho}\bar{\alpha}_1, \tag{8.48}$$
we rewrite (8.47) in the form
$$\dot{z}_1 = \bar{\alpha}_1 + \omega_0 + \bar{\omega}^T\theta + \varepsilon_2 - b_m\left(\dot{y}_r + \bar{\alpha}_1\right)\tilde{\varrho} + b_m z_2. \tag{8.49}$$

Then the choice of the stabilizing function
$$\bar{\alpha}_1 = -c_1 z_1 - d_1 z_1 - \omega_0 - \bar{\omega}^T\hat{\theta} \tag{8.50}$$
results in
$$\dot{z}_1 = -c_1 z_1 - d_1 z_1 + \varepsilon_2 + \bar{\omega}^T\tilde{\theta} - b_m\left(\dot{y}_r + \bar{\alpha}_1\right)\tilde{\varrho} + b_m z_2. \tag{8.51}$$

(Two positive constants c_1 and d_1 are introduced for uniformity with subsequent steps in which d_i is a coefficient of a nonlinear damping term counteracting ε_2.) With (8.46), (8.48), and (8.38), we have
$$\begin{aligned}\bar{\omega}^T\tilde{\theta} + b_m z_2 &= \bar{\omega}^T\tilde{\theta} + \tilde{b}_m z_2 + \hat{b}_m z_2 \\ &= \bar{\omega}^T\tilde{\theta} + \left(v_{m,2} - \hat{\varrho}\dot{y}_r - \alpha_1\right)e_1^T\tilde{\theta} + \hat{b}_m z_2 \\ &= \omega^T\tilde{\theta} - \hat{\varrho}\left(\dot{y}_r + \bar{\alpha}_1\right)e_1^T\tilde{\theta} + \hat{b}_m z_2 \\ &= \left(\omega - \hat{\varrho}\left(\dot{y}_r + \bar{\alpha}_1\right)e_1\right)^T\tilde{\theta} + \hat{b}_m z_2. \end{aligned} \tag{8.52}$$

Substituting (8.52) into (8.51) we get
$$\dot{z}_1 = -c_1 z_1 - d_1 z_1 + \varepsilon_2 + \left(\omega - \hat{\varrho}\left(\dot{y}_r + \bar{\alpha}_1\right)e_1\right)^T\tilde{\theta} - b_m\left(\dot{y}_r + \bar{\alpha}_1\right)\tilde{\varrho} + \hat{b}_m z_2. \tag{8.53}$$

As in Section 4.2.1, the choice of the update law for $\hat{\theta}$ is postponed until the last step and we only select the first tuning function
$$\tau_1 = \left(\omega - \hat{\varrho}\left(\dot{y}_r + \bar{\alpha}_1\right)e_1\right)z_1. \tag{8.54}$$

However, the update law for $\hat{\varrho}$ is designed at the first step as
$$\dot{\hat{\varrho}} = -\gamma\,\text{sgn}(b_m)\left(\dot{y}_r + \bar{\alpha}_1\right)z_1, \qquad \gamma > 0. \tag{8.55}$$

We pause to clarify the arguments of the function α_1. By examining (8.50) along with (8.39) and (8.40), we see that α_1 is a function of $y, \xi, \Xi, \hat{\theta}, \hat{\varrho}, v_{0,2}, \ldots, v_{m-1,2}$ and y_r. For brevity, we denote $X = (y, \xi, \Xi, \hat{\theta}, \hat{\varrho})$. From (8.31) one can show that $v_{i,j}$ can be expressed as
$$v_{i,j} = [*,\ldots,*,1]\begin{bmatrix}\lambda_1 \\ \vdots \\ \lambda_{i+j}\end{bmatrix}, \tag{8.56}$$

where $\lambda_k \triangleq 0$ for $k > n$, and $*$ denotes entries that can have any values. With (8.56) we conclude that α_1 is a function of $y, \xi, \Xi, \hat{\theta}, \hat{\varrho}, \lambda_1, \ldots, \lambda_{m+1}, y_r$. Denoting $\bar{\lambda}_i = (\lambda_1, \ldots, \lambda_i)$, we write $\alpha_1(X, \bar{\lambda}_{m+1}, y_r)$. In the backstepping procedure, α_i is a function of $\left(X, \bar{\lambda}_{m+i}, \bar{y}_r^{(i-1)}\right)$. To make the expositions shorter, this fact is postulated here, and then verified once the α_i's have been selected.

Step $i = 2, \ldots, \rho$. Differentiating (8.46) for $i = 2, \ldots, \rho - 1$, with the help of (8.43) we obtain

$$\begin{aligned}
\dot{z}_i &= \dot{v}_{m,i} - \hat{\varrho} y_r^{(i)} - \dot{\hat{\varrho}} y_r^{(i-1)} - \dot{\alpha}_{i-1}\left(y, \xi, \Xi, \hat{\theta}, \hat{\varrho}, \bar{\lambda}_{m+i-1}, \bar{y}_r^{(i-2)}\right) \\
&= v_{m,i+1} - k_i v_{m,1} - \hat{\varrho} y_r^{(i)} - \dot{\hat{\varrho}} y_r^{(i-1)} - \frac{\partial \alpha_{i-1}}{\partial y}\left(\omega_0 + \omega^T \theta + \varepsilon_2\right) \\
&\quad - \frac{\partial \alpha_{i-1}}{\partial \xi}(A_0 \xi + ky + \phi) - \frac{\partial \alpha_{i-1}}{\partial \Xi}(A_0 \Xi + \Phi) - \sum_{j=1}^{i-1} \frac{\partial \alpha_{i-1}}{\partial y_r^{(j-1)}} y_r^{(j)} \\
&\quad - \sum_{j=1}^{m+i-1} \frac{\partial \alpha_{i-1}}{\partial \lambda_j}(-k_j \lambda_1 + \lambda_{j+1}) - \frac{\partial \alpha_{i-1}}{\partial \hat{\theta}} \dot{\hat{\theta}} - \frac{\partial \alpha_{i-1}}{\partial \hat{\varrho}} \dot{\hat{\varrho}}.
\end{aligned} \quad (8.57)$$

Noting from (8.46) that $v_{m,i+1} - \hat{\varrho} y_r^{(i)} = z_{i+1} + \alpha_i$, we get

$$\begin{aligned}
\dot{z}_i &= z_{i+1} + \alpha_i - k_i v_{m,1} - \frac{\partial \alpha_{i-1}}{\partial y}\left(\omega_0 + \omega^T \theta + \varepsilon_2\right) \\
&\quad - \frac{\partial \alpha_{i-1}}{\partial \xi}(A_0 \xi + ky + \phi) - \frac{\partial \alpha_{i-1}}{\partial \Xi}(A_0 \Xi + \Phi) - \sum_{j=1}^{i-1} \frac{\partial \alpha_{i-1}}{\partial y_r^{(j-1)}} y_r^{(j)} \\
&\quad - \sum_{j=1}^{m+i-1} \frac{\partial \alpha_{i-1}}{\partial \lambda_j}(-k_j \lambda_1 + \lambda_{j+1}) - \frac{\partial \alpha_{i-1}}{\partial \hat{\theta}} \dot{\hat{\theta}} - \left(y_r^{(i-1)} + \frac{\partial \alpha_{i-1}}{\partial \hat{\varrho}}\right) \dot{\hat{\varrho}}.
\end{aligned} \quad (8.58)$$

From (8.44) it follows that the final step $i = \rho$ can be encompassed in these calculations if we define $\alpha_\rho = \sigma(y)u + v_{m,\rho+1} - \hat{\varrho} y_r^{(\rho)}$ and $z_{\rho+1} = 0$. To make the choice of a stabilizing function easier, we add and subtract $-\frac{\partial \alpha_{i-1}}{\partial y} \omega^T \hat{\theta} - \frac{\partial \alpha_{i-1}}{\partial \hat{\theta}} \Gamma \tau_i$ in (8.58) and get

$$\begin{aligned}
\dot{z}_i &= \alpha_i - k_i v_{m,1} - \frac{\partial \alpha_{i-1}}{\partial y}\left(\omega_0 + \omega^T \hat{\theta}\right) \\
&\quad - \frac{\partial \alpha_{i-1}}{\partial \xi}(A_0 \xi + ky + \phi) - \frac{\partial \alpha_{i-1}}{\partial \Xi}(A_0 \Xi + \Phi) - \sum_{j=1}^{i-1} \frac{\partial \alpha_{i-1}}{\partial y_r^{(j-1)}} y_r^{(j)} \\
&\quad - \sum_{j=1}^{m+i-1} \frac{\partial \alpha_{i-1}}{\partial \lambda_j}(-k_j \lambda_1 + \lambda_{j+1}) - \frac{\partial \alpha_{i-1}}{\partial \hat{\theta}} \Gamma \tau_i - \left(y_r^{(i-1)} + \frac{\partial \alpha_{i-1}}{\partial \hat{\varrho}}\right) \dot{\hat{\varrho}} \\
&\quad - \frac{\partial \alpha_{i-1}}{\partial y} \varepsilon_2 - \frac{\partial \alpha_{i-1}}{\partial y} \omega^T \tilde{\theta} - \frac{\partial \alpha_{i-1}}{\partial \hat{\theta}}\left(\dot{\hat{\theta}} - \Gamma \tau_i\right) + z_{i+1}.
\end{aligned} \quad (8.59)$$

The stabilizing function α_i will be chosen to cancel all the terms except those in the last line of (8.59). The potentially destabilizing disturbance $-\frac{\partial \alpha_{i-1}}{\partial y} \varepsilon_2$

8.1 DESIGN WITH K-FILTERS

is counteracted by including the nonlinear damping term $-d_i \left(\frac{\partial \alpha_{i-1}}{\partial y}\right)^2 z_i$ in α_i. Dealing with the remaining three terms in the last line of (8.59) is no different from the way we dealt with similar terms in Section 4.2.1:

- The term $-\frac{\partial \alpha_{i-1}}{\partial y} \omega^T \tilde{\theta}$ determines the ith tuning function

$$\tau_i = \tau_{i-1} - \frac{\partial \alpha_{i-1}}{\partial y} \omega z_i, \qquad i = 2, \ldots, \rho. \tag{8.60}$$

- The term $-\frac{\partial \alpha_{i-1}}{\partial \hat{\theta}} \left(\dot{\hat{\theta}} - \Gamma \tau_i\right)$ represents the mismatch between the actual update law and the tuning function, and it appears because at step i only τ_i can be cancelled in (8.59). As usual in the tuning functions design, our choice of the update law will be $\dot{\hat{\theta}} = \Gamma \tau_\rho$. Therefore, in view of (8.60), we have

$$\dot{\hat{\theta}} - \Gamma \tau_i = -\sum_{j=i+1}^{\rho} \Gamma \frac{\partial \alpha_{j-1}}{\partial y} \omega z_j, \tag{8.61}$$

which yields

$$\begin{aligned} -\frac{\partial \alpha_{i-1}}{\partial \hat{\theta}} \left(\dot{\hat{\theta}} - \Gamma \tau_i\right) &= \sum_{j=i+1}^{\rho} \frac{\partial \alpha_{i-1}}{\partial \hat{\theta}} \Gamma \frac{\partial \alpha_{j-1}}{\partial y} \omega z_j \\ &\overset{\triangle}{=} \sum_{j=i+1}^{\rho} \sigma_{ij} z_j. \end{aligned} \tag{8.62}$$

The stabilizing function α_i will include the term $-\sum_{j=2}^{i-1} \sigma_{ji} z_j$ to achieve skew-symmetry in the error system.

- The 'above-diagonal' term z_{i+1} is compensated by adding the 'below-diagonal' z_{i-1}.

To summarize, our choice of the stabilizing functions for $i = 3, \ldots, \rho$ is

$$\begin{aligned} \alpha_i =& -z_{i-1} - c_i z_i - d_i \left(\frac{\partial \alpha_{i-1}}{\partial y}\right)^2 z_i + k_i v_{m,1} + \frac{\partial \alpha_{i-1}}{\partial y}\left(\omega_0 + \omega^T \hat{\theta}\right) \\ &+ \frac{\partial \alpha_{i-1}}{\partial \xi}(A_0 \xi + ky + \phi) + \frac{\partial \alpha_{i-1}}{\partial \Xi}(A_0 \Xi + \Phi) + \sum_{j=1}^{i-1} \frac{\partial \alpha_{i-1}}{\partial y_r^{(j-1)}} y_r^{(j)} \\ &+ \sum_{j=1}^{m+i-1} \frac{\partial \alpha_{i-1}}{\partial \lambda_j}(-k_j \lambda_1 + \lambda_{j+1}) + \frac{\partial \alpha_{i-1}}{\partial \hat{\theta}} \Gamma \tau_i + \left(y_r^{(i-1)} + \frac{\partial \alpha_{i-1}}{\partial \hat{\varrho}}\right) \dot{\hat{\varrho}} \\ &- \sum_{j=2}^{i-1} \frac{\partial \alpha_{j-1}}{\partial \hat{\theta}} \Gamma \frac{\partial \alpha_{i-1}}{\partial y} z_j. \end{aligned} \tag{8.63}$$

For $i = 2$ the stabilizing function differs because of the term $-\hat{b}_m z_1$ needed to compensate for $\hat{b}_m z_2$ in (8.53):

$$\begin{aligned}\alpha_2 &= -\hat{b}_m z_1 - c_2 z_2 - d_2 \left(\frac{\partial \alpha_1}{\partial y}\right)^2 z_2 + k_2 v_{m,1} + \frac{\partial \alpha_1}{\partial y}\left(\omega_0 + \omega^T \hat{\theta}\right) \\ &+ \frac{\partial \alpha_1}{\partial \xi}(A_0 \xi + ky + \phi) + \frac{\partial \alpha_1}{\partial \Xi}(A_0 \Xi + \Phi) + \frac{\partial \alpha_1}{\partial y_r}\dot{y}_r \\ &+ \sum_{j=1}^{m+1}\frac{\partial \alpha_1}{\partial \lambda_j}(-k_j \lambda_1 + \lambda_{j+1}) + \frac{\partial \alpha_1}{\partial \hat{\theta}}\Gamma \tau_2 + \left(\dot{y}_r + \frac{\partial \alpha_1}{\partial \hat{\varrho}}\right)\dot{\hat{\varrho}}.\end{aligned} \quad (8.64)$$

At the end of the recursive procedure, the last stabilizing function α_ρ is used in the actual control law:

$$u = \frac{1}{\sigma(y)}\left(\alpha_\rho - v_{m,\rho+1} + \hat{\varrho}y_r^{(\rho)}\right). \quad (8.65)$$

and the last tuning function τ_ρ is used in the update law:

$$\dot{\hat{\theta}} = \Gamma \tau_\rho = \Gamma \left\{\omega\left[1, -\frac{\partial \alpha_1}{\partial y}, \ldots, -\frac{\partial \alpha_{\rho-1}}{\partial y}\right]\begin{bmatrix}z_1 \\ z_2 \\ \vdots \\ z_\rho\end{bmatrix} - \hat{\varrho}(\dot{y}_r + \bar{\alpha}_1)e_1 z_1\right\}. \quad (8.66)$$

By substituting (8.63) and (8.64) along with (8.62) into (8.59), the error system becomes

$$\begin{aligned}\dot{z}_1 &= -c_1 z_1 - d_1 z_1 + \hat{b}_m z_2 + \varepsilon_2 + (\omega - \hat{\varrho}(\dot{y}_r + \bar{\alpha}_1)e_1)^T \tilde{\theta} \\ &\quad - b_m(\dot{y}_r + \bar{\alpha}_1)\tilde{\varrho} \end{aligned} \quad (8.67)$$

$$\begin{aligned}\dot{z}_2 &= -c_2 z_2 - d_2\left(\frac{\partial \alpha_1}{\partial y}\right)^2 z_2 - \hat{b}_m z_1 + z_3 + \sum_{j=3}^{\rho}\sigma_{2,j}z_j \\ &\quad -\frac{\partial \alpha_1}{\partial y}\varepsilon_2 - \frac{\partial \alpha_1}{\partial y}\omega^T\tilde{\theta}\end{aligned} \quad (8.68)$$

$$\begin{aligned}\dot{z}_i &= -c_i z_i - d_i\left(\frac{\partial \alpha_{i-1}}{\partial y}\right)^2 z_i - \sum_{j=2}^{i-1}\sigma_{ji}z_j - z_{i-1} + z_{i+1} + \sum_{j=i+1}^{\rho}\sigma_{ij}z_j \\ &\quad -\frac{\partial \alpha_{i-1}}{\partial y}\varepsilon_2 - \frac{\partial \alpha_{i-1}}{\partial y}\omega^T\tilde{\theta}, \qquad i = 3,\ldots,\rho,\end{aligned} \quad (8.69)$$

where $z_{\rho+1} = 0$. This system is compactly written as

$$\dot{z} = A_z(z,t)z + W_\varepsilon(z,t)\varepsilon_2 + W_\theta(z,t)^T\tilde{\theta} - b_m(\dot{y}_r + \bar{\alpha}_1)e_1\tilde{\varrho}, \quad (8.70)$$

8.1 DESIGN WITH K-FILTERS

where the system matrix $A_z(z,t)$ is given by

$$A_z(z,t) =$$

$$\begin{bmatrix} -c_1-d_1 & \hat{b}_m & 0 & \cdots & \cdots & 0 \\ -\hat{b}_m & -c_2-d_2\left(\frac{\partial\alpha_1}{\partial y}\right)^2 & 1+\sigma_{23} & \sigma_{24} & \cdots & \sigma_{2,\rho} \\ 0 & -1-\sigma_{23} & \ddots & \ddots & \ddots & \vdots \\ \vdots & -\sigma_{24} & \ddots & \ddots & \ddots & \sigma_{\rho-2,\rho} \\ \vdots & \vdots & \ddots & \ddots & \ddots & 1+\sigma_{\rho-1,\rho} \\ 0 & -\sigma_{2,\rho} & \cdots & -\sigma_{\rho-2,\rho} & -1-\sigma_{\rho-1,\rho} & -c_\rho-d_\rho\left(\frac{\partial\alpha_{\rho-1}}{\partial y}\right)^2 \end{bmatrix}$$

(8.71)

and $W_\varepsilon(z,t)$ and $W_\theta(z,t)$ are defined as

$$W_\varepsilon(z,t) = \begin{bmatrix} 1 \\ -\frac{\partial\alpha_1}{\partial y} \\ \vdots \\ -\frac{\partial\alpha_{\rho-1}}{\partial y} \end{bmatrix} \in \mathbb{R}^\rho \tag{8.72}$$

$$W_\theta(z,t)^{\mathrm{T}} = W_\varepsilon(z,t)\omega^{\mathrm{T}} - \hat{\varrho}\left(\dot{y}_r + \bar{\alpha}_1\right)e_1 e_1^{\mathrm{T}} \in \mathbb{R}^{\rho\times p}. \tag{8.73}$$

To prepare for the stability analysis in the next subsection, along with the error system (8.70) we consider the error equations for parameter estimators (8.66) and (8.55), as well as the state estimation error (8.18):

$$\dot{\tilde{\theta}} = -\Gamma W_\theta(z,t)z \tag{8.74}$$

$$\dot{\tilde{\varrho}} = \gamma\mathrm{sgn}(b_m)\left(\dot{y}_r + \bar{\alpha}_1\right)e_1^{\mathrm{T}}z \tag{8.75}$$

$$\dot{\varepsilon} = A_0\varepsilon. \tag{8.76}$$

The candidate Lyapunov function for system (8.70), (8.74), (8.75), (8.76) is

$$V = \frac{1}{2}z^{\mathrm{T}}z + \frac{1}{2}\tilde{\theta}^{\mathrm{T}}\Gamma^{-1}\tilde{\theta} + \frac{|b_m|}{2\gamma}\tilde{\varrho}^2 + \sum_{i=1}^{\rho}\frac{1}{4d_i}\varepsilon^{\mathrm{T}}P\varepsilon. \tag{8.77}$$

Recalling (8.12), the derivative of V is

$$\dot{V} = z^{\mathrm{T}}\left(A_z + A_z^{\mathrm{T}}\right)z + z^{\mathrm{T}}W_\varepsilon\varepsilon_2 + z^{\mathrm{T}}W_\theta^{\mathrm{T}}\tilde{\theta} - z^{\mathrm{T}}b_m\left(\dot{y}_r + \bar{\alpha}_1\right)e_1\tilde{\varrho}$$

$$-\tilde{\theta}^{\mathrm{T}}W_\theta z + \tilde{\varrho}b_m\left(\dot{y}_r + \bar{\alpha}_1\right)e_1^{\mathrm{T}}z - \sum_{i=1}^{\rho}\frac{1}{4d_i}\varepsilon^{\mathrm{T}}\varepsilon$$

$$= z^{\mathrm{T}}\left(A_z + A_z^{\mathrm{T}}\right)z - \sum_{i=1}^{\rho}z_i\frac{\partial\alpha_{i-1}}{\partial y}\varepsilon_2 - \sum_{i=1}^{\rho}\frac{1}{4d_i}|\varepsilon|^2, \tag{8.78}$$

where for notational convenience we have introduced $\frac{\partial \alpha_0}{\partial y} \triangleq -1$. The skew-symmetry in (8.71) gives us

$$A_z + A_z^T = -2 \begin{bmatrix} c_1 + d_1 & & & \\ & c_2 + d_2\left(\frac{\partial \alpha_1}{\partial y}\right)^2 & & \\ & & \ddots & \\ & & & c_\rho + d_\rho\left(\frac{\partial \alpha_{\rho-1}}{\partial y}\right)^2 \end{bmatrix}, \quad (8.79)$$

which substituted in (8.78) yields

$$\begin{aligned} \dot{V} &= -\sum_{i=1}^{\rho} c_i z_i^2 - \sum_{i=1}^{\rho} d_i \left(\frac{\partial \alpha_{i-1}}{\partial y}\right)^2 z_i^2 - \sum_{i=1}^{\rho} z_i \frac{\partial \alpha_{i-1}}{\partial y} \varepsilon_2 - \sum_{i=1}^{\rho} \frac{1}{4d_i} |\varepsilon|^2 \\ &= -\sum_{i=1}^{\rho} c_i z_i^2 - \sum_{i=1}^{\rho} d_i \left(z_i \frac{\partial \alpha_{i-1}}{\partial y} + \frac{1}{2d_i}\varepsilon_2\right)^2 - \sum_{i=1}^{\rho} \frac{1}{4d_i}\left(\varepsilon_1^2 + \varepsilon_3^2 + \cdots \varepsilon_n^2\right) \\ &\leq -\sum_{i=1}^{\rho} c_i z_i^2. \end{aligned} \quad (8.80)$$

From this inequality we can conclude that $z, \hat{\theta}, \hat{\varrho}$, and ε are bounded. In the next subsection we use this along with the minimum phase Assumption 8.2 to establish the boundedness of all signals and asymptotic tracking.

To guarantee boundedness without adaptation we add (strong) nonlinear damping terms which counteract the parameter estimation error. To motivate the choice of these terms, we first rewrite the error system (8.70). After replacing (8.67) by (8.51) and adding $\pm b_m z_1$ in (8.68), we arrive at

$$\dot{z} = A_z^*(z,t)z + W_\varepsilon(z,t)\varepsilon_2 + W_\theta^*(z,t)^T\tilde{\theta} - b_m(\dot{y}_r + \bar{\alpha}_1)e_1\tilde{\varrho}, \quad (8.81)$$

where the only difference between A_z and A_z^* is that $\hat{b}_m$ at positions $(1,2)$ and $(2,1)$ is replaced by b_m,

$A_z^*(z,t) =$

$$\begin{bmatrix} -c_1-d_1 & b_m & 0 & \cdots & \cdots & 0 \\ -b_m & -c_2-d_2\left(\frac{\partial \alpha_1}{\partial y}\right)^2 & 1+\sigma_{23} & \sigma_{24} & \cdots & \sigma_{2,\rho} \\ 0 & -1-\sigma_{23} & \ddots & \ddots & \ddots & \vdots \\ \vdots & -\sigma_{24} & \ddots & \ddots & \ddots & \sigma_{\rho-2,\rho} \\ \vdots & \vdots & \ddots & \ddots & \ddots & 1+\sigma_{\rho-1,\rho} \\ 0 & -\sigma_{2,\rho} & \cdots & -\sigma_{\rho-2,\rho} & -1-\sigma_{\rho-1,\rho} & -c_\rho-d_\rho\left(\frac{\partial \alpha_{\rho-1}}{\partial y}\right)^2 \end{bmatrix}$$

$$(8.82)$$

8.1 Design with K-Filters

and W_θ^* is given by

$$W_\theta^*(z,t)^{\mathrm{T}} = \begin{bmatrix} \bar{\omega}^{\mathrm{T}} \\ -\frac{\partial \alpha_1}{\partial y}\omega^{\mathrm{T}} + z_1 e_1^{\mathrm{T}} \\ -\frac{\partial \alpha_2}{\partial y}\omega^{\mathrm{T}} \\ \vdots \\ -\frac{\partial \alpha_{\rho-1}}{\partial y}\omega^{\mathrm{T}} \end{bmatrix} \in \mathbb{R}^\rho. \tag{8.83}$$

While the system (8.70) is more adequate for selecting update laws for $\hat{\theta}$ and $\hat{\varrho}$, the system (8.81)–(8.83) is more adequate for achieving boundedness with constant $\tilde{\theta}$ and $\tilde{\varrho}$. Now, to prevent the destabilizing effect of $\tilde{\theta}$, the nonlinear damping terms are chosen as

$$-\kappa_1 |\bar{\omega}|^2 z_1 \tag{8.84}$$

$$-\kappa_2 \left| \frac{\partial \alpha_1}{\partial y}\omega - z_1 e_1 \right|^2 z_2 \tag{8.85}$$

$$-\kappa_i \left| \frac{\partial \alpha_{i-1}}{\partial y}\omega \right|^2 z_i, \quad i = 3, \ldots, \rho \tag{8.86}$$

and added to (8.50), (8.64), (8.63), respectively. In addition, to counteract $\tilde{\varrho}$, (8.48) is replaced by

$$\alpha_1 = \hat{\varrho}\bar{\alpha}_1 - \kappa_1 \operatorname{sgn}(b_m)(\dot{y}_r + \bar{\alpha}_1)^2 z_1. \tag{8.87}$$

With these terms, the error system becomes

$$\begin{aligned}
\dot{z}_1 &= -c_1 z_1 - d_1 z_1 - \kappa_1|\bar{\omega}|^2 z_1 - \kappa_1|b_m|(\dot{y}_r + \bar{\alpha}_1)^2 z_1 + b_m z_2 \\
&\quad + \varepsilon_2 + \bar{\omega}^{\mathrm{T}}\tilde{\theta} - b_m(\dot{y}_r + \bar{\alpha}_1)\tilde{\varrho}
\end{aligned} \tag{8.88}$$

$$\begin{aligned}
\dot{z}_2 &= -c_2 z_2 - d_2 \left(\frac{\partial \alpha_1}{\partial y}\right)^2 z_2 - \kappa_2 \left|\frac{\partial \alpha_1}{\partial y}\omega - z_1 e_1\right|^2 z_2 - b_m z_1 + z_3 + \sum_{j=3}^{\rho} \sigma_{2,j} z_j \\
&\quad -\frac{\partial \alpha_1}{\partial y}\varepsilon_2 - \left(\frac{\partial \alpha_1}{\partial y}\omega - z_1 e_1\right)^{\mathrm{T}}\tilde{\theta}
\end{aligned} \tag{8.89}$$

$$\begin{aligned}
\dot{z}_i &= -c_i z_i - d_i \left(\frac{\partial \alpha_{i-1}}{\partial y}\right)^2 z_i - \kappa_i \left|\frac{\partial \alpha_{i-1}}{\partial y}\omega\right|^2 z_i \\
&\quad -\sum_{j=2}^{i-1} \sigma_{ji} z_j - z_{i-1} + z_{i+1} + \sum_{j=i+1}^{\rho} \sigma_{ij} z_j \\
&\quad -\frac{\partial \alpha_{i-1}}{\partial y}\varepsilon_2 - \frac{\partial \alpha_{i-1}}{\partial y}\omega^{\mathrm{T}}\tilde{\theta}, \quad i = 3, \ldots, \rho.
\end{aligned} \tag{8.90}$$

The system (8.88)–(8.90) satisfies the Lyapunov inequality (8.80) because the nonlinear damping terms enter it as nonpositive. Moreover, for this system we can also prove that

$$\frac{d}{dt}\left(\frac{1}{2}|z|^2\right) \leq -c_0|z|^2 + \frac{1}{4d_0}\varepsilon_2^2 + \frac{1}{4\kappa_0}\left(|\tilde{\theta}|^2 + |b_m|\tilde{\varrho}^2\right), \tag{8.91}$$

where

$$c_0 = \min_{1\leq i\leq \rho} c_i, \quad d_0 = \left(\sum_{i=1}^{\rho}\frac{1}{d_i}\right)^{-1}, \quad \kappa_0 = \left(\sum_{i=1}^{\rho}\frac{1}{\kappa_i}\right)^{-1}. \tag{8.92}$$

Inequality (8.91) shows that, thanks to the nonlinear damping terms, z remains bounded even when adaptation is turned off, in which case $\hat{\theta}$ and $\hat{\varrho}$ are constant, and ε_2 is bounded and exponentially decaying. In the next subsection we prove that the boundedness of z implies the boundedness of all other signals in the adaptive system.

The complete tuning functions design with K-filters is summarized in Table 8.2. It employs the filters given in Table 8.1.

8.1.3 Stability

For the adaptive scheme developed in the previous subsection, we establish the following result.

Theorem 8.5 (Tuning Functions with K-Filters) *The closed-loop adaptive system consisting of the plant (8.3), the control and update laws in Table 8.2, and the filters in Table 8.1 has the following properties:*

1. *If $\Gamma, \gamma > 0$ and $\kappa_i \geq 0$, $i = 1, \ldots, \rho$, then all the signals are globally uniformly bounded, and asymptotic tracking is achieved:*

$$\lim_{t\to\infty}[y(t) - y_r(t)] = 0. \tag{8.105}$$

2. *If $\Gamma, \gamma = 0$ and $\kappa_i > 0$, $i = 1, \ldots, \rho$, then all the signals are globally uniformly bounded.*

Proof. 1. Due to the piecewise continuity of $y_r(t), \ldots, y_r^{(n)}(t)$ and the smoothness of the nonlinearities in (8.3), the solution of the closed-loop adaptive system exists and is unique. Let its maximum interval of existence be $[0, t_f)$.

Along the solutions of (8.70), (8.74), (8.75), (8.76), we established in (8.80) that

$$\dot{V} \leq -c_0|z|^2. \tag{8.106}$$

By (8.77), this implies that $z, \hat{\theta}, \hat{\varrho}, \varepsilon$ are bounded on $[0, t_f)$. Since z_1 and y_r are bounded, y is also bounded. Then, from (8.28) and (8.29) we conclude that ξ and Ξ are bounded. We have yet to prove the boundedness of λ and x. Our

8.1 Design with K-Filters

Table 8.2: Tuning Functions Design with K-Filters

$$z_1 = y - y_r \qquad (8.93)$$
$$z_i = v_{m,i} - \hat{\varrho} y_r^{(i-1)} - \alpha_{i-1}, \qquad i = 2, \ldots, \rho \qquad (8.94)$$

$$\alpha_1 = \hat{\varrho}\bar{\alpha}_1 - \kappa_1 \mathrm{sgn}(b_m)\left(\dot{y}_r + \bar{\alpha}_1\right)^2 z_1 \qquad (8.95)$$
$$\bar{\alpha}_1 = -\left(c_1 + d_1 + \kappa_1|\bar{\omega}|^2\right)z_1 - \omega_0 - \bar{\omega}^T\hat{\theta} \qquad (8.96)$$
$$\alpha_2 = -\hat{b}_m z_1 - \left[c_2 + d_2\left(\frac{\partial\alpha_1}{\partial y}\right)^2 + \kappa_2\left|\frac{\partial\alpha_1}{\partial y}\omega - z_1 e_1\right|^2\right] z_2$$
$$\qquad + \left(\dot{y}_r + \frac{\partial\alpha_1}{\partial\hat{\varrho}}\right)\dot{\hat{\varrho}} + \frac{\partial\alpha_1}{\partial\hat{\theta}}\Gamma\tau_2 + \beta_2 \qquad (8.97)$$
$$\alpha_i = -z_{i-1} - \left[c_i + \left(d_i + \kappa_i|\omega|^2\right)\left(\frac{\partial\alpha_{i-1}}{\partial y}\right)^2\right] z_i + \left(y_r^{(i-1)} + \frac{\partial\alpha_{i-1}}{\partial\hat{\varrho}}\right)\dot{\hat{\varrho}}$$
$$\qquad + \frac{\partial\alpha_{i-1}}{\partial\hat{\theta}}\Gamma\tau_i - \sum_{j=2}^{i-1}\frac{\partial\alpha_{j-1}}{\partial\hat{\theta}}\Gamma\frac{\partial\alpha_{i-1}}{\partial y}z_j + \beta_i, \qquad i = 3,\ldots,\rho \qquad (8.98)$$
$$\beta_i = \frac{\partial\alpha_{i-1}}{\partial y}\left(\omega_0 + \omega^T\hat{\theta}\right) + \frac{\partial\alpha_{i-1}}{\partial\xi}\left(A_0\xi + ky + \phi\right) + \frac{\partial\alpha_{i-1}}{\partial\Xi}\left(A_0\Xi + \Phi\right)$$
$$\qquad + \sum_{j=1}^{i-1}\frac{\partial\alpha_{i-1}}{\partial y_r^{(j-1)}}y_r^{(j)} + k_i v_{m,1} + \sum_{j=1}^{m+i-1}\frac{\partial\alpha_{i-1}}{\partial\lambda_j}(-k_j\lambda_1 + \lambda_{j+1}) \qquad (8.99)$$

$$\tau_1 = \left(\omega - \hat{\varrho}\left(\dot{y}_r + \bar{\alpha}_1\right)e_1\right)z_1 \qquad (8.100)$$
$$\tau_i = \tau_{i-1} - \frac{\partial\alpha_{i-1}}{\partial y}\omega z_i, \qquad i = 2,\ldots,\rho \qquad (8.101)$$

Adaptive control law:

$$u = \frac{1}{\sigma(y)}\left(\alpha_\rho - v_{m,\rho+1} + \hat{\varrho} y_r^{(\rho)}\right) \qquad (8.102)$$

Parameter update laws:

$$\dot{\hat{\theta}} = \Gamma\tau_\rho \qquad (8.103)$$
$$\dot{\hat{\varrho}} = -\gamma\mathrm{sgn}(b_m)\left(\dot{y}_r + \bar{\alpha}_1\right)z_1 \qquad (8.104)$$

main concern is λ because the boundedness of x follows from the boundedness of ε, ξ, Ξ and λ.

The input filter (8.30) gives

$$\lambda_i = \frac{s^{i-1} + k_1 s^{i-2} + \cdots + k_{i-1}}{K(s)} [\sigma(y)u], \qquad i = 1, \ldots, n, \qquad (8.107)$$

where $K(s) = s^n + k_1 s^{n-1} + \cdots + k_0$. On the other hand, for the plant (8.1) one can show that

$$\frac{d^n y}{dt^n} - \sum_{i=1}^{n} \frac{d^{n-i}}{dt^{n-i}} \left[\varphi_{0,i}(y) + \Phi_{(i)}(y)a\right] = \sum_{i=0}^{m} b_i \frac{d^i}{dt^i} [\sigma(y)u]. \qquad (8.108)$$

Noting that the last sum is $B(s)[\sigma(y)u]$ and substituting (8.108) into (8.107) we get

$$\lambda_i = \frac{s^{i-1} + k_1 s^{i-2} + \cdots + k_{i-1}}{K(s)B(s)} \left\{\frac{d^n y}{dt^n} - \sum_{i=1}^{n} \frac{d^{n-i}}{dt^{n-i}} \left[\varphi_{0,i}(y) + \Phi_{(i)}(y)a\right]\right\},$$

$$i = 1, \ldots, n. \qquad (8.109)$$

The boundedness of y, the smoothness of $\phi(y)$ and $\Phi(y)$, Assumption 8.2, and (8.109) imply that $\lambda_1, \ldots, \lambda_{m+1}$ are bounded. We now return to the coordinate change (8.94), which gives

$$v_{m,i} = z_i + \hat{\varrho} y_r^{(i-1)} + \alpha_{i-1}\left(y, \xi, \Xi, \hat{\theta}, \hat{\varrho}, \bar{\lambda}_{m+i-1}, \bar{y}_r^{(i-2)}\right), \qquad i = 2, \ldots, \rho. \qquad (8.110)$$

Let $i = 2$. The boundedness of $\bar{\lambda}_{m+1}$, along with the boundedness of z_2 and $y, \xi, \Xi, \hat{\theta}, \hat{\varrho}, y_r, \dot{y}_r$, proves that $v_{m,2}$ is bounded. Then from (8.56) it follows that λ_{m+2} is bounded. Continuing in the same fashion, (8.110) and (8.56) recursively establish that λ is bounded. Finally, in view of (8.19), (8.31), (8.32), and the boundedness of ξ, Ξ, λ, and ε, we conclude that x is bounded. Since $\sigma(y)$ is bounded away from zero, u is bounded.

We have thus shown that all of the signals of the closed-loop adaptive system are bounded on $[0, t_f)$ by constants depending only on the initial conditions, design gains, and the external signals $y_r(t), \ldots, y_r^{(n)}(t)$, but not on t_f. This proves that $t_f = \infty$. Hence, all signals are globally uniformly bounded for all $t \geq 0$.

By applying the LaSalle-Yoshizawa theorem (Theorem 2.1) to (8.106), it further follows that $z(t) \to 0$ as $t \to \infty$, which implies that $\lim_{t \to \infty} [y(t) - y_r(t)] = 0$.

2. The boundedness of all the signals without adaptation follows from (8.91), by repeating the argument from point 1. □

Theorem 8.5 established global uniform boundedness of all signals but not global uniform stability of individual solutions. To refer such a stability property to the origin, we now determine an error system such that all of its states

8.1 Design with K-Filters

except the parameter error converge to zero. We start with the subsystem $(z, \varepsilon, \tilde{\theta}, \tilde{\varrho})$ whose $2n + q + 2$ states are encompassed by the Lyapunov function (8.77). Then we derive additional equations to complete the error system.

For filter states we introduce the reference signals ξ^r and Ξ^r defined by

$$\dot{\xi}^r = A_0 \xi^r + k y_r + \phi(y_r) \tag{8.111}$$
$$\dot{\Xi}^r = A_0 \Xi^r + \Phi(y_r). \tag{8.112}$$

This allows us to define the error states $\tilde{\xi} = \xi - \xi^r$ and $\tilde{\Xi}_i = \Xi_i - \Xi_i^r$ governed by

$$\dot{\tilde{\xi}} = A_0 \tilde{\xi} + k z_1 + \tilde{\phi}(z_1, y_r) z_1 \tag{8.113}$$
$$\dot{\tilde{\Xi}} = A_0 \tilde{\Xi} + \tilde{\Phi}(z_1, y_r) z_1, \tag{8.114}$$

where $\tilde{\phi}$ and $\tilde{\Phi}$ are smooth functions defined by the mean-value theorem. The system $(z, \varepsilon, \tilde{\xi}, \tilde{\Xi}, \tilde{\theta}, \tilde{\varrho})$ has $(q+3)n + q + 2$ states, while the original $(x, \xi, \Xi, \lambda, \hat{\theta}, \hat{\varrho})$ system has $(q+3)n + q + m + 2$ states. We recover the missing m error states in the inverse dynamics of (8.3). Let us consider the similarity transformation

$$\begin{bmatrix} x_1 \\ \vdots \\ x_\rho \\ \zeta \end{bmatrix} = \begin{bmatrix} x_1 \\ \vdots \\ x_\rho \\ Tx \end{bmatrix} = \begin{bmatrix} I_\rho & 0_{\rho \times m} \\ & T \end{bmatrix} x, \tag{8.115}$$

where

$$T = [A_b^\rho e_1, \ldots, A_b e_1, I_m] \tag{8.116}$$

$$A_b = \begin{bmatrix} -b_{m-1}/b_m & & \\ \vdots & & I_{m-1} \\ -b_0/b_m & 0 & \cdots & 0 \end{bmatrix}. \tag{8.117}$$

The following two identities are readily verified:

$$T \begin{bmatrix} 0 \\ b \end{bmatrix} = 0, \quad TA = A_b T + T A^\rho \begin{bmatrix} 0 \\ b \end{bmatrix} e_1^T. \tag{8.118}$$

With these identities the inverse dynamics of (8.3) are expressed as

$$\dot{\zeta} = A_b \zeta + T \left(A^\rho \begin{bmatrix} 0 \\ b \end{bmatrix} y + \phi(y) + \Phi(y) a \right). \tag{8.119}$$

Introducing the reference signal ζ^r as

$$\dot{\zeta}^r = A_b \zeta^r + T \left(A^\rho \begin{bmatrix} 0 \\ b \end{bmatrix} y_r + \phi(y_r) + \Phi(y_r) a \right), \tag{8.120}$$

we see that the error state $\tilde{\zeta} = \zeta - \zeta^r$ is governed by

$$\begin{aligned}
\dot{\tilde{\zeta}} &= A_b \tilde{\zeta} + T \left(A^\rho \begin{bmatrix} 0 \\ b \end{bmatrix} z_1 + \tilde{\phi}(z_1, y_r) z_1 + \tilde{\Phi}(z_1, y_r) z_1 a \right) \\
&\triangleq A_b \tilde{\zeta} + \tilde{\varphi}_b(z_1, y_r) z_1 \, .
\end{aligned} \qquad (8.121)$$

We have thus constructed the error system

$$\begin{aligned}
\dot{z} &= A_z(z,t) z + W_\varepsilon(z,t)\varepsilon_2 + W_\theta(z,t)^T \tilde{\theta} - b_m \left(\dot{y}_r + \bar{\alpha}_1 \right) e_1 \tilde{\varrho} & (8.122) \\
\dot{\varepsilon} &= A_0 \varepsilon & (8.123) \\
\dot{\tilde{\zeta}} &= A_b \tilde{\zeta} + \tilde{\varphi}_b(z_1, y_r) z_1 & (8.124) \\
\dot{\tilde{\xi}} &= A_0 \tilde{\xi} + k z_1 + \tilde{\phi}(z_1, y_r) z_1 & (8.125) \\
\dot{\tilde{\Xi}} &= A_0 \tilde{\Xi} + \tilde{\Phi}(z_1, y_r) z_1 & (8.126) \\
\dot{\tilde{\theta}} &= -\Gamma W_\theta(z,t) z & (8.127) \\
\dot{\tilde{\varrho}} &= \gamma \, \mathrm{sgn}(b_m) \left(\dot{y}_r + \bar{\alpha}_1 \right) e_1^T z & (8.128)
\end{aligned}$$

whose stability and regulation properties are established in the following corollary.

Corollary 8.6 *The error system (8.122)–(8.128) has a globally uniformly stable equilibrium at the origin. Its $(q+3)n + q + m + 2$-dimensional state converges to the $q + m + 2$-dimensional manifold*

$$M = \left\{ z = 0, \; \varepsilon = 0, \; \tilde{\zeta} = 0, \; \tilde{\xi} = 0, \; \tilde{\Xi} = 0 \right\}. \qquad (8.129)$$

Proof. From (8.106) and (8.77) it follows that there exists a positive number ν such that

$$\left| \left(z(t), \varepsilon(t), \tilde{\theta}(t), \tilde{\varrho}(t) \right) \right| \leq \nu \left| \left(z(t_0), \varepsilon(t_0), \tilde{\theta}(t_0), \tilde{\varrho}(t_0) \right) \right|, \qquad \forall t \geq t_0 \geq 0 \, . \qquad (8.130)$$

Let us now consider (8.124). Due to the boundedness of y_r and the smoothness of $\tilde{\varphi}_b$, in view of (8.130), there exists a class $\mathcal{K}_\infty$ function γ_ζ such that

$$\left| \tilde{\varphi}_b(z_1(t), y_r(t)) z_1(t) \right| \leq \gamma_\zeta \left(\left| \left(z(t_0), \varepsilon(t_0), \tilde{\theta}(t_0), \tilde{\varrho}(t_0) \right) \right| \right), \qquad \forall t \geq t_0 \geq 0 \, . \qquad (8.131)$$

Then, since A_b is Hurwitz, there exist positive numbers ν_1 and ν_2 such

$$\left| \tilde{\zeta}(t) \right| \leq \nu_1 \left| \tilde{\zeta}(t_0) \right| + \nu_2 \gamma_\zeta \left(\left| \left(z(t_0), \varepsilon(t_0), \tilde{\theta}(t_0), \tilde{\varrho}(t_0) \right) \right| \right), \qquad \forall t \geq t_0 \geq 0 \, . \qquad (8.132)$$

By the same reasoning, we establish bounds for $\tilde{\xi}$ and $\tilde{\Xi}$ analogous to (8.132). Hence, for the complete error state $\mathcal{E} = (z, \varepsilon, \tilde{\zeta}, \tilde{\xi}, \tilde{\Xi}, \tilde{\theta}, \tilde{\varrho})$ we have proven that

$$|\mathcal{E}(t)| \leq \gamma \left(|\mathcal{E}(t_0)| \right), \qquad \forall t \geq t_0 \geq 0, \qquad (8.133)$$

8.1 DESIGN WITH K-FILTERS

where γ is a class $\mathcal{K}_\infty$ function. By Definition A.4, the equilibrium $\mathcal{E} = 0$ is globally uniformly stable.

To establish convergence to the manifold M, we recall that in Theorem 8.5 we showed that z converges to zero, and so does ε because of (8.18). Since the vector $\tilde{\varphi}_b(z_1, y_r)z_1$ in (8.124) converges to zero, then by [28, Theorem IV.1.9] $\tilde{\zeta}$ converges to zero. The same reasoning proves that $\tilde{\xi}, \tilde{\Xi}$ converge to zero. $\square$

Corollary 8.6 establishes global uniform stability, a property stronger than global uniform boundedness. Moreover, it proves the regulation of all the error states except possibly the parameter estimation error.

However, Corollary 8.6 does not establish a correspondence between the two systems—the error system $(z, \varepsilon, \tilde{\zeta}, \tilde{\xi}, \tilde{\Xi}, \tilde{\theta}, \tilde{\varrho})$ and the original system $(x, \lambda, \xi, \Xi, \hat{\theta}, \hat{\varrho})$. We need to show that the coordinate change

$$(x, \lambda, \xi, \Xi, \hat{\theta}, \hat{\varrho}) \mapsto (z, \varepsilon, \tilde{\zeta}, \tilde{\xi}, \tilde{\Xi}, \tilde{\theta}, \tilde{\varrho}) \tag{8.134}$$

is a global C^∞-diffeomorphism for each $t \geq 0$, whenever $B(s)$ and $K(s)$ are coprime. Although the coefficients of $B(s)$ are unknown, this condition is satisfied with probability one.

We only indicate the main idea of the proof that (8.134) is smoothly invertible for each $t \geq 0$. Since $\tilde{\xi} = \xi - \xi^{\mathrm{r}}(t)$, $\tilde{\Xi} = \Xi - \Xi^{\mathrm{r}}(t)$, $\tilde{\theta} = \theta - \hat{\theta}$, and $\tilde{\varrho} = \varrho - \hat{\varrho}$, it is clear that we only need to consider the portions (x, λ) and $(z, \varepsilon, \tilde{\zeta})$ in (8.134). In view of (8.19), (8.31), and (8.32), we have

$$\begin{aligned} x &= \varepsilon + \xi + \Xi a + B(A_0)\lambda \\ &= \varepsilon + \tilde{\xi} + \tilde{\Xi}a + \xi^{\mathrm{r}} + \Xi^{\mathrm{r}}a + B(A_0)\lambda, \end{aligned} \tag{8.135}$$

where $B(A_0) = \sum_{i=0}^{m} b_i A_0^i$. By multiplying (8.135) with T, recalling (8.115), we get

$$TB(A_0)\lambda = \tilde{\zeta} + \zeta^{\mathrm{r}} - T\left(\varepsilon + \tilde{\xi} + \tilde{\Xi}a + \xi^{\mathrm{r}} + \Xi^{\mathrm{r}}a\right). \tag{8.136}$$

It is not hard to prove that

$$TB(A_0) = [T_m \; 0_{m \times \rho}], \tag{8.137}$$

where $T_m \in \mathbb{R}^{m \times m}$. Then, since T is full rank, T_m is nonsingular if and only if $B(A_0)$ is nonsingular. By the fact that the eigenvalues of $B(A_0)$ satisfy $\lambda_i\{B(A_0)\} = B(\lambda_i\{A_0\})$, $i = 1, \ldots, n$, T_m is nonsingular iff $B(s)$ and $K(s)$ are coprime. Consequently, under this condition, $\bar{\lambda}_m$ is a smooth function of $\tilde{\zeta}, \varepsilon, \tilde{\xi}, \tilde{\Xi}$, for each t. Multiplying the identity (8.135) by e_1^{T} from the left and noting that $e_1^{\mathrm{T}} B(A_0) = [*, \ldots, *, 1, 0, \ldots, 0]$, where 1 is the $(m+1)$st entry, one can see that λ_{m+1} is a smooth function of $z_1, \tilde{\zeta}, \varepsilon, \tilde{\xi}, \tilde{\Xi}$, for each t. The rest of the proof exploits (8.110) along with (8.56) to establish that λ is a smooth function of $z, \tilde{\zeta}, \varepsilon, \tilde{\xi}, \tilde{\Xi}$, for each t. Thanks to (8.135), so is x. Thus (8.134) is one-to-one, onto, smooth, and has a smooth inverse for each t, iff $B(s)$ and $K(s)$ are coprime.

Note that the singularity in the coordinate transformation (8.134), which occurs when $B(s)$ and $K(s)$ are coprime, is not in contradiction with the boundedness result of Theorem 8.5 where a different argument was used (cf. (8.107)–(8.109)).

8.1.4 Transient performance

While the stability and convergence results obtained thus far prove that the designed adaptive system has desired asymptotic performance, we now analyze its transient performance by deriving $\mathcal{L}_2$ and $\mathcal{L}_\infty$ transient performance bounds for the error state z. For simplicity, we let $\Gamma = \gamma I$.

Theorem 8.7 (Tuning Functions with K-Filters) *In the adaptive system (8.3), (8.28), (8.29), (8.30), (8.102), (8.103), (8.104), the following inequalities hold:*

$$\|z\|_2 \leq \frac{1}{\sqrt{2c_0}} \left[\frac{1}{\gamma} \left(|\tilde{\theta}(0)|^2 + |b_m| \tilde{\varrho}(0)^2 \right) + \frac{1}{2d_0} |\varepsilon(0)|_P^2 \right]^{1/2}$$
$$+ \frac{1}{\sqrt{2c_0}} |z(0)| \tag{8.138}$$

$$|z(t)| \leq \frac{1}{2\sqrt{c_0}} \left[\frac{1}{\kappa_0} \left(|\tilde{\theta}(0)|^2 + |b_m| \tilde{\varrho}(0)^2 \right) + \frac{1}{d_0} \left(\frac{\gamma}{2\kappa_0} + \frac{1}{\lambda(P)} \right) |\varepsilon(0)|_P^2 \right.$$
$$\left. + \frac{\gamma}{\kappa_0} |z(0)|^2 \right]^{1/2} + |z(0)| e^{-c_0 t}. \tag{8.139}$$

Proof. The bound (8.138) follows immediately by integrating (8.106) over $[0, \infty)$,

$$\|z\|_2^2 = \int_0^\infty |z(\tau)|^2 d\tau \leq -\frac{1}{c_0} \int_0^\infty \dot{V}(\tau) d\tau \leq \frac{1}{c_0} V(0), \tag{8.140}$$

and by recalling that the Lyapunov function V is given by (8.77).

The derivation of the bound (8.139) starts with (8.91). From Lemma C.5(i), it follows that

$$|z(t)|^2 \leq |z(0)|^2 e^{-2c_0 t} + \frac{1}{4c_0} \left[\frac{1}{d_0} \|\varepsilon_2\|_\infty^2 + \frac{1}{\kappa_0} \left\| |\tilde{\theta}|^2 + |b_m| \tilde{\varrho}^2 \right\|_\infty \right]. \tag{8.141}$$

From (8.18) and (8.12) we have $\frac{d}{dt} |\varepsilon|_P^2 \leq -|\varepsilon|^2$, which gives

$$\|\varepsilon_2\|_\infty^2 \leq \frac{1}{\lambda(P)} |\varepsilon(0)|_P^2. \tag{8.142}$$

On the other hand, (8.106) and (8.77) yield

$$\left\| |\tilde{\theta}|^2 + |b_m| \tilde{\varrho}^2 \right\|_\infty \leq 2\gamma V(0). \tag{8.143}$$

8.1 DESIGN WITH K-FILTERS

By substituting (8.142) and (8.143) into (8.141) and expressing $V(0)$ from (8.77), we get (8.139). □

While the $\mathcal{L}_2$ bound (8.138) holds even for $\kappa_0 = 0$, that is, without the nonlinear damping terms (8.84)–(8.87), the $\mathcal{L}_\infty$ bound (8.139) is valid only if $\kappa_0 > 0$.

It is of interest to compare these bounds with the bounds (4.226) and (4.231) for the state feedback case. As expected, the bounds obtained with output feedback have an additional term due to the initial state estimation error $\varepsilon(0)$.

The initial condition $z(0)$ in the above bounds is, in general, dependent on the design parameters $c_0, d_0, \kappa_0, \gamma$. However, as explained in Section 4.3.2 for state feedback, with trajectory initialization we can set $z(0) = 0$. Following (8.93) and (8.94), $z(0)$ is set to zero by selecting

$$y_r(0) = y(0) \tag{8.144}$$

$$y_r^{(i)}(0) = \frac{1}{\hat{\varrho}(0)} \left[v_{m,i+1}(0) - \alpha_i \left(y(0), \xi(0), \Xi(0), \hat{\theta}(0), \hat{\varrho}(0), \bar{\lambda}_{m+i}(0), \bar{y}_r^{(i-1)}(0) \right) \right],$$
$$i = 1, \ldots, \rho - 1. \tag{8.145}$$

Since $b_m \neq 0$, it is reasonable to choose $\hat{b}_m(0) \neq 0$. Then the choice $\hat{\varrho}(0) = 1/\hat{b}_m(0)$ makes (8.145) well-defined. After $z(0)$ is set to zero, the bounds (8.138) and (8.139) become

$$\|z\|_2 \leq \frac{1}{\sqrt{2c_0}} \left[\frac{1}{\gamma} \left(|\tilde{\theta}(0)|^2 + |b_m| \tilde{\varrho}(0)^2 \right) + \frac{1}{2d_0} |\varepsilon(0)|_P^2 \right]^{1/2} \tag{8.146}$$

$$|z(t)| \leq \frac{1}{2\sqrt{c_0}} \left[\frac{1}{\kappa_0} \left(|\tilde{\theta}(0)|^2 + |b_m| \tilde{\varrho}(0)^2 \right) \right.$$
$$\left. + \frac{1}{d_0} \left(\frac{\gamma}{2\kappa_0} + \frac{1}{\underline{\lambda}(P)} \right) |\varepsilon(0)|_P^2 \right]^{1/2}. \tag{8.147}$$

Both of these bounds can be systematically reduced by increasing c_0. Other options for reducing the values of the bounds are as follows. The $\mathcal{L}_2$ bound (8.146) can be reduced by simultaneously increasing γ and d_0. The $\mathcal{L}_\infty$ bound (8.147) can be reduced by simultaneously increasing κ_0 and d_0. In fact, the $\mathcal{L}_\infty$ bound (8.147) can also be systematically reduced by simultaneously increasing γ and d_0. This is easy to see by noting that (8.106) guarantees that $|z(t)|^2 \leq 2V(0)$, which implies

$$|z(t)| \leq \left[\frac{1}{\gamma} \left(|\tilde{\theta}(0)|^2 + |b_m| \tilde{\varrho}(0)^2 \right) + \frac{1}{2d_0} |\varepsilon(0)|_P^2 + |z(0)|^2 \right]^{1/2}, \tag{8.148}$$

where $z(0)$ can be set to zero with trajectory initialization (8.144)–(8.145).

8.2 Design with MT-Filters

8.2.1 Filtered transformations and observer

The design with MT-filters is motivated by the idea of using an adaptive observer for output-feedback control. The main tool in Chapter 5 was the passivity property of the observer error system. This property was easily achieved there because the full state of the plant was measured. When only the output is measured, this property is difficult to achieve.

This difficulty is overcome by filtered transformations introduced in [123], which bring the system (8.7) into a special form. We follow this idea, but we introduce significant modifications which make the filtered transformations easier to understand and implement.

The system (8.7) is rewritten here for convenience:

$$\begin{aligned} \dot{x} &= Ax + \phi(y) + F(y,u)^{\text{T}}\theta \\ y &= x_1, \end{aligned} \qquad (8.149)$$

where

$$\theta = \begin{bmatrix} b \\ a \end{bmatrix} \qquad (8.150)$$

$$F(y,u)^{\text{T}} = \left[\begin{bmatrix} 0_{(\rho-1)\times(m+1)} \\ I_{m+1} \end{bmatrix} \sigma(y)u, \ \Phi(y) \right]. \qquad (8.151)$$

In this design we employ the filters

$$\dot{\xi} = A_l \xi + B_l \phi(y), \qquad \xi \in \mathbb{R}^{n-1} \qquad (8.152)$$
$$\dot{\Omega}^{\text{T}} = A_l \Omega^{\text{T}} + B_l F(y,u)^{\text{T}}, \qquad \Omega \in \mathbb{R}^{p\times(n-1)}, \qquad (8.153)$$

where A_l and B_l are given by

$$A_l = \begin{bmatrix} -\bar{l} & I_{n-2} \\ 0 & \cdots & 0 \end{bmatrix}, \qquad B_l = \begin{bmatrix} -\bar{l}, & I_{n-1} \end{bmatrix}, \qquad (8.154)$$

and the vectors $\bar{l}$ and l are defined via the coefficients of the $(n-1)$-dimensional Hurwitz polynomial $L(s) = s^{n-1} + l_1 s^{n-2} + \cdots + l_{n-1}$:

$$l = \begin{bmatrix} 1 \\ l_0 \\ \vdots \\ l_{n-1} \end{bmatrix} \triangleq \begin{bmatrix} 1 \\ \bar{l} \end{bmatrix}. \qquad (8.155)$$

The filtered transformation, defined as

$$\chi = x - \begin{bmatrix} 0 \\ \xi + \Omega^{\text{T}}\theta \end{bmatrix}, \qquad (8.156)$$

8.2 DESIGN WITH MT-FILTERS

is readily shown to bring the system (8.149) to the *adaptive observer form*

$$\begin{aligned} \dot{\chi} &= A\chi + l\left(\omega_0 + \omega^T\theta\right) \\ y &= \chi_1, \end{aligned} \quad (8.157)$$

where

$$\omega_0 = \varphi_{0,1} + \xi_1 \quad (8.158)$$
$$\omega = F_{(1)} + \Omega_{(1)}. \quad (8.159)$$

The system (8.157) is minimum phase and, considering $\omega_0 + \omega^T\theta$ as its input, its relative degree is one. This "reduction" in relative degree is a property which makes it possible to employ passivity in the design of an adaptive observer.

As before, an important reduction in the dynamic order of the Ω-filter is possible by exploiting the structure of $F(y, u)$ in (8.151). Denote the first $m+1$ columns of Ω^T by $v_m, \ldots, v_1, v_0$. Due to the special dependence of $F(y, u)$ on $\sigma(y)u$, the vectors $v_m, \ldots, v_1, v_0$ are governed by

$$\dot{v}_j = A_l v_j + B_l e_{n-j}\, \sigma(y)u, \quad j = 0, \ldots, m. \quad (8.160)$$

It easy to show that

$$A_l^j e_n = B_l e_{n-j}, \quad j = 0, \ldots, n-1. \quad (8.161)$$

Therefore, the vectors v_l can be obtained from only one implemented filter

$$\dot{\lambda} = A_l \lambda + e_{n-1}\sigma u, \quad \lambda \in \mathbb{R}^{n-1} \quad (8.162)$$

through the algebraic expressions

$$v_j = A_l^j \lambda, \quad j = 0, \ldots, m. \quad (8.163)$$

While we always implement the filter (8.162), for analysis we use the equations (8.160). Now, in view of (8.151) and (8.153), we have

$$\Omega^T = [v_m, \ldots, v_1, v_0, \Xi], \quad (8.164)$$

where

$$\dot{\Xi} = A_0 \Xi + \Phi(y), \quad \Xi \in \mathbb{R}^{(n-1) \times q}. \quad (8.165)$$

The implemented filters are summarized in Table 8.3. We recall that the dynamic order of the K-filters is $n(q+2)$ and can be reduced to $(n-1)(q+2)$ using the reduced-order observer technique. The total dynamic order of the MT-filters is $(n-1)(q+2)$. However, the design with MT-filters also employs an observer of χ, which makes the dynamic order of the complete adaptive scheme with MT-filters higher than with K-filters.

Table 8.3: MT-Filters

MT-filters:		
$\dot{\xi} = A_l \xi + B_l \phi(y)$		(8.166)
$\dot{\Xi} = A_l \Xi + B_l \Phi(y)$		(8.167)
$\dot{\lambda} = A_l \lambda + e_{n-1} \sigma(y) u$		(8.168)
$v_j = A_l^j \lambda, \quad j = 0, \ldots, m$		(8.169)
$\Omega^{\mathrm{T}} = [v_m, \ldots, v_1, v_0, \Xi]$		(8.170)

Observer. The simplest observer for (8.157) is

$$\dot{\hat{\chi}} = A\hat{\chi} + K_o(y - \hat{\chi}_1) + l\left(\omega_0 + \omega^{\mathrm{T}}\hat{\theta}\right), \tag{8.171}$$

where

$$K_o = (A + c_o I) l. \tag{8.172}$$

It is easy to show that, with this choice, $A_o = A - K_o e_1^{\mathrm{T}}$ satisfies $\det(sI - A_o) = (s + c_o) L(s)$, namely,

$$e_1^{\mathrm{T}} (sI - A_o)^{-1} l = \frac{1}{s + c_o}. \tag{8.173}$$

The observer error

$$\varepsilon = \chi - \hat{\chi} \tag{8.174}$$

is governed by

$$\dot{\varepsilon} = A_o \varepsilon + l \omega^{\mathrm{T}} \tilde{\theta}. \tag{8.175}$$

Since $\varepsilon_1 = \dfrac{1}{s + c_o} \left[\omega^{\mathrm{T}} \tilde{\theta}\right]$ is a strictly passive transfer function, the update law

$$\dot{\hat{\theta}} = \Gamma \omega \varepsilon_1 \tag{8.176}$$

guarantees the boundedness of $\tilde{\theta}$ and ε, as well as the square-integrability of ε. This update law will be the starting point in building the tuning functions update law in the next subsection.

To prepare for the backstepping design in the next subsection, we rewrite the output $y = \chi_1$ from (8.157) as

$$\dot{y} = \chi_2 + \omega_0 + \omega^{\mathrm{T}} \hat{\theta}. \tag{8.177}$$

8.2 DESIGN WITH MT-FILTERS

Noting from (8.151) that $F_{(1)} = [\underbrace{0,\ldots,0}_{m+1}, \Phi_{(1)}]^T$, expressions (8.159) and (8.170) yield

$$\omega = [v_{m,1}, v_{m-1,1}, \ldots, v_{0,1},\ \Phi_{(1)} + \Xi_{(1)}]^T. \tag{8.178}$$

Combining (8.177) and (8.178), we obtain the following two important expressions for $\dot{y}$:

$$\dot{y} = \hat{\chi}_2 + \omega_0 + \omega^T\theta + \varepsilon_2 \tag{8.179}$$
$$= b_m v_{m,1} + \hat{\chi}_2 + \omega_0 + \bar{\omega}^T\theta + \varepsilon_2, \tag{8.180}$$

where

$$\bar{\omega} = [0, v_{m-1,1}, \ldots, v_{0,1},\ \Phi_{(1)} + \Xi_{(1)}]^T. \tag{8.181}$$

8.2.2 Adaptive controller design

Equation (8.180) is written in a form to suggest that the filter signal $v_{m,1}$ will play the role of a virtual control. The system to which we apply backstepping is

$$\dot{y} = b_m v_{m,1} + \hat{\chi}_2 + \omega_0 + \bar{\omega}^T\theta + \varepsilon_2 \tag{8.182}$$
$$\dot{v}_{m,i} = v_{m,i+1} - k_i v_{m,1}, \qquad i = 1,\ldots,\rho-1 \tag{8.183}$$
$$\dot{v}_{m,\rho-1} = \sigma(y)u + v_{m,\rho} - k_\rho v_{m,1}. \tag{8.184}$$

The similarity between the systems (8.182)–(8.184) and (8.42)–(8.44) indicates that the backstepping procedure with MT filters will be similar to the backstepping procedure with K-filters. The most important difference is in the the way the state estimation influences the design of the update law. While with the K-filters the system (8.18) is autonomous and exponentially stable, with the MT-filters the observer error system (8.175) is driven by the parameter error. The parameter update law has to take this into account. The strict passivity of $\varepsilon_1 = \dfrac{1}{s+c_o}\left[\omega^T\tilde{\theta}\right]$ dictates the choice of the tuning function

$$\tau_0 = \nu\omega\varepsilon_1, \quad \nu > 0, \tag{8.185}$$

which is the starting point in the design of the update law. The coefficient ν is a weight assigned to $\omega\varepsilon_1$ in the final update law.

Following the idea in Section 4.5.1, for system (8.182)–(8.184) we define

$$z_1 = y - y_r \tag{8.186}$$
$$z_i = v_{m,i-1} - \hat{\varrho} y_r^{(i-1)} - \alpha_{i-1}, \qquad i = 2,\ldots,\rho, \tag{8.187}$$

where $\hat{\varrho}$ is an estimate of $\varrho = 1/b_m$.

Step 1. The equation for the tracking error z_1 obtained from (8.186) and (8.182) is

$$\dot{z}_1 = b_m v_{m,1} + \hat{\chi}_2 + \omega_0 + \bar{\omega}^T \theta + \varepsilon_2 - \dot{y}_r. \tag{8.188}$$

By substituting (8.187) for $i = 2$ into (8.188) and scaling the first stabilizing function α_1 as

$$\alpha_1 = \hat{\varrho}\bar{\alpha}_1, \tag{8.189}$$

(8.188) becomes

$$\dot{z}_1 = \bar{\alpha}_1 + \hat{\chi}_2 + \omega_0 + \bar{\omega}^T \theta + \varepsilon_2 - b_m \left(\dot{y}_r + \bar{\alpha}_1\right) \tilde{\varrho} + b_m z_2. \tag{8.190}$$

Then the choice

$$\bar{\alpha}_1 = -c_1 z_1 - d_1 z_1 - \hat{\chi}_2 - \omega_0 - \bar{\omega}^T \hat{\theta} \tag{8.191}$$

results in

$$\dot{z}_1 = -c_1 z_1 - d_1 z_1 + \varepsilon_2 + \bar{\omega}^T \tilde{\theta} - b_m \left(\dot{y}_r + \bar{\alpha}_1\right) \tilde{\varrho} + b_m z_2. \tag{8.192}$$

Substituting (8.52) into (8.192) we get

$$\dot{z}_1 = -c_1 z_1 - d_1 z_1 + \varepsilon_2 + \left(\omega - \hat{\varrho}\left(\dot{y}_r + \bar{\alpha}_1\right) e_1\right)^T \tilde{\theta} - b_m \left(\dot{y}_r + \bar{\alpha}_1\right) \tilde{\varrho} + \hat{b}_m z_2. \tag{8.193}$$

To select the first tuning function, we recall that the parameter error appearing in the observer error system (8.175) already determined the tuning function τ_0 in (8.185), which we now augment with a term corresponding to (8.193):

$$\tau_1 = \tau_0 + \left(\omega - \hat{\varrho}\left(\dot{y}_r + \bar{\alpha}_1\right) e_1\right) z_1. \tag{8.194}$$

While the choice of the update law for $\hat{\theta}$ is postponed until the last step, the update law for $\hat{\varrho}$ is selected at this step:

$$\dot{\hat{\varrho}} = -\gamma \operatorname{sgn}(b_m) \left(\dot{y}_r + \bar{\alpha}_1\right) z_1, \qquad \gamma > 0. \tag{8.195}$$

We pause to clarify the arguments of α_1. By examining (8.191) along with (8.158) and (8.181), we see that α_1 is a function of $y, \hat{\chi}, \xi, \Xi, \hat{\theta}, \hat{\varrho}, v_{0,1}, \ldots, v_{m-1,1}$ and y_r. For brevity, we denote $X = (y, \hat{\chi}, \xi, \Xi, \hat{\theta}, \hat{\varrho})$. From (8.169) one can show that $v_{i,j}$ can be expressed as

$$v_{i,j} = [*, \ldots, *, 1]\bar{\lambda}_{i+j}, \tag{8.196}$$

where $\lambda_k \triangleq 0$ for $k > n - 1$. With (8.196) we conclude that $\alpha_1(X, \bar{\lambda}_m, y_r)$. Now we present a backstepping procedure in which $\alpha_i\left(X, \bar{\lambda}_{m+i-1}, \bar{y}_r^{(i-1)}\right)$.

8.2 Design with MT-Filters

Step $i = 2, \ldots, \rho$. We now proceed with the design by differentiating (8.187) for $i = 2, \ldots, \rho - 1$ with the help of (8.183):

$$\begin{aligned}
\dot{z}_i &= \dot{v}_{m,i-1} - \hat{\varrho} y_r^{(i)} - \dot{\hat{\varrho}} y_r^{(i-1)} - \dot{\alpha}_{i-1}\left(y, \hat{\chi}, \xi, \Xi, \hat{\theta}, \hat{\varrho}, \bar{\lambda}_{m+i-1}, \bar{y}_r^{(i-2)}\right) \\
&= v_{m,i} - k_i v_{m,1} - \hat{\varrho} y_r^{(i)} - \dot{\hat{\varrho}} y_r^{(i-1)} - \frac{\partial \alpha_{i-1}}{\partial y}\left(\hat{\chi}_2 + \omega_0 + \omega^{\mathrm{T}}\theta + \varepsilon_2\right) \\
&\quad - \frac{\partial \alpha_{i-1}}{\partial \hat{\chi}}\left[A\hat{\chi} + K_o(y - \hat{\chi}_1) + l\left(\omega_0 + \omega^{\mathrm{T}}\hat{\theta}\right)\right] \\
&\quad - \frac{\partial \alpha_{i-1}}{\partial \xi}(A_l \xi + B_l \phi) - \frac{\partial \alpha_{i-1}}{\partial \Xi}(A_l \Xi + B_l \Phi) - \sum_{j=1}^{i-1}\frac{\partial \alpha_{i-1}}{\partial y_r^{(j-1)}} y_r^{(j)} \\
&\quad - \sum_{j=1}^{m+i-2}\frac{\partial \alpha_{i-1}}{\partial \lambda_j}(-k_j \lambda_1 + \lambda_{j+1}) - \frac{\partial \alpha_{i-1}}{\partial \hat{\theta}}\dot{\hat{\theta}} - \frac{\partial \alpha_{i-1}}{\partial \hat{\varrho}}\dot{\hat{\varrho}}. \quad (8.197)
\end{aligned}$$

Noting from (8.187) that $v_{m,i} - \hat{\varrho} y_r^{(i)} = z_{i+1} + \alpha_i$, we get

$$\begin{aligned}
\dot{z}_i &= z_{i+1} + \alpha_i - k_i v_{m,1} - \frac{\partial \alpha_{i-1}}{\partial y}\left(\hat{\chi}_2 + \omega_0 + \omega^{\mathrm{T}}\theta + \varepsilon_2\right) \\
&\quad - \frac{\partial \alpha_{i-1}}{\partial \hat{\chi}}\left[A\hat{\chi} + K_o(y - \hat{\chi}_1) + l\left(\omega_0 + \omega^{\mathrm{T}}\hat{\theta}\right)\right] \\
&\quad - \frac{\partial \alpha_{i-1}}{\partial \xi}(A_l \xi + B_l \phi) - \frac{\partial \alpha_{i-1}}{\partial \Xi}(A_l \Xi + B_l \Phi) - \sum_{j=1}^{i-1}\frac{\partial \alpha_{i-1}}{\partial y_r^{(j-1)}} y_r^{(j)} \\
&\quad - \sum_{j=1}^{m+i-2}\frac{\partial \alpha_{i-1}}{\partial \lambda_j}(-k_j \lambda_1 + \lambda_{j+1}) - \frac{\partial \alpha_{i-1}}{\partial \hat{\theta}}\dot{\hat{\theta}} - \left(y_r^{(i-1)} + \frac{\partial \alpha_{i-1}}{\partial \hat{\varrho}}\right)\dot{\hat{\varrho}}. \quad (8.198)
\end{aligned}$$

From (8.184) it follows that the final step $i = \rho$ can be encompassed in these calculations if we define $\alpha_\rho = \sigma(y)u + v_{m,\rho} - \hat{\varrho} y_r^{(\rho)}$ and $z_{\rho+1} = 0$. To prepare for choosing a stabilizing function we add and subtract $-\frac{\partial \alpha_{i-1}}{\partial y}\omega^{\mathrm{T}}\hat{\theta} - \frac{\partial \alpha_{i-1}}{\partial \hat{\theta}}\Gamma \tau_i$ in (8.198) and get

$$\begin{aligned}
\dot{z}_i &= \alpha_i - k_i v_{m,1} - \frac{\partial \alpha_{i-1}}{\partial y}\left(\hat{\chi}_2 + \omega_0 + \omega^{\mathrm{T}}\hat{\theta}\right) \\
&\quad - \frac{\partial \alpha_{i-1}}{\partial \hat{\chi}}\left[A\hat{\chi} + K_o(y - \hat{\chi}_1) + l\left(\omega_0 + \omega^{\mathrm{T}}\hat{\theta}\right)\right] \\
&\quad - \frac{\partial \alpha_{i-1}}{\partial \xi}(A_l \xi + B_l \phi) - \frac{\partial \alpha_{i-1}}{\partial \Xi}(A_l \Xi + B_l \Phi) - \sum_{j=1}^{i-1}\frac{\partial \alpha_{i-1}}{\partial y_r^{(j-1)}} y_r^{(j)} \\
&\quad - \sum_{j=1}^{m+i-2}\frac{\partial \alpha_{i-1}}{\partial \lambda_j}(-k_j \lambda_1 + \lambda_{j+1}) - \frac{\partial \alpha_{i-1}}{\partial \hat{\theta}}\Gamma \tau_i - \left(y_r^{(i-1)} + \frac{\partial \alpha_{i-1}}{\partial \hat{\varrho}}\right)\dot{\hat{\varrho}} \\
&\quad - \frac{\partial \alpha_{i-1}}{\partial y}\varepsilon_2 - \frac{\partial \alpha_{i-1}}{\partial y}\omega^{\mathrm{T}}\tilde{\theta} - \frac{\partial \alpha_{i-1}}{\partial \hat{\theta}}\left(\dot{\hat{\theta}} - \Gamma \tau_i\right) + z_{i+1}. \quad (8.199)
\end{aligned}$$

The term $-\frac{\partial \alpha_{i-1}}{\partial y}\omega^T\tilde{\theta}$ determines the ith tuning function

$$\tau_i = \tau_{i-1} - \frac{\partial \alpha_{i-1}}{\partial y}\omega z_i, \qquad i = 2,\ldots,\rho. \tag{8.200}$$

Since the final update law will be $\dot{\hat{\theta}} = \Gamma\tau_\rho$, in view of (8.200) we have

$$-\frac{\partial \alpha_{i-1}}{\partial \hat{\theta}}\left(\dot{\hat{\theta}} - \Gamma\tau_i\right) = \sum_{j=i+1}^{\rho} \frac{\partial \alpha_{i-1}}{\partial \hat{\theta}}\Gamma\frac{\partial \alpha_{j-1}}{\partial y}\omega z_j$$

$$\triangleq \sum_{j=i+1}^{\rho} \sigma_{ij}z_j. \tag{8.201}$$

Then our choice of the stabilizing function for $i = 3,\ldots,\rho$ is

$$\begin{aligned}\alpha_i &= -z_{i-1} - c_i z_i - d_i\left(\frac{\partial \alpha_{i-1}}{\partial y}\right)^2 z_i + k_i v_{m,1} + \frac{\partial \alpha_{i-1}}{\partial y}\left(\hat{\chi}_2 + \omega_0 + \omega^T\hat{\theta}\right)\\&\quad + \frac{\partial \alpha_{i-1}}{\partial \hat{\chi}}\left[A\hat{\chi} + K_o(y - \hat{\chi}_1) + l\left(\omega_0 + \omega^T\hat{\theta}\right)\right]\\&\quad + \frac{\partial \alpha_{i-1}}{\partial \xi}(A_l\xi + B_l\phi) + \frac{\partial \alpha_{i-1}}{\partial \Xi}(A_l\Xi + B_l\Phi) + \sum_{j=1}^{i-1}\frac{\partial \alpha_{i-1}}{\partial y_r^{(j-1)}}y_r^{(j)}\\&\quad + \sum_{j=1}^{m+i-2}\frac{\partial \alpha_{i-1}}{\partial \lambda_j}(-k_j\lambda_1 + \lambda_{j+1}) + \frac{\partial \alpha_{i-1}}{\partial \hat{\theta}}\Gamma\tau_i + \left(y_r^{(i-1)} + \frac{\partial \alpha_{i-1}}{\partial \hat{\varrho}}\right)\dot{\hat{\varrho}}\\&\quad - \sum_{j=2}^{i-1}\frac{\partial \alpha_{j-1}}{\partial \hat{\theta}}\Gamma\frac{\partial \alpha_{i-1}}{\partial y}z_j.\end{aligned} \tag{8.202}$$

Only for $i = 2$ the stabilizing function differs from (8.202) because its first term is $-\hat{b}_m z_1$ rather than $-z_1$.

The last stabilizing function α_ρ is used in the actual control law:

$$u = \frac{1}{\sigma(y)}\left(\alpha_\rho - v_{m,\rho} + \hat{\varrho}y_r^{(\rho)}\right). \tag{8.203}$$

and the last tuning function τ_ρ is used in the update law:

$$\dot{\hat{\theta}} = \Gamma\tau_\rho = \Gamma\left\{\omega\left[1, -\frac{\partial \alpha_1}{\partial y}, \cdots, -\frac{\partial \alpha_{\rho-1}}{\partial y}\right]\begin{bmatrix}z_1\\z_2\\\vdots\\z_\rho\end{bmatrix} - \hat{\varrho}(\dot{y}_r + \bar{\alpha}_1)e_1 z_1 + \nu\omega\varepsilon_1\right\}. \tag{8.204}$$

By substituting (8.202) along with (8.201) into (8.199), the error system becomes

$$\dot{z} = A_z(z,t)z + W_\varepsilon(z,t)\varepsilon_2 + W_\theta(z,t)^T\tilde{\theta} - b_m(\dot{y}_r + \bar{\alpha}_1)e_1\tilde{\varrho}, \tag{8.205}$$

with $A_z(z,t)$, $W_\varepsilon(z,t)$ and $W_\theta(z,t)$ defined as in (8.71), (8.72), and (8.73), respectively.

8.2 Design with MT-Filters

Example 8.8 Consider the relative-degree-two nonlinear plant:

$$\begin{aligned} \dot{x}_1 &= x_2 + \theta y^3 \\ \dot{x}_2 &= u \\ y &= x_1, \end{aligned} \tag{8.206}$$

where only y is measured and θ is unknown but constant. In this case only the scalar filters (9.218) and (9.217) are needed:

$$\dot{v} = -lv + u \tag{8.207}$$
$$\dot{\Xi} = -l\Xi - ly^3, \tag{8.208}$$

and the regressor (9.224) is scalar:

$$\omega = y^3 + \Xi. \tag{8.209}$$

The control objective is the regulation of y to zero, namely, $y_r(t) \equiv 0$. The error coordinates (9.238) and (9.239) are

$$\begin{aligned} z_1 &= y \\ z_2 &= v - \alpha_1. \end{aligned} \tag{8.210}$$

The observer (8.171) is given by

$$\dot{\hat{\chi}} = \begin{bmatrix} 0 & 1 \\ 0 & 0 \end{bmatrix} \hat{\chi} + \begin{bmatrix} c_o + l \\ c_o l \end{bmatrix} \varepsilon_1 + \begin{bmatrix} 1 \\ l \end{bmatrix} (v + \omega\hat{\theta}), \tag{8.211}$$

where $\varepsilon_1 = y - \hat{\chi}_1$. The control law is generated via

$$\alpha_1 = -(c+d)z_1 - \hat{\chi}_2 - \omega\hat{\theta} \tag{8.212}$$

$$u = -\left[c + d\left(\frac{\partial \alpha_1}{\partial y}\right)^2\right]z_2 - z_1 + \frac{\partial \alpha_1}{\partial y}(\hat{\chi}_2 + v + \omega\hat{\theta}) - lc_o\varepsilon_1 - \omega\dot{\hat{\theta}}, \tag{8.213}$$

and the parameter update law is

$$\dot{\hat{\theta}} = \gamma\omega\left(z_1 - \frac{\partial \alpha_1}{\partial y}z_2 + \nu\varepsilon_1\right). \tag{8.214}$$

Figure 8.1 shows the system response with $\theta = 2$, $c = c_o = 1$, $d = 0.5$, $l = 1$, $\gamma = 0.4$, $\nu = 1$, and all initial conditions zero except for $x_1(0) = \hat{\chi}_1(0) = 0.8$. The main feature of the responses is that they are fast. The adaptive transient settles in about 0.5 sec, when the parameter error $\tilde{\theta}$ becomes very small. After that, the system is practically feedback linearized, and responses look "linear." The convergence of $\tilde{\theta}$ to zero is fast because the initial swing of the error states and the regressor accelerate the parameter convergence. This is pronounced because of the cubic nonlinearity in the plant which is responsible

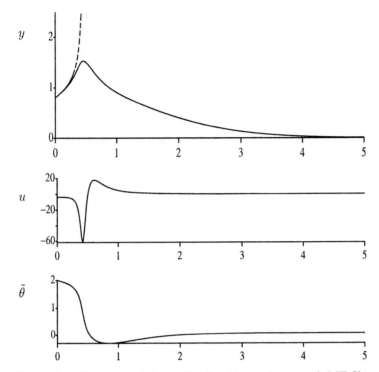

Figure 8.1: Response of the tuning functions scheme with MT-filters.

for the sharp peak in y. (In fact, without adaptation, this nonlinearity would cause an "explosive" escape to infinity; see the dashed curve in Figure 8.1.) With a considerably lower adaptation gain, the parameter estimate still converges fast because the initial swings become larger and further accelerate the convergence. With high values of c, c_o, and/or d, which may increase the control effort, the regulation of y to zero is achieved without $\tilde{\theta}$ converging to zero, that is, without feedback linearization.

We revisit this example to illustrate a modular design in Example 9.23. ◇

To guarantee boundedness without adaptation we add nonlinear damping terms (8.84)–(8.87) to the stabilizing functions, as we did in the design with K-filters. The stabilizing functions are now stabilized in Table 8.4. The resulting error system has the same form (8.88)–(8.90) as in the design with K-filters. Consequently, its state satisfies the inequality

$$\frac{d}{dt}\left(\frac{1}{2}|z|^2\right) \leq -c_0|z|^2 + \frac{1}{4d_0}\varepsilon_2^2 + \frac{1}{4\kappa_0}\left(|\tilde{\theta}|^2 + |b_m|\tilde{\varrho}^2\right), \qquad (8.215)$$

with c_0, d_0, and κ_0 defined in (8.92). When adaptation is turned off, $\tilde{\theta}$ and $\tilde{\varrho}$ are constant and bounded. However, unlike in the design with K-filters,

8.2 DESIGN WITH MT-FILTERS

ε_2 is not guaranteed to be bounded because the input $\omega^T\tilde{\theta}$ to the observer error system (8.175) cannot be guaranteed to be bounded before establishing the boundedness of y. For this reason, we use nonlinear damping to also strengthen the system (8.175). The observer (8.171) is strengthened by adding the nonlinear damping term $\kappa_o|\omega|^2 l(y - \hat{\chi}_1)$:

$$\dot{\hat{\chi}} = A\hat{\chi} + K_o(y - \hat{\chi}_1) + \kappa_o|\omega|^2 l(y - \hat{\chi}_1) + l\left(\omega_0 + \omega^T\hat{\theta}\right). \tag{8.216}$$

Thus the observer error system becomes

$$\dot{\varepsilon} = \left(A_o - \kappa_o|\omega|^2 l e_1^T\right)\varepsilon + l\omega^T\tilde{\theta}. \tag{8.217}$$

The inclusion of the term $\kappa_o|\omega|^2 l(y - \hat{\chi}_1)$ in (8.216) forces us to include this term also in the stabilizing function α_i in (8.202).

The complete tuning functions design with MT-filters is summarized in Tables 8.4 and 8.5. It employs the filters given in Table 8.3.

8.2.3 Stability

The stability properties of the adaptive scheme with MT-filters are as follows.

Theorem 8.9 (Tuning Functions with MT-Filters) *The closed-loop adaptive system consisting of the plant (8.3), the control law, the update laws, the observer in Tables 8.4 and 8.5, and the filters in Table 8.3, has the following properties:*

1. *If $\Gamma, \gamma > 0$, $\kappa_o, \kappa_i \geq 0$, $i = 1, \ldots, \rho$, and*

$$\nu c_o d_0 > l_1^2, \tag{8.218}$$

   then all the signals are globally uniformly bounded, and asymptotic tracking is achieved:

$$\lim_{t \to \infty} [y(t) - y_r(t)] = 0. \tag{8.219}$$

2. *If $\Gamma, \gamma = 0$ and $\kappa_o, \kappa_i > 0$, $i = 1, \ldots, \rho$, then all the signals are globally uniformly bounded.*

Proof. 1. As argued in the proof of Theorem 8.5, the solution of the closed-loop adaptive system exists and is unique on its maximum interval of existence $[0, t_f)$.

Let us introduce the similarity transformation

$$\begin{bmatrix} \varepsilon_1 \\ \eta \end{bmatrix} \triangleq \begin{bmatrix} \varepsilon_1 \\ T\varepsilon \end{bmatrix} = \begin{bmatrix} e_1^T \\ T \end{bmatrix} \varepsilon \tag{8.220}$$

Table 8.4: Tuning Functions Design with MT-Filters (cont'd in Table 8.5)

$$z_1 = y - y_r \tag{8.221}$$

$$z_i = v_{m,i-1} - \hat{\varrho} y_r^{(i-1)} - \alpha_{i-1}, \quad i = 2, \ldots, \rho \tag{8.222}$$

$$\alpha_1 = \hat{\varrho}\bar{\alpha}_1 - \kappa_1 \mathrm{sgn}(b_m)\left(\dot{y}_r + \bar{\alpha}_1\right)^2 z_1 \tag{8.223}$$

$$\bar{\alpha}_1 = -\left(c_1 + d_1 + \kappa_1|\bar{\omega}|^2\right) z_1 - \hat{\chi}_2 - \omega_0 - \bar{\omega}^T \hat{\theta} \tag{8.224}$$

$$\alpha_2 = -\hat{b}_m z_1 - \left[c_2 + d_2 \left(\frac{\partial \alpha_1}{\partial y}\right)^2 + \kappa_2 \left|\frac{\partial \alpha_1}{\partial y}\omega - z_1 e_1\right|^2\right] z_2$$

$$+ \left(\dot{y}_r + \frac{\partial \alpha_1}{\partial \hat{\varrho}}\right)\dot{\hat{\varrho}} + \frac{\partial \alpha_1}{\partial \hat{\theta}}\Gamma \tau_2 + \beta_2 \tag{8.225}$$

$$\alpha_i = -z_{i-1} - \left[c_i + \left(d_i + \kappa_i|\omega|^2\right)\left(\frac{\partial \alpha_{i-1}}{\partial y}\right)^2\right] z_i + \left(y_r^{(i-1)} + \frac{\partial \alpha_{i-1}}{\partial \hat{\varrho}}\right)\dot{\hat{\varrho}}$$

$$+ \frac{\partial \alpha_{i-1}}{\partial \hat{\theta}}\Gamma \tau_i - \sum_{j=2}^{i-1}\frac{\partial \alpha_{j-1}}{\partial \hat{\theta}}\Gamma \frac{\partial \alpha_{i-1}}{\partial y}z_j + \beta_i, \quad i = 3, \ldots, \rho \tag{8.226}$$

$$\beta_i = \frac{\partial \alpha_{i-1}}{\partial y}\left(\hat{\chi}_2 + \omega_0 + \omega^T \hat{\theta}\right) + \frac{\partial \alpha_{i-1}}{\partial \xi}(A_l \xi + B_l \phi) + \frac{\partial \alpha_{i-1}}{\partial \Xi}(A_l \Xi + B_l \Phi)$$

$$+ \frac{\partial \alpha_{i-1}}{\partial \hat{\chi}}\left[A\hat{\chi} + K_o(y - \hat{\chi}_1) + \kappa_o|\omega|^2 l(y - \hat{\chi}_1) + l\left(\omega_0 + \omega^T \hat{\theta}\right)\right]$$

$$+ \sum_{j=1}^{i-1}\frac{\partial \alpha_{i-1}}{\partial y_r^{(j-1)}}y_r^{(j)} + k_i v_{m,1} + \sum_{j=1}^{m+i-2}\frac{\partial \alpha_{i-1}}{\partial \lambda_j}(-k_j \lambda_1 + \lambda_{j+1}) \tag{8.227}$$

$$\tau_0 = \nu \omega \varepsilon_1 \tag{8.228}$$

$$\tau_1 = (\omega - \hat{\varrho}(\dot{y}_r + \bar{\alpha}_1)e_1) z_1 \tag{8.229}$$

$$\tau_i = \tau_{i-1} - \frac{\partial \alpha_{i-1}}{\partial y}\omega z_i, \quad i = 2, \ldots, \rho \tag{8.230}$$

where $T \in \mathbb{R}^{(n-1)\times n}$ is defined as

$$T \triangleq [A_l e_1, \ I_{n-1}] = [A_l, \ e_{n-1}]. \tag{8.231}$$

The following two identities are straightforward to verify:

$$Tl = 0, \quad K_o = c_o l - \begin{bmatrix} A_l e_1 \\ 0 \end{bmatrix} \tag{8.232}$$

8.2 Design with MT-Filters

Table 8.5: Tuning Functions Design w/ MT-Filters (cont'd from Tab. 8.4)

Adaptive control law:
$$u = \frac{1}{\sigma(y)}\left(\alpha_\rho - v_{m,\rho} + \hat{\varrho} y_r^{(\rho)}\right) \qquad (8.233)$$

Parameter update laws:
$$\dot{\hat{\theta}} = \Gamma \tau_\rho \qquad (8.234)$$
$$\dot{\hat{\varrho}} = -\gamma \mathrm{sgn}(b_m)\left(\dot{y}_r + \bar{\alpha}_1\right) z_1 \qquad (8.235)$$

Observer:
$$\dot{\hat{\chi}} = A\hat{\chi} + K_o(y - \hat{\chi}_1) + \kappa_o|\omega|^2 l(y - \hat{\chi}_1) + l\left(\omega_0 + \omega^\mathrm{T}\hat{\theta}\right) \qquad (8.236)$$

(the latter is immediate from (8.172)). Using (8.231)–(8.232), we compute

$$\begin{aligned}
TA_o &= T\left[-c_o l + \begin{bmatrix} A_l e_1 \\ 0 \end{bmatrix} \middle| \begin{bmatrix} I_{n-1} \\ 0 \end{bmatrix}\right] = \left[T\begin{bmatrix} A_l e_1 \\ 0 \end{bmatrix} \middle| T\begin{bmatrix} I_{n-1} \\ 0 \end{bmatrix}\right] \\
&= \left[A_l^2 e_1, \ A_l\right] = A_l\left[A_l e_1, \ I_{n-1}\right] = A_l T \, .
\end{aligned} \qquad (8.237)$$

Combining (8.217) and (8.220) and using (8.237) we have

$$\begin{aligned}
\dot{\eta} &= T\dot{\varepsilon} = TA_o\varepsilon - \kappa_o|\omega|^2 T l e_1^\mathrm{T}\varepsilon + T l \omega^\mathrm{T}\tilde{\theta} \\
&= TA_o\varepsilon = A_l T\varepsilon \, ,
\end{aligned} \qquad (8.238)$$

which because of (8.220) gives

$$\dot{\eta} = A_l \eta \, . \qquad (8.239)$$

This system represents the exponentially stable *inverse dynamics* of (8.217). Since they are not controllable by ε_1, they are also the *zero dynamics* of (8.217). Now we examine the ε_1-equation:

$$\dot{\varepsilon}_1 = -e_1^\mathrm{T} K_o \varepsilon_1 + \varepsilon_2 - \kappa_o|\omega|^2 \varepsilon_1 + \omega^\mathrm{T}\tilde{\theta} \, . \qquad (8.240)$$

The second identity in (8.232) gives $e_1^\mathrm{T} K_o = c_o + l_1$. Therefore

$$\dot{\varepsilon}_1 = -\left(c_o + \kappa_o|\omega|^2\right)\varepsilon_1 + \omega^\mathrm{T}\tilde{\theta} + \varepsilon_2 - l_1\varepsilon_1 \, . \qquad (8.241)$$

Substituting (8.231) into (8.220) we see that

$$\varepsilon_2 - l_1\varepsilon_1 = \eta_1 \qquad (8.242)$$

and obtain
$$\dot{\varepsilon}_1 = -\left(c_o + \kappa_o|\omega|^2\right)\varepsilon_1 + \omega^T\tilde{\theta} + \eta_1. \tag{8.243}$$

Now we are ready for a Lyapunov stability analysis for the closed-loop system consisting of the error system (8.205), the error equations for parameter estimators (8.204) and (8.195), and the observer error system (8.243), (8.239):

$$\dot{z} = A_z(z,t)z + W_\varepsilon(z,t)\varepsilon_2 + W_\theta(z,t)^T\tilde{\theta} - b_m(\dot{y}_r + \bar{\alpha}_1)e_1\tilde{\varrho} \tag{8.244}$$
$$\dot{\tilde{\theta}} = -\Gamma\left(W_\theta(z,t)z + \nu\omega\varepsilon_1\right) \tag{8.245}$$
$$\dot{\tilde{\varrho}} = \gamma\,\text{sgn}(b_m)(\dot{y}_r + \bar{\alpha}_1)e_1^T z \tag{8.246}$$
$$\dot{\varepsilon}_1 = -\left(c_o + \kappa_o|\omega|^2\right)\varepsilon_1 + \omega^T\tilde{\theta} + \eta_1 \tag{8.247}$$
$$\dot{\eta} = A_l\eta. \tag{8.248}$$

(The system matrix A_z in (8.244) is of the form (8.71), but it also incorporates the nonlinear damping terms (8.84)–(8.87).) A candidate Lyapunov function for this system is

$$V = \frac{1}{2}z^T z + \frac{1}{2}\tilde{\theta}^T\Gamma^{-1}\tilde{\theta} + \frac{|b_m|}{2\gamma}\tilde{\varrho}^2 + \frac{\nu}{2}\varepsilon_1^2 + \mu\eta^T P_l\eta, \tag{8.249}$$

where μ is a positive constant to be chosen later, and $P_l = P_l^T > 0$ is the solution to the Lyapunov equation

$$P_l A_l + A_l^T P_l = -I. \tag{8.250}$$

With calculations similar to (8.78)–(8.80), we arrive at

$$\dot{V} \leq -c_0|z|^2 + \frac{1}{4d_0}\varepsilon_2^2 - \nu c_o\varepsilon_1^2 + \nu\varepsilon_1\eta_1 - \mu|\eta|^2. \tag{8.251}$$

By applying Young's inequality to the term $\nu\varepsilon_1\eta_1$, we obtain

$$\dot{V} \leq -c_0|z|^2 + \frac{1}{4d_0}\varepsilon_2^2 - \frac{\nu c_o}{2}\varepsilon_1^2 - \left(\mu - \frac{\nu}{2c_o}\right)|\eta|^2. \tag{8.252}$$

Noting that (8.242) implies $\varepsilon_2^2 \leq 2l_1^2\varepsilon_1^2 + 2\eta_1^2$, we finally get

$$\dot{V} \leq -c_0|z|^2 - \left(\frac{\nu c_o}{2} - \frac{l_1^2}{2d_0}\right)\varepsilon_1^2 - \left(\mu - \frac{\nu}{2c_o} - \frac{1}{2d_0}\right)|\eta|^2. \tag{8.253}$$

By choosing ν as in (8.218), and $\mu = \frac{\nu}{2c_o} + \frac{1}{2d_0}$, we prove that $\dot{V} \leq 0$. This implies that $z, \tilde{\theta}, \tilde{\varrho}, \varepsilon_1, \eta$ are bounded. We now use this to establish the boundedness of all other signals in the adaptive system.

In view of the similarity transformation (8.220), the boundedness of ε_1 and η establish that ε is bounded. The boundedness of y_r and z_1 implies that y is

8.2 Design with MT-Filters

bounded. Therefore ξ and Ξ are bounded. To prove boundedness of $\hat{\chi}$, let us first rewrite (8.236) as

$$\dot{\hat{\chi}} = A_o\hat{\chi} + K_o y + l\left[\kappa_o|\omega|^2(y - \hat{\chi}_1) + \omega_0 + \omega^T\hat{\theta}\right], \qquad (8.254)$$

and note that the boundedness of ε_1 and y implies that $\hat{\chi}_1$ is bounded. To prove that the remaining components of $\hat{\chi}$ are bounded, we employ the similarity transformation

$$\begin{bmatrix} \hat{\chi}_1 \\ \psi \end{bmatrix} \triangleq \begin{bmatrix} \hat{\chi}_1 \\ T\hat{\chi} \end{bmatrix} = \begin{bmatrix} e_1^T \\ T \end{bmatrix} \hat{\chi}, \qquad (8.255)$$

and, from the observer equation (8.254), obtain the system

$$\dot{\psi} = A_l \psi + A_l \bar{l} y, \qquad (8.256)$$

which shows that ψ is independent of the input $\kappa_o|\omega|^2(y - \hat{\chi}_1) + \omega_0 + \omega^T\hat{\theta}$. To arrive at the last equation, we have used the identities (8.232) and (8.237). Because of the boundedness of y and the Hurwitzness of A_l, (8.256) proves that ψ is bounded. By (8.255), the boundedness of $\hat{\chi}_1$ and ψ establishes that $\hat{\chi}$ is bounded. We have yet to prove the boundedness of λ and x. Our main concern is λ because the boundedness of x will follow from the boundedness of $\varepsilon, \hat{\chi}, \xi, \Xi$, and λ. The proof of boundedness of λ is similar to the corresponding part of the proof of Theorem 8.5.

From (8.168) it follows that

$$\lambda_i = \frac{s^{i-1} + l_1 s^{i-2} + \cdots + l_{i-1}}{L(s)} [\sigma(y)u], \qquad i = 1, \ldots, n-1. \qquad (8.257)$$

Substituting (8.108) into (8.257) we get

$$\lambda_i = \frac{s^{i-1} + l_1 s^{i-2} + \cdots + l_{i-1}}{L(s)B(s)} \left\{ \frac{d^n y}{dt^n} - \sum_{i=1}^n \frac{d^{n-i}}{dt^{n-i}} \left[\varphi_{0,i}(y) + \Phi_{(i)}(y)a\right] \right\}$$

$$i = 1, \ldots, n-1. \qquad (8.258)$$

The boundedness of y, the smoothness of $\phi(y)$ and $\Phi(y)$, Assumption 8.2, and (8.258) imply that $\lambda_1, \ldots, \lambda_m$ are bounded. We now return to the coordinate change (8.222) which gives

$$v_{m,i} = z_{i+1} + \hat{\varrho} y_r^{(i)} + \alpha_i \left(y, \hat{\chi}, \xi, \Xi, \hat{\theta}, \hat{\varrho}, \bar{\lambda}_{m+i-1}, \bar{y}_r^{(i-1)}\right), \qquad i = 1, \ldots, \rho - 1. \qquad (8.259)$$

Let $i = 1$. The boundedness of $\bar{\lambda}_{m+1}$, along with the boundedness of z_2 and $y, \hat{\chi}, \xi, \Xi, \hat{\theta}, \hat{\varrho}, y_r, \dot{y}_r$, proves that $v_{m,1}$ is bounded. Then from (8.196) it follows that λ_{m+1} is bounded. Continuing in the same fashion, (8.259) and (8.196) recursively establish that λ is bounded. Since ε and $\hat{\chi}$ are bounded, from (8.174) it follows that χ is bounded. Finally, in view of (8.156), (8.169),

(8.170), and the boundedness of ξ, Ξ, λ and χ, we conclude that x is bounded. Since $\sigma(y)$ is bounded away from zero, u is bounded. By the same argument as in the proof of Theorem 8.5 we conclude that $t_f = \infty$.

By applying the LaSalle-Yoshizawa theorem (Theorem 2.1) to (8.253), it further follows that $z(t) \to 0$ as $t \to \infty$, which implies that $\lim_{t\to\infty}[y(t) - y_r(t)] = 0$.

2. Noting that (8.243) yields

$$\frac{d}{dt}\left(\frac{1}{2}\varepsilon_1^2\right) \leq -\frac{c_o}{2}\varepsilon_1^2 + \frac{1}{4\kappa_o}|\tilde{\theta}|^2 + \frac{1}{2c_o}\zeta_1^2, \qquad (8.260)$$

we conclude that ε_1 is bounded when the adaptation is switched off. In view of the similarity transformation (8.220), the boundedness of ε_1 and η establish that ε is bounded. The boundedness without adaptation now follows from (8.215), by repeating the above argument which deduces the boundedness of all the signals from the boundedness of $z, \hat{\theta}, \hat{\varrho}, \varepsilon_1, \eta$. $\qquad\square$

The requirement $\nu c_o d_o > l_1^2$ given by (8.218), not found in the design with K-filters, is needed in the design with MT-filters because the parameter error $\tilde{\theta}$ appears not only in the z-system (8.205) but also in the observer error system (8.217). The need for each of the factors to be large can be explained as follows:

- ν should be large because the term $\omega\varepsilon_1$ needs to be given a sufficient weight relative to the other terms in the update law (8.204). In other words, the adaptation with respect to the ε-system needs to be sufficiently fast because ε_2 appears as a disturbance in the z-system.

- d_o should be large to prevent the destabilization of the z-system by ε_2 if the adaptation with respect to the ε-system is slow.

- c_o should be large to make the part of the ε-system controllable by $\omega^T\tilde{\theta}$ fast enough if the adaptation with respect to the ε-system is slow.

Hence, if $c_o d_o$ is small, ν should be large enough to satisfy condition (8.218). However, ν must not be too large because, as our performance analysis in Section 8.2.4 shows, an increase in ν may cause a deterioration of performance of z.

As in Corollary 8.6 for the design with K-filters, we can show that the error system

$$\dot{z} = A_z(z,t)z + W_\varepsilon(z,t)\varepsilon_2 + W_\theta(z,t)^T\tilde{\theta} - b_m(\dot{y}_r + \bar{\alpha}_1)e_1\tilde{\varrho} \qquad (8.261)$$
$$\dot{\varepsilon} = \left(A_o - \kappa_o|\omega|^2 l e_1^T\right)\varepsilon + l\omega^T\tilde{\theta} \qquad (8.262)$$
$$\dot{\tilde{\psi}} = A_l\tilde{\psi} + A_l^2 e_1 z_1 \qquad (8.263)$$

8.2 DESIGN WITH MT-FILTERS

$$\dot{\tilde{\zeta}} = A_b\tilde{\zeta} + \tilde{\varphi}_b(z_1, y_r)z_1 \tag{8.264}$$

$$\dot{\tilde{\xi}} = A_l\tilde{\xi} + B_l\tilde{\phi}(z_1, y_r)z_1 \tag{8.265}$$

$$\dot{\tilde{\Xi}} = A_l\tilde{\Xi} + B_l\tilde{\Phi}(z_1, y_r)z_1 \tag{8.266}$$

$$\dot{\tilde{\theta}} = -\Gamma\left(W_\theta(z,t)z + \nu\omega\varepsilon_1\right) \tag{8.267}$$

$$\dot{\tilde{\varrho}} = \gamma \operatorname{sgn}(b_m)\left(\dot{y}_r + \bar{\alpha}_1\right)e_1^\mathrm{T} z \tag{8.268}$$

has a globally uniformly stable equilibrium at the origin, and its $(q+4)n+m$-dimensional state converges to the $q+m+2$-dimensional manifold

$$M = \left\{z = 0,\ \varepsilon = 0,\ \tilde{\psi} = 0,\ \tilde{\zeta} = 0,\ \tilde{\xi} = 0,\ \tilde{\Xi} = 0\right\}. \tag{8.269}$$

The systems (8.263), (8.265), (8.266), are defined from (8.256), (8.166), (8.167), in analogy with Corollary 8.6. One can also show that

$$(x, \hat{\chi}, \lambda, \xi, \Xi, \hat{\theta}, \hat{\varrho}) \mapsto (z, \varepsilon, \tilde{\chi}, \tilde{\zeta}, \tilde{\xi}, \tilde{\Xi}, \tilde{\theta}, \tilde{\varrho}) \tag{8.270}$$

is a global C^∞-diffeomorphism for each $t \geq 0$, whenever $B(s)$ and $L(s)$ are coprime.

8.2.4 Transient performance

We derive $\mathcal{L}_2$ and $\mathcal{L}_\infty$ transient performance bounds for the error state z.

Theorem 8.10 (Tuning Functions with MT-Filters) *In the adaptive system (8.3), (8.166), (8.167), (8.168), (8.233), (8.234), (8.235), (8.236), the following inequalities hold:*

$$\|z\|_2 \leq \frac{1}{\sqrt{2c_0}}\left[\frac{1}{\gamma}\left(|\tilde{\theta}(0)|^2 + |b_m||\tilde{\varrho}(0)|^2\right) + \nu\varepsilon_1(0)^2 + \left(\frac{\nu}{c_0} + \frac{1}{d_0}\right)|\eta(0)|_{P_l}^2\right]^{1/2}$$
$$+ \frac{1}{\sqrt{2c_0}}|z(0)| \tag{8.271}$$

$$|z(t)| \leq \frac{1}{2\sqrt{c_0}}\left\{\left(\frac{1}{\kappa_0} + \frac{2l_1^2}{d_0\nu\gamma}\right)\left(|\tilde{\theta}(0)|^2 + |b_m||\tilde{\varrho}(0)|^2\right) + \left(\frac{\nu\gamma}{\kappa_0} + \frac{2l_1^2}{d_0}\right)\varepsilon_1(0)^2\right.$$
$$+ \left[\left(\frac{\gamma}{\kappa_0} + \frac{2l_1^2}{d_0\nu}\right)\left(\frac{1}{d_0} + \frac{\nu}{c_0}\right) + \frac{2}{d_0\underline{\lambda}(P_l)}\right]|\eta(0)|_{P_l}^2$$
$$\left. + \left(\frac{\gamma}{\kappa_0} + \frac{2l_1^2}{d_0\nu}\right)|z(0)|^2\right\}^{1/2} + |z(0)|e^{-c_0 t}. \tag{8.272}$$

Proof. We first derive the bound (8.271). For ν satisfying condition (8.218) and $\mu = \frac{\nu}{2c_0} + \frac{1}{2d_0}$, inequality (8.253) becomes

$$\dot{V} \leq -c_0|z|^2. \tag{8.273}$$

By integrating (8.273) over $[0, \infty)$,

$$\|z\|_2^2 = \int_0^\infty |z(\tau)|^2 d\tau \leq -\frac{1}{c_0} \int_0^\infty \dot{V}(\tau) d\tau \leq \frac{1}{c_0} V(0), \qquad (8.274)$$

and recalling from (8.249) that the Lyapunov function V is given by

$$V = \frac{1}{2}\left[|z|^2 + \frac{1}{\gamma}\left(|\tilde{\theta}|^2 + |b_m|\tilde{\varrho}^2\right) + \nu\varepsilon_1^2 + \left(\frac{\nu}{c_o} + \frac{1}{d_0}\right)|\eta|_{P_l}^2\right], \qquad (8.275)$$

we arrive at (8.271).

The derivation of the bound (8.272) starts with (8.215). From Lemma C.5(i), it follows that

$$|z(t)|^2 \leq |z(0)|^2 e^{-2c_0 t} + \frac{1}{4c_0}\left[\frac{1}{d_0}\|\varepsilon_2\|_\infty^2 + \frac{1}{\kappa_0}\left\||\tilde{\theta}|^2 + |b_m|\tilde{\varrho}^2\right\|_\infty\right]. \qquad (8.276)$$

From (8.275) and (8.273) we get

$$\left\||\tilde{\theta}|^2 + |b_m|\tilde{\varrho}^2\right\|_\infty \leq 2\gamma V(0). \qquad (8.277)$$

On the other hand, noting that (8.242) implies $\varepsilon_2^2 \leq 2l_1^2 \varepsilon_1^2 + 2\eta_1^2$, we have

$$\|\varepsilon_2\|_\infty^2 \leq 2l_1^2 \|\varepsilon_1\|_\infty^2 + 2\|\eta_1\|_\infty^2. \qquad (8.278)$$

From (8.275) and (8.273) we get

$$\|\varepsilon_1\|_\infty^2 \leq \frac{2}{\nu} V(0), \qquad (8.279)$$

and from (8.239) and (8.250) we have $\frac{d}{dt}|\eta|_{P_l}^2 \leq -|\eta|^2$, which gives

$$\|\eta_1\|_\infty^2 \leq \frac{1}{\underline{\lambda}(P_l)} |\eta(0)|_{P_l}^2. \qquad (8.280)$$

Substituting (8.279) and (8.280) into (8.278), and then along with (8.277) into (8.276) we arrive at the bound (8.272). □

While the $\mathcal{L}_2$ bound (8.271) holds even for $\kappa_0 = 0$, the $\mathcal{L}_\infty$ bound (8.272) is valid only if $\kappa_0 > 0$.

The initial condition $z(0)$ in the bounds in Theorem 8.10 is, in general, dependent on the design parameters $c_0, d_0, \kappa_0, c_o, \gamma, \nu$. However, with trajectory initialization we can set $z(0) = 0$. Following (8.221) and (8.222), $z(0)$ is set to zero by choosing

$$y_r(0) = y(0) \qquad (8.281)$$

$$y_r^{(i)}(0) = \frac{1}{\hat{\varrho}(0)}\big[v_{m,i}(0) \\
-\alpha_i\left(y(0), \hat{\chi}(0), \xi(0), \Xi(0), \hat{\theta}(0), \hat{\varrho}(0), \bar{\lambda}_{m+i-1}(0), \bar{y}_r^{(i-1)}(0)\right)\big] \\
i = 1, \ldots, \rho - 1. \qquad (8.282)$$

After $z(0)$ is set to zero, the bounds (8.271) and (8.272) become

$$\|z\|_2 \leq \frac{1}{\sqrt{2c_0}} \left[\frac{1}{\gamma} \left(|\tilde{\theta}(0)|^2 + |b_m|\tilde{\varrho}(0)^2 \right) + \nu \varepsilon_1(0)^2 + \left(\frac{\nu}{c_o} + \frac{1}{d_0} \right) |\eta(0)|^2_{P_l} \right]^{1/2} \tag{8.283}$$

$$|z(t)| \leq \frac{1}{2\sqrt{c_0}} \left\{ \left(\frac{1}{\kappa_0} + \frac{2l_1^2}{d_0\nu\gamma} \right) \left(|\tilde{\theta}(0)|^2 + |b_m|\tilde{\varrho}(0)^2 \right) + \left(\frac{\nu\gamma}{\kappa_0} + \frac{2l_1^2}{d_0} \right) \varepsilon_1(0)^2 \right.$$
$$\left. + \left[\left(\frac{\gamma}{\kappa_0} + \frac{2l_1^2}{d_0\nu} \right) \left(\frac{1}{d_0} + \frac{\nu}{c_o} \right) + \frac{2}{d_0\underline{\lambda}(P_l)} \right] |\eta(0)|^2_{P_l} \right\}^{1/2}. \tag{8.284}$$

Both of these bounds can be systematically reduced by increasing c_0. Other possibilities for reducing the values of the bounds are easy to find by examining the dependence of the bounds on their arguments. In fact, alternative $\mathcal{L}_\infty$ bounds can be derived which emphasize the role of $\gamma, d_0, c_o, \kappa_o$.

The dependence of the bounds on ν is such that they may increase with ν. On the other hand, ν should be sufficiently large to satisfy the stability condition (8.218). A way to guarantee stability with small ν is to satisfy the condition (8.218) by increasing d_0.

Due to the ν-dependent terms, the bounds with MT-filters are *strictly larger* than the bounds with K-filters. The performance with MT-filters is not as good as with K-filters because the design with MT-filters results in the observer error system (8.217) disturbed by the parameter error, rather than in the autonomous observer error system (8.18). Although this disturbance is easily compensated for in the parameter update law, it causes the performance bounds for the design with MT-filters to be higher.

8.3 Example: Levitated Ball

In this section we consider the magnetic levitation system in Figure 8.2 and design a controller which keeps the iron ball at a desired position y_r. The position of the ball y is sensed by the photoelectric sensor S and fed back to the controller C. The controller is to adjust the electromagnet current I to attract the ball with the force F and thus counteract the gravitation force Mg. Many versions of this system exist in undergraduate control laboratories.

The equation of motion is

$$M\ddot{y} = Mg - F, \tag{8.285}$$

and the key modeling task is to describe the force F as a function of the position y and the current I. We assume that F is proportional to I^2 and that it is a known decaying strictly positive function of y, that is,

$$F = \lambda(y)I^2, \qquad \lambda(y) > 0, \lambda'(y) < 0 \quad \forall y \geq 0. \tag{8.286}$$

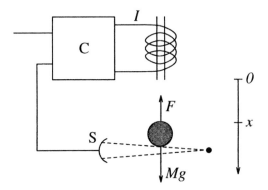

Figure 8.2: Levitated ball system.

The model is meaningful only for $y > 0$, and therefore the design will not yield a global result. To complete the model, its electric part should describe how the electromagnet current I is produced by the control voltage u. Here we will assume a simple RL relationship

$$L\dot{I} = -RI + Gu. \tag{8.287}$$

Thus, our complete third-order model is given by the state equations

$$\begin{aligned} \dot{x}_1 &= x_2 \\ \dot{x}_2 &= g - \theta_1 \lambda(x_1) x_3^2 \\ \dot{x}_3 &= -\theta_2 x_3 + \theta_3 u. \end{aligned} \tag{8.288}$$

with state variables $x_1 = y$, $x_2 = \dot{y}$, and $x_3 = I$ and unknown parameters $\theta_1 = 1/M$, $\theta_2 = R/L$, and $\theta_3 = G/L$. The unknown parameters are all positive. The system (8.288) is not in any of the required forms, but, as we shall see, this does not preclude the backstepping design.

While the states x_1 and x_3 are measured, the velocity $x_2 = \dot{y}$ is not measured. Hence, our design will use partial state feedback. Led by the reduced-order observer idea, we employ the filters

$$\begin{aligned} \dot{\xi} &= -k\xi - k^2 x_1 + g \end{aligned} \tag{8.289}$$
$$\begin{aligned} \dot{\zeta} &= -k\zeta - \lambda(x_1) x_3^2, \end{aligned} \tag{8.290}$$

where $k > 0$, and define the estimate

$$\hat{x}_2 = \xi + \theta_1 \zeta + k x_1, \tag{8.291}$$

so that the estimation error $\varepsilon = x_2 - \hat{x}_2$ satisfies the equation

$$\dot{\varepsilon} = -k\varepsilon. \tag{8.292}$$

8.3 Example: Levitated Ball

Thus, we apply backstepping to the system

$$\begin{aligned} \dot{x}_1 &= \theta_1 \zeta + \xi + kx_1 + \varepsilon \\ \dot{\zeta} &= -k\zeta - \lambda(x_1)x_3^2 \\ \dot{x}_3 &= -\theta_2 x_3 + \theta_3 u. \end{aligned} \quad (8.293)$$

The design objective is to regulate x_1 to a given set point y_r. In the first equation, ζ is considered as virtual control. In the second equation, rather than x_3 alone, the force $\lambda(x_1)x_3^2$ is the virtual control. Thus, our goal is to stabilize system (8.293) in the error coordinates

$$\begin{aligned} z_1 &= x_1 - y_r \\ z_2 &= \zeta - \alpha_1 \\ z_3 &= \lambda(x_1)x_3^2 - \alpha_2. \end{aligned} \quad (8.294)$$

The parameters θ_1 and θ_3 act as unknown virtual control coefficients. Applying the backstepping design with tuning functions, we obtain stabilizing functions

$$\alpha_1 = \hat{\varrho}_1 \left(-c_1 z_1 - d_1 z_1 - kx_1 - \xi \right) \triangleq \hat{\varrho}_1 \bar{\alpha}_1 \quad (8.295)$$

$$\alpha_2 = \hat{\theta}_1 z_1 + c_2 z_2 + d_2 \left(\frac{\partial \alpha_1}{\partial x_1} \right)^2 z_2 - k\zeta - \frac{\partial \alpha_1}{\partial x_1} \left(\xi + \hat{\theta}_1 \zeta + kx_1 \right)$$

$$- \frac{\partial \alpha_1}{\partial \xi} \left(-k\xi - k^2 x_1 + g \right) - \frac{\partial \alpha_1}{\partial \hat{\varrho}_1} \dot{\hat{\varrho}}_1 \quad (8.296)$$

$$\quad (8.297)$$

and the control law

$$u = \frac{\hat{\varrho}_3}{2\lambda(x_1)x_3} \left[z_2 - c_3 z_3 - d_3 \left(\frac{\partial \alpha_2}{\partial x_1} - \lambda'(x_1)x_3 \right)^2 z_3 + 2\hat{\theta}_2 \lambda(x_1)x_3^2 \right.$$

$$+ \left(\frac{\partial \alpha_2}{\partial x_1} - \lambda'(x_1)x_3 \right) \left(\xi + \hat{\theta}_1 \zeta + kx_1 \right) + \frac{\partial \alpha_2}{\partial \xi} \left(\xi + \hat{\theta}_1 \zeta + kx_1 \right)$$

$$\left. + \frac{\partial \alpha_2}{\partial \zeta} \left(-k\zeta - \lambda(x_1)x_3^2 \right) + \frac{\partial \alpha_2}{\partial \hat{\varrho}_1} \dot{\hat{\varrho}}_1 + \frac{\partial \alpha_2}{\partial \hat{\theta}_1} \dot{\hat{\theta}}_1 \right] \triangleq \frac{\hat{\varrho}_3}{2\lambda(x_1)x_3} \bar{\alpha}_3. \quad (8.298)$$

The singularity of this control law at $x_3 = I = 0$ is not an impediment, because the set point value of the current is well above zero. A calculation shows that with the control (8.298) the error system is

$$\begin{bmatrix} \dot{z}_1 \\ \dot{z}_2 \\ \dot{z}_3 \end{bmatrix} = \begin{bmatrix} -c_1 - d_1 & \theta_1 & 0 \\ -\theta_1 & -c_2 - d_2 \left(\frac{\partial \alpha_1}{\partial x_1} \right)^2 & -1 \\ 0 & 1 & -c_3 - d_3 \left(\frac{\partial \alpha_2}{\partial x_1} - \lambda'(x_1)x_3 \right)^2 \end{bmatrix} \begin{bmatrix} z_1 \\ z_2 \\ z_3 \end{bmatrix}$$

$$+ \begin{bmatrix} 1 \\ -\frac{\partial \alpha_1}{\partial x_1} \\ \lambda'(x_1)x_3 - \frac{\partial \alpha_2}{\partial x_1} \end{bmatrix} \varepsilon$$

$$+ \begin{bmatrix} -\theta_1\bar{\alpha}_1 & 0 & 0 & 0 \\ 0 & 0 & z_1 - \frac{\partial \alpha_1}{\partial x_1}\zeta & 0 \\ 0 & -\theta_3\bar{\alpha}_3 & \left(\lambda'(x_1)x_3 - \frac{\partial \alpha_2}{\partial x_1}\right)\zeta & -2\lambda(x_1)x_3^2 \end{bmatrix} \begin{bmatrix} \tilde{\varrho}_1 \\ \tilde{\varrho}_3 \\ \tilde{\theta}_1 \\ \tilde{\theta}_2 \end{bmatrix}. \quad (8.299)$$

The parameter update laws obtained using the tuning functions technique are

$$\dot{\hat{\varrho}}_1 = -\gamma\bar{\alpha}_1 z_1 \qquad (8.300)$$

$$\dot{\hat{\varrho}}_3 = -\gamma\bar{\alpha}_1 z_3 \qquad (8.301)$$

$$\dot{\hat{\theta}}_1 = \gamma\left(z_1 z_2 - \frac{\partial \alpha_1}{\partial x_1}\zeta z_2 - \frac{\partial \alpha_2}{\partial x_1}\zeta z_3 + \lambda'(x_1)x_3^2\zeta z_3\right) \qquad (8.302)$$

$$\dot{\hat{\theta}}_2 = -\gamma 2\lambda(x_1)x_3^2 z_3. \qquad (8.303)$$

The closed-loop adaptive system has an equilibrium at the point

$$\begin{aligned} x_1 &= y_r & \xi &= \frac{g}{k} - ky_r & \hat{\varrho}_1 &= 1/\theta_1 \\ x_2 &= 0 & \zeta &= -\frac{g}{k\theta_1} & \hat{\varrho}_3 &= 1/\theta_3 \\ x_3 &= \sqrt{\frac{g}{\theta_1 \lambda(y_r)}} & & & \hat{\theta}_1 &= \theta_1 \\ & & & & \hat{\theta}_2 &= \theta_2. \end{aligned} \qquad (8.304)$$

Before we discuss the stability properties, let us introduce the signal $\tilde{\xi} = \xi - \frac{g}{k} + ky_r$, which is governed by

$$\dot{\tilde{\xi}} = -k\tilde{\xi} - k^2 z_1. \qquad (8.305)$$

The derivative of the Lyapunov function

$$V = \frac{1}{2}\left(z_1^2 + z_2^2 + z_3^2\right) + \frac{1}{4kd_0}\varepsilon^2 + \frac{c_1}{2k^3}\tilde{\xi}^2 + \frac{1}{2\gamma}\left(\theta_1\tilde{\varrho}_1^2 + \theta_3\tilde{\varrho}_3^2 + \tilde{\theta}_1^2 + \tilde{\theta}_2^2\right) \qquad (8.306)$$

along the solutions of (8.299), (8.292), (8.305), and (8.300)–(8.303) can be easily shown to satisfy

$$\dot{V} \leq -\frac{c_1}{2}z_1^2 - c_2 z_2^2 - c_3 z_3^2 - \frac{1}{4d_0}\varepsilon^2 - \frac{c_1}{2k^2}\tilde{\xi}^2, \qquad (8.307)$$

which means that the equilibrium (8.304) is stable. This property is not global because of the physical constraint confining the solutions to the feasibility set $\mathcal{F}$ defined by $x_1 > 0$ and $x_3 > 0$. The Lyapunov function (8.306) will now be used to estimate the region of attraction within $\mathcal{F}$. Let $\Omega(c)$ be the invariant set defined by $V(x_1, x_2, x_3, \xi, \zeta, \hat{\varrho}_1, \hat{\varrho}_3, \hat{\theta}_1, \hat{\theta}_2) < c$. Then an estimate Ω of the stability region is

$$\Omega = \left\{X = (x_1, x_2, x_3, \xi, \zeta, \hat{\varrho}_1, \hat{\varrho}_3, \hat{\theta}_1, \hat{\theta}_2) \mid V(X) < \arg\sup_{\Omega(c) \subset \mathcal{F}}\{c\}\right\}. \qquad (8.308)$$

From (8.307) we also conclude that all the states except possibly the parameter estimates converge to values given by the equilibrium (8.304). Using LaSalle's invariance theorem one can prove from (8.299) that $\tilde{\varrho}_1$ and $\tilde{\theta}_1$ converge to zero, which means that the estimates of M and $1/M$ converge to their true values. This is due to the gravity g which causes the force of the electromagnet to be nonzero even when $y_r = 0$, thus introducing in the regressor terms which do not vanish at the equilibrium. As we explained in Example 4.13, nonvanishing elements in the regressor contribute to convergence of parameter estimates. From LaSalle's invariance theorem it also follows that a linear combination of $\tilde{\varrho}_3$ and $\tilde{\theta}_2$ converges to zero. Since V converges to a constant, we conclude that $\tilde{\varrho}_3$ and $\tilde{\theta}_2$ converge to constant values.

Notes and References

Both the output-feedback scheme of Marino and Tomei [122, 123] and the modified scheme of Kanellakopoulos, Kokotović, and Morse [72] inherited overparametrization from the original adaptive backstepping design [69].

The designs presented in this chapter avoid overparametrization using the tuning functions technique which was first employed for output-feedback adaptive designs in Krstić, Kanellakopoulos, and Kokotović [95] and in Krstić and Kokotović [99]. While the K-filters employed in this chapter are the same as those in [72] patterned after Kreisselmeier [91], the MT-filters in this chapter are simpler than those proposed for filtered transformations in [122, 123]. The use of nonlinear damping as a tool to improve performance and guarantee boundedness without adaptation was suggested in Kanellakopoulos [64] and in Kanellakopoulos, Krstić, and Kokotović [80].

Teel [189] proposed an output-feedback design where high gain is employed to allow the regressor to depend only on the reference signals, which means that the adaptation is active only in the case of tracking.

Khalil [83] and Janković [57] developed semiglobal adaptive designs for a class of nonlinear systems which includes some systems not transformable into the output feedback form. Their designs employ high-gain observers and control saturation. While Khalil's identifier is Lyapunov-type, Janković employs a passive identifier.

Tao and Kokotović developed adaptive designs for systems with unknown backlash [184] and dead-zone [185].

Chapter 9

Modular Designs

In this chapter we extend the modular approach of Chapters 5 and 6 to the case of output feedback. The output-feedback modular approach results in separation of three design modules: the control law, the identifier, and the state estimator. With the K-filters we design output-feedback control laws which guarantee input-to-state stability with respect to the parameter error, its derivative, and the state estimation error as the inputs. At the end of this chapter we briefly present some schemes with MT-filters.

Following the ideas introduced in Chapters 5 and 6, we develop output-feedback forms of passive and swapping identifiers.

The schemes in this chapter are simpler than the tuning functions schemes because they do not eliminate $\dot{\hat{\theta}}$ from the error system, but instead use stronger ISS-controllers.

As in Chapter 8 we consider the output-feedback systems

$$\begin{aligned} \dot{x} &= Ax + \phi(y) + \Phi(y)a + \begin{bmatrix} 0 \\ b \end{bmatrix} \sigma(y)u \\ y &= e_1^T x\,. \end{aligned} \qquad (9.1)$$

We retain Assumptions 8.2 and 8.3 and modify Assumption 9.1 as follows:

Assumption 9.1 *In addition to* $\mathrm{sgn}(b_m)$, *a positive constant* ς_m *is known such that* $|b_m| \geq \varsigma_m$.

Assumption 9.1, which is stronger than Assumption 8.1, is standard in 'indirect' adaptive control. It allows the control law to contain a division by the estimate $\hat{b}_m$, which is kept away from zero by parameter projection. In the tuning functions designs it was possible to avoid this at the expense of an additional estimate of $\varrho = 1/b_m$.

The K-filters introduced in Section 8.1.1 are repeated for convenience in Table 9.1. With these filters the state x can be represented as

$$x = \xi + \Omega^T \theta + \varepsilon\,, \qquad (9.2)$$

Table 9.1: K-Filters

K-filters:		
$\dot{\xi} = A_0\xi + ky + \phi(y)$		(9.5)
$\dot{\Xi} = A_0\Xi + \Phi(y)$		(9.6)
$\dot{\lambda} = A_0\lambda + e_n\sigma(y)u$		(9.7)
$v_j = A_0^j\lambda,\quad j = 0,\ldots,m$		(9.8)
$\Omega^{\mathrm{T}} = [v_m,\ldots,v_1,v_0,\Xi]$		(9.9)

where

$$\dot{\varepsilon} = A_0\varepsilon. \tag{9.3}$$

We recall that the vector $k = [k_1,\ldots,k_n]^{\mathrm{T}}$ is chosen so that the matrix $A_0 = A - ke_1^{\mathrm{T}}$ is Hurwitz, that is,

$$PA_0 + A_0^{\mathrm{T}}P = -I, \quad P = P^{\mathrm{T}} > 0. \tag{9.4}$$

We also repeat the two crucial expressions for $\dot{y}$:

$$\dot{y} = \omega_0 + \omega^{\mathrm{T}}\theta + \varepsilon_2 \tag{9.10}$$
$$= b_m v_{m,2} + \omega_0 + \bar{\omega}^{\mathrm{T}}\theta + \varepsilon_2, \tag{9.11}$$

where the 'regressor' ω and the 'truncated regressor' $\bar{\omega}$ are defined as

$$\omega = [v_{m,2}, v_{m-1,2},\ldots,v_{0,2},\ \Phi_{(1)} + \Xi_{(2)}]^{\mathrm{T}} \tag{9.12}$$
$$\bar{\omega} = [0, v_{m-1,2},\ldots,v_{0,2},\ \Phi_{(1)} + \Xi_{(2)}]^{\mathrm{T}}, \tag{9.13}$$

and

$$\omega_0 = \varphi_{0,1} + \xi_2. \tag{9.14}$$

9.1 ISS-Controller Design

The output feedback ISS-controller design has many similarities with the state-feedback design in Section 5.3. Also, the technique for dealing with unknown

9.1 ISS-CONTROLLER DESIGN

high frequency gain is the same as in Section 5.9.1. As in Section 8.1.2, our task is to develop a backstepping procedure for the system

$$\dot{y} = b_m v_{m,2} + \omega_0 + \bar{\omega}^T \theta + \varepsilon_2 \tag{9.15}$$
$$\dot{v}_{m,i} = v_{m,i+1} - k_i v_{m,1}, \qquad i = 2, \ldots, \rho - 1 \tag{9.16}$$
$$\dot{v}_{m,\rho} = \sigma(y)u + v_{m,\rho+1} - k_\rho v_{m,1}. \tag{9.17}$$

To achieve modularity, we seek control laws which guarantee input-to-state stability with respect to the parameter error $\tilde{\theta}$, its derivative $\dot{\hat{\theta}}$, and the state estimation error ε. Our main tool is nonlinear damping of Lemma 5.6. Our presentation assumes familiarity with the backstepping procedure of Chapter 8, and the ISS-controller of Section 5.3.

Following Section 5.9.1, for the system (9.15)–(9.17) we define

$$z_1 = y - y_r \tag{9.18}$$
$$z_i = v_{m,i} - \frac{1}{\hat{b}_m} y_r^{(i-1)} - \alpha_{i-1}, \qquad i = 2, \ldots, \rho, \tag{9.19}$$

where the division by the estimate $\hat{b}_m$ poses no problem because $\hat{b}_m$ will be kept away from zero by employing the parameter projection in the update law.

Step 1. The equation for the tracking error z_1 obtained from (9.18) and (9.15) is

$$\dot{z}_1 = b_m v_{m,2} + \omega_0 + \bar{\omega}^T \theta + \varepsilon_2 - \dot{y}_r. \tag{9.20}$$

Substituting (9.19) for $i = 2$ into (9.20), we get

$$\dot{z}_1 = b_m z_2 + b_m \alpha_1 + \omega_0 + \bar{\omega}^T \theta + \varepsilon_2 + \frac{1}{\hat{b}_m} \dot{y}_r \tilde{b}_m. \tag{9.21}$$

In selecting the first stabilizing function α_1 we can disregard the term z_2, but we have to deal with the difficulty created by the unknown b_m multiplying α_1. Writing α_1 in the form

$$\alpha_1 = -\frac{\operatorname{sgn}(b_m)}{\varsigma_m}(c_1 + s_1)z_1 + \frac{1}{\hat{b}_m}\bar{\alpha}_1, \tag{9.22}$$

where s_1 is the first nonlinear damping function, yet to be chosen, from (9.21) we get

$$\dot{z}_1 = -\frac{|b_m|}{\varsigma_m}(c_1 + s_1)z_1 + \bar{\alpha}_1 + \omega_0 + \bar{\omega}^T \theta + \varepsilon_2 + \frac{1}{\hat{b}_m}(\dot{y}_r + \bar{\alpha}_1)\tilde{b}_m + b_m z_2. \tag{9.23}$$

Even though $\bar{\alpha}_1$ appears together with the parameter error $\tilde{b}_m$, it could cancel the term $\omega_0 + \bar{\omega}^T \theta$ if θ were known. This is a convenience provided by (9.22). The choice

$$\bar{\alpha}_1 = -\omega_0 - \bar{\omega}^T \hat{\theta} \tag{9.24}$$

yields

$$\dot{z}_1 = -\frac{|b_m|}{\varsigma_m}(c_1 + s_1)z_1 + \varepsilon_2 + \bar{\omega}^{\mathrm{T}}\tilde{\theta} + \frac{1}{\hat{b}_m}(\dot{y}_r + \bar{\alpha}_1)\tilde{b}_m + b_m z_2 \qquad (9.25)$$

or, noting that $\tilde{b}_m$ is the first component of $\tilde{\theta}$,

$$\dot{z}_1 = -\frac{|b_m|}{\varsigma_m}(c_1 + s_1)z_1 + \varepsilon_2 + \left[\bar{\omega} + \frac{1}{\hat{b}_m}(\dot{y}_r + \bar{\alpha}_1)e_1\right]^{\mathrm{T}}\tilde{\theta} + b_m z_2. \qquad (9.26)$$

In (9.26) we see one more benefit of using $\frac{\mathrm{sgn}(b_m)}{\varsigma_m}$ rather than $\frac{1}{\hat{b}_m}$ in (9.22): the appearance of s_1 in the bracketed term in (9.26) is avoided. Now, to achieve input-to-state stability with respect to ε_2 and $\tilde{\theta}$, the nonlinear damping s_1 is chosen as

$$s_1 = d_1 + \kappa_1 \left|\bar{\omega} + \frac{1}{\hat{b}_m}(\dot{y}_r + \bar{\alpha}_1)e_1\right|^2. \qquad (9.27)$$

As in Section 8.1.2, with (8.56) we conclude that α_1 is a function of $y, \xi, \Xi, \hat{\theta}, \bar{\lambda}_{m+1}, y_r$, and postulate that $\alpha_i\left(y, \xi, \Xi, \hat{\theta}, \bar{\lambda}_{m+i}, \bar{y}_r^{(i-1)}\right)$.

Step 2. Differentiating (9.19), with the help of (9.16), we get

$$\begin{aligned}\dot{z}_2 &= \dot{v}_{m,2} - \frac{1}{\hat{b}_m}\ddot{y}_r + \frac{1}{\hat{b}_m^2}\dot{\hat{b}}_m \dot{y}_r - \dot{\alpha}_1(y, \xi, \Xi, \hat{\theta}, \bar{\lambda}_{m+1}, y_r) \\ &= v_{m,3} - k_2 v_{m,1} - \frac{1}{\hat{b}_m}\ddot{y}_r - \frac{\partial \alpha_1}{\partial y}\left(\omega_0 + \omega^{\mathrm{T}}\theta + \varepsilon_2\right) - \frac{\partial \alpha_1}{\partial \xi}(A_0\xi + ky + \phi) \\ &\quad - \frac{\partial \alpha_1}{\partial \Xi}(A_0\Xi + \Phi) - \frac{\partial \alpha_1}{\partial y_r}\dot{y}_r - \sum_{j=1}^{m+1}\frac{\partial \alpha_1}{\partial \lambda_j}(-k_j\lambda_1 + \lambda_{j+1}) \\ &\quad - \frac{\partial \alpha_1}{\partial \hat{\theta}}\dot{\hat{\theta}} + \frac{1}{\hat{b}_m^2}\dot{y}_r\dot{\hat{b}}_m.\end{aligned} \qquad (9.28)$$

Noting from (9.19) that $v_{m,3} - \frac{1}{\hat{b}_m}\ddot{y}_r = z_3 + \alpha_2$, and by adding and subtracting $\frac{\partial \alpha_1}{\partial y}\omega^{\mathrm{T}}\hat{\theta}$, we get

$$\begin{aligned}\dot{z}_2 &= \alpha_2 - k_2 v_{m,1} - \frac{\partial \alpha_1}{\partial y}\left(\omega_0 + \omega^{\mathrm{T}}\hat{\theta}\right) - \frac{\partial \alpha_1}{\partial \xi}(A_0\xi + ky + \phi) \\ &\quad - \frac{\partial \alpha_1}{\partial \Xi}(A_0\Xi + \Phi) - \frac{\partial \alpha_1}{\partial y_r}\dot{y}_r - \sum_{j=1}^{m+1}\frac{\partial \alpha_1}{\partial \lambda_j}(-k_j\lambda_1 + \lambda_{j+1}) \\ &\quad - \frac{\partial \alpha_1}{\partial y}\varepsilon_2 - \frac{\partial \alpha_1}{\partial y}\omega^{\mathrm{T}}\tilde{\theta} - \left(\frac{\partial \alpha_1}{\partial \hat{\theta}} - \frac{1}{\hat{b}_m^2}\dot{y}_r e_1^{\mathrm{T}}\right)\dot{\hat{\theta}} + z_3.\end{aligned} \qquad (9.29)$$

The stabilizing function α_2 is chosen to cancel all the terms except those in

9.1 ISS-Controller Design

the last line of (9.29):

$$\alpha_2 = -\hat{b}_m z_1 - (c_2 + s_2)z_2 + k_2 v_{m,1} + \frac{\partial \alpha_1}{\partial y}\left(\omega_0 + \omega^T \hat{\theta}\right) + \frac{\partial \alpha_1}{\partial \xi}(A_0 \xi + ky + \phi)$$

$$+ \frac{\partial \alpha_1}{\partial \Xi}(A_0 \Xi + \Phi) + \frac{\partial \alpha_1}{\partial y_r}\dot{y}_r + \sum_{j=1}^{m+1}\frac{\partial \alpha_1}{\partial \lambda_j}(-k_j\lambda_1 + \lambda_{j+1}), \quad (9.30)$$

where the term $-\hat{b}_m z_1$ is added to compensate for $b_m z_2$ in (9.26). Substituting (9.30) into (9.29) we get

$$\dot{z}_2 = -b_m z_1 - (c_2 + s_2)z_2 + z_3$$
$$- \frac{\partial \alpha_1}{\partial y}\varepsilon_2 - \left(\frac{\partial \alpha_1}{\partial y}\omega - z_1 e_1\right)^T \tilde{\theta} - \left(\frac{\partial \alpha_1}{\partial \hat{\theta}} - \frac{1}{\hat{b}_m^2}\dot{y}_r e_1^T\right)\dot{\hat{\theta}}. \quad (9.31)$$

The uncancelled disturbance inputs in the second line of (9.31) are counteracted, term by term, with the nonlinear damping function

$$s_2 = d_2\left(\frac{\partial \alpha_1}{\partial y}\right)^2 + \kappa_2\left|\frac{\partial \alpha_1}{\partial y}\omega - z_1 e_1\right|^2 + g_2\left|\frac{\partial \alpha_1}{\partial \hat{\theta}}^T - \frac{1}{\hat{b}_m^2}\dot{y}_r e_1\right|^2. \quad (9.32)$$

Step $i = 3, \ldots, \rho$. We now proceed with the design by differentiating (9.19) for $i = 3, \ldots, \rho - 1$ with the help of (9.16):

$$\dot{z}_i = \dot{v}_{m,i} - \frac{1}{\hat{b}_m}y_r^{(i)} + \frac{1}{\hat{b}_m^2}\dot{\hat{b}}_m y_r^{(i-1)} - \dot{\alpha}_{i-1}\left(y, \xi, \Xi, \hat{\theta}, \bar{\lambda}_{m+i-1}, \bar{y}_r^{(i-2)}\right)$$

$$= v_{m,i+1} - k_i v_{m,1} - \frac{1}{\hat{b}_m}y_r^{(i)} - \frac{\partial \alpha_{i-1}}{\partial y}\left(\omega_0 + \omega^T \theta + \varepsilon_2\right)$$

$$- \frac{\partial \alpha_{i-1}}{\partial \xi}(A_0\xi + ky + \phi) - \frac{\partial \alpha_{i-1}}{\partial \Xi}(A_0\Xi + \Phi) - \sum_{j=1}^{i-1}\frac{\partial \alpha_{i-1}}{\partial y_r^{(j-1)}}y_r^{(j)}$$

$$- \sum_{j=1}^{m+i-1}\frac{\partial \alpha_{i-1}}{\partial \lambda_j}(-k_j\lambda_1 + \lambda_{j+1}) - \frac{\partial \alpha_{i-1}}{\partial \hat{\theta}}\dot{\hat{\theta}} + \frac{1}{\hat{b}_m^2}y_r^{(i-1)}\dot{\hat{b}}_m. \quad (9.33)$$

Noting from (9.19) that $v_{m,i+1} - \frac{1}{\hat{b}_m}y_r^{(i)} = z_{i+1} + \alpha_i$, we get

$$\dot{z}_i = z_{i+1} + \alpha_i - k_i v_{m,1} - \frac{\partial \alpha_{i-1}}{\partial y}\left(\omega_0 + \omega^T \theta + \varepsilon_2\right)$$

$$- \frac{\partial \alpha_{i-1}}{\partial \xi}(A_0\xi + ky + \phi) - \frac{\partial \alpha_{i-1}}{\partial \Xi}(A_0\Xi + \Phi) - \sum_{j=1}^{i-1}\frac{\partial \alpha_{i-1}}{\partial y_r^{(j-1)}}y_r^{(j)}$$

$$- \sum_{j=1}^{m+i-1}\frac{\partial \alpha_{i-1}}{\partial \lambda_j}(-k_j\lambda_1 + \lambda_{j+1}) - \left(\frac{\partial \alpha_{i-1}}{\partial \hat{\theta}} - \frac{1}{\hat{b}_m^2}y_r^{(i-1)}e_1^T\right)\dot{\hat{\theta}}. \quad (9.34)$$

From (9.17) it follows that the final step $i = \rho$ can be encompassed in these calculations if we define $\alpha_\rho = \sigma(y)u + v_{m,\rho+1} - \frac{1}{b_m}y_r^{(\rho)}$ and $z_{\rho+1} = 0$. To prepare for choosing a stabilizing function we add and subtract $\frac{\partial \alpha_{i-1}}{\partial y}\omega^T\hat{\theta}$ in (9.34) and get

$$\begin{aligned}\dot{z}_i &= \alpha_i - k_i v_{m,1} - \frac{\partial \alpha_{i-1}}{\partial y}\left(\omega_0 + \omega^T\hat{\theta}\right) - \frac{\partial \alpha_{i-1}}{\partial \xi}(A_0\xi + ky + \phi) \\ &\quad - \frac{\partial \alpha_{i-1}}{\partial \Xi}(A_0\Xi + \Phi) - \sum_{j=1}^{i-1}\frac{\partial \alpha_{i-1}}{\partial y_r^{(j-1)}}y_r^{(j)} - \sum_{j=1}^{m+i-1}\frac{\partial \alpha_{i-1}}{\partial \lambda_j}(-k_j\lambda_1 + \lambda_{j+1}) \\ &\quad - \frac{\partial \alpha_{i-1}}{\partial y}\varepsilon_2 - \frac{\partial \alpha_{i-1}}{\partial y}\omega^T\tilde{\theta} - \left(\frac{\partial \alpha_{i-1}}{\partial \hat{\theta}} - \frac{1}{b_m^2}y_r^{(i-1)}e_1^T\right)\dot{\hat{\theta}} + z_{i+1}.\end{aligned} \quad (9.35)$$

The stabilizing function α_i is chosen to cancel all the terms except those in the last line of (9.35):

$$\begin{aligned}\alpha_i &= -z_{i-1} - (c_i + s_i)z_i + k_i v_{m,1} + \frac{\partial \alpha_{i-1}}{\partial y}\left(\omega_0 + \omega^T\hat{\theta}\right) \\ &\quad + \frac{\partial \alpha_{i-1}}{\partial \xi}(A_0\xi + ky + \phi) + \frac{\partial \alpha_{i-1}}{\partial \Xi}(A_0\Xi + \Phi) + \sum_{j=1}^{i-1}\frac{\partial \alpha_{i-1}}{\partial y_r^{(j-1)}}y_r^{(j)} \\ &\quad + \sum_{j=1}^{m+i-1}\frac{\partial \alpha_{i-1}}{\partial \lambda_j}(-k_j\lambda_1 + \lambda_{j+1}).\end{aligned} \quad (9.36)$$

Substituting (9.36) into (9.35) we get

$$\begin{aligned}\dot{z}_i &= -z_{i-1} - (c_i + s_i)z_i + z_{i+1} \\ &\quad - \frac{\partial \alpha_{i-1}}{\partial y}\varepsilon_2 - \frac{\partial \alpha_{i-1}}{\partial y}\omega^T\tilde{\theta} - \left(\frac{\partial \alpha_{i-1}}{\partial \hat{\theta}} - \frac{1}{b_m^2}y_r^{(i-1)}e_1^T\right)\dot{\hat{\theta}}.\end{aligned} \quad (9.37)$$

The uncancelled disturbance inputs in the second line of (9.37) are counteracted, term by term, with the nonlinear damping function

$$s_i = d_i\left(\frac{\partial \alpha_{i-1}}{\partial y}\right)^2 + \kappa_i\left|\frac{\partial \alpha_{i-1}}{\partial y}\omega\right|^2 + g_i\left|\frac{\partial \alpha_{i-1}}{\partial \hat{\theta}}^T - \frac{1}{b_m^2}y_r^{(i-1)}e_1\right|^2. \quad (9.38)$$

The complete design of the control law is summarized in Table 9.2. It employs the filters given in Table 9.1.

The error system (9.26), (9.31), (9.37) is now compactly written as

$$\dot{z} = A_z^*(z,t)z + W_\varepsilon(z,t)\varepsilon_2 + W_\theta^*(z,t)^T\tilde{\theta} + Q(z,t)^T\dot{\hat{\theta}}, \quad z \in \mathbb{R}^\rho, \quad (9.39)$$

9.1 ISS-Controller Design

Table 9.2: ISS-Controller Design with K-Filters

$$z_1 = y - y_r \tag{9.40}$$

$$z_i = v_{m,i} - \frac{1}{\hat{b}_m} y_r^{(i-1)} - \alpha_{i-1}, \qquad i = 2,\ldots,\rho \tag{9.41}$$

$$\alpha_1 = -\frac{\text{sgn}(b_m)}{\varsigma_m}(c_1 + s_1)z_1 + \frac{1}{\hat{b}_m}\bar{\alpha}_1 \tag{9.42}$$

$$\bar{\alpha}_1 = -\omega_0 - \bar{\omega}^T \hat{\theta} \tag{9.43}$$

$$\alpha_2 = -\hat{b}_m z_1 - (c_2 + s_2)z_2 + \beta_2 \tag{9.44}$$

$$\alpha_i = -z_{i-1} - (c_i + s_i)z_i + \beta_i, \qquad i = 3,\ldots,\rho \tag{9.45}$$

$$\beta_i = \frac{\partial \alpha_{i-1}}{\partial y}\left(\omega_0 + \omega^T \hat{\theta}\right) + \frac{\partial \alpha_{i-1}}{\partial \xi}(A_0\xi + ky + \phi) + \frac{\partial \alpha_{i-1}}{\partial \Xi}(A_0\Xi + \Phi)$$

$$+ \sum_{j=1}^{i-1} \frac{\partial \alpha_{i-1}}{\partial y_r^{(j-1)}} y_r^{(j)} + k_i v_{m,1} + \sum_{j=1}^{m+i-1} \frac{\partial \alpha_{i-1}}{\partial \lambda_j}(-k_j \lambda_1 + \lambda_{j+1}) \tag{9.46}$$

$$s_1 = d_1 + \kappa_1 \left|\bar{\omega} + \frac{1}{\hat{b}_m}(\dot{y}_r + \bar{\alpha}_1)e_1\right|^2 \tag{9.47}$$

$$s_2 = d_2\left(\frac{\partial \alpha_1}{\partial y}\right)^2 + \kappa_2\left|\frac{\partial \alpha_1}{\partial y}\omega - z_1 e_1\right|^2 + g_2\left|\frac{\partial \alpha_1}{\partial \hat{\theta}}^T - \frac{1}{\hat{b}_m^2}\dot{y}_r e_1\right|^2 \tag{9.48}$$

$$s_i = d_i\left(\frac{\partial \alpha_{i-1}}{\partial y}\right)^2 + \kappa_i\left|\frac{\partial \alpha_{i-1}}{\partial y}\omega\right|^2 + g_i\left|\frac{\partial \alpha_{i-1}}{\partial \hat{\theta}}^T - \frac{1}{\hat{b}_m^2}y_r^{(i-1)}e_1\right|^2$$
$$i = 3,\ldots,\rho \tag{9.49}$$

Adaptive control law:

$$u = \frac{1}{\sigma(y)}\left(\alpha_\rho - v_{m,\rho+1} + \frac{1}{\hat{b}_m}y_r^{(\rho)}\right) \tag{9.50}$$

where

$$A_z^*(z,t) = \begin{bmatrix} -\frac{|b_m|}{\varsigma_m}(c_1+s_1) & b_m & 0 & \cdots & 0 \\ -b_m & -(c_2+s_2) & 1 & \ddots & \vdots \\ 0 & -1 & \ddots & \ddots & 0 \\ \vdots & & \ddots & \ddots & 1 \\ 0 & \cdots & 0 & -1 & -(c_\rho+s_\rho) \end{bmatrix} \tag{9.51}$$

$$W_\varepsilon(z,t) = \begin{bmatrix} 1 \\ -\frac{\partial \alpha_1}{\partial y} \\ \vdots \\ -\frac{\partial \alpha_{\rho-1}}{\partial y} \end{bmatrix} \in \mathbb{R}^\rho \qquad (9.52)$$

$$W_\theta^*(z,t)^T = \begin{bmatrix} \bar{\omega}^T + \frac{1}{b_m}(\dot{y}_r + \bar{\alpha}_1)e_1^T \\ -\frac{\partial \alpha_1}{\partial y}\omega^T + z_1 e_1^T \\ -\frac{\partial \alpha_2}{\partial y}\omega^T \\ \vdots \\ -\frac{\partial \alpha_{\rho-1}}{\partial y}\omega^T \end{bmatrix} \in \mathbb{R}^{\rho \times p} \qquad (9.53)$$

$$Q(z,t)^T = \begin{bmatrix} 0 \\ -\frac{\partial \alpha_1}{\partial \theta} + \frac{1}{b_m^2}\dot{y}_r e_1^T \\ \vdots \\ -\frac{\partial \alpha_{\rho-1}}{\partial \theta} + \frac{1}{b_m^2}y_r^{(\rho-1)}e_1^T \end{bmatrix} \in \mathbb{R}^{\rho \times p}. \qquad (9.54)$$

We next establish the most important input-to-state properties of the error system (9.39), (9.51)–(9.54), making use of the following constants:

$$c_0 = \min_{1 \le i \le \rho} c_i, \quad d_0 = \left(\sum_{i=1}^{\rho} \frac{1}{d_i}\right)^{-1}, \quad \kappa_0 = \left(\sum_{i=1}^{\rho} \frac{1}{\kappa_i}\right)^{-1}, \quad g_0 = \left(\sum_{i=2}^{\rho} \frac{1}{g_i}\right)^{-1}. \qquad (9.55)$$

Lemma 9.2 *In the error system (9.39), (9.51)–(9.54) with (9.47)–(9.49), the following input-to-state properties hold:*

(i) If $\tilde{\theta}, \dot{\hat{\theta}} \in \mathcal{L}_\infty[0, t_f)$, then $z, x, \xi, \Xi, \lambda \in \mathcal{L}_\infty[0, t_f)$, and

$$|z(t)| \le \frac{1}{2\sqrt{c_0}}\left(\frac{1}{d_0}\|\varepsilon_2\|_\infty^2 + \frac{1}{\kappa_0}\|\tilde{\theta}\|_\infty^2 + \frac{1}{g_0}\|\dot{\hat{\theta}}\|_\infty^2\right)^{\frac{1}{2}} + |z(0)|e^{-c_0 t}. \qquad (9.56)$$

(ii) If $\tilde{\theta} \in \mathcal{L}_\infty[0, t_f)$ and $\dot{\hat{\theta}} \in \mathcal{L}_2[0, t_f)$, then $z, x, \xi, \Xi, \lambda \in \mathcal{L}_\infty[0, t_f)$, and

$$|z(t)| \le \left[\frac{1}{4c_0}\left(\frac{1}{d_0}\|\varepsilon_2\|_\infty^2 + \frac{1}{\kappa_0}\|\tilde{\theta}\|_\infty^2\right) + \frac{1}{2g_0}\|\dot{\hat{\theta}}\|_2^2\right]^{\frac{1}{2}} + |z(0)|e^{-c_0 t}. \qquad (9.57)$$

Proof. Differentiating $\frac{1}{2}|z|^2$ along the solutions of (9.39), (9.51)–(9.51), using the definitions of the nonlinear damping functions (9.47)–(9.49), and noting that $\frac{|b_m|}{\varsigma_m} \ge 1$, we compute

$$\frac{d}{dt}\left(\frac{1}{2}|z|^2\right) = -c_1\frac{|b_m|}{\varsigma_m}z_1^2 - \sum_{i=2}^{\rho} c_i z_i^2$$

9.1 ISS-Controller Design

$$-d_1 \frac{|b_m|}{\varsigma_m} z_1^2 - \sum_{i=2}^{\rho} d_i \left(\frac{\partial \alpha_{i-1}}{\partial y}\right)^2 z_i^2$$

$$-\kappa_1 \frac{|b_m|}{\varsigma_m} \cdot \left|\bar{\omega} + \frac{1}{\hat{b}_m}(\dot{y}_r + \bar{\alpha}_1)e_1\right|^2 z_1^2 - \kappa_2 \left|\frac{\partial \alpha_1}{\partial y}\omega - z_1 e_1\right|^2 z_2^2$$

$$-\sum_{i=3}^{\rho} \kappa_i \left|\frac{\partial \alpha_{i-1}}{\partial y}\omega\right|^2 z_i^2 - \sum_{i=2}^{\rho} g_i \left|\frac{\partial \alpha_{i-1}}{\partial \hat{\theta}}^{\mathrm{T}} - \frac{1}{\hat{b}_m^2} y_r^{(i-1)} e_1\right|^2 z_i^2$$

$$+ z^{\mathrm{T}}\left[W_\varepsilon(z,t)\varepsilon_2 + W_\theta^*(z,t)^{\mathrm{T}}\tilde{\theta} + Q(z,t)^{\mathrm{T}}\dot{\hat{\theta}}\right]$$

$$\leq -\sum_{i=1}^{\rho} c_i z_i^2 - \sum_{i=1}^{\rho} d_i \left(\frac{\partial \alpha_{i-1}}{\partial y}\right)^2 z_i^2$$

$$-\kappa_1 \left|\bar{\omega} + \frac{1}{\hat{b}_m}(\dot{y}_r + \bar{\alpha}_1)e_1\right|^2 z_1^2 - \kappa_2 \left|\frac{\partial \alpha_1}{\partial y}\omega - z_1 e_1\right|^2 z_2^2$$

$$-\sum_{i=3}^{\rho} \kappa_i \left|\frac{\partial \alpha_{i-1}}{\partial y}\omega\right|^2 z_i^2 - \sum_{i=2}^{\rho} g_i \left|\frac{\partial \alpha_{i-1}}{\partial \hat{\theta}}^{\mathrm{T}} - \frac{1}{\hat{b}_m^2} y_r^{(i-1)} e_1\right|^2 z_i^2$$

$$+ z^{\mathrm{T}}\left[W_\varepsilon(z,t)\varepsilon_2 + W_\theta^*(z,t)^{\mathrm{T}}\tilde{\theta} + Q(z,t)^{\mathrm{T}}\dot{\hat{\theta}}\right], \qquad (9.58)$$

where for notational convenience we have defined $\frac{\partial \alpha_0}{\partial y} \triangleq -1$. Now, with (9.52)–(9.54) we get

$$\frac{d}{dt}\left(\frac{1}{2}|z|^2\right) \leq -\sum_{i=1}^{\rho} c_i z_i^2 - \sum_{i=1}^{\rho}\left[d_i\left(\frac{\partial \alpha_{i-1}}{\partial y}\right)^2 z_i^2 + z_i \frac{\partial \alpha_{i-1}}{\partial y}\varepsilon_2\right]$$

$$-\left[\kappa_1 \left|\bar{\omega} + \frac{1}{\hat{b}_m}(\dot{y}_r + \bar{\alpha}_1)e_1\right|^2 z_1^2 - z_1\left(\bar{\omega} + \frac{1}{\hat{b}_m}(\dot{y}_r + \bar{\alpha}_1)e_1\right)^{\mathrm{T}}\tilde{\theta}\right]$$

$$-\left[\kappa_2 \left|\frac{\partial \alpha_1}{\partial y}\omega - z_1 e_1\right|^2 z_2^2 + z_2 \left(\frac{\partial \alpha_1}{\partial y}\omega - z_1 e_1\right)^{\mathrm{T}}\tilde{\theta}\right]$$

$$-\sum_{i=3}^{\rho}\left(\kappa_i \left|\frac{\partial \alpha_{i-1}}{\partial y}\omega\right|^2 z_i^2 + z_i \frac{\partial \alpha_1}{\partial y}\omega^{\mathrm{T}}\tilde{\theta}\right)$$

$$-\sum_{i=2}^{\rho}\left[g_i \left|\frac{\partial \alpha_{i-1}}{\partial \hat{\theta}}^{\mathrm{T}} - \frac{1}{\hat{b}_m^2} y_r^{(i-1)} e_1\right|^2 z_i^2\right]$$

$$+ z_i \left(\frac{\partial \alpha_{i-1}}{\partial \hat{\theta}}^{\mathrm{T}} - \frac{1}{\hat{b}_m^2} y_r^{(i-1)} e_1\right)^{\mathrm{T}}\dot{\hat{\theta}}\right]. \qquad (9.59)$$

By completing the squares in (9.59), we obtain

$$\frac{d}{dt}\left(\frac{1}{2}|z|^2\right) \leq -\sum_{i=1}^{\rho} c_i z_i^2 - \sum_{i=1}^{\rho} d_i \left(\frac{\partial \alpha_{i-1}}{\partial y} z_i + \frac{1}{2d_i}\varepsilon_2\right)^2 + \sum_{i=1}^{\rho} \frac{1}{4d_i}\varepsilon_2^2$$

$$-\kappa_1 \left| \left(\bar{\omega} + \frac{1}{\hat{b}_m}(\dot{y}_r + \bar{\alpha}_1)e_1 \right) z_1 - \frac{1}{2\kappa_1}\dot{\tilde{\theta}} \right|^2 + \frac{1}{4\kappa_1}|\dot{\tilde{\theta}}|^2$$

$$-\kappa_2 \left| \left(\frac{\partial \alpha_1}{\partial y}\omega - z_1 e_1 \right) z_2 + \frac{1}{2\kappa_2}\dot{\tilde{\theta}} \right|^2 + \frac{1}{4\kappa_2}|\dot{\tilde{\theta}}|^2$$

$$-\sum_{i=3}^{\rho} \kappa_i \left| \frac{\partial \alpha_{i-1}}{\partial y}\omega z_i + \frac{1}{2\kappa_i}\dot{\tilde{\theta}} \right|^2 + \sum_{i=3}^{\rho} \frac{1}{4\kappa_i}|\dot{\tilde{\theta}}|^2$$

$$-\sum_{i=2}^{\rho} g_i \left| \left(\frac{\partial \alpha_{i-1}}{\partial \hat{\theta}} \right)^{\mathrm{T}} - \frac{1}{\hat{b}_m^2}y_r^{(i-1)}e_1 \right) z_i + \frac{1}{2g_i}\dot{\hat{\theta}} \right|^2 + \sum_{i=2}^{\rho} \frac{1}{4g_i}|\dot{\hat{\theta}}|^2 . \tag{9.60}$$

which with (9.55) becomes

$$\frac{d}{dt}\left(\frac{1}{2}|z|^2\right) \leq -c_0|z|^2 + \frac{1}{4}\left(\frac{1}{d_0}\varepsilon_2^2 + \frac{1}{\kappa_0}|\tilde{\theta}|^2 + \frac{1}{g_0}|\dot{\hat{\theta}}|^2 \right) \tag{9.61}$$

By applying Lemma C.5 to (9.61) we establish the following two inequalities:

$$|z(t)|^2 \leq |z(0)|^2 \mathrm{e}^{-2c_0 t} + \frac{1}{4c_0}\left(\frac{1}{d_0}\|\varepsilon_2\|_\infty^2 + \frac{1}{\kappa_0}\|\tilde{\theta}\|_\infty^2 + \frac{1}{g_0}\|\dot{\hat{\theta}}\|_\infty^2 \right) \tag{9.62}$$

$$|z(t)|^2 \leq |z(0)|^2 \mathrm{e}^{-2c_0 t} + \frac{1}{4c_0}\left(\frac{1}{d_0}\|\varepsilon_2\|_\infty^2 + \frac{1}{\kappa_0}\|\tilde{\theta}\|_\infty^2 \right) + \frac{1}{2g_0}\|\dot{\hat{\theta}}\|_2^2 \tag{9.63}$$

which, in view of the boundedness of ε, prove that $z \in \mathcal{L}_\infty[0, t_f)$ and (9.56)–(9.57). It remains to prove that the boundedness of z and $\hat{\theta}$ implies that x, ξ, Ξ, and λ are also bounded. A proof of this implication was already given for the tuning functions design in Theorem 8.5 (cf. (8.107)–(8.110)). The same argument is applicable here with (8.110) replaced by

$$v_{m,i} = z_i + \frac{1}{\hat{b}_m}y_r^{(i-1)} + \alpha_{i-1}\left(y, \xi, \Xi, \hat{\theta}, \bar{\lambda}_{m+i-1}, \bar{y}_r^{(i-2)} \right), \qquad i = 2, \ldots, \rho. \tag{9.64}$$

The bounds on z, x, ξ, Ξ, and λ are independent of t_f. □

A consequence of Lemma 9.2 is that even in the absence of adaptation all the closed-loop signals remain bounded, as we stated in Corollary 5.9 for the state feedback case.

With Lemma 9.2 at hand, our next task is to design the identifier module which guarantees that $\tilde{\theta}$ is bounded, and $\dot{\hat{\theta}}$ is either bounded or square-integrable.

9.2 y-Passive Scheme

With the K-filters in Table 9.1, we have obtained the parametric y-model (9.10):

$$\dot{y} = \omega_0 + \omega^{\mathrm{T}}\theta + \varepsilon_2 , \tag{9.65}$$

9.2 y-Passive Scheme

where ω_0 and ω are measured and defined in (9.14) and (9.12), respectively. Except for the state estimation error ε_2, the parametric y-model displays little difference from the parametric x-model (5.128) we used in the state-feedback design. What makes (9.65) desirable is the relative-degree-one property between θ as the input and y as the output.

We introduce the scalar observer

$$\dot{\hat{y}} = -\left(c_0 + \kappa_0|\omega|^2\right)(\hat{y} - y) + \omega_0 + \omega^{\mathrm{T}}\hat{\theta}, \qquad (9.66)$$

where c_0 and κ_0 are as in (9.55). The observer error

$$\epsilon = y - \hat{y} \qquad (9.67)$$

is governed by

$$\dot{\epsilon} = -\left(c_0 + \kappa_0|\omega|^2\right)\epsilon + \omega^{\mathrm{T}}\tilde{\theta} + \varepsilon_2. \qquad (9.68)$$

It can be shown that this system and the system $\dot{\varepsilon} = A_0\varepsilon$ form a system with a strict passivity property from the input $\tilde{\theta}$ to the output $\omega\epsilon$. This determines our choice of the parameter update law:

$$\dot{\hat{\theta}} = \operatorname*{Proj}_{\hat{b}_m}\{\Gamma\omega\epsilon\}, \qquad \hat{b}_m(0)\operatorname{sgn}b_m > \varsigma_m, \qquad \Gamma = \Gamma^{\mathrm{T}} > 0, \qquad (9.69)$$

where the projection operator is employed to guarantee that $|\hat{b}_m(t)| \geq \varsigma_m > 0$, $\forall t \geq 0$. (For a detailed treatment of parameter projection, see Appendix E.)

Lemma 9.3 *Let the maximal interval of existence of solutions of (9.65), (9.66) and (9.69) be $[0, t_f)$. Then the following identifier properties hold:*

$$(o) \quad |\hat{b}_m(t)| \geq \varsigma_m > 0, \ \forall t \in [0, t_f) \qquad (9.70)$$

$$(i) \quad \tilde{\theta} \in \mathcal{L}_\infty[0, t_f) \qquad (9.71)$$

$$(ii) \quad \epsilon \in \mathcal{L}_2[0, t_f) \cap \mathcal{L}_\infty[0, t_f) \qquad (9.72)$$

$$(iii) \quad \dot{\hat{\theta}} \in \mathcal{L}_2[0, t_f). \qquad (9.73)$$

Proof. First, from Lemma E.1 we conclude that the parameter projection guarantees that $|\hat{b}_m(t)| \geq \varsigma_m > 0$, $\forall t \in [0, t_f)$. Let us introduce a Lyapunov-like function

$$V = \frac{1}{2}\left(|\tilde{\theta}|^2_{\Gamma^{-1}} + |\epsilon|^2 + \frac{1}{c_0}|\varepsilon|^2_P\right), \qquad (9.74)$$

whose derivative along the solutions of (9.69), (9.68) and (9.3) is

$$\begin{aligned}
\dot{V} &= -\tilde{\theta}^{\mathrm{T}}\Gamma^{-1}\dot{\hat{\theta}} - c_0\epsilon^2 - \kappa_0|\omega|^2\epsilon^2 + \epsilon\omega^{\mathrm{T}}\tilde{\theta} + \epsilon\varepsilon_2 - \frac{1}{2c_0}|\varepsilon|^2 \\
&\leq -\frac{c_0}{2}\epsilon^2 - \kappa_0|\omega|^2\epsilon^2 + \tilde{\theta}^{\mathrm{T}}\left(\omega\epsilon - \Gamma^{-1}\dot{\hat{\theta}}\right) - \frac{c_0}{2}\epsilon^2 + \epsilon\varepsilon_2 - \frac{1}{2c_0}|\varepsilon_2|^2 \\
&= -\frac{c_0}{2}\epsilon^2 - \kappa_0|\omega|^2\epsilon^2 + \tilde{\theta}^{\mathrm{T}}\left(\omega\epsilon - \Gamma^{-1}\dot{\hat{\theta}}\right) - \frac{c_0}{2}\left(\epsilon - \frac{1}{c_0^2}\varepsilon_2\right)^2 \\
&\leq -\frac{c_0}{2}\epsilon^2 - \kappa_0|\omega|^2\epsilon^2 + \tilde{\theta}^{\mathrm{T}}\left(\omega\epsilon - \Gamma^{-1}\dot{\hat{\theta}}\right). \qquad (9.75)
\end{aligned}$$

Using Lemma E.1, we have

$$-\tilde{\theta}^T \Gamma^{-1} \dot{\hat{\theta}} = -\tilde{\theta}^T \Gamma^{-1} \operatorname{Proj}\{\Gamma \omega \epsilon\} \leq -\tilde{\theta}^T \Gamma^{-1} \Gamma \omega \epsilon = -\tilde{\theta}^T \omega \epsilon, \qquad (9.76)$$

so (9.75) becomes

$$\dot{V} \leq -\frac{c_0}{2}\epsilon^2 - \kappa_0 |\omega|^2 \epsilon^2. \qquad (9.77)$$

The nonpositivity of $\dot{V}$ proves that $\tilde{\theta}, \epsilon \in \mathcal{L}_\infty[0, t_f)$. Now due to Lemma E.1 we have

$$\begin{aligned}
|\dot{\hat{\theta}}|^2 &\leq \frac{1}{\underline{\lambda}(\Gamma^{-1})}|\dot{\hat{\theta}}|^2_{\Gamma^{-1}} \leq \frac{1}{\underline{\lambda}(\Gamma^{-1})}|\Gamma \omega \epsilon|^2_{\Gamma^{-1}} = \frac{1}{\underline{\lambda}(\Gamma^{-1})}|\omega \epsilon|^2_{\Gamma} \\
&\leq \frac{\bar{\lambda}(\Gamma)}{\underline{\lambda}(\Gamma^{-1})}|\omega \epsilon|^2 = \bar{\lambda}(\Gamma)^2 |\omega \epsilon|^2,
\end{aligned} \qquad (9.78)$$

which, substituted into (9.77), yields

$$\dot{V} \leq -\frac{c_0}{2}\epsilon^2 - \frac{\kappa_0}{\bar{\lambda}(\Gamma)^2}|\dot{\hat{\theta}}|^2. \qquad (9.79)$$

Upon integration, we get

$$\begin{aligned}
\frac{c_0}{2}\int_0^t \epsilon(\tau)^2 d\tau + \frac{\kappa_0}{\bar{\lambda}(\Gamma)^2}\int_0^t |\dot{\hat{\theta}}(\tau)|^2 d\tau &\leq -\int_0^t \dot{V} d\tau \leq V(0) - V(t) \\
&\leq V(0) < \infty,
\end{aligned} \qquad (9.80)$$

which proves that $\epsilon, \dot{\hat{\theta}} \in \mathcal{L}_2[0, t_f)$. □

This proof reveals that the nonlinear damping term $-\kappa_0 |\omega|^2 \epsilon$ included in the observer (9.66) is crucial for achieving square-integrability of $\dot{\hat{\theta}}$.

It is important to remember that the bounds in this lemma are all independent of t_f, which will allow us to extend the maximal interval of existence of solutions t_f to ∞ in the proof of the next theorem.

Theorem 9.4 (y-Passive) *All the signals in the closed-loop adaptive system consisting of the plant (9.1), the control law in Table 9.2, the filters in Table 9.1, the observer (9.66), and the update law (9.69), are globally uniformly bounded and global asymptotic tracking is achieved:*

$$\lim_{t \to \infty}[y(t) - y_r(t)] = 0. \qquad (9.81)$$

Proof. The projection operator in Appendix E is locally Lipschitz, as stated in Lemma E.1. Therefore, as argued in the proof of Theorem 8.5, the solution of the closed-loop adaptive system exists and is unique on its maximum interval of existence $[0, t_f)$.

9.2 y-Passive Scheme

From Lemma 9.3 we have $\tilde{\theta} \in \mathcal{L}_\infty[0, t_f)$ and $\dot{\hat{\theta}} \in \mathcal{L}_2[0, t_f)$, which in view of Lemma 9.2(ii) implies that $z, x, \xi, \Xi, \lambda \in \mathcal{L}_\infty[0, t_f)$. Furthermore, Lemma 9.3 guarantees the boundedness of ϵ, which in view of the boundedness of y establishes the boundedness of $\hat{y}$. Since $\hat{b}_m(t)$ is bounded away from zero, the control u is also bounded. By the same argument as in the proof of Theorem 8.5 we conclude that $t_f = \infty$.

Now we set out to prove the tracking. In view of (9.41) and (9.42) we have

$$-\frac{|b_m|}{\varsigma_m}(c_1 + s_1) z_1 + b_m z_2 + \left(\bar{\omega} + \frac{1}{\hat{b}_m}(\dot{y}_r + \bar{\alpha}_1) e_1\right)^{\mathrm{T}} \tilde{\theta}$$

$$= -\frac{\hat{b}_m \operatorname{sgn}(b_m)}{\varsigma_m}(c_1 + s_1) z_1 + \hat{b}_m z_2$$

$$\quad + \left(\bar{\omega} + \left(z_2 - \frac{\operatorname{sgn}(b_m)}{\varsigma_m}(c_1 + s_1) z_1 + \frac{1}{\hat{b}_m}(\dot{y}_r + \bar{\alpha}_1)\right) e_1\right)^{\mathrm{T}} \tilde{\theta}$$

$$= -\frac{\hat{b}_m \operatorname{sgn}(b_m)}{\varsigma_m}(c_1 + s_1) z_1 + \hat{b}_m z_2 + \left(\bar{\omega} + \left(z_2 + \frac{1}{\hat{b}_m}\dot{y}_r + \alpha_1\right) e_1\right)^{\mathrm{T}} \tilde{\theta}$$

$$= -\frac{\hat{b}_m \operatorname{sgn}(b_m)}{\varsigma_m}(c_1 + s_1) z_1 + \hat{b}_m z_2 + (\bar{\omega} + v_{m,2} e_1)^{\mathrm{T}} \tilde{\theta}$$

$$= -\frac{\hat{b}_m \operatorname{sgn}(b_m)}{\varsigma_m}(c_1 + s_1) z_1 + \hat{b}_m z_2 + \omega^{\mathrm{T}} \tilde{\theta}. \tag{9.82}$$

On the other hand,

$$-\hat{b}_m z_1 - \frac{\partial \alpha_1}{\partial y}\omega^{\mathrm{T}}\tilde{\theta} + z_1 e_1^{\mathrm{T}} \tilde{\theta} = -\hat{b}_m z_1 - \frac{\partial \alpha_1}{\partial y}\omega^{\mathrm{T}}\tilde{\theta}. \tag{9.83}$$

With (9.82) and (9.83), the error system (9.39), (9.51)–(9.54) is rewritten as

$$\dot{z} = A_z(z, t) z + W_\varepsilon(z, t)\left(\omega^{\mathrm{T}}\tilde{\theta} + \varepsilon_2\right) + Q(z, t)^{\mathrm{T}} \dot{\hat{\theta}}, \tag{9.84}$$

where A_z (without $*$) is given by

$$A_z = \begin{bmatrix} -\frac{\hat{b}_m \operatorname{sgn}(b_m)}{\varsigma_m}(c_1 + s_1) & \hat{b}_m & 0 & \cdots & 0 \\ -\hat{b}_m & -(c_2 + s_2) & 1 & \ddots & \vdots \\ 0 & -1 & \ddots & \ddots & 0 \\ \vdots & & \ddots & \ddots & 1 \\ 0 & \cdots & 0 & -1 & -(c_\rho + s_\rho) \end{bmatrix} \tag{9.85}$$

and W_ε and Q are as in (9.52) and (9.54). We know that z and $\dot{z}$ are bounded, so to prove that $z(t) \to 0$, we need to show that $z \in \mathcal{L}_2$. Since we also know that $\epsilon \in \mathcal{L}_2$, let us consider

$$\zeta \stackrel{\triangle}{=} z - W_\varepsilon(z, t)\epsilon, \tag{9.86}$$

which, in view of (9.84), (9.68), satisfies

$$\dot{\zeta} = A_z(z,t)\zeta + \left[\left((c_0 + \kappa_0|\omega|^2)I + A_z(z,t)\right)W_\varepsilon(z,t) - \dot{W}_\varepsilon(z,t)\right]\epsilon + Q(z,t)\dot{\hat{\theta}} \quad (9.87)$$

where the bracketed expression and Q are bounded due to the boundedness of all the signals and the smoothness of the plant nonlinearities. Noting that Lemma 9.3 guarantees that $\frac{\hat{b}_m \text{sgn}(b_m)}{\varsigma_m} \geq 1$, with the help of (9.85), it is now straightforward to derive

$$\frac{d}{dt}\left(|\zeta|^2\right) \leq -c_0|\zeta|^2 + \frac{1}{c_0}\left|\left((c_0 + \kappa_0|\omega|^2)I + A_z\right)W_\varepsilon - \dot{W}_\varepsilon\right|^2|\epsilon|^2 + \frac{1}{2g_0}|\dot{\hat{\theta}}|^2. \quad (9.88)$$

Since $\epsilon, \dot{\hat{\theta}} \in \mathcal{L}_2$, it follows by Lemma B.5(ii) that $\zeta \in \mathcal{L}_2$. Thus, by (9.86), $z \in \mathcal{L}_2$. By Barbalat's lemma (Corollary A.7), $z(t) \to 0$ as $t \to \infty$. Since $z_1 = y - y_r$, this proves the tracking. □

The y-passive scheme consisting of the identifier (9.66)–(9.69) and the ISS-controller in Table 9.2 is simpler than the tuning functions scheme in Table 8.2 because the ISS-controller does not incorporate tuning functions terms. The cost of this simplification is very low: only one extra integrator for the observer (9.66).

9.3 y-Swapping Scheme

Since the y-passive scheme is limited to gradient-type update laws we develop a swapping scheme which also allows the least-squares algorithm. We again start with the parametric y-model (9.10)

$$\dot{y} = \omega_0 + \omega^T\theta + \varepsilon_2 \quad (9.89)$$

and introduce the swapping filters

$$\dot{\varpi}_0 = -\left(c_0 + \kappa_0|\omega|^2\right)(\varpi_0 - y) - \omega_0, \qquad \varpi_0 \in \mathbb{R} \quad (9.90)$$
$$\dot{\varpi} = -\left(c_0 + \kappa_0|\omega|^2\right)\varpi + \omega, \qquad \varpi \in \mathbb{R}^p \quad (9.91)$$

and the estimation error

$$\epsilon = y + \varpi_0 - \varpi^T\hat{\theta}. \quad (9.92)$$

Substituting (9.89), (9.90), and (9.91) into (9.92), we get

$$\epsilon = \varpi^T\tilde{\theta} + \tilde{\epsilon}, \quad (9.93)$$

where $\tilde{\epsilon}$ is governed by

$$\dot{\tilde{\epsilon}} = -\left(c_0 + \kappa_0|\omega|^2\right)\tilde{\epsilon} + \varepsilon_2. \quad (9.94)$$

9.3 y-SWAPPING SCHEME

The update law for $\hat{\theta}$ is either the gradient:

$$\dot{\hat{\theta}} = \underset{\hat{b}_m}{\text{Proj}} \left\{ \Gamma \frac{\Omega \epsilon}{1 + \nu |\varpi|^2} \right\}, \qquad \begin{array}{l} \hat{b}_m(0) \operatorname{sgn} b_m > \varsigma_m \\ \Gamma = \Gamma^{\text{T}} > 0, \quad \nu \geq 0 \end{array} \qquad (9.95)$$

or the least-squares:

$$\begin{aligned} \dot{\hat{\theta}} &= \underset{\hat{b}_m}{\text{Proj}} \left\{ \Gamma \frac{\varpi \epsilon}{1 + \nu |\varpi|_\Gamma^2} \right\}, & \hat{b}_m(0) \operatorname{sgn} b_m &> \varsigma_m \\ \dot{\Gamma} &= -\Gamma \frac{\varpi \varpi^{\text{T}}}{1 + \nu |\varpi|_\Gamma^2} \Gamma, & \Gamma(0) &= \Gamma(0)^{\text{T}} > 0, \quad \nu \geq 0 \end{aligned} \qquad (9.96)$$

where the projection operator is employed to guarantee that $|\hat{b}_m(t)| \geq \varsigma_m > 0$, $\forall t \geq 0$, and by allowing $\nu = 0$ we encompass unnormalized gradient and least-squares. (For details of parameter projection, see Appendix E.)

Lemma 9.5 *Let the maximal interval of existence of solutions of (9.89), (9.90)–(9.91) with either (9.95) or (9.96) be $[0, t_f)$. Then for $\nu \geq 0$, the following identifier properties hold:*

$$\begin{aligned} (o)& \quad |\hat{b}_m(t)| \geq \varsigma_m > 0, \quad \forall t \in [0, t_f) & (9.97) \\ (i)& \quad \tilde{\theta} \in \mathcal{L}_\infty[0, t_f) & (9.98) \\ (ii)& \quad \epsilon \in \mathcal{L}_2[0, t_f) \cap \mathcal{L}_\infty[0, t_f) & (9.99) \\ (iii)& \quad \dot{\hat{\theta}} \in \mathcal{L}_2[0, t_f) \cap \mathcal{L}_\infty[0, t_f). & (9.100) \end{aligned}$$

Proof. First, from Lemma E.1 we conclude that $|\hat{b}_m(t)| \geq \varsigma_m > 0$, $\forall t \in [0, t_f)$. Now we consider

$$\begin{aligned} \frac{d}{dt}\left(\frac{1}{2}|\varpi|^2\right) &= -c_0 |\varpi|^2 - \kappa_0 |\omega|^2 |\varpi|^2 + \varpi^{\text{T}} \omega \\ &= -c_0 |\varpi|^2 - \kappa_0 \left(\varpi^{\text{T}} \omega - \frac{1}{2\kappa_0}\right)^2 + \frac{1}{4\kappa_0} \\ &\leq -c_0 |\varpi|^2 + \frac{1}{4\kappa_0}, \end{aligned} \qquad (9.101)$$

which proves that $\varpi \in \mathcal{L}_\infty[0, t_f)$. With (9.4), along the solutions of (9.94) and (9.3) we have

$$\begin{aligned} \frac{1}{2}\frac{d}{dt}\left(\tilde{\epsilon}^2 + \frac{1}{c_0}|\varepsilon|_P^2\right) &= -c_0 \tilde{\epsilon}^2 - \kappa_0 |\omega|^2 \tilde{\epsilon}^2 + \tilde{\epsilon} \varepsilon_2 - \frac{1}{2c_0}|\varepsilon|^2 \\ &\leq -\frac{c_0}{2}\tilde{\epsilon}^2 - \frac{c_0}{2}\left(\tilde{\epsilon} - \frac{1}{c_0}\varepsilon_2\right)^2 + \frac{1}{2c_0}\varepsilon_2^2 - \frac{1}{2c_0}|\varepsilon|^2 \\ &\leq -\frac{c_0}{2}\tilde{\epsilon}^2, \end{aligned} \qquad (9.102)$$

which shows that $\tilde{\epsilon} \in \mathcal{L}_2[0, t_f) \cap \mathcal{L}_\infty[0, t_f)$.

Gradient update law (9.95). We consider the positive definite function

$$V = \frac{1}{2}|\tilde{\theta}|^2_{\Gamma^{-1}} + \frac{1}{2c_0}\left(\tilde{\epsilon}^2 + \frac{1}{c_0}|\varepsilon|^2_P\right). \quad (9.103)$$

Using (9.102), the derivative of V is

$$\dot{V} \leq -\tilde{\theta}^T \Gamma^{-1} \dot{\tilde{\theta}} - \frac{1}{2}\tilde{\epsilon}^2. \quad (9.104)$$

By Lemma E.1, we have

$$-\tilde{\theta}^T \Gamma^{-1} \dot{\tilde{\theta}} = -\tilde{\theta}^T \Gamma^{-1} \text{Proj}\left\{\Gamma \frac{\varpi \epsilon}{1 + \nu|\varpi|^2}\right\} \leq -\tilde{\theta}^T \frac{\varpi \epsilon}{1 + \nu|\varpi|^2}, \quad (9.105)$$

so (9.104) becomes

$$\dot{V} \leq -\tilde{\theta}^T \frac{\varpi \epsilon}{1 + \nu|\varpi|^2} - \frac{1}{2}\tilde{\epsilon}^2. \quad (9.106)$$

In view of (9.93), we have

$$\begin{aligned}
\dot{V} &\leq -\tilde{\theta}^T \frac{\varpi\left(\varpi^T\tilde{\theta} + \tilde{\epsilon}\right)}{1 + \nu|\varpi|^2} - \frac{1}{2}\tilde{\epsilon}^2 = -\frac{\left(\varpi^T\tilde{\theta}\right)^2}{1 + \nu|\varpi|^2} - \frac{\tilde{\theta}^T \varpi \tilde{\epsilon}}{1 + \nu|\varpi|^2} - \frac{1}{2}\tilde{\epsilon}^2 \\
&\leq -\frac{1}{2}\frac{\left(\varpi^T\tilde{\theta}\right)^2}{1 + \nu|\varpi|^2} - \frac{1}{2}\frac{\left(\varpi^T\tilde{\theta}\right)^2}{(1 + \nu|\varpi|^2)^2} - \frac{\tilde{\theta}^T \varpi \tilde{\epsilon}}{1 + \nu|\varpi|^2} - \frac{1}{2}\tilde{\epsilon}^2 \\
&= -\frac{1}{2}\frac{\left(\varpi^T\tilde{\theta}\right)^2}{1 + \nu|\varpi|^2} - \frac{1}{2}\left(\frac{\varpi^T\tilde{\theta}}{1 + \nu|\varpi|^2} + \tilde{\epsilon}\right)^2, \quad (9.107)
\end{aligned}$$

and arrive at

$$\dot{V} \leq -\frac{1}{2}\frac{\epsilon^2}{1 + \nu|\varpi|^2}, \quad (9.108)$$

which implies $\tilde{\theta} \in \mathcal{L}_\infty[0, t_f)$ and $\frac{\epsilon}{\sqrt{1 + \nu|\varpi|^2}} \in \mathcal{L}_2[0, t_f)$. Using this and the boundedness of ϖ and $\tilde{\epsilon}$, we readily show that $\epsilon, \dot{\tilde{\theta}} \in \mathcal{L}_2[0, t_f) \cap \mathcal{L}_\infty[0, t_f)$.

Least-squares update law (9.96). We consider the function

$$V = |\tilde{\theta}|^2_{\Gamma(t)^{-1}} + \frac{1}{c_0}\tilde{\epsilon}^2 + \frac{1}{c_0^2}|\zeta|^2_{P_l}, \quad (9.109)$$

whose derivative is readily shown to be

$$\dot{V} \leq -\frac{\epsilon^2}{1 + \nu|\varpi|^2_\Gamma}, \quad (9.110)$$

which proves that $\tilde{\theta} \in \mathcal{L}_\infty[0, t_f)$ and $\frac{\epsilon}{\sqrt{1 + \nu|\varpi|^2_\Gamma}} \in \mathcal{L}_2[0, t_f)$. Using this and the boundedness of ϖ and $\tilde{\epsilon}$, we show that $\epsilon, \dot{\tilde{\theta}} \in \mathcal{L}_2[0, t_f) \cap \mathcal{L}_\infty[0, t_f)$. $\square$

9.3 y-SWAPPING SCHEME

Theorem 9.6 (*y-Swapping*) *All the signals in the closed-loop adaptive system consisting of the plant (9.1), the control law in Table 9.2, the filters in Table 9.1, and the filters (9.90)–(9.91), with either the gradient (9.95) or the least-squares update law (9.96), are globally uniformly bounded, and global asymptotic tracking is achieved:*

$$\lim_{t\to\infty} [y(t) - y_{\mathrm{r}}(t)] = 0. \tag{9.111}$$

Proof. The projection operator in Appendix E is locally Lipschitz, as stated in Lemma E.1. Therefore, as argued in the proof of Theorem 8.5, the solution of the closed-loop adaptive system exists and is unique on its maximum interval of existence $[0, t_f)$.

From Lemma 9.5 we have $\tilde{\theta}, \dot{\hat{\theta}} \in \mathcal{L}_\infty[0, t_f)$, which in view of Lemma 9.2(i) implies that $z, x, \xi, \Xi, \lambda \in \mathcal{L}_\infty[0, t_f)$. Equations (9.90) and (9.91) imply that ϖ_0 and ϖ are in $\mathcal{L}_\infty[0, t_f)$. Since $\hat{b}_m(t)$ is bounded away from zero, the control u is also bounded. By the same argument as in the proof of Theorem 8.5 we conclude that $t_f = \infty$.

To prove the tracking, let us consider the error system (9.84),

$$\dot{z} = A_z(z,t)z + W_\varepsilon(z,t)\left(\omega^{\mathrm{T}}\tilde{\theta} + \varepsilon_2\right) + Q(z,t)^{\mathrm{T}}\dot{\hat{\theta}}, \tag{9.112}$$

along with the error equation governing ϵ:

$$\dot{\epsilon} = -\left(c_0 + \kappa_0|\omega|^2\right)\epsilon + \omega^{\mathrm{T}}\tilde{\theta} + \varepsilon_2 - \varpi^{\mathrm{T}}\dot{\hat{\theta}}. \tag{9.113}$$

We could finish the proof of convergence of z to zero by an argument similar to (9.86)–(9.88) in the proof of Theorem 9.4. Instead, we present an alternative proof based on the idea in Remark 6.8. As in Remark 6.8, we show that $\dot{\epsilon}(t) \to 0$. Since, by Lemma 9.5, $\dot{\hat{\theta}} \in \mathcal{L}_\infty \cap \mathcal{L}_2$ and $\ddot{\hat{\theta}} \in \mathcal{L}_\infty$, it follows that $\dot{\hat{\theta}}(t) \to 0$. From (9.113) we conclude $\omega(t)^{\mathrm{T}}\tilde{\theta}(t) \to 0$ and therefore, $W_\varepsilon(z(t),t)^{\mathrm{T}}\left(\omega(t)^{\mathrm{T}}\tilde{\theta}(t) + \varepsilon_2(t)\right) \to 0$. Since $\dot{\hat{\theta}}(t) \to 0$, we have $Q(z(t),t)^{\mathrm{T}}\dot{\hat{\theta}}(t) \to 0$. Thus the input $W_\varepsilon\left(\omega^{\mathrm{T}}\tilde{\theta} + \varepsilon_2\right) + Q^{\mathrm{T}}\dot{\hat{\theta}}$ to the error system (9.112) converges to zero. In view of

$$\frac{d}{dt}\left(\frac{1}{2}|z|^2\right) \leq -c_0|z|^2 + z^{\mathrm{T}}\left(W_\varepsilon\left(\omega^{\mathrm{T}}\tilde{\theta} + \varepsilon_2\right) + Q^{\mathrm{T}}\dot{\hat{\theta}}\right)$$

$$\leq -\frac{c_0}{2}|z|^2 + \frac{1}{2c_0}\left|W_\varepsilon\left(\omega^{\mathrm{T}}\tilde{\theta} + \varepsilon_2\right) + Q^{\mathrm{T}}\dot{\hat{\theta}}\right|^2, \tag{9.114}$$

by applying Lemma B.8, we arrive at the conclusion that $z(t) \to 0$. Since $z_1 = y - y_{\mathrm{r}}$, this proves the asymptotic tracking. $\square$

Remark 9.7 In Theorem 6.4 we showed that the state feedback ISS-controller can be simplified by setting $g_i = 0$, $i = 1, \ldots, n$, provided that a particular form of normalization is introduced in the parameter update law. The output feedback case is no different. It is possible to show that the modified *gradient*

$$\dot{\hat{\theta}} = \Proj_{\hat{b}_m} \left\{ \frac{\Gamma}{1+\nu'|Q|_{\mathcal{F}}} \frac{\varpi\epsilon}{1+\nu|\varpi|^2} \right\}, \quad \nu' > 0 \qquad (9.115)$$

and the modified *least-squares*

$$\dot{\hat{\theta}} = \Proj_{\hat{b}_m} \left\{ \frac{\Gamma}{1+\nu'|Q|_{\mathcal{F}}} \frac{\varpi\epsilon}{1+\nu|\varpi|^2_\Gamma} \right\}, \quad \nu' > 0$$

$$\dot{\Gamma} = -\frac{1}{1+\nu'|Q|_{\mathcal{F}}} \Gamma \frac{\varpi\varpi^{\mathrm{T}}}{1+\nu|\varpi|^2_\Gamma} \Gamma \qquad (9.116)$$

update laws, which guarantee the properties

$$(o) \quad |\hat{b}_m(t)| \geq \varsigma_m > 0, \ \forall t \in [0, t_f) \qquad (9.117)$$

$$(i) \quad \tilde{\theta} \in \mathcal{L}_\infty[0, t_f) \qquad (9.118)$$

$$(ii) \quad \epsilon \in \mathcal{L}_\infty[0, t_f), \ \frac{\epsilon}{\sqrt{(1+\nu'|Q|_{\mathcal{F}})}} \in \mathcal{L}_2[0, t_f) \qquad (9.119)$$

$$(iii) \quad Q^{\mathrm{T}}\dot{\hat{\theta}} \in \mathcal{L}_\infty[0, t_f), \ \dot{\hat{\theta}} \in \mathcal{L}_2[0, t_f), \qquad (9.120)$$

allow one to set $g_i = 0$, $i = 1, \ldots, n$ in the ISS-controller in Table 9.2, while retaining the result of Theorem 9.6. ◇

While the y-passive identifier uses an extra integrator to generate $\hat{y}$, the y-swapping identifier employs $p+1$ integrators for ϖ and ϖ_0. Therefore, the price paid for having flexibility in the selection of the update law is the increase of the dynamic order by p.

9.4 x-Swapping Scheme

In this section we design a swapping identifier which uses the K-filters already employed for state estimation. Instead of the parametric y-model we consider the parametric x-model (9.2):

$$x = \xi + \Omega^{\mathrm{T}}\theta + \varepsilon. \qquad (9.121)$$

This parametric model is already in the static form due to the use of the K-filters from Table 9.1. However, this parametric model may seem unimplementable because x is not measured. Only ξ, Ω and the first component of x (the output $y = x_1$) are available. Fortunately, it suffices to consider only the first row of (9.121) where all the signals are measured except for ε_1:

$$y = \xi_1 + \Omega_1^{\mathrm{T}}\theta + \varepsilon_1. \qquad (9.122)$$

9.4 x-Swapping Scheme

It is crucial that ε_1 is bounded and exponentially converging to zero.

We introduce the "prediction" of y as

$$\hat{y} = \xi_1 + \Omega_1^T \hat{\theta}, \tag{9.123}$$

so that the "prediction error" $\epsilon \triangleq y - \hat{y}$ is implemented as

$$\epsilon = y - \xi_1 - \Omega_1^T \hat{\theta} \tag{9.124}$$

and satisfies the following equation linear in the parameter error:

$$\epsilon = \Omega_1^T \tilde{\theta} + \varepsilon_1. \tag{9.125}$$

The update law for $\hat{\theta}$ is either the gradient:

$$\dot{\hat{\theta}} = \operatorname*{Proj}_{\hat{b}_m} \left\{ \Gamma \frac{\Omega_1 \epsilon}{1 + \nu |\Omega_1|^2} \right\}, \qquad \begin{array}{l} \hat{b}_m(0) \operatorname{sgn} b_m > \varsigma_m \\ \Gamma = \Gamma^T > 0, \quad \nu > 0 \end{array} \tag{9.126}$$

or the least-squares:

$$\begin{aligned} \dot{\hat{\theta}} &= \operatorname*{Proj}_{\hat{b}_m} \left\{ \Gamma \frac{\Omega_1 \epsilon}{1 + \nu \Omega_1^T \Gamma_0 \Omega_1} \right\}, & \hat{b}_m(0) \operatorname{sgn} b_m > \varsigma_m \\ \dot{\Gamma} &= -\Gamma \frac{\Omega_1 \Omega_1^T}{1 + \nu \Omega_1^T \Gamma_0 \Omega_1} \Gamma, & \Gamma(0) = \Gamma(0)^T > 0, \quad \nu > 0 \end{aligned} \tag{9.127}$$

where the projection operator is employed to guarantee that $|\hat{b}_m(t)| \geq \varsigma_m > 0$, $\forall t \geq 0$. Matrix Γ_0 in the least-squares update law only indicates that we either use covariance resetting for Γ or let $\Gamma_0 = \Gamma + \gamma_0 I$, $\gamma_0 > 0$, to keep the normalization positive definite.

The above update laws are normalized. With unnormalized update laws we are not able to guarantee boundedness of Ω governed by (8.16),

$$\dot{\Omega}^T = A_0 \Omega^T + F(y, u)^T, \tag{9.128}$$

independently of the boundedness of $F(y, u)$.

Lemma 9.8 *Let the maximal interval of existence of solutions of (9.1), (9.5)–(9.7) with either (9.126) or (9.127) be $[0, t_f)$. Then the following identifier properties hold:*

$$\begin{aligned} (o) & \quad |\hat{b}_m(t)| \geq \varsigma_m > 0, \; \forall t \in [0, t_f) & (9.129) \\ (i) & \quad \tilde{\theta} \in \mathcal{L}_\infty[0, t_f) & (9.130) \\ (ii) & \quad \frac{\epsilon}{\sqrt{1 + \nu |\Omega_1|^2}}, \frac{\epsilon}{\sqrt{1 + \nu \Omega_1^T \Gamma_0 \Omega_1}} \in \mathcal{L}_2[0, t_f) \cap \mathcal{L}_\infty[0, t_f) & (9.131) \\ (iii) & \quad \dot{\hat{\theta}} \in \mathcal{L}_2[0, t_f) \cap \mathcal{L}_\infty[0, t_f). & (9.132) \end{aligned}$$

Proof. First, from Lemma E.1 we conclude that $|\hat{b}_m(t)| \geq \varsigma_m > 0,\ \forall t \in [0, t_f)$.
Gradient update law (9.126). We consider the positive definite function

$$V = \frac{1}{2}|\tilde{\theta}|^2_{\Gamma^{-1}} + \frac{1}{2}|\varepsilon|^2_P. \tag{9.133}$$

By virtue of Lemma E.1 we have $-\tilde{\theta}^T \Gamma^{-1}\dot{\hat{\theta}} \leq -\tilde{\theta}^T \frac{\Omega_1 \epsilon}{1+\nu|\Omega_1|^2}$, which enables us to show that

$$\dot{V} \leq -\frac{1}{2}\frac{\epsilon^2}{1+\nu|\Omega_1|^2}. \tag{9.134}$$

The nonpositivity of $\dot{V}$ proves that $\tilde{\theta} \in \mathcal{L}_\infty[0, t_f)$. Integrating (9.134) we get $\frac{\epsilon}{\sqrt{1+\nu|\Omega_1|^2}} \in \mathcal{L}_2[0, t_f)$. From

$$\frac{|\epsilon|}{\sqrt{1+\nu|\Omega_1|^2}} \leq \frac{|\Omega_1||\tilde{\theta}| + |\varepsilon_1|}{\sqrt{1+\nu|\Omega_1|^2}} \leq \frac{1}{\sqrt{\nu}}|\tilde{\theta}| + |\varepsilon_1|, \tag{9.135}$$

in view of the boundedness of $\tilde{\theta}$ and ε, we establish $\frac{\epsilon}{\sqrt{1+\nu|\Omega_1|^2}} \in \mathcal{L}_\infty[0, t_f)$.

With Lemma E.1 we have

$$|\dot{\hat{\theta}}|^2 \leq \frac{1}{\underline{\lambda}(\Gamma^{-1})}|\dot{\hat{\theta}}|^2_{\Gamma^{-1}} \leq \frac{1}{\underline{\lambda}(\Gamma^{-1})}\frac{\Omega_1^T \Gamma \Omega_1 \epsilon^2}{(1+\nu|\Omega_1|^2)^2} \leq \frac{\bar{\lambda}(\Gamma)}{\underline{\lambda}(\Gamma^{-1})}\frac{|\Omega_1|^2 \epsilon^2}{(1+\nu|\Omega_1|^2)^2}$$
$$\leq \frac{\bar{\lambda}(\Gamma)^2}{\nu}\frac{\epsilon^2}{1+\nu|\Omega_1|^2}, \tag{9.136}$$

which proves that $\dot{\hat{\theta}} \in \mathcal{L}_2[0, t_f) \cap \mathcal{L}_\infty[0, t_f)$.

Least-squares update law (9.127). We consider the function

$$V = |\tilde{\theta}|^2_{\Gamma(t)^{-1}} + |\varepsilon|^2_P, \tag{9.137}$$

which is positive definite because $\Gamma^{-1}(t)$ is positive definite for each t. Using Lemma E.1 and the fact that $\frac{d}{dt}\left(\Gamma^{-1}\tilde{\theta}\right) = -\frac{\Omega_1 \varepsilon_1}{1+\nu \Omega_1^T \Gamma_0 \Omega_1}$, it is straightforward to arrive at

$$\dot{V} \leq -\frac{\epsilon^2}{1+\nu\Omega_1^T \Gamma_0 \Omega_1}. \tag{9.138}$$

In view of the positive definiteness of $\Gamma^{-1}(t)$ this proves that $\tilde{\theta} \in \mathcal{L}_\infty[0, t_f)$. It also proves that $\frac{\epsilon}{\sqrt{1+\nu\Omega_1^T \Gamma_0 \Omega_1}} \in \mathcal{L}_2[0, t_f)$. From

$$\frac{|\epsilon|}{\sqrt{1+\nu|\Omega_1|^2_{\Gamma_0}}} \leq \frac{|\Omega_1||\tilde{\theta}| + |\varepsilon_1|}{\sqrt{1+\nu|\Omega_1|^2_{\Gamma_0}}} \leq \frac{1}{\sqrt{\nu\underline{\lambda}(\Gamma_0)}}|\tilde{\theta}| + |\varepsilon_1|, \tag{9.139}$$

9.4 x-SWAPPING SCHEME

in view of the boundedness of $\tilde{\theta}$ and ε and positive definiteness of Γ_0, we establish $\dfrac{\epsilon}{\sqrt{1+\nu|\Omega_1|_{\Gamma_0}^2}} \in \mathcal{L}_\infty[0,t_f)$. With Lemma E.1 we have

$$|\dot{\tilde{\theta}}|^2 \leq \frac{1}{\underline{\lambda}(\Gamma^{-1})}|\dot{\tilde{\theta}}|^2_{\Gamma^{-1}} \leq \frac{1}{\underline{\lambda}(\Gamma^{-1})}\frac{\Omega_1^{\mathrm{T}}\Gamma\Omega_1\epsilon^2}{\left(1+\nu|\Omega_1|_{\Gamma_0}^2\right)^2} \leq \frac{\bar{\lambda}(\Gamma)}{\underline{\lambda}(\Gamma^{-1})}\frac{|\Omega_1|^2\epsilon^2}{\left(1+\nu|\Omega_1|_{\Gamma_0}^2\right)^2}$$

$$\leq \frac{\bar{\lambda}(\Gamma)^2}{\underline{\lambda}(\Gamma_0)}\frac{|\Omega_1|^2_{\Gamma_0}}{1+\nu|\Omega_1|_{\Gamma_0}^2}\frac{\epsilon^2}{1+\nu|\Omega_1|_{\Gamma_0}^2} \leq \frac{\bar{\lambda}(\Gamma)^2}{\nu\underline{\lambda}(\Gamma_0)}\frac{\epsilon^2}{1+\nu|\Omega_1|_{\Gamma_0}^2}, \tag{9.140}$$

which proves that $\dot{\tilde{\theta}} \in \mathcal{L}_2[0,t_f) \cap \mathcal{L}_\infty[0,t_f)$. □

Combining the x-swapping identifier with the ISS-controller with K-filters, we obtain the following result.

Theorem 9.9 (x-Swapping) *All the signals in the closed-loop adaptive system consisting of the plant (9.1), the control law in Table 9.2, and the filters in Table 9.1, with either the gradient (9.126) or the least-squares update law (9.127), are globally uniformly bounded, and global asymptotic tracking is achieved:*

$$\lim_{t\to\infty}[y(t) - y_{\mathrm{r}}(t)] = 0. \tag{9.141}$$

Proof. The projection operator is locally Lipschitz, so the solution of the closed-loop adaptive system exists and is unique on its maximum interval of existence $[0,t_f)$. From Lemma 9.8 we have $\tilde{\theta}, \dot{\tilde{\theta}} \in \mathcal{L}_\infty[0,t_f)$, which in view of Lemma 9.2(i) implies that $z, x, \xi, \Xi, \lambda \in \mathcal{L}_\infty[0,t_f)$. Hence, $t_f = \infty$.

To prove the tracking, let us consider the error equation (9.125):

$$\epsilon = \Omega_1^{\mathrm{T}}\tilde{\theta} + \varepsilon_1. \tag{9.142}$$

First, from (9.128) we note that

$$\dot{\Omega}_1 = -k_1\Omega_1 + \Omega_2 + F_1(y,u). \tag{9.143}$$

Recalling from (9.9) that $\Omega_2 = [v_{m,2}, v_{m-1,2}, \ldots, v_{0,2}, \Xi_{(2)}]^{\mathrm{T}}$, and from (8.9) that $F_1(y,u) = [0,\ldots,0, \Phi_{(1)}]^{\mathrm{T}}$, we conclude that $\Omega_2 + F_1(y,u) = \omega$, so

$$\dot{\Omega}_1 = -k_1\Omega_1 + \omega. \tag{9.144}$$

With (9.142), (9.144) and (9.3), we now get

$$\dot{\epsilon} = -k_1\epsilon + \omega^{\mathrm{T}}\tilde{\theta} + \varepsilon_2 - k_1\varepsilon_1 - \Omega_1^{\mathrm{T}}\dot{\tilde{\theta}}. \tag{9.145}$$

The rest of the proof follows the lines of (9.112)–(9.114) in the proof of Theorem 9.6. The central part of the argument is to show that (9.145) implies that $\omega^{\mathrm{T}}\tilde{\theta}(t) \to 0$. □

Remark 9.10 As in Remark 9.7 for the y-swapping identifier, we can modify the update laws (9.126) and (9.127) by a normalization with $1+\nu'|Q|_\mathcal{F}$, which guarantees that $Q^\mathrm{T}\dot{\hat{\theta}} \in \mathcal{L}_\infty$, and allows us to set $g_i = 0$ in the ISS-controller in Table 9.2. ◇

The x-swapping scheme is the only modular scheme whose dynamic order is as low as that of the tuning functions scheme. The x-swapping scheme is simpler than the tuning functions scheme, but, as we shall see, its performance properties are a little less strong.

9.5 Schemes with Parametric z-Model

The modular schemes we designed in the last three chapters were based on the parametric y-model (9.10) and the parametric x-model (9.122) rather than on the parametric z-model

$$\dot{z} = A_z(z,t)z + W_\varepsilon(z,t)\left(\omega^\mathrm{T}\tilde{\theta} + \varepsilon_2\right) + Q(z,t)^\mathrm{T}\dot{\hat{\theta}}, \qquad (9.146)$$

with A_z, W_ε, and Q defined in (9.85), (9.52), and (9.54), respectively, and s_i defined in (9.47)–(9.49). (Note that (9.39) cannot be a parametric model because b_m in A_z^* in (9.39) is not known.) Even though the parametric z-model was central in the state-feedback design, our attention was devoted to other parametric models because of the lower dynamic order of the resulting adaptive schemes. For example, while the y-passive scheme employs a scalar observer that generates $\hat{y}$, the z-passive scheme would use an observer of order ρ for $\hat{z}$. The difference is more drastic between the x-swapping scheme which does not use any extra filters, while the z-swapping scheme uses additional filters of total dimension $\rho(p+1)$.

For the sake of completeness and continuity with our state feedback designs, we briefly present two schemes based on the parametric z-model: the z-passive scheme and the z-swapping scheme. We omit the stability proofs.

z-Passive scheme

In analogy with the state-feedback z-passive design in Section 5.5, starting from the parametric model (9.146), we consider the identifier

$$\dot{\hat{z}} = A_z(z,t)\hat{z} + Q(z,t)^\mathrm{T}\dot{\hat{\theta}} \qquad (9.147)$$

$$\epsilon = z - \hat{z} \qquad (9.148)$$

$$\dot{\hat{\theta}} = \operatorname*{Proj}_{\hat{b}_m}\left\{\Gamma\omega W_\varepsilon^\mathrm{T}\epsilon\right\}, \quad \hat{b}_m(0)\operatorname{sgn} b_m > \varsigma_m, \quad \Gamma = \Gamma^\mathrm{T} > 0. \qquad (9.149)$$

If b_m were known, in which case we would not have the estimate $\hat{b}_m$ (and would not use projection), then one could prove the same result as in Theorem 9.4

9.5 Schemes with Parametric z-Model

(cf. [103]). With unknown b_m, the identifier (9.147)–(9.149) is not directly applicable. To see this, recall that the crucial property for establishing global boundedness in passive schemes is $\dot{\hat{\theta}} \in \mathcal{L}_2$. This property was in the state-feedback z-passive design guaranteed automatically by the nonlinear damping terms built into A_z. When b_m is unknown, the nonlinear damping terms (9.47)–(9.49) are capable of guaranteeing input-to-state stability with respect to $\tilde{\theta}$ in the system (9.39), but in the observer error system

$$\dot{\epsilon} = A_z(z,t)\epsilon + W_\varepsilon(z,t)\left(\omega^\mathrm{T}\tilde{\theta} + \varepsilon_2\right) \qquad (9.150)$$

with the parameter update (9.149), they cannot guarantee the square-integrability of $\dot{\hat{\theta}}$ as they would if b_m were known (cf. Lemma 5.10).

To remove this difficulty we strengthen the observer error system (9.150) by including the additional nonlinear damping term $-\mathrm{diag}\left\{\kappa_1|\omega|^2, \kappa_2\left|\frac{\partial\alpha_1}{\partial y}\omega\right|^2, 0, \ldots, 0\right\}(\hat{z} - z)$ in the observer (9.147):

$$\dot{\hat{z}} = A_z(z,t)\hat{z} - \mathrm{diag}\left\{\kappa_1|\omega|^2, \kappa_2\left|\frac{\partial\alpha_1}{\partial y}\omega\right|^2, 0, \ldots, 0\right\}(\hat{z}-z) + Q(z,t)^\mathrm{T}\hat{\theta}. \qquad (9.151)$$

It is possible to show that the adaptive scheme consisting of the control law in Table 9.2, the filters in Table 9.1, the observer (9.151) and the update law (9.149) achieves global uniform boundedness, as well as asymptotic tracking.

z-Swapping scheme

The difficulty with inadequate nonlinear damping terms due to unknown b_m in the z-passive scheme is much easier to deal with in the z-swapping scheme.

A key property of swapping identifiers is that they guarantee that $\dot{\hat{\theta}}$ is bounded. We often achieve this with nonlinear damping terms which guarantee that the filtered regressor is bounded. However, when b_m is not known the nonlinear damping terms built into A_z are not capable of guaranteeing the boundedness of the filtered regressor (cf. Section 6.8). Fortunately, in the swapping design we can use normalization which can guarantee the boundedness of $\dot{\hat{\theta}}$ even when the filtered regressor is growing unbounded (cf. Lemma 6.26).

With this observation one can show that the following z-swapping identifier guarantees the global uniform boundedness and asymptotic tracking:

$$\dot{\mho}^\mathrm{T} = A_z(z,t)\mho^\mathrm{T} + W_\varepsilon(z,t)\omega^\mathrm{T}, \qquad \mho \in \mathbb{R}^{p\times\rho} \qquad (9.152)$$
$$\dot{\mho}_0 = A_z(z,t)\mho_0 + W_\varepsilon(z,t)\omega^\mathrm{T}\hat{\theta} - Q(z,t)^\mathrm{T}\hat{\theta}, \qquad \bar{\mho} \in \mathbb{R}^\rho \qquad (9.153)$$
$$\epsilon = z + \mho_0 - \mho^\mathrm{T}\hat{\theta}, \qquad (9.154)$$

with either the *normalized* gradient:

$$\dot{\hat{\theta}} = \mathop{\mathrm{Proj}}_{\hat{b}_m}\left\{\Gamma\frac{\mho\epsilon}{1+\nu|\mho|_\mathcal{F}^2}\right\}, \qquad \begin{array}{l}\hat{b}_m(0)\,\mathrm{sgn}\,b_m > \varsigma_m \\ \Gamma = \Gamma^\mathrm{T} > 0,\quad \nu > 0\end{array} \qquad (9.155)$$

or the *normalized* least-squares update law:

$$\dot{\hat{\theta}} = \underset{\hat{b}_m}{\text{Proj}}\left\{\Gamma\frac{\mho\epsilon}{1+\nu|\mho|_{\mathcal{F}}^2}\right\}, \qquad \hat{b}_m(0)\,\text{sgn}\,b_m > \varsigma_m$$
$$\dot{\Gamma} = -\Gamma\frac{\mho\mho^T}{1+\nu|\mho|_{\mathcal{F}}^2}\Gamma, \qquad \Gamma(0)=\Gamma(0)^T > 0, \quad \nu > 0. \tag{9.156}$$

9.6 Transient Performance

In this section we derive $\mathcal{L}_2$ and $\mathcal{L}_\infty$ transient performance bounds for the error state z, which include bounds for the tracking error $z_1 = y - y_r$. We first consider the passive schemes (y-passive and z-passive), and then the swapping schemes (y-swapping, x-swapping, and z-swapping). For simplicity, we let $\Gamma = \gamma I$.

9.6.1 Passive schemes

First we derive an $\mathcal{L}_\infty$ performance bound for the y-passive scheme. To eliminate the effect of the initial condition of the estimation error $\epsilon = y - \hat{y}$, we initialize the observer with $\hat{y}(0) = y(0)$, which sets $\epsilon(0) = 0$.

Theorem 9.11 (y-Passive Scheme) *In the adaptive system (9.1), (9.5), (9.6), (9.7), (9.50), (9.66), and (9.69), the following inequality holds:*

$$|z(t)| \leq \frac{1}{\sqrt{c_0}}\left(M|\tilde{\theta}(0)|^2 + N|\varepsilon(0)|_P^2\right)^{1/2} + |z(0)|e^{-c_0t/2}, \tag{9.157}$$

where

$$M = \frac{1}{2\kappa_0}\left(1 + \frac{\gamma^2}{\kappa_0 g_0}\right) \tag{9.158}$$

$$N = \frac{\gamma}{2c_0\kappa_0}\left(1 + \frac{\gamma^2}{\kappa_0 g_0}\right) + \frac{1}{2\lambda(P)}\left(\frac{1}{d_0} + \frac{\gamma^2}{c_0\kappa_0 g_0}\right). \tag{9.159}$$

Proof. To obtain an $\mathcal{L}_\infty$ bound on z, it would seem that inequality (9.56) could be used along with $\mathcal{L}_\infty$ bounds on ε_2, $\tilde{\theta}$ and $\dot{\hat{\theta}}$. However, it is not clear how to obtain a bound on $\|\dot{\hat{\theta}}\|_\infty$ depending only on design parameters and initial conditions. Therefore, we apply a different approach which eliminates the need for $\|\dot{\hat{\theta}}\|_\infty$. First, we note from (9.68) that

$$\frac{d}{dt}\left(\frac{1}{2}\epsilon^2\right) = -c_0\epsilon^2 - \kappa_0|\omega|^2\epsilon^2 + \epsilon\left(\omega^T\tilde{\theta} + \varepsilon_2\right)$$
$$\leq -\frac{c_0}{2}\epsilon^2 - \frac{\kappa_0}{2}|\omega\epsilon|^2 + \frac{1}{2c_0}\varepsilon_2^2 + \frac{1}{2\kappa_0}|\tilde{\theta}|^2. \tag{9.160}$$

9.6 Transient Performance

From Lemma E.1(ii) and (9.69) with $\Gamma = \gamma I$, we have $|\dot{\hat{\theta}}| \leq \gamma |\omega \epsilon|$, which substituted into (9.160) yields

$$\frac{d}{dt}\left(\frac{1}{2}\epsilon^2\right) \leq -\frac{c_0}{2}\epsilon^2 - \frac{\kappa_0}{2\gamma^2}|\dot{\hat{\theta}}|^2 + \frac{1}{2c_0}\varepsilon_2^2 + \frac{1}{2\kappa_0}|\tilde{\theta}|^2. \tag{9.161}$$

In this inequality $|\dot{\hat{\theta}}|^2$ appears with an opposite sign of that in (9.61). Therefore, by adding these two inequalities with an appropriate scaling, we eliminate $|\dot{\hat{\theta}}|^2$:

$$\begin{aligned}
\frac{1}{2}\frac{d}{dt}\left(|z|^2 + \frac{\gamma^2}{2\kappa_0 g_0}\epsilon^2\right) &\leq -c_0|z|^2 + \frac{1}{4}\left(\frac{1}{d_0}\varepsilon_2^2 + \frac{1}{\kappa_0}|\tilde{\theta}|^2 + \frac{1}{g_0}|\dot{\hat{\theta}}|^2\right) \\
&\quad - c_0\frac{\gamma^2}{4\kappa_0 g_0}\epsilon^2 - \frac{\kappa_0}{\gamma^2}\frac{\gamma^2}{4\kappa_0 g_0}|\dot{\hat{\theta}}|^2 \\
&\quad + \frac{1}{c_0}\frac{\gamma^2}{4\kappa_0 g_0}\varepsilon_2^2 + \frac{1}{\kappa_0}\frac{\gamma^2}{4\kappa_0 g_0}|\tilde{\theta}|^2 \\
&\leq -c_0|z|^2 - \frac{c_0\gamma^2}{4\kappa_0 g_0}\epsilon^2 + \frac{1}{4}\left[\left(\frac{1}{d_0} + \frac{\gamma^2}{c_0\kappa_0 g_0}\right)\varepsilon_2^2\right. \\
&\quad \left. + \frac{1}{\kappa_0}\left(1 + \frac{\gamma^2}{\kappa_0 g_0}\right)|\tilde{\theta}|^2\right] + \frac{1}{4g_0}|\dot{\hat{\theta}}|^2 - \frac{\kappa_0}{\gamma^2}\frac{\gamma^2}{4\kappa_0 g_0}|\dot{\hat{\theta}}|^2 \\
&\leq -\frac{c_0}{2}\left(|z|^2 + \frac{\gamma^2}{2\kappa_0 g_0}\epsilon^2\right) \\
&\quad + \frac{1}{4}\left[\left(\frac{1}{d_0} + \frac{\gamma^2}{c_0\kappa_0 g_0}\right)\varepsilon_2^2 + \frac{1}{\kappa_0}\left(1 + \frac{\gamma^2}{\kappa_0 g_0}\right)|\tilde{\theta}|^2\right].
\end{aligned} \tag{9.162}$$

By applying Lemma C.5, we get

$$|z(t)|^2 + \frac{\gamma^2}{2\kappa_0 g_0}\epsilon(t)^2 \leq \frac{1}{2c_0}\left[\left(\frac{1}{d_0} + \frac{\gamma^2}{c_0\kappa_0 g_0}\right)\|\varepsilon_2\|_\infty^2 + \frac{1}{\kappa_0}\left(1 + \frac{\gamma^2}{\kappa_0 g_0}\right)\|\tilde{\theta}\|_\infty^2\right] + \left(|z(0)|^2 + \frac{\gamma^2}{2\kappa_0 g_0}\epsilon(0)^2\right)e^{-c_0 t}. \tag{9.163}$$

Since $\epsilon(0) = 0$, we have

$$|z(t)| \leq \frac{1}{\sqrt{2c_0}}\left[\left(\frac{1}{d_0} + \frac{\gamma^2}{c_0\kappa_0 g_0}\right)\|\varepsilon_2\|_\infty^2 + \frac{1}{\kappa_0}\left(1 + \frac{\gamma^2}{\kappa_0 g_0}\right)\|\tilde{\theta}\|_\infty^2\right]^{1/2} + |z(0)|e^{-c_0 t/2}. \tag{9.164}$$

From (9.3) and (9.4) we have $\frac{d}{dt}|\varepsilon|_P^2 \leq -|\varepsilon|^2$, which gives

$$\|\varepsilon_2\|_\infty^2 \leq \frac{1}{\underline{\lambda}(P)}|\varepsilon(0)|_P^2. \tag{9.165}$$

To obtain a bound on $\|\tilde{\theta}\|_\infty^2$ we recall (9.77) and (9.74), which give

$$\|\tilde{\theta}\|_\infty^2 \leq |\tilde{\theta}(0)|^2 + \frac{\gamma}{c_0}|\varepsilon(0)|_P^2. \tag{9.166}$$

Substituting (9.165) and (9.166) into (9.164), we obtain (9.157) with (9.158)–(9.159). □

The initial condition $z(0)$ in the bound (9.157) is, in general, dependent on the design parameters c_0, d_0, κ_0, g_0. However, as explained in Section 4.3.2 for state feedback, with trajectory initialization we can set $z(0) = 0$. Following (9.40) and (9.41), $z(0)$ is set to zero by selecting

$$y_r(0) = y(0) \tag{9.167}$$
$$y_r^{(i)}(0) = \hat{b}_m(0)\left[v_{m,i+1}(0) - \alpha_i\left(y(0), \xi(0), \Xi(0), \hat{\theta}(0), \bar{\lambda}_{m+i}(0), \bar{y}_r^{(i-1)}(0)\right)\right],$$
$$i = 1, \ldots, \rho - 1. \tag{9.168}$$

Upon setting $z(0)$ to zero, the bound (9.157) can be systematically reduced by increasing c_0. By examining (9.158)–(9.159) one can see that the bound (9.157) can also be systematically reduced by simultaneously increasing κ_0 and d_0.

A careful comparison of (9.157) with (8.139) reveals that the bound for the tuning functions scheme is lower. Also, an advantage of the tuning functions scheme is that an $\mathcal{L}_2$ bound like (8.138) is not available for the y-observer scheme.

We now give performance bounds for the z-passive scheme. The observer initial condition is set to $\hat{z}(0) = z(0)$. For comparison with the y-passive scheme, we select $\kappa_1 = \cdots = \kappa_\rho \stackrel{\triangle}{=} \kappa_0$.

Theorem 9.12 (z-Passive Scheme) *In the adaptive system (9.1), (9.5), (9.6), (9.7), (9.50), (9.149), and (9.151), the following inequality holds:*

$$|z(t)| \leq \frac{1}{\sqrt{c_0}}\left(M_\infty|\tilde{\theta}(0)|^2 + N_\infty|\varepsilon(0)|_P^2\right)^{1/2} + |z(0)|e^{-c_0 t/2}, \tag{9.169}$$

where

$$M_\infty = \frac{1}{2\kappa_0}\left(1 + \frac{\gamma^2}{\kappa_0 g_0}\right) \tag{9.170}$$

$$N_\infty = \frac{1}{8d_0}\left[\left(1 + \frac{\gamma^2}{\kappa_0 g_0}\right)\left(\frac{\gamma}{\kappa_0} + \frac{1}{\lambda(P)}\right) + \frac{1}{\lambda(P)}\right]. \tag{9.171}$$

Moreover, if b_m is known, then the z-passive design with the observer (9.147) results in the $\mathcal{L}_2$ bound

$$\|z\|_2 \leq \frac{1}{\sqrt{c_0}}\left(M_2|\tilde{\theta}(0)|^2 + N_2|\varepsilon(0)|_P^2\right)^{1/2} + \frac{1}{\sqrt{2c_0}}|z(0)|, \tag{9.172}$$

9.6 Transient Performance

where

$$M_2 = \frac{1}{\gamma}\left(1 + \frac{\gamma^2}{4\kappa_0 g_0}\right) \tag{9.173}$$

$$N_2 = \frac{1}{2d_0}\left(1 + \frac{\gamma^2}{4\kappa_0 g_0}\right). \tag{9.174}$$

While the $\mathcal{L}_\infty$ bounds (9.157) and (9.169) are similar, the $\mathcal{L}_2$ bound (9.172) is available only for the z-passive scheme.

9.6.2 Swapping schemes

First we derive an $\mathcal{L}_\infty$ performance bound for the y-swapping scheme. For simplicity, we consider only the gradient update law. To eliminate the effect of the initial condition of the estimation error $\tilde{\epsilon} = y + \varpi_0 - \varpi^T\hat{\theta}$, we initialize it with $\varpi(0) = 0$, $\varpi_0(0) = -y(0)$, which sets $\tilde{\epsilon}(0) = 0$.

Theorem 9.13 (y-Swapping Scheme) *In the adaptive system (9.1), (9.5), (9.6), (9.7), (9.50), (9.90), (9.91), and (9.95), the following inequality holds*

$$|z(t)| \leq \frac{1}{\sqrt{c_0}}\left(M|\tilde{\theta}(0)|^2 + N|\varepsilon(0)|_P^2\right)^{1/2} + |z(0)|e^{-c_0 t}, \tag{9.175}$$

where

$$M = \frac{1}{4\kappa_0}\left(1 + \frac{\gamma^2}{8c_0^2\kappa_0 g_0}\right) \tag{9.176}$$

$$N = \frac{1}{4}\left[\frac{\gamma}{c_0^2\kappa_0}\left(1 + \frac{\gamma^2}{8c_0^2\kappa_0 g_0} + \frac{\gamma}{4g_0}\right) + \frac{1}{d_0\underline{\lambda}(P)}\right]. \tag{9.177}$$

Proof. We derive an $\mathcal{L}_\infty$ bound on z using (9.56). It remains to determine bounds on $\|\varepsilon_2\|_\infty$, $\|\tilde{\theta}\|_\infty$ and $\|\dot{\hat{\theta}}\|_\infty$. First, from (9.3) and (9.4) we have

$$\|\varepsilon_2\|_\infty^2 \leq \frac{1}{\underline{\lambda}(P)}|\varepsilon(0)|_P^2. \tag{9.178}$$

In view of (9.108) and (9.103), using $\tilde{\epsilon}(0) = 0$, we get

$$\|\tilde{\theta}\|_\infty^2 \leq |\tilde{\theta}(0)|^2 + \frac{\gamma}{c_0^2}|\varepsilon(0)|_P^2. \tag{9.179}$$

With the help of (9.95) we write

$$|\dot{\hat{\theta}}|^2 \leq \gamma^2\frac{|\varpi|^2\epsilon^2}{(1+\nu|\varpi|^2)^2} \leq \gamma^2\|\varpi\|_\infty^2\frac{\epsilon^2}{(1+\nu|\varpi|^2)^2} \leq \gamma^2\|\varpi\|_\infty^2\|\epsilon\|_\infty^2 \tag{9.180}$$

and by substituting (9.93) we obtain

$$\|\dot{\tilde{\theta}}\|_\infty^2 \le 2\gamma^2\|\varpi\|_\infty^2 \left(\|\varpi\|_\infty^2\|\tilde{\theta}\|_\infty^2 + \|\tilde{\epsilon}\|_\infty^2\right). \tag{9.181}$$

From (9.101), using $\varpi(0) = 0$, it follows that

$$\|\varpi\|_\infty^2 \le \frac{1}{4c_0\kappa_0}. \tag{9.182}$$

To obtain a bound on $\|\tilde{\epsilon}\|_\infty$, along the solutions of (9.94) and (9.3) we have

$$\frac{d}{dt}\left(\frac{1}{2}|\tilde{\epsilon}|^2 + \frac{1}{4c_0}|\varepsilon|_P^2\right) \le -c_0\tilde{\epsilon}^2 + \tilde{\epsilon}\varepsilon_2 - \frac{1}{4c_0}|\varepsilon|^2 \le 0 \tag{9.183}$$

which yields

$$\|\tilde{\epsilon}\|_\infty^2 \le \frac{1}{2c_0}|\varepsilon(0)|_P^2. \tag{9.184}$$

By substituting (9.179), (9.182), and (9.184) into (9.181), and then, along with (9.178) and (9.179), into (9.56), we arrive at (9.175) with (9.176) and (9.177). □

The initial condition $z(0)$ in the bound (9.175) can be set to zero by the trajectory initialization procedure (9.167)–(9.168). Upon setting $z(0)$ to zero, the bound (9.175) can be systematically reduced by increasing c_0. By examining (9.176)–(9.177) one can see that the bound (9.175) can also be systematically reduced by simultaneously increasing κ_0 and d_0.

The $\mathcal{L}_\infty$ bound (9.175) for the y-swapping scheme is lower that the bound (9.157) for the y-passive scheme but higher than the bound (8.139) for the tuning functions scheme.

Now we derive an $\mathcal{L}_\infty$ performance bound for the x-swapping scheme.

Theorem 9.14 (x-Swapping Scheme) *In the adaptive system (9.1), (9.5), (9.6), (9.7), (9.50), (9.126), the following inequality holds:*

$$|z(t)| \le \frac{1}{\sqrt{c_0}}\left(M|\tilde{\theta}(0)|^2 + N|\varepsilon(0)|_P^2\right)^{1/2} + |z(0)|e^{-c_0 t}, \tag{9.185}$$

where

$$M = \frac{1}{4}\left(\frac{1}{\kappa_0} + \frac{2\gamma^2}{g_0\nu^2}\right) \tag{9.186}$$

$$N = \frac{1}{4}\left[\gamma\left(\frac{1}{\kappa_0} + \frac{2\gamma^2}{g_0\nu^2}\right) + \left(\frac{2\gamma^2}{g_0\nu} + \frac{1}{d_0}\right)\frac{1}{\lambda(P)}\right]. \tag{9.187}$$

9.6 Transient Performance

Proof. We derive an $\mathcal{L}_\infty$ bound on z using (9.56). The bound on $\|\varepsilon_2\|_\infty$ is as in (9.178). It remains to determine bounds on $\|\tilde{\theta}\|_\infty$ and $\|\dot{\hat{\theta}}\|_\infty$. From (9.134) and (9.133) we get

$$\|\tilde{\theta}\|_\infty^2 \leq |\tilde{\theta}(0)|^2 + \gamma|\varepsilon(0)|_P^2 \,. \tag{9.188}$$

By substituting (9.135) into (9.136) we get

$$|\dot{\hat{\theta}}|^2 \leq \frac{2\gamma^2}{\nu}\left(\frac{1}{\nu}|\tilde{\theta}|^2 + |\varepsilon_1|^2\right)\,. \tag{9.189}$$

Since

$$\|\varepsilon_1\|_\infty^2 \leq \frac{1}{\lambda(P)}|\varepsilon(0)|_P^2\,, \tag{9.190}$$

then (9.189) yields

$$\|\dot{\hat{\theta}}\|_\infty^2 \leq \frac{2\gamma^2}{\nu}\left(\frac{1}{\nu}\|\tilde{\theta}\|_\infty^2 + \frac{1}{\lambda(P)}|\varepsilon(0)|_P^2\right)\,. \tag{9.191}$$

By substituting (9.188), (9.178), and (9.191) into (9.56), we arrive at (9.185) with (9.186) and (9.187). □

The initial condition $z(0)$ in the bound (9.185) can be set to zero by the trajectory initialization procedure (9.167)–(9.168). Upon setting $z(0)$ to zero, the bound (9.185) can be systematically reduced by increasing c_0. By examining (9.186)–(9.187) one can see that the bound (9.185) can also be systematically reduced by simultaneously increasing κ_0, g_0, and d_0.

We now give performance bounds for the z-swapping scheme. The filter initial conditions are selected as $\mho(0) = 0$ and $\mho_0(0) = -z(0)$, to set $\tilde{\epsilon}(0) = z(0) + \mho_0(0) - \mho^T(0)\hat{\theta}(0) = 0$.

Theorem 9.15 (z-**Swapping Scheme**) *In the adaptive system (9.1), (9.5), (9.6), (9.7), (9.50), (9.152), (9.153), (9.155), the following inequality holds*

$$|z(t)| \leq \frac{1}{\sqrt{c_0}}\left(M_\infty|\tilde{\theta}(0)|^2 + N_\infty|\varepsilon(0)|_P^2\right)^{1/2} + |z(0)|e^{-c_0 t}\,, \tag{9.192}$$

where

$$M_\infty = \frac{1}{4}\left(\frac{1}{\kappa_0} + \frac{2\gamma^2}{g_0\nu^2}\right) \tag{9.193}$$

$$N_\infty = \frac{1}{4d_0}\left[\frac{\gamma}{2c_0}\left(\frac{1}{\kappa_0} + \frac{2\gamma^2}{g_0\nu^2}\right) + \frac{\gamma^2}{g_0\nu} + \frac{1}{\lambda(P)}\right]\,. \tag{9.194}$$

Moreover, if b_m is known, then the z-swapping scheme results in the $\mathcal{L}_2$ bound

$$\|z\|_2 \leq \frac{1}{\sqrt{c_0}}\left(M_2|\tilde{\theta}(0)|^2 + N_2|\varepsilon(0)|_P^2\right)^{1/2} + \frac{1}{\sqrt{c_0}}|z(0)| + \sqrt{\frac{2}{3\gamma}}|\tilde{\theta}(0)|\,, \tag{9.195}$$

where

$$M_2 = \frac{2}{3}\left[\frac{\nu}{2\gamma\kappa_0} + \frac{\gamma}{\nu}\left(\frac{1}{g_0} + \frac{1}{2c_0^2\kappa_0}\right)\right] \quad (9.196)$$

$$N_2 = \frac{2}{3d_0}\left[1 + \frac{\nu}{4c_0\kappa_0} + \frac{\gamma^2}{2c_0\nu}\left(\frac{1}{g_0} + \frac{1}{2c_0^2\kappa_0}\right)\right]. \quad (9.197)$$

While the $\mathcal{L}_\infty$ bound for the x-swapping is similar to that for the z-swapping scheme, the $\mathcal{L}_2$ bound (9.195) is available only for the z-swapping scheme.

9.7 Swapping Schemes with Weak ISS-Controller

In Chapters 5 and 6 we discussed the possible undesirable effects of the strong nonlinear damping terms. In Section 6.7 we introduced a weaker ISS-controller with a much slower growth of nonlinearities. In this section we develop a weak version of the output-feedback ISS-controller in Table 9.2.

Table 9.3 summarizes the design of the weak ISS-controller. There are two differences between the ISS-controller in Table 9.2 and the weak ISS-controller in Table 9.3:

- Instead of the quadratic nonlinear damping functions in (9.47)–(9.49), the nonlinear damping functions in (9.207)–(9.209) are 'linear-like.'

- The terms $-\hat{b}_m z_1$ and $-z_{i-1}$, present in (9.44) and (9.45), respectively, are eliminated from Table 9.3 to make the analysis simpler.

The resulting closed-loop system has a form similar to (9.39):

$$\dot{z} = A_z^*(z,t)z + W_\varepsilon(z,t)\varepsilon_2 + W_\theta^*(z,t)^\mathrm{T}\tilde{\theta} + Q(z,t)^\mathrm{T}\dot{\hat{\theta}}, \quad (9.198)$$

where W_ε, W_θ^*, and Q are as in (9.52), (9.53), and (9.54), respectively, and A_z^* is without $-b_m, -1, \ldots, -1$ below the diagonal:

$$A_z^*(z,t) = \begin{bmatrix} -\frac{|b_m|}{\varsigma_m}(c_1+s_1) & b_m & & & \\ & -(c_2+s_2) & 1 & & \\ & & \ddots & \ddots & \\ & & & \ddots & 1 \\ & & & & -(c_\rho+s_\rho) \end{bmatrix}. \quad (9.199)$$

Before we select an identifier, we establish an ISS property of the error system (9.198).

9.7 Swapping Schemes with Weak ISS-Controller

Table 9.3: Weak ISS-Controller with K-Filters

$$z_1 = y - y_r \tag{9.200}$$

$$z_i = v_{m,i} - \frac{1}{\hat{b}_m} y_r^{(i-1)} - \alpha_{i-1}, \qquad i = 2, \ldots, \rho \tag{9.201}$$

$$\alpha_1 = -\frac{\text{sgn}(b_m)}{\varsigma_m}(c_1 + s_1) z_1 + \frac{1}{\hat{b}_m} \bar{\alpha}_1 \tag{9.202}$$

$$\bar{\alpha}_1 = -\omega_0 - \bar{\omega}^T \hat{\theta} \tag{9.203}$$

$$\alpha_2 = -(c_2 + s_2) z_2 + \beta_2 \tag{9.204}$$

$$\alpha_i = -(c_i + s_i) z_i + \beta_i, \qquad i = 3, \ldots, \rho \tag{9.205}$$

$$\beta_i = \frac{\partial \alpha_{i-1}}{\partial y}\left(\omega_0 + \omega^T \hat{\theta}\right) + \frac{\partial \alpha_{i-1}}{\partial \xi}(A_0 \xi + ky + \phi) + \frac{\partial \alpha_{i-1}}{\partial \Xi}(A_0 \Xi + \Phi)$$

$$+ \sum_{j=1}^{i-1} \frac{\partial \alpha_{i-1}}{\partial y_r^{(j-1)}} y_r^{(j)} + k_i v_{m,1} + \sum_{j=1}^{m+i-1} \frac{\partial \alpha_{i-1}}{\partial \lambda_j}(-k_j \lambda_1 + \lambda_{j+1}) \tag{9.206}$$

$$s_1 = d_1 + \kappa_1 \sqrt{\left|\bar{\omega} + \frac{1}{\hat{b}_m}(\dot{y}_r + \bar{\alpha}_1) e_1\right|^2 + 1} \tag{9.207}$$

$$s_2 = d_2 \sqrt{\left(\frac{\partial \alpha_1}{\partial y}\right)^2 + 1} + \kappa_2 \sqrt{\left|\frac{\partial \alpha_1}{\partial y}\omega - z_1 e_1\right|^2 + 1}$$

$$+ g_2 \sqrt{\left|\frac{\partial \alpha_1}{\partial \hat{\theta}}^T - \frac{1}{\hat{b}_m^2}\dot{y}_r e_1\right|^2 + 1} \tag{9.208}$$

$$s_i = d_i \sqrt{\left(\frac{\partial \alpha_{i-1}}{\partial y}\right)^2 + 1} + \kappa_i \sqrt{\left|\frac{\partial \alpha_{i-1}}{\partial y}\omega\right|^2 + 1}$$

$$+ g_i \sqrt{\left|\frac{\partial \alpha_{i-1}}{\partial \hat{\theta}}^T - \frac{1}{\hat{b}_m^2} y_r^{(i-1)} e_1\right|^2 + 1}, \qquad i = 3, \ldots, \rho \tag{9.209}$$

Adaptive control law:

$$u = \frac{1}{\sigma(y)}\left(\alpha_\rho - v_{m,\rho+1} + \frac{1}{\hat{b}_m} y_r^{(\rho)}\right) \tag{9.210}$$

Lemma 9.16 *The error system (9.198) is ISS with respect to $\left(\varepsilon_2, \tilde{\theta}, \dot{\hat{\theta}}\right)$.*

Proof. For the states z_i, $i = 3, \ldots, \rho - 1$, of the system (9.198), one readily shows that

$$\begin{aligned}
\frac{d}{dt}\left(\frac{z_i^2}{2}\right) &\leq -\frac{c_i}{2}z_i^2 - \frac{c_i}{2}|z_i|\left(|z_i| - \frac{2}{c_i}|z_{i+1}|\right) \\
&\quad - d_i\left|z_i\frac{\partial \alpha_{i-1}}{\partial y}\right|\left(|z_i| - \frac{1}{d_i}|\varepsilon_2|\right) \\
&\quad - \kappa_i\left|z_i\frac{\partial \alpha_{i-1}}{\partial y}\omega\right|\left(|z_i| - \frac{1}{\kappa_i}|\tilde{\theta}|\right) \\
&\quad - g_i\left|z_i\left(\frac{\partial \alpha_{i-1}}{\partial \hat{\theta}}^{\mathrm{T}} - \frac{1}{\hat{b}_m^2}y_r^{(i-1)}e_1\right)\right|\left(|z_i| - \frac{1}{g_i}|\dot{\hat{\theta}}|\right). \quad (9.211)
\end{aligned}$$

Thus we have the following implication:

$$|z_i| \geq \left(\frac{2}{c_i}|z_{i+1}| + \frac{1}{d_i}|\varepsilon_2| + \frac{1}{\kappa_i}|\tilde{\theta}| + \frac{1}{g_i}|\dot{\hat{\theta}}|\right) \quad\Rightarrow\quad \frac{d}{dt}\left(z_i^2\right) \leq -c_i z_i^2. \quad (9.212)$$

By Theorem C.2, for all $0 \leq s \leq t$,

$$|z_i(t)| \leq |z_i(s)|e^{-c_i t/2} + \sup_{s \leq \tau \leq t}\left\{\frac{2}{c_i}|z_{i+1}(\tau)| + \frac{1}{d_i}|\varepsilon_2(\tau)| + \frac{1}{\kappa_i}|\tilde{\theta}(\tau)| + \frac{1}{g_i}|\dot{\hat{\theta}}(\tau)|\right\}, \quad (9.213)$$

that is, each z_i-equation is ISS with respect to $\left(z_{i+1}, \varepsilon_2, \tilde{\theta}, \dot{\hat{\theta}}\right)$. Inequality (9.213) can also be established for $i = 1, 2$, and ρ, with $z_{\rho+1} \stackrel{\triangle}{=} 0$. Inequalities (9.213) define a set of cascaded ISS inequalities. By repeatedly applying Lemma C.4 we show that there exist positive real numbers β_1, ρ_1, σ independent of initial conditions such that

$$|z(t)| \leq \beta_1 |z(0)|e^{-\sigma t} + \rho_1 \sup_{s \leq \tau \leq t}\left\{|\varepsilon_2(\tau)| + |\tilde{\theta}(\tau)| + |\dot{\hat{\theta}}(\tau)|\right\}, \quad (9.214)$$

which completes the proof. □

Since ε_2 is bounded, we now look for identifiers which guarantee the boundedness of $\tilde{\theta}$ and $\dot{\hat{\theta}}$. Two such identifiers are the y-swapping identifier introduced in Section 9.3 and the x-swapping identifier introduced in Section 9.4.

Theorem 9.17 (Schemes with Weak ISS-Controller) *All the signals in the closed-loop adaptive system consisting of the plant (9.1), the weak ISS-controller in Table 9.3, the filters in Table 9.1, and either the y-swapping or the x-swapping identifier are globally uniformly bounded, and global asymptotic tracking is achieved:*

$$\lim_{t \to \infty}\left[y(t) - y_r(t)\right] = 0. \quad (9.215)$$

9.8 Schemes with MT-Filters

Table 9.4: MT-Filters

MT-filters:		
$\dot{\xi} = A_l \xi + B_l \phi(y)$		(9.216)
$\dot{\Xi} = A_l \Xi + B_l \Phi(y)$		(9.217)
$\dot{\lambda} = A_l \lambda + e_{n-1} \sigma(y) u$		(9.218)
$v_j = A_l^j \lambda, \quad j = 0, \ldots, m$		(9.219)
$\Omega^{\mathrm{T}} = [v_m, \ldots, v_1, v_0, \Xi]$		(9.220)

Proof. With Lemma 9.16 and either Lemma 9.5 or Lemma 9.8, the proof is as in Theorems 9.6 and 9.9. □

Remark 9.18 For a further reduction of the growth of nonlinearities we can eliminate the g_i-terms from the nonlinear damping functions (9.207)–(9.209), provided that we normalize the update laws with $1+\nu'|Q|_{\mathcal{F}}$, as in Remarks 9.7 and 9.10. ◇

9.8 Schemes with MT-Filters

We briefly present some schemes with MT-filters. Most of the technical details are left out.

The MT-filters introduced in Section 8.2.1 are repeated for convenience in Table 9.4. The matrices A_l and B_l are defined in (8.154). With the filters in Table 9.4, the filtered transformation

$$\chi = x - \begin{bmatrix} 0 \\ \xi + \Omega^{\mathrm{T}} \theta \end{bmatrix} \tag{9.221}$$

brings the system (9.1) to the form

$$\begin{aligned} \dot{\chi} &= A\chi + l\left(\omega_0 + \omega^{\mathrm{T}} \theta\right) \\ y &= \chi_1, \end{aligned} \tag{9.222}$$

where l is defined in (8.155), and

$$\omega_0 = \varphi_{0,1} + \xi_1 \tag{9.223}$$

$$\omega = [v_{m,1}, v_{m-1,1}, \ldots, v_{0,1}, \Phi_{(1)} + \Xi_{(1)}]^{\mathrm{T}}. \tag{9.224}$$

9.8.1 ISS-observer

In the modular design we design controllers which guarantee input-to-state stability with respect to the parameter error, its derivative, and the state estimation error ε_2. We also need identifiers which independently guarantee the boundedness of the parameter error and its derivative, as well as observers which guarantee the boundedness of the state estimation error independently of the choice of the identifier.

The observer (8.171), which we used in the tuning functions design, is not strong enough to guarantee the boundedness of the state estimation error. We strengthen it by adding the nonlinear term $\kappa_o|\omega|^2 l(y - \hat{\chi}_1)$:

$$\dot{\hat{\chi}} = A\hat{\chi} + K_o(y - \hat{\chi}_1) + \kappa_o|\omega|^2 l(y - \hat{\chi}_1) + l\left(\omega_0 + \omega^T\hat{\theta}\right). \tag{9.225}$$

In Table 8.5 we have suggested that this strengthened observer may also be used in the tuning functions design. However, while nonlinear damping was not necessary for stabilization in the tuning functions design, and was only used for improving transient performance and guaranteeing boundedness without adaptation, in the modular design the nonlinear damping is indispensable. In the next lemma we show that the observer error system

$$\dot{\varepsilon} = \left(A_o - \kappa_o|\omega|^2 l e_1^T\right)\varepsilon + l\omega^T\tilde{\theta} \tag{9.226}$$

is input-to-state stable with respect to the input $\tilde{\theta}$.

Lemma 9.19 *Let the maximal interval of existence of solutions of (9.226) be $[0, t_f)$. If $\tilde{\theta} \in \mathcal{L}_\infty[0, t_f)$, then $\varepsilon \in \mathcal{L}_\infty[0, t_f)$.*

Proof. Since $e_1^T(sI - A_o)^{-1}l = \dfrac{1}{s + c_o}$ is strictly positive real, then by the Popov-Kalman-Yakubovich lemma (Lemma D.6) there exist $P_o = P_o^T > 0$ and $q_o > 0$ such that
$$\begin{aligned} A_o^T P_o + P_o A_o &\leq -q_o I \\ P_o l &= e_1 \, . \end{aligned} \tag{9.227}$$

Therefore, along the solutions of (9.226) we have

$$\begin{aligned} \frac{d}{dt}\left(|\varepsilon|_{P_o}^2\right) &\leq -q_o|\varepsilon|^2 - 2\kappa_o|\omega|^2 \varepsilon^T P_o l e_1^T \varepsilon + 2\varepsilon^T P_o l \omega^T \tilde{\theta} \\ &= -q_o|\varepsilon|^2 - 2\kappa_o|\omega|^2 \varepsilon_1^2 + 2\varepsilon_1 \omega^T \tilde{\theta} \\ &= -q_o|\varepsilon|^2 - 2\kappa_o\left|\varepsilon_1\omega - \frac{1}{2\kappa_o}\tilde{\theta}\right|^2 + \frac{1}{2\kappa_o}|\tilde{\theta}|^2 \\ &\leq -q_o|\varepsilon|^2 + \frac{1}{2\kappa_o}|\tilde{\theta}|^2 \, , \end{aligned} \tag{9.228}$$

which implies that $\varepsilon \in \mathcal{L}_\infty[0, t_f)$ whenever $\tilde{\theta} \in \mathcal{L}_\infty[0, t_f)$. □

9.8 Schemes with MT-Filters

Before we go on to the design of the controller, we remind the reader of the two crucial expressions for $\dot{y}$ derived in Section 8.2.1:

$$\dot{y} = \hat{\chi}_2 + \omega_0 + \omega^T\theta + \varepsilon_2 \qquad (9.229)$$
$$= b_m v_{m,1} + \hat{\chi}_2 + \omega_0 + \bar{\omega}^T\theta + \varepsilon_2, \qquad (9.230)$$

where

$$\bar{\omega} = [0, v_{m-1,1}, \ldots, v_{0,1},\ \Phi_{(1)} + \Xi_{(1)}]^T. \qquad (9.231)$$

9.8.2 ISS-controller

Applying the backstepping procedure to the system

$$\dot{y} = b_m v_{m,1} + \hat{\chi}_2 + \omega_0 + \bar{\omega}^T\theta + \varepsilon_2 \qquad (9.232)$$
$$\dot{v}_{m,i} = v_{m,i+1} - k_i v_{m,1}, \qquad i = 1, \ldots, \rho-1 \qquad (9.233)$$
$$\dot{v}_{m,\rho-1} = \sigma(y)u + v_{m,\rho} - k_\rho v_{m,1}, \qquad (9.234)$$

with the MT-filters given in Table 9.4, we end up with the design summarized in Table 9.5. One can now show that this results in the same error system (9.39), (9.51)–(9.54) as the ISS-controller with K-filters:

$$\dot{z} = A_z^*(z,t)z + W_\varepsilon(z,t)\varepsilon_2 + W_\theta^*(z,t)^T\tilde{\theta} + Q(z,t)^T\dot{\hat{\theta}} \qquad z \in \mathbb{R}^\rho. \qquad (9.235)$$

Therefore the error system (9.235) possesses the same input-to-state properties as in Lemma 9.2.

Lemma 9.20 *The error system (9.235) satisfies the input-to-state properties stated in Lemma 9.2.*

(i) If $\tilde{\theta}, \dot{\hat{\theta}} \in \mathcal{L}_\infty[0, t_f)$, then $z, x, \hat{\chi}, \xi, \Xi, \lambda \in \mathcal{L}_\infty[0, t_f)$, and

$$|z(t)| \leq \frac{1}{2\sqrt{c_0}}\left(\frac{1}{d_0}\|\varepsilon_2\|_\infty^2 + \frac{1}{\kappa_0}\|\tilde{\theta}\|_\infty^2 + \frac{1}{g_0}\|\dot{\hat{\theta}}\|_\infty^2\right)^{\frac{1}{2}} + |z(0)|e^{-c_0 t}. \qquad (9.236)$$

(ii) If $\tilde{\theta} \in \mathcal{L}_\infty[0, t_f)$ and $\dot{\hat{\theta}} \in \mathcal{L}_2[0, t_f)$, then $z, x, \hat{\chi}, \xi, \Xi, \lambda \in \mathcal{L}_\infty[0, t_f)$, and

$$|z(t)| \leq \left[\frac{1}{4c_0}\left(\frac{1}{d_0}\|\varepsilon_2\|_\infty^2 + \frac{1}{\kappa_0}\|\tilde{\theta}\|_\infty^2\right) + \frac{1}{2g_0}\|\dot{\hat{\theta}}\|_2^2\right]^{\frac{1}{2}} + |z(0)|e^{-c_0 t}. \qquad (9.237)$$

The bounds (9.236) and (9.237) are derived in the proof of Lemma 9.2. The rest of the proof uses Lemma 9.19 and the boundedness argument from the proof of Theorem 8.9.

With Lemmas 9.20 and 9.19 at hand, it remains to design identifiers which guarantee that $\tilde{\theta}$ is bounded, and $\dot{\hat{\theta}}$ is either bounded or square-integrable.

Table 9.5: ISS-Controller Design with MT-Filters

$$z_1 = y - y_r \qquad (9.238)$$

$$z_i = v_{m,i-1} - \frac{1}{\hat{b}_m} y_r^{(i-1)} - \alpha_{i-1}, \qquad i = 2, \ldots, \rho \qquad (9.239)$$

$$\alpha_1 = -\frac{\text{sgn}(b_m)}{\varsigma_m}(c_1 + s_1)z_1 + \frac{1}{\hat{b}_m}\bar{\alpha}_1 \qquad (9.240)$$

$$\bar{\alpha}_1 = -\hat{\chi}_2 - \omega_0 - \bar{\omega}^T\hat{\theta} \qquad (9.241)$$

$$\alpha_2 = -\hat{b}_m z_1 - (c_2 + s_2)z_2 + \beta_2 \qquad (9.242)$$

$$\alpha_i = -z_{i-1} - (c_i + s_i)z_i + \beta_i, \qquad i = 3, \ldots, \rho \qquad (9.243)$$

$$\beta_i = \frac{\partial \alpha_{i-1}}{\partial y}\left(\hat{\chi}_2 + \omega_0 + \omega^T\hat{\theta}\right) + \frac{\partial \alpha_{i-1}}{\partial \xi}(A_l \xi + B_l \phi) + \frac{\partial \alpha_{i-1}}{\partial \Xi}(A_l \Xi + B_l \Phi)$$

$$+ \frac{\partial \alpha_{i-1}}{\partial \hat{\chi}}\left[A\hat{\chi} + K_o(y - \hat{\chi}_1) + \kappa_o|\omega|^2 l(y - \hat{\chi}_1) + l\left(\omega_0 + \omega^T\hat{\theta}\right)\right]$$

$$+ \sum_{j=1}^{i-1} \frac{\partial \alpha_{i-1}}{\partial y_r^{(j-1)}} y_r^{(j)} + k_i v_{m,1} + \sum_{j=1}^{m+i-2} \frac{\partial \alpha_{i-1}}{\partial \lambda_j}(-k_j \lambda_1 + \lambda_{j+1}) \qquad (9.244)$$

$$s_1 = d_1 + \kappa_1 \left|\bar{\omega} + \frac{1}{\hat{b}_m}(\dot{y}_r + \bar{\alpha}_1)e_1\right|^2 \qquad (9.245)$$

$$s_2 = d_2 \left(\frac{\partial \alpha_1}{\partial y}\right)^2 + \kappa_2 \left|\frac{\partial \alpha_1}{\partial y}\omega - z_1 e_1\right|^2 + g_2 \left|\frac{\partial \alpha_1}{\partial \hat{\theta}}^T - \frac{1}{\hat{b}_m^2}\dot{y}_r e_1\right|^2 \qquad (9.246)$$

$$s_i = d_i \left(\frac{\partial \alpha_{i-1}}{\partial y}\right)^2 + \kappa_i \left|\frac{\partial \alpha_{i-1}}{\partial y}\omega\right|^2 + g_i \left|\frac{\partial \alpha_{i-1}}{\partial \hat{\theta}}^T - \frac{1}{\hat{b}_m^2}y_r^{(i-1)}e_1\right|^2$$
$$i = 3, \ldots, \rho \qquad (9.247)$$

Adaptive control law:

$$u = \frac{1}{\sigma(y)}\left(\alpha_\rho - v_{m,\rho} + \frac{1}{\hat{b}_m}y_r^{(\rho)}\right) \qquad (9.248)$$

Observer:

$$\dot{\hat{\chi}} = A\hat{\chi} + K_o(y - \hat{\chi}_1) + \kappa_o|\omega|^2 l(y - \hat{\chi}_1) + l\left(\omega_0 + \omega^T\hat{\theta}\right) \qquad (9.249)$$

9.8.3 χ-Passive scheme

Let us consider the adaptive observer form

$$\dot{\chi} = A\chi + l\left(\omega_0 + \omega^T\theta\right) \qquad (9.250)$$
$$y = \chi_1.$$

This system is a useful parametric model because of the minimum phase and relative-degree-one properties. The observer (9.249) results in the error system (9.226) which possesses a strict passivity property from the input $\tilde{\theta}$ to the output $\omega\varepsilon_1$. Because of this we choose the update law $\dot{\tilde{\theta}} = -\dot{\hat{\theta}}$ in the form

$$\dot{\hat{\theta}} = \Pr_{\hat{b}_m}\{\Gamma\omega\varepsilon_1\}, \qquad \hat{b}_m(0)\operatorname{sgn} b_m > \varsigma_m, \qquad \Gamma = \Gamma^T > 0, \qquad (9.251)$$

where the projection operator is employed to guarantee that $|\hat{b}_m(t)| \geq \varsigma_m > 0$, $\forall t \geq 0$.

Lemma 9.21 *Let the maximal interval of existence of solutions of (9.226) and (9.251) be $[0, t_f)$. Then the following identifier properties hold:*

$$\begin{align}
(o) \quad & |\hat{b}_m(t)| \geq \varsigma_m > 0, \quad \forall t \in [0, t_f) & (9.252)\\
(i) \quad & \tilde{\theta} \in \mathcal{L}_\infty[0, t_f) & (9.253)\\
(ii) \quad & \varepsilon \in \mathcal{L}_2[0, t_f) \cap \mathcal{L}_\infty[0, t_f) & (9.254)\\
(iii) \quad & \dot{\hat{\theta}} \in \mathcal{L}_2[0, t_f). & (9.255)
\end{align}$$

One can use the Lyapunov function

$$V = |\tilde{\theta}|^2_{\Gamma^{-1}} + |\varepsilon|^2_{P_o} \qquad (9.256)$$

along with (9.227) and Lemma E.1 to show that

$$\dot{V} \leq -q_o|\varepsilon|^2 - 2\frac{\kappa_o}{\lambda(\Gamma)^2}|\dot{\hat{\theta}}|^2, \qquad (9.257)$$

and draw the conclusions stated in Lemma 9.21. This lemma guarantees that $\dot{\hat{\theta}}$ is square-integrable, which by Lemma 9.20, is sufficient to establish the boundedness of all signals. The square-integrability of $\dot{\hat{\theta}}$ is due to the nonlinear damping term $-\kappa_o|\omega|^2 l e_1^T$ in (9.226). Hence, this is the second role of nonlinear damping in the observer. Its first role was to guarantee input-to-state stability with respect to $\tilde{\theta}$ in Lemma 9.19.

Theorem 9.22 (χ-Passive) *All the signals in the closed-loop adaptive system consisting of the plant (9.1), the control law in Table 9.5, the filters in Table 9.4, the observer (9.249), and the update law (9.251) are globally uniformly bounded, and global asymptotic tracking is achieved:*

$$\lim_{t\to\infty}[y(t) - y_r(t)] = 0. \qquad (9.258)$$

The boundedness is obtained by combining Lemma 9.21 with Lemma 9.20. One way to establish the convergence of $z(t)$ to zero is to follow the idea in the proof of Theorem 9.4. Instead of (9.86), one should consider

$$\zeta \triangleq z - W_\varepsilon(z,t)\frac{l^{\mathrm{T}}}{|l|^2}\varepsilon \tag{9.259}$$

and combine the systems (9.84) and (9.226).

The χ-passive scheme is the simplest among the schemes with MT-filters. Moreover, its dynamic order is as low as the dynamic order of the tuning functions scheme with MT-filters because the observer (9.249), already used for state estimation, is employed to design an identifier. This is a property the χ-passive scheme has in common with the x-swapping scheme (with K-filters), which employs in the identifier only the filters from Table 9.1, already used for state estimation.

Example 9.23 (Example 8.8, cont'd) For the system (8.206) we now design a χ-passive scheme. The strengthened observer (9.249) is given by

$$\dot{\hat{\chi}} = \begin{bmatrix} 0 & 1 \\ 0 & 0 \end{bmatrix}\hat{\chi} + \begin{bmatrix} c_o + l \\ c_o l \end{bmatrix}\varepsilon_1 + \kappa_o \omega^2 \begin{bmatrix} 1 \\ l \end{bmatrix}\varepsilon_1 + \begin{bmatrix} 1 \\ l \end{bmatrix}(v + \omega\hat{\theta}), \tag{9.260}$$

and the stabilizing function (9.240) and the control law (9.248) are

$$\alpha_1 = -(c + d + \kappa\omega^2)z_1 - \hat{\chi}_2 - \omega\hat{\theta} \tag{9.261}$$

$$u = -\left[c + d\left(\frac{\partial\alpha_1}{\partial y}\right)^2 + \kappa\left(\frac{\partial\alpha_1}{\partial y}\right)^2\omega^2 + g\omega^2\right]z_2 - z_1 + \frac{\partial\alpha_1}{\partial y}(\hat{\chi}_2 + v + \omega\hat{\theta})$$
$$- l(c_o + \kappa_o\omega^2)\varepsilon_1 + 2l\kappa\omega^2 y. \tag{9.262}$$

The update law (9.251) is

$$\dot{\hat{\theta}} = \gamma\omega\varepsilon_1. \tag{9.263}$$

Responses with the χ-passive scheme are shown in Figures 9.1 and 9.2. The values of θ, c, c_o, d, l and all the initial conditions are as for the tuning functions scheme, whereas $\kappa = \kappa_o = g = 1$. Without adaptation and with $\hat{\theta}$ constant and bounded, Lemma 9.20 establishes that all the states of the closed-loop system remain bounded. However, regulation of y to zero is not achieved. Instead, without adaptation the closed-loop system has an asymptotically stable periodic orbit with y-magnitudes as large as ± 1.2. The projection of the periodic orbit to the (ω, y) plane is shown in Figure 9.1a. With slow adaptation, $\gamma = 0.01$, the response is a slowly "shrinking" family of periodic orbits parametrized by $\hat{\theta}$; see Figure 9.1b. When $\hat{\theta}$ reduces enough, the trajectory is attracted by an asymptotically stable equilibrium at the origin. An increase of the adaptation gain to a moderate value of $\gamma = 0.1$ results in the response in

9.8 Schemes with MT-Filters

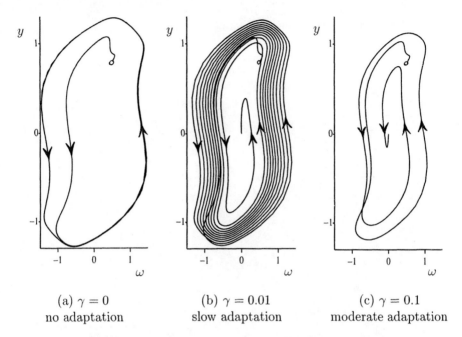

(a) $\gamma = 0$
no adaptation

(b) $\gamma = 0.01$
slow adaptation

(c) $\gamma = 0.1$
moderate adaptation

Figure 9.1: The χ-passive scheme. (a) Without adaptation, the closed-loop system has an asymptotically stable periodic orbit. (b) With slow adaptation, the trajectory passes through a "shrinking" family of periodic orbits parametrized by $\tilde{\theta}$. (c) With moderate adaptation, the trajectory starts by approaching the periodic orbit, but then the adaptation takes over.

Figure 9.1c, which initially approaches the periodic orbit, but when the adaptation takes over, it rapidly converges to the origin. For the same adaptation gain, $\gamma = 0.1$, the time responses of the output y, control u, and parameter error $\tilde{\theta}$ are given in Figure 9.2. They show that the regulation of y is achieved without $\tilde{\theta}$ converging to zero, that is, without feedback linearization. Comparing Figure 9.2 with Figure 8.1, we see that the transient in the χ-passive scheme takes considerably longer to settle than in the tuning functions scheme, but, thanks to nonlinear damping, with much less control effort. ◇

9.8.4 ε-Swapping scheme

Let us consider the observer error system (9.226) as a parametric model. The filters

$$\dot{\mho}^{\mathrm{T}} = \left(A_o - \kappa_o |\omega|^2 l e_1^{\mathrm{T}}\right) \mho^{\mathrm{T}} + l\omega^{\mathrm{T}}, \qquad \mho \in \mathbb{R}^{p \times n} \quad (9.264)$$

$$\dot{\mho}_0 = \left(A_o - \kappa_o |\omega|^2 l e_1^{\mathrm{T}}\right) \mho_0 + l\omega^{\mathrm{T}}\hat{\theta}, \qquad \mho_0 \in \mathbb{R}^n \quad (9.265)$$

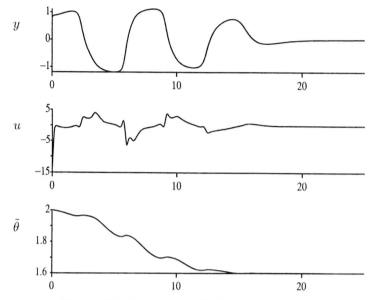

Figure 9.2: Responses with the χ-passive scheme

and the estimation error vector

$$\epsilon = \varepsilon + \mho_0 - \mho^T \hat{\theta} \tag{9.266}$$

result in the static parametric model

$$\epsilon = \mho^T \tilde{\theta} + \tilde{\epsilon}, \tag{9.267}$$

where $\tilde{\epsilon} \triangleq \varepsilon + \mho_0 - \mho^T \theta$ is governed by

$$\dot{\tilde{\epsilon}} = \left(A_o - \kappa_o |\omega|^2 l e_1^T \right) \tilde{\epsilon}. \tag{9.268}$$

Since only $\varepsilon_1 = y - \hat{\chi}_1$ is measured, only the first component of ϵ would be implemented as the estimation error:

$$\epsilon_1 = \varepsilon_1 + \mho_{1,0} - \mho_1^T \hat{\theta}. \tag{9.269}$$

Filters (9.264) and (9.265) have dynamic order $n(p+1)$. Fortunately, the filter dynamic order can be reduced to $p+1$. The reduction is based on recognizing a decomposition of the observer error system (9.226) used in the proof of Theorem 8.9. We remind the reader that the similarity transformation

$$\begin{bmatrix} \varepsilon_1 \\ \eta \end{bmatrix} \triangleq \begin{bmatrix} \varepsilon_1 \\ T\varepsilon \end{bmatrix} = \begin{bmatrix} e_1^T \\ -\bar{l}, \ I_{n-1} \end{bmatrix} \varepsilon \tag{9.270}$$

9.8 Schemes with MT-Filters

converts the system (9.226) into

$$\dot{\varepsilon}_1 = -\left(c_o + \kappa_o|\omega|^2\right)\varepsilon_1 + \omega^\mathrm{T}\tilde{\theta} + \eta_1 \tag{9.271}$$

$$\dot{\eta} = A_l \eta. \tag{9.272}$$

The scalar system (9.271) is a candidate for a parametric model because all the quantities in this system are available, except for the parameter error $\tilde{\theta}$ and the disturbance η_1 which is exponentially decaying.

We introduce the filters

$$\dot{\varpi} = -\left(c_o + \kappa_o|\omega|^2\right)\varpi + \omega, \qquad \varpi \in \mathbb{R}^p \tag{9.273}$$

$$\dot{\varpi}_0 = -\left(c_o + \kappa_o|\omega|^2\right)\varpi_0 + \omega^\mathrm{T}\hat{\theta}, \qquad \varpi_0 \in \mathbb{R} \tag{9.274}$$

and the estimation error

$$\epsilon = \varepsilon_1 + \varpi_0 - \varpi^\mathrm{T}\hat{\theta}. \tag{9.275}$$

Substituting (9.271), (9.273), (9.274) into (9.275), we get

$$\epsilon = \varpi^\mathrm{T}\tilde{\theta} + \tilde{\epsilon}, \tag{9.276}$$

where $\tilde{\epsilon}$ is governed by

$$\dot{\tilde{\epsilon}} = -\left(c_o + \kappa_o|\omega|^2\right)\tilde{\epsilon} + \eta_1. \tag{9.277}$$

The update law for $\hat{\theta}$ is either the gradient:

$$\dot{\hat{\theta}} = \underset{\hat{b}_m}{\operatorname{Proj}}\left\{\Gamma\frac{\varpi\epsilon}{1+\nu|\varpi|^2}\right\}, \qquad \begin{array}{l}\hat{b}_m(0)\operatorname{sgn}b_m > \varsigma_m \\ \Gamma = \Gamma^\mathrm{T} > 0, \quad \nu \geq 0\end{array} \tag{9.278}$$

or the least-quares:

$$\dot{\hat{\theta}} = \underset{\hat{b}_m}{\operatorname{Proj}}\left\{\Gamma\frac{\varpi\epsilon}{1+\nu|\varpi|^2_\Gamma}\right\}, \qquad \hat{b}_m(0)\operatorname{sgn}b_m > \varsigma_m$$

$$\dot{\Gamma} = -\Gamma\frac{\varpi\varpi^\mathrm{T}}{1+\nu|\varpi|^2_\Gamma}\Gamma, \qquad \Gamma(0) = \Gamma(0)^\mathrm{T} > 0, \quad \nu \geq 0, \tag{9.279}$$

where the projection operator is employed to guarantee that $|\hat{b}_m(t)| \geq \varsigma_m > 0$, $\forall t \geq 0$, and by allowing $\nu = 0$ we encompass unnormalized gradient and least-squares.

Lemma 9.24 *Let the maximal interval of existence of solutions of (9.271), (9.273)–(9.274) with either (9.278) or (9.279) be $[0, t_f)$. Then for $\nu \geq 0$ the following identifier properties hold:*

$$(o) \quad |\hat{b}_m(t)| \geq \varsigma_m > 0, \ \forall t \in [0, t_f) \tag{9.280}$$

$$(i) \quad \tilde{\theta} \in \mathcal{L}_\infty[0, t_f) \tag{9.281}$$

$$(ii) \quad \epsilon \in \mathcal{L}_2[0, t_f) \cap \mathcal{L}_\infty[0, t_f) \tag{9.282}$$

$$(iii) \quad \dot{\hat{\theta}} \in \mathcal{L}_2[0, t_f) \cap \mathcal{L}_\infty[0, t_f). \tag{9.283}$$

Theorem 9.25 (ε-Swapping) *All the signals in the closed-loop adaptive system consisting of the plant (9.1), the control law and the observer in Table 9.5, the filters in Table 9.4, the filters (9.273)–(9.274), and either the gradient (9.278) or the least-squares update law (9.279) are globally uniformly bounded, and global asymptotic tracking is achieved:*

$$\lim_{t \to \infty} [y(t) - y_\mathrm{r}(t)] = 0. \tag{9.284}$$

The boundedness is obtained by combining Lemma 9.24 with Lemma 9.20. The convergence of $z(t)$ to zero is proven as in Theorem 9.6, by replacing the estimation error equation (9.113) by

$$\dot{\epsilon} = -\left(c_o + \kappa_o |\omega|^2\right) \epsilon + \omega^\mathrm{T} \tilde{\theta} + \eta_1 - \varpi^\mathrm{T} \dot{\hat{\theta}}. \tag{9.285}$$

As mentioned in Remark 9.7 for the y-swapping scheme, we can eliminate the g_i-terms from the nonlinear damping functions (9.245)–(9.247) in Table 9.5 by normalizing the update laws with $1 + \nu'|Q|_\mathcal{F}$.

The growth of nonlinearities in the controller design in Table 9.5 can be reduced by following the weak ISS approach employed in Section 9.7.

9.8.5 Transient performance

In this section we derive $\mathcal{L}_\infty$ transient performance bounds for the error state z, which include bounds for the tracking error $z_1 = y - y_\mathrm{r}$. $\mathcal{L}_2$ performance bounds are not available. We first consider the χ-passive scheme, and then the ε-swapping scheme. To simplify the analysis, we let $\Gamma = \gamma I$. For the same reason, we implement $\hat{\chi}_1(0) = y(0)$ to get $\varepsilon_1(0) = 0$.

We first consider the χ-passive scheme.

Theorem 9.26 (χ-Passive) *In the adaptive system (9.1), (9.216), (9.217), (9.218), (9.248), (9.249), and (9.251), the following inequality holds*

$$|z(t)| \leq \frac{1}{\sqrt{c}} \left(M|\tilde{\theta}(0)|^2 + N|\eta(0)|^2_{P_l}\right)^{1/2} + |z(0)|e^{-ct/2}, \tag{9.286}$$

where $c = \min\{c_0, c_o\}$ and

$$M = \frac{1}{2}\left(\frac{1}{\kappa_0} + \frac{\gamma^2}{\kappa_o^2 g_0}\right) + \frac{l_1^2}{d_0 \gamma} \tag{9.287}$$

$$N = \frac{\gamma}{2c_o}\left(\frac{1}{\kappa_0} + \frac{\gamma^2}{\kappa_o^2 g_0}\right) + \frac{l_1^2}{c_o d_0} + \frac{1}{\underline{\lambda}(P_l)}\left(\frac{1}{d_0} + \frac{\gamma^2}{2c_o \kappa_o g_0}\right). \tag{9.288}$$

Proof. To obtain an $\mathcal{L}_\infty$ bound on z, it would seem that inequality (9.236) could be used along with $\mathcal{L}_\infty$ bounds on ε_2, $\tilde{\theta}$, and $\dot{\hat{\theta}}$. However, it is not clear

9.8 Schemes with MT-Filters

how to obtain a bound on $\|\dot{\tilde{\theta}}\|_\infty$ depending only on design parameters and initial conditions. Therefore, we apply a different approach which eliminates the need for $\|\dot{\tilde{\theta}}\|_\infty$. Although in Section 9.8.3 we did not use the decomposition (9.270) of (9.226) into (9.271)–(9.272), we here exploit it in order to make $c = \min\{c_0, c_o\}$ appear explicitly in the bounds. First, we note from (9.271) that

$$\frac{d}{dt}\left(\frac{1}{2}\varepsilon_1^2\right) \leq -\frac{c_o}{2}\varepsilon_1^2 - \frac{\kappa_o}{2}|\omega|^2\varepsilon_1^2 + \frac{1}{2c_o}\eta_1^2 + \frac{1}{2\kappa_o}|\tilde{\theta}|^2. \tag{9.289}$$

Combining (9.61) and (9.289) we compute

$$\begin{aligned}\frac{1}{2}\frac{d}{dt}\left(|z|^2 + \frac{\gamma^2}{2\kappa_o g_0}\varepsilon_1^2\right) &\leq -c_0|z|^2 + \frac{1}{4}\left(\frac{1}{d_0}\varepsilon_2^2 + \frac{1}{\kappa_0}|\tilde{\theta}|^2 + \frac{1}{g_0}|\dot{\tilde{\theta}}|^2\right) \\ &\quad -c_o\frac{\gamma^2}{4\kappa_o g_0}\varepsilon_1^2 - \frac{\gamma^2}{4g_0}|\omega\varepsilon_1|^2 + \frac{\gamma^2}{4c_o\kappa_o g_0}\eta_1^2 + \frac{\gamma^2}{4\kappa_o^2 g_0}|\tilde{\theta}|^2 \\ &\leq -\frac{c}{2}\left(|z|^2 + \frac{\gamma^2}{2\kappa_o g_0}\varepsilon_1^2\right) \\ &\quad + \frac{1}{4}\left[\frac{1}{d_0}\varepsilon_2^2 + \left(\frac{1}{\kappa_0} + \frac{\gamma^2}{\kappa_o^2 g_0}\right)|\tilde{\theta}|^2 + \frac{\gamma^2}{c_o\kappa_o g_0}\eta_1^2\right] \\ &\quad + \frac{1}{4g_0}\left(|\dot{\tilde{\theta}}|^2 - |\gamma\omega\varepsilon_1|^2\right), \tag{9.290}\end{aligned}$$

and since by Lemma E.1(ii) and (9.251), $|\dot{\tilde{\theta}}| \leq \gamma|\omega\epsilon|$, with Lemma C.5 we obtain

$$|z(t)| \leq \frac{1}{\sqrt{2c}}\left[\frac{1}{d_0}\|\varepsilon_2\|_\infty^2 + \left(\frac{1}{\kappa_0} + \frac{\gamma^2}{\kappa_o^2 g_0}\right)\|\tilde{\theta}\|_\infty^2 + \frac{\gamma^2}{c_o\kappa_o g_0}\|\eta_1\|_\infty^2\right]^{1/2} + |z(0)|e^{-ct/2}, \tag{9.291}$$

where we have used $\varepsilon_1(0) = 0$. Now we determine bounds on $\|\varepsilon_2\|_\infty$, $\|\tilde{\theta}\|_\infty$ and $\|\eta_1\|_\infty$. First, from (9.272) we have

$$\frac{d}{dt}\left(|\eta|_{P_l}^2\right) = -|\eta|^2, \tag{9.292}$$

which implies that

$$\|\eta_1\|_\infty^2 \leq \frac{1}{\lambda(P_l)}|\eta(0)|_{P_l}^2. \tag{9.293}$$

Along the solutions of (9.271)–(9.272) and (9.251) we have

$$\frac{1}{2}\frac{d}{dt}\left(\varepsilon_1^2 + \frac{1}{c_o}|\eta|_{P_l}^2 + \frac{1}{\gamma}|\tilde{\theta}|^2\right) \leq -\frac{c_o}{2}\varepsilon_1^2 - \frac{\kappa}{\gamma^2}|\tilde{\theta}|^2, \tag{9.294}$$

which yields

$$\|\tilde{\theta}\|_\infty^2 \leq |\tilde{\theta}(0)|^2 + \frac{\gamma}{c_o}|\eta(0)|_{P_l}^2 \tag{9.295}$$

$$\|\varepsilon_1\|_\infty^2 \leq \frac{1}{\gamma}\left(|\tilde{\theta}(0)|^2 + \frac{\gamma}{c_o}|\eta(0)|_{P_l}^2\right). \tag{9.296}$$

To obtain a bound on $\|\varepsilon_2\|_\infty$, from (9.270) we recall that $\varepsilon_2 = l_1\varepsilon_1 + \eta_1$, which by virtue of (9.293) and (9.296), shows that

$$\|\varepsilon_2\|_\infty^2 \leq \frac{2l_1^2}{\gamma}|\tilde{\theta}(0)|^2 + \left(\frac{2l_1^2}{c_o} + \frac{2}{\underline{\lambda}(P_l)}\right)|\eta(0)|_{P_l}^2. \tag{9.297}$$

Substituting (9.293), (9.295), and (9.297) into (9.291), we arrive at (9.286) with (9.287)–(9.288). □

The initial condition $z(0)$ in the bound (9.286) is, in general, dependent on the design parameters c_0, d_0, κ_0, and g_0. However, as explained in Section 4.3.2 for state feedback, with trajectory initialization we can set $z(0) = 0$. Following (9.238) and (9.239), $z(0)$ is set to zero by selecting

$$y_r(0) = y(0) \tag{9.298}$$
$$y_r^{(i)}(0) = \hat{b}_m(0)\left[v_{m,i}(0)\right.$$
$$\left. - \alpha_i\left(y(0), \hat{\chi}(0), \xi(0), \Xi(0), \hat{\theta}(0), \bar{\lambda}_{m+i-1}(0), \bar{y}_r^{(i-1)}(0)\right)\right],$$
$$i = 1, \ldots, \rho - 1. \tag{9.299}$$

Upon setting $z(0)$ to zero, the bound (9.286) can be systematically reduced by increasing c. By examining (9.287)–(9.288) one can see that the bound (9.286) can also be systematically reduced by simultaneously increasing κ_0, κ_o, and d_0.

A comparison of the bound (9.286) with the bound (9.157) for the y-passive scheme (with K-filters) reveals that the latter bound is lower. This is because the design with MT-filters results in an observer error system disturbed by the parameter error, unlike the design with K-filters.

Now we consider the ε-swapping scheme with a gradient update law. It is straightforward to extend this result to a least-squares update law. We initialize $\varpi_0(0) = -\varepsilon_1(0) = 0$ and $\varpi(0) = 0$, which set $\tilde{\epsilon}(0)$ to zero.

Theorem 9.27 (ε-Swapping) *In the adaptive system (9.1), (9.216), (9.217), (9.218), (9.248), (9.249), (9.273), (9.274), and (9.278), the following inequality holds*

$$|z(t)| \leq \frac{1}{\sqrt{c_0}}\left(M|\tilde{\theta}(0)|^2 + N|\eta(0)|_{P_l}^2\right)^{1/2} + |z(0)|e^{-c_0 t}, \tag{9.300}$$

where

$$M = \frac{1}{4}\left(\frac{1}{\kappa_0} + \frac{\gamma^2}{8c_o^2\kappa_o^2 g_0} + \frac{l_1^2}{c_o\kappa_o d_0}\right) \tag{9.301}$$

$$N = \frac{1}{2}\left[\frac{\gamma}{c_o^2}\left(\frac{1}{\kappa_0} + \frac{\gamma^2}{8c_o^2\kappa_o^2 g_0} + \frac{l_1^2}{c_o\kappa_o d_0} + \frac{\gamma}{4\kappa_o g_0}\right)\right.$$
$$\left. + \frac{1}{d_0\underline{\lambda}(P_l)}\left(1 + \frac{l_1^2}{c_o^2}\right)\right]. \tag{9.302}$$

9.8 Schemes with MT-Filters

Proof. We derive an $\mathcal{L}_\infty$ bound on z using (9.236) rewritten as

$$|z(t)| \leq \frac{1}{2\sqrt{c_0}} \left(\frac{1}{d_0} \|\varepsilon_2\|_\infty^2 + \frac{1}{\kappa_0} \|\tilde{\theta}\|_\infty^2 + \frac{1}{g_0} \|\dot{\tilde{\theta}}\|_\infty^2 \right)^{1/2} + |z(0)|e^{-c_0 t}. \quad (9.303)$$

It remains to determine bounds on $\|\varepsilon_2\|_\infty$, $\|\tilde{\theta}\|_\infty$, and $\|\dot{\tilde{\theta}}\|_\infty$. First, along the solutions of (9.278), (9.277), and (9.272), with (8.250) we readily arrive at

$$\frac{d}{dt} \left(\frac{1}{2} |\tilde{\theta}|_{\Gamma^{-1}}^2 + \frac{1}{c_o} \tilde{\epsilon}^2 + \frac{1}{c_o^2} |\eta|_{P_l}^2 \right) \leq -\frac{3}{4} \frac{\epsilon^2}{1 + \nu|\varpi|^2}. \quad (9.304)$$

Then, using $\tilde{\epsilon}(0) = 0$, (9.304) yields

$$\|\tilde{\theta}\|_\infty^2 \leq |\tilde{\theta}(0)|^2 + \frac{2\gamma}{c_o^2} |\eta(0)|_{P_l}^2. \quad (9.305)$$

Noting that (9.271) gives

$$\frac{d}{dt} \left(\frac{1}{2} \varepsilon_1^2 \right) \leq -\frac{c_o}{2} \varepsilon_1^2 + \frac{1}{4\kappa_o} |\tilde{\theta}|^2 + \frac{1}{2c_o} \eta_1^2, \quad (9.306)$$

we obtain

$$\|\varepsilon_1\|_\infty^2 \leq \frac{1}{2c_o \kappa_o} \|\tilde{\theta}\|_\infty^2 + \frac{1}{c_o^2} \|\eta_1\|_\infty^2. \quad (9.307)$$

Recalling that $\varepsilon_2 = l_1 \varepsilon_1 + \eta_1$, using (9.293) and (9.307) we get

$$\|\varepsilon_2\|_\infty^2 \leq \frac{l_1^2}{c_o \kappa_o} \|\tilde{\theta}\|_\infty^2 + \left(1 + \frac{l_1^2}{c_o^2} \right) \frac{2}{\underline{\lambda}(P_l)} |\eta(0)|_{P_l}^2. \quad (9.308)$$

From (9.278) we write

$$|\dot{\tilde{\theta}}|^2 \leq \gamma^2 \frac{|\varpi|^2 \epsilon^2}{(1 + \nu|\varpi|^2)^2} \leq \gamma^2 \|\varpi\|_\infty^2 \frac{\epsilon^2}{(1 + \nu|\varpi|^2)^2} \leq \gamma^2 \|\varpi\|_\infty^2 \|\epsilon\|_\infty^2, \quad (9.309)$$

and by substituting (9.276) we obtain

$$\|\dot{\tilde{\theta}}\|_\infty^2 \leq 2\gamma^2 \|\varpi\|_\infty^2 \left(\|\varpi\|_\infty^2 \|\tilde{\theta}\|_\infty^2 + \|\tilde{\epsilon}\|_\infty^2 \right). \quad (9.310)$$

For the filtered regressor (9.273) we readily show

$$\frac{d}{dt} \left(\frac{1}{2} |\varpi|^2 \right) \leq -c_o |\varpi|^2 + \frac{1}{4\kappa_o}, \quad (9.311)$$

which, using $\varpi(0) = 0$, implies

$$\|\varpi\|_\infty^2 \leq \frac{1}{4c_o \kappa_o}. \quad (9.312)$$

To obtain a bound on $\|\tilde{\epsilon}\|_\infty$, along (9.277) and (9.272), we consider

$$\frac{d}{dt}\left(\frac{1}{2}|\tilde{\epsilon}|^2 + \frac{1}{4c_o}|\eta|^2_{P_l}\right) \leq -c_o\tilde{\epsilon}^2 + \tilde{\epsilon}\eta_1 - \frac{1}{4c_o}|\eta|^2 \leq 0 \qquad (9.313)$$

which yields

$$\|\tilde{\epsilon}\|^2_\infty \leq \frac{1}{2c_o}|\eta(0)|^2_{P_l}. \qquad (9.314)$$

By substituting (9.305) into (9.308), and also (9.305), (9.312), (9.314) into (9.310), and then the two results, along with (9.305), into (9.303), we arrive at (9.300) with (9.301)–(9.302). □

The initial condition $z(0)$ in the bound (9.300) can be set to zero by the trajectory initialization procedure (9.298)–(9.299). Upon setting $z(0)$ to zero, the bound (9.300) can be systematically reduced by increasing c_0. By examining (9.301)–(9.302) one can see that the bound (9.300) can also be systematically reduced by simultaneously increasing κ_0, κ_o, and d_0.

A comparison of the bound (9.300) with the bound (9.175) for the y-swapping scheme (with K-filters) reveals that the latter bound is lower.

Notes and References

The early results on estimation-based output-feedback adaptive control by Kanellakopoulos, Kokotović, and Middleton [67, 68] imposed both structural and growth restrictions on system nonlinearities. Except for the material in Section 9.8 which appeared in Krstić and Kokotović [99], as well as the z-passive scheme presented in Krstić, Kokotović, and Kanellakopoulos [103], the results in this chapter have not been previously published.

Chapter 10

Linear Systems

For more than two decades, adaptive control research has dealt exclusively with linear systems, and has developed controllers which guarantee global boundedness and tracking [5, 44, 51, 142, 165]. How do these traditional adaptive controllers compare with the new controllers developed in this book? How do the controllers developed in Chapters 8 and 9 behave when applied to linear systems? Do they become variants of traditional adaptive controllers or do the resulting feedback systems possess some new properties?

From the preceding two chapters we already know that one of the new controllers' properties is their capability to systematically improve the transient performance. Other new properties include the guaranteed stability without adaptation and passivity of the adaptation loop irrespective of the relative degree. When the controllers designed in this book are applied to linear systems, these new properties come at no increase of the dynamic order of the adaptive schemes.

The task of this chapter is to introduce the new adaptive controllers for linear systems in a self-contained manner, so that a comparison with traditional adaptive control can be made independent of the rest of the book. After introducing the filters used for state estimation in Section 10.1, in Section 10.2 we develop the design procedure, establish the resulting stability and passivity properties, and provide a design example. In Section 10.3 we reveal the structure of the underlying nonadaptive controller and determine conditions for stability without adaptation. In Section 10.4 we derive transient performance bounds, and in Section 10.5 we give a simulation comparison with a traditional linear scheme. Then in Section 10.6 we present modular schemes which are closer to certainty equivalence adaptive linear designs. Finally, in Section 10.7 we summarize the main properties of the new adaptive designs for linear systems.

We consider linear single-input single-output systems

$$y(s) = \frac{B(s)}{A(s)}u(s) = \frac{b_m s^m + \cdots + b_1 s + b_0}{s^n + a_{n-1}s^{n-1} + \cdots + a_1 s + a_0}u(s), \qquad (10.1)$$

where the coefficients a_i's and b_i's are constant but unknown. The control objective is to asymptotically track a reference signal $y_r(t)$ with the output y. Our assumptions about the plant and the reference signal are the same as in traditional model reference adaptive control:

Assumption 10.1 *The plant is minimum phase, i.e., the polynomial $B(s) = b_m s^m + \cdots + b_1 s + b_0$ is Hurwitz.*

Assumption 10.2 *The sign of the high-frequency gain $(\text{sgn}(b_m))$ is known.*

Assumption 10.3 *The relative degree $(\rho = n - m)$ and an upper bound for the plant order (n) are known.*

Assumption 10.4 *The reference signal $y_r(t)$ and its first ρ derivatives are known and bounded, and, in addition, $y_r^{(\rho)}(t)$ is piecewise continuous. In particular, $y_r(t)$ may be the output of a reference model of relative degree $\rho_r \geq \rho$ with a piecewise continuous input $r(t)$.*

10.1 State Estimation Filters

We start by representing the plant (10.1) in the observer canonical form

$$\begin{aligned}
\dot{x}_1 &= x_2 - a_{n-1} y \\
&\vdots \\
\dot{x}_{\rho-1} &= x_\rho - a_{m+1} y \\
\dot{x}_\rho &= x_{\rho+1} - a_m y + b_m u \\
&\vdots \\
\dot{x}_{n-1} &= x_n - a_1 y + b_1 u \\
\dot{x}_n &= -a_0 y + b_0 u \\
y &= x_1
\end{aligned} \qquad (10.2)$$

or, more compactly, as

$$\begin{aligned}
\dot{x} &= Ax - ya + \begin{bmatrix} 0_{(\rho-1)\times 1} \\ b \end{bmatrix} u \\
y &= e_1^T x,
\end{aligned} \qquad (10.3)$$

where

$$A = \begin{bmatrix} 0 & & \\ \vdots & I_{n-1} & \\ 0 & \cdots & 0 \end{bmatrix}, \quad a = \begin{bmatrix} a_{n-1} \\ \vdots \\ a_0 \end{bmatrix}, \quad b = \begin{bmatrix} b_m \\ \vdots \\ b_0 \end{bmatrix}. \qquad (10.4)$$

10.1 STATE ESTIMATION FILTERS

Although the linear system (10.3) is a special case of the nonlinear system (8.3) with $\phi(y) = 0$, $\Phi(y) = -Iy$, and $\sigma(y) = 1$, and the designs from Chapters 8 and 9 are applicable, we proceed with an independent derivation. We first rewrite (10.3) as

$$\begin{aligned} \dot{x} &= Ax + F(y,u)^T \theta \\ y &= e_1^T x, \end{aligned} \quad (10.5)$$

where

$$F(y,u)^T = \left[\begin{bmatrix} 0_{(\rho-1)\times(m+1)} \\ I_{m+1} \end{bmatrix} u, \ -Iy_n \right], \quad (10.6)$$

and the $p = n + m + 1$-dimensional parameter vector θ is defined by

$$\theta = \begin{bmatrix} b \\ a \end{bmatrix}. \quad (10.7)$$

For state estimation we employ the filters

$$\begin{aligned} \dot{\xi} &= A_0 \xi + ky \quad &(10.8) \\ \dot{\Omega}^T &= A_0 \Omega^T + F(y,u)^T, \quad &(10.9) \end{aligned}$$

where the vector $k = [k_1, \ldots, k_n]^T$ is chosen so that the matrix

$$A_0 = A - ke_1^T \quad (10.10)$$

is Hurwitz, and hence P exists such that

$$PA_0 + A_0^T P = -I, \quad P = P^T > 0. \quad (10.11)$$

With so-designed filters our state estimate is

$$\hat{x} = \xi + \Omega^T \theta, \quad (10.12)$$

and it is easy to show that the state estimation error

$$\varepsilon = x - \hat{x} \quad (10.13)$$

vanishes exponentially because it satisfies

$$\dot{\varepsilon} = A_0 \varepsilon. \quad (10.14)$$

What has been achieved thus far is a static relationship between the state x and the unknown parameter θ:

$$x = \xi + \Omega^T \theta + \varepsilon. \quad (10.15)$$

A further practical step is to lower the dynamic order of the Ω-filter by exploiting the structure of $F(y,u)$ in (10.6). We denote the first $m+1$ columns of Ω^T by $v_m, \ldots, v_1, v_0$ and the remaining n columns by Ξ,

$$\Omega^T = [v_m, \ldots, v_1, v_0, \Xi], \quad (10.16)$$

and show that due to the special dependence of $F(y, u)$ on u, the equations for the first $n+1$ columns of Ω^T are governed by

$$\dot{v}_j = A_0 v_j + e_{n-j} u, \qquad j = 0, \ldots, m. \tag{10.17}$$

This means that thanks to the special structure of A_0,

$$A_0^j e_n = e_{n-j}, \qquad j = 0, \ldots, n-1, \tag{10.18}$$

the vectors v_j can be obtained from only one input filter

$$\dot{\lambda} = A_0 \lambda + e_n u, \tag{10.19}$$

through the algebraic expressions

$$v_j = A_0^j \lambda, \qquad j = 0, \ldots, m. \tag{10.20}$$

Similarly, Ξ, governed by

$$\dot{\Xi} = A_0 \Xi - I y, \qquad \Xi \in \mathbb{R}^{n \times n}, \tag{10.21}$$

can be obtained from only one output filter

$$\dot{\eta} = A_0 \eta + e_n y, \tag{10.22}$$

through the algebraic expression

$$\Xi = - \left[A_0^{n-1} \eta, \ldots, A_0 \eta, \eta \right]. \tag{10.23}$$

Finally, with the identity

$$A_0^n e_n = -k, \tag{10.24}$$

the vector ξ in (10.8) can be obtained from the filter (10.22) through the algebraic expression

$$\xi = -A_0^n \eta. \tag{10.25}$$

The implemented input and output filters are summarized in Table 10.1. For comparison with traditional adaptive controllers, we note that the total dynamic order of these K-filters is $2n$ and can further be reduced to $2(n-1)$ by exploiting the fact that u and y are available.

Remark 10.5 From (10.15) and the expressions in Table 10.1 an equivalent expression for the virtual estimate $\hat{x}$ is

$$\begin{aligned} \hat{x} &= -A_0^n \eta - \sum_{i=0}^{n-1} a_i A_0^i \eta + \sum_{i=0}^{m} b_i A_0^i \lambda \\ &= B(A_0) \lambda - A(A_0) \eta, \end{aligned} \tag{10.32}$$

where $A(\cdot)$ and $B(\cdot)$ are matrix-valued polynomial functions. The generation of the virtual estimate $\hat{x}$ is described pictorially in Figure 10.1. With (10.32) we get an explicit relationship among $\lambda, \eta,$ and ε and x:

$$x = B(A_0) \lambda - A(A_0) \eta + \varepsilon. \tag{10.33}$$

◇

10.1 State Estimation Filters

Table 10.1: K-Filters

K-filters:

$$\dot{\eta} = A_0\eta + e_n y \quad (10.26)$$
$$\dot{\lambda} = A_0\lambda + e_n u \quad (10.27)$$

$$\Xi = -\left[A_0^{n-1}\eta, \ldots, A_0\eta, \eta\right] \quad (10.28)$$
$$\xi = -A_0^n \eta \quad (10.29)$$
$$v_j = A_0^j \lambda, \quad j = 0, \ldots, m \quad (10.30)$$
$$\Omega^\mathrm{T} = [v_m, \ldots, v_1, v_0, \Xi] \quad (10.31)$$

The backstepping design for the plant (10.1) starts with its output y, which will be the only plant state allowed to appear in the control law. For this reason, (10.2) is rewritten as:

$$\dot{y} = x_2 - a_{n-1}y = x_2 - y e_1^\mathrm{T} a. \quad (10.34)$$

From the algebraic expression (10.15) we have

$$\begin{aligned} x_2 &= \xi_2 + \Omega_{(2)}^\mathrm{T}\theta + \varepsilon_2 \\ &= \xi_2 + [v_{m,2}, v_{m-1,2}, \ldots, v_{0,2}, \Xi_{(2)}]\theta + \varepsilon_2 \quad (10.35) \\ &= b_m v_{m,2} + \xi_2 + [0, v_{m-1,2}, \ldots, v_{0,2}, \Xi_{(2)}]\theta + \varepsilon_2. \quad (10.36) \end{aligned}$$

Substituting both (10.35) and (10.36) into (10.34), we obtain the following two important expressions for $\dot{y}$:

$$\begin{aligned} \dot{y} &= \xi_2 + \omega^\mathrm{T}\theta + \varepsilon_2 \quad (10.37) \\ &= b_m v_{m,2} + \xi_2 + \bar{\omega}^\mathrm{T}\theta + \varepsilon_2, \quad (10.38) \end{aligned}$$

where the 'regressor' ω and the 'truncated regressor' $\bar{\omega}$ are defined as

$$\omega = \left[v_{m,2}, v_{m-1,2}, \ldots, v_{0,2}, \Xi_{(2)} - y e_1^\mathrm{T}\right]^\mathrm{T} \quad (10.39)$$
$$\bar{\omega} = \left[0, v_{m-1,2}, \ldots, v_{0,2}, \Xi_{(2)} - y e_1^\mathrm{T}\right]^\mathrm{T}. \quad (10.40)$$

We have thus prepared the ground for a backstepping design.

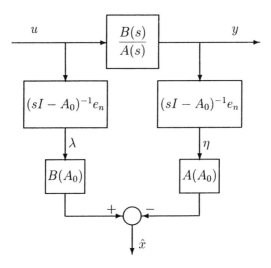

Figure 10.1: Virtual estimate $\hat{x}$ generated with input filter λ and output filter η.

10.2 Tuning Functions Design

10.2.1 Design procedure

We now present the tuning functions design procedure for the case $\rho > 1$. The design for the case $\rho = 1$ can be easily deduced from the first step of the recursive procedure.

Thanks to the minimum phase Assumption 10.1, the design is restricted to the first ρ equations in (10.2):

$$
\begin{aligned}
\dot{x}_1 &= x_2 - a_{n-1}y \\
&\vdots \\
\dot{x}_{\rho-1} &= x_\rho - a_{m+1}y \\
\dot{x}_\rho &= x_{\rho+1} - a_m y + b_m u .
\end{aligned} \tag{10.41}
$$

We will return to the behavior of the last m equations in the stability proof.

In the backstepping approach we view the state variable x_{i+1} as a control input to the subsystem consisting of the states $x_1, \ldots, x_i$, and we design a *stabilizing function* α_i which would achieve the control objective if x_{i+1} were available as a control input. The control law for the actual control input u is obtained at the ρth step of the recursive design. Because only the system output $y = x_1$ is measured, we replace (10.41) with a new system whose states are available. We start with (10.38), which is just an alternative form of the first equation in (10.41). Equation (10.38) suggests that $v_{m,2}$ is chosen instead of the unmeasured x_2 to be the 'virtual control' input for backstepping. The reason for this choice is that both x_2 and $v_{m,2}$ are separated by only $\rho - 1$

10.2 TUNING FUNCTIONS DESIGN

integrators from the actual control u, which is clear from (10.17) for $j = m$:

$$\dot{v}_m = A_0 v_m + e_\rho u. \tag{10.42}$$

A closer examination of the filters in Table 10.1 reveals that more integrators stand in the way of any other variable. Therefore, the design system chosen to replace (10.41) is

$$\begin{aligned}
\dot{y} &= b_m v_{m,2} + \xi_2 + \bar{\omega}^T \theta + \varepsilon_2 \\
\dot{v}_{m,2} &= v_{m,3} - k_i v_{m,1} \\
&\vdots \\
\dot{v}_{m,\rho-1} &= v_{m,\rho} - k_i v_{m,1} \\
\dot{v}_{m,\rho} &= v_{m,\rho+1} - k_\rho v_{m,1} + u.
\end{aligned} \tag{10.43}$$

All of its states are available for feedback. Our design task is to force the output y to asymptotically track the reference output y_r while keeping all the closed-loop signals bounded.

As in the tuning functions design in Chapter 4, we employ the change of coordinates

$$z_1 = y - y_r \tag{10.44}$$

$$z_i = v_{m,i} - \hat{\varrho} y_r^{(i-1)} - \alpha_{i-1}, \quad i = 2, \ldots, \rho, \tag{10.45}$$

where $\hat{\varrho}$ is an estimate of $\varrho = 1/b_m$. Our goal is to regulate $z = [z_1, \ldots, z_\rho]^T$ to zero because by regulating z to zero we will achieve asymptotic tracking of $y_r(t)$ by $y(t)$.

Step 1. We start with the equation for the tracking error z_1 obtained from (10.44) and (10.43):

$$\dot{z}_1 = b_m v_{m,2} + \xi_2 + \bar{\omega}^T \theta + \varepsilon_2 - \dot{y}_r. \tag{10.46}$$

By substituting (10.45) for $i = 2$ into (10.46), we get

$$\begin{aligned}
\dot{z}_1 &= b_m z_2 + b_m \alpha_1 + b_m \hat{\varrho} \dot{y}_r + \xi_2 + \bar{\omega}^T \theta + \varepsilon_2 - \dot{y}_r \\
&= b_m \alpha_1 + \xi_2 + \bar{\omega}^T \theta + \varepsilon_2 - b_m \tilde{\varrho} \dot{y}_r + b_m z_2.
\end{aligned} \tag{10.47}$$

Scaling the first stabilizing function α_1 as

$$\alpha_1 = \hat{\varrho} \bar{\alpha}_1, \tag{10.48}$$

we obtain

$$\dot{z}_1 = \bar{\alpha}_1 + \xi_2 + \bar{\omega}^T \theta + \varepsilon_2 - b_m (\dot{y}_r + \bar{\alpha}_1) \tilde{\varrho} + b_m z_2. \tag{10.49}$$

Then the choice[1]

$$\bar{\alpha}_1 = -c_1 z_1 - d_1 z_1 - \xi_2 - \bar{\omega}^T \hat{\theta} \tag{10.50}$$

---

[1] The reason for using two positive constants c_1 and d_1 is to have uniformity with subsequent steps of the backstepping procedure where d_i is a coefficient of a nonlinear damping term counteracting ε_2.

results in the system

$$\dot{z}_1 = -c_1 z_1 - d_1 z_1 + \varepsilon_2 + \bar{\omega}^T \tilde{\theta} - b_m (\dot{y}_r + \bar{\alpha}_1) \tilde{\varrho} + b_m z_2. \quad (10.51)$$

We stress that (10.51) along with (10.14) would be globally asymptotically stable if $\tilde{\theta}$, $\tilde{\varrho}$, and z_2 were zero. With (10.45), (10.48), and (10.39), we have

$$\begin{aligned}
\bar{\omega}^T \tilde{\theta} + b_m z_2 &= \bar{\omega}^T \tilde{\theta} + \hat{b}_m z_2 + \tilde{b}_m z_2 \\
&= \bar{\omega}^T \tilde{\theta} + (v_{m,2} - \hat{\varrho} \dot{y}_r - \alpha_1) e_1^T \tilde{\theta} + \hat{b}_m z_2 \\
&= \omega^T \tilde{\theta} - \hat{\varrho} (\dot{y}_r + \bar{\alpha}_1) e_1^T \tilde{\theta} + \hat{b}_m z_2 \\
&= \left(\omega - \hat{\varrho} (\dot{y}_r + \bar{\alpha}_1) e_1\right)^T \tilde{\theta} + \hat{b}_m z_2. \quad (10.52)
\end{aligned}$$

Substituting (10.52) into (10.51) we get

$$\dot{z}_1 = -c_1 z_1 - d_1 z_1 + \varepsilon_2 + \left(\omega - \hat{\varrho}(\dot{y}_r + \bar{\alpha}_1) e_1\right)^T \tilde{\theta} - b_m (\dot{y}_r + \bar{\alpha}_1) \tilde{\varrho} + \hat{b}_m z_2. \quad (10.53)$$

This system along with (10.14) is to be stabilized by selecting update laws for the parameter estimates $\hat{\theta}$ and $\hat{\varrho}$. These update laws will be chosen to achieve stability with respect to the Lyapunov function

$$V_1 = \frac{1}{2} z_1^2 + \frac{1}{2} \tilde{\theta}^T \Gamma^{-1} \tilde{\theta} + \frac{|b_m|}{2\gamma} \tilde{\varrho}^2 + \frac{1}{4 d_1} \varepsilon^T P \varepsilon. \quad (10.54)$$

We examine the derivative of V_1:

$$\begin{aligned}
\dot{V}_1 &= z_2 \left[-c_1 z_1 - d_1 z_1 + \varepsilon_2 + (\omega - \hat{\varrho}(\dot{y}_r + \bar{\alpha}_1) e_1)^T \tilde{\theta} - b_m (\dot{y}_r + \bar{\alpha}_1) \tilde{\varrho} + \hat{b}_m z_2 \right] \\
&\quad - \tilde{\theta}^T \Gamma^{-1} \dot{\hat{\theta}} - \frac{|b_m|}{\gamma} \tilde{\varrho} \dot{\hat{\varrho}} - \frac{1}{4 d_1} \varepsilon^T \varepsilon \\
&= -c_1 z_1^2 + \hat{b}_m z_1 z_2 - |b_m| \tilde{\varrho} \frac{1}{\gamma} \left[\gamma \operatorname{sgn}(b_m) (\dot{y}_r + \bar{\alpha}_1) z_1 - \dot{\hat{\varrho}} \right] \\
&\quad + \tilde{\theta}^T \Gamma^{-1} \left[\Gamma (\omega - \hat{\varrho}(\dot{y}_r + \bar{\alpha}_1) e_1) z_1 - \dot{\hat{\theta}} \right] - d_1 z_1^2 + z_1 \varepsilon_2 - \frac{1}{4 d_1} \varepsilon^T \varepsilon. \quad (10.55)
\end{aligned}$$

To eliminate the unknown indefinite $\tilde{\theta}$, $\tilde{\varrho}$-terms in (10.55) we choose

$$\dot{\hat{\varrho}} = -\gamma \operatorname{sgn}(b_m)(\dot{y}_r + \bar{\alpha}_1) z_1, \qquad \gamma > 0 \quad (10.56)$$

and $\dot{\hat{\theta}} = \Gamma \tau_1$, where

$$\tau_1 = (\omega - \hat{\varrho}(\dot{y}_r + \bar{\alpha}_1) e_1) z_1. \quad (10.57)$$

If z_2 where zero, these choices would yield the following expression for the derivative of V_1:

$$\begin{aligned}
\dot{V}_1 &= -c_1 z_1^2 - d_1 \left(z_1 - \frac{1}{2 d_1} \varepsilon_2 \right)^2 - \frac{1}{4 d_1} \left(\varepsilon_1^2 + \varepsilon_3^2 + \cdots + \varepsilon_n^2 \right) \\
&\leq -c_1 z_1^2. \quad (10.58)
\end{aligned}$$

10.2 TUNING FUNCTIONS DESIGN

Since $z_2 \neq 0$, we *do not* use $\dot{\hat{\theta}} = \Gamma \tau_1$ as the update law for $\hat{\theta}$, because θ will reappear in subsequent steps. However, ϱ will not reappear, so we do use (10.56) as the actual update law for $\hat{\varrho}$. We retain (10.57) as our first *tuning function* for $\hat{\theta}$. Substituting (10.56) and (10.57) into (10.55), we obtain

$$\dot{V}_1 \leq -c_1 z_1^2 + \hat{b}_m z_1 z_2 + \tilde{\theta}^T \left(\tau_1 - \Gamma^{-1}\dot{\hat{\theta}}\right). \tag{10.59}$$

We pause to determine the arguments of the function α_1. By examining (10.50) along with (10.40), we see that α_1 is a function of $y, \eta, \hat{\theta}, \hat{\varrho}, v_{0,2}, \ldots, v_{m-1,2}$, and y_r. In view of (10.30), $v_{i,j}$ can be expressed as

$$v_{i,j} = [*, \ldots, *, 1]\bar{\lambda}_{i+j}, \tag{10.60}$$

where $\bar{\lambda}_k \triangleq [\lambda_1, \ldots, \lambda_k]^T$ and $\lambda_k \triangleq 0$ for $k > n$. With (10.60) we conclude that α_1 is a function of $y, \eta, \hat{\theta}, \hat{\varrho}, \bar{\lambda}_{m+1}$, and y_r. As we shall see in the subsequent steps, α_i is a function of $y, \eta, \hat{\theta}, \hat{\varrho}, \bar{\lambda}_{m+i}$, and $\bar{y}_r^{(i-1)}$.

Step 2. Differentiating (10.45) for $i = 2$, with the help of the second equation in (10.43) we obtain

$$\begin{aligned}
\dot{z}_2 &= \dot{v}_{m,2} - \hat{\varrho}\ddot{y}_r - \dot{\hat{\varrho}}\dot{y}_r - \dot{\alpha}_1(y, \eta, \hat{\theta}, \hat{\varrho}, \bar{\lambda}_{m+1}, y_r) \\
&= v_{m,3} - k_2 v_{m,1} - \hat{\varrho}\ddot{y}_r - \dot{\hat{\varrho}}\dot{y}_r - \frac{\partial \alpha_1}{\partial y}\left(\xi_2 + \omega^T \theta + \varepsilon_2\right) \\
&\quad - \frac{\partial \alpha_1}{\partial \eta}(A_0 \eta + e_n y) - \frac{\partial \alpha_1}{\partial y_r}\dot{y}_r - \sum_{j=1}^{m+1} \frac{\partial \alpha_1}{\partial \lambda_j}(-k_j \lambda_1 + \lambda_{j+1}) \\
&\quad - \frac{\partial \alpha_1}{\partial \hat{\theta}}\dot{\hat{\theta}} - \frac{\partial \alpha_1}{\partial \hat{\varrho}}\dot{\hat{\varrho}} \\
&\triangleq v_{m,3} - \beta_2 - \hat{\varrho}\ddot{y}_r - \frac{\partial \alpha_1}{\partial y}\left(\omega^T \tilde{\theta} + \varepsilon_2\right) - \frac{\partial \alpha_1}{\partial \hat{\theta}}\dot{\hat{\theta}},
\end{aligned} \tag{10.61}$$

where β_2 is a function of available signals:

$$\begin{aligned}
\beta_2 &= k_2 v_{m,1} + \frac{\partial \alpha_1}{\partial y}\left(\xi_2 + \omega^T \hat{\theta}\right) + \frac{\partial \alpha_1}{\partial \eta}(A_0 \eta + e_n y) + \frac{\partial \alpha_1}{\partial y_r}\dot{y}_r \\
&\quad + \sum_{j=1}^{m+1} \frac{\partial \alpha_1}{\partial \lambda_j}(-k_j \lambda_1 + \lambda_{j+1}) + \left(\ddot{y}_r + \frac{\partial \alpha_1}{\partial \hat{\varrho}}\right)\dot{\hat{\varrho}}.
\end{aligned} \tag{10.62}$$

Noting from (10.45) that $v_{m,3} - \hat{\varrho}\ddot{y}_r = z_3 + \alpha_2$, we get

$$\dot{z}_2 = \alpha_2 - \beta_2 - \frac{\partial \alpha_1}{\partial y}\left(\omega^T \tilde{\theta} + \varepsilon_2\right) - \frac{\partial \alpha_1}{\partial \hat{\theta}}\dot{\hat{\theta}} + z_3. \tag{10.63}$$

Since our system is augmented by the new state z_2, we augment the Lyapunov function (10.54) as

$$V_2 = V_1 + \frac{1}{2}z_2^2 + \frac{1}{4d_2}\varepsilon^T P \varepsilon, \tag{10.64}$$

where another ε-term was included to account for the presence of ε_2 in (10.63). In view of (10.59), (10.63), and (10.14), the derivative of V_2 is

$$\begin{aligned}\dot{V}_2 &\leq -c_1 z_1^2 + \hat{b}_m z_1 z_2 + \tilde{\theta}^{\mathrm{T}}\left(\tau_1 - \Gamma^{-1}\dot{\hat{\theta}}\right) \\ &\quad + z_2\left[\alpha_2 - \beta_2 - \frac{\partial \alpha_1}{\partial y}\left(\omega^{\mathrm{T}}\tilde{\theta} + \varepsilon_2\right) - \frac{\partial \alpha_1}{\partial \hat{\theta}}\dot{\hat{\theta}} + z_3\right] - \frac{1}{4d_2}\varepsilon^{\mathrm{T}}\varepsilon \\ &\leq -c_1 z_1^2 + z_2 z_3 + \tilde{\theta}^{\mathrm{T}}\Gamma^{-1}\left(\tau_1 - \frac{\partial \alpha_1}{\partial y}\omega z_2 - \Gamma^{-1}\dot{\hat{\theta}}\right) \\ &\quad + z_2\left[\alpha_2 + \hat{b}_m z_1 - \beta_2 - \frac{\partial \alpha_1}{\partial \hat{\theta}}\dot{\hat{\theta}}\right] - z_2\frac{\partial \alpha_1}{\partial y}\varepsilon_2 - \frac{1}{4d_2}\varepsilon_2^2 \\ &= -c_1 z_1^2 + z_2 z_3 + \tilde{\theta}^{\mathrm{T}}\Gamma^{-1}\left(\tau_1 - \frac{\partial \alpha_1}{\partial y}\omega z_2 - \Gamma^{-1}\dot{\hat{\theta}}\right) \\ &\quad + z_2\left[\alpha_2 + \hat{b}_m z_1 - \beta_2 - \frac{\partial \alpha_1}{\partial \hat{\theta}}\dot{\hat{\theta}}\right] \\ &\quad + d_2\left(\frac{\partial \alpha_1}{\partial y}\right)^2 z_2^2 - d_2\left(z_2\frac{\partial \alpha_1}{\partial y} + \frac{1}{2d_2}\varepsilon_2\right)^2. \end{aligned} \quad (10.65)$$

The elimination of the unknown indefinite $\tilde{\theta}$-term from (10.65) can be achieved with the update law $\dot{\hat{\theta}} = \Gamma\tau_2$, where

$$\tau_2 = \tau_1 - \frac{\partial \alpha_1}{\partial y}\omega z_2. \quad (10.66)$$

Then, if z_3 were zero, the stabilizing function

$$\alpha_2 = -c_2 z_2 - d_2\left(\frac{\partial \alpha_1}{\partial y}\right)^2 z_2 - \hat{b}_m z_1 + \beta_2 + \frac{\partial \alpha_1}{\partial \hat{\theta}}\Gamma\tau_2 \quad (10.67)$$

would yield

$$\dot{V}_2 \leq -c_1 z_1^2 - c_2 z_2^2 - d_2\left(z_2\frac{\partial \alpha_1}{\partial y} + \frac{1}{2d_2}\varepsilon_2\right)^2 \leq -c_1 z_1^2 - c_2 z_2^2. \quad (10.68)$$

However, since $z_3 \neq 0$, we do not use $\dot{\hat{\theta}} = \Gamma\tau_2$ as an update law. Instead, we retain τ_2 as our second tuning function and α_2 as our second stabilizing function. Upon the substitution into (10.65), we obtain

$$\dot{V}_2 \leq -c_1 z_1^2 - c_2 z_2^2 + z_2 z_3 + \tilde{\theta}^{\mathrm{T}}\left(\tau_2 - \Gamma^{-1}\dot{\hat{\theta}}\right) + z_2\frac{\partial \alpha_1}{\partial \hat{\theta}}\left(\Gamma\tau_2 - \dot{\hat{\theta}}\right). \quad (10.69)$$

The mismatch term $z_2\frac{\partial \alpha_1}{\partial \hat{\theta}}\left(\Gamma\tau_2 - \dot{\hat{\theta}}\right)$ will be dealt with in subsequent steps.

10.2 TUNING FUNCTIONS DESIGN

Step 3. This step is crucial for understanding the tuning functions technique. By differentiating (10.45) for $i = 3$, with the help of the third equation in (10.43), we have

$$\begin{aligned}\dot{z}_3 &= \dot{v}_{m,3} - \hat{\varrho}y_{\mathrm{r}}^{(3)} - \dot{\hat{\varrho}}\ddot{y}_{\mathrm{r}} - \dot{\alpha}_2(y, \eta, \hat{\theta}, \hat{\varrho}, \bar{\lambda}_{m+2}, y_{\mathrm{r}}, \dot{y}_{\mathrm{r}}) \\ &= v_{m,4} - \beta_3 - \hat{\varrho}y_{\mathrm{r}}^{(3)} - \frac{\partial \alpha_2}{\partial y}\left(\omega^{\mathrm{T}}\tilde{\theta} + \varepsilon_2\right) - \frac{\partial \alpha_2}{\partial \hat{\theta}}\dot{\hat{\theta}},\end{aligned} \qquad (10.70)$$

where

$$\begin{aligned}\beta_3 &= k_3 v_{m,1} + \frac{\partial \alpha_2}{\partial y}\left(\xi_2 + \omega^{\mathrm{T}}\hat{\theta}\right) + \frac{\partial \alpha_2}{\partial \eta}\left(A_0\eta + e_n y\right) + \frac{\partial \alpha_2}{\partial y_{\mathrm{r}}}\dot{y}_{\mathrm{r}} + \frac{\partial \alpha_2}{\partial \dot{y}_{\mathrm{r}}}\ddot{y}_{\mathrm{r}} \\ &\quad + \sum_{j=1}^{m+2}\frac{\partial \alpha_1}{\partial \lambda_j}(-k_j\lambda_1 + \lambda_{j+1}) + \left(\ddot{y}_{\mathrm{r}} + \frac{\partial \alpha_2}{\partial \hat{\varrho}}\right)\dot{\hat{\varrho}}.\end{aligned} \qquad (10.71)$$

Noting from (10.45) that $v_{m,4} - \hat{\varrho}y_{\mathrm{r}}^{(3)} = z_4 + \alpha_3$, we get

$$\dot{z}_3 = \alpha_3 - \beta_3 - \frac{\partial \alpha_2}{\partial y}\left(\omega^{\mathrm{T}}\tilde{\theta} + \varepsilon_2\right) - \frac{\partial \alpha_2}{\partial \hat{\theta}}\dot{\hat{\theta}} + z_4. \qquad (10.72)$$

Since our system is augmented by the new state z_3, we also augment the Lyapunov function (10.64):

$$V_3 = V_2 + \frac{1}{2}z_3^2 + \frac{1}{4d_3}\varepsilon^{\mathrm{T}}P\varepsilon. \qquad (10.73)$$

In view of (10.69), (10.72), and (10.14), the derivative of V_3 is

$$\begin{aligned}\dot{V}_3 &\leq -c_1 z_1^2 - c_2 z_2^2 + z_3 z_4 \\ &\quad + \tilde{\theta}^{\mathrm{T}}\Gamma^{-1}\left(\tau_2 - \frac{\partial \alpha_2}{\partial y}\omega z_3 - \Gamma^{-1}\dot{\hat{\theta}}\right) + z_2\frac{\partial \alpha_1}{\partial \hat{\theta}}\left(\Gamma \tau_2 - \dot{\hat{\theta}}\right) \\ &\quad + z_3\left[\alpha_3 + z_2 - \beta_3 - \frac{\partial \alpha_2}{\partial \hat{\theta}}\dot{\hat{\theta}}\right] - z_3\frac{\partial \alpha_2}{\partial y}\varepsilon_2 - \frac{1}{4d_3}\varepsilon_2^2.\end{aligned} \qquad (10.74)$$

As in the previous steps, for the elimination of the unknown indefinite $\tilde{\theta}$-term from (10.74), we can choose the update law $\dot{\hat{\theta}} = \Gamma\tau_3$, where

$$\tau_3 = \tau_2 - \frac{\partial \alpha_2}{\partial y}\omega z_3. \qquad (10.75)$$

Noting that

$$\Gamma\tau_2 - \dot{\hat{\theta}} = \Gamma\tau_2 - \Gamma\tau_3 + \Gamma\tau_3 - \dot{\hat{\theta}} = \Gamma\frac{\partial \alpha_2}{\partial y}\omega z_3 + \left(\Gamma\tau_3 - \dot{\hat{\theta}}\right), \qquad (10.76)$$

(10.74) becomes

$$\begin{aligned}
\dot{V}_3 &\leq -c_1 z_1^2 - c_2 z_2^2 + z_3 z_4 \\
&\quad + \tilde{\theta}^T \Gamma^{-1}\left(\tau_3 - \Gamma^{-1}\dot{\hat{\theta}}\right) + z_2 \frac{\partial \alpha_1}{\partial \hat{\theta}}\left(\Gamma \tau_3 - \dot{\hat{\theta}}\right) + z_2 \frac{\partial \alpha_1}{\partial \hat{\theta}} \Gamma \frac{\partial \alpha_2}{\partial y} \omega z_3 \\
&\quad + z_3 \left[\alpha_3 + z_2 - \beta_3 - \frac{\partial \alpha_2}{\partial \hat{\theta}}\dot{\hat{\theta}}\right] \\
&\quad + d_3 \left(\frac{\partial \alpha_2}{\partial y}\right)^2 z_3^2 - d_3 \left(z_3 \frac{\partial \alpha_2}{\partial y} + \frac{1}{2d_3}\varepsilon_2\right)^2 \\
&\leq -c_1 z_1^2 - c_2 z_2^2 + z_3 z_4 + \tilde{\theta}^T \Gamma^{-1}\left(\tau_3 - \Gamma^{-1}\dot{\hat{\theta}}\right) + z_2 \frac{\partial \alpha_1}{\partial \hat{\theta}}\left(\Gamma \tau_3 - \dot{\hat{\theta}}\right) \\
&\quad + z_3 \left[\alpha_3 + z_2 - \beta_3 - \frac{\partial \alpha_2}{\partial \hat{\theta}}\dot{\hat{\theta}} + z_2 \frac{\partial \alpha_1}{\partial \hat{\theta}} \Gamma \frac{\partial \alpha_2}{\partial y} \omega\right] \\
&\quad + d_3 \left(\frac{\partial \alpha_2}{\partial y}\right)^2 z_3^2. \quad (10.77)
\end{aligned}$$

Then, if z_4 were zero, and if $\dot{\hat{\theta}}$ were chosen to be $\Gamma \tau_3$, the stabilizing function

$$\alpha_3 = -c_3 z_3 - d_3 \left(\frac{\partial \alpha_2}{\partial y}\right)^2 z_3 - z_2 + \beta_3 + \frac{\partial \alpha_2}{\partial \hat{\theta}} \Gamma \tau_3 - z_2 \frac{\partial \alpha_1}{\partial \hat{\theta}} \Gamma \frac{\partial \alpha_2}{\partial y} \omega \quad (10.78)$$

would yield

$$\dot{V}_3 \leq -c_1 z_1^2 - c_2 z_2^2 - c_3 z_3^2. \quad (10.79)$$

However, $z_4 \neq 0$, and we do not use $\dot{\hat{\theta}} = \Gamma \tau_3$ as an update law. Instead, we retain τ_3 as our third tuning function and α_3 as our third stabilizing function. Upon the substitution into (10.77), we obtain

$$\begin{aligned}
\dot{V}_3 &\leq -c_1 z_1^2 - c_2 z_2^2 - c_3 z_3^2 + z_3 z_4 \\
&\quad + \tilde{\theta}^T \left(\tau_3 - \Gamma^{-1}\dot{\hat{\theta}}\right) + \left(z_2 \frac{\partial \alpha_1}{\partial \hat{\theta}} + z_3 \frac{\partial \alpha_2}{\partial \hat{\theta}}\right)\left(\Gamma \tau_3 - \dot{\hat{\theta}}\right). \quad (10.80)
\end{aligned}$$

As we pointed out, this step is crucial for understanding the general design procedure. Unlike the first two stabilizing functions, the third stabilizing function α_3 in (10.78) contains the term $-z_2 \frac{\partial \alpha_1}{\partial \hat{\theta}} \Gamma \frac{\partial \alpha_2}{\partial y} \omega$. The role of this term is to cancel the indefinite term $z_2 \frac{\partial \alpha_1}{\partial \hat{\theta}} \Gamma (\tau_2 - \tau_3)$ in the Lyapunov function derivative $\dot{V}_3$. This is achieved by recognizing in (10.76) that $\tau_2 - \tau_3$ has z_3 as a factor. We provide additional interpretation after equation (10.118). For further insight, the reader is referred to Sections 4.1 and 4.2.

Step $i = 4, \ldots, \rho$. We now proceed fast in presenting the remaining steps of the design procedure. By differentiating (10.45) for $i = 4, \ldots, \rho - 1$, with the

10.2 TUNING FUNCTIONS DESIGN

help of (10.43) we obtain

$$\begin{aligned}\dot{z}_i &= \dot{v}_{m,i} - \hat{\varrho} y_{\mathrm{r}}^{(i)} - \dot{\hat{\varrho}} y_{\mathrm{r}}^{(i-1)} - \dot{\alpha}_{i-1}\left(y, \eta, \hat{\theta}, \hat{\varrho}, \bar{\lambda}_{m+i-1}, \bar{y}_{\mathrm{r}}^{(i-2)}\right)\\ &= v_{m,i+1} - \beta_i - \hat{\varrho} y_{\mathrm{r}}^{(i)} - \frac{\partial \alpha_{i-1}}{\partial y}\left(\omega^{\mathrm{T}}\tilde{\theta} + \varepsilon_2\right) - \frac{\partial \alpha_{i-1}}{\partial \hat{\theta}}\dot{\hat{\theta}},\end{aligned} \quad (10.81)$$

where

$$\begin{aligned}\beta_i &= k_i v_{m,1} + \frac{\partial \alpha_{i-1}}{\partial y}\left(\xi_2 + \omega^{\mathrm{T}}\hat{\theta}\right) + \frac{\partial \alpha_{i-1}}{\partial \eta}(A_0\eta + e_n y) + \sum_{j=1}^{i-1}\frac{\partial \alpha_{i-1}}{\partial y_{\mathrm{r}}^{(j-1)}}y_{\mathrm{r}}^{(j)}\\ &\quad + \sum_{j=1}^{m+i-1}\frac{\partial \alpha_{i-1}}{\partial \lambda_j}(-k_j\lambda_1 + \lambda_{j+1}) + \left(y_{\mathrm{r}}^{(i-1)} + \frac{\partial \alpha_{i-1}}{\partial \hat{\varrho}}\right)\dot{\hat{\varrho}}.\end{aligned} \quad (10.82)$$

Noting from (10.45) that $v_{m,i+1} - \hat{\varrho} y_{\mathrm{r}}^{(i)} = z_{i+1} + \alpha_i$, we get

$$\dot{z}_i = \alpha_i - \beta_i - \frac{\partial \alpha_{i-1}}{\partial y}\left(\omega^{\mathrm{T}}\tilde{\theta} + \varepsilon_2\right) - \frac{\partial \alpha_{i-1}}{\partial \hat{\theta}}\dot{\hat{\theta}} + z_{i+1}. \quad (10.83)$$

From (10.43) it follows that the final step $i = \rho$ can be encompassed in these calculations if we define

$$\alpha_\rho = u + v_{m,\rho+1} - \hat{\varrho} y_{\mathrm{r}}^{(\rho)}, \qquad z_{\rho+1} = 0. \quad (10.84)$$

Now consider the Lyapunov function

$$\begin{aligned}V_i &= V_{i-1} + \frac{1}{2}z_i^2 + \frac{1}{4d_i}\varepsilon^{\mathrm{T}}P\varepsilon\\ &= \frac{1}{2}\sum_{k=1}^{i}\left(z_k^2 + \frac{1}{2d_k}\varepsilon^{\mathrm{T}}P\varepsilon\right) + \frac{1}{2}\tilde{\theta}^{\mathrm{T}}\Gamma^{-1}\tilde{\theta} + \frac{|b_m|}{2\gamma}\tilde{\varrho}^2,\end{aligned} \quad (10.85)$$

where, in analogy with (10.80), V_{i-1} satisfies

$$\dot{V}_{i-1} \le -\sum_{k=1}^{i-1}c_k z_k^2 + z_{i-1}z_i + \tilde{\theta}^{\mathrm{T}}\left(\tau_{i-1} - \Gamma^{-1}\dot{\hat{\theta}}\right) + \left(\sum_{k=2}^{i-1}z_k\frac{\partial \alpha_{k-1}}{\partial \hat{\theta}}\right)\left(\Gamma\tau_{i-1} - \dot{\hat{\theta}}\right). \quad (10.86)$$

With (10.86), (10.83), and (10.14), the derivative of V_i is

$$\begin{aligned}\dot{V}_i &\le -\sum_{k=1}^{i-1}c_k z_k^2 + z_i z_{i+1}\\ &\quad + \tilde{\theta}^{\mathrm{T}}\left(\tau_{i-1} - \frac{\partial \alpha_{i-1}}{\partial y}\omega z_i - \Gamma^{-1}\dot{\hat{\theta}}\right) + \left(\sum_{k=2}^{i-1}z_k\frac{\partial \alpha_{k-1}}{\partial \hat{\theta}}\right)\left(\Gamma\tau_{i-1} - \dot{\hat{\theta}}\right)\\ &\quad + z_i\left[\alpha_i + z_{i-1} - \beta_i - \frac{\partial \alpha_{i-1}}{\partial \hat{\theta}}\dot{\hat{\theta}}\right] - z_i\frac{\partial \alpha_{i-1}}{\partial y}\varepsilon_2 - \frac{1}{4d_i}\varepsilon_2^2.\end{aligned} \quad (10.87)$$

We can now eliminate the indefinite $\tilde{\theta}$-term from (10.87) by choosing the update law $\dot{\hat{\theta}} = \Gamma \tau_i$, where

$$\tau_i = \tau_{i-1} - \frac{\partial \alpha_{i-1}}{\partial y} \omega z_i. \tag{10.88}$$

Noting that

$$\Gamma \tau_{i-1} - \dot{\hat{\theta}} = \Gamma \tau_{i-1} - \Gamma \tau_i + \Gamma \tau_i - \dot{\hat{\theta}} = \Gamma \frac{\partial \alpha_{i-1}}{\partial y} \omega z_i + \left(\Gamma \tau_i - \dot{\hat{\theta}} \right), \tag{10.89}$$

(10.87) becomes

$$\begin{aligned}
\dot{V}_i &\leq -\sum_{k=1}^{i-1} c_k z_k^2 + z_i z_{i+1} + \tilde{\theta}^{\mathrm{T}} \left(\tau_i - \Gamma^{-1} \dot{\hat{\theta}} \right) \\
&\quad + \left(\sum_{k=2}^{i-1} z_k \frac{\partial \alpha_{k-1}}{\partial \hat{\theta}} \right) \left(\Gamma \tau_i - \dot{\hat{\theta}} \right) + \left(\sum_{k=2}^{i-1} z_k \frac{\partial \alpha_{k-1}}{\partial \hat{\theta}} \right) \Gamma \frac{\partial \alpha_{i-1}}{\partial y} \omega z_i \\
&\quad + z_i \left[\alpha_i + z_{i-1} - \beta_i - \frac{\partial \alpha_{i-1}}{\partial \hat{\theta}} \dot{\hat{\theta}} \right] \\
&\quad + d_i \left(\frac{\partial \alpha_{i-1}}{\partial y} \right)^2 z_i^2 - d_i \left(z_i \frac{\partial \alpha_{i-1}}{\partial y} + \frac{1}{2 d_i} \varepsilon_2 \right)^2 \\
&\leq -\sum_{k=1}^{i-1} c_k z_k^2 + z_i z_{i+1} + \tilde{\theta}^{\mathrm{T}} \left(\tau_i - \Gamma^{-1} \dot{\hat{\theta}} \right) + \left(\sum_{k=2}^{i-1} z_k \frac{\partial \alpha_{k-1}}{\partial \hat{\theta}} \right) \left(\Gamma \tau_i - \dot{\hat{\theta}} \right) \\
&\quad + z_i \left[\alpha_i + z_{i-1} - \beta_i - \frac{\partial \alpha_{i-1}}{\partial \hat{\theta}} \dot{\hat{\theta}} + \left(\sum_{k=2}^{i-1} z_k \frac{\partial \alpha_{k-1}}{\partial \hat{\theta}} \right) \Gamma \frac{\partial \alpha_{i-1}}{\partial y} \omega \right] \\
&\quad + d_i \left(\frac{\partial \alpha_{i-1}}{\partial y} \right)^2 z_i^2. \tag{10.90}
\end{aligned}$$

Then, if z_{i+1} were zero and if $\dot{\hat{\theta}}$ were chosen as $\Gamma \tau_i$, the stabilizing function

$$\alpha_i = -c_i z_i - d_i \left(\frac{\partial \alpha_{i-1}}{\partial y} \right)^2 z_i - z_{i-1} + \beta_i + \frac{\partial \alpha_{i-1}}{\partial \hat{\theta}} \Gamma \tau_i - \left(\sum_{k=2}^{i-1} z_k \frac{\partial \alpha_{k-1}}{\partial \hat{\theta}} \right) \Gamma \frac{\partial \alpha_{i-1}}{\partial y} \omega \tag{10.91}$$

would yield

$$\dot{V}_i \leq -\sum_{k=1}^{i} c_k z_k^2 + z_i z_{i+1}. \tag{10.92}$$

However, $z_{i+1} \neq 0$, and we do not use $\dot{\hat{\theta}} = \Gamma \tau_i$ as an update law. Instead, we retain τ_i in (10.88) as our ith tuning function and α_i in (10.91) as our ith stabilizing function. Upon the substitution into (10.90), we obtain

$$\dot{V}_i \leq -\sum_{k=1}^{i} c_k z_k^2 + z_i z_{i+1} + \tilde{\theta}^{\mathrm{T}} \left(\tau_i - \Gamma^{-1} \dot{\hat{\theta}} \right) + \left(\sum_{k=2}^{i} z_k \frac{\partial \alpha_{k-1}}{\partial \hat{\theta}} \right) \left(\Gamma \tau_i - \dot{\hat{\theta}} \right). \tag{10.93}$$

10.2 TUNING FUNCTIONS DESIGN

Comparing (10.93) with (10.86), we see that the design at step i has resulted in the Lyapunov derivative of the same form as in step $i-1$.

At the end of the design procedure, that is, when $i = \rho$, we select the actual update law as

$$\dot{\hat{\theta}} = \Gamma\tau_\rho. \tag{10.94}$$

This choice makes $\dot{V}_\rho$ negative semidefinite:

$$\dot{V}_\rho \leq -\sum_{k=1}^{\rho} c_k z_k^2, \tag{10.95}$$

because from (10.84) we have $z_{\rho+1} = 0$. The control law (10.84) which has helped us to achieve (10.95) is our actual control law:

$$u = \alpha_\rho - v_{m,\rho+1} + \hat{\varrho}y_{\rm r}^{(\rho)}. \tag{10.96}$$

The complete tuning functions design employing the filters in Table 10.1 is summarized in Table 10.2.

The resulting error system is

$$\begin{aligned}
\dot{z}_1 &= -c_1 z_1 - d_1 z_1 + \hat{b}_m z_2 + \varepsilon_2 + (\omega - \hat{\varrho}(\dot{y}_{\rm r} + \bar{\alpha}_1)e_1)^{\rm T}\tilde{\theta} \\
&\quad - b_m(\dot{y}_{\rm r} + \bar{\alpha}_1)\tilde{\varrho}
\end{aligned} \tag{10.109}$$

$$\begin{aligned}
\dot{z}_2 &= -c_2 z_2 - d_2 \left(\frac{\partial \alpha_1}{\partial y}\right)^2 z_2 - \hat{b}_m z_1 + z_3 - \frac{\partial \alpha_1}{\partial \hat{\theta}}\left(\dot{\hat{\theta}} - \Gamma\tau_2\right) \\
&\quad - \frac{\partial \alpha_1}{\partial y}\varepsilon_2 - \frac{\partial \alpha_1}{\partial y}\omega^{\rm T}\tilde{\theta}
\end{aligned} \tag{10.110}$$

$$\begin{aligned}
\dot{z}_i &= -c_i z_i - d_i \left(\frac{\partial \alpha_{i-1}}{\partial y}\right)^2 z_i - z_{i-1} + z_{i+1} \\
&\quad - \sum_{j=2}^{i-1}\frac{\partial \alpha_{j-1}}{\partial \hat{\theta}}\Gamma\frac{\partial \alpha_{i-1}}{\partial y}z_j - \frac{\partial \alpha_{i-1}}{\partial \hat{\theta}}\left(\dot{\hat{\theta}} - \Gamma\tau_i\right) \\
&\quad - \frac{\partial \alpha_{i-1}}{\partial y}\varepsilon_2 - \frac{\partial \alpha_{i-1}}{\partial y}\omega^{\rm T}\tilde{\theta}, \qquad i = 3,\ldots,\rho,
\end{aligned} \tag{10.111}$$

where $z_{\rho+1} = 0$. In view of (10.105) we have

$$\dot{\hat{\theta}} - \Gamma\tau_i = -\sum_{j=i+1}^{\rho}\Gamma\frac{\partial \alpha_{j-1}}{\partial y}\omega z_j. \tag{10.112}$$

Defining

$$\sigma_{ij} \triangleq \frac{\partial \alpha_{i-1}}{\partial \hat{\theta}}\Gamma\frac{\partial \alpha_{j-1}}{\partial y}\omega, \tag{10.113}$$

(10.112) yields

$$-\frac{\partial \alpha_{i-1}}{\partial \hat{\theta}}\left(\dot{\hat{\theta}} - \Gamma\tau_i\right) = \sum_{j=i+1}^{\rho}\sigma_{ij}z_j. \tag{10.114}$$

Table 10.2: Tuning Functions Design for Linear Systems

$$z_1 = y - y_r \qquad (10.97)$$
$$z_i = v_{m,i} - \hat{\varrho} y_r^{(i-1)} - \alpha_{i-1}, \qquad i = 2, \ldots, \rho \qquad (10.98)$$

$$\alpha_1 = \hat{\varrho} \bar{\alpha}_1 \qquad (10.99)$$
$$\bar{\alpha}_1 = -(c_1 + d_1) z_1 - \xi_2 - \bar{\omega}^T \hat{\theta} \qquad (10.100)$$
$$\alpha_2 = -\hat{b}_m z_1 - \left[c_2 + d_2 \left(\frac{\partial \alpha_1}{\partial y} \right)^2 \right] z_2 + \beta_2 + \frac{\partial \alpha_1}{\partial \hat{\theta}} \Gamma \tau_2 \qquad (10.101)$$
$$\alpha_i = -z_{i-1} - \left[c_i + d_i \left(\frac{\partial \alpha_{i-1}}{\partial y} \right)^2 \right] z_i + \beta_i + \frac{\partial \alpha_{i-1}}{\partial \hat{\theta}} \Gamma \tau_i$$
$$\quad - \sum_{j=2}^{i-1} \frac{\partial \alpha_{j-1}}{\partial \hat{\theta}} \Gamma \frac{\partial \alpha_{i-1}}{\partial y} z_j, \qquad i = 3, \ldots, \rho \qquad (10.102)$$
$$\beta_i = \frac{\partial \alpha_{i-1}}{\partial y} \left(\xi_2 + \omega^T \hat{\theta} \right) + \frac{\partial \alpha_{i-1}}{\partial \eta} (A_0 \eta + e_n y) + \sum_{j=1}^{i-1} \frac{\partial \alpha_{i-1}}{\partial y_r^{(j-1)}} y_r^{(j)} + k_i v_{m,1}$$
$$\quad + \sum_{j=1}^{m+i-1} \frac{\partial \alpha_{i-1}}{\partial \lambda_j} (-k_j \lambda_1 + \lambda_{j+1}) + \left(y_r^{(i-1)} + \frac{\partial \alpha_{i-1}}{\partial \hat{\varrho}} \right) \dot{\hat{\varrho}} \qquad (10.103)$$

$$\tau_1 = (\omega - \hat{\varrho} (\dot{y}_r + \bar{\alpha}_1) e_1) z_1 \qquad (10.104)$$
$$\tau_i = \tau_{i-1} - \frac{\partial \alpha_{i-1}}{\partial y} \omega z_i, \qquad i = 2, \ldots, \rho \qquad (10.105)$$

Adaptive control law:
$$u = \alpha_\rho - v_{m,\rho+1} + \hat{\varrho} y_r^{(\rho)} \qquad (10.106)$$

Parameter update laws:
$$\dot{\hat{\theta}} = \Gamma \tau_\rho \qquad (10.107)$$
$$\dot{\hat{\varrho}} = -\gamma \, \text{sgn}(b_m) (\dot{y}_r + \bar{\alpha}_1) z_1 \qquad (10.108)$$

10.2 TUNING FUNCTIONS DESIGN

By substituting (10.114), we bring the error system (10.109)–(10.111) into the compact form

$$\dot{z} = A_z(z,t)z + W_\varepsilon(z,t)\varepsilon_2 + W_\theta(z,t)^{\mathrm{T}}\tilde{\theta} - b_m(\dot{y}_{\mathrm{r}} + \bar{\alpha}_1)e_1\tilde{\varrho}, \qquad (10.115)$$

where the system matrix $A_z(z,t)$ is given by

$$A_z(z,t) =$$

$$\begin{bmatrix} -c_1-d_1 & \hat{b}_m & 0 & \cdots & \cdots & 0 \\ -\hat{b}_m & -c_2-d_2\left(\frac{\partial\alpha_1}{\partial y}\right)^2 & 1+\sigma_{23} & \sigma_{24} & \cdots & \sigma_{2,\rho} \\ 0 & -1-\sigma_{23} & \ddots & \ddots & \ddots & \vdots \\ \vdots & -\sigma_{24} & \ddots & \ddots & \ddots & \sigma_{\rho-2,\rho} \\ \vdots & \vdots & \ddots & \ddots & \ddots & 1+\sigma_{\rho-1,\rho} \\ 0 & -\sigma_{2,\rho} & \cdots & -\sigma_{\rho-2,\rho} & -1-\sigma_{\rho-1,\rho} & -c_\rho-d_\rho\left(\frac{\partial\alpha_{\rho-1}}{\partial y}\right)^2 \end{bmatrix}$$

$$(10.116)$$

and $W_\varepsilon(z,t)$ and $W_\theta(z,t)$ are

$$W_\varepsilon(z,t) = \begin{bmatrix} 1 \\ -\frac{\partial\alpha_1}{\partial y} \\ \vdots \\ -\frac{\partial\alpha_{\rho-1}}{\partial y} \end{bmatrix} \in \mathbb{R}^\rho \qquad (10.117)$$

$$W_\theta(z,t)^{\mathrm{T}} = W_\varepsilon(z,t)\omega^{\mathrm{T}} - \hat{\varrho}(\dot{y}_{\mathrm{r}} + \bar{\alpha}_1)e_1 e_1^{\mathrm{T}} \in \mathbb{R}^{\rho\times p}. \qquad (10.118)$$

The structure of the matrix $A_z(z,t)$ is very important. The terms σ_{ij} above the diagonal are due to the terms $-\frac{\partial\alpha_{i-1}}{\partial\hat{\theta}}\Gamma(\tau_j-\tau_{j-1})$ in the z_i-equation of the error system. Since $i < j$, these terms cannot be eliminated from the error system. However, by introducing their negative images $-\sigma_{ij}$ below the diagonal in the matrix $A_z(z,t)$, we achieve *skew-symmetry* off its negative diagonal. This makes the homogeneous part of the error system (10.115) exponentially stable and yields nonpositivity of the Lyapunov derivative (10.95).

The closed-loop system which satisfies (10.95) consists of (10.115) along with

$$\dot{\hat{\theta}} = -\Gamma W_\theta(z,t)z \qquad (10.119)$$
$$\dot{\hat{\varrho}} = \gamma\,\mathrm{sgn}(b_m)(\dot{y}_{\mathrm{r}} + \bar{\alpha}_1)e_1^{\mathrm{T}}z \qquad (10.120)$$
$$\dot{\varepsilon} = A_0\varepsilon, \qquad (10.121)$$

summarized here from (10.107), (10.108), and (10.14), respectively.

10.2.2 Stability analysis

For the adaptive scheme developed in the previous subsection, we establish the following result.

Theorem 10.6 (Tuning Functions) *All the signals in the closed-loop adaptive system consisting of the plant (10.1), the control and update laws in Table 10.2, and the filters in Table 10.1 are globally uniformly bounded, and asymptotic tracking is achieved:*

$$\lim_{t \to \infty} [y(t) - y_{\rm r}(t)] = 0. \qquad (10.122)$$

Proof. Due to the piecewise continuity of $y_{\rm r}(t), \ldots, y_{\rm r}^{(n)}(t)$ and the smoothness of the control law, the update laws, and the filters, the solution of the closed-loop adaptive system exists and is unique. Let its maximum interval of existence be $[0, t_f)$.

Let us consider the Lyapunov function

$$V_\rho = \frac{1}{2} z^{\rm T} z + \frac{1}{2} \sum_{k=1}^{\rho} \frac{1}{4 d_k} \varepsilon^{\rm T} P \varepsilon + \frac{1}{2} \tilde{\theta}^{\rm T} \Gamma^{-1} \tilde{\theta} + \frac{|b_m|}{2\gamma} \tilde{\varrho}^2. \qquad (10.123)$$

In (10.95) we established that $V_\rho(t)$ is nonincreasing. Hence, $z, \hat{\theta}, \hat{\varrho}$, and ε are bounded on $[0, t_f)$. Since z_1 and $y_{\rm r}$ are bounded, y is also bounded. Then, from (10.26) we conclude that η is bounded. We have yet to prove the boundedness of λ and x. Our main concern is λ because the boundedness of x will be immediate from the boundedness of ε, η, and λ.

From (10.27) it follows that

$$\lambda_i(s) = \frac{s^{i-1} + k_1 s^{i-2} + \cdots + k_{i-1}}{K(s)} u(s), \qquad i = 1, \ldots, n, \qquad (10.124)$$

where $K(s) = s^n + k_1 s^{n-1} + \cdots + k_0$. By substituting (10.1) we get

$$\lambda_i(s) = \frac{(s^{i-1} + k_1 s^{i-2} + \cdots + k_{i-1}) A(s)}{K(s) B(s)} y(s), \qquad i = 1, \ldots, n. \qquad (10.125)$$

In view of the boundedness of y and Assumption 10.1, the last expression proves that $\lambda_1, \ldots, \lambda_{m+1}$ are bounded. We now return to the coordinate change (10.98) which gives

$$v_{m,i} = z_i + \hat{\varrho} y_{\rm r}^{(i-1)} + \alpha_{i-1}\left(y, \eta, \hat{\theta}, \hat{\varrho}, \bar{\lambda}_{m+i-1}, \bar{y}_{\rm r}^{(i-2)}\right), \qquad i = 2, \ldots, \rho. \qquad (10.126)$$

Let $i = 2$. The boundedness of $\bar{\lambda}_{m+1}$, along with the boundedness of z_2 and $y, \eta, \hat{\theta}, \hat{\varrho}, y_{\rm r}$, and $\dot{y}_{\rm r}$, proves that $v_{m,2}$ is bounded. Then, from (10.60) it follows that λ_{m+2} is bounded. Continuing in the same fashion, (10.126) and (10.60)

10.2 TUNING FUNCTIONS DESIGN

recursively establish that λ is bounded. Finally, in view of (10.33) and the boundedness of η, λ, and ε, we conclude that x is bounded.

We have thus shown that all of the signals of the closed-loop adaptive system are bounded on $[0, t_f)$ by constants depending only on the initial conditions, design gains, and the external signals $y_r(t), \ldots, y_r^{(n)}(t)$, but not on t_f. The independence of the bound of t_f proves that $t_f = \infty$. Hence, all signals are globally uniformly bounded on $[0, \infty)$.

By applying the LaSalle-Yoshizawa theorem (Theorem 2.1) to (10.95), it further follows that $z(t) \to 0$ as $t \to \infty$, which implies that $\lim_{t \to \infty} [y(t) - y_r(t)] = 0$. $\square$

Theorem 10.6 establishes global uniform boundedness of all signals but not global uniform stability of individual trajectories. We now determine an error system which translates the investigated system to the origin. Then we prove that the equilibrium at the origin is globally uniformly stable, and all the error states except the parameter error are regulated to zero.

We start with the subsystem $(z, \varepsilon, \tilde{\theta}, \tilde{\varrho})$ whose $3n+2$ states are encompassed by the Lyapunov function (10.123), and construct additional equations to form a complete error system. We first introduce the equation for the reference signal η^r

$$\dot{\eta}^r = A_0 \eta^r + e_n y_r, \qquad (10.127)$$

so that the error state $\tilde{\eta} = \eta - \eta^r$ is governed by

$$\dot{\tilde{\eta}} = A_0 \tilde{\eta} + e_n z_1. \qquad (10.128)$$

The system $(z, \varepsilon, \tilde{\eta}, \tilde{\theta}, \tilde{\varrho})$ has $4n + 2$ states, while the original $(x, \eta, \lambda, \hat{\theta}, \hat{\varrho})$ system has $4n + m + 2$ states. We find the remaining m error states in the inverse dynamics of (10.3). Let us consider the similarity transformation

$$\begin{bmatrix} x_1 \\ \vdots \\ x_\rho \\ \zeta \end{bmatrix} = \begin{bmatrix} x_1 \\ \vdots \\ x_\rho \\ Tx \end{bmatrix} = \begin{bmatrix} I_\rho & 0_{\rho \times m} \\ & T \end{bmatrix} x, \qquad (10.129)$$

where

$$T = [A_b^\rho e_1, \ldots, A_b e_1, I_m] \qquad (10.130)$$

$$A_b = \begin{bmatrix} -b_{m-1}/b_m & & & \\ \vdots & & I_{m-1} & \\ -b_0/b_m & 0 & \cdots & 0 \end{bmatrix}. \qquad (10.131)$$

With the help of the readily verifiable identities

$$T \begin{bmatrix} 0 \\ b \end{bmatrix} = 0, \qquad TA = A_b T + T A^\rho \begin{bmatrix} 0 \\ b \end{bmatrix} e_1^T, \qquad (10.132)$$

we represent the inverse dynamics of (10.3) as

$$\dot{\zeta} = A_b \zeta + T\left(A^\rho \begin{bmatrix} 0 \\ b \end{bmatrix} - a\right) y$$
$$\triangleq A_b \zeta + b_b y \,. \tag{10.133}$$

The equation for the corresponding reference signal ζ^r is

$$\dot{\zeta}^r = A_b \zeta^r + b_b y_r \,, \tag{10.134}$$

so that the error state $\tilde{\zeta} = \zeta - \zeta^r$ is governed by

$$\dot{\tilde{\zeta}} = A_b \tilde{\zeta} + b_b z_1 \,. \tag{10.135}$$

We have now characterized the error system

$$\dot{z} = A_z(z,t)z + W_\varepsilon(z,t)\varepsilon_2 + W_\theta(z,t)^\mathrm{T}\tilde{\theta} - b_m(\dot{y}_r + \bar{\alpha}_1)e_1\tilde{\varrho} \tag{10.136}$$
$$\dot{\varepsilon} = A_0 \varepsilon \tag{10.137}$$
$$\dot{\tilde{\zeta}} = A_b \tilde{\zeta} + b_b z_1 \tag{10.138}$$
$$\dot{\tilde{\eta}} = A_0 \tilde{\eta} + e_n z_1 \tag{10.139}$$
$$\dot{\tilde{\theta}} = -\Gamma W_\theta(z,t) z \tag{10.140}$$
$$\dot{\tilde{\varrho}} = \gamma \operatorname{sgn}(b_m)(\dot{y}_r + \bar{\alpha}_1)e_1^\mathrm{T} z \tag{10.141}$$

which possesses the desired stability and regulation properties.

Corollary 10.7 *The error system (10.137)–(10.141) has a globally uniformly stable equilibrium at the origin. Moreover, its $4n + m + 2$-dimensional state converges to the $n + m + 2$-dimensional manifold*

$$M = \left\{ z = 0,\ \varepsilon = 0,\ \tilde{\zeta} = 0,\ \tilde{\eta} = 0 \right\}. \tag{10.142}$$

Proof. We augmented the Lyapunov function V_ρ by the terms quadratic in $\tilde{\eta}$ and $\tilde{\zeta}$:

$$\begin{aligned} V &= V_\rho + \frac{1}{k_\eta} \tilde{\eta}^\mathrm{T} P \tilde{\eta} + \frac{1}{k_\zeta} \tilde{\zeta}^\mathrm{T} P_b \tilde{\zeta} \\ &= \frac{1}{2} z^\mathrm{T} z + \sum_{j=1}^{\rho} \frac{1}{4d_j} \varepsilon^\mathrm{T} P \varepsilon + \frac{|b_m|}{2\gamma} \tilde{\varrho}^2 + \frac{1}{2} \tilde{\theta}^\mathrm{T} \Gamma^{-1} \tilde{\theta} + \frac{1}{k_\eta} \tilde{\eta}^\mathrm{T} P \tilde{\eta} + \frac{1}{k_\zeta} \tilde{\zeta}^\mathrm{T} P_b \tilde{\zeta} \,, \end{aligned} \tag{10.143}$$

where P_b satisfies

$$P_b A_b + A_b^\mathrm{T} P_b = -I \,, \qquad P_b = P_b^\mathrm{T} > 0 \,, \tag{10.144}$$

10.2 TUNING FUNCTIONS DESIGN

and k_η, k_ζ are positive constants to be chosen. In view of (10.95), (10.138), (10.139), (10.11), and (10.144), the derivative of V is given by

$$\begin{aligned}
\dot{V} &\leq -\sum_{j=1}^{\rho} c_j z_j^2 - \frac{1}{k_\eta}|\tilde{\eta}|^2 - \frac{1}{k_\zeta}|\tilde{\zeta}|^2 + \frac{2}{k_\eta}\tilde{\eta}^T P_0 e_n z_1 + \frac{2}{k_\zeta}\tilde{\zeta}^T P_b b_b z_1 \\
&= -\sum_{j=1}^{\rho} c_j z_j^2 - \frac{1}{2k_\eta}|\tilde{\eta}|^2 - \frac{1}{2k_\eta}\left|\tilde{\eta} - 2P e_n z_1\right|^2 + \frac{2}{k_\eta}|P e_n|^2 z_1^2 \\
&\quad -\frac{1}{2k_\zeta}|\tilde{\zeta}|^2 - \frac{1}{2k_\zeta}\left|\tilde{\zeta} - 2P_b b_b z_1\right|^2 + \frac{2}{k_\zeta}|P_b b_b|^2 z_1^2 \\
&\leq -\sum_{j=2}^{\rho} c_j z_j^2 - \left(c_1 - \frac{2}{k_\eta}|P e_n|^2 - \frac{2}{k_\zeta}|P_b b_b|^2\right) z_1^2 \\
&\quad -\frac{1}{2k_\eta}|\tilde{\eta}|^2 - \frac{1}{2k_\zeta}|\tilde{\zeta}|^2.
\end{aligned} \qquad (10.145)$$

Thus, if we choose k_η and k_ζ as

$$k_\eta \geq \frac{4|P e_n|^2}{c_1}, \qquad k_\zeta \geq \frac{4|P_b b_b|^2}{c_1}, \qquad (10.146)$$

the derivative of V will be nonpositive,

$$\dot{V} \leq -\frac{1}{2k_\eta}|\tilde{\eta}|^2 - \frac{1}{2k_\zeta}|\tilde{\zeta}|^2, \qquad (10.147)$$

which proves the global uniform stability. By applying the LaSalle-Yoshizawa theorem (Theorem 2.1) to (10.147), it further follows that $\tilde{\eta}(t), \tilde{\zeta}(t) \to 0$ as $t \to \infty$. The convergence of $z(t)$ to zero follows from (10.95), and the convergence of ε follows from the exponential stability of the system (10.137). □

Corollary 10.7 establishes stability properties which are stronger than the properties guaranteed by traditional certainty equivalence adaptive controllers [5, 44, 51, 142, 165]. In addition to the usual global uniform boundedness and asymptotic tracking, the error system possesses a stronger property of global uniform stability. Moreover, all the states converge to zero, except possibly the parameter estimation errors $\tilde{\theta}$ and $\tilde{\varrho}$. As usual in adaptive control, in order to also have the convergence of $\tilde{\theta}$ and $\tilde{\varrho}$ to zero, one needs to assume some form of persistency of excitation (PE). Thus, the achieved stability and convergence properties are as strong as possible without PE.

Corollary 10.7 has not dealt with a correspondence between the original system $(x, \lambda, \eta, \hat{\theta}, \hat{\varrho})$ and the error system $(z, \varepsilon, \tilde{\zeta}, \tilde{\eta}, \tilde{\theta}, \tilde{\varrho})$, which can be done by analyzing the coordinate change

$$(x, \lambda, \eta, \hat{\theta}, \hat{\varrho}) \mapsto (z, \varepsilon, \tilde{\zeta}, \tilde{\eta}, \tilde{\theta}, \tilde{\varrho}). \qquad (10.148)$$

It can be shown that whenever $B(s)$ and $K(s)$ are coprime, this coordinate change is a global C^∞-diffeomorphism for each $t \geq 0$. The proof of a similar claim is given in Section 8.1.3. Although the coprimeness condition cannot be guaranteed by design because the coefficients of $B(s)$ are unknown, it is satisfied with probability one. We stress that the singularity of the coordinate transformation (10.148) is not in contradiction with the boundedness result of Theorem 10.6 (cf. (10.124)–(10.125)) whose proof remains valid when $B(s)$ and $K(s)$ are coprime.

10.2.3 Passivity

An additional novelty of the tuning functions design is a passivity property for any relative degree of the plant transfer function. At each step of the backstepping procedure, the corresponding error system is strictly passive from $\tilde{\theta}$ as the input to the tuning function as the output. We will show this assuming, for simplicity, that the high-frequency gain b_m is known (in which case $\hat{b}_m(t) = 1/\hat{\varrho}(t) \equiv b_m$). We consider the error system at step i, obtained by setting $z_{i+1} = 0$ and $\dot{\hat{\theta}} = \Gamma \tau_i$:

$$\begin{bmatrix} \dot{\bar{z}}_i \\ \dot{\varepsilon} \end{bmatrix} = \begin{bmatrix} A_z^i(z,t) & W_\varepsilon^i(z,t)e_2^T \\ A_0 & 0 \end{bmatrix} \begin{bmatrix} \bar{z}_i \\ \varepsilon \end{bmatrix} + \begin{bmatrix} W_\varepsilon^i(z,t)\omega^T \\ 0 \end{bmatrix} \tilde{\theta}$$

$$\tau_i = \begin{bmatrix} W_\varepsilon^i(z,t)\omega^T \\ 0 \end{bmatrix}^T \begin{bmatrix} \bar{z}_i \\ \varepsilon \end{bmatrix},$$
(10.149)

where

$$A_z^i = \begin{bmatrix} -c_1 - d_1 & b_m & 0 & \cdots & 0 \\ -b_m & -c_2 - d_2\left(\frac{\partial \alpha_1}{\partial y}\right)^2 & 1 + \sigma_{23} & \cdots & \sigma_{2,i} \\ 0 & -1 - \sigma_{23} & \ddots & \ddots & \vdots \\ \vdots & \vdots & \ddots & \ddots & 1 + \sigma_{i-1,i} \\ 0 & -\sigma_{2,i} & \cdots & -1 - \sigma_{i-1,i} & -c_i - d_i\left(\frac{\partial \alpha_{i-1}}{\partial y}\right)^2 \end{bmatrix}$$

$$W_\varepsilon^i = \begin{bmatrix} 1 \\ -\frac{\partial \alpha_1}{\partial y} \\ \vdots \\ -\frac{\partial \alpha_{i-1}}{\partial y} \end{bmatrix}.$$
(10.150)

To show that this system is strictly passive from the input $\tilde{\theta}$ to the output τ_i, by Definition D.2, we need to find a nonnegative storage function $U_i(\bar{z}_i, \varepsilon)$ and a positive definite dissipation rate $\psi_i(\bar{z}_i, \varepsilon)$ such that

$$\int_0^t \tau_i^T \tilde{\theta} d\sigma \geq U_i(t) - U_i(0) + \int_0^t \psi_i(\sigma) d\sigma.$$
(10.151)

10.2 TUNING FUNCTIONS DESIGN

We employ the positive definite and radially unbounded functions

$$U_i(\bar{z}_i, \varepsilon) = \frac{1}{2} \sum_{k=1}^{i} \left(z_k^2 + \frac{1}{d_k} \varepsilon^T P \varepsilon \right) \tag{10.152}$$

$$\psi_i(\bar{z}_i, \varepsilon) = \sum_{k=1}^{i} \left(c_k z_k^2 + \frac{1}{4d_k} \varepsilon^T \varepsilon \right). \tag{10.153}$$

Along the solutions of the system (10.149), the derivative of U_i is

$$\begin{aligned}
\dot{U}_i &= -\sum_{k=1}^{i} c_k z_k^2 - \sum_{k=1}^{i} d_k \left(\frac{\partial \alpha_{k-1}}{\partial y} \right)^2 z_k^2 - \sum_{k=1}^{i} z_k \frac{\partial \alpha_{k-1}}{\partial y} \varepsilon_2 - \sum_{k=1}^{i} \frac{1}{2d_k} \varepsilon^T \varepsilon \\
&\quad + \bar{z}_i^T W_\varepsilon^i \omega^T \tilde{\theta} \\
&\leq -\sum_{k=1}^{i} c_k z_k^2 - \sum_{k=1}^{i} d_k \left(\frac{\partial \alpha_{k-1}}{\partial y} z_k - \frac{1}{2d_k} \varepsilon_2 \right)^2 - \sum_{k=1}^{i} \frac{1}{4d_k} \varepsilon^T \varepsilon + \tau_i^T \tilde{\theta} \\
&\leq -\sum_{k=1}^{i} \left(c_k z_k^2 + \frac{1}{4d_k} \varepsilon^T \varepsilon \right) + \tau_i^T \tilde{\theta} \\
&= -\psi_i + \tau_i^T \tilde{\theta}. \tag{10.154}
\end{aligned}$$

By integrating over $[0, t]$, we arrive at the inequality (10.151), which proves the strict passivity between $\tilde{\theta}$ and τ_i.

The strict passivity results from a special choice of the output (tuning function) τ_i and the stabilizing function α_i. At the end of the recursive procedure, for $i = \rho$, we obtain the error system

$$\begin{bmatrix} \dot{z} \\ \dot{\varepsilon} \end{bmatrix} = \begin{bmatrix} A_z(z,t) & W_\varepsilon(z,t) e_2^T \\ A_0 & 0 \end{bmatrix} \begin{bmatrix} z \\ \varepsilon \end{bmatrix} + \begin{bmatrix} W_\varepsilon(z,t) \omega^T \\ 0 \end{bmatrix} \tilde{\theta}$$

$$\tau_\rho = \begin{bmatrix} W_\varepsilon(z,t) \omega^T \\ 0 \end{bmatrix}^T \begin{bmatrix} z \\ \varepsilon \end{bmatrix}. \tag{10.155}$$

Because of the strict passivity from $\tilde{\theta}$ to τ_ρ, by Theorem D.4, the feedback system will be stable if the feedback path from τ_i to $-\tilde{\theta}$ is passive. The simplest implementable choice is an integrator:

$$-\tilde{\theta} = \frac{\Gamma}{s} \tau_\rho. \tag{10.156}$$

The resulting feedback system, shown in Figure 10.2, is

$$\begin{bmatrix} \dot{z} \\ \dot{\varepsilon} \end{bmatrix} = \begin{bmatrix} A_z(z,t) & W_\varepsilon(z,t) e_2^T \\ A_0 & 0 \end{bmatrix} \begin{bmatrix} z \\ \varepsilon \end{bmatrix} + \begin{bmatrix} W_\varepsilon(z,t) \omega^T \\ 0 \end{bmatrix} \tilde{\theta}$$

$$\dot{\tilde{\theta}} = -\begin{bmatrix} W_\varepsilon(z,t) \omega^T \\ 0 \end{bmatrix}^T \begin{bmatrix} z \\ \varepsilon \end{bmatrix}. \tag{10.157}$$

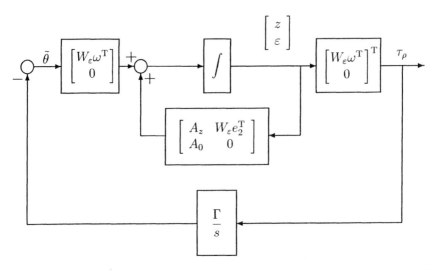

Figure 10.2: The feedback connection of the strictly passive (z, ε)-system with a passive update law.

By Theorem D.4, the system (10.157) has a globally uniformly stable equilibrium $z = 0, \varepsilon = 0$, and $\tilde{\theta} = 0$, and, in addition, $z(t), \varepsilon(t) \to 0$ as $t \to 0$. The boundedness of all the signals is argued as in the proof of Theorem 10.6.

The fact that the passivity property of the error system is achieved for plants of higher relative degree does not contradict the relative degree restriction: From $\tilde{\theta}$ to τ_ρ the relative degree is still one.

10.2.4 Design example

We illustrate the new design procedure on an unstable relative-degree-three plant

$$y(s) = \frac{1}{s^2(s-a)} u(s), \qquad (10.158)$$

where $a = 3$ is considered to be unknown. Simulations of this example are presented in Sections 10.4.2, 10.5, and 10.6.4.

The relative-degree-three design contains the main features of the general design procedure. The control objective is to asymptotically track the output of the reference model

$$y_r(s) = \frac{1}{(s+1)^3} r(s). \qquad (10.159)$$

To derive the adaptive controller resulting from our nonlinear design, the

10.2 Tuning Functions Design

plant (10.158) is first rewritten in the state-space form (10.2):

$$\begin{aligned} \dot{x}_1 &= x_2 + ax_1 \\ \dot{x}_2 &= x_3 \\ \dot{x}_3 &= u \\ y &= x_1 . \end{aligned} \tag{10.160}$$

The filters from Table 10.1 are implemented as

$$\dot{\eta} = A_0\eta + e_3 y, \quad \Xi = -\begin{bmatrix} A_0^2\eta, & A_0\eta, & \eta \end{bmatrix}, \quad \xi = -A_0^3\eta \tag{10.161}$$

$$\dot{\lambda} = A_0\lambda + e_3 u, \quad v = \lambda, \tag{10.162}$$

$$A_0 = \begin{bmatrix} -k_1 & 1 & 0 \\ -k_2 & 0 & 1 \\ -k_3 & 0 & 0 \end{bmatrix} . \tag{10.163}$$

The signals $y_r, \dot{y}_r, \ddot{y}_r$, and $y_r^{(3)}$ are implemented from the reference model (10.159) as

$$\begin{aligned} y_r &= r_1 \\ \dot{y}_r &= r_2 \\ \ddot{y}_r &= r_3 \\ y_r^{(3)} &= -3r_3 - 3r_2 - r_1 + r, \end{aligned} \tag{10.164}$$

by writing the reference model (10.159) as follows:

$$\begin{aligned} \dot{r}_1 &= r_2 \\ \dot{r}_2 &= r_3 \\ \dot{r}_3 &= -3r_3 - 3r_2 - r_1 + r . \end{aligned} \tag{10.165}$$

Since in this example the high-frequency gain is known, in the first step we can directly treat v_2 as a virtual control and do not need the additional parameter estimate $\hat{\varrho}$. The virtual estimate (10.12) is $\hat{x} = -(A_0^3 - aA_0^2)\eta + \lambda$, and by defining

$$\omega = \Xi_{2,1} - y = -e_2^T A_0^2 \eta - y, \tag{10.166}$$

the results of the three steps of our design procedure are:

Step 1.

$$z_1 = y - y_r \tag{10.167}$$
$$\tau_1 = \omega z_1 \tag{10.168}$$
$$\alpha_1 = -(c_1 + d_1)z_1 - \xi_2 - \omega\hat{a} . \tag{10.169}$$

Step 2.

$$z_2 = v_2 - \alpha_1 - \dot{y}_r \tag{10.170}$$

$$\tau_2 = \tau_1 - \frac{\partial \alpha_1}{\partial y}\omega z_2 \tag{10.171}$$

$$\alpha_2 = -z_1 - \left[c_2 + d_2\left(\frac{\partial \alpha_1}{\partial y}\right)^2\right]z_2 + k_2 v_1 + \frac{\partial \alpha_1}{\partial y}(v_2 + \xi_2 + \omega \hat{a})$$

$$+ \frac{\partial \alpha_1}{\partial y_r}\dot{y}_r + \frac{\partial \alpha_1}{\partial \eta}(A_0 \eta + e_3 y) + \frac{\partial \alpha_1}{\partial \hat{a}}\gamma \tau_2. \tag{10.172}$$

Step 3.

$$z_3 = v_3 - \alpha_2 - \ddot{y}_r \tag{10.173}$$

$$\tau_3 = \tau_2 - \frac{\partial \alpha_2}{\partial y}\omega z_3 \tag{10.174}$$

$$\alpha_3 = -z_2 - \left[c_3 + d_3\left(\frac{\partial \alpha_2}{\partial y}\right)^2\right]z_3 + k_3 v_1 + \frac{\partial \alpha_2}{\partial y}(v_2 + \xi_{3,2} + \omega \hat{a})$$

$$+ \frac{\partial \alpha_2}{\partial y_r}\dot{y}_r + \frac{\partial \alpha_2}{\partial \dot{y}_r}\ddot{y}_r + \frac{\partial \alpha_2}{\partial \eta}(A_0 \eta + e_3 y) + \frac{\partial \alpha_2}{\partial \lambda}(A_0 \lambda + e_3 u)$$

$$+ \frac{\partial \alpha_2}{\partial \hat{a}}\gamma \tau_3 - \gamma z_2 \frac{\partial \alpha_1}{\partial \hat{a}}\frac{\partial \alpha_2}{\partial y}\omega. \tag{10.175}$$

The adaptive control law (10.106) and the parameter update law (10.107) are

$$u = \alpha_3 + y_r^{(3)} \tag{10.176}$$

$$\dot{\hat{a}} = \gamma \tau_3. \tag{10.177}$$

The matrix form of the $(z_1, z_2, z_3, \tilde{a})$-system is

$$\begin{bmatrix} \dot{z}_1 \\ \dot{z}_2 \\ \dot{z}_3 \end{bmatrix} = \begin{bmatrix} -c_1 - d_1 & 1 & 0 \\ -1 & -c_2 - d_2\left(\frac{\partial \alpha_1}{\partial y}\right)^2 & 1 + \sigma \\ 0 & -1 - \sigma & -c_3 - d_3\left(\frac{\partial \alpha_2}{\partial y}\right)^2 \end{bmatrix} \begin{bmatrix} z_1 \\ z_2 \\ z_3 \end{bmatrix}$$

$$+ \begin{bmatrix} 1 \\ -\frac{\partial \alpha_1}{\partial y} \\ -\frac{\partial \alpha_2}{\partial y} \end{bmatrix}(\tilde{a}\omega + \varepsilon_2) \tag{10.178}$$

$$\dot{\tilde{a}} = -\gamma\left[1, -\frac{\partial \alpha_1}{\partial y}, -\frac{\partial \alpha_2}{\partial y}\right]\omega z, \tag{10.179}$$

where $\sigma = \gamma \frac{\partial \alpha_1}{\partial \hat{a}}\frac{\partial \alpha_2}{\partial y}\omega$. Note again the skew-symmetry of the off-diagonal entries and the stabilizing role of the diagonal entries.

The block diagram in Figure 10.3 shows that the overall structure of the new adaptive system has the familiar form of the input and output filters

10.3 PROPERTIES OF THE NONADAPTIVE SYSTEM

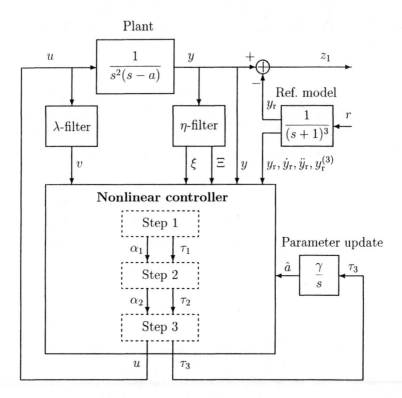

Figure 10.3: The distinguishing feature of the tuning functions design is the "Nonlinear controller" block. In contrast to certainty-equivalence designs which result in linear-like control laws, the tuning functions design produces a control law in which both parameter estimates and filter signals enter nonlinearly.

feeding into an estimator/controller block. The fundamental difference, however, is that this block is now a nonlinear controller. Whereas in traditional schemes this block would be a "certainty-equivalence" linear controller, the tuning functions design produces the control law (10.176) in which both parameter estimates and filter signals enter nonlinearly.

10.3 Properties of the Nonadaptive System

10.3.1 Underlying linear controller

The underlying nonadaptive 'detuned' controller is obtained by setting the adaptation gain Γ to zero, that is, by freezing the parameter estimates at some 'detuned' values.

In traditional certainty equivalence adaptive control, there is a clear separation between the controller and the identifier. The identifier 'tunes' the parameters of the *linear* controller. Although the control law is a nonlinear function of the parameter estimates, it is linear in the output y and the filter states.

In the tuning functions design the control law is a nonlinear function of both the parameter estimates, and the output and filter states. The nonlinearity in the output and filter states comes from including the parameter update law in the control law. The resulting error system is also nonlinear. One would expect that the controller and the error system obtained after setting $\Gamma = 0$ are also nonlinear. We show that this is not so: The underlying nonadaptive controller used in the tuning functions design is *linear*.

To make the analysis simpler, we assume that the high-frequency gain b_m is known,[2] and without loss of generality, let $b_m = 1$. With b_m known, the first components of the unknown parameter vector θ and the filtered regressor ω are dropped from (10.7) and (10.39), respectively:

$$\theta = [b_{m-1}, \ldots, b_0, a_{n-1}, \ldots, a_0]^T \tag{10.180}$$

$$\omega = \left[v_{m-1,2}, \ldots, v_{0,2}, \Xi_{(2)} - ye_1^T\right]^T \tag{10.181}$$

Thus, the closed-loop system (10.115)–(10.121) becomes

$$\dot{z} = A_z(z,t)z + W_\varepsilon(z,t)\left(\omega^T\tilde{\theta} + \varepsilon_2\right) \tag{10.182}$$

$$\dot{\varepsilon} = A_0\varepsilon \tag{10.183}$$

$$\dot{\tilde{\theta}} = -\Gamma\omega W_\varepsilon^T(z,t)z, \tag{10.184}$$

where the system matrix $A_z(z,t)$ is given by

$A_z(z,t) =$

$$\begin{bmatrix} -c_1-d_1 & 1 & 0 & \cdots & \cdots & 0 \\ -1 & -c_2-d_2\left(\frac{\partial\alpha_1}{\partial y}\right)^2 & 1+\sigma_{23} & \sigma_{24} & \cdots & \sigma_{2,\rho} \\ 0 & -1-\sigma_{23} & \ddots & \ddots & \ddots & \vdots \\ \vdots & -\sigma_{24} & \ddots & \ddots & \ddots & \sigma_{\rho-2,\rho} \\ \vdots & \vdots & \ddots & \ddots & \ddots & 1+\sigma_{\rho-1,\rho} \\ 0 & -\sigma_{2,\rho} & \cdots & -\sigma_{\rho-2,\rho} & -1-\sigma_{\rho-1,\rho} & -c_\rho-d_\rho\left(\frac{\partial\alpha_{\rho-1}}{\partial y}\right)^2 \end{bmatrix}$$

$$\tag{10.185}$$

and $W_\varepsilon(z,t)$ and σ_{ij} are defined as

$$W_\varepsilon(z,t) = \begin{bmatrix} 1 \\ -\frac{\partial\alpha_1}{\partial y} \\ \vdots \\ -\frac{\partial\alpha_{\rho-1}}{\partial y} \end{bmatrix} \tag{10.186}$$

---

[2] When b_m is unknown, a slight modification in the design leads to the same conclusions.

10.3 Properties of the Nonadaptive System

$$\sigma_{ij} = \frac{\partial \alpha_{i-1}}{\partial \hat{\theta}} \Gamma \frac{\partial \alpha_{j-1}}{\partial y} \omega. \tag{10.187}$$

Now we set out to determine the detuned nonadaptive controller. With $\Gamma = 0$, the z-system (10.182) becomes

$$\dot{z} = A_z z + b_z \left(\tilde{\theta}^{\mathrm{T}} \omega + \varepsilon_2 \right) \tag{10.188}$$

where $A_z = A_z(z,t)\mid_{\Gamma=0}$ and $b_z = W_\varepsilon(z,t)\mid_{\Gamma=0}$, that is,

$$A_z = \begin{bmatrix} -c_1 - d_1 & 1 & 0 & \cdots & 0 \\ -1 & -c_2 - d_2 \left(\frac{\partial \alpha_1}{\partial y}\right)^2 & 1 & \ddots & \vdots \\ 0 & -1 & \ddots & \ddots & 0 \\ \vdots & \ddots & \ddots & \ddots & 1 \\ 0 & \cdots & 0 & -1 & -c_\rho - d_\rho \left(\frac{\partial \alpha_{\rho-1}}{\partial y}\right)^2 \end{bmatrix} \tag{10.189}$$

$$b_z = \begin{bmatrix} 1 \\ -\frac{\partial \alpha_1}{\partial y} \\ \vdots \\ -\frac{\partial \alpha_{\rho-1}}{\partial y} \end{bmatrix} \tag{10.190}$$

The nonlinear terms σ_{ij} have disappeared from the matrix A_z because they have Γ as a factor (cf. (10.187)). Examining the expressions for $\alpha_1, \ldots, \alpha_{\rho-1}$ in Table 10.2, it can be established that when $\Gamma = 0$, all the derivatives $\frac{\partial \alpha_i}{\partial y}$ are known constants depending on c_i, d_i, and $\hat{\theta}$ only. Hence, the matrix A_z and the vector b_z are constant. Since A_z is Hurwitz (as a sum of a skew-symmetric matrix and a negative diagonal matrix), the transfer function from $\tilde{\theta}^{\mathrm{T}}\omega$ to z_1,

$$\frac{\beta_z(s)}{\alpha_z(s)} = e_1^{\mathrm{T}}(sI - A_z)^{-1} b_z, \tag{10.191}$$

is stable, its relative degree is one, and both $\beta_z(s)$ and $\alpha_z(s)$ are monic polynomials.

Disregarding the initial conditions and the exponentially decaying input ε_2, the linear error system (10.188) is rewritten as

$$z_1 = y - y_r = \frac{\beta_z(s)}{\alpha_z(s)} \tilde{\theta}^{\mathrm{T}} \omega, \tag{10.192}$$

which is one part of the detuned feedback system. However, this system alone does not reveal the underlying linear controller because u does not appear in (10.192). In order to have u appear, let us examine $\tilde{\theta}^{\mathrm{T}}\omega$ along with (10.180),

(10.181), and the filter expressions from Table 10.1:

$$\tilde{\theta}^{\mathrm{T}}\omega = \begin{bmatrix} \tilde{b}_{m-1}, \ldots, \tilde{b}_0, \tilde{a}_{n-1}, \ldots, \tilde{a}_0 \end{bmatrix} \begin{bmatrix} v_{m-1,2} \\ \vdots \\ v_{0,2} \\ \Xi_{(2)}^{\mathrm{T}} - ye_1 \end{bmatrix}$$

$$= \begin{bmatrix} \tilde{b}_{m-1}, \ldots, \tilde{b}_0 \end{bmatrix} \frac{s+k_1}{s^n + k_1 s^{n-1} + \cdots + k_n} \begin{bmatrix} s^{m-1} \\ \vdots \\ s \\ 1 \end{bmatrix} u$$

$$+ \begin{bmatrix} \tilde{a}_{n-1}, \ldots, \tilde{a}_0 \end{bmatrix} \frac{s+k_1}{s^n + k_1 s^{n-1} + \cdots + k_n} \begin{bmatrix} s^{n-1} \\ \vdots \\ s \\ 1 \end{bmatrix} y$$

$$\triangleq \frac{s+k_1}{K(s)} \left(-\tilde{A}(s)y + \tilde{B}(s)u\right), \qquad (10.193)$$

where $\tilde{A}(s) = A(s) - \hat{A}(s)$, $\hat{A}(s) = s^n + \hat{a}_{n-1}s^{n-1} + \cdots + \hat{a}_0$, $\tilde{B}(s) = B(s) - \hat{B}(s)$, and $\hat{B}(s) = s^m + \hat{b}_{m-1}s^{m-1} + \cdots + \hat{b}_0$.

Combining (10.192) and (10.193) we obtain the system

$$y - y_{\mathrm{r}} = \frac{\beta_z(s)}{\alpha_z(s)} \tilde{\theta}^{\mathrm{T}}\omega \qquad (10.194)$$

$$\tilde{\theta}^{\mathrm{T}}\omega = \frac{s+k_1}{K(s)} \left(-\tilde{A}(s)y + \tilde{B}(s)u\right). \qquad (10.195)$$

Substituting (10.195) into (10.194) we get

$$y - y_{\mathrm{r}} = \frac{\beta_z(s)}{\alpha_z(s)} \frac{(s+k_1)}{K(s)} \left(-\tilde{A}(s)y + \tilde{B}(s)u\right) = \frac{\beta_z(s)}{\alpha_z(s)} \frac{(s+k_1)}{K(s)} \left(\hat{A}(s)y - \hat{B}(s)u\right), \qquad (10.196)$$

which shows that the underlying linear controller is defined by

$$(s+k_1)\beta_z(s)\hat{B}(s)u = \left((s+k_1)\beta_z(s)\hat{A}(s) - \alpha_z(s)K(s)\right)y + \alpha_z(s)K(s)y_{\mathrm{r}}. \qquad (10.197)$$

In the block diagram of the underlying linear system in Figure 10.4 the transfer function $\frac{(s+k_1)\beta_z\hat{B}-K}{K}$ is strictly proper because $\beta_z(s)$, $K(s)$, and $\hat{B}(s)$ are monic.

To prove that the transfer function $\frac{\alpha_z K - (s+k_1)\beta_z\hat{A}}{K}$ is proper, we examine the control law $u\left(y, \eta, \lambda, \bar{y}_{\mathrm{r}}^{(\rho)}, \hat{\theta}\right)$. Since u is linear in $y, \eta, \lambda, \bar{y}_{\mathrm{r}}^{(\rho)}$, from the definitions of the filter states η and λ we conclude that

$$u = \bar{q}y + \frac{q_{n-1}(s)}{K(s)}y + \frac{p_{n-1}(s)}{K(s)}u + r_\rho(s)y_{\mathrm{r}}, \qquad (10.198)$$

10.3 Properties of the Nonadaptive System

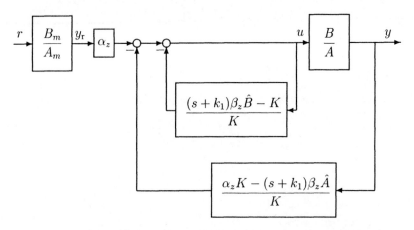

Figure 10.4: The underlying detuned linear controller.

where $\deg q_{n-1}(s) \leq n-1$, $\deg p_{n-1}(s) \leq n-1$, and $\deg r_\rho(s) \leq \rho$. This shows that (10.197) has the following form:

$$\frac{K(s) - p_{n-1}(s)}{K(s)} u = r_\rho(s) y_r + \frac{q_n(s)}{K(s)} y, \quad (10.199)$$

where

$$q_n(s) = q_{n-1}(s) + \bar{q} K(s) = -\alpha_z(s) K(s) + (s + k_1)\beta_z(s) \hat{A}(s), \quad (10.200)$$

and $\deg q_n(s) = \deg[q_{n-1}(s) + \bar{q} K(s)] = n$. Hence, from (10.200) we conclude that $\deg\left[\alpha_z(s) K(s) - (s+k_1)\beta_z(s)\hat{A}(s)\right] = n$. Thus, the transfer function $\frac{\alpha_z K - (s+k_1)\beta_z \hat{A}}{K}$ is proper and the system in Figure 10.4 is implementable.

From the block diagram in Figure 10.4 we obtain the transfer function of the detuned system from the reference output y_r to the output y:

$$y = \frac{\alpha_z K B}{\alpha_z K B + (s+k_1)\beta_z\left(\hat{B}A - \hat{A}B\right)} y_r. \quad (10.201)$$

The matching condition for $y(s) = y_r(s)$ is

$$\frac{\hat{B}(s)}{\hat{A}(s)} = \frac{B(s)}{A(s)}, \quad (10.202)$$

which becomes

$$\hat{A}(s) = A(s), \quad \hat{B}(s) = B(s) \quad (10.203)$$

when $A(s)$ and $B(s)$ are coprime. When the matching is achieved, the characteristic polynomial of (10.201) is $\alpha_z(s) K(s) B(s)$, so that the closed-loop poles consist of:

1. the roots of $\alpha_z(s)$, i.e., the eigenvalues of the error system matrix A_z,

2. the roots of $K(s)$, i.e., the eigenvalues of the filter matrix A_0,

3. the roots of $B(s)$, i.e., the zeros of the plant.

It is of interest to compare these closed-loop poles with those of the traditional MRAC whose characteristic polynomial is $A_m(s)K(s)B(s)$. The design here replaces the reference model denominator polynomial $A_m(s)$ with the error system denominator polynomial $\alpha_z(s)$. In this way, our controller allows the closed-loop poles to be placed independently of the reference model poles.

10.3.2 Parametric robustness and nonadaptive performance

We now show that the design parameters can be chosen to make the detuned closed-loop system stable, that is, the underlying nonadaptive controller has a *parametric robustness* property. In addition, the design parameters can be used for systematic performance improvement. Traditional certainty equivalence adaptive linear controllers do not possess these properties.

We express the regressor ω as the response to the output y:

$$\begin{aligned}
\omega &= \begin{bmatrix} v_{m-1,2}, \ldots, v_{0,2}, \Xi_{(2)} - y e_1^T \end{bmatrix}^T \\
&= \begin{bmatrix} \dfrac{(s+k_1)}{K(s)}[s^{m-1}, \ldots, s, 1]u, \; \dfrac{s+k_1}{K(s)}[s^{n-1}, \ldots, s, 1]y \end{bmatrix}^T \\
&= \dfrac{s+k_1}{K(s)} \begin{bmatrix} [s^{m-1}, \ldots, s, 1]\dfrac{A(s)}{B(s)}, \; [s^{n-1}, \ldots, s, 1] \end{bmatrix}^T y \\
&\triangleq H_\omega(s) y.
\end{aligned} \qquad (10.204)$$

The $(n+m) \times 1$ transfer matrix $H_\omega(s)$ is proper and stable, and its coefficients depend only on the plant parameters θ and the filter coefficients $k_1, \ldots, k_n$.

The expressions (10.192) and (10.204), rewritten here as

$$z_1 = \frac{\beta_z(s)}{\alpha_z(s)} \tilde{\theta}^T \omega \qquad (10.205)$$

$$\tilde{\theta}^T \omega = \tilde{\theta}^T H_\omega(s) z_1 + \tilde{\theta}^T H_\omega(s) y_r, \qquad (10.206)$$

define the feedback systems shown in Figure 10.5.

The closed-loop transfer function of the detuned system in Figure 10.5 is

$$\frac{y(s)}{y_r(s)} = G_c(s) = \frac{1}{1 - \dfrac{\beta_z(s)}{\alpha_z(s)} \tilde{\theta}^T H_\omega(s)}. \qquad (10.207)$$

10.3 Properties of the Nonadaptive System

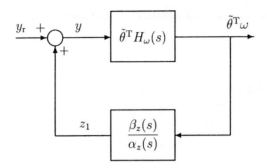

Figure 10.5: The detuned linear feedback system.

The system is tuned when the parameter error is zero, $\tilde{\theta} = 0$, in which case the transfer function is $\frac{y(s)}{y_r(s)} = 1$.

We will establish the parametric robustness property by showing that the controller parameters can be selected to guarantee stability when $\tilde{\theta} \neq 0$. Both blocks in Figure 10.5 are stable, and the transfer function $\frac{\beta_z(s)}{\alpha_z(s)}\tilde{\theta}^T H_\omega(s)$ is strictly proper. Thus, the feedback system is well-defined. Our proof of parametric robustness will proceed using the small gain theorem.

In addition to the parametric robustness property, we will establish transient performance bounds for the state of the closed-loop system expressed in the error coordinates:

$$\dot{z} = A_z z + b_z \tilde{\theta}^T \omega, \qquad z(0) = 0 \qquad (10.208)$$

$$\dot{\zeta} = A_b \zeta + b_b z_1, \qquad \tilde{\eta}(0) = 0 \qquad (10.209)$$

$$\dot{\tilde{\eta}} = A_0 \tilde{\eta} + e_n z_1, \qquad \tilde{\zeta}(0) = 0. \qquad (10.210)$$

The system $\dot{\varepsilon} = A_0 \varepsilon$ is not included because $\varepsilon(0) = 0$. The initial conditions $\tilde{\eta}(0)$ and $\tilde{\zeta}(0)$ can always be set to zero because the initial conditions of the conceptual filters (10.127) and (10.134) are free for us to chose. From (10.209) and (10.210) we define transfer functions

$$W_{\tilde{\eta}}(s) \triangleq (sI - A_0)^{-1} e_n, \qquad (10.211)$$

$$W_{\tilde{\zeta}}(s) \triangleq (sI - A_b)^{-1} b_b, \qquad (10.212)$$

and denote the respective impulse responses by $w_{\tilde{\eta}}(t), w_{\tilde{\zeta}}(t)$.

In the sequel, we will make use of the following equivalent gains:

$$c_0 = \min_{1 \leq i \leq p} c_i, \qquad d_0 = \left(\sum_{i=1}^{p} \frac{1}{d_i}\right)^{-1}. \qquad (10.213)$$

Theorem 10.8 (Stability and $\mathcal{L}_\infty$ Performance) *The nonadaptive system (10.208)–(10.210) is asymptotically stable for*

$$2\sqrt{c_0 d_0} > |\tilde{\theta}| \|h_\omega\|_1. \qquad (10.214)$$

The $\mathcal{L}_\infty$ norms of the states of this system are bounded by

$$\|z\|_\infty \leq \frac{|\tilde{\theta}|\|h_\omega\|_1}{2\sqrt{c_0 d_0} - |\tilde{\theta}|\|h_\omega\|_1}\|y_r\|_\infty \tag{10.215}$$

$$\|\tilde{\eta}\|_\infty \leq \frac{|\tilde{\theta}|\|h_\omega\|_1}{2\sqrt{c_0 d_0} - |\tilde{\theta}|\|h_\omega\|_1}\|y_r\|_\infty \|w_{\tilde{\eta}}\|_1 \tag{10.216}$$

$$\|\tilde{\zeta}\|_\infty \leq \frac{|\tilde{\theta}|\|h_\omega\|_1}{2\sqrt{c_0 d_0} - |\tilde{\theta}|\|h_\omega\|_1}\|y_r\|_\infty \|w_{\tilde{\zeta}}\|_1 . \tag{10.217}$$

Proof. Differentiating $\frac{1}{2}|z|^2$ along the solutions of (10.208), (10.189), and (10.190), we get

$$\begin{aligned}\frac{d}{dt}\left(\frac{1}{2}|z|^2\right) &= -\sum_{k=1}^{\rho} c_k z_k^2 - \sum_{k=1}^{\rho} d_k z_k^2 \left(\frac{\partial \alpha_{k-1}}{\partial y}\right)^2 - \sum_{k=1}^{\rho} z_k \frac{\partial \alpha_{k-1}}{\partial y}\tilde{\theta}^T \omega \\ &= -\sum_{k=1}^{\rho} c_k z_k^2 - \sum_{k=1}^{\rho} d_k \left[\frac{\partial \alpha_{k-1}}{\partial y} z_k + \frac{1}{2d_k}\tilde{\theta}^T\omega\right]^2 + \left(\sum_{k=1}^{\rho}\frac{1}{4d_k}\right)\left(\tilde{\theta}^T\omega\right)^2 \\ &\leq -c_0|z|^2 + \frac{1}{4d_0}\left(\tilde{\theta}^T\omega\right)^2 . \end{aligned} \tag{10.218}$$

By applying Lemma C.5(i), we obtain

$$|z(t)| \leq \frac{1}{2\sqrt{c_0 d_0}}\|\tilde{\theta}^T\omega\|_\infty \leq \frac{1}{2\sqrt{c_0 d_0}}|\tilde{\theta}|\|\omega\|_\infty . \tag{10.219}$$

On the other hand, since $H_\omega(s)$, as defined in (10.204), is a stable proper transfer matrix, then by Theorem B.2(i) we have

$$\begin{aligned}\|\omega\|_\infty &\leq \|h_\omega\|_1 \|y\|_\infty \\ &\leq \|h_\omega\|_1 \|z_1\|_\infty + \|h_\omega\|_1 \|y_r\|_\infty \\ &\leq \|h_\omega\|_1 \|z\|_\infty + \|h_\omega\|_1 \|y_r\|_\infty . \end{aligned} \tag{10.220}$$

To apply the small gain theorem (Theorem B.10) to (10.219)–(10.220), we note that $\frac{1}{2\sqrt{c_0 d_0}}$ in (10.219) can be made arbitrarily small by a choice of c_0 and d_0. Since $\|h_\omega\|_1$ is finite and independent of c_0 and d_0, the loop gain $\frac{1}{2\sqrt{c_0 d_0}}|\tilde{\theta}|\|h_\omega\|_1$ can be made less than one. Thus, by Theorem B.10, the $\mathcal{L}_\infty$-stability of the feedback system in Figure 10.5 is guaranteed. Next, we show that the $\mathcal{L}_\infty$-stability also guarantees the internal asymptotic stability of this system. Consider the closed-loop system (10.201):

$$y = \frac{\alpha_z K B}{\alpha_z K B + (s+k_1)\beta_z\left(\hat{B}A - \hat{A}B\right)} y_r . \tag{10.221}$$

Since $\alpha_z(s)$, $K(s)$, and $B(s)$ are all Hurwitz, if there are cancellations in the transfer function in (10.221), they are all in the open left half-plane, so that

10.3 PROPERTIES OF THE NONADAPTIVE SYSTEM

the denominator in (10.221) is also Hurwitz. We have thus shown that for sufficiently large $c_0 d_0$ the linear system (10.221) is asymptotically stable. The bound (10.215) follows by substituting (10.220) into (10.219):

$$\|z\|_\infty \leq \frac{|\tilde{\theta}|\|h_\omega\|_1}{2\sqrt{c_0 d_0}}\|z\|_\infty + \frac{|\tilde{\theta}|\|h_\omega\|_1}{2\sqrt{c_0 d_0}}\|y_r\|_\infty. \qquad (10.222)$$

Finally, from (10.209)–(10.210) and (10.211)–(10.212), by Theorem B.2(i) we have

$$\|\tilde{\eta}\|_\infty \leq \|w_{\tilde{\eta}}\|_1 \|z_1\|_\infty \qquad (10.223)$$
$$\|\tilde{\zeta}\|_\infty \leq \|w_{\tilde{\zeta}}\|_1 \|z_1\|_\infty, \qquad (10.224)$$

which, in view of (10.215), proves (10.216) and (10.217). □

Theorem 10.8 is a high-gain result: The coefficients c_i and d_i should be high not only to improve performance, but also to satisfy the stability condition (10.214). However, this result is significant because it shows that the controller employed in the tuning functions design can be used both as an adaptive controller and as a parametrically robust high-gain controller. Traditional adaptive linear controllers do not possess this property. They can be used only as adaptive, and parameter adaptation is the only tool they can use to guarantee stability in the presence of unknown parameters.

This nonadaptive controller does not, in general, achieve asymptotic tracking, so we cannot talk about its $\mathcal{L}_2$ performance. However, it is possible to prove that mean-square performance can be made arbitrarily good. In addition to this, the following theorem also provides an alternative stability condition.

Theorem 10.9 (Stability and Mean-Square Performance) *The nonadaptive system (10.208)–(10.210) is asymptotically stable for*

$$2\sqrt{c_0 d_0} > |\tilde{\theta}| \|H_\omega\|_\infty. \qquad (10.225)$$

The mean-square values of $z, \tilde{\eta}, \tilde{\zeta}$ are bounded by

$$\left(\frac{1}{t}\int_0^t |z(\tau)|^2 d\tau\right)^{\frac{1}{2}} \leq \frac{|\tilde{\theta}| \|H_\omega\|_\infty}{2\sqrt{c_0 d_0} - |\tilde{\theta}| \|H_\omega\|_\infty} \|y_r\|_\infty \qquad (10.226)$$

$$\left(\frac{1}{t}\int_0^t |\tilde{\eta}(\tau)|^2 d\tau\right)^{\frac{1}{2}} \leq \frac{|\tilde{\theta}| \|H_\omega\|_\infty}{2\sqrt{c_0 d_0} - |\tilde{\theta}| \|H_\omega\|_\infty} \|y_r\|_\infty \|W_{\tilde{\eta}}\|_\infty \qquad (10.227)$$

$$\left(\frac{1}{t}\int_0^t |\tilde{\zeta}(\tau)|^2 d\tau\right)^{\frac{1}{2}} \leq \frac{|\tilde{\theta}| \|H_\omega\|_\infty}{2\sqrt{c_0 d_0} - |\tilde{\theta}| \|H_\omega\|_\infty} \|y_r\|_\infty \|W_{\tilde{\zeta}}\|_\infty. \qquad (10.228)$$

Proof. By applying Lemma B.5 to (10.218), we get

$$|z(t)|^2 \leq \frac{1}{2d_0} \int_0^t e^{-2c_0(t-\tau)} \left(\tilde{\theta}^T \omega(\tau)\right)^2 d\tau, \qquad (10.229)$$

which, upon integration over $[0,t]$ becomes

$$\int_0^t |z(\tau)|^2 d\tau \le \frac{1}{2d_0} \int_0^t \left[\int_0^\tau e^{-2c_0(\tau-s)} \left(\tilde{\theta}^{\mathrm{T}} \omega(s)\right)^2 ds \right] d\tau. \quad (10.230)$$

Changing the order of integration, (10.230) becomes

$$\int_0^t |z(\tau)|^2 d\tau \le \frac{1}{2d_0} \int_0^t e^{2c_0 s} \left(\tilde{\theta}^{\mathrm{T}} \omega(s)\right)^2 \left(\int_s^t e^{-2c_0 \tau} d\tau\right) ds$$

$$\le \frac{1}{2d_0} \int_0^t e^{2c_0 s} \left(\tilde{\theta}^{\mathrm{T}} \omega(s)\right)^2 \frac{1}{2c_0} e^{-2c_0 s} ds \quad (10.231)$$

because

$$\int_s^t e^{-2c_0 \tau} d\tau = \frac{1}{2c_0} \left(e^{-2c_0 s} - e^{-2c_0 t}\right) \le \frac{1}{2c_0} e^{-2c_0 s}. \quad (10.232)$$

Now, the cancellation $e^{2c_0 s} e^{-2c_0 s} = 1$ in (10.231) yields

$$\int_0^t |z(\tau)|^2 d\tau \le \frac{|\tilde{\theta}|^2}{4c_0 d_0} \int_0^t |\omega(\tau)|^2 d\tau. \quad (10.233)$$

On the other hand, since $H_\omega(s)$ is stable and proper, then by Theorem B.2(ii) the truncated $\mathcal{L}_2$ norms of ω and z are related as

$$\|\omega\|_{2,t} \le \|H_\omega\|_\infty \|y\|_{2,t} \le \|H_\omega\|_\infty \left(\|z\|_{2,t} + \|y_r\|_{2,t}\right). \quad (10.234)$$

From (10.233) and (10.234), by the small gain theorem (Theorem B.10), $\mathcal{L}_2$ stability is guaranteed for $2\sqrt{c_0 d_0} > |\tilde{\theta}| \|H_\omega\|_\infty$, and asymptotic stability follows as in the proof of Theorem 10.8. Substituting (10.234) into (10.233) and solving for $\|z\|_{2,t} = \left(\int_0^t |z(\tau)|^2 d\tau\right)^{\frac{1}{2}}$, we get

$$\left(\int_0^t |z(\tau)|^2 d\tau\right)^{\frac{1}{2}} \le \frac{|\tilde{\theta}| \|H_\omega\|_\infty}{2\sqrt{c_0 d_0} - |\tilde{\theta}| \|H_\omega\|_\infty} \|y_r\|_{2,t}. \quad (10.235)$$

and (10.226) follows because $\|y_r\|_{2,t}^2 = \int_0^t |y_r(\tau)|^2 d\tau \le \|y_r\|_\infty^2 t$. From (10.209)–(10.210) and (10.211)–(10.212), by Theorem B.2(ii) we have

$$\|\tilde{\eta}_t\|_2 \le \|W_{\tilde{\eta}}\|_\infty \|(z_1)_t\|_2 \quad (10.236)$$

$$\|\tilde{\zeta}_t\|_2 \le \|W_{\tilde{\zeta}}\|_\infty \|(z_1)_t\|_2, \quad (10.237)$$

which, in view of (10.226), proves inequalities (10.227) and (10.228). □

Theorems 10.8 and 10.9 provide two different stability conditions, (10.188) and (10.225), of which (10.225) is directly computable [16] and less conservative because $\|H_\omega\|_\infty \le \|h_\omega\|_1$ (see Theorem B.2(iii)).

Another way of expressing the performance properties is by comparing the detuned closed-loop transfer function (10.207) with the desired transfer function $\frac{y(s)}{y_r(s)} = 1$.

10.4 TRANSIENT PERFORMANCE WITH TUNING FUNCTIONS

Theorem 10.10 (Frequency Domain Performance) *In the nonadaptive system (10.207), the design parameters c_i and d_i, $1 \leq i \leq \rho$, can be chosen to satisfy the following tracking performance specification for any $\delta_c > 0$:*

$$|G_c(j\omega) - 1| < \delta_c, \qquad \forall \omega \in \mathbb{R}. \tag{10.238}$$

Proof. By setting $t = \infty$ in (10.233) we see that the induced $\mathcal{L}_2$ norm of $\frac{\beta_z(s)}{\alpha_z(s)}$ is no greater than $\frac{1}{2\sqrt{c_0 d_0}}$. This, in turn, means that $\left\|\frac{\beta_z}{\alpha_z}\right\|_\infty \leq \frac{1}{2\sqrt{c_0 d_0}}$, which implies

$$\left|\frac{\beta_z(j\omega)}{\alpha_z(j\omega)}\right| \leq \frac{1}{2\sqrt{c_0 d_0}}, \qquad \forall \omega \in \mathbb{R}. \tag{10.239}$$

From (10.207) we now have

$$|G_c(j\omega) - 1| = \left|\frac{\frac{\beta_z(j\omega)}{\alpha_z(j\omega)} \tilde{\theta}^T H_\omega(j\omega)}{1 - \frac{\beta_z(j\omega)}{\alpha_z(j\omega)} \tilde{\theta}^T H_\omega(j\omega)}\right| \leq \frac{\frac{1}{2\sqrt{c_0 d_0}} |\tilde{\theta}| \|H_\omega\|_\infty}{1 - \frac{1}{2\sqrt{c_0 d_0}} |\tilde{\theta}| \|H_\omega\|_\infty} \tag{10.240}$$

which is less than any given δ_c provided that $c_0 d_0$ is sufficiently large:

$$2\sqrt{c_0 d_0} > \left(1 + \frac{1}{\delta_c}\right) |\tilde{\theta}| \|H_\omega\|_\infty. \tag{10.241}$$

□

As expected, the tracking condition (10.241) is more stringent than the corresponding stability condition (10.225). The required value of $2\sqrt{c_0 d_0}$ is increased by the factor $1 + \frac{1}{\delta_c}$ and tends to infinity as $\delta_c \to 0$. In this sense the underlying linear controller is a "high-gain" controller which achieves a good tracking performance at the expense of an increase of the bandwidth.

10.4 Transient Performance with Tuning Functions

In the absence of disturbances and unmodeled dynamics, the tracking error of most adaptive control schemes converges to zero, that is, they meet the stated asymptotic performance objective. In applications, however, the system's transient performance is often more important.

Analytical quantification and improvement of transient performance have been long-standing open problems in adaptive control. For traditional adaptive linear controllers there are virtually no results which allow the designer to a priori compute bounds on the transient behavior, let alone to meet a given transient performance specification.

Transient responses of traditional adaptive schemes suffer from large initial swings [202] because the certainty equivalence controller does not take into account the parameter estimation transients. In addition, the identifier, which is designed separately from the controller, is driven by an 'estimation error' signal that is unrelated to the control objective. We will show that the transient performance can be improved by letting the controller and the identifier exchange information during the operation of the adaptive system.

The tuning functions controller takes into account the effect of the parameter estimation transients by incorporating the parameter update law $\dot{\hat{\theta}} = \Gamma\tau_\rho$, and its fast unnormalized update law is driven by the tracking error.

10.4.1 Transient performance of the adaptive system

We now derive computable bounds on both $\mathcal{L}_2$ and $\mathcal{L}_\infty$ norms of the states $z, \tilde{\eta}$, and $\tilde{\zeta}$ of the adaptive system, and we show how they can be made arbitrarily small by a choice of the design parameters c_i, d_i, and Γ.

Theorem 10.11 ($\mathcal{L}_2$ **performance**) *The $\mathcal{L}_2$ norms of the states $z, \tilde{\eta}, \tilde{\zeta}$ of the adaptive system (10.182)–(10.184), (10.209), (10.210), are bounded by*

$$\|z\|_2 \leq \frac{1}{\sqrt{c_0}}\sqrt{V_\rho(0)} \tag{10.242}$$

$$\|\tilde{\eta}\|_2 \leq \frac{1}{\sqrt{c_0}}\sqrt{V_\rho(0)}\|W_{\tilde{\eta}}\|_\infty \tag{10.243}$$

$$\|\tilde{\zeta}\|_2 \leq \frac{1}{\sqrt{c_0}}\sqrt{V_\rho(0)}\|W_{\tilde{\zeta}}\|_\infty, \tag{10.244}$$

where $\|W_{\tilde{\eta}}\|_\infty$ and $\|W_{\tilde{\zeta}}\|_\infty$ are independent of c_0, d_0, and Γ.

Proof. As shown in (10.95), the derivative of V_ρ along the solutions of (10.182)–(10.184) is

$$\dot{V}_\rho \leq -c_0|z|^2. \tag{10.245}$$

Since V_ρ is nonincreasing, we have

$$\|z\|_2^2 = \int_0^\infty |z(\tau)|^2 d\tau \leq \frac{1}{c_0}(V_\rho(0) - V_\rho(\infty)) \leq \frac{1}{c_0}V_\rho(0), \tag{10.246}$$

which implies (10.242). From (10.209) and (10.211), by Theorem B.2(ii), we get

$$\|\tilde{\eta}\|_2 \leq \|W_{\tilde{\eta}}\|_\infty \|z_1\|_2 \leq \frac{1}{\sqrt{c_0}}\sqrt{V_\rho(0)}\|W_{\tilde{\eta}}\|_\infty \tag{10.247}$$

and, from (10.210) and (10.212) we get

$$\|\tilde{\zeta}\|_2 \leq \|W_{\tilde{\zeta}}\|_\infty \|z_1\|_2 \leq \frac{1}{\sqrt{c_0}}\sqrt{V_\rho(0)}\|W_{\tilde{\zeta}}\|_\infty. \tag{10.248}$$

□

10.4 Transient Performance with Tuning Functions

The initial value of the Lyapunov function is

$$V_\rho(0) = \frac{1}{2}|z(0)|^2 + \frac{1}{4d_0}|\varepsilon(0)|_P^2 + \frac{1}{2}|\tilde{\theta}(0)|_{\Gamma^{-1}}^2. \qquad (10.249)$$

From (10.242) and (10.249) it may appear that by increasing c_0 we reduce the bound on $\|z\|_2$. This would be so only if $\varepsilon(0)$, $\tilde{\theta}(0)$, and $z(0)$ were independent of c_0. While $\varepsilon(0)$, $\tilde{\theta}(0)$, and $z_1(0) = y(0) - y_r(0)$ are clearly independent of c_i, d_i, and Γ, the initial values $z_2(0), \ldots, z_\rho(0)$ depend on c_i, d_i, and Γ. Fortunately, we can set $z(0)$ to zero by appropriately *initializing the reference trajectory*. Following (10.97) and (10.98), $z(0)$ is set to zero by selecting

$$y_r(0) = y(0) \qquad (10.250)$$

$$y_r^{(i)}(0) = \frac{1}{\hat{\varrho}(0)}\left[v_{m,i+1}(0) - \alpha_i\left(y(0), \eta(0), \hat{\theta}(0), \hat{\varrho}(0), \bar{\lambda}_{m+i}(0), \bar{y}_r^{(i-1)}(0)\right)\right]$$
$$i = 1, \ldots, \rho - 1. \qquad (10.251)$$

Since $b_m \neq 0$, it is reasonable to choose $\hat{b}_m(0) \neq 0$. Then the choice $\hat{\varrho}(0) = 1/\hat{b}_m(0)$ makes (10.251) well-defined. A detailed presentation of the initialization procedure, given in Section 4.3.2 for the case of state feedback, explains how to modify a prespecified reference trajectory, so that the implemented trajectory is properly initialized.

Thus, by setting $z(0) = 0$, we make

$$V_\rho(0) = \frac{1}{4d_0}|\varepsilon(0)|_P^2 + \frac{1}{2}|\tilde{\theta}(0)|_{\Gamma^{-1}}^2 \qquad (10.252)$$

a decreasing function of d_0 and Γ, independent of c_0. This means that the bounds resulting from (10.242)–(10.244) and (10.252) for $\Gamma = \gamma I$,

$$\|z\|_2 \leq \frac{1}{\sqrt{2c_0}}\left(\frac{1}{\gamma}|\tilde{\theta}(0)|^2 + \frac{1}{4d_0}|\varepsilon(0)|_P^2\right)^{1/2} \qquad (10.253)$$

$$\|\tilde{\eta}\|_2 \leq \frac{1}{\sqrt{2c_0}}\left(\frac{1}{\gamma}|\tilde{\theta}(0)|^2 + \frac{1}{4d_0}|\varepsilon(0)|_P^2\right)^{1/2}\|W_{\tilde{\eta}}\|_\infty \qquad (10.254)$$

$$\|\tilde{\zeta}\|_2 \leq \frac{1}{\sqrt{2c_0}}\left(\frac{1}{\gamma}|\tilde{\theta}(0)|^2 + \frac{1}{4d_0}|\varepsilon(0)|_P^2\right)^{1/2}\|W_{\tilde{\zeta}}\|_\infty, \qquad (10.255)$$

can be systematically reduced either by increasing c_0 or by simultaneously increasing d_0 and γ. The possibility to improve performance with the adaptation gain γ is particularly clear in the case $\varepsilon(0) = 0$, when the $\mathcal{L}_2$ bounds of Theorem 10.11 become

$$\|z\|_2 \leq \frac{1}{\sqrt{2c_0\gamma}}|\tilde{\theta}(0)| \qquad (10.256)$$

$$\|\tilde{\eta}\|_2 \leq \frac{1}{\sqrt{2c_0\gamma}}|\tilde{\theta}(0)|\|W_{\tilde{\eta}}\|_\infty \qquad (10.257)$$

$$\|\tilde{\zeta}\|_2 \leq \frac{1}{\sqrt{2c_0\gamma}}|\tilde{\theta}(0)|\|W_{\tilde{\zeta}}\|_\infty. \qquad (10.258)$$

These expressions provide insight, although, of course, $\varepsilon(0)$ cannot be set to zero by design because it depends on the initial value of the unmeasured state $x(0)$.

Another advantage of the derived bounds is that they are computable. The bound for $\|z\|_2$ is explicit, while the bound for $\|\tilde{\eta}\|_2$ involves $\|W_{\tilde{\eta}}\|_\infty$ which is known from (10.211). Only the factor $\|W_{\tilde{\zeta}}\|_\infty$ in the bound for the zero dynamics $\|\tilde{\zeta}\|_2$ depends on the unknown parameters $b_0, \ldots, b_{m-1}$. When these parameters belong to known intervals, $\|W_{\tilde{\zeta}}\|_\infty$ can be computed using [16].

For a further characterization of the achieved performance, we proceed to derive $\mathcal{L}_\infty$ norm bounds for the states of the adaptive system (10.182)–(10.184), (10.209), (10.210). These bounds are also useful for a comparison with nonadaptive systems.

We first give simple bounds on $\|z\|_\infty$ and $\|\tilde{\theta}\|_\infty$:

$$\|z\|_\infty \leq \sqrt{2V_\rho(0)} \qquad (10.259)$$

$$\|\tilde{\theta}\|_\infty \leq \sqrt{\bar{\lambda}(\Gamma)}\sqrt{2V_\rho(0)}. \qquad (10.260)$$

Since $\dot{V}_\rho \leq 0$, the bound (10.259) follows immediately from

$$2V_\rho(t) = |z(t)|^2 + \frac{1}{2d_0}|\varepsilon(t)|_P^2 + |\tilde{\theta}(t)|_{\Gamma^{-1}}^2 \leq 2V_\rho(0), \qquad (10.261)$$

and the bound (10.260) is obtained by noting that

$$\frac{1}{\bar{\lambda}(\Gamma)}|\tilde{\theta}|^2 \leq |\tilde{\theta}|_{\Gamma^{-1}}^2 \leq 2V_\rho(0). \qquad (10.262)$$

For $\Gamma = \gamma I$, it further follows from (10.260)–(10.261) that

$$\|\tilde{\theta}\|_\infty \leq \sqrt{\gamma}|z(0)| + \sqrt{\frac{\gamma}{2d_0}}|\varepsilon(0)|_P + |\tilde{\theta}(0)|. \qquad (10.263)$$

In this way, $\|\tilde{\theta}\|_\infty$ is explicitly related to initial conditions and design parameters.

Theorem 10.12 ($\mathcal{L}_\infty$ Performance) *The states $z, \tilde{\eta}$, and $\tilde{\zeta}$ of the adaptive system (10.182)–(10.184), (10.209), (10.210) are bounded by*

$$|z(t)| \leq \frac{1}{\sqrt{c_0 d_0}}M + |z(0)|e^{-c_0 t} \qquad (10.264)$$

$$|\tilde{\eta}(t)| \leq \left(\frac{1}{\sqrt{c_0 d_0}}M + |z(0)|\right)\|w_{\tilde{\eta}}\|_1 \qquad (10.265)$$

$$|\tilde{\zeta}(t)| \leq \left(\frac{1}{\sqrt{c_0 d_0}}M + |z(0)|\right)\|w_{\tilde{\zeta}}\|_1, \qquad (10.266)$$

10.4 Transient Performance with Tuning Functions

where

$$M \triangleq \frac{1}{2}\left\{\sqrt{\bar{\lambda}(\Gamma)}\sqrt{2V_\rho(0)}\left[\|h_\omega\|_1\left(\sqrt{2V_\rho(0)}+\|y_r\|_\infty\right)+\kappa_\omega\right]+\frac{1}{\sqrt{\underline{\lambda}(P)}}|\varepsilon(0)|_P\right\}, \tag{10.267}$$

and $\|w_{\tilde{\eta}}\|_1, \|w_{\tilde{\zeta}}\|_1, \|h_\omega\|_1$, and κ_ω are independent of c_0, d_0, and Γ.

Proof. Differentiating $\frac{1}{2}|z|^2$ along the solutions of (10.182) and (10.185), we get

$$\frac{d}{dt}\left(\frac{1}{2}|z|^2\right) = -\sum_{k=1}^{\rho}c_k z_k^2 - \sum_{k=1}^{\rho}d_k z_k^2 \left(\frac{\partial \alpha_{k-1}}{\partial y}\right)^2 - \sum_{k=1}^{\rho}z_k\frac{\partial \alpha_{k-1}}{\partial y}\left(\tilde{\theta}^T\omega+\varepsilon_2\right)$$

$$\leq -c_0|z|^2 + \frac{1}{4d_0}\left(\tilde{\theta}^T\omega+\varepsilon_2\right)^2. \tag{10.268}$$

By applying Lemma C.5(i), we obtain

$$|z(t)|^2 \leq |z(0)|^2 e^{-2c_0 t} + \frac{1}{4c_0 d_0}\|\tilde{\theta}^T\omega+\varepsilon_2\|_\infty^2. \tag{10.269}$$

From (10.14) and (10.11) we have $\frac{d}{dt}|\varepsilon|_P^2 \leq -|\varepsilon|^2$, which gives

$$\|\varepsilon_2\|_\infty^2 \leq \frac{1}{\underline{\lambda}(P)}|\varepsilon(0)|_P^2. \tag{10.270}$$

With (10.269) and (10.270) we obtain

$$|z(t)| \leq \frac{1}{2\sqrt{c_0 d_0}}\left(\|\tilde{\theta}\|_\infty\|\omega\|_\infty + \frac{1}{\sqrt{\underline{\lambda}(P)}}|\varepsilon(0)|_P\right) + |z(0)|e^{-c_0 t}. \tag{10.271}$$

It was shown in (10.204) that

$$\omega = H_\omega(s)y + \omega_0(t), \tag{10.272}$$

where $|\omega_0(t)| \leq \kappa_\omega e^{-\sigma t}$ is the response due to the initial conditions of $\eta(0)$ and $\lambda(0)$, and κ_ω and σ depend only on the plant and filter parameters and not on c_0, d_0, and Γ. Now, using $y = z_1 + y_r$ and (10.259), we get

$$\|\omega\|_\infty \leq \|h_\omega\|_1(\|z_1\|_\infty + \|y_r\|_\infty) + \kappa_\omega e^{-\sigma t} \leq \|h_\omega\|_1\left(\sqrt{2V_\rho(0)}+\|y_r\|_\infty\right)+\kappa_\omega. \tag{10.273}$$

Substituting (10.273) into (10.271) and using (10.260) we obtain

$$|z(t)| \leq \frac{1}{\sqrt{c_0 d_0}}\frac{1}{2}\left\{\sqrt{\bar{\lambda}(\Gamma)}\sqrt{2V_\rho(0)}\left[\|h_\omega\|_1\left(\sqrt{2V_\rho(0)}+\|y_r\|_\infty\right)+\kappa_\omega\right]\right.$$

$$\left.+\frac{1}{\sqrt{\underline{\lambda}(P)}}|\varepsilon(0)|_P\right\} + |z(0)|e^{-c_0 t}$$

$$= \frac{1}{\sqrt{c_0 d_0}}M + |z(0)|e^{-c_0 t} \tag{10.274}$$

From (10.209) and (10.211), by Theorem B.2(i), we get

$$|\tilde{\eta}(t)| \leq \|w_{\tilde{\eta}}\|_1 \|z_1\|_\infty \leq \left(\frac{1}{\sqrt{c_0 d_0}} M + |z(0)|\right) \|w_{\tilde{\eta}}\|_1, \qquad (10.275)$$

and from (10.210) and (10.212) we get

$$|\tilde{\zeta}(t)| \leq \|w_{\tilde{\zeta}}\|_1 \|z_1\|_\infty \leq \left(\frac{1}{\sqrt{c_0 d_0}} M + |z(0)|\right) \|w_{\tilde{\zeta}}\|_1. \qquad (10.276)$$

$\square$

With the initialization $z(0) = 0$, the expression (10.252) for $V_\rho(0)$ and (10.267) show that M is a decreasing function of d_0 independent of c_0.

With the bounds on $z, \tilde{\eta}$, and $\tilde{\zeta}$ that can be reduced by design parameters, and fixed bounds on ε and $\tilde{\theta}$, the entire state of the adaptive system is guaranteed to have a good $\mathcal{L}_\infty$ performance.

Since M in (10.267) depends on $\|h_\omega\|_1$, the bounds (10.264)–(10.266) require computation of $\|h_\omega\|_1$, $\|w_{\tilde{\eta}}\|_1$, and $\|w_{\tilde{\zeta}}\|_1$. Although $\|h_\omega\|_1$ and $\|w_{\tilde{\zeta}}\|_1$ depend on uncertain parameters, we can employ the procedure of [16] to compute their $\mathcal{H}_\infty$ norms and then apply the well-known inequality $\|g\|_1 \leq (2n + 1)\|G\|_\infty$, where $G(s)$ is a stable transfer function, n is its McMillan degree, and $g(t)$ is its impulse response (see Theorem B.2(iii)).

A special form of the above $\mathcal{L}_\infty$ bounds is more revealing.

Corollary 10.13 *In the case $z(0) = 0$, $\varepsilon(0) = \eta(0) = \lambda(0) = 0$, and $\Gamma = \gamma I$, the $\mathcal{L}_\infty$ bounds of Theorem 10.12 become*

$$\|z\|_\infty \leq \frac{|\tilde{\theta}(0)| \|h_\omega\|_1}{2\sqrt{c_0 d_0}} \left(\|y_r\|_\infty + \frac{1}{\sqrt{\gamma}} |\tilde{\theta}(0)|\right) \qquad (10.277)$$

$$\|\tilde{\eta}\|_\infty \leq \frac{|\tilde{\theta}(0)| \|h_\omega\|_1}{2\sqrt{c_0 d_0}} \left(\|y_r\|_\infty + \frac{1}{\sqrt{\gamma}} |\tilde{\theta}(0)|\right) \|w_{\tilde{\eta}}\|_1 \qquad (10.278)$$

$$\|\tilde{\zeta}\|_\infty \leq \frac{|\tilde{\theta}(0)| \|h_\omega\|_1}{2\sqrt{c_0 d_0}} \left(\|y_r\|_\infty + \frac{1}{\sqrt{\gamma}} |\tilde{\theta}(0)|\right) \|w_{\tilde{\zeta}}\|_1. \qquad (10.279)$$

The assumption $z(0) = 0$, $\varepsilon(0) = \eta(0) = \lambda(0) = 0$ is satisfied in the particular case where $x(0) = \eta(0) = \lambda(0) = 0$ and the trajectory initialization is performed. In this case the system is driven only by the reference trajectory.

The form of bounds in Corollary 10.13 clarifies the dependence of the $\mathcal{L}_\infty$ performance on the parameter uncertainty $|\tilde{\theta}(0)|$ and the design parameters c_0, d_0, and γ. Any increase in those parameters results in an improvement of the $\mathcal{L}_\infty$ performance. It is of interest to observe that d_0, present in the $\mathcal{L}_\infty$ bounds (10.277)–(10.279), is absent from the $\mathcal{L}_2$ bounds (10.256)–(10.258).

10.4 TRANSIENT PERFORMANCE WITH TUNING FUNCTIONS

Remark 10.14 The bound (10.260) can be useful for ensuring stability of the system in case of an accidental interruption of adaptation. By substituting (10.252) into (10.260) we get

$$\|\tilde{\theta}\|_\infty \leq \left(|\tilde{\theta}(0)|^2 + \frac{\gamma}{2d_0}|\varepsilon(0)|_P^2\right)^{1/2} \tag{10.280}$$

Suppose that the design parameters c_0 and d_0 are chosen so that

$$2\sqrt{c_0 d_0} > |\tilde{\theta}(0)|\|H_\omega\|_\infty, \tag{10.281}$$

which, by Theorem 10.9, means that without adaptation the resulting closed-loop linear system would be stable. Let us now suppose that this system is running *with* adaptation until time T, when the adaptation is disconnected. The bound (10.280) indicates that the parameter error $|\tilde{\theta}(T)|$ may be larger than the initial value $|\tilde{\theta}(0)|$. Therefore, $|\tilde{\theta}(T)|$ may violate the stability condition (10.281), and the resulting linear system may be unstable. It may therefore appear that the parametric robustness results of Theorems 10.8 and 10.9 hold only if adaptation is disconnected at $T = 0$. Fortunately, the bound (10.280) shows that adaptation can be disconnected at any time $T \geq 0$ without destroying the system stability provided that c_0 and d_0 are chosen so that

$$2\sqrt{c_0 d_0} > \left(|\tilde{\theta}(0)|^2 + \frac{\gamma}{2d_0}|\varepsilon(0)|_P^2\right)^{1/2} \|H_\omega\|_\infty. \tag{10.282}$$

Suppose, for simplicity, that the filter initial conditions are $\eta(0) = \lambda(0) = 0$, which, in view of (10.33), means that $\varepsilon(0) = x(0)$. It is reasonable to assume that, even though x is not measured, we know a bound on its initial condition $x(0)$. This bound, along with a bound on the initial parameter error, can be used in (10.282) to select c_0 and d_0 which guarantee stability. In other words, with sufficiently high c_0 and d_0, each 'frozen controller' is stabilizing. ◇

10.4.2 Performance improvement due to adaptation

In the literature, robust and adaptive designs coexist as two separate approaches with little quantitative evidence for performance comparison. One would expect adaptive controllers to perform better because they are using additional knowledge about the uncertainty acquired on-line, while the robust controller design is based only on a priori knowledge. However, for traditional adaptive controllers this is true only asymptotically because they go through big initial transient swings.

With the performance bounds derived in Sections 10.3.2 and 10.4.1, we have assembled a data base for a quantitative performance comparison of the nonlinear adaptive system and its linear nonadaptive counterpart.

Before we present a comparison of transient performance, we review the basic differences in boundedness properties and asymptotic performance between the adaptive and the nonadaptive controllers:

- *Boundedness* of the adaptive system is guaranteed to be global for any positive values of the design parameters c_0, d_0, and Γ. No a priori information is required about the parameter uncertainty. In contrast, the linear controller guarantees boundedness only if a bound on the parameter uncertainty is known so that the value of $c_0 d_0$ can be set large enough to satisfy the stability condition (10.225).

- *Asymptotic tracking* is achieved by the adaptive controller for any parameter uncertainty and any positive c_0, d_0, and Γ. The tracking error of the linear system can be reduced but, in general, does not converge to zero. To make the tracking error small, the value of $c_0 d_0$ is required to be high. It can be shown that the increase of $c_0 d_0$ increases the bandwidth, which may be undesirable.

We show now that the *transient performance* of the adaptive system can be improved over that of the nonadaptive system without an increase of $c_0 d_0$. We use the superscripts A and N to denote the quantities in the adaptive and in the nonadaptive system, respectively. For convenience, we repeat the nonadaptive bound (10.215):

$$\|z^N\|_\infty \leq \frac{|\tilde{\theta}^N| \| h_\omega \|_1}{2\sqrt{c_0^N d_0^N} - |\tilde{\theta}^N| \| h_\omega \|_1} \|y_r\|_\infty \triangleq B_N . \tag{10.283}$$

Under the conditions of Corollary 10.13, because $\|\tilde{\theta}^A\|_\infty \leq |\tilde{\theta}^A(0)|$, the same type of bound holds for the adaptive system:

$$\|z^A\|_\infty \leq \frac{|\tilde{\theta}^A(0)| \| h_\omega \|_1}{2\sqrt{c_0^A d_0^A} - |\tilde{\theta}^A(0)| \| h_\omega \|_1} \|y_r\|_\infty . \tag{10.284}$$

In addition, for the adaptive system we have the bound (10.259),

$$\|z^A\|_\infty \leq \frac{1}{\sqrt{\gamma}} |\tilde{\theta}^A(0)| , \tag{10.285}$$

and the bound (10.277),

$$\|z^A\|_\infty \leq \frac{|\tilde{\theta}^A(0)| \| h_\omega \|_1}{2\sqrt{c_0^A d_0^A}} \left(\|y_r\|_\infty + \frac{1}{\sqrt{\gamma}} |\tilde{\theta}^A(0)| \right) . \tag{10.286}$$

Thus, the tightest adaptive bound we have is the smallest of the above three bounds, (10.284), (10.285), and (10.286):

$$\|z^A\|_\infty \leq \min \left\{ \frac{|\tilde{\theta}^A(0)| \| h_\omega \|_1}{2\sqrt{c_0^A d_0^A}} \left(\|y_r\|_\infty + \frac{1}{\sqrt{\gamma}} |\tilde{\theta}^A(0)| \right) , \frac{1}{\sqrt{\gamma}} |\tilde{\theta}^A(0)| , \right.$$

$$\left. \frac{|\tilde{\theta}^A(0)| \| h_\omega \|_1}{2\sqrt{c_0^A d_0^A} - |\tilde{\theta}^A(0)| \| h_\omega \|_1} \|y_r\|_\infty \right\} \triangleq B_A . \tag{10.287}$$

10.4 TRANSIENT PERFORMANCE WITH TUNING FUNCTIONS

A good measure of the performance improvement due to adaptation is the performance ratio

$$R_{\mathcal{L}_\infty} \triangleq \frac{B_{\mathrm{A}}}{B_{\mathrm{N}}} \qquad (10.288)$$

between the $\mathcal{L}_\infty$ bounds (10.287) and (10.283). The improvement is achieved if the performance ratio is small: $R_{\mathcal{L}_\infty} \leq \bar{R}_{\mathcal{L}_\infty} < 1$. For the same parameter uncertainty in the adaptive and the nonadaptive cases, $\tilde{\theta}^{\mathrm{N}} = \tilde{\theta}^{\mathrm{A}}(0) \triangleq \tilde{\theta}$, the following corollary is established by direct calculation.

Corollary 10.15 *Let the initial conditions of $z, \varepsilon, \eta, \lambda$ be zero. Then with adaptation gain*

$$\gamma \geq \left(\frac{2\sqrt{c_0^{\mathrm{N}} d_0^{\mathrm{N}}} - |\tilde{\theta}| \|h_w\|_1}{\max\left\{|\tilde{\theta}|\|h_w\|_1 + 2\sqrt{c_0^{\mathrm{A}} d_0^{\mathrm{A}}}\, \bar{R}_{\mathcal{L}_\infty} - 2\sqrt{c_0^{\mathrm{N}} d_0^{\mathrm{N}}},\ |\tilde{\theta}|\|h_w\|_1 \bar{R}_{\mathcal{L}_\infty}\right\}} \right)^2 \frac{|\tilde{\theta}|^2}{\|y_r\|_\infty^2} \qquad (10.289)$$

and $|\tilde{\theta}|\|h_w\|_1 + 2\sqrt{c_0^{\mathrm{A}} d_0^{\mathrm{A}}}\, \bar{R}_{\mathcal{L}_\infty} > 2\sqrt{c_0^{\mathrm{N}} d_0^{\mathrm{N}}} > |\tilde{\theta}|\|h_w\|_1$, the performance ratio $R_{\mathcal{L}_\infty}$ is no greater than $\bar{R}_{\mathcal{L}_\infty} < 1$.

From this corollary we can deduce two further advantages of the adaptive controller.

- First, the adaptation gain γ provides an additional degree of freedom with which the performance can be improved when the adaptive and the nonadaptive gains are the same, $c_0^{\mathrm{A}} d_0^{\mathrm{A}} = c_0^{\mathrm{N}} d_0^{\mathrm{N}} \triangleq c_0 d_0$, and satisfy the stability condition $2\sqrt{c_0 d_0} - |\tilde{\theta}|\|h_w\|_1 > 0$. In this case the adaptive bound is lower than the nonadaptive bound provided that

$$\gamma > \left(\frac{2\sqrt{c_0 d_0} - |\tilde{\theta}|\|h_w\|_1}{\|h_w\|_1 \|y_r\|_\infty} \right)^2 \triangleq \gamma^*, \qquad (10.290)$$

and the bounds are the same when $\gamma \leq \gamma^*$. Figure 10.6 shows a qualitative plot of the quantity

$$Q(\gamma) = \ln \frac{B_{\mathrm{N}}}{B_{\mathrm{A}}} = \ln \frac{1}{R_{\mathcal{L}_\infty}} \qquad (10.291)$$

obtained using the bounds (10.283) and (10.287). While Corollary 10.15 demonstrates a performance improvement due to adaptation only for $\gamma > \gamma^*$, the simulations, some of which are shown in Example 10.16, exhibit a performance improvement for all $\gamma > 0$.

- Second, and more important, performance improvement can be achieved even with $c_0^{\mathrm{A}} d_0^{\mathrm{A}}$ smaller than $c_0^{\mathrm{N}} d_0^{\mathrm{NA}}$. In the presence of a large parameter

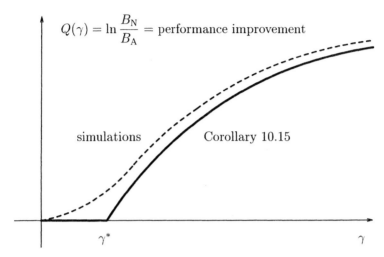

Figure 10.6: Performance improvement due to adaptation.

uncertainty $\tilde{\theta}$, the nonadaptive controller must use $c_0^N d_0^{NA}$ sufficiently large to satisfy $2\sqrt{c_0^N d_0^N} - |\tilde{\theta}|\|h_\omega\|_1 > 0$, thus increasing the bandwidth. From Corollary 10.15 it is clear that with the adaptive controller such an undesirable bandwidth increase can be avoided, because when both $\tilde{\theta}$ and $c_0^N d_0^N$ are large, the condition $2\sqrt{c_0^A d_0^A} \bar{R}_{\mathcal{L}_\infty} > 2\sqrt{c_0^N d_0^N} - |\tilde{\theta}|\|h_\omega\|_1$ can be satisfied with $c_0^A d_0^A$ much smaller than $c_0^N d_0^N$.

This analytically confirms that adaptation is an efficient tool for reducing the effects of large parametric uncertainty without unacceptable widening of system bandwidth. For small parametric uncertainty, the linear controller is effective.

Example 10.16 The improvement of performance due to adaptation is now briefly illustrated with the example introduced in Section 10.2.4. We consider the unstable relative-degree-three plant

$$y(s) = \frac{1}{s^2(s-a)}u(s), \qquad a > 0 \quad \text{unknown} . \qquad (10.292)$$

The control objective is to asymptotically track the output of the reference model

$$y_r(s) = \frac{1}{(s+1)^3}r(s). \qquad (10.293)$$

The tuning functions design for this problem was shown in detail in Section 10.5.

To illustrate the parametric robustness (Theorems 10.8 and 10.9), we switch off the adaptation ($\gamma = 0$) at a constant estimate $\hat{a} = 1$, when

10.4 TRANSIENT PERFORMANCE WITH TUNING FUNCTIONS

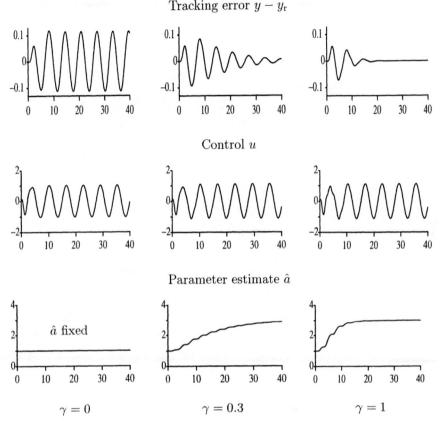

Figure 10.7: Adaptation improves the tracking error transients without an increase in control effort. The plant is driven by $r(t) = \sin t$, and the plant parameter is $a = 3$.

the parameter error $\tilde{a} = 2$ is significant. With $c_1 = c_2 = c_3 = 3$ and $d_1 = d_2 = d_3 = 0.1$, the resulting detuned linear system is unstable. With an increase to $c_1 = c_2 = c_3 = 5$, the system is stabilized. However, without adaptation, the tracking error, shown in Figure 10.7, is about 12% of the reference input, which is not acceptable in most applications.

The adaptive controller is simulated with the same coefficients $c_1 = c_2 = c_3 = 5$ and $d_1 = d_2 = d_3 = 0.1$. The effectiveness of the adaptive scheme is demonstrated by the fact that even with slow adaptation ($\gamma = 0.3$), the tracking error is reduced to zero after a few periods of the reference input, as shown in Figure 10.7. It is remarkable that even during the adaptation transients, the tracking error is smaller than in the nonadaptive system, while the control effort is about the same. When the adaptation gain is increased to $\gamma = 1$, the tracking performance is further improved with about the same control effort.

While Corollary 10.15 shows the performance improvement only beyond a certain γ, the simulations indicate that the performance improvement is present for any $\gamma \geq 0$. $\diamond$

As a conclusion to this section, we point out that the improvement of performance due to adaptation is the first such result in the literature. There are two reasons for this. First, the traditional certainty equivalence adaptive controllers do not possess the parametric robustness property, so they do not have nonadaptive counterparts which can achieve stability, let alone a given level of performance. Second, even if they stabilize the plant with some constant estimate, the adaptation is likely to make the performance worse during the transient because it is not based on the control objective (the identifier is not driven by the tracking error), and the controller does not account for the parameter estimation transients.

10.5 Comparison with a Traditional Scheme

The tuning functions scheme is now compared using simulations with a standard certainty-equivalence scheme on the basis of transient performance and control effort. The comparison is made for the relative-degree-three unstable system from Section 10.2.4.

10.5.1 Choice of a traditional scheme

The comparison with a direct MRAC scheme is not pursued because such a scheme updates *at least three* parameters. This is clear from its control law

$$u = r + \left[\frac{\theta_0 s^2 + \theta_1 s + \theta_2}{s^2 + m_1 s + m_2}\right] y + \left[\frac{\theta_3 s + \theta_4}{s^2 + m_1 s + m_2}\right] u, \qquad (10.294)$$

where $s^2 + m_1 s + m_2$ is a Hurwitz polynomial. A calculation using the Bezout identity gives

$$s^5 + s^4[m_1 - \theta_3 - a] + s^3[m_2 - \theta_4 - a(m_1 - \theta_3)] - s^2[\theta_0 + (m_2 - \theta_4)a]$$
$$= (s+1)^3(s^2 + m_1 s + m_2) + \theta_1 s + \theta_2, \qquad (10.295)$$

which shows that θ_0, θ_3, and θ_4 have to be updated, while θ_1 and θ_2 can be fixed at $\theta_1 = -m_1 - 3m_2$, $\theta_2 = -m_2$. Simulations showed that the update of three parameters results in transient performance inferior to indirect linear schemes which update only one parameter estimate.

Therefore, we compare our new controller to a standard indirect scheme [43, 129], in which the plant equation $s^2(s-a)y(s) = u(s)$ is filtered by a Hurwitz

10.5 COMPARISON WITH A TRADITIONAL SCHEME

observer polynomial $s^3 + k_1 s^2 + k_2 s + k_3$ to obtain the estimation equation

$$\phi = \psi a$$
$$\phi = \frac{s^3}{s^3 + k_1 s^2 + k_2 s + k_3} y(s) - \frac{1}{s^3 + k_1 s^2 + k_2 s + k_3} u(s) \quad (10.296)$$
$$\psi = \frac{s^2}{s^3 + k_1 s^2 + k_2 s + k_3} y(s),$$

and the parameter update law is a normalized gradient[3]:

$$\dot{\hat{a}} = \gamma \frac{\psi e}{1 + \psi^2}, \quad e = \phi - \psi \hat{a}. \quad (10.297)$$

The control law (10.294) is implemented by replacing a with $\hat{a}$ in (10.295) and then solving it for the controller parameters: $\theta_3 = -(3+\hat{a})$, $\theta_4 = -[3+3m_1+\hat{a}(m_1-\theta_3)]$, $\theta_0 = -[1+3m_1+3m_2+(m_2-\theta_4)\hat{a}]$, $\theta_1 = -m_1 - 3m_2$, $\theta_2 = -m_2$.

The indirect adaptive linear scheme and the tuning functions scheme were applied to the plant (10.158) with the true parameter $a = 3$. In all tests the initial parameter estimate was $\hat{a}(0) = 0$, so that, with the adaptation switched off, both closed-loop systems were unstable. The reference input was $r(t) = \sin t$.

For a fair comparison, our first task was to adjust the design parameters of the indirect scheme to achieve the best transient performance with a prescribed control effort. This was done in detail in [95, Section VII]. The trade-off between transient performance and control effort was examined for various initial conditions. To reduce the transients due to the mismatch of initial conditions, the initial condition of the reference model output was set in all tests to be equal to the initial value of the plant output. The available design constants were the adaptation gain γ and the coefficients of the observer polynomial $s^3 + k_1 s^2 + k_2 s + k_3$ and of the controller polynomial $s^2 + m_1 s + m_2$. All the roots of the observer polynomial were placed at $s = -2$ with $k_1 = 6, k_2 = 12, k_3 = 8$, while the roots of the controller polynomial were placed in a Butterworth configuration of radius 3 with $m_1 = 4.2426, m_2 = 9$. These were judged to yield the best trade-off between transient performance and control effort for different initial conditions. On the basis of the simulation results shown in [95, Figures 2–4] the best compromise between transient performance and control effort was judged to be for the value of the adaptation gain $\gamma = 1000$.

10.5.2 Comparison of the schemes

For a comparison of transient performance, the tuning functions scheme was adjusted to employ about the same control effort as the indirect linear scheme.

---

[3]The simulation results with a least-squares update law were virtually identical and are therefore omitted.

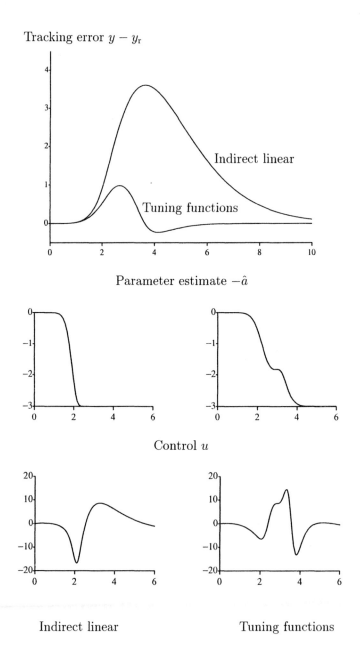

Figure 10.8: Comparison for $y(0) = 0$. The tuning functions controller improves performance with about the same control effort by incorporating the update law $\dot{\hat{a}}$.

10.5 COMPARISON WITH A TRADITIONAL SCHEME

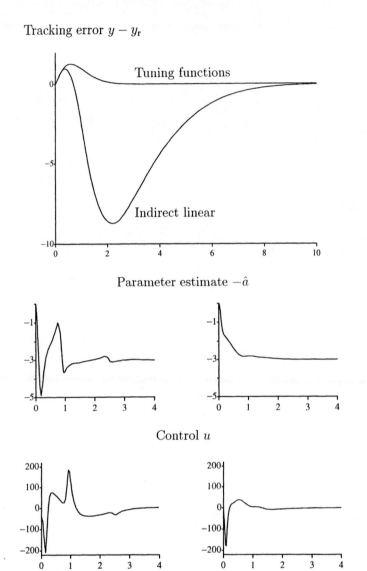

Figure 10.9: Comparison for $y(0) = 1$. The parameter estimate in the tuning functions scheme is smoother because its update law is driven by the state of the error system $z(t)$.

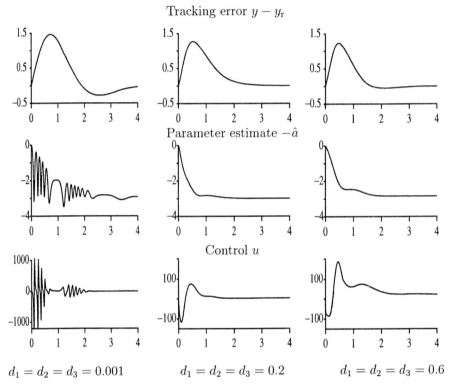

Figure 10.10: Nonlinear damping for $y(0) = 1$. The effect of state estimation error is attenuated.

This was achieved with $k_1 = 6, k_2 = 12, k_3 = 8, c_1 = c_2 = c_3 = 1, d_1 = d_2 = d_3 = 0.1$, and the adaptation gain $\gamma = 0.5$. The plots in Figures 10.8 and 10.9 show that the transient performance of the tuning functions scheme was far superior for both sets of initial conditions. Measured by any norm, the tracking error with the tuning functions scheme is only a fraction of the indirect linear scheme error.

The simulations presented here confirm the strong transient performance properties of the tuning functions design derived in Section 10.4.2. As we explained, the distinctive feature of the tuning functions design is that the controller incorporates the parameter update law $\dot{\hat{a}} = \gamma \tau_3$, with which it accounts for parameter estimation transients. The effect of this additional information about $\hat{a}$ is that the settling time of the tracking error is much shorter for the tuning functions scheme. Figures 10.8 and 10.9 show that the settling time of the tracking error is closely coupled to that of the parameter error. In contrast, the tracking error of the indirect linear scheme continues to grow even after the parameter estimate has converged to its true value.

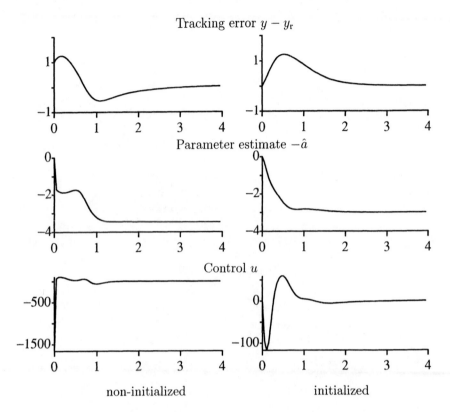

Figure 10.11: Reference model initialization for $y(0) = 1$. The large initial value of control is eliminated, and the parameter transient is made almost monotonic.

Two other most important factors which contributed to the superior performance of the tuning functions scheme are nonlinear damping and reference model initialization.

Nonlinear damping. The nonlinear damping terms $-d_i \left(\frac{\partial \alpha_{i-1}}{\partial y}\right)^2 z_i$ contributed to a significant reduction of the effect of initial conditions on the new adaptive system. Its attenuating effect is displayed in Figure 10.10. If the damping is increased over an optimum rate, the tracking error continues to decrease, but the control effort increases.

Reference model initialization. In contrast to the indirect scheme, the new tuning functions scheme provides clear guidelines for reference model initialization, which follow from the design objective of driving the z-variables to zero. According to (10.167), (10.170), and (10.173), the initial values of z-variables are set to zero by choosing $r_1(0) = y(0)$, $r_2(0) = v_2(0) - \alpha_1(0)$, and $r_3(0) = v_3(0) - \alpha_2(0)$. In general, it is always possible to set $z(0) = 0$ by expressions (10.250)–(10.251). In all tests, the reference model initialization was found to significantly improve both the transient performance and the control effort. A typical example is Figure 10.11.

10.6 Modular Designs

The tuning functions controller and update law are interlaced in an intricate fashion, which makes the design complex. In Chapter 9 we introduced modular output-feedback schemes with independently designed controllers and identifiers. We pursue here the same idea and design modular backstepping schemes for *linear* systems. However, for linear systems we do not use the strong ISS-controllers because their underlying nonadaptive controllers are nonlinear even for linear systems. Instead, we employ an *SG-controller* which is a certainty equivalence version of the nonadaptive linear controller in Section 10.3.1. This controller is different from traditional certainty equivalence controllers because its backstepping structure endows it with design coefficients which are useful in shaping the transient behavior. Because of their certainty equivalence nature, the modular schemes will serve nicely for a qualitative comparison between the traditional certainty equivalence schemes and the tuning functions scheme.

In addition to Assumptions 10.1–10.4, we make the following assumption about a lower bound on the high-frequency gain, standard in 'indirect' adaptive control.

Assumption 10.17 *In addition to* $\text{sgn}(b_m)$, *a positive constant* ς_m *is known such that* $|b_m| \geq \varsigma_m$.

Assumption 10.17 strengthens Assumption 10.2 in the tuning functions design. It allows the control law to contain a division by the estimate $\hat{b}_m$, which is kept away from zero by parameter projection. In the tuning functions designs it was possible to avoid this assumption by introducing an additional estimate of $\varrho = 1/b_m$.

This section is organized as follows. In Section 10.6.1 we present the SG-controller design using the knowledge of the tuning functions controller from Section 10.2.1. Then in Sections 10.6.2 and 10.6.3 we present two identifiers and stability analyses for corresponding closed-loop adaptive systems. Finally in Section 10.6.4 we compare the modular designs with the tuning functions design.

10.6.1 SG-controller

The SG-controller, given in Table 10.3, is a modification of the tuning functions controller in Table 10.2.

We briefly discuss the modifications leading to the controller in Table 10.3.

10.6 MODULAR DESIGNS

Table 10.3: SG-Controller for Linear Systems

$$z_1 = y - y_r \tag{10.298}$$

$$z_i = v_{m,i} - \frac{1}{\hat{b}_m} y_r^{(i-1)} - \alpha_{i-1}, \qquad i = 2, \ldots, \rho \tag{10.299}$$

$$\alpha_1 = \frac{1}{\hat{b}_m} \bar{\alpha}_1 \tag{10.300}$$

$$\bar{\alpha}_1 = -(c_1 + d_1) z_1 - \xi_2 - \bar{\omega}^T \hat{\theta} \tag{10.301}$$

$$\alpha_2 = -\hat{b}_m z_1 - \left[c_2 + d_2 \left(\frac{\partial \alpha_1}{\partial y} \right)^2 \right] z_2 + \beta_2 \tag{10.302}$$

$$\alpha_i = -z_{i-1} - \left[c_i + d_i \left(\frac{\partial \alpha_{i-1}}{\partial y} \right)^2 \right] z_i + \beta_i, \qquad i = 3, \ldots, \rho \tag{10.303}$$

$$\beta_i = \frac{\partial \alpha_{i-1}}{\partial y} \left(\xi_2 + \omega^T \hat{\theta} \right) + \frac{\partial \alpha_{i-1}}{\partial \eta} (A_0 \eta + e_n y) + \sum_{j=1}^{i-1} \frac{\partial \alpha_{i-1}}{\partial y_r^{(j-1)}} y_r^{(j)}$$

$$+ k_i v_{m,1} + \sum_{j=1}^{m+i-1} \frac{\partial \alpha_{i-1}}{\partial \lambda_j} (-k_j \lambda_1 + \lambda_{j+1}) \tag{10.304}$$

Adaptive control law:

$$u = \alpha_\rho - v_{m,\rho+1} + \frac{1}{\hat{b}_m} y_r^{(\rho)} \tag{10.305}$$

- The adaptation gains Γ and γ are set to zero, which means that we eliminate $\frac{\partial \alpha_1}{\partial \hat{\theta}} \Gamma \tau_2$ from (10.101), $\frac{\partial \alpha_{i-1}}{\partial \hat{\theta}} \Gamma \tau_i - \sum_{j=2}^{i-1} \frac{\partial \alpha_{j-1}}{\partial \hat{\theta}} \Gamma \frac{\partial \alpha_{i-1}}{\partial y} z_j$ from (10.102), and $\left(y_r^{(i-1)} + \frac{\partial \alpha_{i-1}}{\partial \hat{\varrho}} \right) \dot{\hat{\varrho}}$ from (10.103).

- With Assumption 10.17, $\hat{\varrho}$ is replaced by $1/\hat{b}_m$, and $\hat{b}_m(t)$ is kept from crossing zero by parameter projection.

Straightforward but lengthy calculations show that the resulting error system is

$$\dot{z} = A_z^*(\hat{\theta}) z + W_\varepsilon(\hat{\theta}) \varepsilon_2 + W_\theta^*(z,t)^T \tilde{\theta} + Q(z,t)^T \dot{\hat{\theta}}, \quad z \in \mathbb{R}^\rho, \tag{10.306}$$

where

$$A_z^* = \begin{bmatrix} -c_1-d_1 & b_m & 0 & \cdots & & 0 \\ -b_m & -c_2-d_2\left(\frac{\partial \alpha_1}{\partial y}\right)^2 & 1 & \ddots & & \vdots \\ 0 & -1 & \ddots & \ddots & & 0 \\ \vdots & & \ddots & \ddots & & 1 \\ 0 & \cdots & & 0 & -1 & -c_\rho-d_\rho\left(\frac{\partial \alpha_{\rho-1}}{\partial y}\right)^2 \end{bmatrix} \quad (10.307)$$

$$W_\varepsilon = \begin{bmatrix} 1 \\ -\frac{\partial \alpha_1}{\partial y} \\ \vdots \\ -\frac{\partial \alpha_{\rho-1}}{\partial y} \end{bmatrix} \in \mathbb{R}^\rho \quad (10.308)$$

$$W_\theta^{*\mathrm{T}} = \begin{bmatrix} \bar{\omega}^{\mathrm{T}} + \frac{1}{\hat{b}_m}(\dot{y}_r + \bar{\alpha}_1)e_1^{\mathrm{T}} \\ -\frac{\partial \alpha_1}{\partial y}\omega^{\mathrm{T}} + z_1 e_1^{\mathrm{T}} \\ -\frac{\partial \alpha_2}{\partial y}\omega^{\mathrm{T}} \\ \vdots \\ -\frac{\partial \alpha_{\rho-1}}{\partial y}\omega^{\mathrm{T}} \end{bmatrix} \in \mathbb{R}^{\rho \times p} \quad (10.309)$$

$$Q^{\mathrm{T}} = \begin{bmatrix} 0 \\ -\frac{\partial \alpha_1}{\partial \hat{\theta}} + \frac{1}{\hat{b}_m^2}\ddot{y}_r e_1^{\mathrm{T}} \\ \vdots \\ -\frac{\partial \alpha_{\rho-1}}{\partial \hat{\theta}} + \frac{1}{\hat{b}_m^2}y_r^{(\rho-1)}e_1^{\mathrm{T}} \end{bmatrix} \in \mathbb{R}^{\rho \times p} \quad (10.310)$$

By examining the expressions for α_i in Table 10.3, it is not hard to see that $\frac{\partial \alpha_i}{\partial y}$ is a function of only $\hat{\theta}$, c_i, and d_i, which means that A_z^* and W_ε depend on $\hat{\theta}$, but not on other states of the closed-loop system. Thus, in contrast to the tuning functions design and all the other output-feedback designs so far (cf. Chapters 8 and 9), the matrix A_z^* and the vector W_ε are functions of $\hat{\theta}$ only. This is due not only to the SG-controller design, but also to the linearity of the plant. In view of (10.98) and (10.99) we have

$$\begin{aligned} b_m z_2 + \left(\bar{\omega} + \frac{1}{\hat{b}_m}(\dot{y}_r + \bar{\alpha}_1)e_1\right)^{\mathrm{T}}\tilde{\theta} &= \hat{b}_m z_2 + \left(\bar{\omega} + \left(z_2 + \frac{1}{\hat{b}_m}(\dot{y}_r + \bar{\alpha}_1)\right)e_1\right)^{\mathrm{T}}\tilde{\theta} \\ &= \hat{b}_m z_2 + \left(\bar{\omega} + \left(z_2 + \frac{1}{\hat{b}_m}\dot{y}_r + \alpha_1\right)e_1\right)^{\mathrm{T}}\tilde{\theta} \\ &= \hat{b}_m z_2 + (\bar{\omega} + v_{m,2}e_1)^{\mathrm{T}}\tilde{\theta} \\ &= \hat{b}_m z_2 + \omega^{\mathrm{T}}\tilde{\theta}. \quad (10.311) \end{aligned}$$

10.6 MODULAR DESIGNS

On the other hand,

$$-\hat{b}_m z_1 - \frac{\partial \alpha_1}{\partial y}\omega^T\tilde{\theta} + z_1 e_1^T\tilde{\theta} = -\hat{b}_m z_1 - \frac{\partial \alpha_1}{\partial y}\omega^T\tilde{\theta}. \tag{10.312}$$

With (10.311) and (10.312), the error system (10.306)–(10.310) is rewritten as

$$\dot{z} = A_z(\hat{\theta})z + W_\varepsilon(\hat{\theta})\left(\omega^T\tilde{\theta} + \varepsilon_2\right) + Q(z,t)^T\dot{\hat{\theta}}, \tag{10.313}$$

where

$$A_z(\hat{\theta}) = \begin{bmatrix} -c_1 - d_1 & \hat{b}_m & 0 & \cdots & & 0 \\ -\hat{b}_m & -c_2 - d_2\left(\frac{\partial \alpha_1}{\partial y}\right)^2 & 1 & \ddots & & \vdots \\ 0 & -1 & \ddots & \ddots & & 0 \\ \vdots & & \ddots & \ddots & & 1 \\ 0 & \cdots & & 0 & -1 & -c_p - d_p\left(\frac{\partial \alpha_{p-1}}{\partial y}\right)^2 \end{bmatrix}. \tag{10.314}$$

The error system (10.313) is similar in form to the error system (9.84) in the modular nonlinear design. However, because of the absence of the κ_i and g_i nonlinear damping terms, (10.313) is not input-to-state stable with respect to $\tilde{\theta}$ and $\dot{\hat{\theta}}$.

For a brief comment on the underlying nonadaptive SG-controller, we let $\dot{\hat{\theta}} \equiv 0$, so that the error system (10.313) becomes

$$\dot{z} = A_z(\hat{\theta})z + W_\varepsilon(\hat{\theta})\left(\omega^T\tilde{\theta} + \varepsilon_2\right), \qquad \hat{\theta} = \text{const}. \tag{10.315}$$

Let, for simplicity, $b_m = \hat{b}_m = 1$. The detuned error system (10.188) in the tuning functions design and the detuned error system (10.315) in the modular design are identical. Therefore, the parametric robustness and nonadaptive performance properties established in Theorems 10.8–10.10 for the tuning functions design hold also for the modular designs. These properties distinguish our modular designs from the traditional estimation-based certainty equivalence designs.

The backstepping approach results in a nonadaptive controller which has parametric robustness and performance properties, and it can be made adaptive in two different ways—using either a Lyapunov methodology (tuning functions) or a modular input-output methodology with identifiers designed separately from the SG-controller.

10.6.2 y-Passive scheme

Consider the parametric y-model (10.37)

$$\dot{y} = \xi_2 + \omega^T\theta + \varepsilon_2, \tag{10.316}$$

where ξ_2 and ω are available and defined in (10.29) and (10.39), respectively. We introduce the simple scalar observer

$$\dot{\hat{y}} = -\left(c_0 + \kappa_0|\omega|^2\right)(\hat{y} - y) + \omega_0 + \omega^T\hat{\theta}, \qquad (10.317)$$

where κ_0 is a positive constant, and c_0 is as defined in (10.213). The observer error

$$\epsilon = y - \hat{y} \qquad (10.318)$$

is governed by the system

$$\dot{\epsilon} = -\left(c_0 + \kappa_0|\omega|^2\right)\epsilon + \omega^T\tilde{\theta} + \varepsilon_2. \qquad (10.319)$$

The parameter update law is chosen to be

$$\dot{\hat{\theta}} = \operatorname*{Proj}_{\hat{b}_m}\{\Gamma\omega\epsilon\}, \qquad \begin{array}{c}\hat{b}_m(0)\operatorname{sgn}b_m > \varsigma_m \\ \Gamma = \Gamma^T > 0\end{array} \qquad (10.320)$$

where the projection operator is employed to guarantee that $|\hat{b}_m(t)| \geq \varsigma_m > 0$, $\forall t \geq 0$, (see Appendix E).

Lemma 10.18 *Let the maximal interval of existence of solutions of (10.316), (10.317), and (10.320) be $[0, t_f)$. Then the following identifier properties hold:*

$$\begin{align}
(o) \quad & |\hat{b}_m(t)| \geq \varsigma_m > 0, \; \forall t \in [0, t_f) & (10.321) \\
(i) \quad & \tilde{\theta} \in \mathcal{L}_\infty[0, t_f) & (10.322) \\
(ii) \quad & \epsilon \in \mathcal{L}_2[0, t_f) \cap \mathcal{L}_\infty[0, t_f) & (10.323) \\
(iii) \quad & \omega\epsilon, \dot{\hat{\theta}} \in \mathcal{L}_2[0, t_f). & (10.324)
\end{align}$$

The above identifier is the same as the identifier used in Section 9.2. Therefore, the proof of Lemma 10.18 is identical to that of Lemma 9.3. However, the stability proof for the resulting error system is different from that of Theorem 9.4 because the SG-controller makes the proof of the following theorem considerably more involved.

Theorem 10.19 (y-Passive) *All the signals in the closed-loop adaptive system consisting of the plant (10.1), the control law in Table 10.3, the filters in Table 10.1, the observer (10.317), and the update law (10.320) are globally uniformly bounded, and asymptotic tracking is achieved:*

$$\lim_{t\to\infty}[y(t) - y_r(t)] = 0. \qquad (10.325)$$

Proof. The projection operator in Appendix E is locally Lipschitz, as stated in Lemma E.1. Therefore, as argued in the proof of Theorem 10.6, the solution of

10.6 MODULAR DESIGNS

the closed-loop adaptive system exists and is unique on its maximum interval of existence $[0, t_f)$.

Let us consider the systems (10.313) and (10.319),

$$\dot{z} = A_z(\hat{\theta})z + W_\varepsilon(\hat{\theta})\left(\omega^T\tilde{\theta} + \varepsilon_2\right) + Q(z,t)^T\dot{\hat{\theta}} \qquad (10.326)$$

$$\dot{\epsilon} = -c_0\epsilon + \left(\omega^T\tilde{\theta} + \varepsilon_2\right) - \kappa_0\omega^T\omega\epsilon, \qquad (10.327)$$

and define the signal

$$\zeta \triangleq z - W_\varepsilon(\hat{\theta})\epsilon. \qquad (10.328)$$

Equations (10.326), (10.327), and (10.328) yield the system

$$\dot{\zeta} = A_z(\hat{\theta})\zeta + \left[A_z(\hat{\theta}) + c_0 I\right] W_\varepsilon(\hat{\theta})\epsilon - \frac{\partial W_\varepsilon(\hat{\theta})}{\partial \hat{\theta}}\dot{\hat{\theta}}\epsilon$$
$$+ Q(z,t)^T\dot{\hat{\theta}} + \kappa_0 W_\varepsilon(\hat{\theta})\omega^T\omega\epsilon. \qquad (10.329)$$

From Lemma 10.18, $\hat{\theta} \in \mathcal{L}_\infty[0, t_f)$. Since $\frac{\partial \alpha_i}{\partial y}$ are smooth functions of $\hat{\theta}$ whenever $\hat{b}_m \neq 0$, and Lemma 10.18 guarantees that $\hat{b}_m(t) \not\equiv 0$, then W_ε, A_z and $\frac{\partial W_\varepsilon}{\partial \hat{\theta}}$ are bounded. Since, by Lemma 10.18, $\epsilon, \dot{\hat{\theta}} \in \mathcal{L}_2[0, t_f)$ and $\epsilon \in \mathcal{L}_\infty[0, t_f)$, we conclude that

$$\left[A_z(\hat{\theta}) + c_0 I\right] W_\varepsilon(\hat{\theta})\epsilon - \frac{\partial W_\varepsilon(\hat{\theta})}{\partial \hat{\theta}}\dot{\hat{\theta}}\epsilon \triangleq L_1 \in \mathcal{L}_2[0, t_f). \qquad (10.330)$$

Thus, system (10.329) becomes

$$\dot{\zeta} = A_z(\hat{\theta})\zeta + L_1 + Q(z,t)^T\underbrace{\dot{\hat{\theta}}}_{\in \mathcal{L}_2} + \kappa_0 W_\varepsilon(\hat{\theta})\omega^T\underbrace{\omega\epsilon}_{\in \mathcal{L}_2}. \qquad (10.331)$$

Since $A_z(\hat{\theta})$ is exponentially stable uniformly in $\hat{\theta}$, we view (10.331) as a perturbed linear time-varying system which cannot be destabilized by the square-integrable disturbance L_1. So we focus our attention to $Q(z,t)^T\dot{\hat{\theta}} + \kappa_0 W_\varepsilon(\hat{\theta})\omega^T\omega\epsilon$.

Claim 10.20 *There exist functions $p_k^{ij} \in \mathcal{L}_2[0, t_f)$ and $q_k^{ij} \in \mathcal{L}_2[0, t_f)$, $i, j = 1, \ldots, \rho$, $k = 0, \ldots, n+m$, such that*

$$Q(z,t)^T\dot{\hat{\theta}} + \kappa_0 W_\varepsilon(\hat{\theta})\omega^T\omega\epsilon = H(t,s)[z] + G(t,s)\left[\bar{y}_r^{(\rho-1)}\right], \qquad (10.332)$$

where the linear operators H and G are

$$H(t,s) = \{h_{ij}(t,s)\}_{\rho \times \rho}, \quad h_{ij}(t,s) = \sum_{k=0}^{n+m} p_k^{ij}(t)\frac{s^k}{K(s)B(s)} \qquad (10.333)$$

$$G(t,s) = \{g_{ij}(t,s)\}_{\rho \times \rho}, \quad g_{ij}(t,s) = \sum_{k=0}^{n+m} q_k^{ij}(t)\frac{s^k}{K(s)B(s)}. \qquad (10.334)$$

The proof of this claim is lengthy but straightforward. It relies on the fact that the α_i's are linear in y, η, λ and $\bar{y}_r^{(\rho-1)}$ and nonlinear only in $\hat{\theta}$.

Substituting (10.328) into (10.332) and then into (10.331), one obtains

$$\dot{\zeta} = A_z(\hat{\theta})\zeta + L_1 + H(t,s)[\zeta] + H(t,s)\left[W_\varepsilon(\hat{\theta})\epsilon\right] + G(t,s)\left[\bar{y}_r^{(\rho-1)}\right]. \quad (10.335)$$

Since $W_\varepsilon(\hat{\theta})\epsilon$ and $\bar{y}_r^{(\rho-1)}$ are bounded and the coefficients in $H(t,s)$ and $G(t,s)$ are square-integrable, then

$$L_1 + H(t,s)\left[W_\varepsilon(\hat{\theta})\epsilon\right] + G(t,s)\left[\bar{y}_r^{(\rho-1)}\right] \triangleq L_2 \in \mathcal{L}_2[0, t_f). \quad (10.336)$$

Hence, our problem is reduced to the study of the system

$$\dot{\zeta} = A_z(\hat{\theta})\zeta + H(t,s)[\zeta] + L_2. \quad (10.337)$$

Let us denote by l_1 a generic function in $\mathcal{L}_1[0, t_f)$. (Note that this notation allows us to say $kl_1 \leq l_1$ for any finite k.) In view of (10.333), we have

$$\begin{aligned}\dot{V}_h &\leq -\delta V_h + |\zeta|^2 \\ |H(t,s)[\zeta]|^2 &\leq l_1|\zeta|^2 + l_1 V_h,\end{aligned} \quad (10.338)$$

for some $\delta > 0$. The derivative of $|\zeta|^2/2$ along the solutions of (10.337) with (10.314) is

$$\begin{aligned}\frac{d}{dt}\left(\frac{1}{2}|\zeta|^2\right) &\leq -c_0|\zeta|^2 + \zeta^T H(t,s)[\zeta] + \zeta^T L_2 \\ &\leq -\frac{c_0}{2}|\zeta|^2 + \frac{1}{c_0}|H(t,s)[\zeta]|^2 + \frac{1}{c_0}|L_2|^2. \end{aligned} \quad (10.339)$$

Combining (10.339) with (10.338), we get

$$\frac{d}{dt}\left(|\zeta|^2\right) \leq -(c_0 - l_1)|\zeta|^2 + l_1 V_h + l_1 \quad (10.340)$$

$$\dot{V}_h \leq -\delta V_h + |\zeta|^2. \quad (10.341)$$

These two differential inequalities are interconnected and form a loop with small gain because V_h appears multiplied by l_1 in (10.340). To finish the proof we define the "superstate"

$$X \triangleq V_h + \frac{2}{c_0}|\zeta|^2, \quad (10.342)$$

differentiate it and substitute (10.340)–(10.341),

$$\begin{aligned}\dot{X} &\leq -\delta V_h - 2|\zeta|^2 + l_1\left(\frac{2}{c_0}|\zeta|^2 + \frac{2}{c_0}V_h\right) + |\zeta|^2 + l_1 \\ &\leq -\delta V_h - |\zeta|^2 + l_1 X + l_1, \end{aligned} \quad (10.343)$$

10.6 MODULAR DESIGNS

so we get

$$\dot{X} \leq -\left(\min\left\{\delta, \frac{c_0}{2}\right\} - l_1\right) X + l_1. \tag{10.344}$$

By applying Lemma B.6 we conclude that X is bounded and integrable on $[0, t_f)$. The boundedness of X implies the boundedness of ζ, which, in turn, implies that z is bounded. It remains to prove that the boundedness of z, $\hat{\theta}$ and $\bar{y}_r^{(\rho)}$ implies that x, η and λ are also bounded. A proof of this implication was already given for the tuning functions design in Theorem 10.6 (cf. (10.124)–(10.126)). The same argument is applicable here with (10.126) replaced by

$$v_{m,i} = z_i + \frac{1}{b_m} y_r^{(i-1)} + \alpha_{i-1}\left(y, \eta, \hat{\theta}, \bar{\lambda}_{m+i-1}, \bar{y}_r^{(i-2)}\right), \qquad i = 2, \ldots, \rho. \tag{10.345}$$

All the bounds are independent of t_f. By the same argument as in the proof of Theorem 10.6 we conclude that $t_f = \infty$.

To prove the tracking, we first note that the integrability of X implies that ζ is square-integrable. Thus $z \in \mathcal{L}_2$. The boundedness of all the signals and (10.313) prove that $\dot{z}$ is also bounded. By Barbalat's lemma (Corollary A.7) $z(t) \to 0$ as $t \to \infty$, and therefore asymptotic tracking is achieved. □

The simple update law (10.320) is based on passivity of the ϵ-system (10.319). The term $\kappa_0 |\omega|^2$ in (10.317) serves as a form of 'normalization' which slows down the adaptation and achieves $\dot{\hat{\theta}} \in \mathcal{L}_2$. The proof of Theorem 10.19 reveals that this property is crucial for stability.

It is easy to see that Theorem 10.19 holds even when $d_1 = \cdots = d_\rho = 0$. We have chosen to propose the design with $d_0 > 0$ because it can also guarantee stability without adaptation.

10.6.3 x-Swapping scheme

Of the two swapping identifiers presented in Chapter 9, we choose the x-swapping identifier because it uses only the K-filters with no additional swapping filters.

Consider the parametric x-model obtained by substituting (10.12) into (10.13):

$$x = \xi + \Omega^T \theta + \varepsilon. \tag{10.346}$$

In the first row of (10.346),

$$y = \xi_1 + \Omega_1^T \theta + \varepsilon_1, \tag{10.347}$$

all the signals are measured except for the bounded, exponentially decaying ε_1. The "prediction error" is implemented as

$$\epsilon = y - \xi_1 - \Omega_1^T \hat{\theta}, \tag{10.348}$$

and its relationship with the parameter error is linear:

$$\epsilon = \Omega_1^T \tilde{\theta} + \varepsilon_1. \qquad (10.349)$$

The update law for $\hat{\theta}$ is either the gradient:

$$\dot{\hat{\theta}} = \operatorname*{Proj}_{\hat{b}_m}\left\{\Gamma \frac{\Omega_1 \epsilon}{1 + \nu |\Omega_1|^2}\right\}, \qquad \hat{b}_m(0)\operatorname{sgn} b_m > \varsigma_m \qquad (10.350)$$
$$\Gamma = \Gamma^T > 0, \quad \nu > 0$$

or the least-squares:

$$\dot{\hat{\theta}} = \operatorname*{Proj}_{\hat{b}_m}\left\{\Gamma \frac{\Omega_1 \epsilon}{1 + \nu |\Omega_1|^2}\right\}, \qquad \hat{b}_m(0)\operatorname{sgn} b_m > \varsigma_m$$
$$\dot{\Gamma} = -\Gamma \frac{\Omega_1 \Omega_1^T}{1 + \nu |\Omega_1|^2}\Gamma, \qquad \Gamma(0) = \Gamma(0)^T > 0, \quad \nu > 0, \qquad (10.351)$$

where the projection operator is employed to guarantee that $|\hat{b}_m(t)| \geq \varsigma_m > 0$, $\forall t \geq 0$.

The proof of the following lemma is the same as that of Lemma 9.8.

Lemma 10.21 *Let the maximal interval of existence of solutions of (10.2) and (10.26)–(10.27) with either (10.350) or (10.351) be $[0, t_f)$. Then the following identifier properties hold:*

$$(o) \quad |\hat{b}_m(t)| \geq \varsigma_m > 0, \ \forall t \in [0, t_f) \qquad (10.352)$$
$$(i) \quad \tilde{\theta} \in \mathcal{L}_\infty[0, t_f) \qquad (10.353)$$
$$(ii) \quad \frac{\epsilon}{\sqrt{1 + \nu |\Omega_1|^2}} \in \mathcal{L}_2[0, t_f) \cap \mathcal{L}_\infty[0, t_f) \qquad (10.354)$$
$$(iii) \quad \dot{\hat{\theta}} \in \mathcal{L}_2[0, t_f) \cap \mathcal{L}_\infty[0, t_f). \qquad (10.355)$$

In contrast to the y-passive identifier, the normalization is employed here in the update law. The normalization makes $\dot{\hat{\theta}}$ not only square-integrable, but also bounded. It slows down the adaptation sufficiently so that the following stability result holds.

Theorem 10.22 (x-Swapping) *All the signals in the closed-loop adaptive system consisting of the plant (10.1), the control law in Table 10.3, and the filters in Table 10.1, with either the gradient (10.350) or the least-squares update law (10.351), are globally uniformly bounded, and asymptotic tracking is achieved:*

$$\lim_{t \to \infty} [y(t) - y_r(t)] = 0. \qquad (10.356)$$

10.6 MODULAR DESIGNS

Proof. The projection operator in Appendix E is locally Lipschitz, so the solution of the closed-loop adaptive system exists and is unique on its maximum interval of existence $[0, t_f)$.

In the proof of Theorem 9.9, we showed that

$$\dot{\Omega}_1 = -k_1 \Omega_1 + \omega, \tag{10.357}$$

which, along with (10.349) and (10.14), means that

$$\dot{\epsilon} = -k_1 \epsilon + \omega^T \tilde{\theta} + \varepsilon_2 - k_1 \varepsilon_1 - \Omega_1^T \dot{\hat{\theta}}. \tag{10.358}$$

Let us now define the signal

$$\zeta \triangleq z - W_\varepsilon(\hat{\theta}) \epsilon. \tag{10.359}$$

Combining (10.359) with (10.313) and (10.358), we get

$$\dot{\zeta} = A_z(\hat{\theta})\zeta + \underbrace{k_1 W_\varepsilon(\hat{\theta})\varepsilon_1}_{\triangleq L_1 \in \mathcal{L}_2} + \left[Q(z,t)^T + W_\varepsilon(\hat{\theta})\Omega_1^T\right] \underbrace{\dot{\hat{\theta}}}_{\in \mathcal{L}_2}$$

$$+ \underbrace{\left\{\underbrace{\left[A_z(\hat{\theta}) + c_0 I\right] W_\varepsilon(\hat{\theta}) - \frac{\partial W_\varepsilon(\hat{\theta})}{\partial \hat{\theta}} \dot{\hat{\theta}}}_{\in \mathcal{L}_\infty}\right\} \underbrace{\frac{\epsilon}{\sqrt{1+\nu|\Omega_1|^2}}}_{\in \mathcal{L}_2} \sqrt{1 + \nu|\Omega_1|^2},}_{\triangleq L_2 \in \mathcal{L}_2}$$

$$\tag{10.360}$$

where the $\mathcal{L}_2$ and $\mathcal{L}_\infty$ signal properties are established using Lemma 10.21.

Claim 10.23 *There exist functions $p_k^{ij} \in \mathcal{L}_2[0, t_f)$ and $q_k^{ij} \in \mathcal{L}_2[0, t_f)$, $i,j = 1, \ldots, \rho$, $k = 0, \ldots, n+m$, such that*

$$\left[Q(z,t)^T + W_\varepsilon(\hat{\theta})\Omega_1^T\right] \dot{\hat{\theta}} = H(t,s)[z] + G(t,s)\left[\bar{y}_r^{(\rho-1)}\right], \tag{10.361}$$

where the linear operators H and G are

$$H(t,s) = \{h_{ij}(t,s)\}_{\rho \times \rho}, \quad h_{ij}(t,s) = \sum_{k=0}^{n+m} p_k^{ij}(t) \frac{s^k}{K(s)B(s)} \tag{10.362}$$

$$G(t,s) = \{g_{ij}(t,s)\}_{\rho \times \rho}, \quad g_{ij}(t,s) = \sum_{k=0}^{n+m} q_k^{ij}(t) \frac{s^k}{K(s)B(s)}. \tag{10.363}$$

The proof of this claim is omitted. With (10.361) and (10.359), system (10.360) becomes

$$\dot{\zeta} = A_z(\hat{\theta})\zeta + L_1 + H(t,s)[\zeta] + H(t,s)\left[W_\varepsilon(\hat{\theta})\epsilon\right] + G(t,s)\left[\bar{y}_r^{(\rho-1)}\right]$$
$$+ L_2 \sqrt{1 + \nu|\Omega_1|^2}. \tag{10.364}$$

Since $\bar{y}_r^{(\rho-1)}$ is bounded and the coefficients in $G(t,s)$ are square-integrable, then $L_1 + G(t,s)\left[\bar{y}_r^{(\rho-1)}\right] \triangleq L_3 \in \mathcal{L}_2[0,t_f)$. Thus (10.364) becomes

$$\dot{\zeta} = A_z(\hat{\theta})\zeta + L_3 + H(t,s)[\zeta] + H(t,s)\left[W_\varepsilon(\hat{\theta})\frac{\epsilon}{\sqrt{1+\nu|\Omega_1|^2}}\sqrt{1+\nu|\Omega_1|^2}\right]$$
$$+ L_2\sqrt{1+\nu|\Omega_1|^2}. \qquad (10.365)$$

We note that $W_\varepsilon(\hat{\theta})\dfrac{\epsilon}{\sqrt{1+\nu|\Omega_1|^2}} \in \mathcal{L}_\infty[0,t_f)$, which implies that there exists a function $B_1 \in \mathcal{L}_\infty[0,t_f)$ and functions $r_k^{ij} \in \mathcal{L}_2[0,t_f)$, $i,j = 1,\ldots,\rho$, $k = 0,\ldots,n+m$, such that

$$H(t,s)\left[W_\varepsilon(\hat{\theta})\frac{\epsilon}{\sqrt{1+\nu|\Omega_1|^2}}\sqrt{1+\nu|\Omega_1|^2}\right] + L_2\sqrt{1+\nu|\Omega_1|^2}$$
$$= F(t,s)\left[B_1\sqrt{1+\nu|\Omega_1|^2}\right], \qquad (10.366)$$

where the linear operator F is

$$F(t,s) = \{f_i(t,s)\}_{\rho \times 1}, \qquad f_i(t,s) = \sum_{k=0}^{n+m} r_k^i(t)\frac{s^k}{K(s)B(s)}. \qquad (10.367)$$

With (10.366), we write (10.365) in the form

$$\dot{\zeta} = A_z(\hat{\theta})\zeta + L_3 + H(t,s)[\zeta] + F(t,s)\left[B_1\sqrt{1+\nu|\Omega_1|^2}\right]. \qquad (10.368)$$

By combining (10.357) with (10.204), we get

$$\begin{aligned}
\Omega_1 &= \frac{1}{s+k_1}H_\omega(s)[y] \\
&= \frac{1}{s+k_1}H_\omega(s)[z_1 + y_r] \\
&= \frac{1}{s+k_1}H_\omega(s)[\zeta_1 - \epsilon + y_r] \\
&= \frac{1}{s+k_1}H_\omega(s)\left[\zeta_1 - \frac{\epsilon}{\sqrt{1+\nu|\Omega_1|^2}}\sqrt{1+\nu|\Omega_1|^2} + y_r\right], \quad (10.369)
\end{aligned}$$

where $\dfrac{\epsilon}{\sqrt{1+\nu|\Omega_1|^2}}$ is square-integrable. Since $\dfrac{1}{s+k_1}H_\omega(s)$ is stable and proper, we have

$$\frac{d}{dt}\left(|\Omega_1|^2\right) \leq -\delta_\Omega|\Omega_1|^2 + k_\Omega\left(\zeta_1^2 + \frac{\epsilon^2}{1+\nu|\Omega_1|^2}\left(1+\nu|\Omega_1|^2\right) + y_r^2\right), \qquad (10.370)$$

10.6 MODULAR DESIGNS

where δ_Ω and k_Ω are positive constants. Let k denote a generic positive constant, and let l_1 denote a generic function in $\mathcal{L}_1[0, t_f)$. Then (10.370) is rewritten as

$$\frac{d}{dt}\left(|\Omega_1|^2\right) \leq -(\delta_\Omega - l_1)|\Omega_1|^2 + k_\Omega|\varsigma|^2 + k. \tag{10.371}$$

In view of (10.362), we have

$$\begin{aligned} \dot{V}_h &\leq -\delta V_h + |\varsigma|^2 \\ |H(t,s)[\varsigma]|^2 &\leq l_1 V_h + l_1|\varsigma|^2, \end{aligned} \tag{10.372}$$

and in view of (10.367), we have

$$\begin{aligned} \dot{V}_f &\leq -\delta V_f + B_1^2\left(1 + \nu|\Omega_1|^2\right) \\ \left|F(t,s)\left[B_1\sqrt{1 + \nu|\Omega_1|^2}\right]\right|^2 &\leq l_1 V_f + l_1 B_1^2\left(1 + \nu|\Omega_1|^2\right). \end{aligned} \tag{10.373}$$

The derivative of $|\varsigma|^2/2$ along the solutions of (10.368) and (10.314) is

$$\begin{aligned}\frac{d}{dt}\left(\frac{1}{2}|\varsigma|^2\right) &\leq -c_0|\varsigma|^2 + \varsigma^T H(t,s)[\varsigma] + \varsigma^T F(t,s)\left[B_1\sqrt{1+\nu|\Omega_1|^2}\right] + \varsigma^T L_3 \\ &\leq -\frac{c_0}{4}|\varsigma|^2 + \frac{1}{c_0}|H(t,s)[\varsigma]|^2 + \frac{1}{c_0}\left|F(t,s)\left[B_1\sqrt{1+\nu|\Omega_1|^2}\right]\right|^2 \\ &\quad + \frac{1}{c_0}|L_3|^2. \end{aligned} \tag{10.374}$$

Thus (10.371)–(10.374) define a system of differential inequalities

$$\begin{aligned} \frac{d}{dt}\left(|\varsigma|^2\right) &\leq -\left(\frac{c_0}{2} - l_1\right)|\varsigma|^2 + l_1|\Omega_1|^2 + l_1 V_h + l_1 V_f + l_1 \\ \frac{d}{dt}\left(|\Omega_1|^2\right) &\leq -(\delta_\Omega - l_1)|\Omega_1|^2 + k_\Omega|\varsigma|^2 + k \\ \dot{V}_h &\leq -\delta V_h + |\varsigma|^2 \\ \dot{V}_f &\leq -\delta V_f + k|\Omega_1|^2 + k. \end{aligned} \tag{10.375}$$

These four inequalities are interconnected in several loops with small gain l_1. To finish the proof we define the "superstate"

$$X \triangleq |\varsigma|^2 + \frac{c_0}{8}\left(\frac{1}{k_\Omega}|\Omega_1|^2 + V_h + \frac{\delta_\Omega}{2kk_\Omega}V_f\right). \tag{10.376}$$

Differentiating X along the solutions of (10.375), after arranging the terms, we get

$$\dot{X} \leq -(\delta_X - l_1)X + l_1 + k, \tag{10.377}$$

where
$$\delta_X = \min\left\{\frac{c_0}{8}, \frac{\delta_\Omega}{2}, \frac{8\delta}{c_0}, \delta\right\} > 0. \tag{10.378}$$

By applying the Lemma B.6 we conclude that X is bounded on $[0, t_f)$. The boundedness of X implies the boundedness of ζ and Ω_1, which, in turn, implies that ϵ and z are bounded. The boundedness of x, η and λ is argued as in Theorem 10.19. All the bounds are independent of t_f, so $t_f = \infty$.

To prove the tracking, we first note from (10.360) that, due to the boundedness of all the signals, all the inputs to the ζ-system are square-integrable. Then, with Lemma B.5 we show that ζ itself is square-integrable. Due to the boundedness of Ω_1, $\epsilon \in \mathcal{L}_2$, which establishes that $z \in \mathcal{L}_2$. The boundedness of all the signals and (10.313) prove that $\dot{z}$ is also bounded, so by Barbalat's lemma (Corollary A.7) $z(t) \to 0$ as $t \to \infty$. Hence, asymptotic tracking is achieved. $\square$

The stability proof for the x-swapping scheme is more involved than for the y-passive scheme. However, the x-swapping scheme is of lower dynamic order because it uses only the K-filters in the identifier, as opposed to an additional observer in the y-observer scheme. The dynamic order of the x-swapping scheme is the same as in the tuning functions scheme.

Like Theorem 10.19, Theorem 10.22 also holds even when $d_1 = \cdots = d_\rho = 0$, but we have presented the design with $d_0 > 0$ because it can also guarantee stability without adaptation.

Our presentation in this subsection was with normalized update laws. It is also possible, however, to prove the result of Theorem 10.22 with the *unnormalized least-squares* update law. The proof exploits the properties of the unnormalized least-squares algorithm: $\dot{\hat{\theta}} \in \mathcal{L}_1, \epsilon \in \mathcal{L}_2$ (see, e.g., [157]).

10.6.4 Comparison: modular vs. tuning functions design

Even though the tuning functions design and the modular design both use the backstepping approach with the same underlying nonadaptive controller, they result in fundamentally different adaptive schemes.

The tuning functions scheme is designed using a single Lyapunov function, and results in a simple and direct stability analysis. The stability analysis of modular schemes is far more involved. Even though one can derive transient performance bounds for the modular schemes, they are neither as simple nor as insightful as the tuning functions performance bounds.

We now illustrate the difference in performance behavior between the tuning functions and modular schemes on the example introduced in Section 10.2.4. Of the two modular schemes, we present simulations only for the x-swapping scheme. The responses with the y-passive scheme are qualitatively

10.6 MODULAR DESIGNS

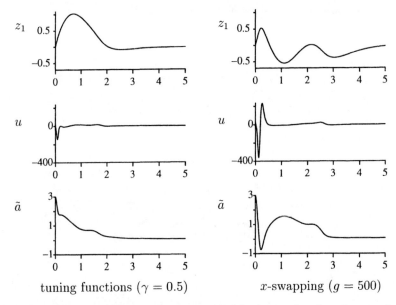

Figure 10.12: Comparison of responses with the tuning functions and the x-swapping schemes. While the tuning functions update law (driven by the tracking error z_1) monotonically reduces the parameter error $\tilde{a}$, the x-swapping update law generates an 'overshoot' in $\tilde{a}$, which results in a considerably higher peak in the response of the control u.

similar. The SG-controller used in the modular schemes is obtained by setting $\gamma = 0$ in the tuning functions controller (10.167)–(10.176). The x-swapping scheme employs the update law

$$\dot{\hat{a}} = -g\frac{\Xi_{1,1}\epsilon}{1+\Xi_{1,1}^2}, \qquad \epsilon = y - \xi_1 + \hat{a}\Xi_{1,1} - v_1. \tag{10.379}$$

For both schemes we use the design parameter values $c_1 = c_2 = c_3 = 1$, $d_1 = d_2 = d_3 = 1$, $k_1 = 6, k_2 = 12$, and $k_3 = 8$.

A comparison of responses with the tuning functions and the x-swapping schemes is given in Figure 10.12. While the tuning functions update law (driven by the tracking error z_1) reduces the parameter error $\tilde{a}$ almost monotonically, the x-swapping update law generates an 'overshoot' in $\tilde{a}$, which results in a considerably higher peak in the control u. It is important to stress that the responses given in Figure 10.12 are not the best possible with either of the schemes. They are only chosen to illustrate a fundamental difference in transient properties between the tuning functions and modular schemes.

Even though the modular schemes may not have performance properties as strong as the tuning functions scheme, they are simpler to design and offer flexibility in the selection of an update law.

While the modular schemes presented in this chapter have a lot in common with the certainty equivalence adaptive linear schemes, they remove some of the shortcomings of the certainty equivalence schemes. As shown in Section 10.3.2, their underlying nonadaptive controller can achieve stability even without adaptation, and their design parameters can be used for systematic improvement of nonadaptive performance.

10.7 Summary

We have presented two classes of adaptive designs for linear systems: tuning functions and modular. These designs have the same underlying nonadaptive linear controller based on backstepping. This nonadaptive controller can guarantee stability without adaptation when a bound on the parametric uncertainty is known, and in addition, achieve a prescribed level of tracking performance. This is an improvement over traditional adaptive designs which cannot guarantee stability when the adaptation is disconnected.

The tuning functions design removes several other obstacles from adaptive linear control. Since the design is based on a single Lyapunov function incorporating both the state of the error system and the update law, the proof of global uniform stability is direct and simple. Moreover, all the error states except for the parameter error converge to zero. This is the strongest convergence property without persistency of excitation.

The main advantage of the tuning functions design over traditional certainty equivalence adaptive designs is in transient performance. The nonlinear control law which incorporates the parameter update law keeps the parameter estimation transients from causing bad tracking transients. The performance bounds obtained for the tuning functions scheme are computable and can be used for systematic improvement of transient performance.

The modular schemes are simpler to design than the tuning functions scheme, but do not possess as strong performance properties (although their design parameters c_i and d_i can be used to influence the performance). Their underlying nonadaptive controller offers an advantage over traditional certainty equivalence schemes in guaranteeing stability without adaptation.

The dynamic order of the adaptive schemes in this chapter can be as low as $3n + m$, which is no higher (and in most cases is significantly lower) than with any other scheme that achieves tracking.

Notes and References

Adaptive control of linear systems is thoroughly covered in the books of Goodwin and Sin [44], Åström and Wittenmark [5], Narendra and Annaswamy [142], Sastry and Bodson [165], and Ioannou and Sun [51] and in the research monographs by Egardt [31], Landau [109], Ioannou and Kokotović [50], and Anderson et al. [2].

Transient performance of adaptive systems has recently received a great deal of attention in the works of Ioannou and Tao [52], Zang and Bitmead [202], Krause, Khargonekar, and Stein [90], Ydstie [200], Naik, Kumar, and Ydstie [140], Ortega [146], and Miller and Davison [131]. Sun, Olbrot, and Polis [181] developed nonadaptive controllers which guarantee stability in the presence of uncertain parameters. Sun [180] used these controllers with adaptation, and added a modifying signal to improve transient performance. This led to model reference adaptive controllers of Datta and Ioannou [24] which can reduce both the $\mathcal{L}_\infty$ norm and the mean-square value of the tracking error. The $\mathcal{L}_\infty$ and $\mathcal{L}_2$ performance bounds in Krstić, Kokotović, and Kanellakopoulos [102] are computable and show that the tuning functions controller outperforms its nonadaptive counterpart.

Passivity of the tuning functions design was studied by Krstić, Kanellakopoulos, and Kokotović [96] and Brogliato, Lozano, and Landau [11]. Parameter convergence was addressed first by Ortega and Fradkov [147] and then solved by Zhang, Ioannou, and Chien [203] who showed that the tuning functions design has parameter convergence properties stronger than the traditional adaptive schemes. Initial progress on developing a robust adaptive design with tuning functions has been reported by Li, Wen, and Soh [111, 112]. Wen [196] solved the decentralized adaptive regulation problem by employing the tuning functions design.

Sections 10.1–10.5 are based on Krstić, Kanellakopoulos, and Kokotović [95, 96] and Krstić, Kokotović, and Kanellakopoulos [102]. The results of Section 10.6 have not been previously published.

Appendices

Appendix A

Lyapunov Stability and Convergence

In addition to the main Lyapunov stability notions and theorem reviewed in Section 2.1, we now briefly review the main invariance theorems. For completeness and convenience, we repeat some of the definitions and theorems already presented in Section 2.1.

Consider the nonautonomous system

$$\dot{x} = f(x, t) \tag{A.1}$$

where $f : \mathbb{R}^n \times \mathbb{R}_+ \to \mathbb{R}^n$ is locally Lipschitz in x and piecewise continuous in t.

Definition A.1 *The origin $x = 0$ is the equilibrium point for (A.1) if*

$$f(0, t) = 0, \quad \forall t \geq 0. \tag{A.2}$$

Scalar *comparison functions* are important stability tools frequently used in this book.

Definition A.2 *A continuous function $\gamma : [0, a) \to \mathbb{R}_+$ is said to belong to class $\mathcal{K}$ if it is strictly increasing and $\gamma(0) = 0$. It is said to belong to class $\mathcal{K}_\infty$ if $a = \infty$ and $\gamma(r) \to \infty$ as $r \to \infty$.*

Definition A.3 *A continuous function $\beta : [0, a) \times \mathbb{R}_+ \to \mathbb{R}_+$ is said to belong to class $\mathcal{KL}$ if for each fixed s the mapping $\beta(r, s)$ belongs to class $\mathcal{K}$ with respect to r, and for each fixed r the mapping $\beta(r, s)$ is decreasing with respect to s and $\beta(r, s) \to 0$ as $s \to \infty$. It is said to belong to class $\mathcal{KL}_\infty$ if, in addition, for each fixed s the mapping $\beta(r, s)$ belongs to class $\mathcal{K}_\infty$ with respect to r.*

Using comparison functions we can restate the stability definitions from Section 2.1.1 in a more practical form:

Definition A.4 *The equilibrium point $x = 0$ of (A.1) is*

- *uniformly stable, if there exists a class $\mathcal{K}$ function $\gamma(\cdot)$ and a positive constant c, independent of t_0, such that*

$$|x(t)| \leq \gamma(|x(t_0)|), \quad \forall t \geq t_0 \geq 0, \quad \forall x(t_0) \mid |x(t_0)| < c; \quad \text{(A.3)}$$

- *uniformly asymptotically stable, if there exists a class $\mathcal{KL}$ function $\beta(\cdot,\cdot)$ and a positive constant c, independent of t_0, such that*

$$|x(t)| \leq \beta(|x(t_0)|, t - t_0), \quad \forall t \geq t_0 \geq 0, \quad \forall x(t_0) \mid |x(t_0)| < c; \quad \text{(A.4)}$$

- *exponentially stable, if (A.4) is satisfied with $\beta(r,s) = kre^{-\alpha s}$, $k > 0$, $\alpha > 0$;*

- *globally uniformly stable, if (A.3) is satisfied with $\gamma \in \mathcal{K}_\infty$ for any initial state $x(t_0)$;*

- *globally uniformly asymptotically stable, if (A.4) is satisfied with $\beta \in \mathcal{KL}_\infty$ for any initial state $x(t_0)$; and*

- *globally exponentially stable, if (A.4) is satisfied for any initial state $x(t_0)$ and with $\beta(r,s) = kre^{-\alpha s}$, $k > 0$, $\alpha > 0$.*

The main Lyapunov stability theorem is then formulated as follows (see, for example, [81, Theorem 4.1, Corollaries 4.1 and 4.2]):

Theorem A.5 (Uniform Stability) *Let $x = 0$ be an equilibrium point of (A.1) and $D = \{x \in \mathbb{R}^n \mid |x| < r\}$. Let $V : D \times \mathbb{R}^n \to \mathbb{R}_+$ be a continuously differentiable function such that $\forall t \geq 0$, $\forall x \in D$,*

$$\gamma_1(|x|) \leq V(x,t) \leq \gamma_2(|x|) \quad \text{(A.5)}$$

$$\frac{\partial V}{\partial t} + \frac{\partial V}{\partial x} f(x,t) \leq -\gamma_3(|x|), \quad \text{(A.6)}$$

Then the equilibrium $x = 0$ is

- *uniformly stable, if γ_1 and γ_2 are class $\mathcal{K}$ functions on $[0, r)$ and $\gamma_3(\cdot) \geq 0$ on $[0, r)$;*

- *uniformly asymptotically stable, if γ_1, γ_2 and γ_3 are class $\mathcal{K}$ functions on $[0, r)$;*

- *exponentially stable, if $\gamma_i(\rho) = k_i \rho^\alpha$ on $[0, r)$, $k_i > 0$, $\alpha > 0$, $i = 1, 2, 3$;*

- *globally uniformly stable, if $D = \mathbb{R}^n$, γ_1 and γ_2 are class $\mathcal{K}_\infty$ functions, and $\gamma_3(\cdot) \geq 0$ on $\mathbb{R}_+$;*

- globally uniformly asymptotically stable, *if $D = \mathbb{R}^n$, γ_1 and γ_2 are class $\mathcal{K}_\infty$ functions, and γ_3 is a class $\mathcal{K}$ function on $\mathbb{R}_+$; and*

- globally exponentially stable, *if $D = \mathbb{R}^n$ and $\gamma_i(\rho) = k_i \rho^\alpha$ on $\mathbb{R}_+$, $k_i > 0$, $\alpha > 0$, $i = 1, 2, 3$.*

In general, our goal is to achieve convergence to a set. For time-invariant systems, the main convergence tool is LaSalle's Invariance Theorem (Theorem 2.2). For time-varying systems, a more refined tool was developed by LaSalle [110] and Yoshizawa [201]. For pedagogical reasons, we will introduce it via a technical lemma due to Barbalat [155]. These key results and their proofs are of utmost importance in guaranteeing that an adaptive system will fulfill its tracking task.

Lemma A.6 (Barbalat) *Consider the function $\phi : \mathbb{R}_+ \to \mathbb{R}$. If ϕ is uniformly continuous and $\lim_{t \to \infty} \int_0^\infty \phi(\tau) d\tau$ exists and is finite, then*

$$\lim_{t \to \infty} \phi(t) = 0. \tag{A.7}$$

Proof. Suppose that (A.7) does not hold; that is, either the limit does not exist or it is not equal to zero. Then there exists $\varepsilon > 0$ such that for every $T > 0$ one can find $t_1 \geq T$ with $|\phi(t_1)| > \varepsilon$. Since ϕ is uniformly continuous, there is a positive constant $\delta(\varepsilon)$ such that $|\phi(t) - \phi(t_1)| < \varepsilon/2$ for all $t_1 \geq 0$ and all t such that $|t - t_1| \leq \delta(\varepsilon)$. Hence, for all $t \in [t_1, t_1 + \delta(\varepsilon)]$, we have

$$\begin{aligned} |\phi(t)| &= |\phi(t) - \phi(t_1) + \phi(t_1)| \\ &\geq |\phi(t_1)| - |\phi(t) - \phi(t_1)| \\ &> \varepsilon - \frac{\varepsilon}{2} = \frac{\varepsilon}{2}, \end{aligned} \tag{A.8}$$

which implies that

$$\left| \int_{t_1}^{t_1 + \delta(\varepsilon)} \phi(\tau) d\tau \right| = \int_{t_1}^{t_1 + \delta(\varepsilon)} |\phi(\tau)| d\tau > \frac{\varepsilon \delta(\varepsilon)}{2}, \tag{A.9}$$

where the first equality holds since $\phi(t)$ does not change sign on $[t_1, t_1 + \delta(\varepsilon)]$. Noting that $\int_0^{t_1 + \delta(\varepsilon)} \phi(\tau) d\tau = \int_0^{t_1} \phi(\tau) d\tau + \int_{t_1}^{t_1 + \delta(\varepsilon)} \phi(\tau) d\tau$, we conclude that $\int_0^t \phi(\tau) d\tau$ cannot converge to a finite limit as $t \to \infty$, which contradicts the assumption of the lemma. Thus, $\lim_{t \to \infty} \phi(t) = 0$. $\square$

Corollary A.7 *Consider the function $\phi : \mathbb{R}_+ \to \mathbb{R}$. If $\phi, \dot{\phi} \in \mathcal{L}_\infty$, and $\phi \in \mathcal{L}_p$ for some $p \in [1, \infty)$, then*

$$\lim_{t \to \infty} \phi(t) = 0. \tag{A.10}$$

Theorem A.8 (LaSalle-Yoshizawa) *Let $x = 0$ be an equilibrium point of (A.1) and suppose f is locally Lipschitz in x uniformly in t. Let $V : \mathbb{R}^n \times \mathbb{R}_+ \to \mathbb{R}_+$ be a continuously differentiable function such that*

$$\gamma_1(|x|) \leq V(x, t) \leq \gamma_2(|x|) \tag{A.11}$$

$$\dot{V} = \frac{\partial V}{\partial t} + \frac{\partial V}{\partial x} f(x, t) \leq -W(x) \leq 0 \tag{A.12}$$

$\forall t \geq 0$, $\forall x \in \mathbb{R}^n$, *where γ_1 and γ_2 are class $\mathcal{K}_\infty$ functions and W is a continuous function. Then, all solutions of (A.1) are globally uniformly bounded and satisfy*

$$\lim_{t \to \infty} W(x(t)) = 0. \tag{A.13}$$

In addition, if $W(x)$ is positive definite, then the equilibrium $x = 0$ is globally uniformly asymptotically stable.

Proof. Since $\dot{V} \leq 0$, V is nonincreasing. Thus, in view of the first inequality in (A.11), we conclude that x is globally uniformly bounded, that is, $|x(t)| \leq B$, $\forall t \geq 0$. Since $V(x(t), t)$ is nonincreasing and bounded from below by zero, we conclude that it has a limit V_∞ as $t \to \infty$. Integrating (A.12), we have

$$\begin{aligned} \lim_{t \to \infty} \int_{t_0}^{t} W(x(\tau))d\tau &\leq -\lim_{t \to \infty} \int_{t_0}^{t} \dot{V}(x(\tau), \tau)d\tau \\ &= \lim_{t \to \infty} \{V(x(t_0), t_0) - V(x(t), t)\} \\ &= V(x(t_0), t_0) - V_\infty, \end{aligned} \tag{A.14}$$

which means that $\int_{t_0}^{\infty} W(x(\tau))d\tau$ exists and is finite. Now we show that $W(x(t))$ is also uniformly continuous. Since $|x(t)| \leq B$ and f is locally Lipschitz in x uniformly in t, we see that for any $t \geq t_0 \geq 0$,

$$\begin{aligned} |x(t) - x(t_0)| &= \left| \int_{t_0}^{t} f(x(\tau), \tau)d\tau \right| \leq L \int_{t_0}^{t} |x(\tau)|d\tau \\ &\leq LB|t - t_0|, \end{aligned} \tag{A.15}$$

where L is the Lipschitz constant of f on $\{|x| \leq B\}$. Choosing $\delta(\varepsilon) = \frac{\varepsilon}{LB}$, we have

$$|x(t) - x(t_0)| < \varepsilon, \quad \forall \, |t - t_0| \leq \delta(\varepsilon), \tag{A.16}$$

which means that $x(t)$ is uniformly continuous. Since W is continuous, it is uniformly continuous on the compact set $\{|x| \leq B\}$. From the uniform continuity of $W(x)$ and $x(t)$, we conclude that $W(x(t))$ is uniformly continuous. Hence, it satisfies the conditions of Lemma A.6, which then guarantees that $W(x(t)) \to 0$ as $t \to \infty$.

If, in addition, $W(x)$ is positive definite, there exists a class $\mathcal{K}$ function $\gamma_3(\cdots)$ such that $W(x) \geq \gamma_3(|x|)$. Using Theorem A.5, we conclude that $x = 0$ is globally uniformly asymptotically stable. □

In applications, we usually have $W(x) = x^T Q x$, where Q is a symmetric positive semidefinite matrix. For this case, the proof of Theorem A.8 simplifies using Corollary A.7 with $p = 1$.

Appendix B
Input-Output Stability

In addition to a review of basic input-output stability results, we give several technical lemmas used in the book.

For a function $x : \mathbb{R}_+ \to \mathbb{R}^n$ we define the $\mathcal{L}_p$ norm, $p \in [1, \infty]$, as

$$\|x\|_p = \begin{cases} \left(\int_0^\infty |x(t)|^p dt\right)^{1/p} & p \in [1, \infty) \\ \sup_{t \geq 0} |x(t)| & p = \infty \end{cases} \tag{B.1}$$

and the $\mathcal{L}_{p,e}$ norm (truncated $\mathcal{L}_p$ norm) as

$$\|x_t\|_p = \begin{cases} \left(\int_0^t |x(\tau)|^p d\tau\right)^{1/p} & p \in [1, \infty) \\ \sup_{\tau \in [0,t]} |x(\tau)| & p = \infty \end{cases} \tag{B.2}$$

We consider an LTI causal system described by the convolution

$$y(t) = h * u = \int_0^t h(t-\tau) u(\tau) d\tau, \tag{B.3}$$

where $u : \mathbb{R}_+ \to \mathbb{R}$ is the input, $y : \mathbb{R}_+ \to \mathbb{R}$ is the output, and $h : \mathbb{R} \to \mathbb{R}$ is the system's impulse response, which is defined to be zero for negative values of its argument. Let $H(s)$ denote the Laplace transform of $h(t)$.

Theorem B.1 [28] *A strictly proper rational transfer function $H(s)$ is analytic in $\Re\{s\} \geq 0$ if and only if $h \in \mathcal{L}_1$.*

The quantity

$$\|H\|_\infty = \sup_{\omega \in \mathbb{R}} |H(j\omega)| \tag{B.4}$$

is referred to as the $\mathcal{H}_\infty$ norm of the transfer function $H(s)$.

The following theorem is known from [10, 23, 28, 30].

Theorem B.2 *For system (B.3), if $h \in \mathcal{L}_1$ and $H \in \mathcal{H}_\infty$, then*

$$\text{(i)} \quad \|y_t\|_\infty \leq \|h\|_1 \|u_t\|_\infty \tag{B.5}$$
$$\text{(ii)} \quad \|y_t\|_2 \leq \|H\|_\infty \|u_t\|_2 \tag{B.6}$$
$$\text{(iii)} \quad \|H\|_\infty \leq \|h\|_1 \leq 2n\|H\|_\infty, \tag{B.7}$$

where n is the McMillan degree of $H(s)$.

Lemma B.3 (Hölder's Inequality) *If $p, q \in [1, \infty]$ and $\frac{1}{p} + \frac{1}{q} = 1$, then*

$$\|(fg)_t\|_1 \leq \|f_t\|_p \|g_t\|_q, \qquad \forall t \geq 0. \tag{B.8}$$

The following theorem is referred to as Young's convolution theorem [197].

Theorem B.4 *If $h \in \mathcal{L}_{1,e}$, then*

$$\|(h*u)_t\|_p \leq \|h_t\|_1 \|u_t\|_p, \qquad p \in [1, \infty]. \tag{B.9}$$

Proof. Let $y = h * u$. Then, for $p \in [1, \infty)$, we have

$$
\begin{aligned}
|y(t)| &\leq \int_0^t |h(t-\tau)|\,|u(\tau)|\,d\tau \\
&= \int_0^t |h(t-\tau)|^{\frac{p-1}{p}} |h(t-\tau)|^{\frac{1}{p}} |u(\tau)|\,d\tau \\
&\leq \left(\int_0^t |h(t-\tau)|\,d\tau\right)^{\frac{p-1}{p}} \left(\int_0^t |h(t-\tau)|\,|u(\tau)|^p\,d\tau\right)^{\frac{1}{p}} \\
&= \|h_t\|_1^{\frac{p-1}{p}} \left(\int_0^t |h(t-\tau)|\,|u(\tau)|^p\,d\tau\right)^{\frac{1}{p}}, \tag{B.10}
\end{aligned}
$$

where the second inequality is obtained by applying Hölder's inequality. Raising (B.10) to power p and integrating from 0 to t, we get

$$
\begin{aligned}
\|y_t\|_p^p &\leq \int_0^t \|h_t\|_1^{p-1} \left(\int_0^\tau |h(\tau-s)|\,|u(s)|^p\,ds\right) d\tau \\
&= \|h_t\|_1^{p-1} \int_0^t \left(\int_s^t |h(\tau-s)|\,|u(s)|^p\,d\tau\right) ds \\
&= \|h_t\|_1^{p-1} \int_0^t \left(\int_0^t |h(\tau-s)|\,|u(s)|^p\,d\tau\right) ds \\
&= \|h_t\|_1^{p-1} \int_0^t |u(s)|^p \left(\int_0^t |h(\tau-s)|\,d\tau\right) ds \\
&\leq \|h_t\|_1^{p-1} \int_0^t |u(s)|^p \left(\int_0^t |h(\tau)|\,d\tau\right) ds \\
&\leq \|h_t\|_1^{p-1} \|h\|_1^1 \|u_t\|_1^p \\
&\leq \|h_t\|_1^p \|u_t\|_p^p, \tag{B.11}
\end{aligned}
$$

where the second line is obtained by changing the sequence of integration, and the third line by using the causality of h. The proof for the case $p = \infty$ is immediate by taking a supremum of u over $[0, t]$ in the convolution. □

Input-Output Stability

Lemma B.5 *Let v and ρ be real-valued functions defined on $\mathbb{R}_+$, and let b and c be positive constants. If they satisfy the differential inequality*

$$\dot{v} \leq -cv + b\rho(t)^2, \qquad v(0) \geq 0, \tag{B.12}$$

(i) then the following integral inequality holds:

$$v(t) \leq v(0)e^{-ct} + b\int_0^t e^{-c(t-\tau)}\rho(\tau)^2 d\tau. \tag{B.13}$$

(ii) If, in addition, $\rho \in \mathcal{L}_2$, then $v \in \mathcal{L}_1$ and

$$\|v\|_1 \leq \frac{1}{c}\left(v(0) + b\|\rho\|_2^2\right). \tag{B.14}$$

Proof. *(i)* Upon multiplication of (B.12) by e^{ct}, it becomes

$$\frac{d}{dt}\left(v(t)e^{ct}\right) \leq b\rho(t)^2 e^{ct}. \tag{B.15}$$

Integrating (B.15) over $[0, t]$, we arrive at (B.13).
(ii) By integrating (B.13) over $[0, t]$, we get

$$\int_0^t v(\tau)d\tau \leq \int_0^t v(0)e^{-c\tau}d\tau + b\int_0^t \left[\int_0^\tau e^{-c(\tau-s)}\rho(s)^2 ds\right] d\tau$$
$$\leq \frac{1}{c}v(0) + b\int_0^t \left[\int_0^\tau e^{-c(\tau-s)}\rho(s)^2 ds\right] d\tau. \tag{B.16}$$

Noting that the second term is $b\|(h*\rho^2)_t\|_1$ where $h(t) = e^{-ct}$, $t \geq 0$, we apply Theorem B.4. Since $\|h\|_1 = \frac{1}{c}$, we obtain (B.14). □

Lemma B.6 *Let v, l_1, and l_2 be real-valued functions defined on $\mathbb{R}_+$, and let c be a positive constant. If l_1 and l_2 are nonnegative and in $\mathcal{L}_1$ and satisfy the differential inequality*

$$\dot{v} \leq -cv + l_1(t)v + l_2(t), \qquad v(0) \geq 0 \tag{B.17}$$

then $v \in \mathcal{L}_\infty \cap \mathcal{L}_1$ and

$$v(t) \leq \left(v(0)e^{-ct} + \|l_2\|_1\right)e^{\|l_1\|_1} \tag{B.18}$$

$$\|v\|_1 \leq \frac{1}{c}\left(v(0) + \|l_2\|_1\right)e^{\|l_1\|_1}. \tag{B.19}$$

Proof. Using the fact that fact that $w(t) \leq v(t)$, $\dot{w} = -cw + l_1(t)w + l_2(t)$, $w(0) = v(0)$ (the comparison principle; see, for example, [108, 132]), and applying the variation of constants formula, the differential inequality (B.17) is rewritten as

$$\begin{aligned} v(t) &\leq v(0)e^{\int_0^t [-c+l_1(s)]ds} + \int_0^t e^{\int_\tau^t [-c+l_1(s)]ds} l_2(\tau)d\tau \\ &\leq v(0)e^{-ct}e^{\int_0^\infty l_1(s)ds} + \int_0^t e^{-c(t-\tau)} l_2(\tau)d\tau e^{\int_0^\infty l_1(s)ds} \\ &\leq \left[v(0)e^{-ct} + \int_0^t e^{-c(t-\tau)} l_2(\tau)d\tau \right] e^{\|l_1\|_1} . \end{aligned} \quad (\text{B.20})$$

By taking a supremum of $e^{-c(t-\tau)}$ over $[0,\infty]$, we obtain (B.18). Integrating (B.20) over $[0,\infty]$, we get

$$\int_0^t v(\tau)d\tau \leq \left(\frac{1}{c} v(0) + \int_0^t \left[\int_0^\tau e^{-c(\tau-s)} l_2(s)ds \right] d\tau \right) e^{\|l_1\|_1} . \quad (\text{B.21})$$

Applying Theorem B.4 to the double integral, we arrive at (B.19). □

Remark B.7 An alternative proof that $v \in \mathcal{L}_\infty \cap \mathcal{L}_1$ in Lemma B.6 is using the Gronwall lemma (Lemma B.11). However, with the Gronwall lemma, the estimates of the bounds (B.18) and (B.19) are more conservative:

$$v(t) \leq \left(v(0)e^{-ct} + \|l_2\|_1 \right) \left(1 + \|l_1\|_1 e^{\|l_1\|_1} \right) \quad (\text{B.22})$$

$$\|v\|_1 \leq \frac{1}{c} \left(v(0) + \|l_2\|_1 \right) \left(1 + \|l_1\|_1 e^{\|l_1\|_1} \right), \quad (\text{B.23})$$

because $e^x < (1 + xe^x)$, $\forall x > 0$. Note that the ratio between the bounds (B.22) and (B.18), and the ratio between the bounds (B.23) and (B.19), are of order $\|l_1\|_1$ when $\|l_1\|_1 \to \infty$. ◇

For cases where l_1 and l_2 are functions of time that converge to zero but are not in $\mathcal{L}_p$ for any $p \in [1,\infty)$ we have the following lemma.

Lemma B.8 *Consider the differential inequality*

$$\dot{v} \leq -[c - \beta_1(r_0,t)]v + \beta_2(r_0,t) + \rho, \qquad v(0) = v_0 \geq 0, \quad (\text{B.24})$$

where $c > 0$ and $r_0 \geq 0$ are constants, and β_1 and β_2 are class $\mathcal{KL}$ functions. Then there exists a class $\mathcal{KL}$ function β_v and a class $\mathcal{K}$ function γ_v such that

$$v(t) \leq \beta_v(v_0 + r_0, t) + \gamma_v(\rho), \qquad \forall t \geq 0. \quad (\text{B.25})$$

Moreover, if $\beta_i(r,t) = \alpha_i(r)e^{-\sigma_i t}$, $i = 1,2$, where $\alpha_i \in \mathcal{K}$ and $\sigma_i > 0$, then there exists $\alpha_v \in \mathcal{K}$ and $\sigma_v > 0$ such that $\beta_v(r,t) = \alpha_v(r)e^{-\sigma_v t}$.

INPUT-OUTPUT STABILITY

Proof. We start by introducing $\tilde{v} = v - \frac{\rho}{c}$ and rewriting (B.24) as

$$\dot{\tilde{v}} \leq -[c - \beta_1(r_0, t)]\tilde{v} + \frac{\rho}{c}\beta_1(r_0, t) + \beta_2(r_0, t). \tag{B.26}$$

It then follows that

$$v(t) \leq v_0 e^{\int_0^t [\beta_1(r_0, s) - c] ds} + \int_0^t \left[\frac{\rho}{c}\beta_1(r_0, \tau) + \beta_2(r_0, \tau)\right] e^{\int_\tau^t [\beta_1(r_0, s) - c] ds} d\tau + \frac{\rho}{c}. \tag{B.27}$$

We recall the standard result (see, for example, [81, Lemma 4.6]):

$$e^{\int_\tau^t [\beta_1(r_0, s) - c] ds} \leq k(r_0) e^{-\frac{c}{2}(t - \tau)}, \quad \forall \tau \in [0, t], \tag{B.28}$$

where k is a positive, continuous, increasing function. To get an estimate of the overshoot coefficient $k(r_0)$, we provide a proof of (B.28). For each c there exists a class $\mathcal{K}$ function $T_c : \mathbb{R}_+ \to \mathbb{R}_+$ such that

$$\beta_1(r_0, s) \leq \frac{c}{2}, \quad \forall s \geq T_c(r_0). \tag{B.29}$$

Therefore, for $0 \leq \tau \leq T_c(r_0) \leq t$, we have

$$\begin{aligned}\int_\tau^t [\beta_1(r_0, s) - c] ds &\leq \int_\tau^{T_c(r_0)} [\beta_1(r_0, s) - c] ds + \int_{T_c(r_0)}^t \left(-\frac{c}{2}\right) ds \\ &\leq (\beta_1(r_0, 0) - c)(T_c(r_0) - \tau) - \frac{c}{2}(t - T_c(r_0)) \\ &\leq T_c(r_0)\beta_1(r_0, 0) - \frac{c}{2}(t - \tau),\end{aligned} \tag{B.30}$$

so the overshoot coefficient in (B.28) is given by

$$k(r_0) \triangleq e^{T_c(r_0)\beta_1(r_0, 0)}. \tag{B.31}$$

For the other two cases, $t \leq T_c(r_0)$ and $T_c(r_0) \leq \tau$, getting (B.28) with $k(r_0)$ as in (B.31) is immediate. Now substituting (B.28) into (B.27), we get

$$v(t) \leq v_0 k(r_0) e^{-\frac{c}{2}t} + k(r_0) \int_0^t \left[\frac{\rho}{c}\beta_1(r_0, \tau) + \beta_2(r_0, \tau)\right] e^{-\frac{c}{2}(t - \tau)} d\tau + \frac{\rho}{c}. \tag{B.32}$$

To complete the proof, we show that a class $\mathcal{KL}$ function β convolved with an exponentially decaying kernel is bounded by another class $\mathcal{KL}$ function:

$$\begin{aligned}\int_0^t e^{-\frac{c}{2}(t-\tau)} \beta(r_0, \tau) d\tau &= \int_0^{t/2} e^{-\frac{c}{2}(t-\tau)} \beta(r_0, \tau) d\tau + \int_{t/2}^t e^{-\frac{c}{2}(t-\tau)} \beta(r_0, \tau) d\tau \\ &\leq \beta(r_0, 0) \int_0^{t/2} e^{-\frac{c}{2}(t-\tau)} d\tau + \beta(r_0, t/2) \int_{t/2}^t e^{-\frac{c}{2}(t-\tau)} d\tau \\ &\leq \frac{2}{c} \left[\beta(r_0, 0) e^{-\frac{c}{4}t} + \beta(r_0, t/2)\right].\end{aligned} \tag{B.33}$$

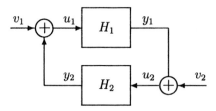

Figure B.1: Feedback connection.

Thus, (B.32) becomes

$$v(t) \leq k(r_0)\left\{\left[v_0 + \frac{2\rho}{c^2}\beta_1(r_0,0) + \frac{2}{c}\beta_2(r_0,0)\right]e^{-\frac{c}{4}t}\right.$$
$$\left. + \frac{2\rho}{c^2}\beta_1(r_0,t/2) + \frac{2}{c}\beta_2(r_0,t/2)\right\} + \frac{\rho}{c}. \tag{B.34}$$

By applying Young's inequality to the terms $k(r_0)\frac{2\rho}{c^2}\beta_1(r_0,0)e^{-\frac{c}{4}t}$ and $k(r_0)\frac{2\rho}{c^2}\beta_1(r_0,t/2)$, we obtain (B.25) with

$$\beta_v(r,t) = k(r)\left\{\left[r + \frac{k(r)}{c^2}\beta_1(r,0)^2 + \frac{2}{c}\beta_2(r,0)\right]e^{-\frac{c}{4}t}\right.$$
$$\left. + \frac{k(r)}{c^2}\beta_1(r,t/2)^2 + \frac{2}{c}\beta_2(r,t/2)\right\} \tag{B.35}$$

$$\gamma_v(r) = \frac{r}{c} + \frac{r^2}{c^2}. \tag{B.36}$$

The last statement of the lemma is immediate by substitution into (B.35). □

All the above results describe input-output properties of individual systems. To prepare for a basic result on feedback connections of systems, we first give a definition of $\mathcal{L}_p$ stability.

Definition B.9 *A mapping $H : \mathcal{L}_{p,e} \to \mathcal{L}_{p,e}$ is said to be $\mathcal{L}_p$ stable if there exist finite positive numbers γ and β such that $\|(Hu)_t\|_p \leq \gamma\|u_t\|_p + \beta$ for all $u \in \mathcal{L}_{p,e}$ and all $t \in [0,\infty)$.*

The following theorem, known as the *small gain theorem* [28], gives sufficient conditions for $\mathcal{L}_p$ stability of the feedback connection in Figure B.1.

Theorem B.10 *Consider the system in Figure B.1. Let $H_1, H_2 : \mathcal{L}_{p,e} \to \mathcal{L}_{p,e}$, $p \in [1,\infty]$, be two $\mathcal{L}_{p,e}$ stable operators with finite gains γ_1, γ_2 and associated constants β_1, β_2. Let the operator $H_1 H_2$ be strictly causal. If*

$$\gamma_1\gamma_2 < 1, \tag{B.37}$$

INPUT-OUTPUT STABILITY

then for all $v_1, v_2 \in \mathcal{L}_{p,e}$,

$$\|u_{1t}\|_p \leq \frac{1}{1-\gamma_1\gamma_2}(\|v_{1t}\|_p + \gamma_2\|v_{2t}\|_p + \beta_2 + \gamma_2\beta_1) \qquad (B.38)$$

$$\|u_{2t}\|_p \leq \frac{1}{1-\gamma_1\gamma_2}(\|v_{2t}\|_p + \gamma_1\|v_{1t}\|_p + \beta_1 + \gamma_1\beta_2) \qquad (B.39)$$

for all $t \in [0, \infty)$. If, in addition, $v_1, v_2 \in \mathcal{L}_p$, then (B.38)–(B.39) hold with all subscripts t dropped, which implies that u_1, u_2, y_1, and y_2 have bounded $\mathcal{L}_p$ norms.

Now we give a version of Gronwall's lemma. For the proof, the reader is referred to [51]. Similar lemmas can be found, among other references, in [28, 81, 165].

Lemma B.11 (Gronwall) *Consider the continuous functions* $\lambda : \mathbb{R}_+ \to \mathbb{R}$, $\mu : \mathbb{R}_+ \to \mathbb{R}_+$, *and* $\nu : \mathbb{R}_+ \to \mathbb{R}_+$, *where μ and ν are also nonnegative. If a continuous function* $y : \mathbb{R}_+ \to \mathbb{R}$ *satisfies the inequality*

$$y(t) \leq \lambda(t) + \mu(t)\int_{t_0}^{t} \nu(s)y(s)ds, \qquad \forall t \geq t_0 \geq 0, \qquad (B.40)$$

then

$$y(t) \leq \lambda(t) + \mu(t)\int_{t_0}^{t} \lambda(s)\nu(s)e^{\int_s^t \mu(\tau)\nu(\tau)d\tau}ds, \qquad \forall t \geq t_0 \geq 0. \qquad (B.41)$$

In particular, if $\lambda(t) \equiv \lambda$ is a constant and $\mu(t) \equiv 1$, then

$$y(t) \leq \lambda e^{\int_{t_0}^{t} \nu(\tau)d\tau}, \qquad \forall t \geq t_0 \geq 0. \qquad (B.42)$$

Appendix C

Input-to-State Stability

Input-to-state stability introduced by Sontag [173] plays a crucial role in our modular adaptive nonlinear designs. We extend Sontag's definition to time-varying systems:

Definition C.1 *The system*

$$\dot{x} = f(t, x, u), \qquad (C.1)$$

where f is piecewise continuous in t and locally Lipschitz in x and u, is said to be input-to-state stable (ISS) if there exist a class $\mathcal{KL}$ function β and a class $\mathcal{K}$ function γ, such that, for any $x(0)$ and for any input $u(\cdot)$ continuous and bounded on $[0, \infty)$ the solution exists for all $t \geq 0$ and satisfies

$$|x(t)| \leq \beta(|x(t_0)|, t - t_0) + \gamma\left(\sup_{t_0 \leq \tau \leq t} |u(\tau)|\right) \qquad (C.2)$$

for all t_0 and t such that $0 \leq t_0 \leq t$.

The following theorem establishes the connection between the existence of a Lyapunov-like function and the input-to-state stability.

Theorem C.2 [173, Claim on p. 441] *Suppose that for the system (C.1) there exists a C^1 function $V : \mathbb{R}_+ \times \mathbb{R}^n \to \mathbb{R}_+$ such that for all $x \in \mathbb{R}^n$ and $u \in \mathbb{R}^m$,*

$$\gamma_1(|x|) \leq V(t, x) \leq \gamma_2(|x|) \qquad (C.3)$$

$$|x| \geq \rho(|u|) \quad \Rightarrow \quad \frac{\partial V}{\partial t} + \frac{\partial V}{\partial x} f(t, x, u) \leq -\gamma_3(|x|), \qquad (C.4)$$

where γ_1, γ_2, and ρ are class $\mathcal{K}_\infty$ functions and γ_3 is a class $\mathcal{K}$ function. Then the system (C.1) is ISS with $\gamma = \gamma_1^{-1} \circ \gamma_2 \circ \rho$.

Proof (Outline). If $x(t_0)$ is in the set

$$R_{t_0} = \left\{ x \in \mathbb{R}^n \,\bigg|\, |x| \leq \rho\left(\sup_{\tau \geq t_0} |u(\tau)|\right) \right\}, \tag{C.5}$$

then $x(t)$ remains within the set

$$S_{t_0} = \left\{ x \in \mathbb{R}^n \,\bigg|\, |x| \leq \gamma_1^{-1} \circ \gamma_2 \circ \rho\left(\sup_{\tau \geq t_0} |u(\tau)|\right) \right\} \tag{C.6}$$

for all $t \geq t_0$. Define $B = [t_0, T)$ as the time interval before $x(t)$ enters R_{t_0} for the first time. In view of the definition of R_{t_0}, we have

$$\dot{V} \leq -\gamma_3 \circ \gamma_2^{-1}(V), \qquad \forall t \in B. \tag{C.7}$$

Then, by [173, Lemma 6.1], there exists a class $\mathcal{KL}$ function β_V such that

$$V(t) \leq \beta_V(V(t_0), t - t_0), \qquad \forall t \in B, \tag{C.8}$$

which implies

$$|x(t)| \leq \gamma_1^{-1}(\beta_V(\gamma_2(|x(t_0)|), t - t_0)) \triangleq \beta(|x(t_0)|, t - t_0), \qquad \forall t \in B. \tag{C.9}$$

On the other hand, by (C.6), we conclude

$$|x(t)| \leq \gamma_1^{-1} \circ \gamma_2 \circ \rho\left(\sup_{\tau \geq t_0} |u(\tau)|\right) \triangleq \gamma\left(\sup_{\tau \geq t_0} |u(\tau)|\right), \qquad \forall t \in [t_0, \infty] \setminus B. \tag{C.10}$$

Then, by (C.9) and (C.10),

$$|x(t)| \leq \beta(|x(t_0)|, t - t_0) + \gamma\left(\sup_{\tau \geq t_0} |u(\tau)|\right), \qquad \forall t \geq t_0 \geq 0. \tag{C.11}$$

By causality, it follows that

$$|x(t)| \leq \beta(|x(t_0)|, t - t_0) + \gamma\left(\sup_{t_0 \leq \tau \leq t} |u(\tau)|\right), \qquad \forall t \geq t_0 \geq 0. \tag{C.12}$$

$\square$

A function V satisfying conditions of Theorem C.2 is called an *ISS-Lyapunov function*. Sontag, Wang, and Lin recently proved that the inverse of Theorem C.2 is also true. They also introduced an equivalent dissipativity-type characterization of ISS.

Theorem C.3 [115, 177] *For the system*

$$\dot{x} = f(x, u),$$

the following properties are equivalent:

1. *the system is ISS,*

2. *there exists a smooth ISS-Lyapunov function,*

3. *there exists a smooth positive definite radially unbounded function V and class $\mathcal{K}_\infty$ functions ρ_1 and ρ_2 such that the following dissipativity inequality is satisfied:*

$$\frac{\partial V}{\partial x} f(x, u) \leq -\rho_1(|x|) + \rho_2(|u|).$$

The following lemma establishes a useful property that a cascade of two ISS systems is itself ISS.

Lemma C.4 *Suppose that in the system*

$$\begin{aligned}
\dot{x}_1 &= f_1(t, x_1, x_2, u) & \text{(C.13)} \\
\dot{x}_2 &= f_2(t, x_2, u) & \text{(C.14)}
\end{aligned}$$

the x_1-subsystem is ISS with respect to x_2 and u, and the x_2-subsystem is ISS with respect to u, that is,

$$|x_1(t)| \leq \beta_1(|x_1(s)|, t-s) + \gamma_1\left(\sup_{s \leq \tau \leq t} \{|x_2(\tau)| + |u(\tau)|\}\right) \quad \text{(C.15)}$$

$$|x_2(t)| \leq \beta_2(|x_2(s)|, t-s) + \gamma_2\left(\sup_{s \leq \tau \leq t} |u(\tau)|\right), \quad \text{(C.16)}$$

where β_1 and β_2 are class $\mathcal{KL}$ functions and γ_1 and γ_2 are class $\mathcal{K}$ functions. Then the complete $x = (x_1, x_2)$-system is ISS with

$$|x(t)| \leq \beta(|x(s)|, t-s) + \gamma\left(\sup_{s \leq \tau \leq t} |u(\tau)|\right), \quad \text{(C.17)}$$

where

$$\begin{aligned}
\beta(r, t) &= \beta_1(2\beta_1(r, t/2) + 2\gamma_1(2\beta_2(r, 0)), t/2) \\
&\quad + \gamma_1(2\beta_2(r, t/2)) + \beta_2(r, t), & \text{(C.18)} \\
\gamma(r) &= \beta_1(2\gamma_1(2\gamma_2(r) + 2r), 0) + \gamma_1(2\gamma_2(r) + 2r) + \gamma_2(r). & \text{(C.19)}
\end{aligned}$$

Proof. With $(s,t) = (t/2, t)$, (C.15) is rewritten as

$$|x_1(t)| \leq \beta_1(|x_1(t/2)|, t/2) + \gamma_1 \left(\sup_{t/2 \leq \tau \leq t} \{|x_2(\tau)| + |u(\tau)|\} \right). \quad \text{(C.20)}$$

From (C.16) we have

$$\sup_{t/2 \leq \tau \leq t} |x_2(\tau)| \leq \sup_{t/2 \leq \tau \leq t} \left\{ \beta_2(|x_2(0)|, \tau) + \gamma_2 \left(\sup_{0 \leq \sigma \leq \tau} |u(\sigma)| \right) \right\}$$

$$\leq \beta_2(|x_2(0)|, t/2) + \gamma_2 \left(\sup_{0 \leq \tau \leq t} |u(\tau)| \right), \quad \text{(C.21)}$$

and from (C.15) we obtain

$$|x_1(t/2)| \leq \beta_1(|x_1(0)|, t/2) + \gamma_1 \left(\sup_{0 \leq \tau \leq t/2} \{|x_2(\tau)| + |u(\tau)|\} \right)$$

$$\leq \beta_1(|x_1(0)|, t/2)$$

$$+ \gamma_1 \left(\sup_{0 \leq \tau \leq t/2} \left\{ \beta_2(|x_2(0)|, \tau) + \gamma_2 \left(\sup_{0 \leq \sigma \leq \tau} |u(\sigma)| \right) + |u(\tau)| \right\} \right)$$

$$\leq \beta_1(|x_1(0)|, t/2)$$

$$+ \gamma_1 \left(\beta_2(|x_2(0)|, 0) + \sup_{0 \leq \tau \leq t/2} \{\gamma_2(|u(\tau)|) + |u(\tau)|\} \right)$$

$$\leq \beta_1(|x_1(0)|, t/2) + \gamma_1 (2\beta_2(|x_2(0)|, 0))$$

$$+ \gamma_1 \left(2 \sup_{0 \leq \tau \leq t/2} \{\gamma_2(|u(\tau)|) + |u(\tau)|\} \right), \quad \text{(C.22)}$$

where in the last inequality we have used the fact that $\delta(a+b) \leq \delta(2a) + \delta(2b)$, for any class $\mathcal{K}$ function δ and any nonnegative a and b. Then, substituting (C.21) and (C.22) into (C.20) we get

$$|x_1(t)| \leq \beta_1 \left(\beta_1(|x_1(0)|, t/2) + \gamma_1(2\beta_2(|x_2(0)|, 0)) \right.$$

$$\left. + \gamma_1 \left(2 \sup_{0 \leq \tau \leq t/2} \{\gamma_2(|u(\tau)|) + |u(\tau)|\} \right) \right)$$

$$+ \gamma_1 \left(\beta_2(|x_2(0)|, t/2) + \gamma_2 \left(\sup_{0 \leq \tau \leq t} |u(\tau)| \right) + \sup_{t/2 \leq \tau \leq t} \{|u(\tau)|\} \right)$$

$$\leq \beta_1(2\beta_1(|x_1(0)|, t/2) + 2\gamma_1(2\beta_2(|x_2(0)|, 0)), t/2)$$

$$+ \gamma_1(2\beta_2(|x_2(0)|, t/2))$$

$$+ \beta_1 \left(2\gamma_1 \left(2 \sup_{0 \leq \tau \leq t} \{\gamma_2(|u(\tau)|) + |u(\tau)|\} \right), 0 \right)$$

$$+ \gamma_1 \left(2 \sup_{0 \leq \tau \leq t} \{\gamma_2(|u(\tau)|) + |u(\tau)|\} \right). \quad \text{(C.23)}$$

Input-to-State Stability

Combining (C.23) and (C.16) we arrive at (C.17) with (C.18)–(C.19). □

The proof of Lemma C.4 follows the lines of the proof of [173, Proposition 7.2]. An alternative proof has been given in [176] by using part 3 of Theorem C.3. Since (C.13) and (C.14) are ISS, then there exist ISS-Lyapunov functions V_1 and V_2 and class $\mathcal{K}_\infty$ functions $\alpha_1, \rho_1, \alpha_2,$ and ρ_2 such that

$$\frac{\partial V_1}{\partial x_1} f_1(t, x_1, x_2, u) \leq -\alpha_1(|x_1|) + \rho_1(|x_2|) + \rho_1(|u|) \qquad (C.24)$$

$$\frac{\partial V_2}{\partial x_2} f_2(t, x_2, u) \leq -\alpha_2(|x_2|) + \rho_2(|u|). \qquad (C.25)$$

It is shown in [176] that $V_1, V_2, \alpha_1, \rho_1, \alpha_2, \rho_2$ can be found such that

$$\rho_1 = \alpha_2/2. \qquad (C.26)$$

Then the ISS-Lyapunov function for the complete system (C.13)–(C.14) can be defined as

$$V(x) = V_1(x_1) + V_2(x_2), \qquad (C.27)$$

and its derivative

$$\dot{V} \leq -\alpha_1(|x_1|) - \frac{1}{2}\alpha_2(|x_2|) + \rho_1(|u|) + \rho_2(|u|) \qquad (C.28)$$

establishes the ISS property of (C.13)–(C.14) by part 3 of Theorem C.3.

In many of our applications of input-to-state stability, we use the following lemma which is much simpler than Theorem C.2.

Lemma C.5 *Let v and ρ be real-valued functions defined on $\mathbb{R}_+$, and let b and c be positive constants. If they satisfy the differential inequality*

$$\dot{v} \leq -cv + b\rho(t)^2, \qquad v(0) \geq 0, \qquad (C.29)$$

then the following holds:

(i) If $\rho \in \mathcal{L}_\infty$, then $v \in \mathcal{L}_\infty$ and

$$v(t) \leq v(0)e^{-ct} + \frac{b}{c}\|\rho\|_\infty^2. \qquad (C.30)$$

(ii) If $\rho \in \mathcal{L}_2$, then $v \in \mathcal{L}_\infty$ and

$$v(t) \leq v(0)e^{-ct} + b\|\rho\|_2^2. \qquad (C.31)$$

Proof. *(i)* From Lemma B.5, we have

$$\begin{aligned}
v(t) &\leq v(0)e^{-ct} + b\int_0^t e^{-c(t-\tau)}\rho(\tau)^2 d\tau \\
&\leq v(0)e^{-ct} + b\sup_{\tau\in[0,t]}\{\rho(\tau)^2\}\int_0^t e^{-c(t-\tau)}d\tau \\
&\leq v(0)e^{-ct} + b\|\rho\|_\infty^2 \frac{1}{c}\left(1 - e^{-ct}\right) \\
&\leq v(0)e^{-ct} + \frac{b}{c}\|\rho\|_\infty^2.
\end{aligned} \qquad (C.32)$$

(ii) From (B.13) we have

$$\begin{aligned}
v(t) &\leq v(0)e^{-ct} + b\sup_{\tau\in[0,t]}\left\{e^{-c(t-\tau)}\right\}\int_0^t \rho(\tau)^2 d\tau \\
&= v(0)e^{-ct} + b\|\rho\|_2^2.
\end{aligned} \qquad (C.33)$$

□

Remark C.6 From Lemma C.5, it follows that if

$$\dot{v} \leq -cv + b_1\rho_1(t)^2 + b_2\rho_2(t)^2, \qquad v(0) \geq 0 \qquad (C.34)$$

and $\rho_1 \in \mathcal{L}_\infty$ and $\rho_2 \in \mathcal{L}_2$, then $v \in \mathcal{L}_\infty$ and

$$v(t) \leq v(0)e^{-ct} + \frac{b_1}{c}\|\rho_1\|_\infty^2 + b_2\|\rho_2\|_2^2. \qquad (C.35)$$

This, in particular, implies the input-to-state stability with respect to two inputs: ρ_1 and $\|\rho_2\|_2$. ◇

Appendix D

Passivity

Now we briefly review some basic passivity results. The concept of passivity, which was first used in network synthesis, became a fundamental feedback control concept in the seminal work of Popov [155]. Its applications to adaptive control were pioneered by Parks [150] and Landau [109], and to other areas of systems and control theory by Willems [198] and Hill and Moylan [45]. We use the passivity definitions of Byrnes, Isidori, and Willems [14] extended to time-varying nonlinear systems. Consider systems of the form

$$\begin{aligned} \dot{x} &= f(x,t) + g(x,t)u \\ y &= h(x,t), \end{aligned} \qquad \text{(D.1)}$$

with $x \in \mathbb{R}^n$, $y \in \mathbb{R}^m$, $u \in \mathbb{R}^m$, and f, g, h continuous in t and smooth in x. Suppose $f(0,t) = 0$ and $h(0,t) = 0$ for all $t \geq 0$.

Definition D.1 *The system (D.1) is said to be* passive *if there exists a continuous nonnegative ("storage") function* $V : \mathbb{R}^n \times \mathbb{R}_+ \to \mathbb{R}_+$, *which satisfies* $V(0,t) = 0$, $\forall t \geq 0$, *such that for all* $u \in C^0$, $x(0) \in \mathbb{R}^n$, $t \geq t_0 \geq 0$

$$\int_{t_0}^{t} y^{\mathrm{T}}(\sigma) u(\sigma) d\sigma \geq V(x(t),t) - V(x(t_0),t_0). \qquad \text{(D.2)}$$

Definition D.2 *The system (D.1) is said to be* strictly passive *if there exist a continuous nonnegative (storage) function* $V : \mathbb{R}^n \times \mathbb{R}_+ \to \mathbb{R}_+$, *which satisfies* $V(0,t) = 0$, $\forall t \geq 0$, *and a positive definite function (dissipation rate)* $\psi : \mathbb{R}^n \to \mathbb{R}_+$, *such that for all* $u \in C^0$, $x(0) \in \mathbb{R}^n$, $t \geq t_0 \geq 0$

$$\int_{t_0}^{t} y^{\mathrm{T}}(\sigma) u(\sigma) d\sigma \geq V(x(t),t) - V(x(t_0),t_0) + \int_{t_0}^{t} \psi(x(\sigma)) d\sigma. \qquad \text{(D.3)}$$

Passivity and Lyapunov stability are closely related concepts.

Lemma D.3 *Suppose the system (D.1) is (strictly) passive. If V is positive definite, radially unbounded, and decrescent, that is, if there exist class $\mathcal{K}_\infty$*

functions γ_1 and γ_2 such that $\gamma_1(|x|) \leq V(x,t) \leq \gamma_2(|x|)$, $\forall (x,t) \in \mathbb{R}^n \times \mathbb{R}_+$, then, for $u \equiv 0$, the equilibrium $x = 0$ of (D.1) is globally uniformly (asymptotically) stable.

Proof. When $u \equiv 0$, in the case of strict passivity, differentiating (D.3), we have
$$\dot{V} \leq -\psi(x). \tag{D.4}$$
Thus, the equilibrium $x = 0$ is globally uniformly asymptotically stable. The case of passivity is analogous. □

Many problems in parameter identification and adaptive control can be studied as feedback interconnections of passive systems (see Figure D.1):

$$\Sigma_1 : \quad \begin{aligned} \dot{x}_1 &= f_1(x,t) + g_1(x,t)u_1 \\ y_1 &= h_1(x,t) \end{aligned} \tag{D.5}$$

$$\Sigma_2 : \quad \begin{aligned} \dot{x}_2 &= f_2(x,t) + g_2(x,t)u_2 \\ y_2 &= h_2(x,t) \end{aligned} \tag{D.6}$$

connected by the relations

$$u_1 = -y_2 + v_1 \tag{D.7}$$
$$u_2 = y_1, \tag{D.8}$$

where v_1 is a new input to the system.

Theorem D.4 *Suppose the system Σ_1 is (strictly) passive with storage function V_1 (and dissipation rate ψ_1) independent of x_2. Likewise, suppose the system Σ_2 is (strictly) passive with storage function V_2 (and dissipation rate ψ_2) independent of x_1. Then the interconnected system (D.5)–(D.8) with input v_1 and output y_1 is*

1. *strictly passive if both Σ_1 and Σ_2 are strictly passive,*

2. *passive if at least one of the systems Σ_1 and Σ_2 is passive but not strictly passive.*

Moreover, when $v_1 \equiv 0$, if Σ_1 is strictly passive and Σ_2 is passive, then the equilibrium $x = 0$ is globally uniformly stable and $\lim_{t \to \infty} x_1(t) = 0$.

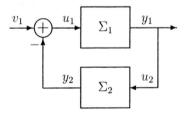

Figure D.1: Feedback interconnection of two passive systems.

Passivity

Proof. Let us first assume that Σ_1 and Σ_2 are both strictly passive. Then, in view of (D.7)–(D.8) we have

$$\int_{t_0}^{t} y_1^\mathrm{T}[v_1 - y_2]d\sigma \geq V_1(x_1(t),t) - V_1(x_1(t_0),t_0) + \int_{t_0}^{t} \psi_1(x_1)d\sigma \quad (\text{D.9})$$

$$\int_{t_0}^{t} y_2^\mathrm{T} y_1 d\sigma \geq V_2(x_2(t),t) - V_2(x_2(t_0),t_0) + \int_{t_0}^{t} \psi_2(x_2)d\sigma. \quad (\text{D.10})$$

Adding inequalities (D.9) and (D.10), we obtain

$$\int_{t_0}^{t} y_1^\mathrm{T}(\sigma)v_1(\sigma)d\sigma \geq V(x(t),t) - V(x(t_0),t_0) + \int_{t_0}^{t} \psi(x(\sigma))d\sigma, \quad (\text{D.11})$$

where the storage function V and the dissipation rate ψ for the complete x-system are defined as

$$V(t,x) = V_1(x_1,t) + V_2(x_2,t) \quad (\text{D.12})$$
$$\psi(x) = \psi_1(x_1) + \psi_2(x_2). \quad (\text{D.13})$$

Since V is positive definite, radially unbounded and decrescent, and ψ is positive definite, this proves the strict passivity. If at least one of the systems Σ_1 and Σ_2 is passive but not strictly passive, then its dissipation rate ψ_i is at best positive semidefinite but not positive definite, and the overall system is only passive. Finally, when $v_1 \equiv 0$, if Σ_1 is strictly passive and Σ_2 is passive, then ψ_2 is positive semidefinite, and by differentiating (D.11) we get

$$\dot{V} \leq -\psi_1(x_1). \quad (\text{D.14})$$

Thus, by Theorem 2.1, $x = 0$ is globally uniformly stable and $\lim_{t\to\infty} x_1(t) = 0$. $\square$

Now we turn our attention to linear time-invariant passive systems.

Definition D.5 *A rational transfer function $G(s)$ is said to be* positive real *if $G(s)$ is real for all real s, and $\mathfrak{Re}\{G(s)\} \geq 0$ for all $\mathfrak{Re}\{s\} \geq 0$. If, in addition, $G(s-\epsilon)$ is positive real for some $\epsilon > 0$, then $G(s)$ is said to be* strictly positive real.

For completeness, we quote the celebrated Popov-Kalman-Yakubovich lemma. A recent version of its proof can be found in Tao and Ioannou [183].

Lemma D.6 *Let the strictly positive real transfer function $G(s)$ have the state-space representation (A,b,c,d), $d \geq 0$. Then, for any given $L = L^\mathrm{T} > 0$, there exists a scalar $\nu > 0$, a vector q, and a $P = P^\mathrm{T} > 0$ such that*

$$A^\mathrm{T} P + PA = -qq^\mathrm{T} - \nu L \quad (\text{D.15})$$
$$Pb - c^\mathrm{T} = q\sqrt{2d}. \quad (\text{D.16})$$

With this lemma a Lyupunov function $V = x^T P x$ can be constructed such that P satisfies not only the Lyapunov equation (D.15) but also the input-output condition (D.16) from which the restriction to relative degree zero ($d > 0$) or one ($d = 0$, $cb > 0$) is apparent. The main utility of this special Lyapunov function for adaptive and cascade designs is that the indefinite term in its derivative $\dot{V}$ depends on the output y and not on the whole state x.

Appendix E

Parameter Projection

The modular adaptive controllers have a point of singularity $\hat{b}_m = 0$, where $\hat{b}_m$ is the estimate of the high-frequency gain (virtual control coefficient) b_m. In order to prevent $\hat{b}_m$ from taking the value zero, we use the parameter projection in our identifiers. For this, we need to know the sign of the actual high-frequency gain b_m. We first give a treatment of projection for a general convex parameter set and then specialize to the case where only the high-frequency gain is constrained.

Let us define the following convex set

$$\Pi = \left\{ \hat{\theta} \in \mathbb{R}^p \,\middle|\, \mathcal{P}(\hat{\theta}) \leq 0 \right\}, \tag{E.1}$$

where by assuming that the convex function $\mathcal{P} : \mathbb{R}^p \to \mathbb{R}$ is smooth, we assure that the boundary $\partial \Pi$ of Π is smooth. Let us denote the interior of Π by $\overset{\circ}{\Pi}$ and observe that $\nabla_{\hat{\theta}} \mathcal{P}$ represents an outward normal vector at $\hat{\theta} \in \partial \Pi$. The standard projection operator is

$$\text{Proj}\{\tau\} = \begin{cases} \tau, & \hat{\theta} \in \overset{\circ}{\Pi} \text{ or } \nabla_{\hat{\theta}} \mathcal{P}^\text{T} \tau \leq 0 \\ \left(I - \Gamma \dfrac{\nabla_{\hat{\theta}} \mathcal{P} \, \nabla_{\hat{\theta}} \mathcal{P}^\text{T}}{\nabla_{\hat{\theta}} \mathcal{P}^\text{T} \Gamma \nabla_{\hat{\theta}} \mathcal{P}} \right) \tau, & \hat{\theta} \in \partial \Pi \text{ and } \nabla_{\hat{\theta}} \mathcal{P}^\text{T} \tau > 0, \end{cases} \tag{E.2}$$

where Γ belongs to the set $\mathcal{G}$ of all positive definite symmetric $p \times p$ matrices. Although Proj is a function of three arguments, τ, $\hat{\theta}$ and Γ, for compactness of notation we write only $\text{Proj}\{\tau\}$.

The meaning of (E.2) is that, when $\hat{\theta}$ is in the interior of Π or at the boundary with τ pointing inward, then $\text{Proj}\{\tau\} = \tau$. When $\hat{\theta}$ is at the boundary with τ pointing outward, then Proj projects τ on the hyperplane tangent to $\partial \Pi$ at $\hat{\theta}$.

In general, the mapping (E.2) is discontinuous. This is undesirable for two reasons. First, the discontinuity represents a difficulty for implementation in continuous time. Second, since the Lipschitz continuity is violated, we cannot use standard theorems for existence of solutions. Therefore, we need to smooth the projection operator. Let us consider the following convex set

$$\Pi_\varepsilon = \left\{\hat{\theta} \in \mathbb{R}^p \,\middle|\, \mathcal{P}(\hat{\theta}) \leq \varepsilon\right\}, \tag{E.3}$$

which is a union of the set Π and an $O(\varepsilon)$-boundary layer around it. We now modify (E.2) to achieve continuity of the transition from the vector field τ on the boundary of Π to the vector field $\left(I - \Gamma \frac{\nabla_{\hat{\theta}}\mathcal{P}\, \nabla_{\hat{\theta}}\mathcal{P}^{\mathrm{T}}}{\nabla_{\hat{\theta}}\mathcal{P}^{\mathrm{T}}\Gamma\nabla_{\hat{\theta}}\mathcal{P}}\right)\tau$ on the boundary of Π_ε:

$$\mathrm{Proj}\{\tau\} = \begin{cases} \tau, & \hat{\theta} \in \overset{\circ}{\Pi} \text{ or } \nabla_{\hat{\theta}}\mathcal{P}^{\mathrm{T}}\tau \leq 0 \\ \left(I - c(\hat{\theta})\Gamma\frac{\nabla_{\hat{\theta}}\mathcal{P}\,\nabla_{\hat{\theta}}\mathcal{P}^{\mathrm{T}}}{\nabla_{\hat{\theta}}\mathcal{P}^{\mathrm{T}}\Gamma\nabla_{\hat{\theta}}\mathcal{P}}\right)\tau, & \hat{\theta} \in \Pi_\varepsilon \setminus \overset{\circ}{\Pi} \text{ and } \nabla_{\hat{\theta}}\mathcal{P}^{\mathrm{T}}\tau > 0 \end{cases} \tag{E.4}$$

$$c(\hat{\theta}) = \min\left\{1, \frac{\mathcal{P}(\hat{\theta})}{\varepsilon}\right\}. \tag{E.5}$$

It is helpful to note that $c(\partial\Pi) = 0$ and $c(\partial\Pi_\varepsilon) = 1$.

In the proofs of stability of identifiers we need the following technical properties of the projection operator (E.4).

Lemma E.1 (Projection Operator) *The following are the properties of the projection operator (E.4):*

(i) The mapping $\mathrm{Proj} : \mathbb{R}^p \times \Pi_\varepsilon \times \mathcal{G} \to \mathbb{R}^p$ *is locally Lipschitz in its arguments* $\tau, \hat{\theta}, \Gamma$.

(ii) $\mathrm{Proj}\{\tau\}^{\mathrm{T}}\Gamma^{-1}\mathrm{Proj}\{\tau\} \leq \tau^{\mathrm{T}}\Gamma^{-1}\tau$, $\forall \hat{\theta} \in \Pi_\varepsilon$.

(iii) Let $\Gamma(t), \tau(t)$ *be continuously differentiable and*

$$\dot{\hat{\theta}} = \mathrm{Proj}\{\tau\}, \qquad \hat{\theta}(0) \in \Pi_\varepsilon.$$

Then, on its domain of definition, the solution $\hat{\theta}(t)$ *remains in* Π_ε.

(iv) $-\tilde{\theta}^{\mathrm{T}}\Gamma^{-1}\mathrm{Proj}\{\tau\} \leq -\tilde{\theta}^{\mathrm{T}}\Gamma^{-1}\tau$, $\forall \hat{\theta} \in \Pi_\varepsilon, \theta \in \Pi$.

Proof. *(i)* The proof of this point is lengthy but straightforward. The reader is referred to [157, Lemma (103)].

(ii) For $\hat{\theta} \in \overset{\circ}{\Pi}$ or $\nabla_{\hat{\theta}}\mathcal{P}^{\mathrm{T}}\tau \leq 0$, we have $\mathrm{Proj}\{\tau\} = \tau$ and *(ii)* trivially holds with equality. Otherwise, a direct computation gives

$$\begin{aligned}
\mathrm{Proj}\{\tau\}^{\mathrm{T}}\Gamma^{-1}\mathrm{Proj}\{\tau\} &= \tau^{\mathrm{T}}\Gamma^{-1}\tau - 2c(\hat{\theta})\frac{\left(\nabla_{\hat{\theta}}\mathcal{P}^{\mathrm{T}}\tau\right)^2}{\nabla_{\hat{\theta}}\mathcal{P}^{\mathrm{T}}\Gamma\nabla_{\hat{\theta}}\mathcal{P}} + c(\hat{\theta})^2\frac{\left|\nabla_{\hat{\theta}}\mathcal{P}\nabla_{\hat{\theta}}\mathcal{P}^{\mathrm{T}}\tau\right|^2_\Gamma}{(\nabla_{\hat{\theta}}\mathcal{P}^{\mathrm{T}}\Gamma\nabla_{\hat{\theta}}\mathcal{P})^2} \\
&= \tau^{\mathrm{T}}\Gamma^{-1}\tau - c(\hat{\theta})\left(2 - c(\hat{\theta})\right)\frac{\left(\nabla_{\hat{\theta}}\mathcal{P}^{\mathrm{T}}\tau\right)^2}{\nabla_{\hat{\theta}}\mathcal{P}^{\mathrm{T}}\Gamma\nabla_{\hat{\theta}}\mathcal{P}} \\
&\leq \tau^{\mathrm{T}}\Gamma^{-1}\tau, \tag{E.6}
\end{aligned}$$

Parameter Projection

where the last inequality follows by noting that $c(\hat{\theta}) \in [0,1]$ for $\hat{\theta} \in \Pi_\varepsilon \setminus \overset{\circ}{\Pi}$.

(iii) Using the definition of the Proj operator, we get

$$\nabla_{\hat{\theta}} \mathcal{P}^{\mathrm{T}} \operatorname{Proj}\{\tau\} = \begin{cases} \nabla_{\hat{\theta}} \mathcal{P}^{\mathrm{T}} \tau, & \hat{\theta} \in \overset{\circ}{\Pi} \text{ or } \nabla_{\hat{\theta}} \mathcal{P}^{\mathrm{T}} \tau \leq 0 \\ \left(1 - c(\hat{\theta})\right) \nabla_{\hat{\theta}} \mathcal{P}^{\mathrm{T}} \tau, & \hat{\theta} \in \Pi_\varepsilon \setminus \overset{\circ}{\Pi} \text{ and } \nabla_{\hat{\theta}} \mathcal{P}^{\mathrm{T}} \tau > 0, \end{cases} \quad \text{(E.7)}$$

which, in view of the fact that $c(\hat{\theta}) \in [0,1]$ for $\hat{\theta} \in \Pi_\varepsilon \setminus \overset{\circ}{\Pi}$, implies that

$$\nabla_{\hat{\theta}} \mathcal{P}^{\mathrm{T}} \operatorname{Proj}\{\tau\} \leq 0 \quad \text{whenever } \hat{\theta} \in \partial \Pi_\varepsilon, \tag{E.8}$$

that is, the vector $\operatorname{Proj}\{\tau\}$ either points inside Π_ε or is tangential to the hyperplane of $\partial \Pi_\varepsilon$ at $\hat{\theta}$. Since $\hat{\theta}(0) \in \Pi_\varepsilon$, it follows that $\hat{\theta}(t) \in \Pi_\varepsilon$ as long as the solution exists.

(iv) For $\hat{\theta} \in \overset{\circ}{\Pi}$, *(iv)* trivially holds with equality. For $\hat{\theta} \in \Pi_\varepsilon \setminus \overset{\circ}{\Pi}$, since $\theta \in \Pi$ and $\mathcal{P}$ is a convex function, we have

$$(\theta - \hat{\theta})^{\mathrm{T}} \nabla_{\hat{\theta}} \mathcal{P} \leq 0 \quad \text{whenever } \hat{\theta} \in \Pi_\varepsilon \setminus \overset{\circ}{\Pi}. \tag{E.9}$$

With (E.9) we now calculate

$$\begin{aligned} -\tilde{\theta}^{\mathrm{T}} \Gamma^{-1} \operatorname{Proj}\{\tau\} &= -\tilde{\theta}^{\mathrm{T}} \Gamma^{-1} \tau \\ &\quad + \begin{cases} 0, & \hat{\theta} \in \overset{\circ}{\Pi} \text{ or } \nabla_{\hat{\theta}} \mathcal{P}^{\mathrm{T}} \tau \leq 0 \\ c(\hat{\theta}) \dfrac{(\tilde{\theta}^{\mathrm{T}} \nabla_{\hat{\theta}} \mathcal{P})(\nabla_{\hat{\theta}} \mathcal{P}^{\mathrm{T}} \tau)}{\nabla_{\hat{\theta}} \mathcal{P}^{\mathrm{T}} \Gamma \nabla_{\hat{\theta}} \mathcal{P}}, & \hat{\theta} \in \Pi_\varepsilon \setminus \overset{\circ}{\Pi} \text{ and } \nabla_{\hat{\theta}} \mathcal{P}^{\mathrm{T}} \tau > 0 \end{cases} \\ &\leq -\tilde{\theta}^{\mathrm{T}} \Gamma^{-1} \tau, \end{aligned} \tag{E.10}$$

which completes the proof. $\square$

Since we intend to use the projection operator only to keep the estimate $\hat{b}_m$ of the high-frequency gain b_m from becoming zero, we now specialize the projection operator for this case. We assume that $|b_m| \geq \varsigma_m > 0$, where $\operatorname{sgn} b_m$ and ς_m are known. Recalling that b_m is the first element of the parameter vector θ, i.e., $\theta = [b_m, \theta_2, \ldots, \theta_p]^{\mathrm{T}}$, we define $\mathcal{P}(\hat{\theta}) = \varsigma_m - \hat{b}_m \operatorname{sgn} b_m$ and note that $\nabla_{\hat{\theta}} \mathcal{P} = -\operatorname{sgn} b_m e_1^{\mathrm{T}}$. Let us denote the nominal vector field for the parameter update law by $\tau = [\tau_1, \tau_2, \ldots, \tau_p]^{\mathrm{T}}$ and choose $\varepsilon \in (0, \varsigma_m)$. The update law of the form $\dot{\hat{\theta}} = \operatorname{Proj}\{\tau\}$ using the projection operator (E.4)–(E.5) is given by

$$\dot{\hat{b}}_m = \tau_1 \begin{cases} 1, & \hat{b}_m \operatorname{sgn} b_m > \varsigma_m \text{ or } \tau_1 \operatorname{sgn} b_m \geq 0 \\ \max\left\{0, \dfrac{\varepsilon - \varsigma_m + \hat{b}_m \operatorname{sgn} b_m}{\varepsilon}\right\}, & \hat{b}_m \operatorname{sgn} b_m \leq \varsigma_m \text{ and } \tau_1 \operatorname{sgn} b_m < 0 \end{cases} \tag{E.11}$$

$$\dot{\hat{\theta}}_i = \begin{cases} \tau_i, & \hat{b}_m \operatorname{sgn} b_m > \varsigma_m \text{ or } \tau_1 \operatorname{sgn} b_m \geq 0 \\ \tau_i - \tau_1 \dfrac{\Gamma_{1i}}{\Gamma_{11}} \min\left\{1, \dfrac{\varsigma_m - \hat{b}_m \operatorname{sgn} b_m}{\varepsilon}\right\}, & \hat{b}_m \operatorname{sgn} b_m \leq \varsigma_m \text{ and } \tau_1 \operatorname{sgn} b_m < 0 \end{cases} \tag{E.12}$$

where $i = 2, \ldots, p$, and Γ_{1i} is the $(1, i)$ element of the positive definite symmetric matrix Γ. This update law achieves $\hat{b}_m(t)\,\text{sgn}\,b_m \geq \varsigma_m - \varepsilon, \forall t \geq 0$ whenever $\hat{b}_m(0)\,\text{sgn}\,b_m \geq \varsigma_m - \varepsilon$.

From expression (E.12) one should observe that when the update law is gradient it can be simplified to $\dot{\hat{\theta}}_i = \tau_i$ by the choice of the adaptation gain matrix $\Gamma = \text{diag}\{\gamma_1, \Gamma_2\}$, where γ_1 is a positive scalar and Γ_2 is a symmetric positive definite $(p-1) \times (p-1)$ matrix. Thus, projection is applied only to $\hat{b}_m$. This simplification is not possible with the least-squares update law because $\Gamma(t)$ is not a constant matrix, so it does not maintain a block-diagonal structure even if $\Gamma(0)$ is block-diagonal.

By setting $\varepsilon = 0$ in (E.11) and (E.12), we obtain update laws with the standard discontinuous projection, which in the case of gradient estimation simplifies to

$$\dot{\hat{b}}_m = \begin{cases} \tau_1 & \hat{b}_m\,\text{sgn}\,b_m > \varsigma_m \text{ or } \tau_1\,\text{sgn}\,b_m \geq 0 \\ 0 & \hat{b}_m\,\text{sgn}\,b_m \leq \varsigma_m \text{ and } \tau_1\,\text{sgn}\,b_m < 0 \end{cases} \quad \text{(E.13)}$$

For the most part, the properties of the projection operator given in this appendix are recapitulated from [157] (see also [43, 51, 165]).

Appendix F

Nonlinear Swapping

The well-known Swapping Lemma [138] is ubiquitous in adaptive linear control. Here we provide its nonlinear counterpart.

Lemma F.1 (Nonlinear Swapping Lemma) *Consider the nonlinear time-varying system*

$$\Sigma_1 : \quad \begin{aligned} \dot{z} &= A(z,t)z + g(z,t)W(z,t)^{\mathrm{T}}\tilde{\theta} - Q(z,t)^{\mathrm{T}}\dot{\tilde{\theta}} + E(z,t)e \\ y_1 &= h(z,t)z + l(z,t)W(z,t)^{\mathrm{T}}\tilde{\theta} \end{aligned} \quad \text{(F.1)}$$

where $\tilde{\theta} : \mathbb{R}_+ \to \mathbb{R}^p$ *is differentiable,* $e : \mathbb{R}_+ \to \mathbb{R}$ *is continuous and* $\lim_{t\to\infty} e(t) = 0$, *the matrix-valued functions* $A : \mathbb{R}^n \times \mathbb{R}_+ \to \mathbb{R}^{n\times n}$, $g : \mathbb{R}^n \times \mathbb{R}_+ \to \mathbb{R}^{n\times m}$, $W : \mathbb{R}^n \times \mathbb{R}_+ \to \mathbb{R}^{p\times m}$, $Q : \mathbb{R}^n \times \mathbb{R}_+ \to \mathbb{R}^{p\times n}$, $E : \mathbb{R}^n \times \mathbb{R}_+ \to \mathbb{R}^n$, *and* $l : \mathbb{R}^n \times \mathbb{R}_+ \to \mathbb{R}^{r\times m}$ *are locally Lipschitz in* z *and continuous and bounded in* t, *and* $h : \mathbb{R}^n \times \mathbb{R}_+ \to \mathbb{R}^{r\times n}$ *is bounded in* z *and* t. *Along with (F.1) consider the linear time-varying systems*

$$\Sigma_2 : \quad \begin{aligned} \dot{\Omega}^{\mathrm{T}} &= A(z,t)\Omega^{\mathrm{T}} + g(z,t)W(z,t)^{\mathrm{T}} \\ y_2 &= h(z,t)\Omega^{\mathrm{T}} + l(z,t)W(z,t)^{\mathrm{T}} \end{aligned} \quad \text{(F.2)}$$

$$\Sigma_3 : \quad \begin{aligned} \dot{\psi} &= A(z,t)\psi + \Omega^{\mathrm{T}}\dot{\tilde{\theta}} + Q(z,t)^{\mathrm{T}}\dot{\tilde{\theta}} \\ y_3 &= -h(z,t)\psi . \end{aligned} \quad \text{(F.3)}$$

Assume that $z(t)$ *is continuous on* $[0,\infty)$ *and there exists a continuously differentiable function* $V : \mathbb{R}^n \times \mathbb{R}_+ \to \mathbb{R}_+$ *such that*

$$\alpha_1|\zeta|^2 \leq V(\zeta,t) \leq \alpha_2|\zeta|^2, \quad \text{(F.4)}$$

and for each $z \in C^0$,

$$\frac{\partial V}{\partial \zeta}A(z,t)\zeta + \frac{\partial V}{\partial t} \leq -\alpha_3|\zeta|^2 - \alpha_4\left(\frac{\partial V}{\partial \zeta}E(z,t)\right)^2 \quad \text{(F.5)}$$

$\forall t \geq 0$, $\forall \zeta \in \mathbb{R}^n$, $\alpha_1, \alpha_2, \alpha_3, \alpha_4 > 0$. Then for $\forall z(0), \psi(0) \in \mathbb{R}^n$, $\forall \Omega(0) \in \mathbb{R}^{p \times n}$, $\forall t \geq 0$ the outputs of systems (F.1)-(F.3) are related by

$$y_1 = y_2 \tilde{\theta} + y_3 + y_\epsilon \tag{F.6}$$

where y_ϵ is uniformly bounded and $y_\epsilon(t) \to 0$ as $t \to \infty$.

Proof. Due to the continuity of $z(t)$, the matrix-valued functions $g(z(t),t)$, $W(z(t),t)$, $Q(z(t),t)$, and $E(z(t),t)$ are continuous in t. Since the input gW^T to the linear time-varying system Σ_2 is continuous on $[0,\infty)$, then $\Omega \in \mathcal{L}_{\infty e}$. Since also $\dot{\tilde{\theta}} \in \mathcal{L}_{\infty e}$ and system Σ_3 is linear time-varying, then $\psi \in \mathcal{L}_{\infty e}$. Hence the solution of the nonlinear system (F.1)-(F.3) is defined on $[0,\infty)$. Differentiating $\tilde{\epsilon} = z + \psi - \Omega^T \tilde{\theta}$, we compute

$$\begin{aligned}
\dot{\tilde{\epsilon}} &= \dot{z} + \dot{\psi} - \dot{\Omega}^T \tilde{\theta} - \Omega^T \dot{\tilde{\theta}} \\
&= Az + gW^T \tilde{\theta} - Q^T \tilde{\theta} + Ee + A\psi + \Omega^T \tilde{\theta} + Q^T \tilde{\theta} \\
&\quad - A\Omega^T \tilde{\theta} - gW^T \tilde{\theta} - \Omega^T \dot{\tilde{\theta}} \\
&= A(z + \psi - \Omega^T \tilde{\theta}) + Ee
\end{aligned} \tag{F.7}$$

and obtain

$$\dot{\tilde{\epsilon}} = A(z,t)\tilde{\epsilon} + E(z,t)e. \tag{F.8}$$

In view of (F.5) we have

$$\begin{aligned}
\dot{V}(\tilde{\epsilon},t) &= \frac{\partial V}{\partial \tilde{\epsilon}}[A(z,t)\tilde{\epsilon} + E(z,t)e] + \frac{\partial V}{\partial t} \\
&\leq -\alpha_3 |\tilde{\epsilon}|^2 - \alpha_4 \left(\frac{\partial V}{\partial \tilde{\epsilon}} E(z,t)\right)^2 + \frac{\partial V}{\partial \tilde{\epsilon}} E(z,t)e \\
&= -\alpha_3 |\tilde{\epsilon}|^2 - \alpha_4 \left(\frac{\partial V}{\partial \tilde{\epsilon}} E(z,t) - \frac{1}{2\alpha_4}\right)^2 + \frac{1}{4\alpha_4} e^2 \\
&\leq -\alpha_3 |\tilde{\epsilon}|^2 + \frac{1}{4\alpha_4} e^2,
\end{aligned} \tag{F.9}$$

which, according to (F.4), yields

$$\dot{V} \leq -\frac{\alpha_3}{\alpha_2} V + \frac{1}{4\alpha_4} e^2. \tag{F.10}$$

By Lemma B.8 we conclude that V is uniformly bounded and $V(t) \to 0$ as $t \to \infty$. In view of (F.4) we have $|\tilde{\epsilon}(t)| \leq \sqrt{V(t)/\alpha_1}$, which implies that $\tilde{\epsilon}$ is uniformly bounded and converges to zero. Noting that

$$\begin{aligned}
y_\epsilon(t) &= y_1 - y_3 - y_2 \tilde{\theta} \\
&= hz + h\psi + lW^T \tilde{\theta} - h\Omega^T \tilde{\theta} - lW^T \tilde{\theta} + h\psi \\
&= h(z + \psi - \Omega^T \tilde{\theta}) = h(z,t)\tilde{\epsilon},
\end{aligned} \tag{F.11}$$

since $h(z,t)$ is bounded, y_ϵ is bounded and converges to zero. $\square$

Nonlinear Swapping

Remark F.2 If instead of $[0, \infty)$, the maximal interval of existence of $z(t)$ is $[0, t_f)$, then the lemma holds on this interval, stating that y_ϵ is uniformly bounded on $[0, t_f)$. ◇

Remark F.3 When $Q(z,t) \equiv 0$ and $E(z,t) \equiv 0$, the result of Lemma F.1 is reminiscent of Morse's linear Swapping Lemma [138]. To see this, we rewrite (F.6) as

$$T_z[W^T\tilde{\theta}] = T[W^T]\tilde{\theta} + T_h\left[T_g[W^T]\dot{\tilde{\theta}}\right] + y_\epsilon. \quad (F.12)$$

In this notation $T_z : W^T\tilde{\theta} \mapsto y_1$ is the nonlinear operator defined by (F.1) with $D(z,t) \equiv 0$, while the system

$$\begin{aligned} \dot{\xi} &= A(z(t),t)\xi + g(z(t),t)u \\ y &= h(z(t),t)\xi + l(z(t),t)u \end{aligned} \quad (F.13)$$

is used to define the linear time-varying operators: $T : u \mapsto y$, $T_g : u \mapsto y$ for $h = I$ and $l = 0$, $T_h : u \mapsto y$ for $g = I$ and $l = 0$. When A, g, h and l are constant, then the operator $T_z(s) = T(s) = h(sI - A)^{-1}g + l$ is a proper stable rational transfer function, $T_g(s) = (sI - A)^{-1}g$, $T_h(s) = -h(sI - A)^{-1}$, and Lemma F.1 reduces to Lemma 3.6.5 from [165]. ◇

In some texts on adaptive linear control, an extended result which guarantees that $\dot{\tilde{\theta}} \in \mathcal{L}_2 \Rightarrow T_z[W^T\tilde{\theta}] - T[W^T]\tilde{\theta} \in \mathcal{L}_2$ is also referred to as Swapping Lemma. Our next lemma is a nonlinear time-varying generalization of this result.

Lemma F.4 *Consider systems (F.1)–(F.3) with the same set of assumptions as in Lemma F.1. Furthermore, assume that $z \in \mathcal{L}_\infty$ and $e \in \mathcal{L}_2$. If $\dot{\tilde{\theta}} \in \mathcal{L}_2$, then*

$$y_1 - y_2\tilde{\theta} \in \mathcal{L}_2. \quad (F.14)$$

If $\dot{\tilde{\theta}} \in \mathcal{L}_2 \cap \mathcal{L}_\infty$, then

$$\lim_{t \to \infty}\left[y_1(t) - y_2(t)\tilde{\theta}(t)\right] = 0. \quad (F.15)$$

Proof. Since $z \in \mathcal{L}_\infty$, then $gW^T \in \mathcal{L}_\infty$ and $Q \in \mathcal{L}_\infty$. Due to the exponential stability of $A(z,t)$, it follows that $\Omega \in \mathcal{L}_\infty$. We need to prove that $y_3 \in \mathcal{L}_2$ and $y_\epsilon \in \mathcal{L}_2$. The solution of (F.3) is

$$\psi(t) = \Phi_z(t,0)\psi(0) + \int_0^t \Phi_z(t,\tau)[\Omega(\tau) + Q(z(\tau),\tau)]^T\dot{\tilde{\theta}}(\tau)d\tau \quad (F.16)$$

where (F.4)–(F.5) guarantee that the state transition matrix $\Phi_z : \mathbb{R}_+ \times \mathbb{R}_+ \to \mathbb{R}^{n\times n}$ is such that $|\Phi_z(t,\tau)|_2 \leq k e^{-\alpha(t-\tau)}$, $k, \alpha > 0$. Since Ω and Q are bounded, then

$$\begin{aligned}|\psi(t)| &\leq k e^{-\alpha t}|\psi(0)| + k\|\Omega + Q\|_\infty \int_0^t e^{-\alpha(t-\tau)}|\dot{\tilde{\theta}}(\tau)|d\tau \\ &\leq k e^{-\alpha t}|\psi(0)| + k\|\Omega + Q\|_\infty \left(\int_0^t e^{-\alpha(t-\tau)}d\tau\right)^{\frac{1}{2}} \left(\int_0^t e^{-\alpha(t-\tau)}|\dot{\tilde{\theta}}(\tau)|^2 d\tau\right)^{\frac{1}{2}} \\ &\leq k e^{-\alpha t}|\psi(0)| + k\|\Omega + Q\|_\infty \frac{1}{\sqrt{\alpha}} \left(\int_0^t e^{-\alpha(t-\tau)}|\dot{\tilde{\theta}}(\tau)|^2 d\tau\right)^{\frac{1}{2}}, \quad (F.17)\end{aligned}$$

where the second inequality is obtained using the Schwartz inequality. By squaring (F.17) and integrating over $[0, t]$ we obtain

$$\int_0^t |\psi(\tau)|^2 d\tau \leq \frac{k^2}{2\alpha}|\psi(0)|^2 + \frac{k^2}{\alpha}\|\Omega + Q\|_\infty^2 \int_0^t \left[\int_0^\tau e^{-\alpha(\tau-s)}|\dot{\tilde{\theta}}(s)|^2 ds\right] d\tau. \quad (F.18)$$

Changing the sequence of integration, (F.18) becomes

$$\begin{aligned}\int_0^t |\psi(\tau)|^2 d\tau &\leq \frac{k^2}{2\alpha}|\psi(0)|^2 + \frac{k^2}{\alpha}\|\Omega + Q\|_\infty^2 \int_0^t e^{\alpha s}|\dot{\tilde{\theta}}(s)|^2 \left(\int_s^t e^{-\alpha \tau}d\tau\right) ds \\ &\leq \frac{k^2}{2\alpha}|\psi(0)|^2 + \frac{k^2}{\alpha}\|\Omega + Q\|_\infty^2 \int_0^t e^{\alpha s}|\dot{\tilde{\theta}}(s)|^2 \frac{1}{\alpha}e^{-\alpha s}ds \quad (F.19)\end{aligned}$$

because $\int_s^t e^{-\alpha\tau}d\tau = \frac{1}{\alpha}(e^{-\alpha s} - e^{-\alpha t}) \leq \frac{1}{\alpha}e^{-\alpha s}$. Now, the cancellation $e^{\alpha s}e^{-\alpha s} = 1$ in (F.19) yields

$$\|\psi\|_2 \leq \frac{k}{\sqrt{2\alpha}}|\psi(0)| + \frac{k}{\alpha}\|\Omega + Q\|_\infty \|\dot{\tilde{\theta}}\|_2 < \infty, \quad (F.20)$$

which proves $\psi \in \mathcal{L}_2$. Due to the uniform boundedness of h, it follows that $y_3 \in \mathcal{L}_2$. As for y_ϵ, by applying Lemma B.5 to (F.10), we arrive at

$$\|V\|_1 \leq \frac{\alpha_2}{\alpha_3}V(0) + \frac{\alpha_2}{4\alpha_3\alpha_4}\|e\|_2^2 < \infty. \quad (F.21)$$

Because $h \in \mathcal{L}_\infty$, then in view of (F.4) we have

$$\|y_\epsilon\|_2^2 = \|h\tilde{e}\|_2^2 \leq \left\||h|\sqrt{\frac{V}{\alpha_1}}\right\|_2^2 \leq \|h\|_\infty^2 \frac{1}{\alpha_1}\|V\|_1 < \infty, \quad (F.22)$$

and hence $y_\epsilon \in \mathcal{L}_2$. This completes the proof of (F.14). When $\dot{\tilde{\theta}} \in \mathcal{L}_2 \cap \mathcal{L}_\infty$, then $\psi \in \mathcal{L}_\infty \cap \mathcal{L}_2$ and $\dot{\psi} \in \mathcal{L}_\infty$. Thus, by Barbalat's lemma, $\psi(t) \to 0$ as $t \to \infty$. Therefore $y_3(t) \to 0$ as $t \to \infty$. This proves (F.15) because $y_\epsilon(t) \to 0$ as $t \to \infty$. □

Remark F.5 When $D(z,t) \equiv 0$, we rewrite (F.14) as

$$T_z[W^{\mathrm{T}}\tilde{\theta}] - T[W^{\mathrm{T}}]\tilde{\theta} \in \mathcal{L}_2 \qquad (\text{F.23})$$

and (F.15) as

$$\lim_{t\to\infty}\left\{T_z[W^{\mathrm{T}}\tilde{\theta}](t) - \left(T[W^{\mathrm{T}}]\tilde{\theta}\right)(t)\right\} = 0 \qquad (\text{F.24})$$

with T_z and T as in Remark 4.2. For constant A, g, h, and l, the operator $T_z = T$ is a proper stable rational transfer function, and Lemma F.4 reduces to Lemma 2.11 from [142]. ◇

Comparing the z-parametric model (6.24) to (F.1), we see that Lemma F.1 is directly applicable as a design tool. By filtering, we can transform the dynamic parametric model (F.1) into the static parametric model (F.6), where $y_\epsilon(t)$ converges to zero and y_1, y_2 and y_3 are available.

Hence, with the Nonlinear Swapping Lemma at hand, the design of swapping-based identifiers for the z-parametric models is easy. A natural question arises: How to use this lemma, for example, for the x-parametric model

$$\dot{x} = f(x,u) + F(x,u)^{\mathrm{T}}\theta, \qquad (\text{F.25})$$

where the complete state x is available? This system obviously does not match the form of the system (F.1) in Lemma F.1. What is needed to bring the x-model into the form (F.1)? First, we need a presence of $\hat{\theta}$ instead of θ, we need an exponentially stable homogeneous part, and we must remove $f(x,u)$ because only "disturbances" converging to zero (represented by e) are allowed in (F.1). Namely, we would prefer to have the model

$$\dot{\tilde{x}} = A(x,t)\tilde{x} + F(x,u)^{\mathrm{T}}\tilde{\theta}, \qquad (\text{F.26})$$

where $A(x,t)$ is exponentially stable for each x continuous in t. To obtain this, we define $\tilde{x} = x - \hat{x}$ and introduce

$$\dot{\hat{x}} = A(x,t)(\hat{x} - x) + f(x,u) + F(x,u)^{\mathrm{T}}\hat{\theta}. \qquad (\text{F.27})$$

Now with system (F.26) in the form (F.1) we can perform our identifier design according to the filters and the static parametric model in the Nonlinear Swapping Lemma.

Remark F.6 Lemma F.1 is given in the form convenient for analysis. For design, instead of the ψ-filter we use

$$\Sigma_4: \quad \begin{aligned} \dot{\Omega}_0 &= A(z,t)\Omega_0 + g(z,t)W(z,t)^{\mathrm{T}}\hat{\theta} \\ y_4 &= h(z,t)\Omega_0 + l(z,t)W(z,t)^{\mathrm{T}}\hat{\theta}, \end{aligned} \qquad (\text{F.28})$$

and instead of (F.6) we use

$$y_1 + y_4 - y_2\hat{\theta} = y_2\tilde{\theta} + y_\epsilon. \qquad (\text{F.29})$$

◇

The following lemma is essential in proofs of convergence in swapping-based schemes where the identifier is not designed directly from the error system but from the plant model.

Lemma F.7 *Let $T_i : u \mapsto \zeta_i$, $i = 1, 2$ be linear time-varying operators defined by*
$$\dot{\zeta}_i = A_i(t)\zeta_i + u, \tag{F.30}$$
where $A_i : \mathbb{R}_+ \to \mathbb{R}^{n \times n}$ are continuous, bounded and exponentially stable. Suppose $\tilde{\theta} : \mathbb{R}_+ \to \mathbb{R}^p$ is differentiable, $\phi : \mathbb{R}_+ \to \mathbb{R}^{p \times m}$ is piecewise continuous and bounded, and $M : \mathbb{R}_+ \to \mathbb{R}^{n \times n}$ is bounded and has a bounded derivative on $\mathbb{R}_+$. If $\dot{\tilde{\theta}} \in \mathcal{L}_2$, then
$$T_1[\phi^T]\tilde{\theta} \in \mathcal{L}_2 \;\Rightarrow\; T_2[M\phi^T]\tilde{\theta} \in \mathcal{L}_2. \tag{F.31}$$
If, moreover, $M(t)$ is nonsingular $\forall t$ and M^{-1} is bounded and has a bounded derivative on $\mathbb{R}_+$, then (F.31) holds in both directions.

Proof. Suppose that $T_1[\phi^T]\tilde{\theta} \in \mathcal{L}_2$. By Lemma F.4, $T_1[\phi^T\tilde{\theta}] - T_1[\phi^T]\tilde{\theta} \in \mathcal{L}_2$ and therefore $\zeta_1 \triangleq T_1[\phi^T\tilde{\theta}] \in \mathcal{L}_2$. We will show first that $\zeta_2 \triangleq T_2[M\phi^T\tilde{\theta}] \in \mathcal{L}_2$. By substituting $\phi^T\tilde{\theta} = \dot{\zeta}_1 - A_1(t)\zeta_1$ into the variation of constants formula and applying partial integration, we calculate

$$\begin{aligned}\zeta_2(t) &= \Phi_2(t,0)\zeta_2(0) + \int_0^t \Phi_2(t,\tau)M(\tau)\phi^T(\tau)\tilde{\theta}(\tau)d\tau \\ &= \Phi_2(t,0)\zeta_2(0) + \int_0^t \Phi_2(t,\tau)M(\tau)\left[\dot{\zeta}_1(\tau) - A_1(\tau)\zeta_1(\tau)\right]d\tau \\ &= \Phi_2(t,0)\zeta_2(0) + M(t)\zeta_1(t) - \Phi_2(t,0)M(0)\zeta_1(0) \\ &\quad + \int_0^t \Phi_2(t,\tau)\left[\dot{M}(\tau) + A_2(\tau)M(\tau) - M(\tau)A_1(\tau)\right]\zeta_1(\tau)d\tau \quad (\text{F.32})\end{aligned}$$

where $\Phi_2(t,\tau)$ is the state transition matrix of $A_2(t)$ that satisfies $|\Phi_2(t,\tau)|_2 \le ke^{-\alpha(t-\tau)}$, $k, \alpha > 0$. It is clear that $\Phi_2(t,0)\zeta_2(0) + M(t)\zeta_1(t) - \Phi_2(t,0)M(0)\zeta_1(0) \in \mathcal{L}_2$ because $\Phi_2(t,0)$ is exponentially decaying, $M(t)$ is bounded and $\zeta_1 \in \mathcal{L}_2$. Since

$$\left|\int_0^t \Phi_2(t,\tau)\left[\dot{M}(\tau) + A_2(\tau)M(\tau) - M(\tau)A_1(\tau)\right]\zeta_1(\tau)d\tau\right|^2$$
$$\le \|\dot{M} + A_2M - MA_1\|_\infty^2 k^2 \int_0^t e^{-2\alpha(t-\tau)}|\zeta_1(\tau)|^2 d\tau, \quad (\text{F.33})$$

then similarly to (F.18)–(F.20) from the proof of Lemma F.4, we can show that the expression (F.33) is in $\mathcal{L}_2$. Thus $\zeta_2 = T_2[M\phi^T\tilde{\theta}] \in \mathcal{L}_2$. By Lemma F.4, $T_2[M\phi^T\tilde{\theta}] - T_2[M\phi^T]\tilde{\theta} \in \mathcal{L}_2$ and therefore $T_2[M\phi^T\tilde{\theta}] \in \mathcal{L}_2$. The proof of the other direction of (F.31) is identical (when $M(t)$ is nonsingular $\forall t$ and M^{-1} is bounded and has a bounded derivative on $\mathbb{R}_+$). □

Appendix G

Differential Geometric Conditions

For completeness, we now derive necessary and sufficient conditions under which a nonlinear system can be transformed into one of the canonical forms considered in this book. These conditions represent coordinate-free characterizations of classes of nonlinear systems suitable for our backstepping designs. The differential geometry background required for a full understanding of this appendix is contained in Isidori [53] and Nijmeijer and van der Schaft [144].

G.1 Partial-State-Feedback Forms

We first give geometric conditions which are necessary and sufficient for single-input single-output nonlinear systems of the form

$$\begin{aligned} \dot{\zeta} &= f_0(\zeta) + \sum_{j=1}^{p} \theta_j f_j(\zeta) + \left[g_0(\zeta) + \sum_{j=1}^{p} \theta_j g_j(\zeta) \right] u \\ y &= h(\zeta) \end{aligned} \qquad \text{(G.1)}$$

to be transformable via a parameter-independent diffeomorphism into the canonical forms of Chapters 3, 4 and 7. The first form considered is obtained by setting $k = 1$ in (7.180) (so that $x^m = [x_1, \ldots, x_{m_1}]$):

$$\begin{aligned}
\dot{x}_1 &= x_2 + \varphi_{0,1}(x_1) + \sum_{j=1}^{p} \theta_j \varphi_{j,1}(x_1) \\
\dot{x}_2 &= x_3 + \varphi_{0,2}(x_1, x_2) + \sum_{j=1}^{p} \theta_j \varphi_{j,2}(x_1, x_2) \\
&\vdots \\
\dot{x}_{m_1} &= x_{m_1+1} + \varphi_{0,m_1}(x_1, \ldots, x_{m_1}) + \sum_{j=1}^{p} \theta_j \varphi_{j,m_1}(x_1, \ldots, x_{m_1}) \\
\dot{x}_{m_1+1} &= x_{m_1+2} + \varphi_{0,m_1+1}(x^m) + \sum_{j=1}^{p} \theta_j \varphi_{j,m_1+1}(x^m) \quad\text{(G.2)} \\
&\vdots \\
\dot{x}_{\rho-1} &= x_\rho + \varphi_{0,\rho-1}(x^m) + \sum_{j=1}^{p} \theta_j \varphi_{j,\rho-1}(x^m) \\
\dot{x}_\rho &= x_{\rho+1} + \varphi_{0,\rho}(x^m) + \sum_{j=1}^{p} \theta_j \varphi_{j,\rho}(x^m) + b_m \beta(x^m) u \\
&\vdots \\
\dot{x}_n &= \varphi_{0,n}(x^m) + \sum_{j=1}^{p} \theta_j \varphi_{j,n}(x^m) + b_0 \beta(x^m) u \,.
\end{aligned}$$

We start with a result for the case when $\theta = 0$ and $u = 0$.

Proposition G.1 *The system*

$$\begin{aligned}
\dot{\zeta} &= f_0(\zeta) \\
y &= h(\zeta)
\end{aligned} \quad\text{(G.3)}$$

can be transformed via a global diffeomorphism $x = \phi(\zeta)$ into the partial-state-feedback form

$$\begin{aligned}
\dot{x}_1 &= x_2 \\
&\vdots \\
\dot{x}_{m_1-1} &= x_{m_1} \\
\dot{x}_{m_1} &= x_{m_1+1} + \varphi_{0,m_1}(x_1, \ldots, x_{m_1}) \\
&\vdots \\
\dot{x}_n &= \varphi_{0,n}(x_1, \ldots, x_{m_1}) \\
y &= x_1
\end{aligned} \quad\text{(G.4)}$$

if and only if the following conditions are globally satisfied:

Differential Geometric Conditions

(i) rank $\left\{ dh, d(L_{f_0}h), \ldots, d\left(L_{f_0}^{n-1}h\right) \right\} = n$,

(ii) $\left[\operatorname{ad}_{f_0}^i r, \operatorname{ad}_{f_0}^{i+1} r\right] = 0$, $\quad 0 \leq i \leq n - m_1 - 1$, and

(iii) the vector fields $r, \operatorname{ad}_{f_0} r, \ldots, \operatorname{ad}_{f_0}^{n-m_1} r, \operatorname{ad}_{\bar{f}_0}^{n-m_1+1} r, \ldots, \operatorname{ad}_{\bar{f}_0}^{n-1} r$ are complete,

where r and $\bar{f}_0$ are the vector fields defined by

$$L_r L_{f_0}^i h = \begin{cases} 0, & i = 0 \ldots, n-2 \\ 1, & i = n-1, \end{cases} \qquad \bar{f}_0 = f_0 - \left(L_{f_0}^n h\right) r. \tag{G.5}$$

Proof. *Sufficiency.* Condition (i) implies that the change of coordinates $\chi_i = L_{f_0}^{i-1} h(\zeta)$, $1 \leq i \leq n$, is a local diffeomorphism transforming the system (G.3) into the system

$$\begin{aligned}
\dot{\chi}_1 &= \chi_2 \\
&\vdots \\
\dot{\chi}_{n-1} &= \chi_n \\
\dot{\chi}_n &= \mu(\chi) \\
y &= \chi_1,
\end{aligned} \tag{G.6}$$

where $\mu(\chi)$ is the function $L_{f_0}^n h$ expressed in the coordinates of (G.6). The definitions (G.5) imply that in these coordinates the vector fields r and $\bar{f}_0$ are expressed as

$$r = \frac{\partial}{\partial \chi_n}, \qquad \bar{f}_0 = \chi_2 \frac{\partial}{\partial \chi_1} + \cdots + \chi_n \frac{\partial}{\partial \chi_{n-1}}. \tag{G.7}$$

Hence, the vector fields $\operatorname{ad}_{\bar{f}_0}^i r$, $0 \leq i \leq n-1$, become

$$\operatorname{ad}_{\bar{f}_0}^i r = (-1)^i \frac{\partial}{\partial \chi_{n-i}}, \qquad 0 \leq i \leq n-1. \tag{G.8}$$

We now show by induction that condition (ii) implies that $\chi_{m_1+1}, \ldots, \chi_n$ can be replaced by new coordinates $x_{m_1+1}, \ldots, x_n$ such that in the x-coordinates (with $x_1 = \chi_1, \ldots, x_{m_1} = \chi_{m_1}$) the system (G.6) takes on the form (G.4) and, moreover, the vector fields $\operatorname{ad}_{f_0}^i r$, $0 \leq i \leq n - m_1$, become

$$\operatorname{ad}_{f_0}^i r = (-1)^i \frac{\partial}{\partial x_{n-i}}, \qquad 0 \leq i \leq n - m_1. \tag{G.9}$$

• *First induction step* ($i = n$): In the coordinates of (G.6), the vector field $\operatorname{ad}_{f_0} r$ is expressed as

$$\operatorname{ad}_{f_0} r = \left[\sum_{j=1}^{n-1} \chi_{j+1} \frac{\partial}{\partial \chi_j} + \mu(\chi) \frac{\partial}{\partial \chi_n}, \frac{\partial}{\partial \chi_n} \right] = -\frac{\partial}{\partial \chi_{n-1}} - \frac{\partial \mu}{\partial \chi_n}(\chi) \frac{\partial}{\partial \chi_n}. \tag{G.10}$$

Then, the condition $[r, \text{ad}_{f_0} r] = 0$ implies that

$$\left[\frac{\partial}{\partial \chi_n}, -\frac{\partial}{\partial \chi_{n-1}} - \frac{\partial \mu}{\partial \chi_n}(\chi)\frac{\partial}{\partial \chi_n}\right] = -\frac{\partial^2 \mu}{\partial \chi_n^2}(\chi)\frac{\partial}{\partial \chi_n} = 0. \quad (G.11)$$

Hence, the function $\mu(\chi)$ can be expressed as

$$\mu(\chi) = \mu_1(\chi_1, \ldots, \chi_{n-1}) + \mu_2(\chi_1, \ldots, \chi_{n-1})\chi_n. \quad (G.12)$$

Let us then define the new coordinate $\bar{\chi}_n$ as

$$\bar{\chi}_n = \chi_n - \int_0^{\chi_{n-1}} \mu_2(\chi_1, \ldots, \chi_{n-2}, s)ds \stackrel{\triangle}{=} \chi_n - \bar{\mu}(\chi_1, \ldots, \chi_{n-1}). \quad (G.13)$$

In the coordinates $(\chi_1, \ldots, \chi_{n-1}, \bar{\chi}_n)$ the system (G.3) becomes

$$\begin{aligned}
\dot{\chi}_j &= \chi_{j+1}, \quad 1 \leq j \leq n-2 \\
\dot{\chi}_{n-1} &= \bar{\chi}_n + \bar{\mu}(\chi_1, \ldots, \chi_{n-1}) \\
\dot{\bar{\chi}}_n &= \mu_1(\chi_1, \ldots, \chi_{n-1}) \\
y &= \chi_1,
\end{aligned} \quad (G.14)$$

and the vector fields $r, \text{ad}_{f_0} r$ become

$$r = \frac{\partial}{\partial \bar{\chi}_n}, \quad \text{ad}_{f_0} r = -\frac{\partial}{\partial \chi_{n-1}}. \quad (G.15)$$

- **Induction hypothesis** ($i = k+1$, $m_1 + 1 \leq k \leq n-1$): Assume that we have replaced $\chi_{k+1} \ldots, \chi_n$ by new coordinates $\xi_{k+1}, \ldots, \xi_n$ such that in the $(\chi_1, \ldots, \chi_k, \xi_{k+1}, \ldots, \xi_n)$-coordinates the system (G.3) is expressed as

$$\begin{aligned}
\dot{\chi}_j &= \chi_{j+1}, \quad 1 \leq j \leq k-1 \\
\dot{\chi}_k &= \xi_{k+1} + \nu_k(\chi_1, \ldots, \chi_k) \\
\dot{\xi}_{k+1} &= \xi_{k+2} + \nu_{k+1}(\chi_1, \ldots, \chi_k) \\
&\vdots \\
\dot{\xi}_n &= \nu_n(\chi_1, \ldots, \chi_k) \\
y &= \chi_1,
\end{aligned} \quad (G.16)$$

and, moreover, the vector fields $\text{ad}_{f_0}^j r$, $0 \leq j \leq n-k$, are expressed as

$$\text{ad}_{f_0}^j r = (-1)^j \frac{\partial}{\partial \xi_{n-j}}, \quad 0 \leq j \leq n-k-1, \quad \text{ad}_{f_0}^{n-k} r = (-1)^{n-k}\frac{\partial}{\partial \chi_k}. \quad (G.17)$$

DIFFERENTIAL GEOMETRIC CONDITIONS

• *Induction proof* ($i = k$, $m_1 + 1 \leq k \leq n - 1$): In the coordinates of (G.16), the vector field $\mathrm{ad}_{f_0}^{n-k+1} r$ is expressed as

$$\begin{aligned}
\mathrm{ad}_{f_0}^{n-k+1} r &= \left[f_0, \mathrm{ad}_{f_0}^{n-k} r \right] \\
&= \left[\sum_{j=1}^{k-1} \chi_{j+1} \frac{\partial}{\partial \chi_j} + (\xi_{k+1} + \nu_k) \frac{\partial}{\partial \chi_k} \right. \\
&\quad \left. + \sum_{j=k+1}^{n-1} (\xi_{j+1} + \nu_j) \frac{\partial}{\partial \xi_j} + \nu_n \frac{\partial}{\partial \xi_n} \,,\, (-1)^{n-k} \frac{\partial}{\partial \chi_k} \right] \\
&= (-1)^{n-k+1} \left(\frac{\partial}{\partial \chi_{k-1}} + \frac{\partial \nu_k}{\partial \chi_k} \frac{\partial}{\partial \chi_k} + \sum_{j=k+1}^{n} \frac{\partial \nu_j}{\partial \chi_k} \frac{\partial}{\partial \xi_j} \right). \quad \text{(G.18)}
\end{aligned}$$

Then, the condition $[\mathrm{ad}_{f_0}^{n-k} r, \mathrm{ad}_{f_0}^{n-k+1} r] = 0$ implies that

$$\frac{\partial^2 \nu_j}{\partial \chi_k^2} = 0 \;\Rightarrow\; \nu_j(\chi_1, \ldots, \chi_k) = \nu_{j,1}(\chi_1, \ldots, \chi_{k-1})$$
$$+ \nu_{j,2}(\chi_1, \ldots, \chi_{k-1}) \chi_k\,, \quad k \leq j \leq n. \quad \text{(G.19)}$$

Let us then define the new coordinates $\bar{\xi}_k, \ldots, \bar{\xi}_n$ as

$$\begin{aligned}
\bar{\xi}_k &= \chi_k - \int_0^{\chi_{k-1}} \nu_{k,2}(\chi_1, \ldots, \chi_{k-2}, s) ds \triangleq \chi_k - \bar{\nu}_{k-1}(\chi_1, \ldots, \chi_{k-1}) \\
\bar{\xi}_j &= \xi_j - \int_0^{\chi_{k-1}} \nu_{j,2}(\chi_1, \ldots, \chi_{k-2}, s) ds\,, \quad k+1 \leq j \leq n.
\end{aligned} \quad \text{(G.20)}$$

In the coordinates $(\chi_1, \ldots, \chi_{k-1}, \bar{\xi}_k, \ldots, \bar{\xi}_n)$ the system (G.3) becomes

$$\begin{aligned}
\dot\chi_j &= \chi_{j+1}, \quad 1 \leq j \leq k-2 \\
\dot\chi_{k-1} &= \bar{\xi}_k + \bar{\nu}_{k-1}(\chi_1, \ldots, \chi_{k-1}) \\
\dot{\bar{\xi}}_k &= \bar{\xi}_{k+1} + \bar{\nu}_k(\chi_1, \ldots, \chi_{k-1}) \\
&\vdots \\
\dot{\bar{\xi}}_n &= \bar{\nu}_n(\chi_1, \ldots, \chi_{k-1}) \\
y &= \chi_1,
\end{aligned} \quad \text{(G.21)}$$

and, moreover, the vector fields $\mathrm{ad}_{f_0}^j r$, $0 \leq j \leq n - k + 1$, become

$$\mathrm{ad}_{f_0}^j r = (-1)^j \frac{\partial}{\partial \bar{\xi}_{n-j}}, \quad 0 \leq j \leq n-k, \quad \mathrm{ad}_{f_0}^{n-k+1} r = (-1)^{n-k+1} \frac{\partial}{\partial \chi_{k-1}}. \quad \text{(G.22)}$$

Thus, we have shown that conditions (i) and (ii) guarantee the local existence of a diffeomorphism $x = \phi(\zeta)$ transforming (G.3) into (G.4). Furthermore, (G.8) and (G.22) imply that in the x-coordinates we have

$$\begin{aligned}
\mathrm{ad}_{f_0}^i r &= (-1)^i \frac{\partial}{\partial x_{n-i}}, \quad 0 \leq i \leq n - m_1, \\
\mathrm{ad}_{f_0}^i r &= (-1)^i \frac{\partial}{\partial x_{n-i}}, \quad n - m_1 + 1 \leq i \leq n - 1.
\end{aligned} \quad \text{(G.23)}$$

Then, from condition (iii) and [162, Corollary 2.4] we conclude that this diffeomorphism is global.

Necessity. If there exists a diffeomorphism $x = \phi(\zeta)$ that transforms (G.3) into (G.4), one can directly verify that the coordinate-free conditions (i)–(iii) are satisfied for the system (G.4), and hence for the system (G.3). □

Theorem G.2 *The system (G.1) can be transformed via a global parameter-independent diffeomorphism $x = \phi(\zeta)$ into the partial-state-feedback form (G.2) if and only if, in addition to the conditions of Proposition G.1, the following conditions are globally satisfied:*

(iv) $\left[f_j, ad_{f_0}^i r\right] = 0$, $0 \leq i \leq n - m_1 - 1$, $1 \leq j \leq p$,

(v) $d\left(L_{f_j} L_{f_0}^i h\right) \in \text{span}\left\{dh, \ldots, d\left(L_{f_0}^j h\right)\right\}$, $0 \leq i \leq m_1 - 2$, $1 \leq j \leq p$,

(vi) $\left[g_j, ad_{f_0}^i r\right] = 0$, $0 \leq i \leq n - m_1 - 1$, $0 \leq j \leq p$, *and*

(vii) $g_0 + \sum_{j=1}^{p} \theta_j g_j = \beta(\cdot) \sum_{i=0}^{m} b_i(-1)^i ad_{f_0}^i r$,

where $\beta(\cdot)$ is a smooth nonlinear function and r is the vector field defined by (G.5).

Proof. *Sufficiency.* In the x-coordinates of Proposition G.1, condition (iv) becomes (cf. (G.23)):

$$\left[f_j, (-1)^i \frac{\partial}{\partial x_{n-i}}\right] = 0, \quad 0 \leq i \leq n - m_1 - 1, \ 1 \leq j \leq p. \quad \text{(G.24)}$$

Hence, the vector fields f_j are expressed in the x-coordinates as

$$f_j = \sum_{i=1}^{n} \varphi_{j,i}(x_1, \ldots, x_{m_1}) \frac{\partial}{\partial x_i}, \quad 1 \leq j \leq p. \quad \text{(G.25)}$$

Furthermore, since in the x-coordinates we have

$$\begin{aligned} x_{i+1} &= L_{f_0}^i h, \quad 0 \leq i \leq m_1 - 1 \\ \varphi_{j,i+1}(x_1, \ldots, x_{m_1}) &= L_{f_j} L_{f_0}^i h, \quad 0 \leq i \leq m_1 - 1, \ 1 \leq j \leq p, \end{aligned} \quad \text{(G.26)}$$

condition (v) becomes

$$d\varphi_{j,i} \in \text{span}\{dx_1, \ldots, dx_i\}, \quad 1 \leq i \leq m_1 - 1, \ 1 \leq j \leq p. \quad \text{(G.27)}$$

Combining (G.25) with (G.27), we obtain

$$f_j = \sum_{i=1}^{m_1-1} \varphi_{j,i}(x_1, \ldots, x_i) \frac{\partial}{\partial x_i} + \sum_{i=m_1}^{n} \varphi_{j,i}(x_1, \ldots, x_{m_1}) \frac{\partial}{\partial x_i}, \quad 1 \leq j \leq p. \quad \text{(G.28)}$$

Differential Geometric Conditions

Similarly, conditions (vi) and (vii) imply that in the x-coordinates we have

$$g_0 + \sum_{j=1}^{p} \theta_j g_j = \beta(x_1, \ldots, x_{m_1}) \sum_{i=0}^{m} b_i \frac{\partial}{\partial x_{n-i}}. \qquad (G.29)$$

From (G.4), (G.28) and (G.29) we conclude that in the x-coordinates, which are globally defined, the system (G.1) is expressed as (G.2) with $\varphi_{0,i} \equiv 0$, $1 \leq i \leq m_1 - 1$.

Necessity. Again, it is straightforward to directly verify that conditions (iv)–(vii) are satisfied for the system (G.2). □

We now proceed to another partial-state-feedback form, in which the measured variables are $x^m = [x_1, \ldots, x_{m_1}, x^r]$:

$$\dot{x}_1 = x_2 + \varphi_{0,1}(x_1) + \sum_{j=1}^{p} \theta_j \varphi_{j,1}(x_1)$$

$$\dot{x}_2 = x_3 + \varphi_{0,2}(x_1, x_2) + \sum_{j=1}^{p} \theta_j \varphi_{j,2}(x_1, x_2)$$

$$\vdots$$

$$\dot{x}_{q-1} = x_q + \varphi_{0,q-1}(x_1, \ldots, x_{q-1}) + \sum_{j=1}^{p} \theta_j \varphi_{j,q-1}(x_1, \ldots, x_{q-1})$$

$$\dot{x}_q = x_{q+1} + \varphi_{0,q}(x_1, \ldots, x_q, x^r) + \sum_{j=1}^{p} \theta_j \varphi_{j,q}(x_1, \ldots, x_q, x^r)$$

$$\vdots$$

$$\dot{x}_{m_1} = x_{m_1+1} + \varphi_{0,m_1}(x_1, \ldots, x_{m_1}, x^r) + \sum_{j=1}^{p} \theta_j \varphi_{j,m_1}(x_1, \ldots, x_{m_1}, x^r)$$

$$\dot{x}_{m_1+1} = x_{m_1+2} + \varphi_{0,m_1+1}(x^m) + \sum_{j=1}^{p} \theta_j \varphi_{j,m_1+1}(x^m) \qquad (G.30)$$

$$\vdots$$

$$\dot{x}_{\rho-1} = x_\rho + \varphi_{0,\rho-1}(x^m) + \sum_{j=1}^{p} \theta_j \varphi_{j,\rho-1}(x^m)$$

$$\dot{x}_\rho = \varphi_{0,\rho}(x^m) + \sum_{j=1}^{p} \theta_j \varphi_{j,\rho}(x^m) + b_m \beta(x^m) u$$

$$\dot{x}^r = \Phi_0(x_1, \ldots, x_q, x^r) + \sum_{j=1}^{p} \theta_j \Phi_j(x_1, \ldots, x_q, x^r)$$

$$y = x_1.$$

Comparing (G.30) with (G.2), we note several differences: In (G.30) the zero dynamics subsystem (x^r) is nonlinear, its states are measured, and they enter the nonlinearities of the $\dot{x}_q,\ldots,\dot{x}_\rho$ equations. In contrast, the states $x_{\rho+1},\ldots,x_n$ of the linear zero dynamics subsystem of (G.2) are not measured and do not enter the $\dot{x}_1,\ldots,\dot{x}_\rho$ equations.

Proposition G.3 *The system*

$$\begin{aligned} \dot{\zeta} &= f_0(\zeta) + g_0(\zeta)u \\ y &= h(\zeta) \end{aligned} \qquad (G.31)$$

can be transformed via a diffeomorphism $x = \phi(\zeta)$, which is satisfied in a neighborhood $\mathcal{U}$ of a point ζ_0, into the partial-state-feedback form

$$\begin{aligned} \dot{x}_1 &= x_2 \\ &\vdots \\ \dot{x}_{m_1-1} &= x_{m_1} \\ \dot{x}_{m_1} &= x_{m_1+1} + \varphi_{0,m_1}(x_1,\ldots,x_{m_1},x^r) \\ &\vdots \\ \dot{x}_{\rho-1} &= \varphi_{0,\rho-1}(x_1,\ldots,x_{m_1},x^r) \\ \dot{x}_\rho &= \varphi_{0,\rho}(x_1,\ldots,x_{m_1},x^r) + \beta(x_1,\ldots,x_{m_1},x^r)u \\ \dot{x}^r &= \Phi_0(x_1,\ldots,x_q,x^r) \\ y &= x_1, \end{aligned} \qquad (G.32)$$

where $q \leq m_1$, if and only if the following conditions are valid in a neighborhood $\mathcal{U}_1 \supseteq \mathcal{U}$:

(i) $L_{g_0}L_{f_0}^i h \equiv 0$, $0 \leq i \leq \rho - 2$, $L_{g_0}L_{f_0}^{\rho-1}h \neq 0$,

(ii) the distribution $\mathcal{G}^{\rho-q} = \mathrm{span}\left\{g_0, \mathrm{ad}_{f_0}g_0, \ldots, \mathrm{ad}_{f_0}^{\rho-q}g_0\right\}$ is involutive and of constant rank $\rho - q + 1$,

(iii) $\left[\mathrm{ad}_{f_0}^i \bar{g}_0, \mathrm{ad}_{f_0}^{i+1}\bar{g}_0\right] = 0$, $0 \leq i \leq \rho - m_1 - 1$, *and*

(iv) $\left[g_0, \mathrm{ad}_{f_0}^i \bar{g}_0\right] = 0$, $0 \leq i \leq \rho - m_1 - 1$,

where the vector field $\bar{g}_0$ is defined by

$$\bar{g}_0 = \frac{1}{L_{g_0}L_{f_0}^{\rho-1}h} g_0. \qquad (G.33)$$

Proof. *Sufficiency.* It was proved in [118, Proposition 10] and in [13] that conditions (i) and (ii) guarantee the existence of a local diffeomorphism $\chi =$

Differential Geometric Conditions

$\bar{\phi}(\zeta)$, with $\chi_i = L_{f_0}^{i-1} h(\zeta), 1 \leq i \leq \rho$, which transforms the system (G.31) into the system

$$
\begin{aligned}
\dot{\chi}_1 &= \chi_2 \\
&\vdots \\
\dot{\chi}_{\rho-1} &= \chi_\rho \\
\dot{\chi}_\rho &= \mu(\chi) + \beta(\chi) u \\
\dot{\chi}^r &= \Phi_0(\chi_1, \ldots, \chi_q, \chi^r) \\
y &= \chi_1,
\end{aligned}
\quad (G.34)
$$

where $\chi^r = [\chi_{\rho+1}, \ldots, \chi_n]^T$ and $\mu(\chi), \beta(\chi)$ are the functions $L_{f_0}^\rho h$ and $L_{g_0} L_{f_0}^{\rho-1} h$ expressed in the coordinates of (G.34). The definition (G.33) implies that in these coordinates the vector fields g_0 and $\bar{g}_0$ are expressed as

$$
g_0 = \beta(\chi) \frac{\partial}{\partial \chi_\rho}, \quad \bar{g}_0 = \frac{\partial}{\partial \chi_\rho}. \quad (G.35)
$$

We now show by induction that conditions (iii) and (iv) imply that $\chi_{m_1+1}, \ldots, \chi_n$ can be replaced by new coordinates $x_{m_1+1}, \ldots, x_n$ such that in the x-coordinates (with $x_1 = \chi_1, \ldots, x_{m_1} = \chi_{m_1}, x^r = \chi^r$) the system (G.34) takes on the form (G.32) and, moreover, the vector fields $\text{ad}_{f_0}^i g_0$, $0 \leq i \leq \rho - m_1$, become

$$
\text{ad}_{f_0}^i \bar{g}_0 = (-1)^i \frac{\partial}{\partial x_{\rho-i}}, \quad 0 \leq i \leq \rho - m_1. \quad (G.36)
$$

- *First induction step* ($i = \rho$): In the coordinates of (G.34) the vector field $\text{ad}_{f_0} \bar{g}_0$ is expressed as

$$
\begin{aligned}
\text{ad}_{f_0} \bar{g}_0 &= \left[\sum_{j=1}^{\rho-1} \chi_{j+1} \frac{\partial}{\partial \chi_j} + \mu(\chi) \frac{\partial}{\partial \chi_\rho} + \sum_{j=\rho+1}^{n} \Phi_{0,j}(\chi_1, \ldots, \chi_q, \chi^r) \frac{\partial}{\partial \chi_j}, \frac{\partial}{\partial \chi_\rho} \right] \\
&= -\frac{\partial}{\partial \chi_{\rho-1}} - \frac{\partial \mu}{\partial \chi_\rho}(\chi) \frac{\partial}{\partial \chi_\rho}.
\end{aligned}
\quad (G.37)
$$

Then, the conditions $[\bar{g}_0, \text{ad}_{f_0} \bar{g}_0] = 0$ and $[g_0, \bar{g}_0] = 0$ imply that

$$
\left[\frac{\partial}{\partial \chi_\rho}, -\frac{\partial}{\partial \chi_{\rho-1}} - \frac{\partial \mu}{\partial \chi_\rho}(\chi) \frac{\partial}{\partial \chi_\rho} \right] = -\frac{\partial^2 \mu}{\partial \chi_\rho^2}(\chi) \frac{\partial}{\partial \chi_\rho} = 0 \quad (G.38)
$$

$$
\left[\beta(\chi) \frac{\partial}{\partial \chi_\rho}, \frac{\partial}{\partial \chi_\rho} \right] = -\frac{\partial \beta}{\partial \chi_\rho}(\chi) \frac{\partial}{\partial \chi_\rho} = 0. \quad (G.39)
$$

Hence, $\beta(\cdot)$ is independent of χ_ρ and the function $\mu(\chi)$ can be expressed as

$$
\mu(\chi) = \mu_1(\chi_1, \ldots, \chi_{\rho-1}, \chi^r) + \mu_2(\chi_1, \ldots, \chi_{\rho-1}, \chi^r) \chi_\rho. \quad (G.40)
$$

Let us then define the new coordinate $\bar{\chi}_\rho$ as

$$\bar{\chi}_\rho = \chi_\rho - \int_0^{\chi_{\rho-1}} \mu_2(\chi_1, \ldots, \chi_{\rho-2}, s, \chi^r) ds \triangleq \chi_\rho - \bar{\mu}(\chi_1, \ldots, \chi_{\rho-1}, \chi^r). \quad (G.41)$$

In the coordinates $(\chi_1, \ldots, \chi_{\rho-1}, \bar{\chi}_\rho, \chi^r)$, the system (G.31) becomes

$$\begin{aligned}
\dot{\chi}_j &= \chi_{j+1}, \quad 1 \leq j \leq \rho - 2 \\
\dot{\chi}_{\rho-1} &= \bar{\chi}_\rho + \bar{\mu}(\chi_1, \ldots, \chi_{\rho-1}, \chi^r) \\
\dot{\bar{\chi}}_\rho &= \mu_1(\chi_1, \ldots, \chi_{\rho-1}, \chi^r) + \beta(\chi_1, \ldots, \chi_{\rho-1}, \chi^r) u \\
\dot{\chi}^r &= \Phi_0(\chi_1, \ldots, \chi_q, \chi^r) \\
y &= \chi_1,
\end{aligned} \quad (G.42)$$

and the vector fields $\bar{g}_0, \mathrm{ad}_{f_0}\bar{g}_0$ become

$$\bar{g}_0 = \frac{\partial}{\partial \bar{\chi}_\rho}, \quad \mathrm{ad}_{f_0}\bar{g}_0 = -\frac{\partial}{\partial \chi_{\rho-1}}. \quad (G.43)$$

- *Induction hypothesis* ($i = k+1, m_1 + 1 \leq k \leq \rho - 1$): Assume that we have replaced $\chi_{k+1} \ldots, \chi_\rho$ by new coordinates $\xi_{k+1}, \ldots, \xi_\rho$ such that in the $(\chi_1, \ldots, \chi_k, \xi_{k+1}, \ldots, \xi_\rho, \chi^r)$-coordinates the system (G.31) is expressed as

$$\begin{aligned}
\dot{\chi}_j &= \chi_{j+1}, \quad 1 \leq j \leq k-1 \\
\dot{\chi}_k &= \xi_{k+1} + \nu_k(\chi_1, \ldots, \chi_k, \chi^r) \\
\dot{\xi}_{k+1} &= \xi_{k+2} + \nu_{k+1}(\chi_1, \ldots, \chi_k, \chi^r) \\
&\vdots \\
\dot{\xi}_\rho &= \nu_\rho(\chi_1, \ldots, \chi_k, \chi^r) + \beta(\chi_1, \ldots, \chi_k, \chi^r) u \\
\dot{\chi}^r &= \Phi_0(\chi_1, \ldots, \chi_q, \chi^r) \\
y &= \chi_1,
\end{aligned} \quad (G.44)$$

and, moreover, the vector fields $\mathrm{ad}_{f_0}^j \bar{g}_0$, $0 \leq j \leq \rho - k$, are expressed as

$$\mathrm{ad}_{f_0}^j \bar{g}_0 = (-1)^j \frac{\partial}{\partial \xi_{\rho-j}}, \quad 0 \leq j \leq \rho - k - 1, \quad \mathrm{ad}_{f_0}^{\rho-k} r = (-1)^{\rho-k} \frac{\partial}{\partial \chi_k}. \quad (G.45)$$

- *Induction proof* ($i = k$, $m_1 + 1 \leq k \leq \rho - 1$): In the coordinates of (G.44), the vector field $\mathrm{ad}_{f_0}^{\rho-k+1} \bar{g}_0$ is expressed as

$$\begin{aligned}
\mathrm{ad}_{f_0}^{\rho-k+1} \bar{g}_0 &= \left[f_0, \mathrm{ad}_{f_0}^{\rho-k} \bar{g}_0 \right] \\
&= \left[\sum_{j=1}^{k-1} \chi_{j+1} \frac{\partial}{\partial \chi_j} + (\xi_{k+1} + \nu_k) \frac{\partial}{\partial \chi_k} + \sum_{j=k+1}^{\rho-1} (\xi_{j+1} + \nu_j) \frac{\partial}{\partial \xi_j} \right. \\
&\quad + \nu_\rho \frac{\partial}{\partial \xi_\rho} + \sum_{j=\rho+1}^{n} \Phi_{0,j} \frac{\partial}{\partial \chi_j}, (-1)^{\rho-k} \frac{\partial}{\partial \chi_k} \Bigg] \\
&= (-1)^{\rho-k+1} \left(\frac{\partial}{\partial \chi_{k-1}} + \frac{\partial \nu_k}{\partial \chi_k} \frac{\partial}{\partial \chi_k} + \sum_{j=k+1}^{\rho} \frac{\partial \nu_j}{\partial \chi_k} \frac{\partial}{\partial \xi_j} \right). \quad (G.46)
\end{aligned}$$

DIFFERENTIAL GEOMETRIC CONDITIONS

Then, the conditions $[\mathrm{ad}_{f_0}^{\rho-k}\bar{g}_0, \mathrm{ad}_{f_0}^{\rho-k+1}\bar{g}_0] = 0$ and $[g_0, \mathrm{ad}_{f_0}^{\rho-k}\bar{g}_0] = 0$ imply that

$$\frac{\partial^2 \nu_j}{\partial \chi_k^2} = 0 \Rightarrow \nu_j(\chi_1, \ldots, \chi_k, \chi^r) = \nu_{j,1}(\chi_1, \ldots, \chi_{k-1}, \chi^r)$$
$$+ \nu_{j,2}(\chi_1, \ldots, \chi_{k-1}, \chi^r)\chi_k, \quad k \leq j \leq \rho \quad (G.47)$$

$$\frac{\partial \beta}{\partial \chi_k}(\chi_1, \ldots, \chi_k, \chi^r) = 0. \quad (G.48)$$

Let us then define the new coordinates $\bar{\xi}_k, \ldots, \bar{\xi}_\rho$ as

$$\bar{\xi}_k = \chi_k - \int_0^{\chi_{k-1}} \nu_{k,2}(\chi_1, \ldots, \chi_{k-2}, s, \chi^r) ds \triangleq \chi_k - \bar{\nu}_{k-1}(\chi_1, \ldots, \chi_{k-1}, \chi^r)$$
$$\bar{\xi}_j = \xi_j - \int_0^{\chi_{k-1}} \nu_{j,2}(\chi_1, \ldots, \chi_{k-2}, s, \chi^r) ds, \quad k+1 \leq j \leq n. \quad (G.49)$$

In the coordinates $(\chi_1, \ldots, \chi_{k-1}, \bar{\xi}_k, \ldots, \bar{\xi}_\rho, \chi^r)$, the system (G.31) is expressed as

$$\begin{aligned}
\dot{\chi}_j &= \chi_{j+1}, \quad 1 \leq j \leq k-2 \\
\dot{\chi}_{k-1} &= \bar{\xi}_k + \bar{\nu}_{k-1}(\chi_1, \ldots, \chi_{k-1}, \chi^r) \\
\dot{\bar{\xi}}_k &= \bar{\xi}_{k+1} + \bar{\nu}_k(\chi_1, \ldots, \chi_{k-1}, \chi^r) \\
&\vdots \\
\dot{\bar{\xi}}_\rho &= \bar{\nu}_\rho(\chi_1, \ldots, \chi_{k-1}, \chi^r) + \beta(\chi_1, \ldots, \chi_{k-1}, \chi^r)u \\
\dot{\chi}^r &= \Phi_0(\chi_1, \ldots, \chi_q, \chi^r) \\
y &= \chi_1,
\end{aligned} \quad (G.50)$$

and, moreover, the vector fields $\mathrm{ad}_{f_0}^j \bar{g}_0$, $0 \leq j \leq \rho - k + 1$, become

$$\mathrm{ad}_{f_0}^j \bar{g}_0 = (-1)^j \frac{\partial}{\partial \bar{\xi}_{\rho-j}}, \quad 0 \leq j \leq \rho - k, \quad \mathrm{ad}_{f_0}^{\rho-k+1} \bar{g}_0 = (-1)^{\rho-k+1} \frac{\partial}{\partial \chi_{k-1}}. \quad (G.51)$$

The necessity is again straightforward. □

The conditions of Proposition G.3 are necessary and sufficient only for the *local* existence of a diffeomorphism transforming (G.31) into (G.32). At this time there are no necessary and sufficient conditions for the *global* existence of such a diffeomorphism. Of course, the global validity of conditions (i)–(iv) of Proposition G.3 is necessary, as are the completeness of the vector fields $\mathrm{ad}_{f_0}^i \bar{g}_0$, $0 \leq i \leq \rho - m_1 - 1$, and the connectedness of the manifold $M = \{\zeta \in \mathbb{R}^n : h(\zeta) = L_{f_0} h(\zeta) = \cdots = L_{f_0}^{\rho-1} h(\zeta) = 0\}$, as proved in [13]. Therefore, to formulate the counterpart of Theorem G.2, we make the following assumption:

Assumption G.4 *The system (G.31) can be transformed via a global diffeomorphism $x = \phi(\zeta)$ into (G.32).*

Theorem G.5 *Under Assumption G.4, the system (G.1) can be transformed via a global parameter-independent diffeomorphism $x = \phi(\zeta)$ into the partial-state-feedback form (G.30) if and only if the following conditions are globally satisfied:*

(i) $\left[f_j, \mathrm{ad}_{f_0}^i \bar{g}_0\right] = 0$, $0 \leq i \leq \rho - m_1 - 1$, $1 \leq j \leq p$,

(ii) $d\left(L_{f_j} L_{f_0}^i h\right) \in \mathrm{span}\left\{dh, \ldots, d\left(L_{f_0}^j h\right)\right\}$, $0 \leq i \leq q - 2$, $1 \leq j \leq p$,

(iii) $\left[f_j, \mathrm{ad}_{f_0}^i g_0\right] \in \mathcal{G}^i = \mathrm{span}\left\{g_0, \mathrm{ad}_{f_0} g_0, \ldots, \mathrm{ad}_{f_0}^i g_0\right\}$, $0 \leq i \leq \rho - q - 1$, $1 \leq j \leq p$,

(iv) $\left[g_j, \mathrm{ad}_{f_0}^i \bar{g}_0\right] = 0$, $0 \leq i \leq \rho - m_1 - 1$, $0 \leq j \leq p$, and

(v) $\sum_{j=1}^{p} \theta_j g_j = (b_m - 1) g_0$,

where $\bar{g}_0$ is the vector field defined in (G.33).

Proof. *Sufficiency.* In the x-coordinates, which are defined globally by Assumption G.4, we have

$$\mathrm{ad}_{f_0}^i \bar{g}_0 = (-1)^i \frac{\partial}{\partial x_{\rho - i}}, \quad 0 \leq i \leq \rho - m_1 \tag{G.52}$$

$$\mathcal{G}^i = \mathrm{span}\left\{\frac{\partial}{\partial x_\rho}, \ldots, \frac{\partial}{\partial x_{\rho - i}}\right\}, \quad 0 \leq i \leq \rho - q. \tag{G.53}$$

Hence, condition (i) becomes

$$\left[f_j, (-1)^i \frac{\partial}{\partial x_{\rho - i}}\right] = 0, \quad 0 \leq i \leq \rho - m_1 - 1, \ 1 \leq j \leq p, \tag{G.54}$$

which implies that the vector fields f_j are expressed in the x-coordinates as

$$f_j = \sum_{i=1}^{n} \varphi_{j,i}(x_1, \ldots, x_{m_1}, x^r) \frac{\partial}{\partial x_i}, \quad 1 \leq j \leq p. \tag{G.55}$$

Since in the x-coordinates we have

$$\begin{aligned} x_{i+1} &= L_{f_0}^i h, \quad 0 \leq i \leq m_1 - 1 \\ \varphi_{j,i+1}(x_1, \ldots, x_{m_1}, x^r) &= L_{f_j} L_{f_0}^i h, \quad 0 \leq i \leq m_1 - 1, \ 1 \leq j \leq p, \end{aligned} \tag{G.56}$$

condition (ii) becomes

$$d\varphi_{j,i} \in \mathrm{span}\{dx_1, \ldots, dx_i\}, \quad 1 \leq i \leq q - 1, \ 1 \leq j \leq p, \tag{G.57}$$

Differential Geometric Conditions 533

or, equivalently,

$$\frac{\partial \varphi_{j,i}}{\partial x_k} = 0, \quad i+1 \le k \le n, \ 1 \le i \le q-1, \ 1 \le j \le p. \tag{G.58}$$

From (G.53) and (G.55) we see that condition (iii) is equivalently expressed in the x-coordinates as

$$\left[\frac{\partial}{\partial x_{\rho-i}}, f_j\right] \in \text{span}\left\{\frac{\partial}{\partial x_\rho}, \ldots, \frac{\partial}{\partial x_{\rho-i}}\right\}, \quad 0 \le i \le \rho - q, \ 1 \le j \le p, \tag{G.59}$$

which implies that

$$\begin{aligned}
\frac{\partial \varphi_{j,i}}{\partial x_k} &= 0, \quad i+1 \le k \le \rho, \ q \le i \le \rho - 1, \ 1 \le j \le p \\
\frac{\partial \varphi_{j,i}}{\partial x_k} &= 0, \quad q+1 \le k \le \rho, \ \rho+1 \le i \le n, \ 1 \le j \le p.
\end{aligned} \tag{G.60}$$

Combining (G.55), (G.58), and (G.60), we see that the vector fields f_j, $1 \le j \le p$, are expressed in the x-coordinates as

$$\begin{aligned}
f_j &= \sum_{i=1}^{q} \varphi_{j,i}(x_1, \ldots, x_i) \frac{\partial}{\partial x_i} + \sum_{i=q+1}^{m_1} \varphi_{j,i}(x_1, \ldots, x_i, x^{\mathrm{r}}) \frac{\partial}{\partial x_i} \\
&\quad + \sum_{i=m_1+1}^{\rho} \varphi_{j,i}(x_1, \ldots, x_{m_1}, x^{\mathrm{r}}) \frac{\partial}{\partial x_i} + \sum_{i=\rho+1}^{n} \varphi_{j,i}(x_1, \ldots, x_q, x^{\mathrm{r}}) \frac{\partial}{\partial x_i}.
\end{aligned} \tag{G.61}$$

Similarly, conditions (iv) and (v) imply that in the x-coordinates we have

$$g_0 + \sum_{j=1}^{p} \theta_j g_j = b_m \beta(x_1, \ldots, x_{m_1}, x^{\mathrm{r}}) \frac{\partial}{\partial x_\rho}. \tag{G.62}$$

From Assumption G.4, (G.61), and (G.62), we conclude that in the x-coordinates the system (G.1) is expressed as (G.30) with $\varphi_{0,i} \equiv 0$, $1 \le i \le m_1 - 1$. The proof of necessity is straightforward. □

G.2 Output-Feedback Forms

Setting $m_1 = 1$ in (G.2), in Proposition G.1, and in Theorem G.2, we obtain the following corollary, which was proved in [122, 121]:

Corollary G.6 *The system (G.3) can be transformed via a global parameter-independent diffeomorphism $x = \phi(\zeta)$ into the parametric output-feedback form (7.101) if and only if the following conditions hold globally:*

(i) $\text{rank}\left\{dh, d(L_{f_0}h), \ldots, d\left(L_{f_0}^{n-1}h\right)\right\} = n$,

(ii) $\left[\mathrm{ad}_{f_0}^i r, \mathrm{ad}_{f_0}^{i+1} r\right] = 0, \quad 0 \leq i \leq n-2$,

(iii) $\left[f_j, \mathrm{ad}_{f_0}^i r\right] = 0, \quad 0 \leq i \leq n-2, \ 1 \leq j \leq p$,

(iv) $\left[g_j, \mathrm{ad}_{f_0}^i r\right] = 0, \quad 0 \leq i \leq n-2, \ 1 \leq j \leq p$,

(vi) $g_0 + \sum_{j=1}^{p} \theta_j g_j = \beta(\cdot) \sum_{i=0}^{m} b_i (-1)^i \mathrm{ad}_{f_0}^i r$, and

(vii) the vector fields $r, \mathrm{ad}_{f_0} r, \ldots, \mathrm{ad}_{f_0}^{n-1} r$ are complete,

where β is a smooth nonlinear function and r is the vector field defined by

$$L_r L_{f_0}^i h = \begin{cases} 0, & i = 0, \ldots, n-2 \\ 1, & i = n-1. \end{cases} \tag{G.63}$$

Corollary G.6 gives necessary and sufficient conditions for (G.3) to be globally transformable into the parametric output-feedback canonical form (7.101) via a *parameter-independent* diffeomorphism. However, this would unnecessarily exclude many systems such as the robotic example of Section 7.3.3, for which a *parameter-dependent* diffeomorphism is needed to go from the physical coordinates into the output-feedback form. In the full-state feedback case, we need parameter-independent diffeomorphisms, because we want to be able to calculate the new state variables from the measurements of the original ones. When only the output is measured, the dependence of the diffeomorphism on the unknown parameters is acceptable because the states do not appear in the control law. Therefore, we now give necessary and sufficient conditions for the system

$$\begin{aligned} \dot{\zeta} &= f(\zeta; \bar{\theta}) + g(\zeta; \bar{\theta}) u \\ y &= h(\zeta; \bar{\theta}), \end{aligned} \tag{G.64}$$

where $\bar{\theta}$ is a vector of unknown parameters, to be globally transformable into (7.101) via diffeomorphism which is allowed to depend on the unknown parameters. The following result was first given in [72]:

Corollary G.7 *The system (G.64) can be transformed via a global diffeomorphism $x = \phi(\zeta; \bar{\theta})$ into the output-feedback canonical form (7.101) if and only if the following conditions are satisfied for all $\zeta \in \mathbb{R}^n$ and for the true value of the parameter vector $\bar{\theta}$:*

(i) $\mathrm{rank}\left\{dh, d(L_f h), \ldots, d\left(L_f^{n-1} h\right)\right\} = n$,

(ii) $\left[\mathrm{ad}_f^i r, \mathrm{ad}_f^{i+1} r\right] = 0, \quad 0 \leq i \leq n-2$,

Differential Geometric Conditions

(iii) $\mathrm{ad}_f^n r = \sum_{i=0}^{n-1} \left[\varphi'_{0,n-i}(y) + \sum_{j=1}^{p} \theta_j \varphi'_{j,n-i}(y) \right] (-1)^{n-i} \mathrm{ad}_f^i r,$

where $\varphi_{j,n-i}(y) = \int_0^y \varphi'_{j,n-i}(s)ds,\ 0 \le i \le n-1,\ 0 \le j \le p,$

(iv) $\left[g, \mathrm{ad}_f^i r \right] = 0,\ 0 \le i \le n-2,$

(v) $g = \beta(\cdot) \sum_{i=0}^{m} b_i (-1)^i \mathrm{ad}_f^i r,$ and

(vi) the vector fields $r, \mathrm{ad}_f r, \ldots, \mathrm{ad}_f^{n-1} r$ are complete,

where β is a smooth nonlinear function and r is the vector field defined by

$$L_r L_f^i h = \begin{cases} 0, & i = 0, \ldots, n-2 \\ 1, & i = n-1 \end{cases} \quad (G.65)$$

G.3 Full-State-Feedback Forms

In this section, we consider the full-state feedback case and, hence, we require the diffeomorphisms to be parameter-independent. The results in this section were first obtained in [69], except for Theorem G.9, which was first given in [1].

Setting $k = 1$, $m_1 = \rho$ in (G.30)–(G.32), in Proposition G.3, and in Theorem G.5, we obtain the following corollary:

Corollary G.8 *There exists a parameter-independent diffeomorphism $x = \phi(\zeta)$, satisfied in a neighborhood $\mathcal{U}$ of a point ζ_0, which transforms (G.1) into the form*

$$\begin{aligned}
\dot{x}_1 &= x_2 + \varphi_1^\mathrm{T}(x_1)\theta \\
\dot{x}_2 &= x_3 + \varphi_2 \theta^\mathrm{T}(x_1, x_2)\theta \\
&\vdots \\
\dot{x}_{q-1} &= x_q + \varphi_{q-1}^\mathrm{T}(x_1, \ldots, x_{q-1})\theta \\
\dot{x}_q &= x_{q+1} + \varphi_q^\mathrm{T}(x_1, \ldots, x_q, x^\mathrm{r})\theta \\
&\vdots \\
\dot{x}_{p-1} &= x_p + \varphi_{p-1}^\mathrm{T}(x_1, \ldots, x_{p-1}, x^\mathrm{r})\theta \\
\dot{x}_p &= \varphi_{0,p}(x) + \varphi_p^\mathrm{T}(x)\theta + \beta(x)u \\
\dot{x}^\mathrm{r} &= \Phi_0(x_1, \ldots, x_q, x^\mathrm{r}) + \sum_{j=1}^{p} \theta_j \Phi_j(x_1, \ldots, x_q, x^\mathrm{r}) \\
y &= x_1,
\end{aligned} \quad (G.66)$$

if and only if the following conditions are valid in a neighborhood $\mathcal{U}_1 \supseteq \mathcal{U}$:

(i) $L_{g_0}L_{f_0}^i h \equiv 0$, $0 \leq i \leq \rho - 2$, $L_{g_0}L_{f_0}^{\rho-1} h \neq 0$,

(ii) the distribution $\mathcal{G}^{\rho-q} = \text{span}\left\{g_0, \text{ad}_{f_0} g_0, \ldots, \text{ad}_{f_0}^{\rho-q} g_0\right\}$ is involutive and of constant rank $\rho - q + 1$,

(iii) $d\left(L_{f_j} L_{f_0}^i h\right) \in \text{span}\left\{dh, \ldots, d\left(L_{f_0}^j h\right)\right\}$, $0 \leq i \leq q - 2$, $1 \leq j \leq p$,

(iv) $\left[f_j, \text{ad}_{f_0}^i g_0\right] \in \mathcal{G}^i = \text{span}\left\{g_0, \text{ad}_{f_0} g_0, \ldots, \text{ad}_{f_0}^i g_0\right\}$, $0 \leq i \leq \rho - q - 1$, $1 \leq j \leq p$,

(v) $g_j \equiv 0$, $1 \leq j \leq p$.

For the diffeomorphism of Corollary G.8 to be globally valid, it is *necessary* that the above conditions (i)–(v) be globally valid and that the manifold $M = \left\{\zeta \in \mathbb{R}^n : h(\zeta) = L_{f_0} h(\zeta) = \cdots = L_{f_0}^{\rho-1} h(\zeta) = 0\right\}$ be connected. As can be shown using the results of [13], these conditions, together with the completeness of the vector fields $\bar{g}_0, \text{ad}_{\bar{f}_0} \bar{g}_0, \ldots, \text{ad}_{\bar{f}_0}^{\rho-1} \bar{g}_0$, where $\bar{g}_0$ is defined in (G.33) and $\bar{f}_0 = f_0 - L_{f_0}^\rho \bar{g}_0$, are *sufficient* for $x = \phi(\zeta)$ to be a global diffeomorphism. In the case $q = 1$, these conditions are actually *necessary and sufficient* [13, Corollary 5.7]. However, for $q > 1$, the completeness of $\text{ad}_{\bar{f}_0} \bar{g}_0, \ldots, \text{ad}_{\bar{f}_0}^{\rho-1} \bar{g}_0$ is not necessary. For example, consider the system

$$\begin{aligned} \dot{x}_1 &= x_2 \\ \dot{x}_2 &= u \\ \dot{x}_3 &= -x_3^3 - x_2 x_3^2 \\ y &= x_1. \end{aligned} \qquad (G.67)$$

This system is already in the form (G.66), but the vector field

$$\begin{aligned} \text{ad}_{\bar{f}_0} \bar{g}_0 &= \left[x_2 \frac{\partial}{\partial x_1} - \left(x_3^3 + x_2 x_3^2\right) \frac{\partial}{\partial x_3}, \frac{\partial}{\partial x_2}\right] \\ &= -\frac{\partial}{\partial x_1} + x_3^2 \frac{\partial}{\partial x_3} \end{aligned} \qquad (G.68)$$

is not complete, since the solutions of the system

$$\begin{aligned} \dot{x}_1 &= -1 \\ \dot{x}_2 &= 0 \\ \dot{x}_3 &= x_3^2 \end{aligned} \qquad (G.69)$$

starting from any point with $x_3(0) > 0$ escape to infinity in finite time.

We now turn our attention to nonlinear systems of the form

$$\dot{\zeta} = f_0(\zeta) + \sum_{j=1}^{p} \theta_j f_j(\zeta) + \left[g_0(\zeta) + \sum_{j=1}^{p} \theta_j g_j(\zeta)\right] u, \qquad (G.70)$$

DIFFERENTIAL GEOMETRIC CONDITIONS

where f_j, g_j, $0 \leq j \leq p$, are smooth vector fields in a neighborhood of the origin $\zeta = 0$ with $f_j(0) = 0$, $0 \leq j \leq p$, $g_0(0) \neq 0$, and give necessary and sufficient conditions for (G.70) to be locally transformable via a parameter-independent diffeomorphism $x = \phi(\zeta)$ into the *parametric pure-feedback form* (4.285), which is repeated here for convenience:

$$\begin{aligned}
\dot{x}_1 &= x_2 + \varphi_1^T(x_1, x_2)\theta \\
\dot{x}_2 &= x_3 + \varphi_2^T(x_1, x_2, x_3)\theta \\
&\vdots \\
\dot{x}_{n-1} &= x_n + \varphi_n^T(x_1, \ldots, x_n)\theta \\
\dot{x}_n &= \varphi_{0,n}(x) + \varphi_n^T(x)\theta + [\beta_0(x) + \beta^T(x)\theta]u,
\end{aligned} \qquad (G.71)$$

where

$$\varphi_{0,n}(0) = 0, \; \varphi_1(0) = \cdots = \varphi_n(0) = 0, \; \beta_0(0) \neq 0. \qquad (G.72)$$

Theorem G.9 *A parameter-independent diffeomorphism $x = \phi(\zeta)$, with $\phi(0) = 0$, transforming (G.70) into (G.71), exists in a neighborhood $\bar{B}_x \subset \mathcal{U}$ of the origin if and only if the following conditions are satisfied in $\mathcal{U}$:*

(i) Feedback linearization condition. The distributions

$$\mathcal{G}^i = \operatorname{span}\left\{g_0, \operatorname{ad}_{f_0} g_0, \ldots, \operatorname{ad}_{f_0}^i g_0\right\}, \quad 0 \leq i \leq n-1 \qquad (G.73)$$

are involutive and of constant rank $i+1$.

(ii) Parametric pure-feedback condition.

$$\begin{aligned}
g_j &\in \mathcal{G}^0, \quad 1 \leq j \leq p \\
[X, f_j] &\in \mathcal{G}^{i+1}, \quad \forall X \in \mathcal{G}^i, \quad 0 \leq i \leq n-3, \; 1 \leq j \leq p.
\end{aligned} \qquad (G.74)$$

Proof. *Sufficiency.* As proved in [56], condition (i) is sufficient for the existence of a diffeomorphism $x = \phi(\zeta)$ with $\phi(0) = 0$ which transforms the system

$$\dot{\zeta} = f_0(\zeta) + g_0(\zeta)u, \; f_0(0) = 0, \; g_0(0) \neq 0 \qquad (G.75)$$

into the system

$$\begin{aligned}
\dot{x}_i &= x_{i+1}, \quad 1 \leq i \leq n-1 \\
\dot{x}_n &= \varphi_{0,n}(x) + \beta_0(x)u,
\end{aligned} \qquad (G.76)$$

with

$$\varphi_{0,n}(0) = 0, \; \beta_0(0) \neq 0. \qquad (G.77)$$

Hence, in the coordinates of (G.76) we have

$$f_0 = x_2 \frac{\partial}{\partial x_1} + \cdots + x_n \frac{\partial}{\partial x_{n-1}} + \varphi_{0,n}(x) \frac{\partial}{\partial x_n} \tag{G.78}$$

$$g_0 = \beta_0(x) \frac{\partial}{\partial x_n} \tag{G.79}$$

$$\mathcal{G}^i = \text{span}\left\{\frac{\partial}{\partial x_n}, \ldots, \frac{\partial}{\partial x_{n-i}}\right\}, \quad 0 \leq i \leq n-1. \tag{G.80}$$

Because of (G.80), the parametric pure-feedback condition (G.74), expressed in the x-coordinates, states that

$$\begin{aligned} g_j &\in \text{span}\left\{\frac{\partial}{\partial x_n}\right\}, \quad 1 \leq j \leq p \\ \left[\frac{\partial}{\partial x_i}, f_j\right] &\in \text{span}\left\{\frac{\partial}{\partial x_n}, \ldots, \frac{\partial}{\partial x_{i-1}}\right\}, \quad 3 \leq i \leq n, \ 1 \leq j \leq p. \end{aligned} \tag{G.81}$$

But (G.81) can be equivalently rewritten as

$$\begin{aligned} g_j &= \beta_j(x) \frac{\partial}{\partial x_n}, \quad 1 \leq j \leq p \\ f_j &= \varphi_{j,1}(x_1, x_2) \frac{\partial}{\partial x_1} + \varphi_{j,2}(x_1, x_2, x_3) \frac{\partial}{\partial x_2} + \cdots \\ &\quad + \varphi_{j,n-1}(x_1, \ldots, x_n) \frac{\partial}{\partial x_{n-1}} + \varphi_{j,n}(x_1, \ldots, x_n) \frac{\partial}{\partial x_n}, \quad 1 \leq j \leq p. \end{aligned} \tag{G.82}$$

Furthermore, since $\phi(0) = 0$ and $f_j(0) = 0, 1 \leq j \leq p$, we conclude from (G.82) that

$$\varphi_1(0) = \cdots = \varphi_n(0) = 0. \tag{G.83}$$

Combining (G.78), (G.79), (G.82) and (G.83), we see that in the x-coordinates the system (G.70) becomes (G.71). The necessity is straightforward. □

Remark G.10 The design of Section 4.5.3 can be applied to the system (G.70), after using the diffeomorphism of Theorem G.9 to transform it into (G.71). Then, the feasibility region $\mathcal{F} = B_x \times B_\theta$ of Proposition 4.24 must be a subset of the region on which the diffeomorphism exists. This can be ensured by selecting B_x to be a subset of $\bar{B}_x \subset \mathcal{U}$. ◇

Remark G.11 A special case of the parametric pure-feedback condition (G.74) is the extended-matching condition of [65]:

$$g_j \in \mathcal{G}^0, \ f_j \in \mathcal{G}^1, \ 1 \leq j \leq p. \tag{G.84}$$

Differential Geometric Conditions

This is clear from the proof of Theorem G.9: if (G.74) is replaced by (G.84), then (G.82) still holds, but with $\varphi_1 \equiv 0, \ldots, \varphi_{n-2} \equiv 0$. Then, the system (G.70) is expressed in the x-coordinates as

$$\begin{aligned}
\dot{x}_1 &= x_2 \\
\dot{x}_2 &= x_3 \\
&\vdots \\
\dot{x}_{n-2} &= x_{n-1} \\
\dot{x}_{n-1} &= x_n + \varphi_{n-1}^{\mathrm{T}}(x_1, \ldots, x_n)\theta \\
\dot{x}_n &= \varphi_{0,n}(x) + \varphi_n^{\mathrm{T}}(x)\theta + \left[\beta_0(x) + \beta^{\mathrm{T}}(x)\theta\right] u\,.
\end{aligned} \qquad (\text{G.85})$$

$\diamond$

Remark G.12 The expressions given in (G.73) and (G.74) for the feedback linearization and parametric pure-feedback conditions are convenient for the proof of Theorem G.9, but they are *not minimal*. As shown in [178, 49], the equivalent minimal form of (G.73) is

$$\mathcal{G}^{n-2} \text{ is involutive and } \mathcal{G}^{n-1} \text{ has constant rank } n\,. \qquad (\text{G.86})$$

The minimal form of (G.74) is

$$\left[\mathrm{ad}_{f_0}^i g_0, f_j\right] \in \mathcal{G}^{i+1}, \quad 0 \leq i \leq n-3, \quad 1 \leq j \leq p\,. \qquad (\text{G.87})$$

The equivalence of (G.74) and (G.87) follows from the involutivity of $\mathcal{G}^{n-2}$. $\diamond$

Finally, we consider the class of systems of the form (G.70) which can be transformed into the *parametric strict-feedback form* (3.70):

$$\begin{aligned}
\dot{x}_1 &= x_2 + \varphi_1(x_1)^{\mathrm{T}}\theta \\
\dot{x}_2 &= x_3 + \varphi_2(x_1, x_2)^{\mathrm{T}}\theta \\
&\vdots \\
\dot{x}_{n-1} &= x_n + \varphi_{n-1}^{\mathrm{T}}(x_1, \ldots, x_{n-1})\theta \\
\dot{x}_n &= \varphi_{0,n}(x) + \varphi_n(x)^{\mathrm{T}}\theta + \beta_0(x)u\,.
\end{aligned} \qquad (\text{G.88})$$

To characterize this class of systems, we use the following assumption about the part of the system (G.70) that does not contain unknown parameters:

Assumption G.13 *There exists a global diffeomorphism* $x = \phi(\zeta)$, *with* $\phi(0) = 0$, *transforming the system*

$$\dot{\zeta} = f_0(\zeta) + g_0(\zeta)u\,, \qquad (\text{G.89})$$

into the system

$$\begin{aligned}
\dot{x}_i &= x_{i+1}, \quad 1 \leq i \leq n-1 \\
\dot{x}_n &= \varphi_{0,n}(x) + \beta_0(x)u\,,
\end{aligned} \qquad (\text{G.90})$$

with

$$\varphi_{0,n}(0) = 0, \ \beta_0(x) \neq 0 \ \forall x \in \mathbb{R}^n\,. \qquad (\text{G.91})$$

Remark G.14 The *local* existence of such a diffeomorphism is equivalent to the feedback linearization condition (G.73). At present, there are no necessary and sufficient conditions verifying the *global* validity of this assumption. Sufficient conditions for Assumption G.13 are given in [48], while necessary and sufficient conditions for the case in which $\beta_0(x)$ is constant can be found in [27, 162]. ◇

Theorem G.15 *Under Assumption G.13, the system (G.70) is globally diffeomorphically equivalent through $x = \phi(\zeta)$ to (G.88) if and only if the following condition holds globally:*

(i) Parametric strict-feedback condition.

$$g_j \equiv 0, \quad 1 \leq j \leq p$$
$$[X, f_j] \in \mathcal{G}^i, \quad \forall X \in \mathcal{G}^i, \quad 0 \leq i \leq n-2, \quad 1 \leq j \leq p, \quad (G.92)$$

with $\mathcal{G}^i$, $0 \leq i \leq n-1$, as defined in (G.73).

Proof. The proof is similar to that of Theorem G.9. First, because of the assumptions that the diffeomorphism $x = \phi(\zeta)$ is global and $\beta_0(x) \neq 0 \; \forall x \in \mathbb{R}^n$, the distributions $\mathcal{G}^i$, $0 \leq i \leq n-1$, are globally defined and can be expressed in the x-coordinates as

$$\mathcal{G}^i = \text{span}\left\{\frac{\partial}{\partial x_n}, \cdots, \frac{\partial}{\partial x_{n-i}}\right\}, \quad 0 \leq i \leq n-1. \quad (G.93)$$

To prove sufficiency, note that if the parametric pure-feedback condition (G.74) of Theorem G.9 is replaced by the parametric strict-feedback condition (G.92), then (G.81) is replaced by

$$g_j \equiv 0, \quad 1 \leq j \leq p$$
$$\left[\frac{\partial}{\partial x_i}, f_j\right] \in \text{span}\left\{\frac{\partial}{\partial x_n}, \cdots, \frac{\partial}{\partial x_i}\right\}, \quad 2 \leq i \leq n, \quad 1 \leq j \leq p. \quad (G.94)$$

Thus, the expression for f_j in (G.82) becomes

$$f_j = \varphi_{j,1}(x_1)\frac{\partial}{\partial x_1} + \varphi_{j,2}(x_1, x_2)\frac{\partial}{\partial x_2} + \cdots$$
$$+ \varphi_{j,n-1}(x_1, \ldots, x_{n-1})\frac{\partial}{\partial x_{n-1}} + \varphi_{j,n}(x_1, \ldots, x_n)\frac{\partial}{\partial x_n}, \quad \leq j \leq p. \quad (G.95)$$

□

Bibliography

[1] O. Akhrif and G. L. Blankenship, "Robust stabilization of feedback linearizable systems," *Proceedings of the 27th IEEE Conference on Decision and Control*, Austin, TX, 1988, pp. 1714–1719.

[2] B. D. O. Anderson, R. R. Bitmead, C. R. Johnson, Jr., P. V. Kokotović, R. L. Kosut, I. Mareels, L. Praly, and B. D. Riedle, *Stability of Adaptive Systems: Passivity and Averaging Analysis*, Cambridge, MA: MIT Press, 1986.

[3] A. M. Annaswamy, D. Seto, and J. Baillieul, "Adaptive control of a class of nonlinear systems," *Proceedings of the 7th Yale Workshop on Adaptive and Learning Systems*, New Haven, CT, 1992.

[4] Z. Artstein, "Stabilization with relaxed controls," *Nonlinear Analysis*, vol. TMA-7, pp. 1163–1173, 1983.

[5] K. J. Åström and B. Wittenmark, *Adaptive Control*, second edition, New York: Addison-Wesley, 1995.

[6] B. R. Barmish, M. Corless, and G. Leitmann, "A new class of stabilizing controllers for uncertain dynamical systems," *SIAM Journal on Control and Optimization*, vol. 21, pp. 246–255, 1983.

[7] G. Bastin, "Adaptive nonlinear control of fed-batch stirred tank reactors," *International Journal of Adaptive Control and Signal Processing*, vol. 6, pp. 273–284, 1992.

[8] G. Bastin and G. Campion, "Indirect adaptive control of linearly parametrized nonlinear systems," *Proceedings of the 3rd IFAC Symposium on Adaptive Systems in Control, and Signal Processing*, Glasgow, UK, 1989.

[9] F. Blaschke, "The principle of field orientation applied to the new transvector closed-loop control system for rotating field machines," *Siemens Review*, vol. 39, pp. 217–220, 1972.

[10] S. Boyd and J. C. Doyle, "Comparison of peak and RMS gains for discrete systems," *Systems & Control Letters*, vol. 9, pp. 1–6, 1987.

[11] B. Brogliato, R. Lozano, and I. Landau, "New relationships between Lyapunov functions and the passivity theorem," *International Journal of Adaptive Control and Signal Processing*, vol. 7, pp. 353–365, 1993.

[12] C. I. Byrnes and A. Isidori, "New results and examples in nonlinear feedback stabilization," *Systems & Control Letters*, vol. 12, pp. 437–442, 1989.

[13] C. I. Byrnes and A. Isidori, "Asymptotic stabilization of minimum phase nonlinear systems," *IEEE Transactions on Automatic Control*, vol. 36, pp. 1122–1137, 1991.

[14] C.I. Byrnes, A. Isidori, and J.C. Willems, "Passivity, feedback equivalence, and the global stabilization of minimum phase nonlinear systems," *IEEE Transactions on Automatic Control*, vol. 36, pp. 1228–1240, 1991.

[15] G. Campion and G. Bastin, "Indirect adaptive state-feedback control of linearly parametrized nonlinear systems," *International Journal of Adaptive Control and Signal Processing*, vol. 4, pp. 345–358, 1990.

[16] H. Chapellat, M. Dahleh, and S. P. Bhattacharyya, "Robust stability under structured and unstructured perturbations," *IEEE Transactions on Automatic Control*, vol. 35, pp. 1100–1108, 1990.

[17] C. Chen and M. Tomizuka, "Steering and braking control of tractor-semitrailer vehicles in automated highway systems," *Proceedings of the 1995 American Control Conference*, Seattle, WA, to appear.

[18] Y. H. Chen, "A new matching condition for robust control design," *Proceedings of the 1993 American Control Conference*, San Francisco, CA, pp. 122–126, 1993.

[19] Y. H. Chen and G. Leitmann, "Robustness of uncertain systems in the absence of matching assumptions," *International Journal of Control*, vol. 45, pp. 1527–1542, 1987.

[20] M. Corless and G. Leitmann, "Continuous state feedback guaranteeing uniform ultimate boundedness for uncertain dynamical systems," *IEEE Transactions on Automatic Control*, vol. 26, pp. 1139–1144, 1981.

[21] J.-M. Coron and L. Praly, "Adding an integrator for the stabilization problem," *Systems & Control Letters*, vol. 17, pp. 89–104, 1991.

[22] J. J. Craig, *Adaptive Control of Mechanical Manipulators*, Reading, MA: Addison-Wesley, 1988.

[23] M. A. Dahleh and I. J. Diaz-Bobillo, *Control of Uncertain Systems: A Linear Programming Approach*, Englewood Cliffs, NJ: Prentice-Hall, 1995.

[24] A. Datta and P. Ioannou, "Performance improvement versus robust stability in model reference adaptive control," *IEEE Transactions on Automatic Control*, vol. 39, pp. 2370–2387, 1994.

[25] D. M. Dawson, J. J. Carroll and M. Schneider, "Integrator backstepping control of a brushed DC motor turning a robotic load," *IEEE Transactions on Control Systems Technology*, vol. 2, pp. 233–244, 1994.

[26] D. M. Dawson, J. Hu, and T. Burg, *Nonlinear Control of Electric Machinery*, Piscataway, NJ: IEEE Press, to appear, 1995.

[27] W. Dayawansa, W. M. Boothby and D. L. Elliott, "Global state and feedback equivalence of nonlinear systems," *Systems & Control Letters*, vol. 6, pp. 229–234, 1985.

[28] C. A. Desoer, and M. Vidyasagar, *Feedback Systems: Input-Output Properties*, New York: Academic Press, 1975.

[29] S. Dharmasena, and P. V. Kokotović, "Robustness of the adaptive backstepping design for unmodeled dynamics — regulation," submitted to *International Journal of Adaptive Control and Signal Processing*, 1994.

[30] J. C. Doyle, B. A. Francis, and A. R. Tannenbaum, *Feedback Control Theory*, New York: Macmillan, 1992.

[31] B. Egardt, *Stability of Adaptive Controllers*, New York: Springer-Verlag, 1979.

[32] J. M. Elzebda, A. H. Nayfeh, and D. T. Mook, "Development of an analytical model of wing rock for slender delta wings," *AIAA Journal of Aircraft*, vol. 26, pp. 737–743, 1989.

[33] G. Espinosa and R. Ortega, "State observers are unnecessary for induction motor control," *Systems & Control Letters*, vol. 23, pp. 315–323, 1994.

[34] K. M. Eveker, C. N. Nett, "Control of compression system surge and rotating stall: a laboratory-based 'hands-on' introduction," *Proceedings of the 1994 American Control Conference*, Baltimore, MD, pp. 1307–1311.

[35] A. Feuer, A. S. Morse, "Adaptive control of single-input single-output linear systems," *IEEE Transactions on Automatic Control*, vol. 23, pp. 557–569, 1978.

[36] A. Feuer, A. S. Morse, "Local stability of parameter adaptive control systems," *Proceedings of the 1978 Conference on Information Sciences and Systems*, Johns Hopkins, Baltimore, MD, pp. 107–111, 1978.

[37] R. A. Freeman and P. V. Kokotović, "Backstepping design of robust controllers for a class of nonlinear systems," *Proceedings of the IFAC Nonlinear Control Systems Design Symposium*, Bordeaux, France, June 1992, pp. 307–312.

[38] R. A. Freeman and P. V. Kokotović, "Design of 'softer' robust nonlinear control laws," *Automatica*, vol. 29, pp. 1425–1437, 1993.

[39] R. A. Freeman and P. V. Kokotović, "Inverse optimality in robust stabilization," *SIAM Journal on Control and Optimization*, to appear.

[40] R. A. Freeman and P. V. Kokotović, "Tracking controllers for systems linear in the unmeasured states," submitted to *Automatica*, 1994.

[41] R. Ghanadan and G. L. Blankenship, "Adaptive control of nonlinear systems via approximate linearization," *IEEE Transactions on Automatic Control*, to appear.

[42] R. Ghanadan, "Computer algebra tools for recursive design of robust and adaptive nonlinear controllers," Technical Report CCEC-06-09-94, University of California, Santa Barbara, 1994.

[43] G. C. Goodwin and D. Q. Mayne, "A parameter estimation perspective of continuous time model reference adaptive control," *Automatica*, vol. 23, pp. 57–70, 1987.

[44] G. C. Goodwin and K. S. Sin, *Adaptive Filtering Prediction and Control*. Englewood Cliffs, NJ: Prentice-Hall, 1984.

[45] D. Hill and P. Moylan, "Dissipative Dynamical Systems," *J. Franklin Inst.*, 1980, vol. 309, pp. 327–357.

[46] J. Hu, D. M. Dawson, and Y. Qian, "Position tracking control for robot manipulators driven by induction motors without flux measurements," *IEEE Transactions on Robotics and Automation*, to appear, 1995.

[47] C. H. Hsu and E. Lan, "Theory of wing rock," *AIAA Journal of Aircraft*, vol. 22, pp. 920–924, 1985.

[48] L. R. Hunt, R. Su, and G. Meyer, "Global transformations of nonlinear systems," *IEEE Transactions on Automatic Control*, vol. 28, pp. 24–31, 1983.

[49] L. R. Hunt, R. Su, and G. Meyer, "Design for multi-input nonlinear systems," in *Differential Geometric Control Theory*, R. W. Brockett, R. S. Millman, and H. S. Sussman, Eds., Boston: Birkhäuser, 1983.

[50] P. A. Ioannou and P. V. Kokotović, *Adaptive Systems with Reduced Models*, New York: Springer-Verlag, 1983.

[51] P. A. Ioannou and J. Sun, *Stable and Robust Adaptive Control*, Englewood Cliffs, NJ: Prentice-Hall, 1995.

[52] P. A. Ioannou and G. Tao, "Dominant richness and improvement of performance of robust adaptive control," *Automatica*, vol. 25, pp. 287–291, March 1989.

[53] A. Isidori, *Nonlinear Control Systems*, Berlin: Springer-Verlag, 1989.

[54] D. H. Jacobson, *Extensions of Linear-Quadratic Control, Optimization and Matrix Theory*, New York: Academic Press, 1977.

[55] S. Jain and F. Khorrami, "Global decentralized adaptive control of large scale nonlinear systems without strict matching," *Proceedings of the 1995 American Control Conference*, Seattle, WA, to appear.

[56] B. Jakubczyk, W. Respondek, "On linearization of systems," *Bulletin of the Polish Academy of Science, Series on Mathematical Science*, vol. 28, no. 9-10, pp. 517–522, 1980.

[57] M. Janković, "Adaptive output feedback control of nonlinear feedback-linearizable systems," *International Journal of Adaptive Control and Signal Processing*, to appear.

[58] Z. P. Jiang and J.-B. Pomet, "Combining backstepping and time-varying techniques for a new set of adaptive controllers," *Proceedings of the 33rd IEEE Conference on Decision and Control*, Lake Buena Vista, FL, December 1994, pp. 2207–2212.

[59] Z. P. Jiang and L. Praly, "Iterative designs of adaptive controllers for systems with nonlinear integrators," *Proceedings of the 30th IEEE Conference on Decision and Control*, Brighton, UK, December 1991, pp. 2482–2487.

[60] Z. P. Jiang and L. Praly, "Preliminary results about robust Lagrange stability in adaptive nonlinear regulation," *International Journal on Adaptive Control and Signal Processing*, vol. 6, pp. 285–307, 1992.

[61] Z. P. Jiang, A. R. Teel, and L. Praly, "Small-gain theorem for ISS systems and applications," *Mathematics of Control, Signals, and Systems*, vol. 7, pp. 95–120, 1995.

[62] V. Jurdjevic and J. P. Quinn, "Controllability and Stability," *Journal of Differential Equations*, vol. 28, pp. 381–389, 1978.

[63] I. Kanellakopoulos, *Adaptive Control of Nonlinear Systems*, Ph.D. Dissertation, University of Illinois, Urbana, 1991.

[64] I. Kanellakopoulos, "Passive adaptive control of nonlinear systems," *International Journal of Adaptive Control and Signal Processing*, vol. 7, pp. 339–352, 1993.

[65] I. Kanellakopoulos, P. V. Kokotović, and R. Marino, "An extended direct scheme for robust adaptive nonlinear control," *Automatica*, vol. 27, pp. 247–255, 1991.

[66] I. Kanellakopoulos, P. V. Kokotović, R. Marino, and P. Tomei, "Adaptive control of nonlinear systems with partial state feedback," *Proceedings of the 1991 European Control Conference*, Grenoble, France, July 1991, pp. 1322–1327.

[67] I. Kanellakopoulos, P. V. Kokotović, and R. H. Middleton, "Observer-based adaptive control of nonlinear systems under matching conditions," *Proceedings of the 1990 American Control Conference*, San Diego, CA, pp. 549–552.

[68] I. Kanellakopoulos, P. V. Kokotović, and R. H. Middleton, "Indirect adaptive output-feedback control of a class of nonlinear systems," *Proceedings of the 29th IEEE Conference on Decision and Control*, Honolulu, HI, December 1990, pp. 2714–2719.

[69] I. Kanellakopoulos, P. V. Kokotović, and A. S. Morse, "Systematic design of adaptive controllers for feedback linearizable systems," *IEEE Transactions on Automatic Control*, vol. 36, pp. 1241–1253, 1991.

[70] I. Kanellakopoulos, P. V. Kokotović, and A. S. Morse, "Adaptive output-feedback control of systems with output nonlinearities," pp. 495–525 in [86].

[71] I. Kanellakopoulos, P. V. Kokotović, and A. S. Morse, "Adaptive output-feedback control of systems with output nonlinearities," *IEEE Transactions on Automatic Control*, vol. 37, pp. 1266–1282, 1992.

[72] I. Kanellakopoulos, P. V. Kokotović, and A. S. Morse, "Adaptive output-feedback control of a class of nonlinear systems," *Proceedings of the 30th IEEE Conference on Decision and Control*, Brighton, UK, December 1991, pp. 1082–1087.

[73] I. Kanellakopoulos, P. V. Kokotović, and A. S. Morse, "A toolkit for nonlinear feedback design," *Systems & Control Letters*, vol. 18, pp. 83–92, 1992.

[74] I. Kanellakopoulos, P. V. Kokotović, and A. S. Morse, "Adaptive nonlinear control with incomplete state information," *International Journal of Adaptive Control and Signal Processing*, vol. 6, pp. 367–394, 1992.

[75] I. Kanellakopoulos and P. T. Krein, "Integral-action nonlinear control of induction motors," *Proceedings of the 12th IFAC World Congress*, Sydney, Australia, July 1993, pp. 251–254.

[76] I. Kanellakopoulos, P. T. Krein, and F. Disilvestro, "Nonlinear flux-observer-based control of induction motors," *Proceedings of the 1992 American Control Conference*, Chicago, IL, June 1992, pp. 1700–1704.

[77] I. Kanellakopoulos, P. T. Krein, and F. Disilvestro, "A new controller-observer design for induction motor control," DSC-vol. 43, pp. 43–47, *ASME Winter Annual Meeting*, Anaheim, CA, November 1992.

[78] I. Kanellakopoulos, M. Krstić, and P. V. Kokotović, "Interlaced controller-observer design for adaptive nonlinear control," *Proceedings of the 1992 American Control Conference*, Chicago, IL, June 1992, pp. 1337–1342.

[79] I. Kanellakopoulos, M. Krstić and P. V. Kokotović, "Trajectory initialization in adaptive nonlinear control," *Proceedings of the 1993 IEEE Mediterranean Symposium on New Directions in Control Theory*, Chania, Greece.

[80] I. Kanellakopoulos, M. Krstić, and P. V. Kokotović, "κ-adaptive control of output-feedback nonlinear systems," *Proceedings of the 32nd IEEE Conference on Decision and Control*, San Antonio, TX, pp. 1061–1066, 1993.

[81] H. K. Khalil, *Nonlinear Systems*, New York: Macmillan, 1992.

[82] H. Khalil, "Robust servomechanism output feedback controllers for a class of feedback linearizable systems," *Proceedings of the 12th IFAC World Congress*, Sydney, Australia, vol. 8, pp. 35–38, 1993.

[83] H. Khalil, "Adaptive output-feedback control of nonlinear systems represented by input-output models," *Proceedings of the 33rd IEEE Conference on Decision and Control*, Lake Buena Vista, FL, December 1994, pp. 199–204, also submitted to *IEEE Transactions on Automatic Control*.

[84] D. E. Koditschek, "Adaptive techniques for mechanical systems," *Proceedings of the 5th Yale Workshop on Adaptive Systems*, New Haven, CT, 1987, pp. 259–265.

[85] P. V. Kokotović and H. J. Sussmann, "A positive real condition for global stabilization of nonlinear systems," *Systems & Control Letters*, vol. 13, pp. 125–133, 1989.

[86] P. V. Kokotović, Ed., *Foundations of Adaptive Control*, Berlin: Springer-Verlag, 1991.

[87] P. V. Kokotović, I. Kanellakopoulos, and A. S. Morse, "Adaptive feedback linearization of nonlinear systems," pp. 311–346 in [86].

[88] P. V. Kokotović, "The joy of feedback: nonlinear and adaptive," 1991 Bode Prize Lecture, *Control Systems Magazine*, vol. 12, pp. 7–17.

[89] A. A. Kolesnikov, "Analytical construction of nonlinear aggregated regulators for a given set of invariant manifolds," (in Russian), *Electromechanika, Izv. VTU*, Novocherkask, 1987, Part I, No. 3, pp. 100–109, Part II, No. 5, pp. 58–66.

[90] J. M. Krause, P. P. Khargonekar, and G. Stein, "Robust adaptive control: stability and asymptotic performance," *IEEE Transactions on Automatic Control*, vol. 37, pp. 316–332, 1992.

[91] G. Kreisselmeier, "Adaptive observers with exponential rate of convergence," *IEEE Transactions on Automatic Control*, vol. 22, pp. 2–8, 1977.

[92] M. Krstić, *Adaptive Nonlinear Control*, Ph.D. Dissertation, University of California, Santa Barbara, 1994.

[93] M. Krstić, "Asymptotic properties of adaptive nonlinear stabilizers," *Proceedings of the 1995 American Control Conference*, Seattle, WA, to appear, also submitted to *IEEE Transactions on Automatic Control*.

[94] M. Krstić, I. Kanellakopoulos, and P. V. Kokotović, "Adaptive nonlinear control without overparametrization," *Systems & Control Letters*, vol. 19, pp. 177–185, 1992.

[95] M. Krstić, I. Kanellakopoulos, and P. V. Kokotović, "Nonlinear design of adaptive controllers for linear systems," *IEEE Transactions on Automatic Control*, vol. 39, pp. 783–752, 1994.

[96] M. Krstić, I. Kanellakopoulos, and P. V. Kokotović, "Passivity and parametric robustness of a new class of adaptive systems," *Automatica*, vol. 30, pp. 1703–1716, 1994.

[97] M. Krstić and P. V. Kokotović, "Observer-based schemes for adaptive nonlinear state-feedback control," *International Journal of Control*, vol. 59, pp. 1373–1381, 1994.

[98] M. Krstić and P. V. Kokotović, "Adaptive nonlinear design with controller-identifier separation and swapping," *IEEE Transactions on Automatic Control*, vol. 40, pp. 426–440, 1995.

[99] M. Krstić and P. V. Kokotović, "Adaptive nonlinear output-feedback schemes with Marino-Tomei controller," *Proceedings of the 1994 American Control Conference*, Baltimore, MD, pp. 861–866, 1994, also to appear in *IEEE Transactions on Automatic Control*.

[100] M. Krstić and P. V. Kokotović, "Control Lyapunov functions for adaptive nonlinear stabilization," *Systems & Control Letters*, to appear, 1995.

[101] M. Krstić and P. V. Kokotović, "Modular approach to adaptive nonlinear stabilization," submitted to *Automatica*, 1994.

[102] M. Krstić, P. V. Kokotović, and I. Kanellakopoulos, "Transient performance improvement with a new class of adaptive controllers," *Systems & Control Letters*, vol. 21, pp. 451–461, 1993.

[103] M. Krstić, P. V. Kokotović, and I. Kanellakopoulos, "Adaptive nonlinear output-feedback control with an observer-based identifier," *Proceedings of the 1993 American Control Conference*, San Francisco, CA, pp. 2821–2825, 1993.

[104] M. Krstić, J. Sun, and P. V. Kokotović, "Robust control of nonlinear systems with input unmodeled dynamics," submitted to *IEEE Transactions on Automatic Control*, 1994.

[105] M. Krstić and P. V. Kokotović, "Avoiding cancellation of nonlinearities: a jet engine example," submitted to *The 4th IEEE Conference on Control Applications*, 1995.

[106] Z. Krzeminski, "Nonlinear control of induction motor," *Proceedings of the 10th IFAC World Congress*, Munich, Germany, 1987, pp. 349–354.

[107] P. Kudva and K. S. Narendra, "Synthesis of an adaptive observer using Lyapunov's direct method," *International Journal of Control*, vol. 18, pp. 1201–1210, 1973.

[108] V. Lakshmikantham and S. Leela, *Differential and Integral Inequalities*, New York: Academic Press, 1969.

[109] Y. D. Landau, *Adaptive Control*, New York: Marcel Dekker, 1979.

[110] J. P. LaSalle, "Stability theory for ordinary differential equations," *Journal of Differential Equations*, vol. 4, pp. 57–65, 1968.

[111] Z. Li, C. Wen, and C.-B. Soh, "Robustness of Krstic's new adaptive control scheme," *Proceedings of the IFAC Symposium on Nonlinear Control System Design*, Tahoe City, CA, 1995, to appear.

[112] Z. Li, C. Wen, and C.-B. Soh, "A new robust adaptive scheme using auxiliary errors," preprint, 1994.

[113] D. C. Liaw and E. H. Abed, "Stability analysis and control of rotating stall," *Proceedings of the IFAC Nonlinear Control Systems Design Symposium*, Bordeaux, France, June 1992.

[114] Y. Lin, *Lyapunov Function Techniques for Stabilization*, Ph.D. Dissertation, Rutgers University, New Brunswick, NJ, 1992.

[115] Y. Lin, E. Sontag, and Y. Wang, "Recent results on Lyapunov-theoretic techniques for nonlinear stability," *Proceedings of the 1994 American Control Conference*, Baltimore, MD, pp. 1771–1775.

[116] R. Lozano, B. Brogliato, and I. D. Landau, "Passivity and global stabilization of cascaded nonlinear systems," *IEEE Transactions on Automatic Control*, vol. 37, pp. 1386–1388, 1992.

[117] R. Marino, "On the largest feedback linearizable subsystem," *Systems & Control Letters*, vol. 6, pp. 345–351, 1986.

[118] R. Marino, W. M. Boothby, and D. L. Elliot, "Geometric properties of linearizable control systems," *Mathematical Systems Theory*, vol. 18, pp. 97–123, 1985.

[119] R. Marino, S. Peresada, and P. Valigi, "Adaptive input-output linearizing control of induction motors," *IEEE Transactions on Automatic Control*, vol. 38, pp. 208–221, 1993.

[120] R. Marino and P. Tomei, "Dynamic output-feedback linearization and global stabilization," *Systems & Control Letters*, vol. 17, pp. 115–121, 1991.

[121] R. Marino and P. Tomei, "Global adaptive observers for nonlinear systems via filtered transformations," *IEEE Transactions on Automatic Control*, vol. 37, pp. 1239–1245, 1992.

[122] R. Marino and P. Tomei, "Global adaptive observers and output-feedback stabilization for a class of nonlinear systems," pp. 455–493 in [86].

Bibliography

[123] R. Marino and P. Tomei, "Global adaptive output-feedback control of nonlinear systems, Part I: linear parametrization," *IEEE Transactions on Automatic Control*, vol. 38, pp. 17–32, 1993.

[124] R. Marino and P. Tomei, "Global adaptive output-feedback control of nonlinear systems, Part II: nonlinear parametrization," *IEEE Transactions on Automatic Control*, vol. 38, pp. 33–49, 1993.

[125] R. Marino and P. Tomei, "Robust stabilization of feedback linearizable time-varying uncertain nonlinear systems," *Automatica*, vol. 29, pp. 181–189, 1993.

[126] F. Mazenc and L. Praly, "Adding an integration and global asymptotic stabilization of feedforward systems," submitted to *IEEE Transactions on Automatic Control*.

[127] F. Mazenc, L. Praly, and W. P. Dayawansa, "Global stabilization by output feedback: examples and counterexamples," *Systems & Control Letters*, vol. 23, pp. 119–125, 1994.

[128] F. E. McCaughan, "Bifurcation analysis of axial flow compressor stability," *SIAM Journal of Applied Mathematics*, vol. 20, pp. 1232–1253, 1990.

[129] R. H. Middleton, "Indirect continuous time adaptive control," *Automatica*, vol. 23, pp. 793–795, 1987.

[130] R. H. Middleton and G. C. Goodwin, "Adaptive computed torque control for rigid link manipulators," *Systems & Control Letters*, vol. 10, pp. 9–16, 1988.

[131] D. E. Miller and E. J. Davison, "An Adaptive Controller which provides an arbitrarily good transient and steady state response," *IEEE Transactions on Automatic Control*, vol. 36, pp. 68–81, 1991.

[132] R. K. Miller and A. N. Michel, *Ordinary Differential Equations*, New York: Academic Press, 1982.

[133] M. M. Monahemi, J. B. Barlow, and M. Krstić, "Control of wing rock motion of slender delta wings using adaptive feedback linearization," submitted to *Journal of Guidance, Control, and Dynamics*, 1994.

[134] M. M. Monahemi, J. B. Barlow, and M. Krstić, "Adaptive nonlinear control of aircraft wing rock," in preparation, 1995.

[135] J. Monod, "La technique de cultures continues: théorie et applications," *Ann. Inst. Pasteur*, tome. 79, nr. 4, 1950.

[136] R. V. Monopoli, "Model reference adaptive control with an augmented error signal," *IEEE Transactions on Automatic Control*, vol. 19, pp. 474–484, 1974.

[137] F. K. Moore and E. M. Greitzer, "A theory of post-stall transients in axial compression systems—Part I: Development of equations," *Journal of Turbomachinery*, vol. 108, pp. 68–76, 1986.

[138] A. S. Morse, "Global stability of parameter-adaptive control systems," *IEEE Transactions on Automatic Control*, vol. 25, pp. 433–439, 1980.

[139] A. S. Morse, "High-order parameter tuners for the adaptive control of linear and nonlinear systems," *Proceedings of the US-Italy joint seminar "Systems, models and feedback: theory and application"*, Capri, Italy, 1992.

[140] S. M. Naik, P. R. Kumar and B. E. Ydstie, "Robust continuous-time adaptive control by parameter projection," *IEEE Transactions on Automatic Control*, vol. 37, pp. 182–197, 1992.

[141] K. Nam and A. Arapostathis, "A model-reference adaptive control scheme for pure-feedback nonlinear systems," *IEEE Transactions on Automatic Control*, vol. 33, pp. 803–811, 1988.

[142] K. S. Narendra and A. M. Annaswamy, *Stable Adaptive Systems*, Englewood Cliffs, NJ: Prentice-Hall, 1989.

[143] L. T. Nguyen, D. Whipple, and J. M. Brandon, "Recent experiences of unsteady aerodynamic effect on aircraft dynamics at high angle of attack," *AGARD Conference Proceedings No. 386*, Göttingen, Federal Republic of Germany, May 1985, Paper No. 28.

[144] H. Nijmeijer and A. van der Schaft, *Nonlinear Dynamical Control Systems*, New York: Springer-Verlag, 1990.

[145] R. Ortega, "Passivity properties for stabilization of cascaded nonlinear systems," *Automatica*, vol. 27, pp. 423–424, 1991.

[146] R. Ortega, "On Morse's new adaptive controller: parameter convergence and transient performance," *IEEE Transactions on Automatic Control*, vol. 38, pp. 1191–1202, 1993.

[147] R. Ortega and A. Fradkov, "Asymptotic stability of a class of adaptive systems," *International Journal of Adaptive Control and Signal Processing*, vol. 7, pp. 255–260, 1993.

[148] R. Ortega and M.W. Spong, "Adaptive motion control of rigid robots: a tutorial," *Automatica*, vol. 25, pp. 877–888, 1989.

[149] J. D. Paduano, L. Valavani, A. H. Epstein, E. M. Greitzer, and G. R. Guenette. "Modeling for control of rotating stall," *Automatica*, vol. 30, pp. 1357–1373, 1966.

[150] P.C. Parks, "Lyapunov redesign of model reference adaptive control systems," *IEEE Transactions on Automatic Control*, vol. 11, pp. 362–367, 1966.

[151] M. M. Polycarpou and P. A. Ioannou, "A robust adaptive nonlinear control design," *Proceedings of the 1993 American Control Conference*, San Francisco, CA, pp. 1365–1369, 1993.

[152] J.-B. Pomet and L. Praly, "Indirect adaptive nonlinear control," *Proceedings of the 27th IEEE Conference on Decision and Control*, Austin, TX, December 1988, pp. 2414–2415.

[153] J.-B. Pomet and L. Praly, "Adaptive nonlinear control: an estimation-based algorithm," in *New Trends in Nonlinear Control Theory*, J. Descusse, M. Fliess, A. Isidori, and D. Leborgne, Eds., Springer-Verlag, New York, 1989.

[154] J.-B. Pomet and L. Praly, "Adaptive nonlinear regulation: estimation from the Lyapunov equation," *IEEE Transactions on Automatic Control*, vol. 37, pp. 729–740, 1992.

[155] V.M. Popov, *Hyperstability of Automatic Control Systems*, Editura Academiei Republicii Socialiste România, Bucharest, 1966 (in Romanian).

[156] L. Praly, "Adaptive regulation: Lyapunov design with a growth condition," *International Journal of Adaptive Control and Signal Processing*, vol. 6, pp. 329–351, 1992.

[157] L. Praly, G. Bastin, J.-B. Pomet and Z. P. Jiang, "Adaptive stabilization of nonlinear systems," pp. 347–434 in [86].

[158] L. Praly, B. d'Andréa-Novel, and J.-M. Coron, "Lyapunov design of stabilizing controllers for cascaded systems," *IEEE Transactions on Automatic Control*, vol. 36, pp. 1177–1181, 1991.

[159] L. Praly and Z. P. Jiang, "Stabilization by output-feedback for systems with ISS inverse dynamics," *Systems and Control Letters*, vol. 21, pp. 19–33, 1993.

[160] Z. Qu, "Robust control of nonlinear uncertain systems under generalized matching conditions," *Automatica*, vol. 29, pp. 985–998, 1993.

[161] Z. Qu, "A new systematic procedure of designing robust control for nonlinear uncertain systems," submitted to *IEEE Transactions on Automatic Control*, 1993.

[162] W. Respondek, "Global aspects of linearization, equivalence to polynomial forms and decomposition of nonlinear control systems," in *Algebraic and Geometric Methods in Nonlinear Control Theory*, M. Fliess and M. Hazewinkel, Eds., D. Reidel Publishing Co., Dordrecht, 1986.

[163] A. Saberi, P.V. Kokotović, and H.J. Sussmann, "Global stabilization of partially linear composite systems," *SIAM J. Control Opt.*, 1990, vol. 28, pp. 1491–1503.

[164] P. Sannuti and A. Saberi, "Special coordinate basis for multivariable linear systems—finite and infinite zero structure, squaring down and decoupling," *International Journal of Control*, vol. 45, pp. 1655–1704, 1987.

[165] S. S. Sastry and M. Bodson, *Adaptive Control: Stability, Convergence and Robustness*, Englewood Cliffs, NJ: Prentice-Hall, 1989.

[166] S. S. Sastry and A. Isidori, "Adaptive control of linearizable systems," *IEEE Transactions on Automatic Control*, vol. 34, pp. 1123–1131, 1989.

[167] D. Seto, A. M. Annaswamy, and J. Baillieul, "Adaptive control of a class of nonlinear systems with a triangular structure," *IEEE Transactions on Automatic Control*, vol. 39, pp. 1411–1428, 1994.

[168] J.-J. E. Slotine and J. K. Hedrick, "Robust input-output feedback linearization," *International Journal of Control*, vol. 57, pp. 1133–1139, 1993.

[169] J.-J. E. Slotine and W. Li, "On the adaptive control of robot manipulators," *International Journal of Robotics Research*, vol. 6, pp. 49–59, 1987.

[170] J.-J. E. Slotine and W. Li, "Adaptive manipulator control: a case study," *IEEE Transactions on Automatic Control*, vol. 33, pp. 995–1003, 1988.

[171] E. D. Sontag, "A 'universal' construction of Artstein's theorem on nonlinear stabilization," *Systems & Control Letters*, vol. 13, pp. 117–123, 1989.

[172] E. D. Sontag, "A Lyapunov-like characterization of asymptotic controllability," *SIAM Journal of Control and Optimization*, vol. 21, pp. 462–471, 1983.

[173] E. D. Sontag, "Smooth stabilization implies coprime factorization," *IEEE Transactions on Automatic Control*, vol. 34, pp. 435–443, 1989.

[174] E. D. Sontag, "Further facts about input to state stabilization," *IEEE Transactions on Automatic Control*, vol. 35, pp. 473–476, 1990.

[175] E. D. Sontag and H. J. Sussmann, "Further comments on the stabilizability of the angular velocity of a rigid body," *Systems & Control Letters*, vol. 12, pp. 437–442, 1988.

[176] E. D. Sontag and A. R. Teel, "Changing supply functions in input/state stable systems," submitted to *IEEE Transactions on Automatic Control*, 1995.

[177] E. D. Sontag and Y. Wang, "On characterizations of input-to-state stability property," *Systems & Control Letters*, to appear, 1994.

[178] R. Su, "On the linear equivalents of nonlinear systems," *Systems & Control Letters*, vol. 2, pp. 48–52, 1982.

[179] R. Su and L. R. Hunt, "A canonical expansion for nonlinear systems," *IEEE Transactions on Automatic Control*, vol. 31, pp. 670–673, 1986.

[180] J. Sun, "A modified model reference adaptive control scheme for improved transient performance," *IEEE Transactions on Automatic Control*, vol. 38, pp. 1255–1259, 1993.

[181] J. Sun, A. Olbrot and M. Polis, "Robust stabilization and robust performance using model reference control and modeling error compensation," *IEEE Transactions on Automatic Control*, vol. 39, pp. 630–635, 1994.

[182] H. J. Sussmann and P. V. Kokotović, "The peaking phenomenon and the global stabilization of nonlinear systems," *IEEE Transactions on Automatic Control*, vol. 36, pp. 424–440, 1991.

[183] G. Tao and P. A. Ioannou, "Strictly positive real matrices and the Lefschetz-Kalman-Yakubovich lemma," *IEEE Transactions on Automatic Control*, vol. 33, pp. 1183–1185, 1988.

[184] G. Tao and P. V. Kokotović, "Adaptive control of systems with backlash," *Automatica*, vol. 29, pp. 323–335, 1993.

[185] G. Tao and P. V. Kokotović, "Adaptive control of plants with unknown dead-zones," *IEEE Transactions on Automatic Control*, vol. 39, pp. 59–68, 1994.

[186] D. Taylor, P. V. Kokotović, R. Marino and I. Kanellakopoulos, "Adaptive regulation of nonlinear systems with unmodeled dynamics," *IEEE Transactions on Automatic Control*, vol. 34, pp. 405–412, 1989.

[187] A. R. Teel, "Using saturation to stabilize a class of single-input partially linear composite systems," *Proceedings of the IFAC Nonlinear Control Systems Design Symposium*, Bordeaux, France, June 1992, pp. 24–26.

[188] A. R. Teel, "Error-based adaptive non-linear control and regions of feasibility," *International Journal of Adaptive Control and Signal Processing*, vol. 6, pp. 319–327, 1992.

[189] A. R. Teel, "Adaptive tracking with robust stability," *Proceedings of the 32nd IEEE Conference on Decision and Control*, San Antonio, TX, December 1993, pp. 570–575.

[190] A. R. Teel, R. R. Kadiyala, P. V. Kokotović and S. S. Sastry, "Indirect techniques for adaptive input-output linearization of non-linear systems," *International Journal of Control*, vol. 53, pp. 193–222, 1991.

[191] A. R. Teel and L. Praly, "On output-feedback stabilization for systems with ISS inverse dynamics and uncertainties," *Proceedings of the 32nd IEEE Conference on Decision and Control*, San Antonio, TX, pp. 1942–1947, 1993.

[192] A. R. Teel and L. Praly, "Tools for semiglobal stabilization by partial state and output feedback," *SIAM Journal of Control and Optimization*, to appear.

[193] J. Tsinias, "Sufficient Lyapunov-like conditions for stabilization," *Mathematics of Control, Signals and Systems*, vol. 2, pp. 343–357, 1989.

[194] G. Verghese and S. Sanders, "Observers for flux estimation in induction machines," *IEEE Transactions on Industrial Electronics*, vol. 35, pp. 85–94, 1988.

[195] M. Vidyasagar, *Nonlinear systems analysis*, New York: Marcel Dekker, 1977.

[196] C. Wen, "Decentralized adaptive regulation," *IEEE Transactions on Automatic Control*, vol. 39, pp. 2163–2166, 1994.

[197] R. L. Wheeden and A. Zygmund, *Measure and Integral: An Introduction to Real Analysis*, Englewood Cliffs, NJ : Prentice-Hall, 1993.

[198] J. C. Willems, "Dissipative dynamical systems—Part I: general theory," *Arch. Rational Mechanics and Analysis*, vol. 45, pp. 321–351, 1972.

[199] B. Yao and M. Tomizuka, "Robust adaptive nonlinear control with guaranteed transient performance," *Proceedings of the 1995 American Control Conference*, Seattle, WA, to appear.

[200] B. E. Ydstie, "Transient performance and robustness of direct adaptive control," *IEEE Transactions on Automatic Control*, vol. 37, 1091–1106, 1992.

[201] T. Yoshizawa, *Stability Theory by Lyapunov's Second Method*, The Mathematical Society of Japan, 1966.

[202] Z. Zang and R. R. Bitmead, "Transient bounds for adaptive control systems," *IEEE Transactions on Automatic Control*, vol. 39, 171–175, 1994.

[203] Y. Zhang, P. A. Ioannou, and C.-C. Chien, "Parameter convergence of a new class of adaptive controllers," submitted to *IEEE Transactions on Automatic Control*.

Index

Italic numbers denote pages where a given concept is either introduced or presented in the most detail.

active suspension, 37, 55
adaptation as dynamic feedback, 88
adaptation gain, 157
adaptive backstepping, 11, 13, 87, 92, *96*, 121
adaptive backstepping via aclf, 134
adaptive block-backstepping, 98
adaptive control Lyapunov function (aclf), *129*, 132, 134, 136
adaptive linear control, 2, 17, 484
adaptive observer backstepping, 301
adaptive observer form, 349, 407
aircraft wing rock, 180
approximately feedback linearizable systems, 234
asymptotic stability, 22
attractivity, 22
augmented error, 239
avoiding cancellations, 21, 28, 32, *70*

backstepping, 21, 29, *33*, 82, 86
bandwidth, 462
Barbalat's lemma, 491
benchmark example, 128
biochemical process, 113
block backstepping, 49, 51, 54
block-strict-feedback systems, 64
bounded-input bounded-state (BIBS), 174
boundedness, 22, 75, 82
boundedness without adaptation, 105, 118, 160, 275

cascade connection, 41
cascade systems, 42, 86
cascade of ISS systems, 503
certainty equivalence, 2, 14, 124, 131, 186, 222
chain of integrators, 37
class-$\mathcal{K}$ ($\mathcal{K}_\infty$, $\mathcal{KL}$, $\mathcal{KL}_\infty$) functions, 489
closed-loop poles, 448
comparison functions, 489
comparison principle, 495
compression system, 67
control Lyapunov function (clf), 26
controllable, 35
convergence, 151, 344, 436, 484, 489
correction term, 144, 146
χ-passive identifier, 407

decentralized design, 183
detuned linear system, 449
detuned nonadaptive controller, 445
differential-geometric conditions, 521
differential-geometric control theory, 10
direct adaptive control, 2
dissipation rate, 507
dissipativity characterization of ISS, 196, 503
dynamic order, 13, 16, 123, 349, 417, 484

electric machines, 86

559

equation-error filtering identifiers, 233
equilibrium, 23, 489
equilibrium manifold, 152
error variable, 30
error system, 148, 203, 336, 344, 362, 433, 436, 471
estimation-based design, 5, 14, 233
estimation error, 240, 384
existence of solutions, 211, 215, 293
exponential stability, 490
extended matching, 10, 110, 121, 126, 538
ϵ-swapping identifier, 409

fast adaptation, 159, 264
feasibility region, 121, 178, 368
feedback connection, 498, 508
feedback linearization, 10, 21, 27, 35, *39*, 72
feedback passive, 46
feedback positive real (FPR), 43
feedback strictly passive, 46
fed-batch process, 113
filtered transformations, 11, 348
finite escape time, 8, 78, 188, 288
flexibility of backstepping, 32, 36
flux observer, 296
full-state feedback linearization, 42

high-frequency gain, 169, 418, 470, 511
high-gain controller, 451
Hölder's inequality, 494

global exponential stability, 490
global stability, 23
global uniform asymptotic stability, 490
global uniform stability, 490
globally adaptively quadratically stabilizable, 132, 134

globally adaptively stabilizable, 128, 189
gradient update law, 230, 249, 385, 478, 514
Gronwall's lemma, 499
growth conditions, 10, 186, 232, 282, 324, 416

indirect adaptive control, 2
indirect linear scheme, 464
induction motor, 294, 325
input-output feedback linearization, 42
input-output stability, 22, *493*
input-to-state stability (ISS), 25, 68, 77, 98, 106, 174, 189, *501*
input-to-state stabilization, 190
instability, 22
integrator backstepping
interlaced controller-observer design, 315
interlacing functions, 315
invariant manifolds, 183
invariant set, 24
inverse dynamics, 343, 436
ISS (*see* input-to-state stability)
ISS-backstepping, 195
ISS-control Lyapunov function (ISS-clf), 190
ISS-controller, 198, *202*, 235
ISS-controller with K-filters, 377
ISS-controller with MT-filters, 406
ISS-Lyapunov function, 190, 502
ISS-observer, 403

jet engine, 66

κ-terms, 115
K-filters, 331, 372, 421
Kalman-Yakubovich lemma (*see* Popov-Kalman-Yakubovich lemma)

INDEX 561

$\mathcal{L}_p$ stability, 498
LaSalle's invariance theorem, 25
LaSalle-Yoshizawa theorem, 24, *492*
least-squares update law, 240, 250, 385, 478, 514
levitated ball, 365
linear systems, 15, 417
locally Lipschitz, 211, 233, 512
Lyapunov-based design, 4, 9
Lyapunov stability, 22, 489
Lyapunov stability theorem, 490
Lyapunov-type update law, 185

magnetic levitation, 365
matching condition, 10, 86, 90, 121
MIMO design, 294
minimum phase, *42*, 51, 291, 418
model reference adaptive control, 17
modified system, 129
modular design, 14, 185, 233, 235, 371
modular linear design, 470
Moore-Greitzer model, 67
MT-filters, 350, 403
multi-input systems, 103
multi-input parametric-strict-feedback systems (*see* multi-input systems)

nonadaptive system, 443
nonholonomic systems, 183
nonlinear damping, 73, *78*, 157, 166, *199*, 224, 246, 264, 272, 289, 376, 469
nonlinear observer, 291
nonlinear swapping, 515
nonminimum phase, 46
normalization, 477, 478
normalized update law, 246, 249, 266, 280
norms
    Euclidean, 17

Frobenius, 17, 240
induced 2-norm, 17
$\mathcal{L}_1, \mathcal{L}_2, \mathcal{L}_\infty$, 17
$\mathcal{L}_p, \mathcal{L}_{p,e}, \mathcal{H}_\infty$, 493
notation, 17

observer backstepping, 285
observer-based identifiers, 233
observer canonical form, 418
observers, 206, 291
operator gain, 77
optimality, 86
output-feedback design, 15, 285
output-feedback systems, 291, 327, 371, 533
overparametrization, 11, 101, 110, 152, 183, 325, 369

parameter projection, 232, 278, 381, 385, 407, 473, *511*
parametric x-model, 207, 388, 477
parametric y-model, 380, 384, 473
parametric z-model, 206, 392
parametric-block-strict-feedback systems, 105, 173
parametric-pure-feedback systems, 113, 121, *175*, 537
parametric output-feedback systems, 307, 533
parametric robustness, 448, 473
parametric-strict-feedback systems, 99, 121, 139, 198, 539
partial-state-feedback systems, 320, 521
passive identifiers, 185, 233
passivity, 132, 154, 206, 437, *507*
periodic orbits, 408
persistency of excitation, 437
Popov-Kalman-Yakubovich lemma, 509
positive real (PR), 43, 509
prediction error, 238, 249, 389
projection operator, 511

pure-feedback systems, 61

quadratic Lyapunov function, 70

radial unboundedness, 23
reduced-order observer, 331, 350, 366
reference model, 156, 418
reference model initialization, 469
reference trajectory, 162
region of attraction, 23, 180, 368
regressor, 145, 331, 421
regulation, 23, 192
relative degree, 40, 43
robotic manipulator, 313
robust control Lyapunov function (rclf), 234
robust strict-feedback systems, 84
robustness, 17, 18
rotating stall, 67

semiglobal adaptive designs, 369
separation principle, 285
set-point regulation, 139
SG-controller, 223
SG-controller for linear systems, 470
SG-scheme, 222, 265
single link flexible robot, 313
singularities, 35
skew-symmetry, 32, 95, *144*, 149, 433
slow adaptation, 246, 252, 280
small control property, 27, 130, 192
small-gain theorem, 498
Sontag's formula, 26, 28, 129, 136, 191
static parametric models, 239, 388
stability, 22, 489
stability without adaptation, 160, 484
stabilizability

stabilizing function, 30, 422
steering and braking, 86
storage function, 507
strengthened passive identifier, 225
strict-feedback systems, 58
strict passivity, 507
strictly positive real (SPR), 509
surge, 67
swapping identifiers, 235, 282
swapping lemma, 515
swapping technique, 238
symbolic software, 183

tracking, 24, 156
traditional scheme, 464
trajectory initialization, 118, *162*, 165, 220, 254, 259, 347, 364, 396, 414, 455, 458
transient performance, 16, 118, 157, 218, 254, 346, 363, 394, 412, 453, 484, 485
 x-passive, 221
 x-swapping, 258, 398
 y-passive, 394
 y-swapping, 396
 z-passive, 219, 396
 z-swapping, 256, 399
 ϵ-swapping, 414
 χ-passive, 412
 improvement due to adaptation, 168, 459
 nonadaptive, 448, 473
  frequency domain, 453
  $\mathcal{L}_\infty$, 449
  mean-square, 451
 tuning functions, 165
  $\mathcal{L}_2$, 165, 454
  $\mathcal{L}_\infty$, 166, 456
 tuning functions with K-filters, 346
 tuning functions with MT-filters, 363

tuning function, 123, *131*, 136, 150, 155, 425
tuning functions design, 11, 13, 123, 327
  linear systems, 432
  state-feedback, 158
  vs. modular design 263, 482
  with K-filters, 341
  with MT-filters, 358, 359

uncertainty
  matched, 73
  backstepping with, 80, 86
underlying linear controller, 443, 447, 484
underlying nonadaptive controller, 224
uniform asymptotic stability, 490
uniform stability, 23, 490
unnormalized least-squares, 482
unnormalized update law, 240, 246, 249
unobservable, 42
unstable (*see* instability)
update law, 87

virtual control, 30, 286, 422
virtual control coefficients, 168, 229, 277, 511
virtual estimate, 304, 308

weak ISS-controller, 271
weak ISS-controller with K-filters, 401
weak minimum phase, 45, 49

x-passive identifier, 212
x-swapping identifier
  linear systems, 477
  output-feedback, 388
  state-feedback 248, 277

y-passive identifier, 380, 473
y-swapping identifier, 384
Young's convolution theorem, 494
Young's inequality, 75

z-passive identifier, 209, 392
z-swapping identifier, 239, 279, 393
zero dynamics, *42*, 291